Cardiac Imaging in Electrophysiology

Angelo Auricchio • Jagmeet Singh
Frank E. Rademakers
Editors

Cardiac Imaging in Electrophysiology

Springer

Editors
Angelo Auricchio, M.D., Ph.D.
Director, Electrophysiology Program
Division of Cardiology
Fondazione Cardiocentro Ticino
Lugano
Switzerland

Frank Rademakers, M.D., Ph.D.
Medical Director
Division of Cardiology
University Hospitals Leuven
Leuven
Belgium

Jagmeet Singh, M.D., Ph.D.
Director of Cardiac Resynchronization
Therapy Program
Massachusetts General Hospital
Heart Center, Boston
Associate Professor of Medicine
Harvard Medical School
Boston, MA
USA

ISBN 978-1-4471-6894-2 ISBN 978-1-84882-486-7 (eBook)
DOI 10.1007/978-1-84882-486-7
Springer London Dordrecht Heidelberg New York

British Library Cataloguing in Publication Data
A catalogue record for this book is available from the British Library

Printed on acid-free paper

Springer is part of Springer Science+Business Media (www.springer.com)

To my family for their support, love and patience.

Angelo Auricchio

To Noopur, Ashan and Ahaana for always being there.

Jagmeet Singh

To my wife and children for their love, support and understanding.

Frank Rademakers

Foreword

Cardiovascular imaging has become a subspecialty in cardiology on its own merit. It is central to every diagnostic pathway from detection to treatment of patients with all sorts of cardiac conditions. The advances and sophistication of all imaging modalities we have in our possession have progressed so much that close collaboration between imaging experts and clinicians is more essential than ever. Such is the level of sophistication of each individual imaging modality that it has become virtually impossible for any imager alone to be expert in all modalities so that individual expertise is required. At the same time, the cost of the technology has increased and a rational approach is required in order to avoid duplication of information. A careful balance between the diagnostic yield with the cost and potential safety concerns (radiation and contrast) is important. The real value of imaging in relation to improved outcomes however remains subjective, which explains, at least in part, the inconsistent utilisation of those technologies around the world.

Imaging has also become essential for the diagnosis and treatment of electrical disorders and the wealth of imaging options for the practicing electrophysiologist has exploded over the last decade, led perhaps by the ability to effectively ablate patients with atrial fibrillation and device implantation for treating heart failure patients. This book, by experts in electrophysiology and cardiovascular imaging, offers an essential guide to advanced imaging in patients with electrical disorders from a critical appraisal of technology through ablation and device therapies and the latest advances in magnetic and robotic navigation. It is ultimately, however, in the hands of the clinician to make the best use of the available technology for the patient's best benefit and avoid duplication and overlap, and this book offers exactly that.

London, UK Petros Nihoyannopoulos

Preface

Major technological, basic, and clinical research breakthroughs have led to impressive changes in both diagnostic and therapeutic capabilities for the management of cardiac rhythm disorders. Over the last three decades, the diagnostic part has been enriched by sophisticated ECG recording systems and automatic analysis techniques, by the development of 3-dimensional (3D) electroanatomical mapping systems and advances in novel implantable devices capable of recording electrical signals as well as intracardiac and intravascular pressures. On the other hand, the most striking progress in the therapeutic arena has been in catheter-based ablation of different rhythm disorders, fast-paced progress in device therapy inclusive of implantable cardioverter-defibrillators, and cardiac resynchronization therapy.

Consequently, the role of the electrophysiologist in the field of cardiology has significantly expanded and the role of cardiac imaging has become a crucial and integral part of diagnostic and therapeutic electrophysiological procedures. This book attempts to cover the increased role of cardiac imaging in diagnosis and electrophysiological procedures. Various cardiac imaging modalities are now available to guide the electrophysiologist, from the first encounter with the patient and subsequently throughout the procedural and postprocedural care. These modalities have come to play an important role in appropriately selecting patients, guiding therapy, thereby reducing complications, and enabling us to closely monitor the effects of device therapy or catheter ablation procedures.

The advances in imaging modalities have been immense, and their use is no longer confined to the domain of the noninvasive cardiologist, but has expanded into the realm of image-guided interventions. This trend is clearly appreciated in transcatheter-based treatment of atrial and ventricular arrhythmias as well as in device implantation and management. Some of the potential advantages of imaging during invasive diagnostic and curative procedures include easier navigation, greater precision in targeting the region of interest, better catheter stability, minimizing collateral damage, and a reduction in radiation exposure to both patient and physician.

It is probably true that the diagnosis and treatment of atrial arrhythmias represent the best example of cardiac rhythm disorders in which imaging-assisted therapy delivery has played a major role. In particular, the demands created by the complexity of treating atrial fibrillation by catheter ablation has forced the paradigm shift from a pure electrophysiological based approach, in which intracardiac signal recording and fluoroscopy had originally founded the basis of therapy delivery, to the more contemporary approach involving 3D electroanatomical mapping integrated with either rotational angiography, cardiac tomography (CT), or magnetic resonance imaging (MRI). Image-guided atrial fibrillation ablation has facilitated the accuracy in localizing the anatomical target, enhanced safety while significantly reducing the procedure, and fluoroscopy time. At present, catheter ablation for AF is considered a reasonable option when antiarrhythmic drugs have failed. The cornerstone for most AF ablation procedures is the electrical isolation of pulmonary veins. The role of imaging in catheter ablation for AF encompasses (1) segmentation and integration of left atrium (LA) 3D anatomy into navigation systems, (2) intraprocedural visualization of the pulmonary veins, (3) delineating macro-reentrant circuits via activation maps, (4) mapping complex fractionated electrograms, and (5) most importantly manual as well as robotic-guidance of the ablation catheter to the specific area of interest.

More recently, the need to better define the substrate and more precisely define the anatomical characteristics of the chamber wall has come to be considered important for individualizing the ablative approach. A large number of imaging modalities have become potentially available for visualization of the LA wall, its thickness, and related tissue characterization. Rather than the conventional nomenclature of paroxysmal or persistent atrial fibrillation dictating the approach and extent of the ablation, it has become clear that more attention needs to be paid to structural characteristics such as atrial fibrosis. It is noteworthy that fibrosis may result from different disease processes that either directly alter the atrial wall 3D architecture and anatomical characteristics, or as a consequence of thermal damage occurring during isolation of the AF circuits. By analyzing fibrosis with delayed enhancement MRI, one may be able to better describe predictors of success and also help plan the ablation approach. Interestingly, recent work has shown that patients who had recurrent AF showed delayed enhancement in all portions of the LA, whereas patients free of AF-recurrence showed delayed enhancement confined to the posterior wall and septum. Image-guided ablation of these areas of contention appears to be related to improved ablation outcomes. Targeted and individualized approaches will help limit the extent of the ablation while enhancing success. Imminently, the future approaches will consist of real-time 3D imaging, thereby eliminating the errors and limitations of integration techniques.

Similar to catheter ablation of atrial arrhythmias, there is growing evidence that cardiac imaging can now better define the anatomic substrate and ablation targets for ventricular tachycardia (VT). CT and MRI strategies have become effective in complementing the conventional electroanatomical voltage map, providing an anatomic correlate to the underlying electrophysiological data during substrate ablation of scar-related VT. Novel multi-array mapping catheters specifically designed for VT ablation together with innovative postprocessing algorithms capable of high spatial resolution of scar tissue identified by MRI will continue to increase the accuracy to specifically localize the reentry isthmus and potentially improving the effectiveness of ablative therapy. Whether functional and metabolic information, derived either by nuclear medicine techniques or by MRI, will further help in defining the area of abnormal tissue activity is an area of intense investigation.

A novel concept within the constantly evolving imaging world involves interventional MRI, where the electrophysiologist in conjunction with 3D electroanatomical mapping performs real-time MR tracking of a deflectable MR compatible catheter. At present, this method involves real-time MRI positioning of an electrophysiologic catheter, which is overlaid onto high-resolution, time-resolved images. This technique provides real-time visualization of the anatomical substrate identical to the exposure of tissues obtained during surgery. Future development faces challenges such as catheter device improvement and acoustic noise reduction. Despite these challenges, MRI could revolutionize image-guided ablations above and beyond what can be currently provided by conventional imaging approaches. Another innovative technology, still in its preliminary stages, involves the use of real-time intracardiac 3D and transesophageal 4D ultrasound imaging probes to guide catheter ablation. The advantage of intracardiac or transesophageal ultrasound imaging is intrinsically related to the widespread utilization of the technique, the easy accessibility of the technology compared to MRI, the limited amount of human and finance resource utilization required for running it, as well as the modest initial investment.

A significant problem in the use of cardiac imaging for both device and ablation procedures by electrophysiologists is the limited training that most fellows in electrophysiology usually receive. Imaging techniques such as standard echocardiography or nuclear imaging are usually learned during cardiology training whereas more sophisticated techniques such as CT, MRI, or 3D/4D echocardiography are usually neglected. On the other hand, most imaging cardiologists or cardiac radiologists have little exposure to electrophysiological procedures or device implantation; therefore, for the purpose of guiding complex electrophysiological procedures, acquired images may be of suboptimal quality or image resolution and consequently integration into 3D electroanatomical systems may be inadequate. This gap between the community of electrophysiologists and cardiac imagers should be bridged by more extensive common educational programs offered by scientific societies or dedicated training and educational programs.

Still questionable is whether the procedural efficiency will grow, and if the learning curve will get shortened. Evidently, the future of imaging modalities for treating arrhythmias and implementing device therapy will be determined by the overall clinical utility and cost-effectiveness of their use. At present, the appropriateness criterion for cardiac CT and cardiac MRI allows electrophysiologists to be formally guided for the use of imaging modalities pre- and post-procedurally. There are however limited data about the value of intra-procedural cardiac imaging in most electrophysiological procedures, inclusive of device implantation. The complexity in quantifying the real value of cardiac imaging is intertwined with the need for a corresponding advancement in the efficacy of the delivery of electrophysiologic therapy. Also of concern is that the imaging-enhanced guidance of electrophysiological procedures may have a significant impact on workflow and increase the absolute procedural costs.

In conclusion, the benefit of noninvasive and invasive imaging methods during electrophysiology procedures is as important as the benefits obtained pre- and post-procedurally. Identification of anatomical landmarks and prediction of possible postprocedural complications is crucial to deliver appropriate and safe therapy. The choice of imaging methods varies between institutions and is reflected not only by patient population, but also by the experience, expertise, and technological availability at each institution. The patient's age, patient's co-morbidities, biological risk of exposure to radiation, and economics of each method should be weighed when evaluating patients. In order to improve cost-effectiveness, future studies should be focused on decreasing the percentage of nonresponders to therapies, reduction of failure rate and recurrence rate, while measuring success in terms of patient-centered end-points. The future of imaging in electrophysiology not only depends on the physician's skill and clinical knowledge, but also on a willingness to explore new frontiers. Ultimately, this will improve the efficiency, better individualize our treatment strategies while enhancing the safety profile of these procedures.

Angelo Auricchio
Jagmeet Singh
Frank Rademakers

Contents

Part III Device Therapy

Part IV Advances in Navigation and Integration

Contributors

Theodore P. Abraham, M.D. Division of Cardiology, Department of Medicine, Johns Hopkins University, Baltimore, MD, USA

José Alberto de Agustín, M.D. Unidad de Imagen Cardiovascular, Hospital Clínico San Carlos, Madrid, Spain

Amin Al-Ahmad, M.D. Cardiac Arrhythmia Service and Cardiac Electrophysiology Laboratory, Department of Cardiovascular Medicine, Stanford University, Stanford, CA, USA

Jeroen J. Bax, M.D., Ph.D. Department of Cardiology, Leiden University Medical Center, Leiden, ZA, The Netherlands

Paolo Della Bella, M.D. Arrhythmia Unit and of the Electrophysiology Laboratory of the Centro Cardiologico Monzino, Institute of Cardiology, University of Milan, Milan, Italy

Matteo Bertini, M.D. Department of Cardiology, Leiden University Medical Center, Leiden, ZA, The Netherlands

Kenneth Bilchick, M.D. Department of Medicine, Cardiovascular Division, University of Virginia Health System, Charlottesville, VA, USA

Rutger J. van Bommel, M.D. Department of Cardiology, Leiden University Medical Center, Leiden, ZA, The Netherlands

Anthony G. Brooks, Ph.D. Department of Cardiac Electrophysiology, Centre for Heart Rhythm Disorders, University of Adelaide and Royal Adelaide Hospital, Adelaide, SA, Australia

Corrado Carbucicchio, M.D. Arrhythmia Department and Clinical Electrophysiology Laboratory, Centro Cardiologico Monzino, Milan, Italy

Andre D'Avila, M.D., Ph.D. Department of Medicine, The Mount Sinai Heart Centre, Mount Sinai Medical Center, New York, NY, USA

Stephan Danik, M.D. Experimental Electrophysiology Laboratory, Department of Medicine, Massachusetts General Hospital, Boston, MA, USA

Victoria Delgado, M.D. Department of Cardiology, Leiden University Medical Center, Leiden, ZA, The Netherlands

Veronica Lea J. Dimaano, M.D. Division of Cardiology, Johns Hopkins University, Baltimore, MD, USA

Sabine Ernst, M.D. Royal Brompton and Harefield Hospital, London, UK

Christian S. Eversull, M.D. Department of Emergency Medicine, VA Palo Alto Health Care Services, Palo Alto, CA, USA

Francesco Fulvio Faletra, M.D. Division of Cardiology, Cardiocentro Ticino, Lugano, Switzerland

Amparo L. Figueroa, M.D. Department of Radiology, Massachusetts General Hospital, Boston, MA, USA

Ann C. Garlitski, M.D. Department of Medicine, Tufts Medical Center, Boston, MA, USA

John Gorcsan, M.D. Department of Medicine, Cardiovascular Institute, University of Pittsburgh, Pittsburgh, PA, USA

Udo Hoffmann, M.D., M.P.H. Department of Radiology, Massachusetts General Hospital and Harvard Medical School, Boston, MA, USA

Henry H. Hsia, M.D. Cardiac Arrhythmia Service and Cardiac Electrophysiology Laboratory, Department of Cardiovascular Medicine, Stanford University, Stanford, CA, USA

Pierre Jaïs, M.D. Department of Cardiology, Hôpital Cardiologique du Haut- Lévêque, CHU Bordeaux, Pessac, France

C. Jan Willem Borleffs, M.D. Department of Cardiology, Leiden University Medical Center, Leiden, ZA, The Netherlands

Subodh B. Joshi, M.B.B.S. Department of Radiology, Massachusetts General Hospital and Harvard Medical School, Boston, MA, USA

Josef Kautzner, M.D., Ph.D. Department of Cardiology, Institute for Clinical and Experimental Medicine, Prague, Czech Republic

Sébastien Knecht, M.D. Cardiology Service, CHU Brugmann, Brussels, France

Juhani Knuuti, M.D. Turku PET Centre, Turku University Hospital, Turku, Finland

Albert Lardo, Ph.D. Department of Medicine and Biomedical Engineering, Heart and Vascular Institute, Johns Hopkins University, Baltimore, MD, USA

Bruce D. Lindsay, M.D. Department of Cardiovascular Medicine, Cleveland Clinic, Cleveland, OH, USA

Giuseppe Maccabelli, M.D. Department of Arrhythmia, Electrophysiology and Electrostimulation Unit, Fondazione Centro San Raffaele del Monte Tabor, Milano, Italy

Nassir F. Marrouche, M.D. Division of Cardiology, Comprehensive Arrhythmia Research & Management Center, University of Utah Health Sciences Center, Salt Lake City, UT, USA

Eliot McVeigh, Ph.D. Department of Biomedical Engineering, Johns Hopkins University School of Medicine, Baltimore, MD, USA

Narayanan Namboodiri, M.D. Department of Cardiac Electrophysiology, Centre for Heart Rhythm Disorders, University of Adelaide and Royal Adelaide Hospital, Adelaide, SA, Australia

Natesa G. Pandian, M.D. Heart Valve Center, Tufts University School of Medicine, Boston, MA, USA
Cardiovascular Imaging Center, Tufts University School of Medicine, Boston, MA, USA
Cardiovascular Ultrasound Research, Department of Medicine, Tufts University School of Medicine, Boston, MA, USA

Petr Peichl, M.D., Ph.D. Department of Cardiology, Institute for Clinical and Experimental Medicine, Prague, Czech Republic

Marco Perez, M.D. Department of Cardiovascular Medicine, Stanford University, Stanford, CA, USA

Aurélio C. Pinheiro, M.D., Ph.D. Division of Cardiology, Johns Hopkins University, Baltimore, MD, USA

François Regoli, M.D. Division of Cardiology, Fondazione Cardiocentro Ticino, Lugano, Switzerland

Prashanthan Sanders, M.B.B.S., Ph.D. Department of Cardiac Electrophysiology, Centre for Heart Rhythm Disorders, University of Adelaide and Royal Adelaide Hospital, Adelaide, SA, Australia

Martin J. Schalij, M.D., Ph.D. Department of Cardiology, Leiden University Medical Center, Leiden, ZA, The Netherlands

Jagmeet Singh, M.D., Ph.D. Director of Cardiac Resynchronization Therapy Program, Massachusetts General Hospital Heart Center, Boston, Associate Professor of Medicine, Harvard Medical School, Boston, MA, USA

Sheldon M. Singh, M.D. Department of Medicine, Schulich Heart Center, Sunnybrook Health Sciences Center, University of Toronto, Toronto, ON, Canada

Ahmed Tawakol, M.D. Department of Medicine, Division of Cardiology, Massachusetts General Hospital and Harvard Medical School, Boston, MA, USA

John Triedman, M.D. Department of Pediatrics, Harvard Medical School, Boston, MA, USA and Department of Cardiology, Children's Hospital Boston, Boston, MA, USA

Quynh A. Truong, M.D. Cardiac Division, Department of Radiology, Massachusetts General Hospital, Boston, MA, USA

Mintu Turakhia, M.D., M.A.S. Department of Cardiovascular Medicine, Stanford University, Stanford, CA, USA

Gaston R. Vergara, M.D. Division of Cardiology, Department of Medicine, University of Utah School of Medicine, Salt Lake City, UT, USA

Paul J. Wang, M.D. Cardiac Arrhythmia Service and Cardiac Electrophysiology Laboratory, Department of Cardiovascular Medicine, Stanford University, Stanford, CA, USA

Oussama Wazni, M.D. Department of Cardiovascular Medicine, Cleveland Clinic, Cleveland, OH, USA

Matthew Wright, M.B.B.S., Ph.D. Department of Cardiology, St. Thomas' Hospital, London, United Kingdom

José Zamorano, M.D. Unidad de Imagen Cardiovascular, Hospital Clínico San Carlos, Madrid, Spain

Paul C. Zei, M.D., Ph.D. Cardiac Arrhythmia Service and Cardiac Electrophysiology Laboratory, Department of Cardiovascular Medicine, Stanford University, Stanford, CA, USA

Part I

Technology Evaluation

Imaging-Based Heart Anatomy

1

Francesco Fulvio Faletra, Ann C. Garlitski,
François Regoli, and Natesa G. Pandian

Abstract

Improvements in imaging technology have led to a better understanding of dynamic morphology of the heart and resulted in improved safety and efficacy of invasive electrophysiologic procedures. Delineation of anatomic structures and morphologic variants have allowed the cardiac electrophysiologist to perform increasing complex ablations and implantation of sophisticated devices. Novel tools have also reduced the reliance on fluoroscopy, minimizing radiation exposure to the patient and to the clinician.

Keywords

Image-based heart anatomy • Heart anatomy based on imaging • Right atrium anatomy • Right ventricle anatomy • Fluoroscopy for cardiac anatomy

Over the last few decades, there have been revolutionary advances in the understanding of the pathophysiology of cardiac arrhythmias. In parallel, technological advancements have led to new therapeutic modalities. Novel catheters, energy sources, and modes of imaging have allowed for the mapping and ablation of abnormal endocardial and epicardial supraventricular and ventricular circuits. In addition, our contemporary treatment armamentarium includes pacemakers, defibrillators, and cardiac resynchronization therapy devices in order to treat potentially life-threatening bradycardias and tachycardias as well as congestive heart failure symptoms. The ability to perform such a variety of procedures with high success and low complication rates involves an intimate knowledge of cardiac anatomy. An array of sophisticated imaging techniques have developed over the years which allow the electrophysiologist to ablate integral parts of abnormal circuits, rendering the patient cured of the arrhythmia, and to implant devices in patients with cardiomyopathies, both acquired and congenital. Such rapid developments in electrophysiologic interventions have triggered a renewed interest in the anatomy of the heart. In order for the clinician to be successful, it is important to understand the advantages of each imaging modality and how these tools can be used alone or in combination in order to define intracardiac and extracardiac structures.

The basic imaging approach to guide catheter placement and manipulation in the electrophysiology lab is fluoroscopy.[1] Fluoroscopy is routinely used to localize anatomic landmarks during electrophysiologic procedures. However, the only anatomic reference with fluoroscopy is the heart shadow and its relation to catheters positioned at certain fixed locations. Figures 1.1–1.6 show the differences in imaging representation between fluoroscopy and 3D multislice computed tomography (CT) volume rendering, CT slice, and real-time 3D transesophageal echocardiography (RT 3D TEE), respectively. Moreover, fluoroscopy is

N.G. Pandian (✉)
Heart Valve Center, Tufts University School of Medicine,
Boston, MA, USA

Cardiovascular Imaging Center, Tufts University School of Medicine,
Boston, MA, USA

Cardiovascular Ultrasound Research, Department of Medicine,
Tufts University School of Medicine, Boston, MA, USA
e-mail: npandian@tuftsmedicalcenter.org

A. Auricchio et al. (eds.), *Cardiac Imaging in Electrophysiology*,
DOI 10.1007/978-1-84882-486-7_1, © Springer-Verlag London Limited 2012

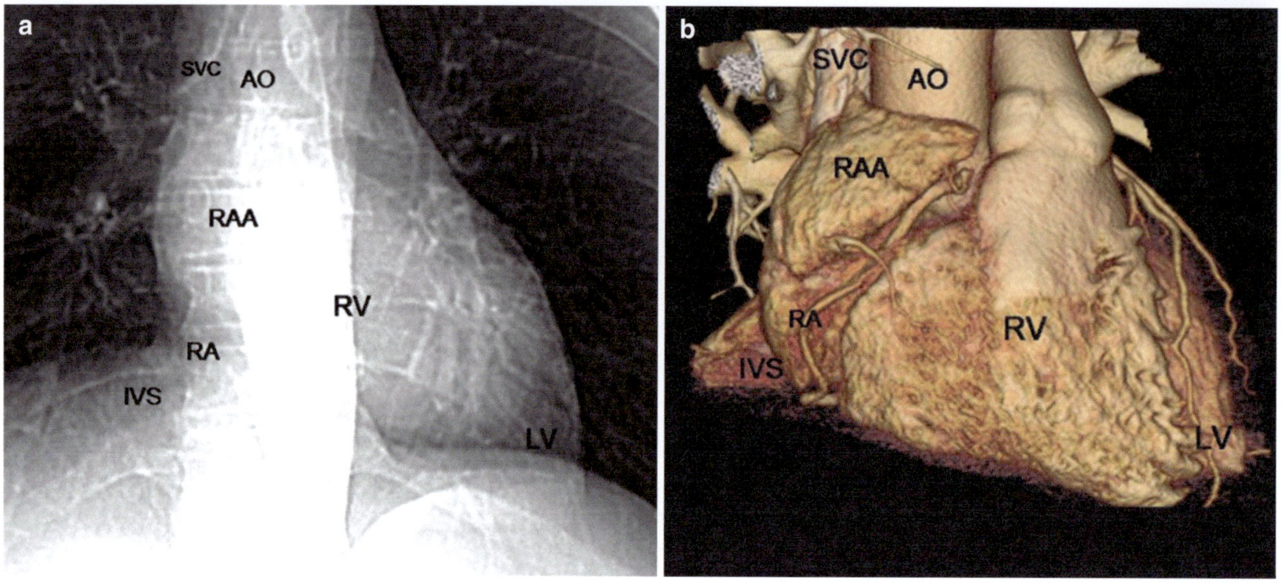

Fig. 1.1 (**a**) Fluoroscopic image in antero-posterior projection of the heart and (**b**) the corresponding 3D CT volume rendering. *IVS* inferior vena cava, *SVC* superior vena cava, *RA* Right atrium, *RAA* right atrial appendage, *RV* right ventricle, *LV* left ventricle, *Ao* Aorta

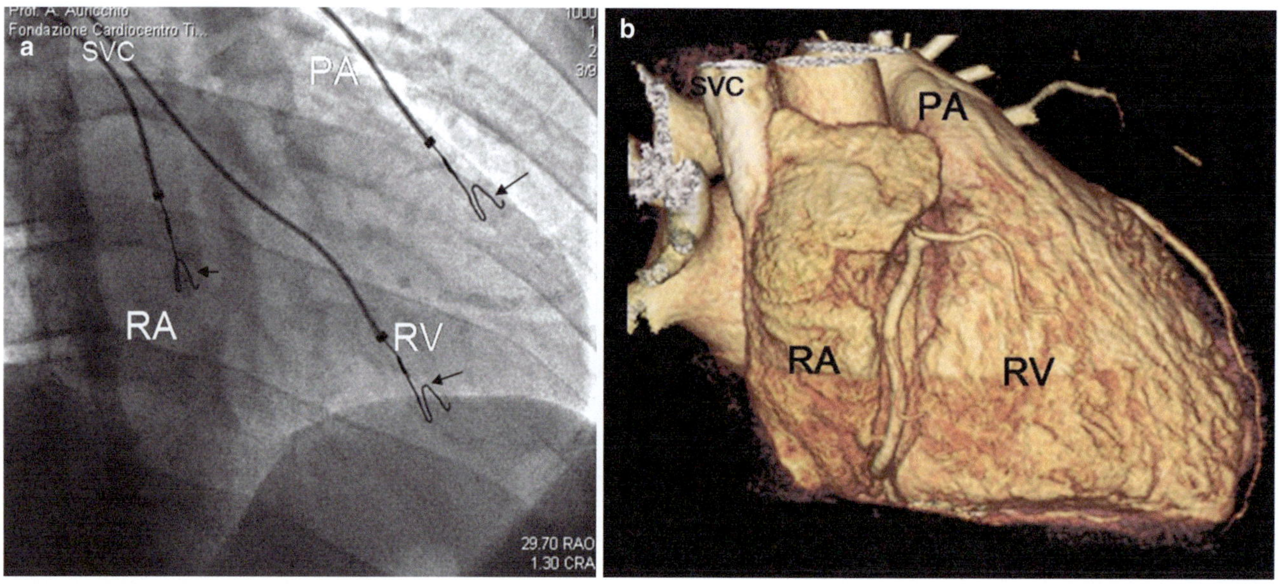

Fig. 1.2 (**a**) Fluoroscopic image in right anterior oblique projection and (**b**) the corresponding 3D CT volume rendering. *RA* Right atrium, *RV* right ventricle, *PA* pulmonary artery, *SVC* superior vena cava. Because of the limitation of imaging complex three-dimensional structures on two-dimensional projections, the surface electrodes (*arrows*) seem to be into the heart cavities

limited by its two-dimensional projection of complex 3D structures that limit interpretation and analysis, and various cardiac structures, such as fossa ovalis, crista terminalis, Eustachian valve, coronary sinus, and pulmonary vein ostia, are difficult to precisely define at fluoroscopy. Accurate pre-procedural imaging of these structures might become a pre-requisite for successful electrophysiological procedures, providing an anatomic "road map" to allow faster and more precise placement of intracardiac catheters. Sophisticated imaging technologies such as magnetic resonance imaging (MRI), CT, and RT 3D TEE provide relevant anatomic information in exquisite detail. These techniques have the potential to give pre-procedural anatomic information for locating important landmarks relative to current electrophysiological interventions or cardiac resynchronization therapy.

In this chapter, we describe the normal cardiac anatomy and anatomic landmarks of interest to electrophysiologists,

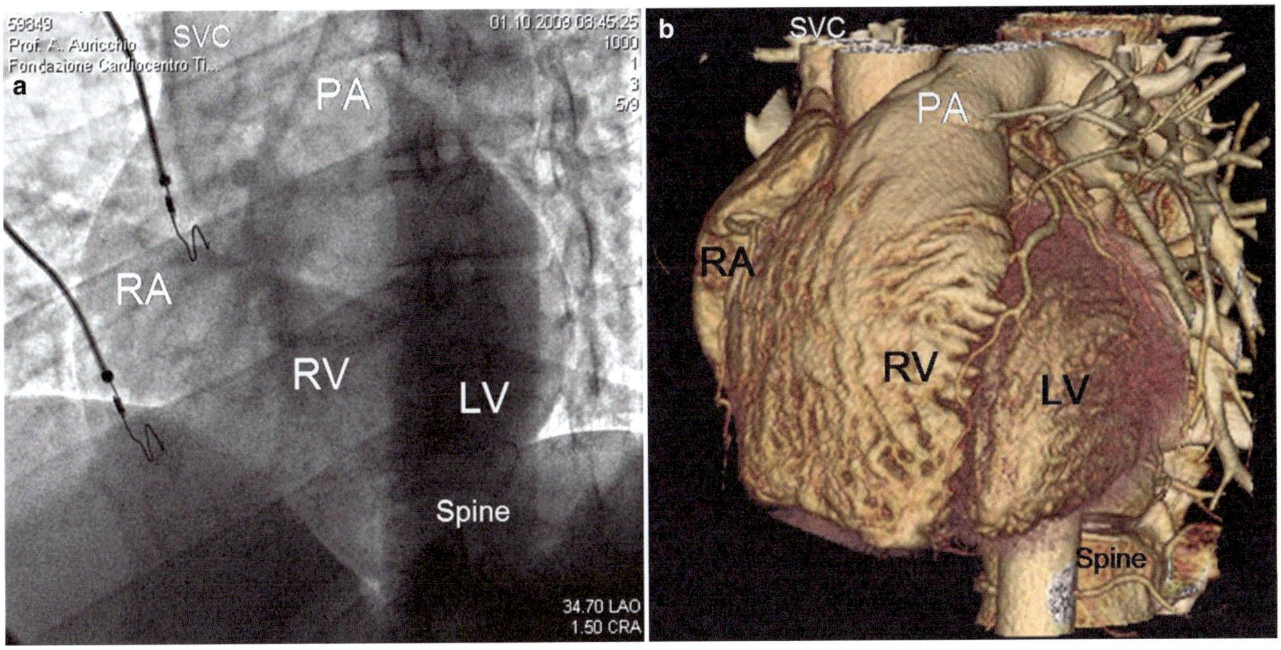

Fig. 1.3 (**a**) Fluoroscopic image in left anterior oblique projection and (**b**) the corresponding 3D CT volume rendering. *RA* Right atrium, *RV* right ventricle, *PA* pulmonary artery, *SVC* superior vena cava

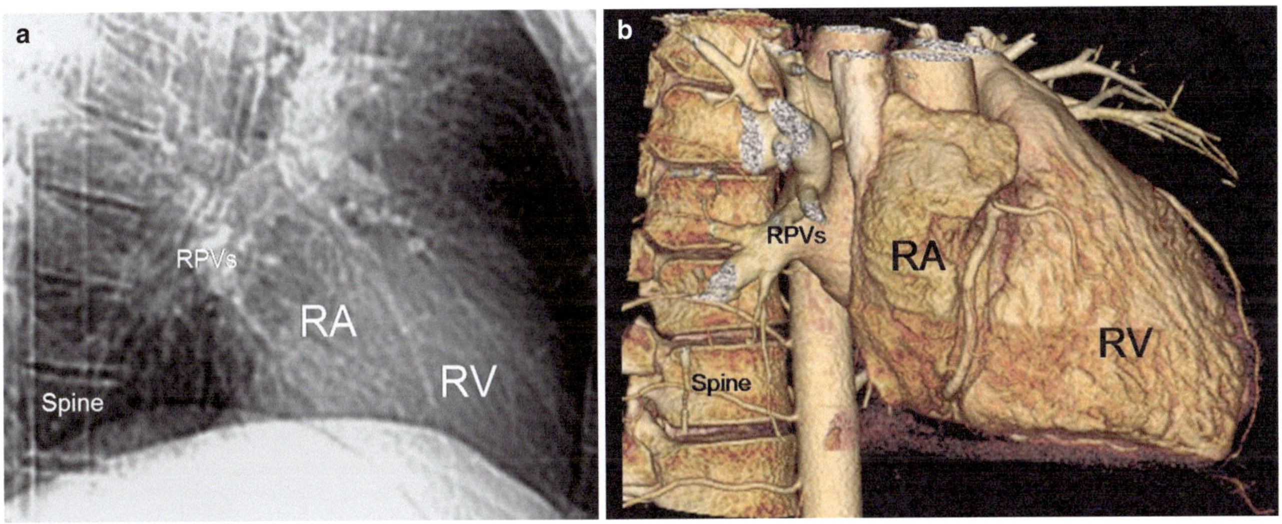

Fig. 1.4 (**a**) Fluoroscopic image in right lateral projection and (**b**) the corresponding 3D CT volume rendering. *RA* Right atrium, *RV* right ventricle, *RPVs* right pulmonary veins

discussing and illustrating these landmarks in terms of their localization, their relationships with other structures, and their anatomic variants. Images shown herein are obtained with CT, RT 3D TEE, and MRI.

1.1 The Right Atrium

The right atrium consists of four components: the venous component (sinus venous), the right atrial appendage, the vestibulum, and the atrial septum which is shared by the two atria.[2]

1.1.1 The Sinus Venosus

The sinus receives the superior and the inferior vena cavae and forms the posterior and the rightward part of the atrium. It is characterized by a smooth wall. The internal anatomical features of the sinus venosus can be appreciated by 3D volume rendering obtained with CT, while the internal features can be best appreciated by RT 3D TEE (Figs. 1.7 and 1.8). The superior vena cava opens to the top of the venous component looking toward the tricuspid orifice, while the inferior vena cava opens to the bottom of the sinus and looks toward the fossa ovalis (Fig. 1.9).

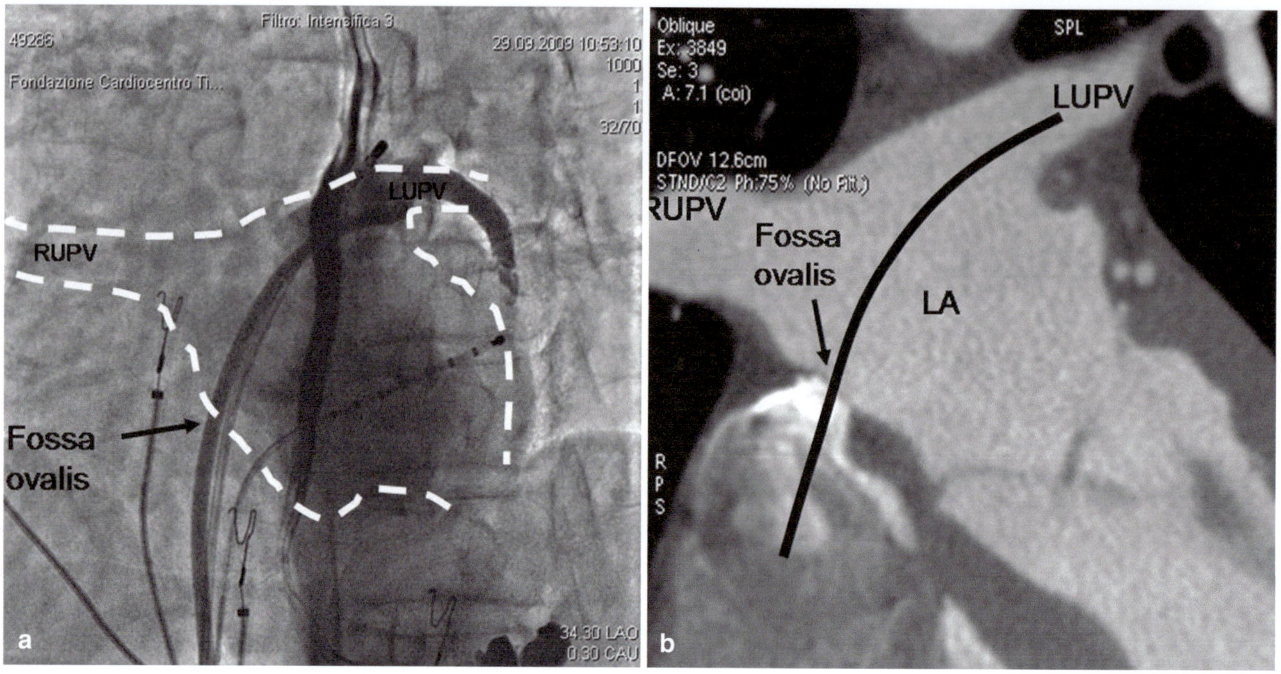

Fig. 1.5 (**a**) Fluoroscopic image in LAO projection of the heart and (**b**) the corresponding MSCT slice showing the catheter (represented by a *black line*) in left upper pulmonary vein (*LUPV*) after having crossed the fossa ovalis. In fluoroscopy, the fossa ovalis is not visible. *RUPV* right upper pulmonary vein

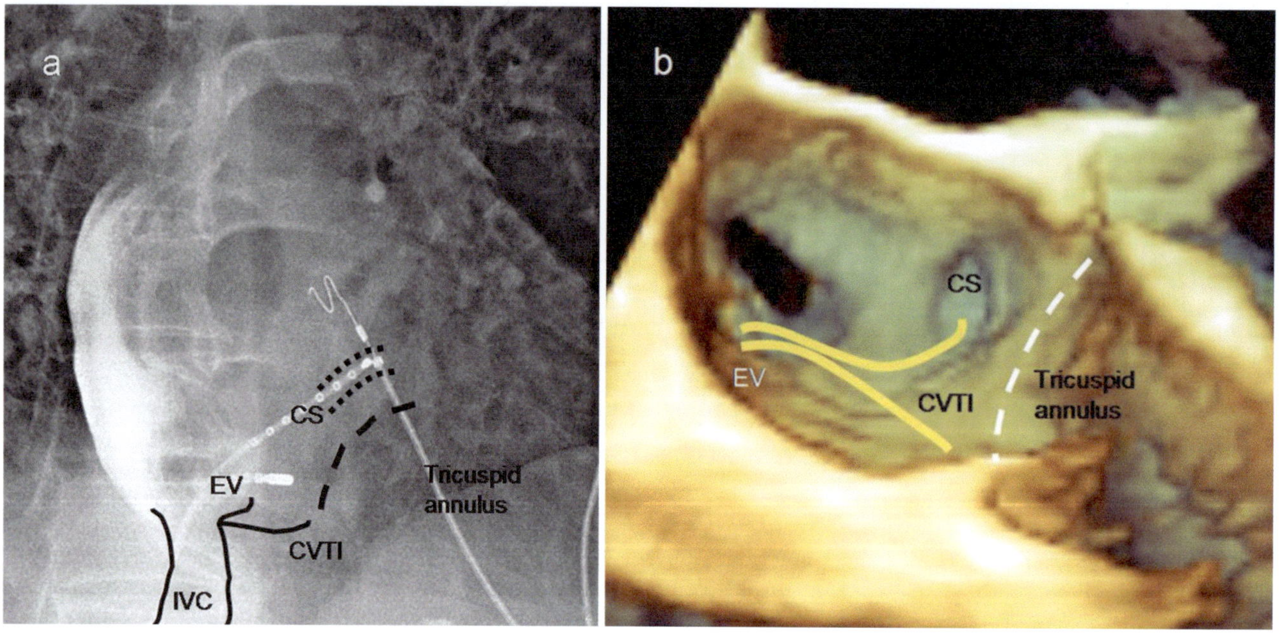

Fig. 1.6 (**a**) Fluoroscopic image in right oblique view showing one catheter in coronary sinus (*CS*) and one lying on the cavo-tricuspid isthmus (*CVTI*). (**b**) RT 3D TEE image in the same orientation. Catheters are represented as *yellow lines*. Fluoroscopic image does not allow to recognition of structures such as the cavo-tricuspid isthmus (*CVTI*), the Eustachian valve (*EV*), or the coronary sinus (*CS*)

1.1.2 The Right Atrial Appendage

The right atrial appendage (RAA) is roughly triangular and has a broad junction with the atrial chamber. Antero-medially the RAA protrudes from the right atrium and overlaps the aortic root forming nearly the entirety of the anterior wall of the atrium. An excellent external appearance of RAA can be obtained by CT using 3D volume rendering modality (Fig. 1.10) or MRI. The internal appearance of the appendage can be shown either by RT 3D TEE or by CT with the

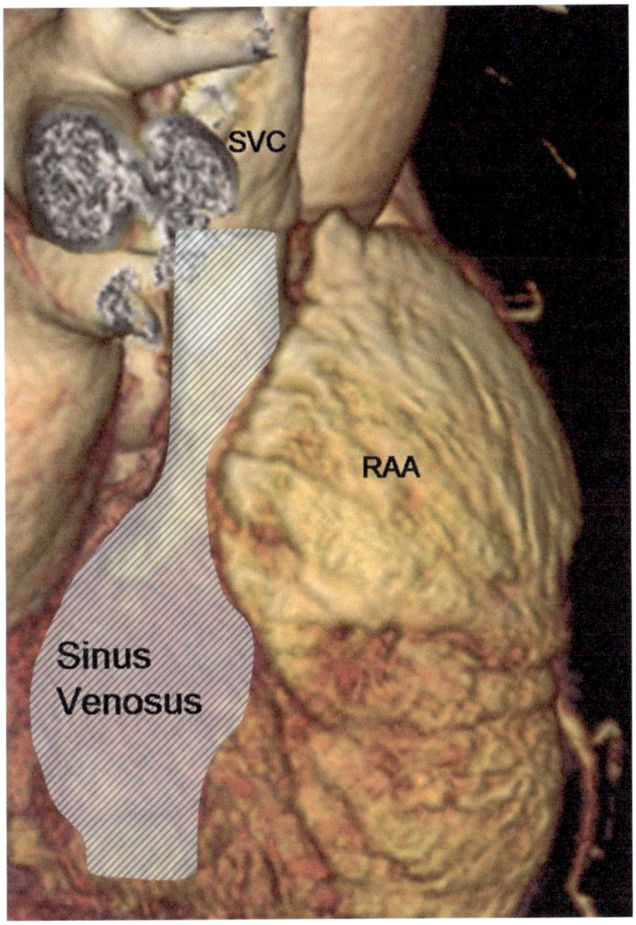

Fig. 1.7 3D image obtained with MSCT from a posterior perspective showing the anatomic boundaries of the sinus venous of the right atrium (*shaded area*)

"virtual endoscopy" modality (Fig. 1.11). The most peculiar aspect of the RAA is the fact that it is lined with pectinate muscles emerging in branch fashion from the crista terminalis and terminating at the vestibule (see also the paragraph of the crista terminalis). In between the ridges of pectinate muscles, the wall is very thin and almost parchment-like. Although extensively arranged, the pectinate muscles never reach the orifice of the tricuspid valve. Several different patterns of arrangement of pectinate muscles have been described: quite often pectinate muscles are oriented perpendicularly to the crista with uniform spacing and lack of crossover, less frequently they originate from a common solitary trunk in an harborizing fashion, or have a haphazard, trabecular arrangement with multiple crossovers.[3]

1.1.3 The Vestibule

The vestibule is a smooth muscular rim surrounding the orifice of the tricuspid valve. The vestibule forms the leftward margin of the right atrium, its musculature inserting into the leaflets of the tricuspid valve.

1.1.4 Anatomical Landmarks Relevant to Electrophysiologists

Two anatomical landmarks of the right atrium are relevant for electrophysiologists being target for catheter-based ablation for many tachyarrhythmias: the *crista terminalis* and the *cavo-tricuspid isthmus*.

The crista terminalis. An external fat-filled groove, the sulcus terminalis, divides the venous part of the atrium from the right atrial appendage (the true primitive atrium) (Fig. 1.12). The sulcus terminalis corresponds internally to the *crista terminalis*. The crista is a roughly C-shaped muscular band that separates the smooth wall of the venous component from the rough wall of the appendage.[2,3] Planar images of the crista terminalis can be obtained with CT slices (Fig. 1.13a) while RT 3D TEE and CT in virtual endoscopy modality provide excellent images of these structures in 3D format (Fig. 1.11–1.13b). Superiorly, the crista terminalis arches anterior to the orifice of the superior vena cava, extends to the area of the anterior interatrial groove, and merges with the interatrial bundle (Bachmann bundle). Inferiorly, it ends near the orifice of the inferior vena cava, merging with the fine trabeculations of the cavo-tricuspid isthmus. An extensive array of pectinate muscles arise from the crista spreading throughout the entire wall of the appendage (Fig. 1.11). The morphology of the crista with the corresponding pectinate muscles is subject to a wide range of variability. The crista may vary in size and thickness appearing as a small, thin valve-like or a broad-based structure. A very large crista terminalis mimicking a mass can be found in the so-called lipomatous hypertrophy of atrial septum and it is caused by an extensive fatty infiltration of terminalis sulcus. The nonuniform architecture of myofibers at the junctions of the crista with pectinate muscle is most likely the cause of the marked anisotropy creating the substrate for reentry. Indeed, more than two-thirds of right-sided atrial tachycardias originate from the crista terminalis.[4] In addition, the crista terminalis is a natural posterior barrier to transverse conduction of typical atrial flutter as evidenced by widely spaced double potentials that can be recorded along its length.[5,6]

Cavo-tricuspid isthmus The CVTI is a roughly quadrilateral-shaped endocardial surface bordered by the tricuspid hingeline (annulus) anteriorly and by the Eustachian valve and the Eustachian ridge posteriorly. The superior (medial) border is lined by the inferior border of coronary sinus ostium. Finally, the inferior-lateral border is lined by the final ramification of the crista terminalis.[7] Planar images of this structure can be obtained by angiography,[8] CT[9,10] (Fig. 1.14) and MRI.[11,12] Three-dimensional images can be obtained either by RT 3D TEE or CT in virtual endoscopy modality (Fig. 1.15). CTVI is a critical component of slow conduction of counterclockwise and clockwise macro-reentrant atrial flutter, and it is the target of ablation.[13]

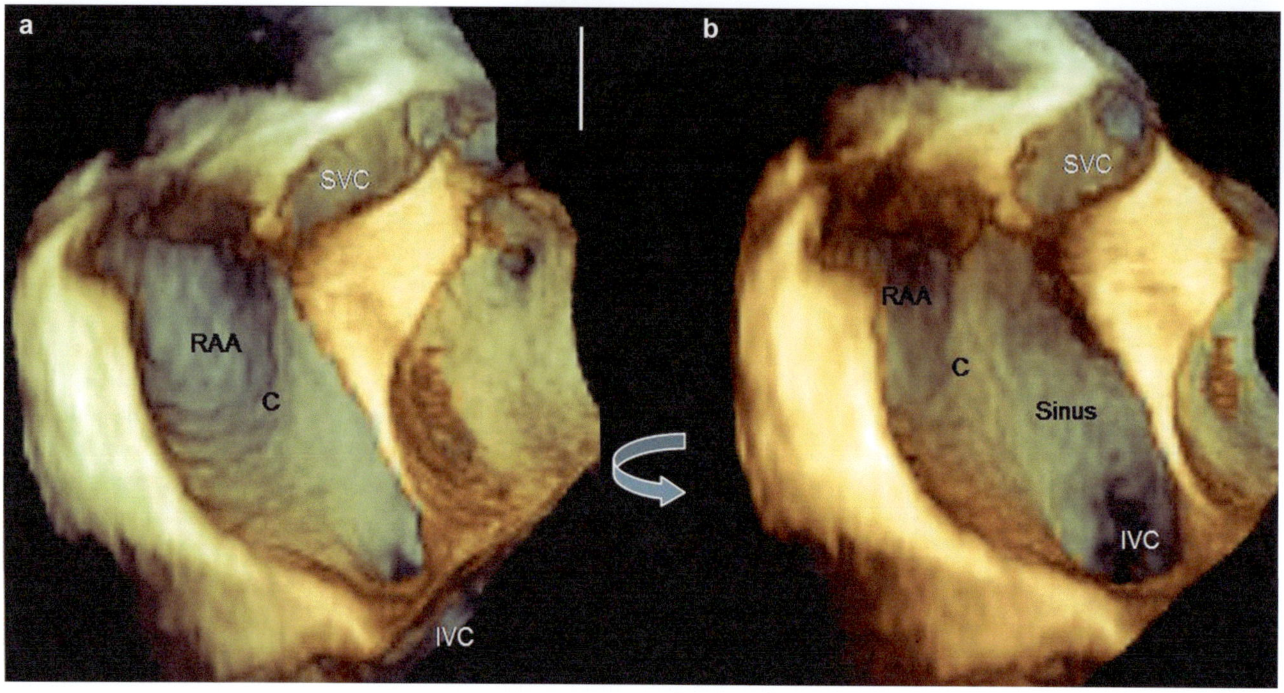

Fig. 1.8 RT 3D TEE images of the internal aspect of the right atrium seen from (**a**) right perspective. The *curved arrow* indicates the counterclockwise rotation of image to reveal (**b**) the sinus. The crista terminalis (*C*) separates the smooth wall of the sinus from the rough wall of the right atrial appendage (*RAA*). *SVC* superior vena cava, *IVC* inferior vena cava

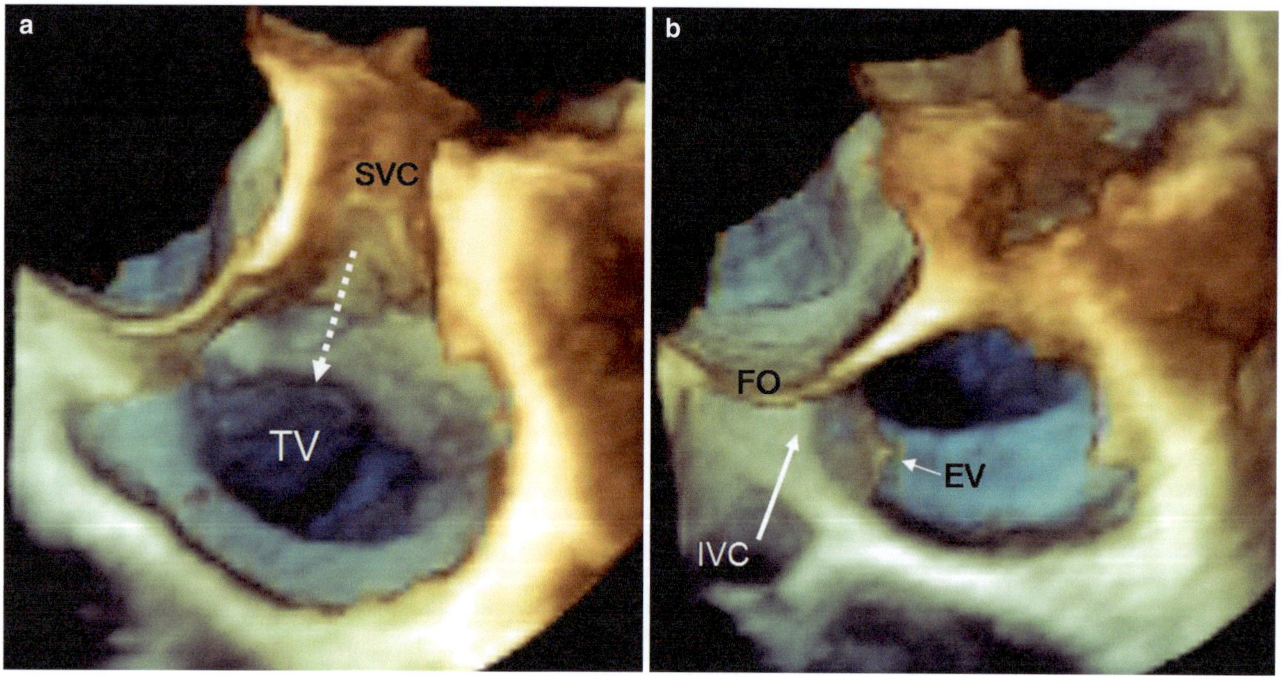

Fig. 1.9 (**a**) RT 3D TEE image of the superior vena cava (*SVC*) looking toward the tricuspid valve (*TV*) and (**b**) of the inferior vena cava (*IVC*) looking toward the fossa ovalis (*FO*). The Eustachian valve (*EV*) guards the entrance of IVC. During the fetal life, the EV directs the blood into the left atrium throughout an open fossa ovalis

Right atrial conventional angiography and autopsy reports reveal a high variability of isthmian anatomy. Recognizing these anatomic variants and, consequently, adapting ablation approaches to them may contribute to successful procedures. *A deep sub-Eustachian pouch* is one of the main sources of procedural difficulty.[14] A suboptimal

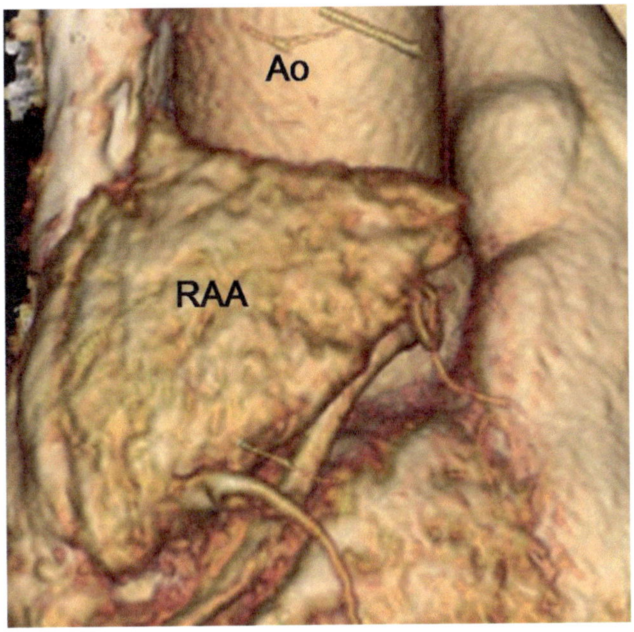

Fig. 1.10 3D volume rendering obtained by CT. Anterior perspective of the right atrium showing the right atrial appendage (*RAA*) overlapping the Aorta (*Ao*)

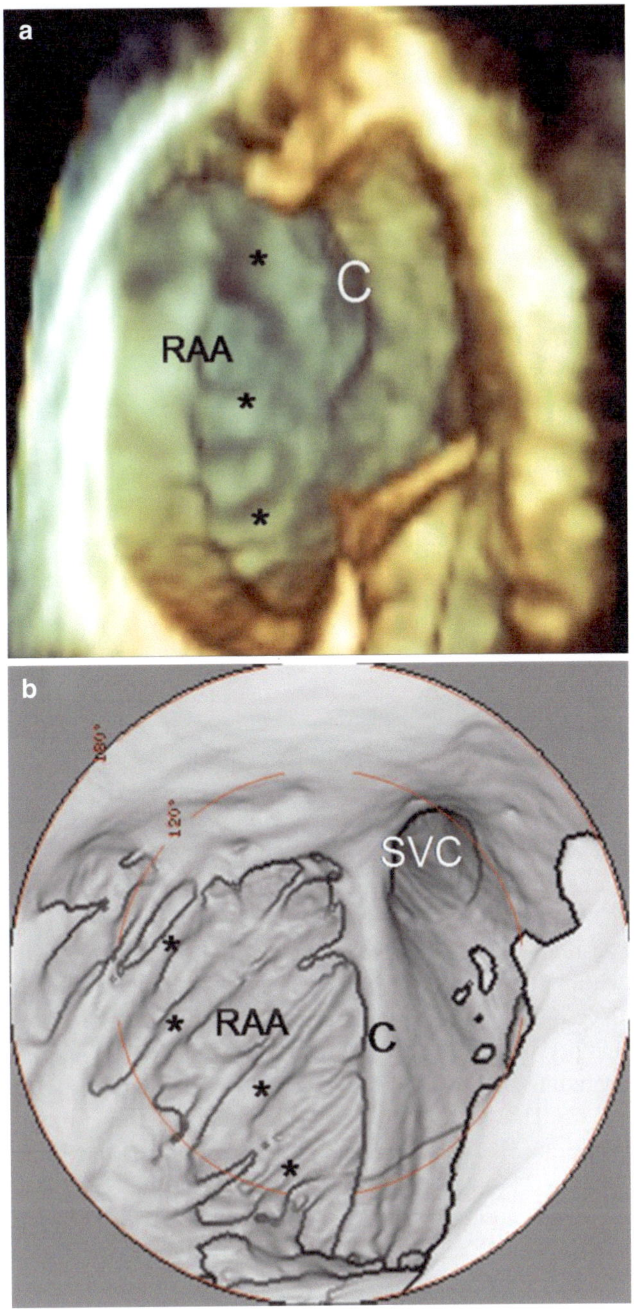

Fig. 1.11 (**a**) RT 3D TEE and (**b**) MSCT "virtual endoscopy" images of the right atrial appendage. Both images show pectinate muscles (*asterisks*) lining the wall of the right atrial appendage (*RAA*). *SVC* superior vena cava. *C* crista terminalis

power delivery due to poor blood flow within the pouch and a difficult catheter manipulation may explain gaps in the line of ablation when drawing back the catheter from the tricuspid valve to the inferior vena cava and inadequate ablation.[15] Currently fluoroscopy, CT, and MRI may discover this anatomical variant (Fig. 1.16). In 3D format, a deep pouch can be appreciated by RT 3D TEE (Fig. 1.17). *Occasionally pectinate muscles may encroach* onto the cavo-tricuspid isthmus rendering a transmural ablation lesion difficult.[15] The presence of anatomically distinct pectinate muscles traversing the isthmus can be recognized with intracardiac ultrasound or RT 3D TEE (Fig. 1.18). *The Eustachian valve* guarding the entrance of the inferior caval vein is usually a triangular flap of fibrous or fibrous-muscular tissue that inserts medially to the border of the fossa ovalis and the coronary sinus (Eustachian Ridge or sinus septum). Both Eustachian valve and ridge can be well imaged by 3D TEE (Fig. 1.19). Occasionally the valve is perforated, or even takes form of a delicate filigreed mesh. When it is extensive, it is described as Chiari's network. However, in some cases, the valve is particularly large and muscular, posing an obstacle to the catheter passed from the inferior vena cava to the CVTI[15] (Fig. 1.20).

1.1.4.1 The Coronary Sinus

With the establishment of cardiac resynchronization therapy (CRT) as a non-pharmacological treatment modality for heart failure patients, there is a great deal of interest in characterizing coronary sinus anatomy. This includes identifying the location, presence, and anatomic characteristics of venous side branches (first- and second-order branches) especially their take-off, anatomical course, and relationship with adjacent anatomical structures (myocardial scar, phrenic nerve, and diaphragm). Moreover, for the novice or low-volume operators, cannulation of the CS is very challenging. Anatomy of coronary sinus and its side branches can be evaluated at the time of CRT implantation by retro-

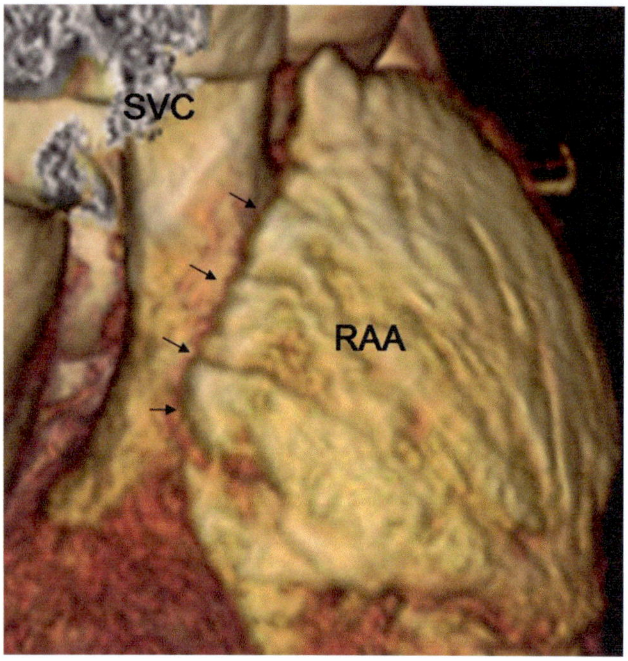

Fig. 1.12 3D volume rendering obtained by CT. Postero-lateral perspective of the right atrium showing the superior vena cava (*SVC*), the sulcus terminalis (*arrows*), and the right atrial appendage (*RAA*)

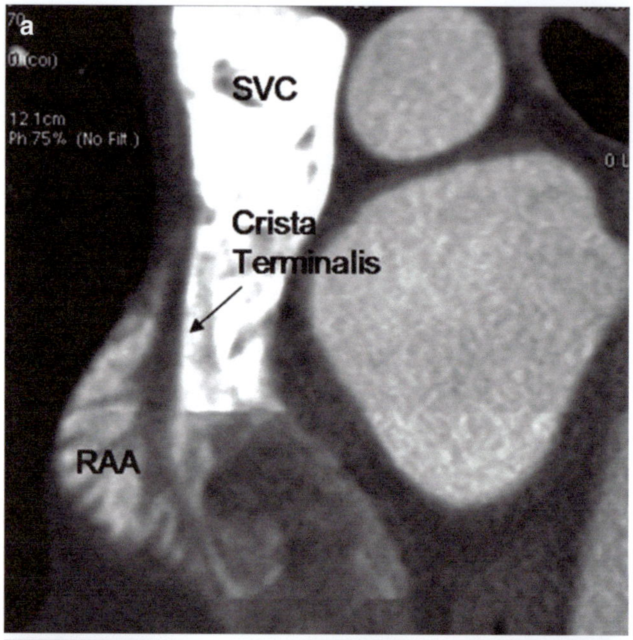

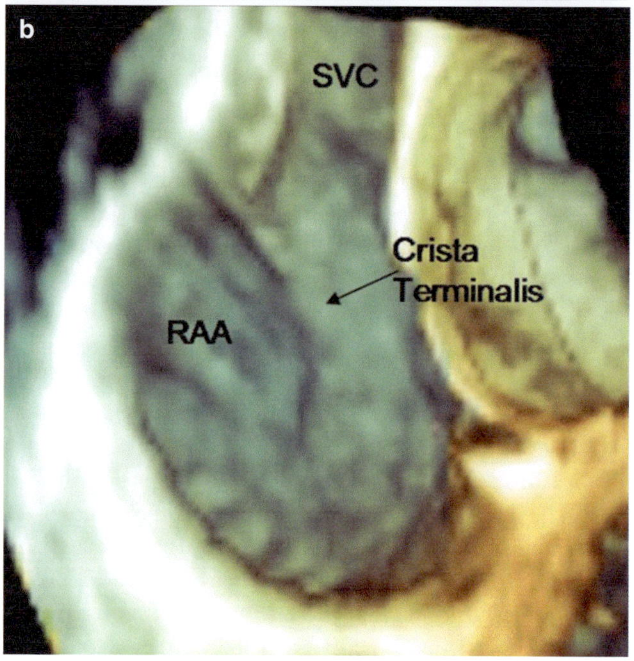

Fig. 1.13 (**a**) Slice image obtained with MSCT and (**b**) 3D image obtained with RT 3D TEE of the right atrium showing the crista terminalis. *SVC* superior vena cava, *RAA* right atrial appendage

grade venography (Fig. 1.21). However, because of the extreme variability of coronary venous anatomy, pre-procedural assessment of coronary sinus and target veins (instead of "on the table" assessment) may become an important pre-requisite before proceeding to CRT. Finally, ablation within the coronary sinus may be indicated in the treatment of ectopic atrial tachycardias, atrioventricular bypass tracts, and persistent AF. Several noninvasive imaging techniques, including MRI, CT, and echocardiography, can visualize the CS. However, for imaging the entire course of coronary sinus and its side branches both in planar image or in 3D format, CT is probably the best imaging technique.

In the normal heart, the CS is a wide venous channel often situated slightly proximal to the atrioventricular groove and usually runs along the inferior wall of the left atrium rather than in the atrioventricular groove. It is surrounded to a greater or lesser extent by a muscular coat that often has continuity with left atrial myocardium.[16] The tributaries are the great, small and middle cardiac veins, the posterior vein of the left ventricle, and the oblique vein of the left atrium (Marshall's vein), all of which except the vein of Marshall may have valves at their ostia.[17] In patients with depressed left ventricular function, there may be dilatation of the CS. The most significant dilatation occurs in patients with a persistent left superior vena cava (SVC), which most often courses between the LAA and the left PV and then drains into the CS. Therefore, if a pre-procedure echocardiogram reveals a dilated CS ostium, the presence of a persistent left SVC, should be suspected. This finding is of particular importance as it increases the technical complexity of lead placement for an intracardiac device. The presence of a coronary sinus diverticulum, although uncommon, is most often reported in association with posteroseptal accessory pathways[18] and is usually located at the junction of the middle cardiac vein[19] Of note, a CS diverticulum should be distinguished from a subthebesian pouch which is located in the right atrium, below the CS ostium. Anatomic observations have shown several degrees of elevation and variable "arched" courses of coronary sinus (Fig. 1.22). A "normal" position of the coronary

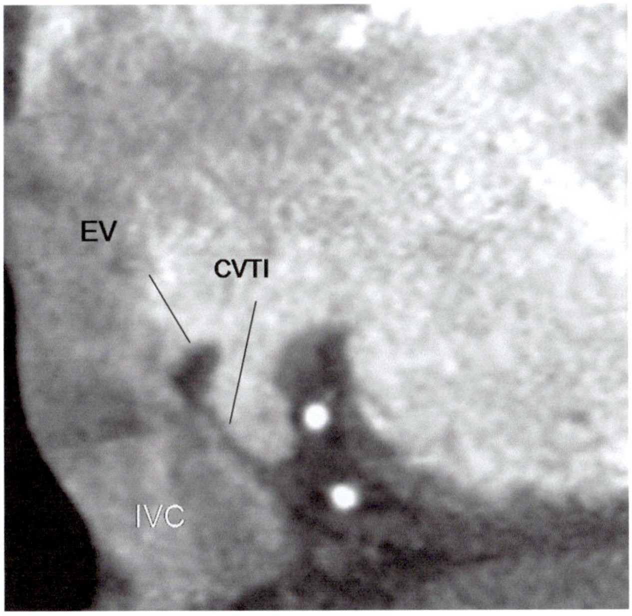

Fig. 1.14 Slice image in a simulated right oblique anterior view showing the cavo-tricuspid isthmus (*CVTI*).*IVC* inferior vena cava, *EV* Eustachian valve

sinus in the left posterior atrioventricular groove can be found in only 16% of cases, a slight elevation (1–3 mm) in 12%, a moderate elevation (4–7 mm) in 50%, and an extreme elevation (8–15 mm) in 22%. This fact is well known by electrophysiologists because of the large near-field atrial signal noted on catheters placed within the coronary sinus. This is due to embryological reasons since both the postero-inferior left atrium and coronary sinus derive from the left horn of the sinus venous. Viewed from above, the coronary sinus describes a gentle curve in the left inferior coronary sulcus (or, when it is elevated as previously described) on the inferior left atrial wall. Recently, we observed that CS is shifted toward the posterior part of the MVA when the left cardiac chambers and mitral annulus dilate[20] (Fig. 1.23). The average length of coronary sinus is 40–45 mm and the average diameter is of 10 mm. With regard to CRT, several variations in coronary venous anatomy exist, which can render an endocardial approach of the coronary sinus difficult or impossible or can lead to suboptimal resynchronization. Specific examples include small or absent lateral branches, veins with acute branch angles, compression of the coronary sinus against the spine because of cardiomegaly, prominent Thebesian valves blocking the coronary sinus ostium, coronary sinus atresia, and persistent left superior vena cava (Fig. 1.24). The coronary sinus ostium opens in the right atrium between the opening of the inferior vena cava and the tricuspid orifice at the inferior border of the triangle of Koch. In nearly 80% of cases, its orifice is guarded by a semilunar valve, the valve of the coronary sinus (valve of Thebesius). Thebesian valve can be appreciated either with CT or with RT 3D TEE (Figs. 1.25 and 1.26).

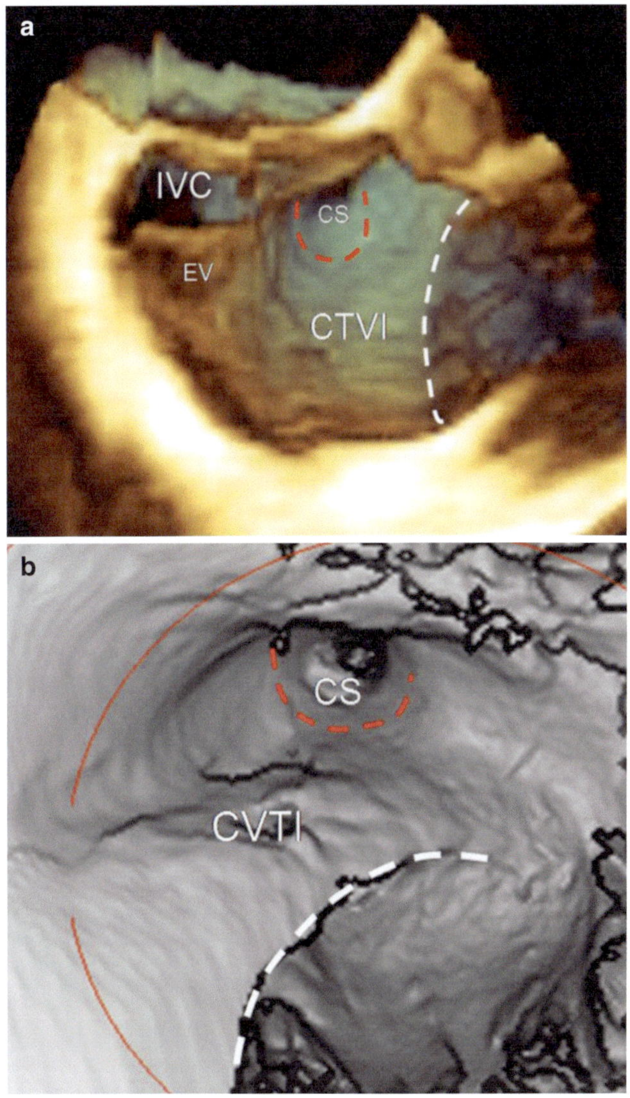

Fig. 1.15 (**a**) RT 3D TEE showing the cavo-tricuspid isthmus (*CTVI*) bordered posteriorly by the inferior vena cava (*IVC*) and Eustachian valve (*EV*), anteriorly by the hinge of tricuspid valve (*white dotted line*), and medially by the inferior border of the ostium of coronary sinus (*CS*) (*red dotted line*). (**b**) Similar images obtained with MSCT in virtual endoscopy modality

The proximity of coronary sinus to the circumflex artery and its side branches may be relevant when ablation is performed in coronary sinus. The left circumflex artery may be acutely occluded during ablation within the coronary sinus.[21] Prior to the application of radiofrequency energy, the electrophysiologist should therefore consider the relationship of the coronary sinus to the left circumflex artery. Reduced power settings have been proposed for ablation within the coronary sinus in order to decrease the risk of such complications. Thus a pre-procedural assessment of these relationships may have important implications for safety.

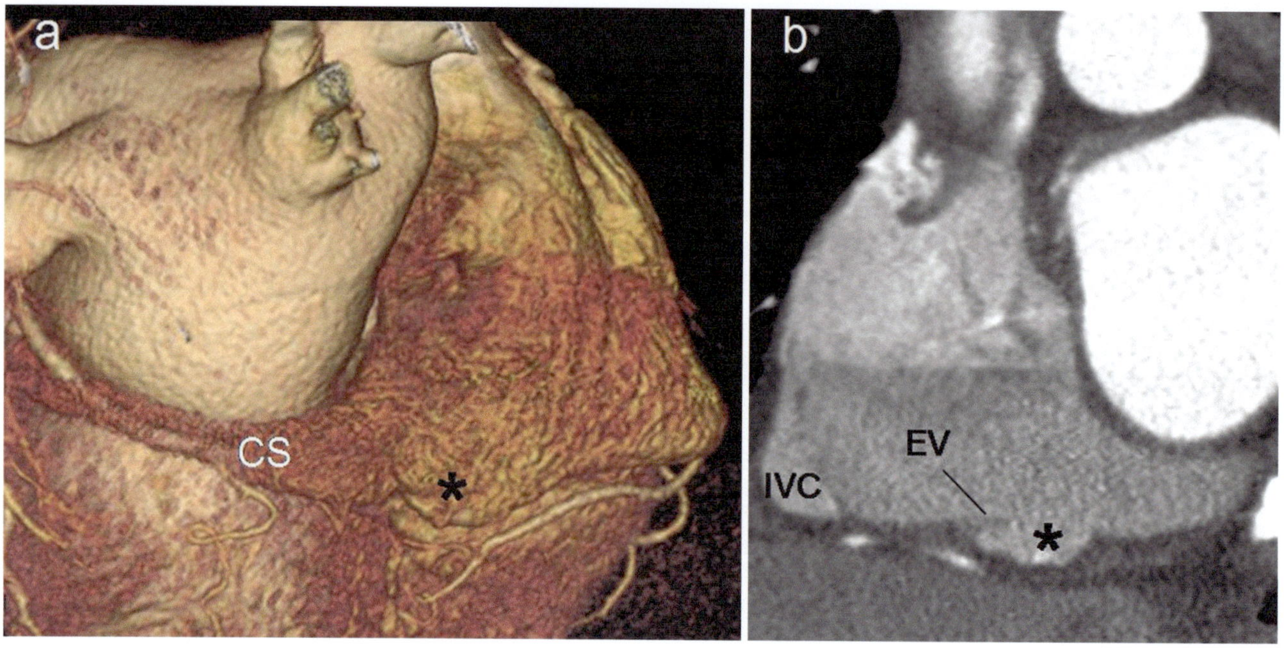

Fig. 1.16 (**a**) MSCT 3D volume rendering and (**b**) MSCT slice obtained at level of coronary sinus (*CS*) showing a relatively large sub-Eustachian pouch (*asterisk*). *IVC* Inferior vena cava. *CS* coronary sinus

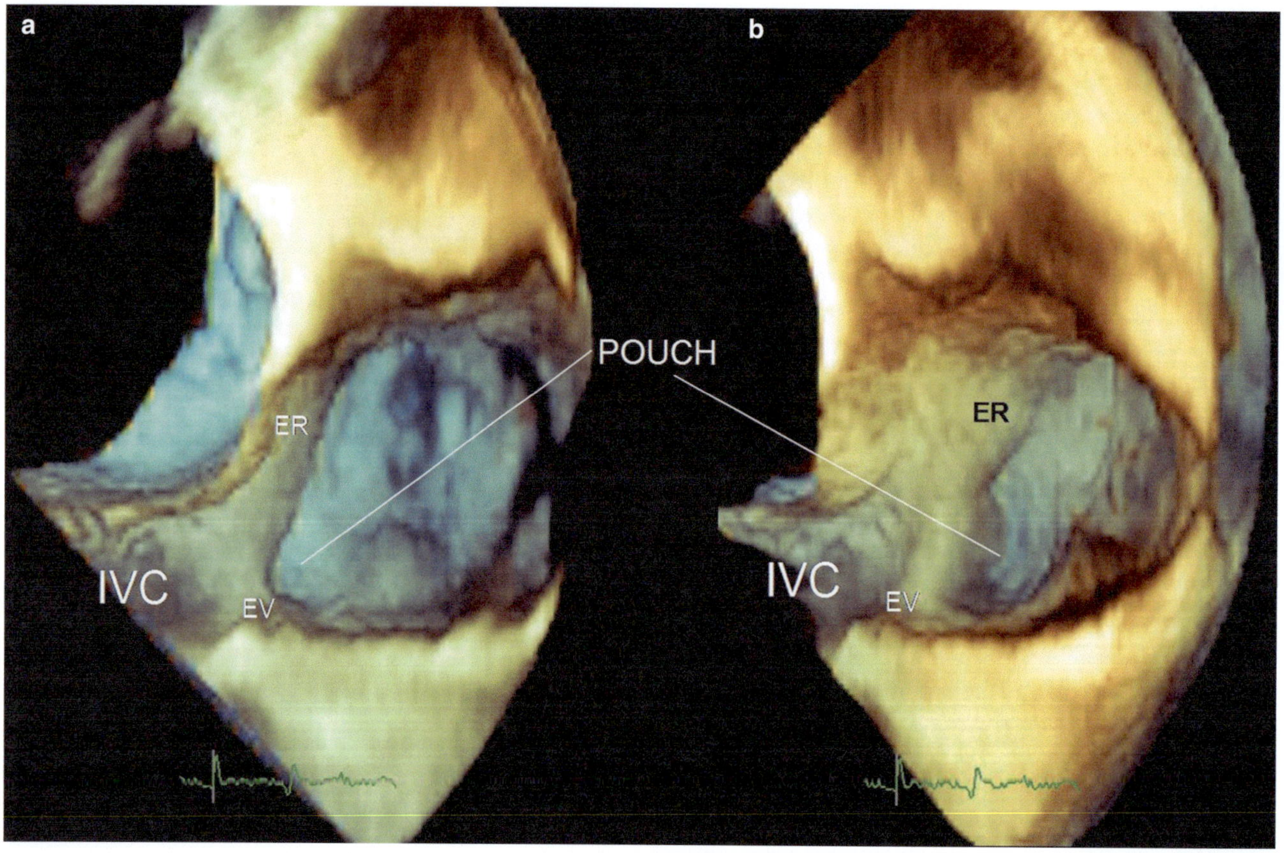

Fig. 1.17 Images RT 3D TEE with two slightly different perspectives (**a** and **b**) showing a deep sub-Eustachian pouch. *IVC* inferior vena cava, *EV* Eustachian valve, *ER* Eustachian ridge

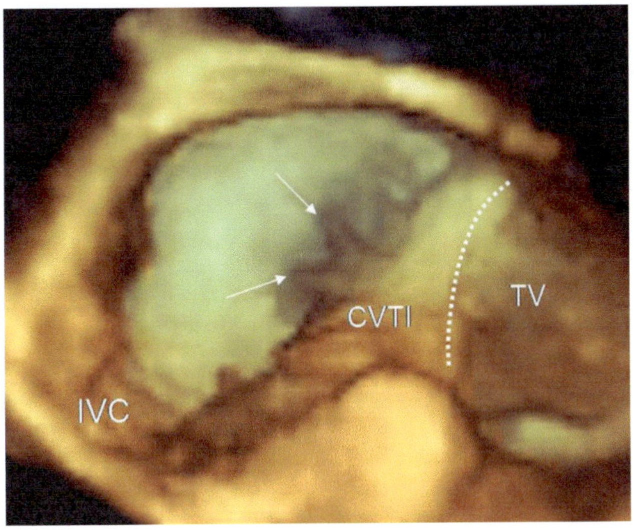

Fig. 1.18 RT 3D TEE image showing pectinate muscle (*arrows*) traversing the cavo-tricuspid isthmus (*CVTI*). The *dotted line* marks the tricuspid (*TV*) annulus. *IVC* inferior vena cava

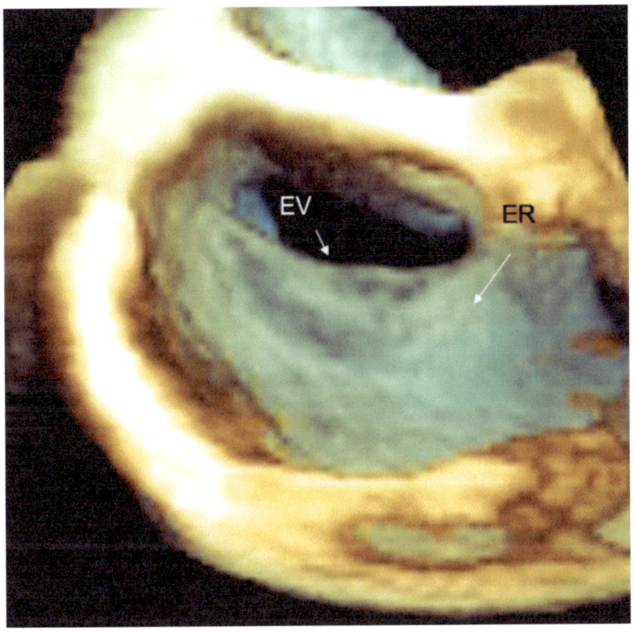

Fig. 1.19 Images RT 3D TEE from a frontal perspective showing a thin semilunar-shaped Eustachian valve (*EV*). Medially the EV is inserted along the margin of the Eustachian ridge (*ER*). The tendon of Todaro courses deep into the crest on the ridge. The ridge itself is an elevation of cavo-tricuspid isthmus and borders the fossa ovalis inferiorly and the ostium of coronary sinus superiorly

Left phrenic nerve stimulation after CRT is a well-recognized complication.[22] Usually the left phrenic nerve descends along the fibrous pericardium in close relation to the lateral coronary veins. Given the anatomic variability of lateral coronary veins and the proximity of the phrenic nerve to them, it is important to understand their relationships. CT has the potential to detect the phrenic nerve as it passes near the lateral veins[23] (Fig. 1.27).

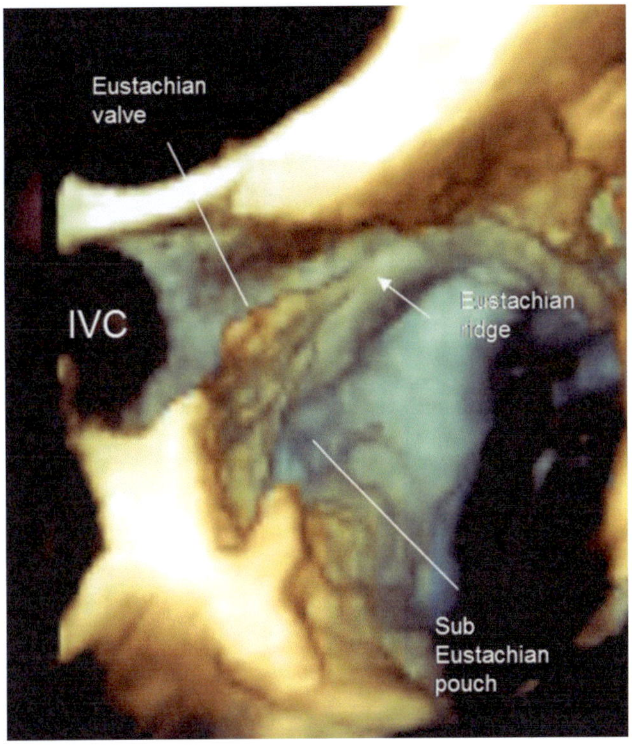

Fig. 1.20 RT 3D TEE mimics the right anterior oblique projection showing a prominent Eustachian valve and a sub-Eustachian pouch

1.1.4.2 The Atrial Septum

The atrial septum can be defined as the medial wall of the atria that can be removed without exposing the atrial cavities to the extracardiac structures. Thus the "true" septum is limited to the flap valve of the fossa ovalis (septum primum) and part of its anterior and inferior muscular rim. The flap of the foramen ovale closes against the atrial septum, with fusion usually occurring within the first 2 years of life. Fusion is incomplete in about 25% of the population, resulting in probe patent defect, or patent foramen ovale.[24] Because of this arrangement, the right side of the fossa ovalis appears as a crater-like structure; on the contrary, the left side is rather indistinguishable from the parietal atrial wall.[2] These anatomic features can be easily imaged by RT 3D TEE which provides an excellent "en face" view of the septal surfaces both from left and right perspective (Fig. 1.28). The extensive muscular surface seen antero-medially relative to the fossa ovalis is not septum, being the wall of the atrium lying immediately behind the aorta and the vena cava (Fig. 1.29). The superior rim of the fossa (septum secundum) is the infolded wall between the superior vena cava and the right upper pulmonary vein. The planar image with CT scan (or MRI) shows this area filled by fat (Fig. 1.30a). In 3D format, this area can be hardly appreciated by 3D TEE as fat because of the low difference in acoustic impedance between fat and wall (Fig. 1.30b). In 3D CT volume rendering, the area appears

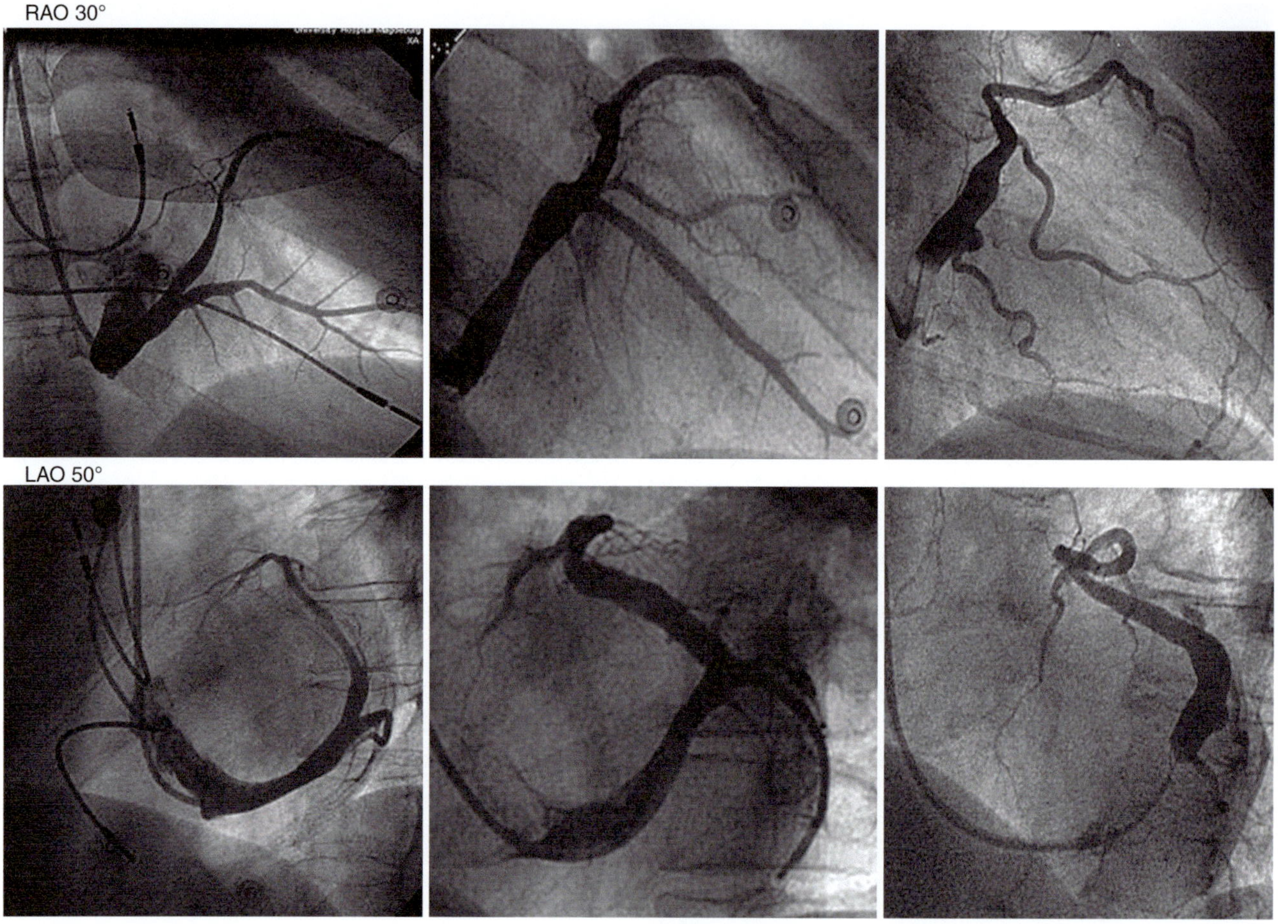

Fig. 1.21 Multiple projections of retrograde coronary venous angiography showing the anatomy of coronary sinus and its side branches

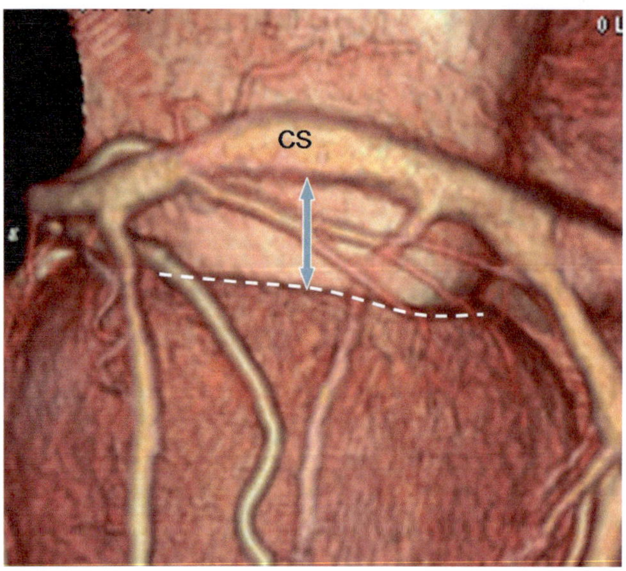

Fig. 1.22 3D MSCT volume rendering showing as the course of coronary sinus (*CS*) in normal heart does not usually lie on the atrioventricular groove (*dotted line*)

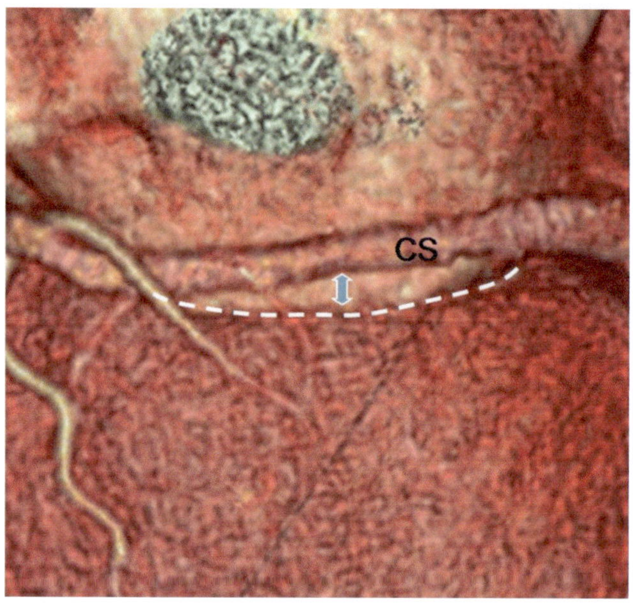

Fig. 1.23 3D MSCT volume rendering showing the course of coronary sinus (*CS*) in patients with dilated left atrium, left ventricle, and mitral annulus. In comparison with Fig. 1.25, the coronary sinus (*CS*) lies closer to the atrioventricular groove (*dotted line*)

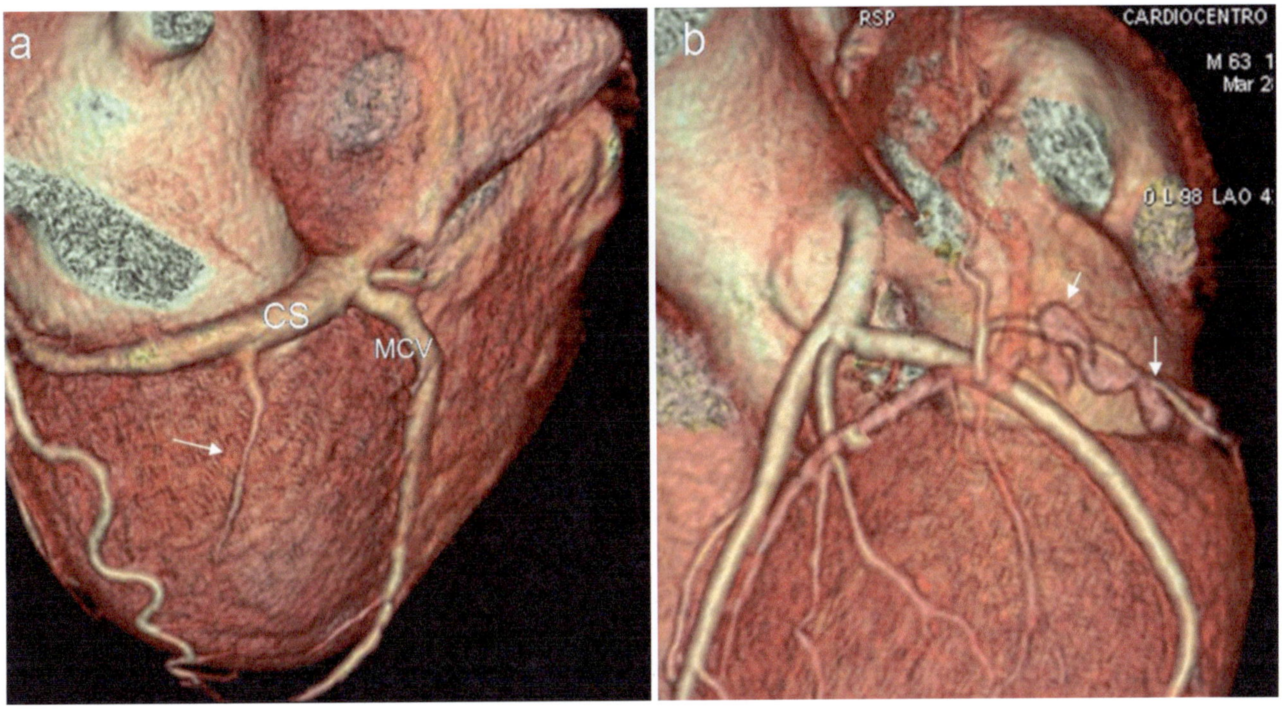

Fig. 1.24 3D MSCT volume rendering showing (**a**) a small lateral vein (*arrow*) and (**b**) a diverticular arrangement of coronary sinus (*arrows*)

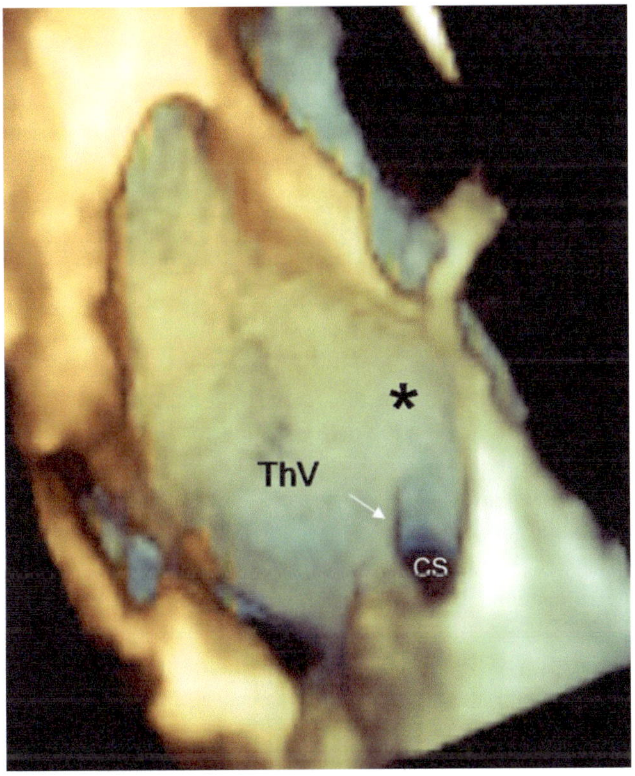

Fig. 1.25 RT 3D TEE showing the coronary sinus ostium (*CS*) and the Thebesian valve (*ThV*). The *asterisk* marks the point where it is supposed to be the atrioventricular node

unfilled (Fig. 1.31). The membranous fossa ovalis is the portion of the interatrial septum which is of great interest to electrophysiologists, because it is the ideal site for transseptal puncture. Transseptal puncture is an important technique in the field of electrophysiology as a variety of tachycardias such as atrial fibrillation (AF), atrial tachycardias, and left sided atrioventricular accessory pathways are ablated in the left atrium. Transseptal access to the left atrium allows for superior catheter manipulation as compared to a retrograde aortic approach to the left atrium. In addition, the percutaneous placement of left atrial appendage occlusion devices, which are currently under study, is delivered via a transseptal approach.

1.2 The Left Atrium

As with the right atrium, the left atrium possesses a venous component, a vestibule a prominent body, and an appendage. Most of the left atrium (i.e., the vestibule, the posterior and superior wall that make the body of the atrium, the pulmonary component, and the atrial septum) is smooth walled.[25] Characteristically, the left atrial appendage (LAA) appears as a small finger-like extension from the atrial chamber. Pectinate muscles are exclusively confined to the LAA. Because of its narrow junction with the rest of the atrium and the complex network formed by pectinate muscles, the left atrial appendage is the potential site for deposition of thrombi.

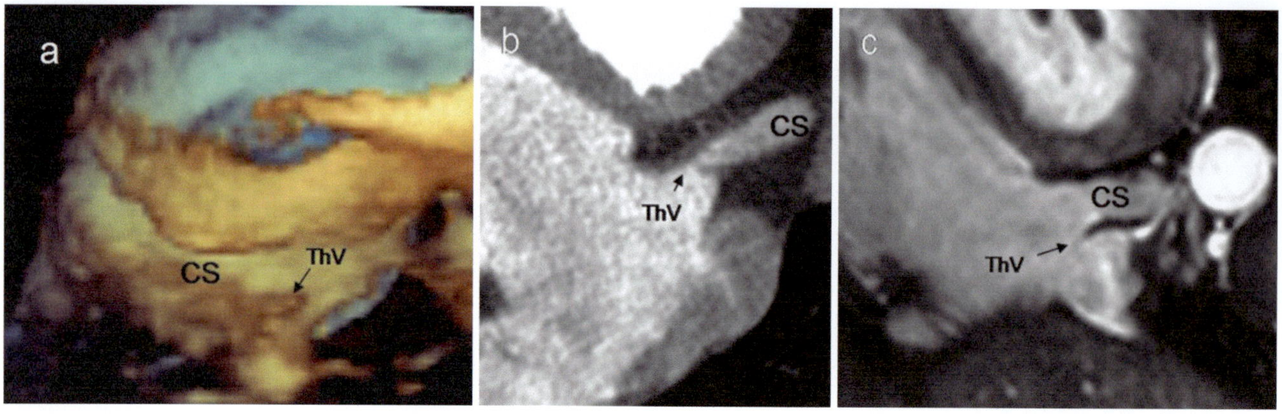

Fig. 1.26 (**a**) RT 3D TEE image, (**b**) CT, and (**c**) MRI slices showing the coronary sinus (*CS*) and the Thebesian valve (*ThV*)

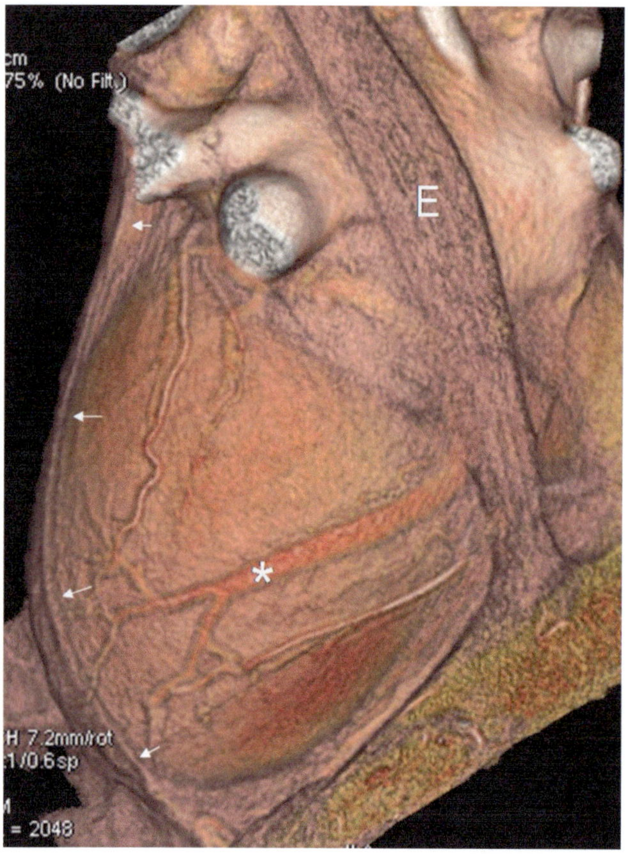

Fig. 1.27 3D volume rendering image by MSCT showing the course of Phrenic nerve (*arrows*) and its close proximity to the lateral vein (*asterisk*). *E* esophagus

1.2.1 The Vestibule

The vestibule surrounds the mitral valve orifice and has no anatomical distinctive characteristics. Part of the vestibule is the left atrial isthmus which is the area between the left lower pulmonary vein and the mitral hinge line. This area is one of the targets for a linear ablation in atrial fibrillation. The length of the left atrial isthmus is highly variable ranging from 17 to 51 mm.[26] Real-time 3D TEE, as well as CT and MRI, can visualize the left atrial isthmus (Fig. 1.32). One of the complications of ablation in this area is injury to the adjacent vessels, including the circumflex artery. 3D volume rendering by CT can demonstrate the relationships between the coronary sinus, circumflex artery, and atrial wall, thereby providing pre-procedural data for a safer approach in this type of ablation (Fig. 1.33)

1.2.1.1 Pulmonary Veins

There are sleeves of myocardial tissue with varying orientations which reflect back into the pulmonary veins (PVs).[25–30] This atrial myocardium is found between the adventitia and the media of the venous wall and may possess electrical ectopic activity that acts as a trigger of atrial fibrillation. These structures are therefore the targets of PV isolation ablation.[31,32] Regardless the strategy used to achieve PV isolation, knowledge of the PV anatomy and of the exact location of the junction of the left atrium and PVs are important guides to the electrophysiologist, providing pre-procedural anatomic details noninvasively. With CT and MRI, information regarding the size, number, location, and anatomic variants of the PVs can be easily obtained before performing the procedure.[33]

The most common pattern of the entry of pulmonary veins into the left atrium is two veins from the hilum of each lung. The superior veins project in a forward and upward fashion; whereas, the inferior veins project in a backward and downward fashion.[34] The external view of pulmonary veins is best obtained by CT and MRI (Fig. 1.34).[35,36] The internal aspect (the ostia) of PVs can be imaged in 3D format within virtual endoscopy modality (Fig. 1.35). Because of its limited angle of scanning, 3D TEE cannot visualize the entire roof with all four pulmonary veins; however, it can provide high quality images focused on one or two pulmonary veins (Fig. 1.36).

Anatomic studies have reported significant variability in dimension, shape, and branching patterns of the PVs. The most frequent anatomical variations are: (1) the early

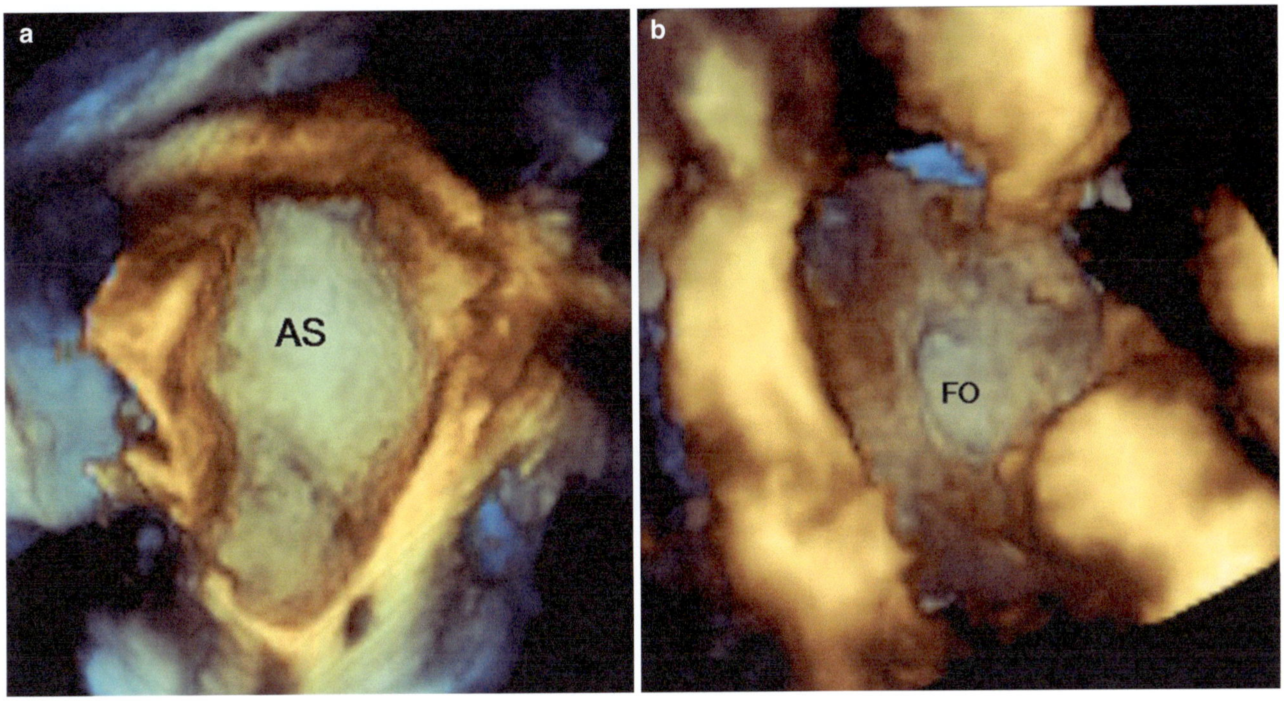

Fig. 1.28 (**a**) RT 3D TEE showing the atrial septum from left and (**b**) right perspective. In (**a**) the septum is rather featureless while (**b**) from right-hand perspective, the position of the fossa ovalis (*FO*) is clearly visible. The extensive muscular surface seen antero-medially relative to the fossa ovalis is not septum, being the wall of the atrium lying immediately behind the aorta and the vena cava

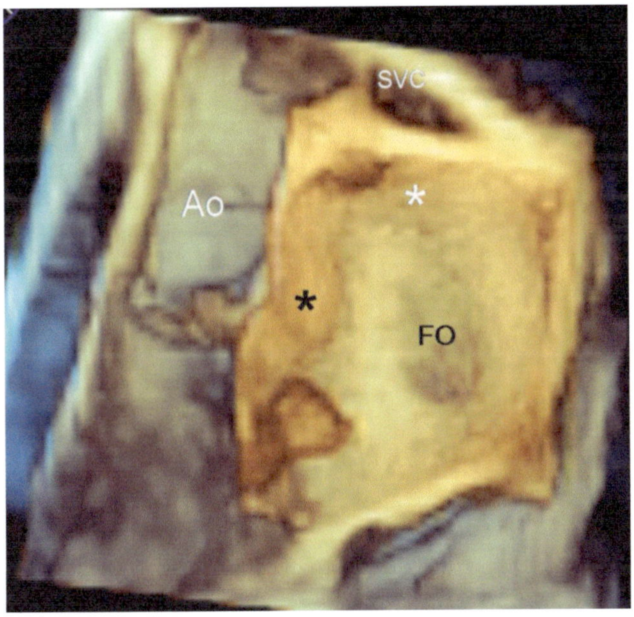

Fig. 1.29 Atrial septum from left perspective. The antero-medial wall relative to the fossa is not septum but the atrial wall behind the aorta (*black asterisk*) and superior vena cava (*white asterisk*). *Ao* aorta, *SVC* superior vena cava, *FO* fossa ovalis

branching of right lower pulmonary vein (70–90% of the cases); (2) the presence of a short or long common left trunk (up to 35% of cases); (3) right supranumerary veins (20–30% of the cases) (Figs. 1.37 and 1.38). In some patients, the supranumerary pulmonary veins insert with a perpendicular orientation to the posterior wall (Fig. 1.37c, d).[25,37]

The pulmonary vein ostia are usually oval. Pulmonary vein diameters and area can be obtained with MSCT, RT 3D TEE, and MRI. Usually the pulmonary veins ostia are larger (19–20 mm) than the inferior (16–17 mm). Similarly the superior pulmonary veins tend to have a longer trunk (i.e., the distance from the ostium to the first-order branches) than the inferior (22 ± 7 mm versus 14 ± 6 mm).[35]

1.2.2 Anatomical Relationship Between Pulmonary Veins and Adjacent Structures

The right superior PV lies just behind the superior vena cava. The relationship between superior vena cava and right upper pulmonary vein can be demonstrated by RT 3D TEE from an internal perspective (Fig. 1.39). An excellent external view of the relationships between the superior vena cava and the pulmonary veins can be obtained by CT or MRI in 3D format (Fig. 1.40). Left PVs are positioned between the left atrial appendage (LAA) and descending aorta. The external view of this anatomical relationship can be imaged by 3D CT volume rendering (Fig. 1.41). The upper left pulmonary vein is separated from the left appendage by a fold in the atrial wall. The infolded atrial wall appears like a ridge in the endocardial surface. The width of this ridge is not uniform being narrower

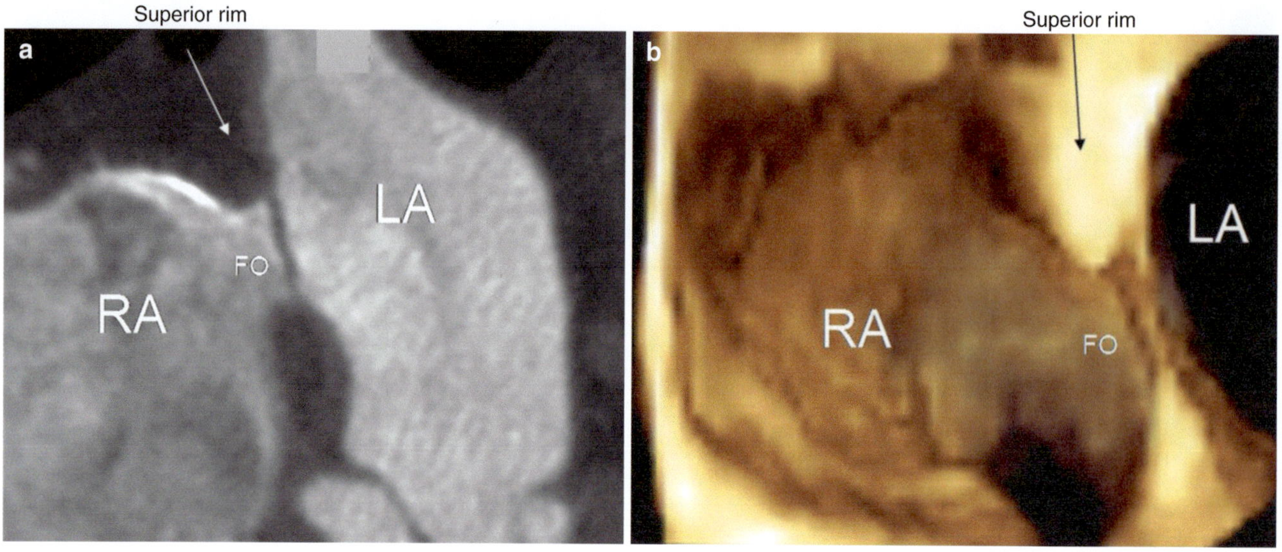

Fig. 1.30 (**a**) MSCT slice and (**b**) magnified image of the atrial septum by RT 3D TEE showing how the true septum is formed by the valve of foramen ovalis (*FO*) and its immediate muscular rim. The *arrows* mark the infolded atrial wall filled by fat

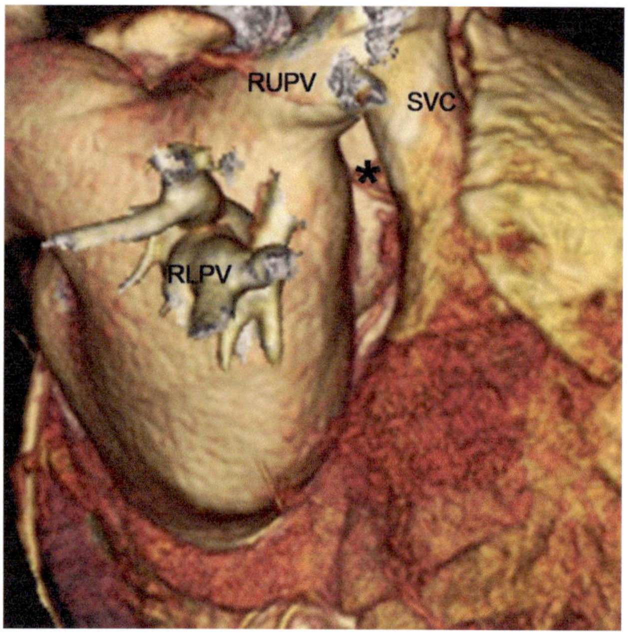

Fig. 1.31 3D volume rendering by CT. The *asterisk* indicates the infolding between the superior vena cava (*SVC*) and the right upper pulmonary artery (*RUPV*). *RLPV* right lower pulmonary vein

at the antero-superior level. It ranges from 3 to 6 mm at its narrowest point at the level of left upper and left lower pulmonary vein, respectively.[37] Its shape and size are of relevance during catheter ablation of atrial fibrillation when encircling the orifices of the left PVs or during ablation of extrapulmonary vein triggers arising around or inside the LAA. It is also the site that is most challenging for accomplishing electrical isolation of the left pulmonary veins. An external perspective of the infolded atrial wall can be obtained by 3D volume rendering CT or MRI, while an internal perspective of the LAA and the ridge can be obtained by CT "virtual endoscopy" (Fig. 1.35) or by RT 3D TEE (Fig. 1.42).

1.2.3 Left Atrial Appendage

Characteristically the left atrial appendage (LAA) is a finger-like extension from the atrial chamber and usually points anteriorly and superiorly overlying the left ventricular wall. Its ostium (i.e., the orifice opening into the atrium) lies immediately above the level of the atrioventricular groove.[27] Anatomically the LAA is likely the most variable structure of the entire heart with one, two, or more lobes spreading in different directions (Fig. 1.43). Unlike the RAA, the LAA has a narrow junction with the venous component of the atrium and the junction with the left atrium in not externally marked either internally or externally by a crest or groove. As in the right appendage, the LAA is lined by small ridges (pectinate muscles). Planar imaging of the internal appearance of the left atrial appendage can be obtained by CT and MRI slices (Fig. 1.44). The internal surface of the appendage can be appreciated in 3D configuration by CT in virtual endoscopy modality or by 3D TEE (Fig. 1.45).

1.2.4 Anatomic Relations Between Left Atrium and Esophagus

A critical clinical issue is the relation of the posterior wall of the left atrium and the esophagus because of a possible rare, but dreadful complication of catheter ablation, namely, atrio-esophageal fistula.[28] Understanding the anatomic

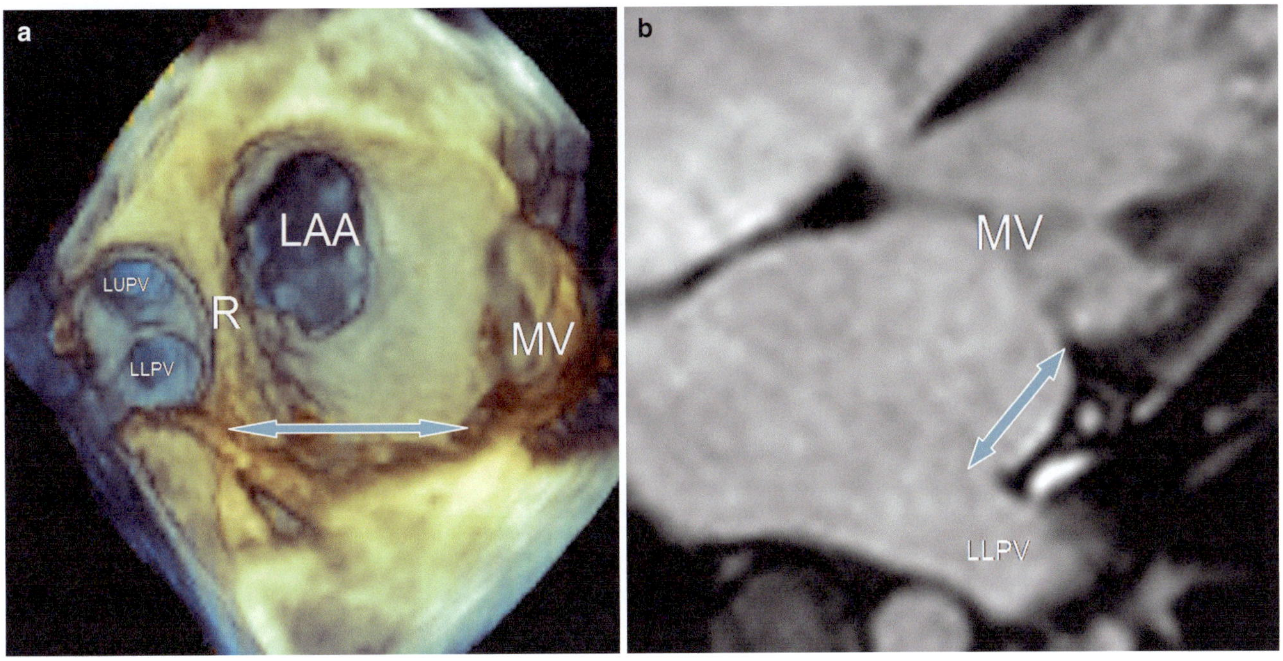

Fig. 1.32 (**a**) RT 3D TEE images showing the left upper (*LUPV*) and lower (*LLPV*) pulmonary veins, the ridge (*R*) between pulmonary veins and left atrial appendage (*LAA*), and the mitral valve (*MV*). The *double-headed arrow* marks the area between LLPV and the postero-inferior hinge line of the mitral valve (*left atrial isthmus*). Pre-procedure imaging may help to evaluate length and morphological variant of this region. (**b**) Slice image by MRI. The *double-headed arrow* marks the left atrial isthmus

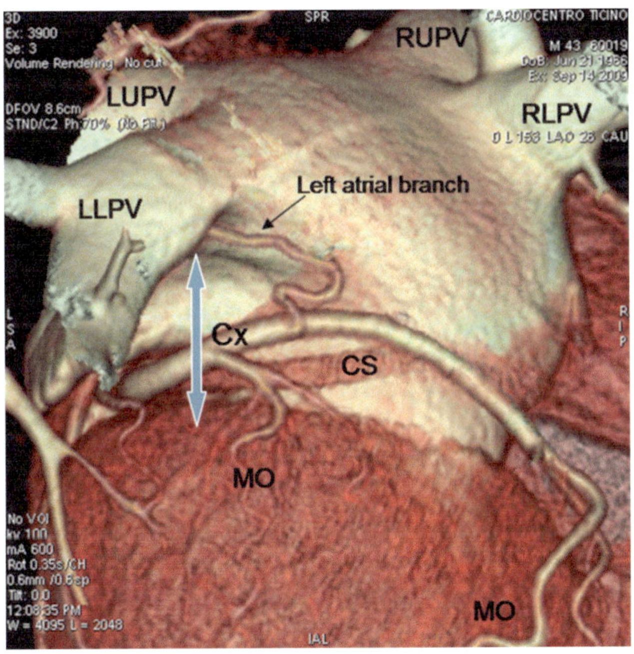

Fig. 1.33 CT volume rendering showing the external aspect of the *left isthmus*. The relationship between the ablation line and the adjacent vessels is clearly demonstrated. *LUPV* left upper pulmonary vein, *LLPV* left lower pulmonary vein, *RUPV* right upper pulmonary vein, *RLPV* right lower pulmonary vein, *CS* coronary sinus, *MO* marginal obtuse branch, *Cx* circumflex artery

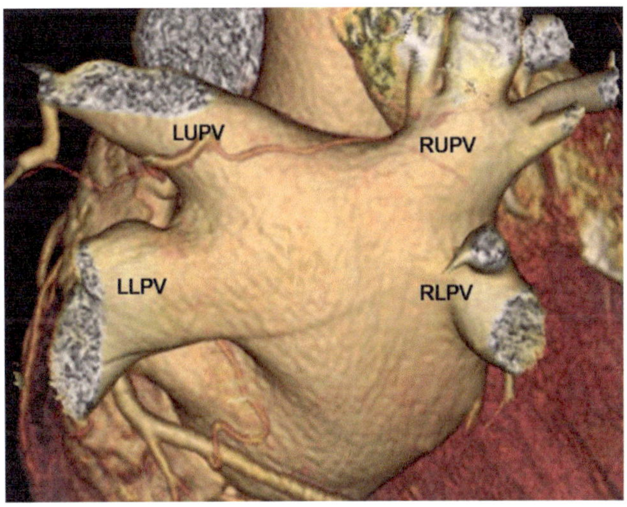

Fig. 1.34 3D CT volume rendering showing four pulmonary veins entering into the left atrium. *LUPV* Left upper pulmonary vein, *LLPV* left lower pulmonary vein, *RUPV* right upper pulmonary vein, *RLPV* right lower pulmonary vein

relationship between the esophagus and pulmonary veins/left atrium provides useful information for avoiding esophageal injury during catheter ablation.[29] The anterior aspect of the esophagus is always in direct contact with the posterior wall of the LA. In general, the esophagus lies closer to

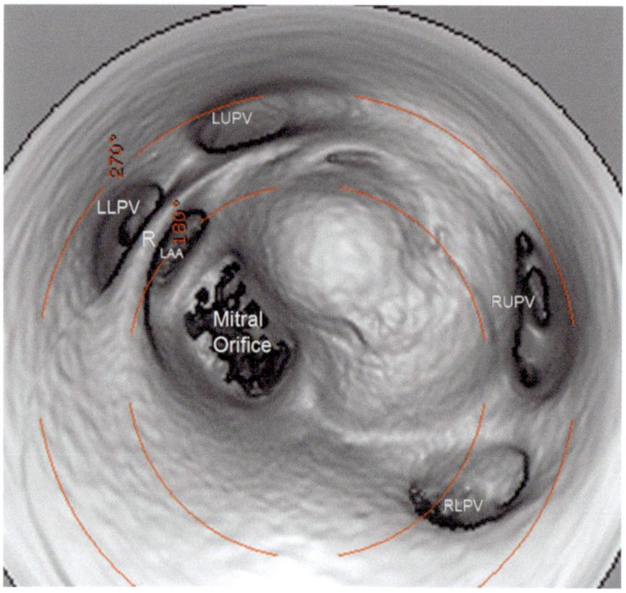

Fig. 1.35 CT virtual endoscopy modality showing a comprehensive view of the roof of the *left atrium*. The left upper (*LUPV*) and lower (*LLPV*) pulmonary veins, the ridge (*R*) between left pulmonary veins and left atrial appendage (*LAA*), the mitral orifice and the right upper (*RUPV*) and right lower (*RLPV*) are imaged in one image. Interestingly, because this image is obtained by making transparent the contrast and opaque the wall, mitral leaflets cannot be seen. The distortion of the image depends on the specific software used in this modality. Rather than being parallel, the projected light rays are focused to converge on the viewpoint. The resulting distortion of the anatomical structures allows the perception of the distance not on the basis of color (or gray scale as in 3D echocardiography), but on the basis of the object size (simulating the natural light convergence of human retina). Thus, objects near the viewpoint appear larger, whereas objects farther away appear smaller

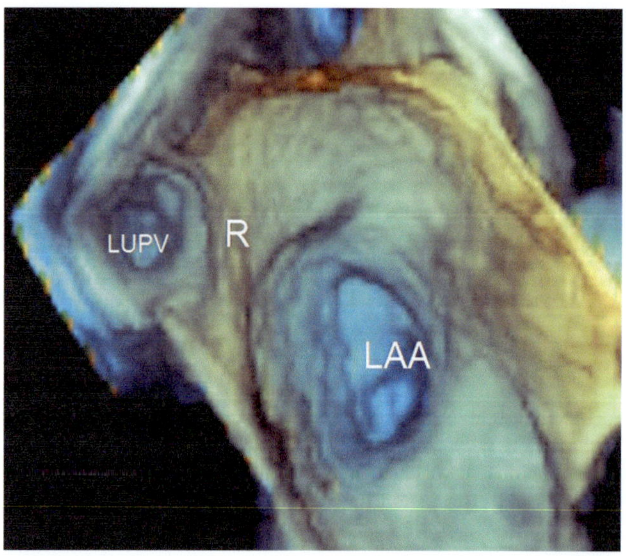

Fig. 1.36 RT 3D TEE showing the left atrial appendage (*LAA*) and the left upper pulmonary vein (*LUPV*) separated by the ridge (*R*)

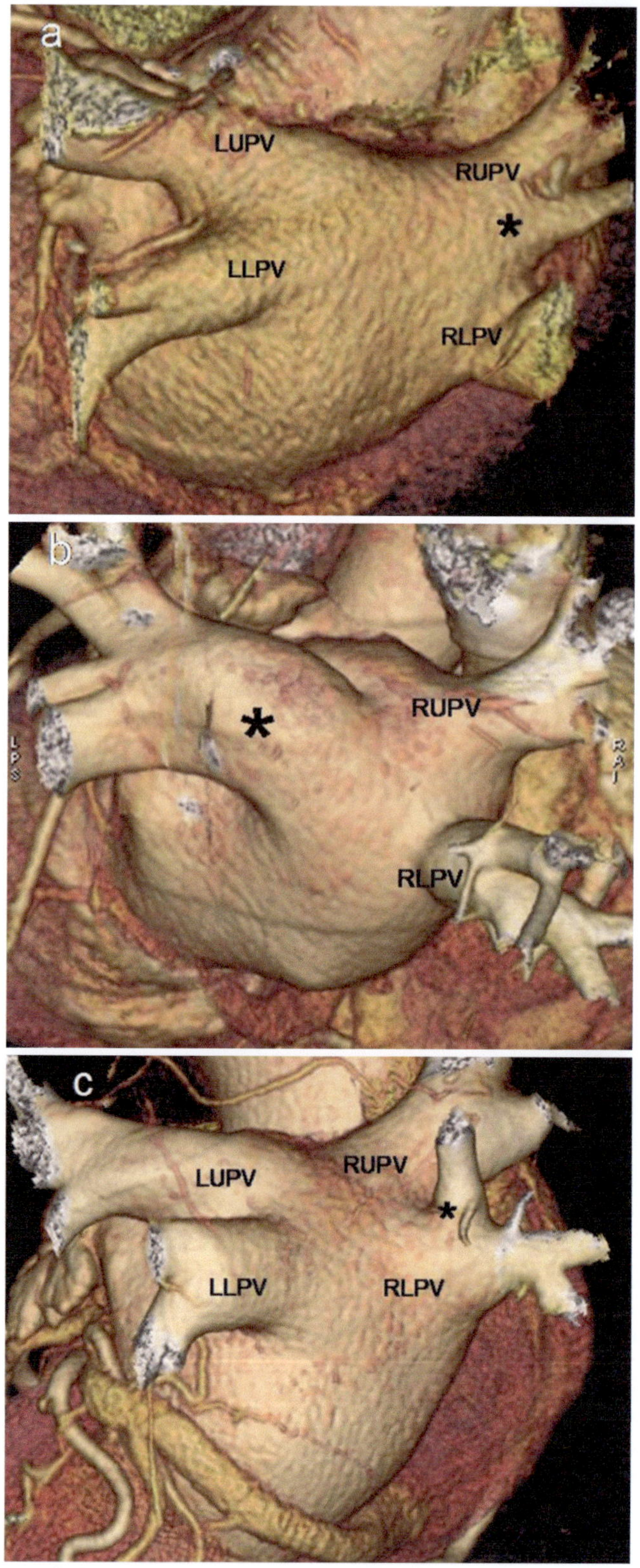

Fig. 1.37 3D volume rendering image obtained by CT showing several different anatomical patterns of pulmonary vein entry into the left atrium. (**a**) Early branching of right upper pulmonary vein (*asterisk*), (**b**) a common left trunk (*asterisk*), (**c**) and (**d**) right supranumerary veins (*asterisks*). *LUPV* Left upper pulmonary vein, *LLPV* left lower pulmonary vein, *RUPV* right upper pulmonary vein, *RLPV* right lower pulmonary vein

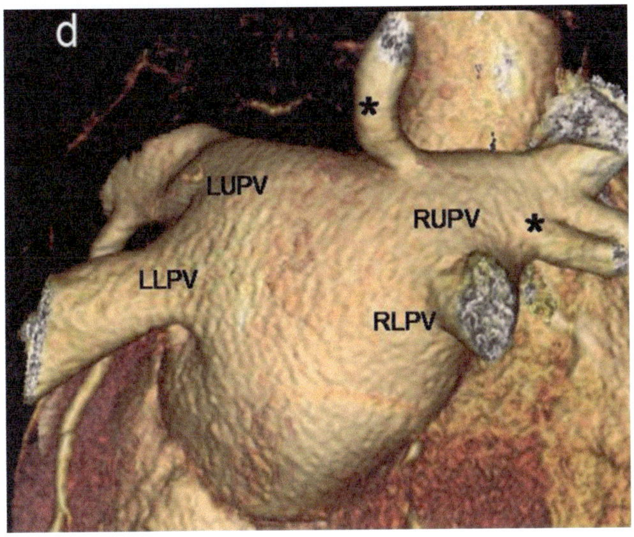

Fig. 1.37 (continued)

the ostia of the left PVs than the right PVs. However, in any given patient, the position of the esophagus can vary in the posterior mediastinum and therefore is unpredictable. Intervening fat between the LA and the anterior aspect of the esophagus was seen only in the most superior and inferior aspects of the LA. Although the position of the esophagus is influenced by peristalsis and dynamic movement, the pre-procedural assessment of the relationship between these structures may be useful in determining the location of the ablation lesions in the LA and to understand the possible risk of esophageal injury[29] (Fig. 1.46).

1.3 The Right Ventricle

The right ventricle (RV) is the most anteriorly situated cardiac chamber, lying directly behind the sternum. The RV cavity is a complex, crescent-shaped structure wrapped around the left ventricle.[38] The external appearance of the RV and its relationships with surrounding structures is nicely imaged by MSCT in 3D volume rendering (Fig. 1.47).

In the normal heart, the wall of the RV is considerably thinner than that of the left ventricle. It ranges in thickness from 3 to 7 mm. At the tip of the apex, however, the wall may be particularly thin. Planar slices obtained with MSCT or MRI provide reliable measurement of RV wall thickness (Fig. 1.48). The internal appearance of the RV is typical. The shape of the cavity can be imaged as an open "V" with a wide muscular separation between tricuspid and pulmonary valve. The right ventricular outflow tract (RVOT) is positioned in a leftward direction, whereas the left ventricular outflow tract (LVOT) is directed rightward and passes under the right outlet. The septum of the RVOT is convex, and the portion of the outflow tract that borders the pulmonary truck

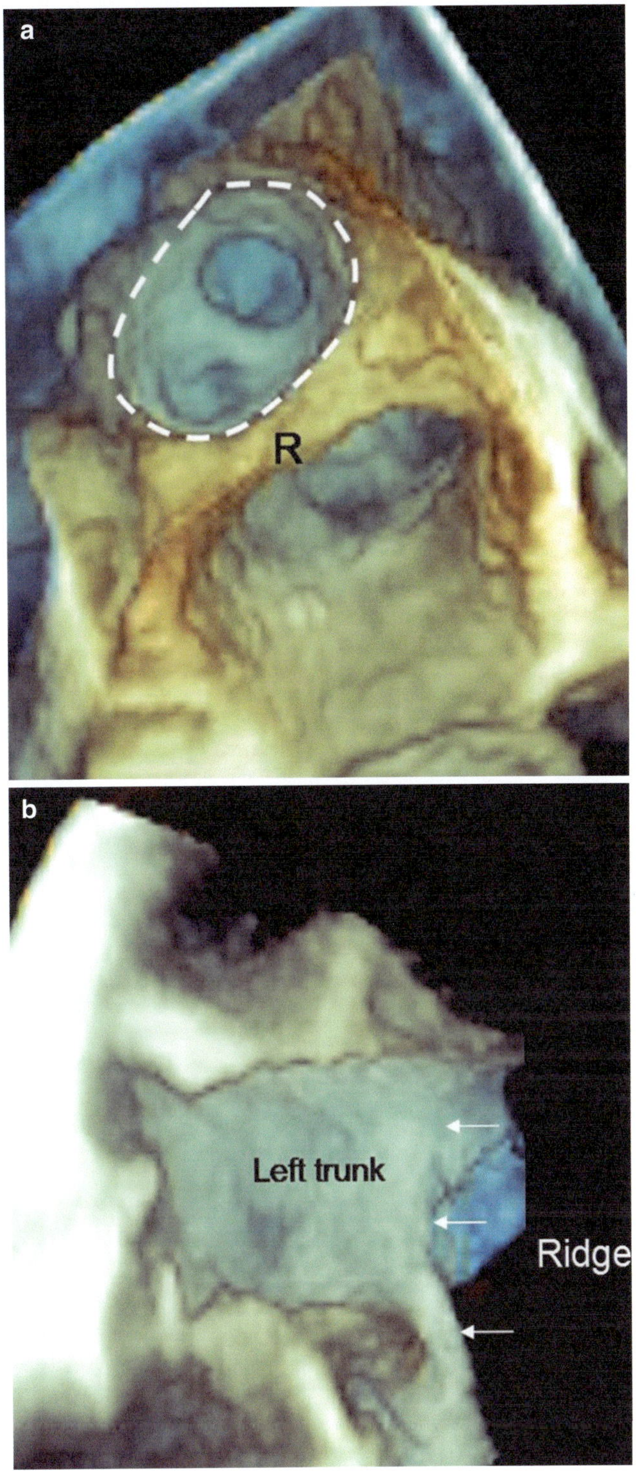

Fig. 1.38 Real-time 3D TEE showing a left common trunk in (**a**) en face perspective and (**b**) in a long axis cut. *R* Ridge

is the infundibulum where the wall thickness is 1 to 2 mm.[39] This area is of interest to the electrophysiologist as it is the site where idiopathic RVOT, a monomorphic ventricular tachycardia with a left bundle branch block and inferior axis, can be ablated.

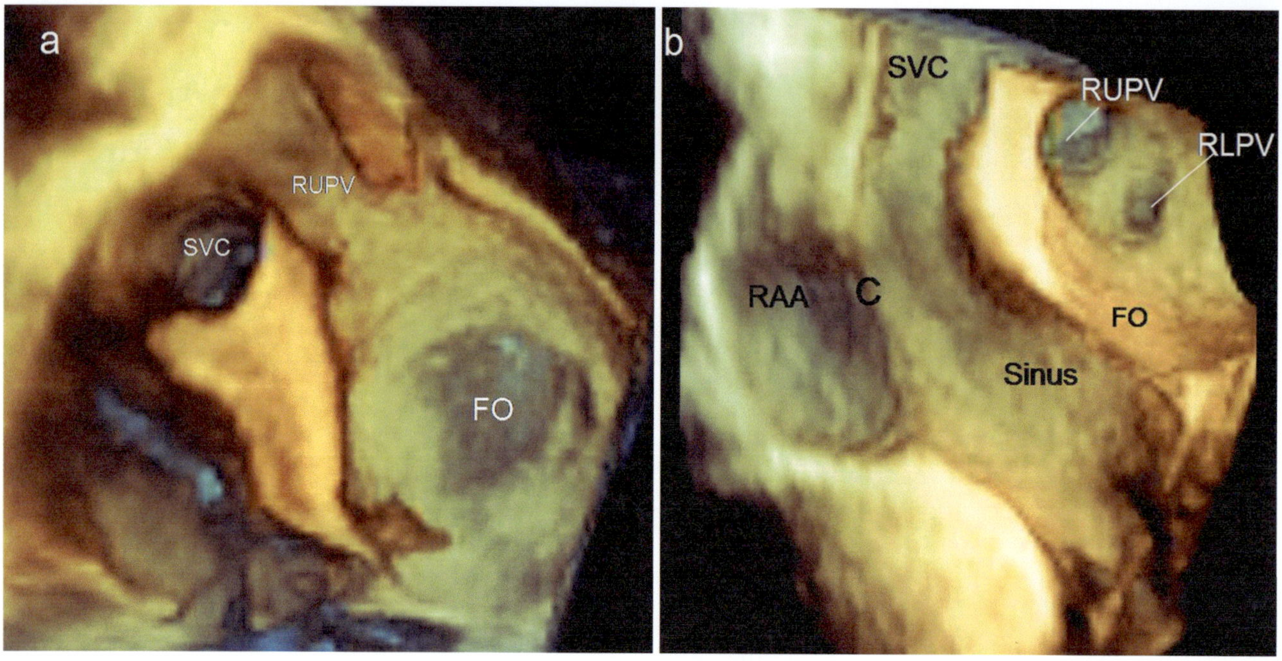

Fig. 1.39 3D TEE showing the relationship between superior vena cava (*SVC*) and right upper pulmonary vein (*RUPV*).(**a**) A proper cut shows the right upper pulmonary vein (*RUPV*) in long axis view. Because of the different long axis orientations, the superior vena cava is imaged in a short axis view. (**b**) The image is cut to show the *SVC* in long axis view; as a consequence, both the right pulmonary vein ostia are seen "en face"

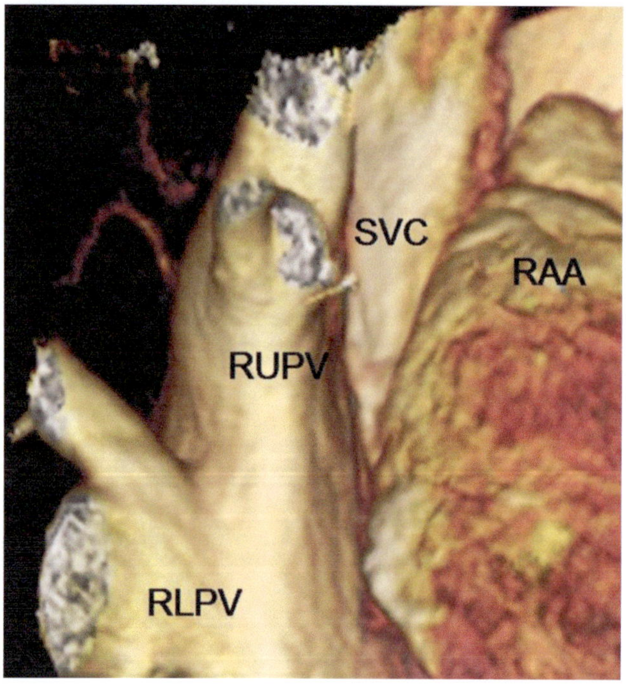

Fig. 1.40 3D volume rendering by CT showing the relationship between the superior vena cava (*SVC*) and the right upper pulmonary vein (*RUPV*). *RLPV* right lower pulmonary vein. *RAA* right atrial appendage

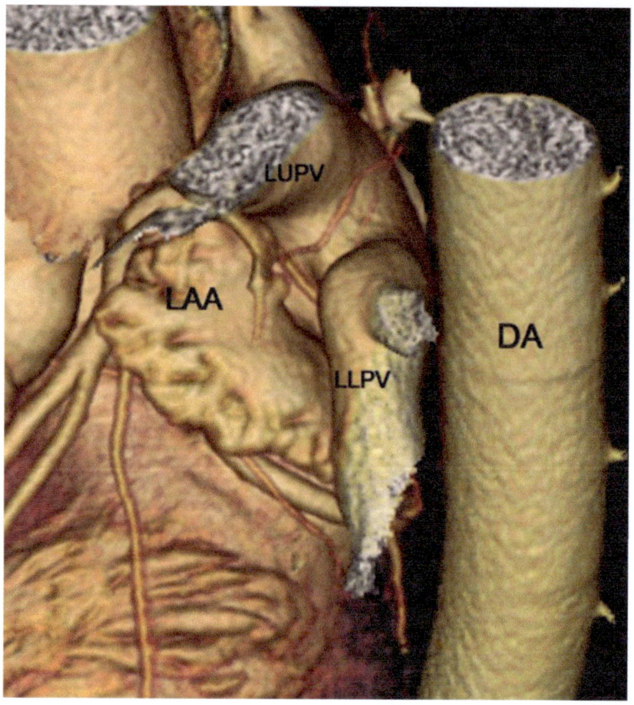

Fig. 1.41 3D Volume rendering by CT showing the relationship between the left atrial appendage (*LAA*), the left upper (*LUPV*) and lower (*LLPV*) pulmonary veins, and the descending aorta (*DA*)

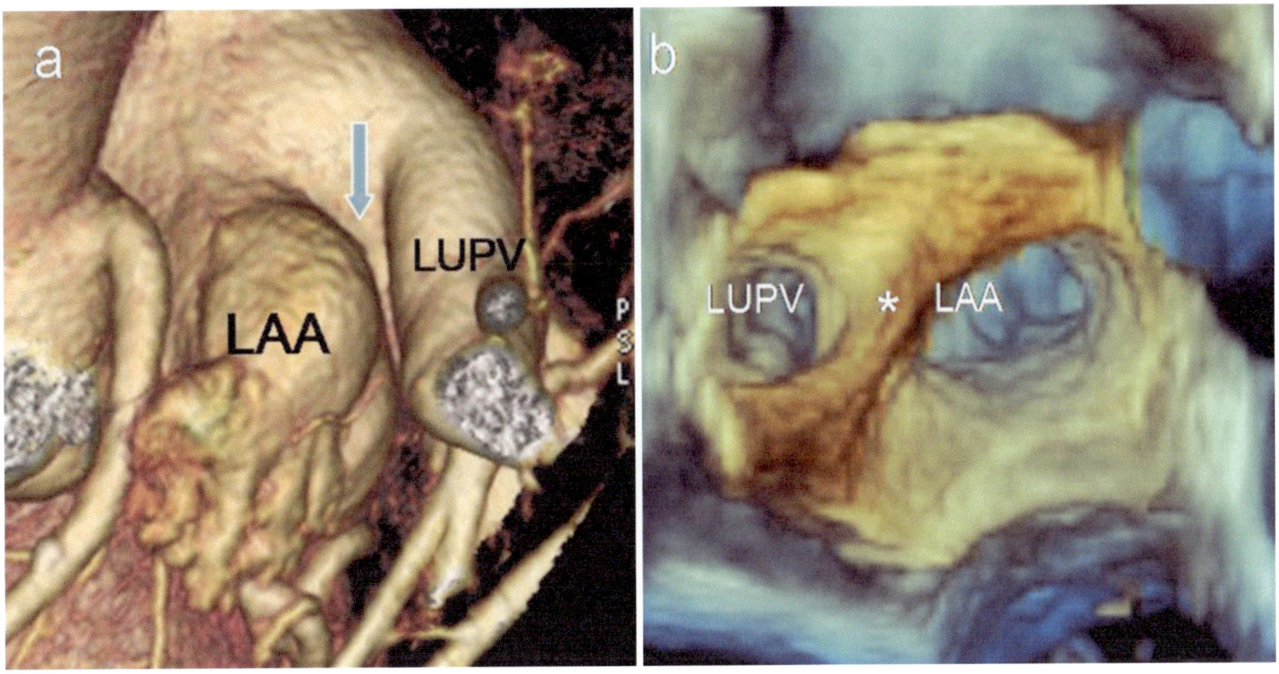

Fig. 1.42 (**a**) 3D volume rendering CT image showing the infolding of the wall of the left atrium between the left atrial appendage (*LAA*) and left upper pulmonary vein (*LUPV*), this infolding is filled by pericardial fat (*arrow*). (**b**) RT 3D TEE image showing the endocardial surface of the infolding which appears as a ridge (*asterisk*) between *LUPV* and *LAA*

The RV can be divided into three components: the inlet, the apical, and the outlet components (Fig. 1.49a). The most constant characteristic feature of the RV is the presence of coarse apical trabeculations compared to fine trabeculations that are found in the left ventricle (Fig. 1.49b). The RV is characterized by the presence of specific anatomical arrangements such as (a) the crista supraventricular (or supraventricular crest), (b) the moderator band, and (c) the trabecula septomarginalis (or septomarginal trabeculation).

The crista supraventricularis is the prominent muscular structure which separates tricuspid from pulmonary valves.[40] It comprises of the ventriculo-infundibular fold and its septal insertion into the trabecular septomarginalis. The external aspect of the ventriculo-infundibular fold can be imaged by MSCT 3D volume rendering, while the internal features are best imaged by RT 3D TEE (Fig. 1.50).

The trabecula septomarginalis is a Y-shaped muscle band that appears like a column supporting the ventriculo-infundibular fold in between its arms.[40] When abnormally formed or hypertrophied, it can be a substrate for dividing the ventricular cavity into two chambers. The body of the "trabeculation" ends near the apex, splitting into several smaller muscle bundles. One of these usually takes a characteristic course crossing the right ventricular cavity. This branch has been named the moderator band (Fig. 1.51). *The moderator band* runs toward the anterior wall, and joins the base of anterior papillary muscle (Fig. 1.51).

The outlet component, or infundibulum, is a freestanding and completely muscular structure. The pulmonary valve leaflets are supported entirely by this freestanding musculature. There is an extensive external tissue plane between the walls of the aorta and the pulmonary infundibulum (Fig. 1.52).

Arrhythmogenic right ventricular dysplasia/cardiomyopathy (ARVD) is a desmosomal cardiomyopathy resulting in progressive fibro-fatty infiltration of the myocardium, and it may be characterized by ventricular tachycardia or ventricular fibrillation. This disease affects particular areas of the right ventricle, specifically the subtricuspid portion, RVOT, and apex of the RV – known as the "triangle of dysplasia".[41] Of note, right ventricular aneurysms, often found in the thinnest portion of the RV myocardium, are considered pathognomonic and are seen in approximately 50% of autopsy cases.[42] Cardiac MRI, which is able to characterize fibro-fatty infiltration and abrupt wall thinning, offers higher reproducibility and lower observer variability than echocardiography[43] Endocardial voltage mapping allows for the identification of areas of myocardial atrophy and fibro-fatty substitution which demonstrate low amplitude electrograms. The resultant three-dimensional electroanatomic map may be used to facilitate substrate-based linear ablation of macroreentrant ventricular tachycardia in these patients. In addition, a recent study showed that RVOT and early forms of ARVD could be differentiated with the use of voltage mapping.[44]

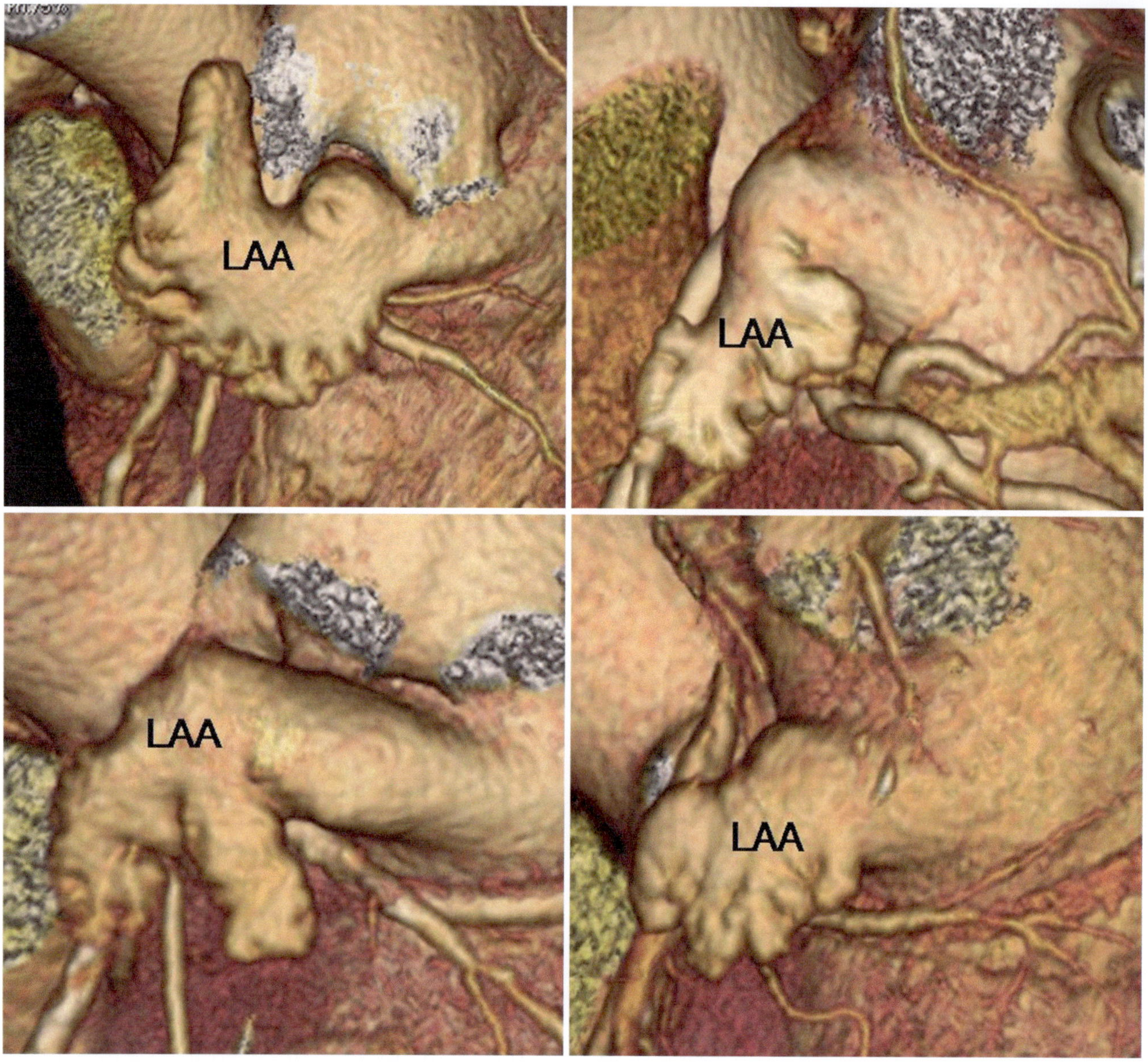

Fig. 1.43 3D volume rendering CT showing the left atrial appendage (*LAA*) of four consecutive patients. Note the extreme variability of *LAA* morphology

1.4 The Left Ventricle

The morphological left ventricle forms the apex and the lower part of the left heart border. It is shaped like a cone (ellipsoid of revolution) with its long axis directed from the apex to the base.[45] Short axis cross sections, perpendicular to long axis, reveal a roughly circular geometry. 3D images obtained by MSCT or MRI volume rendering or by RT 3D TEE can beautifully show the LV shape (Fig. 1.53 and 1.54).

The endocardial surface is irregular relative to the epicardial surface due to the two groups of papillary muscles and trabeculations. The papillary muscles are protuberences of the LV musculature that anchor the cords to the left ventricular wall. Rupture of one of their heads results in severe valvular regurgitation. The two groups are located beneath the commissures, occupying anterolateral and posteromedial positions (Fig. 1.55).

Tomographic imaging has consistently shown that the bases of the papillary muscles are not solid. Instead, they are composed of muscular continuations from the trabeculations that line the ventricular cavity (Fig. 1.56).

Like the right ventricle, the left ventricle possesses an inlet, an apical trabecular component, and an outlet. The inlet

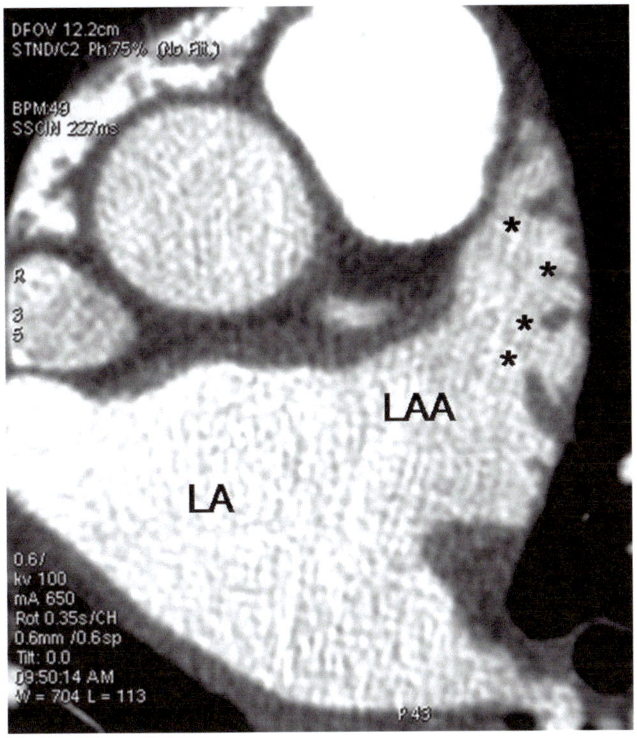

Fig. 1.44 MSCT axial slice showing the LAA appendage (*LAA*) and pectinate muscles (*asterisks*) inside. *LA* left atrium

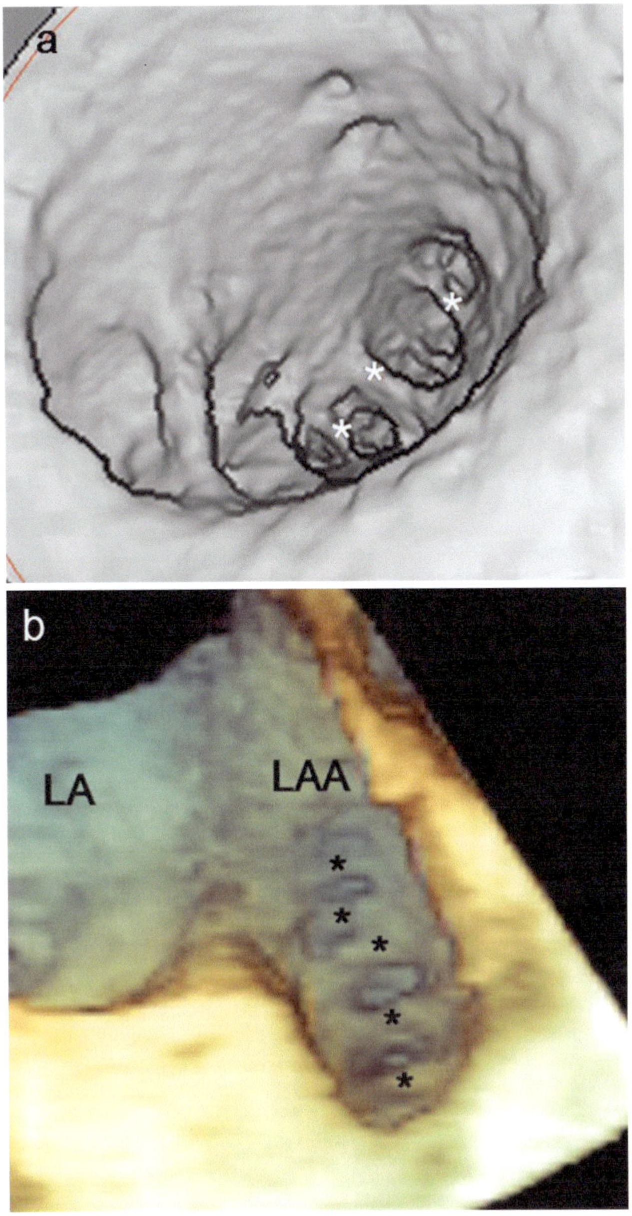

Fig. 1.45 (**a**) MSCT in virtual endoscopy modality and (**b**) RT 3D TEE showing the left atrial appendage (*LAA*) and the small irregular pectinate muscles (*asterisks*) which divide the *LAA* in small lobes

component contains the mitral valve and extends from the atrioventricular junction to the attachment of the papillary muscles (Fig. 1.57).

The apical trabecular portion is the most characteristic feature of the morphological left ventricle and contains fine trabeculations. This part helps identification since the left ventricle never possesses a septomarginal trabeculation or a moderator band (Fig. 1.58).

The LVOT component supports the aortic valve and consists of both muscular and fibrous portions. This is in contrast to the infundibulum of the right ventricle, which is comprised entirely of muscle (Fig. 1.59). The LVOT arrhythmias may arise from left and right aortic cusps, the mitral valve including the aortomitral continuity, or the base of the left ventricle. The septal portion of the LVOT, although primarily muscular, also includes the membranous portion of the ventricular septum (Fig. 1.60a). The anterolateral quadrant of the outflow tract is again muscular and consists of the lateral margin of the inner curvature of the heart (Fig. 1.60b). The posterior quadrant of the outflow tract consists of an extensive fibrous curtain that extends from the fibrous skeleton of the heart across the aortic leaflet of the mitral valve, and supports the leaflets of the aortic valve in the area of aortic-mitral fibrous continuity (Fig. 1.60c).

During systole, the ventricle thickens radially (from the epicardium to the center of cavity, in the direction of radii perpendicular to the long axis) and shortens from the base to apex along meridians (curved lines parallel to long axis). Moreover, shortening occurs also circumferentially along curved lines in short axis planes and, finally, the apex twists relative to the base (Fig. 1.61). These simultaneous movements in different directions can be explained by the complex architecture of fibers orientation. In the midwall (half away between the epicardium and the endocardium), the fibers lie in the circumferential plane, aligned with the short axis section. In the subendocardium (approximately 10% of

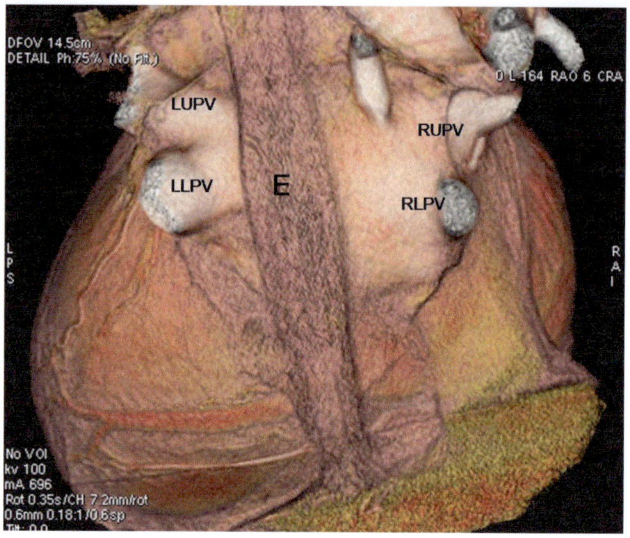

Fig. 1.46 3D volume rendering obtained by CT showing the relations between the esophagus (*E*) and the posterior roof of the left atrium. *LUPV* Left upper pulmonary vein, *LLPV* left lower pulmonary vein, *RUPV* right upper pulmonary vein, *RLPV* right lower pulmonary vein

wall thickness from the endocardium surface), the fibers course upward and to the right averaging 60° oblique to the circumferential plane. Finally, in the subepicardium (10% of wall thickness closest to epicardial surface), the fibers course downward to the right averaging 60° oblique to the circumferential plane, thus overlapping the subendocardial fibers at 120°.

1.5 Summary

Improvements in imaging technology have led to a better understanding of dynamic morphology of the heart and resulted in improved safety and efficacy of invasive electrophysiologic procedures. Three-dimensional delineation of anatomic structures and morphologic variants has allowed the cardiac electrophysiologist to perform increasing complex ablations and implantation of sophisticated devices. Novel tools have also reduced the reliance on fluoroscopy, minimizing radiation exposure to the patient and to the clinician.

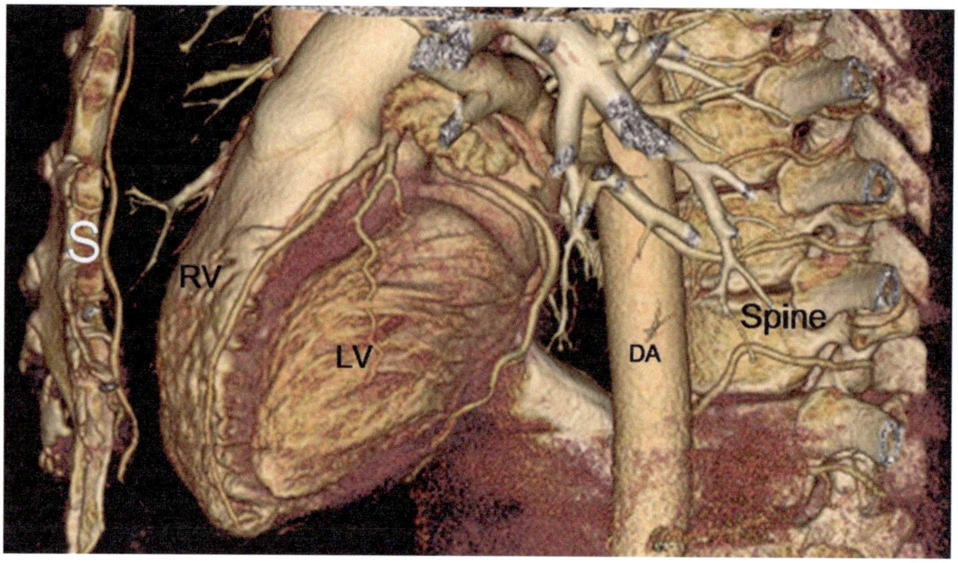

Fig. 1.47 3D volume rendering CT showing the anterior position of the right ventricle (*RV*) behind the sternum (*S*). The right ventricle overlaps most of the left ventricle (*LV*). *DA* descending aorta

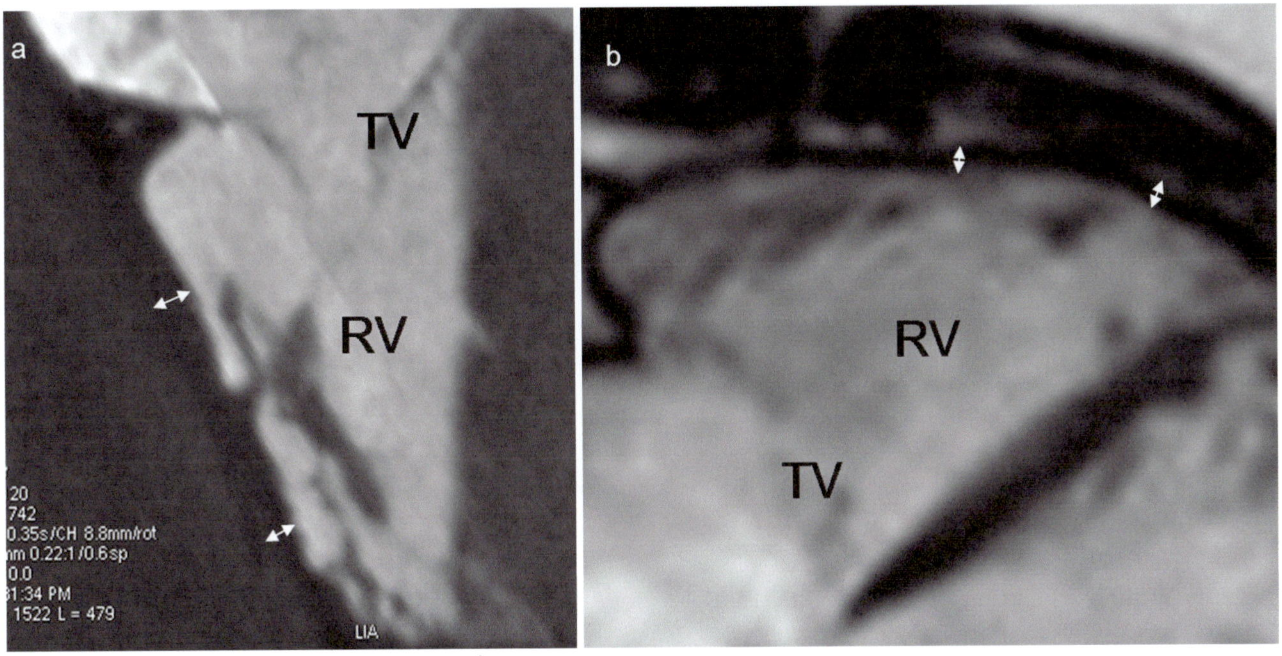

Fig. 1.48 (**a**) Slice image by CT and (**b**) MRI showing the remarkable thin wall of the right ventricle (*RV*). *TV* tricuspid valve

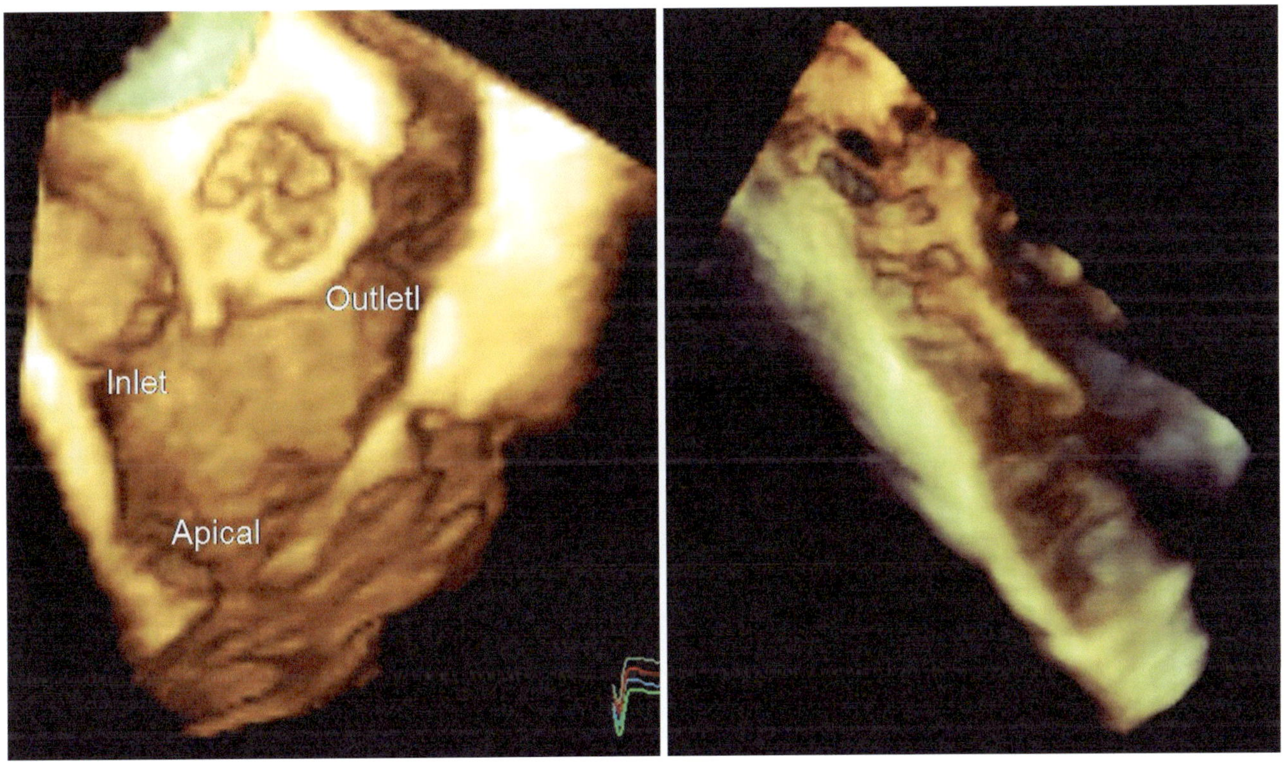

Fig. 1.49 RT 3D TEE showing (**a**) the three component of the right ventricle from a right perspective and (**b**) the coarse trabeculations near the apex

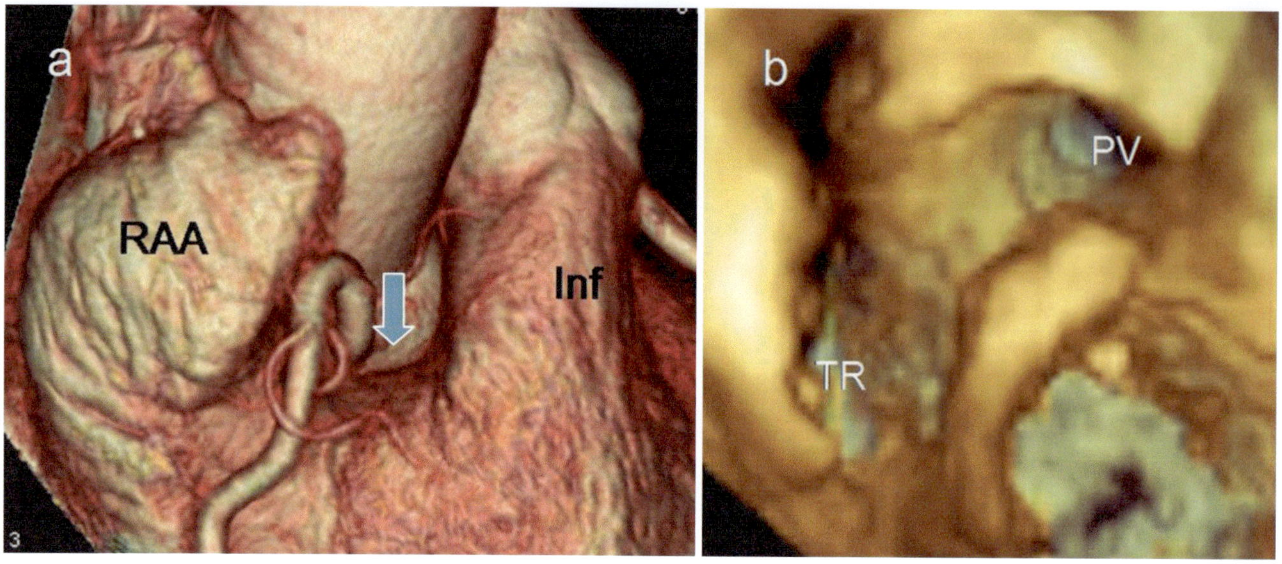

Fig. 1.50 (**a**) 3D volume rendering MSCT and (**b**) RT 3D TEE showing the crista supraventricularis from an external and internal perspective respectively

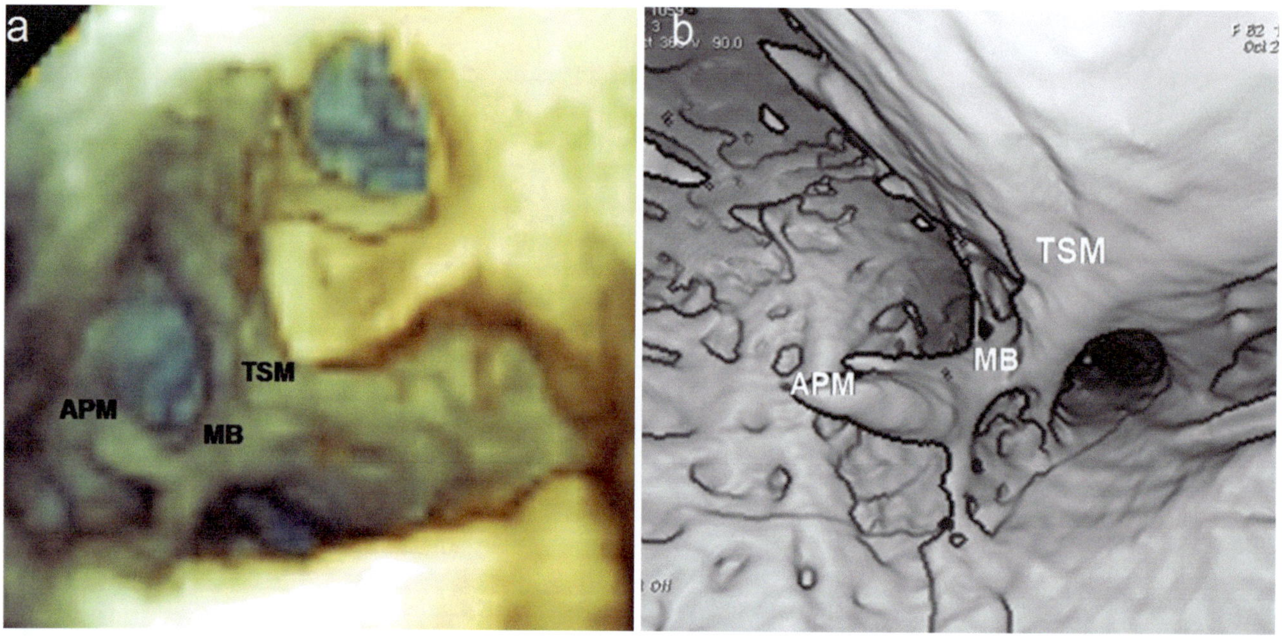

Fig. 1.51 (**a**) RT 3D TEE and (**b**) virtual endoscopy by MSCT showing the trabecula septomarginalis (*TSM*), the moderator band (*MB*), and the anterior papillary muscle (*APM*) from an above perspective

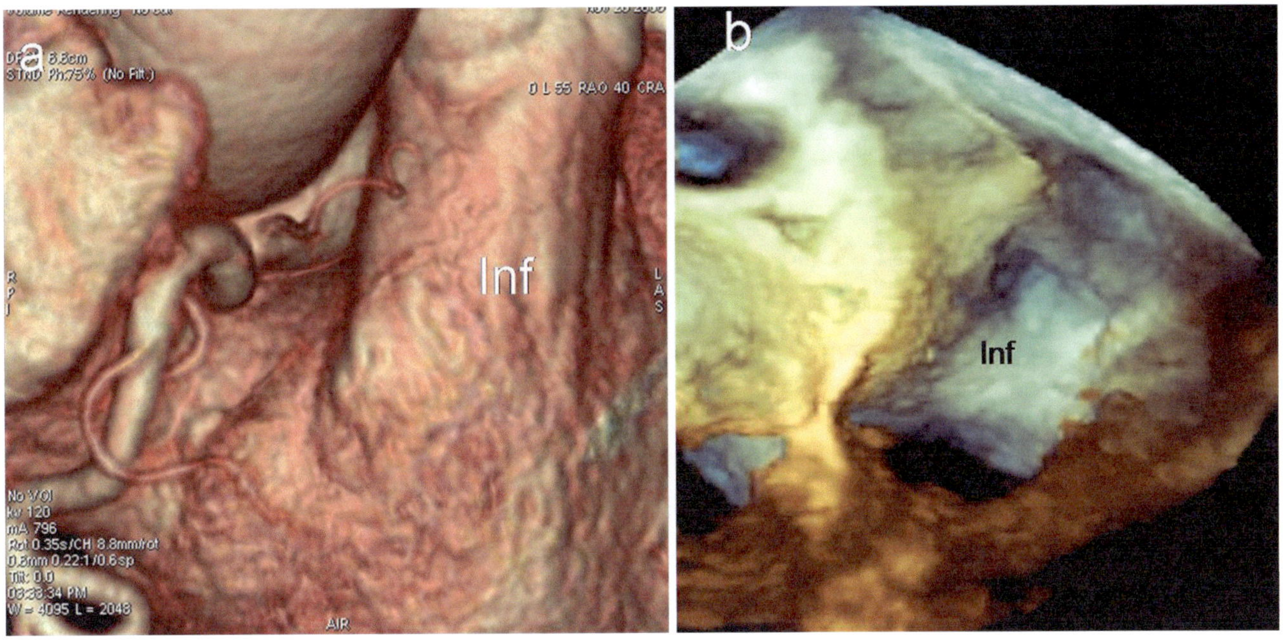

Fig. 1.52 (**a**) 3D MSCT volume rendering and (**b**) RT 3D TEE showing the right infundibulum from an external and internal perspective respectively

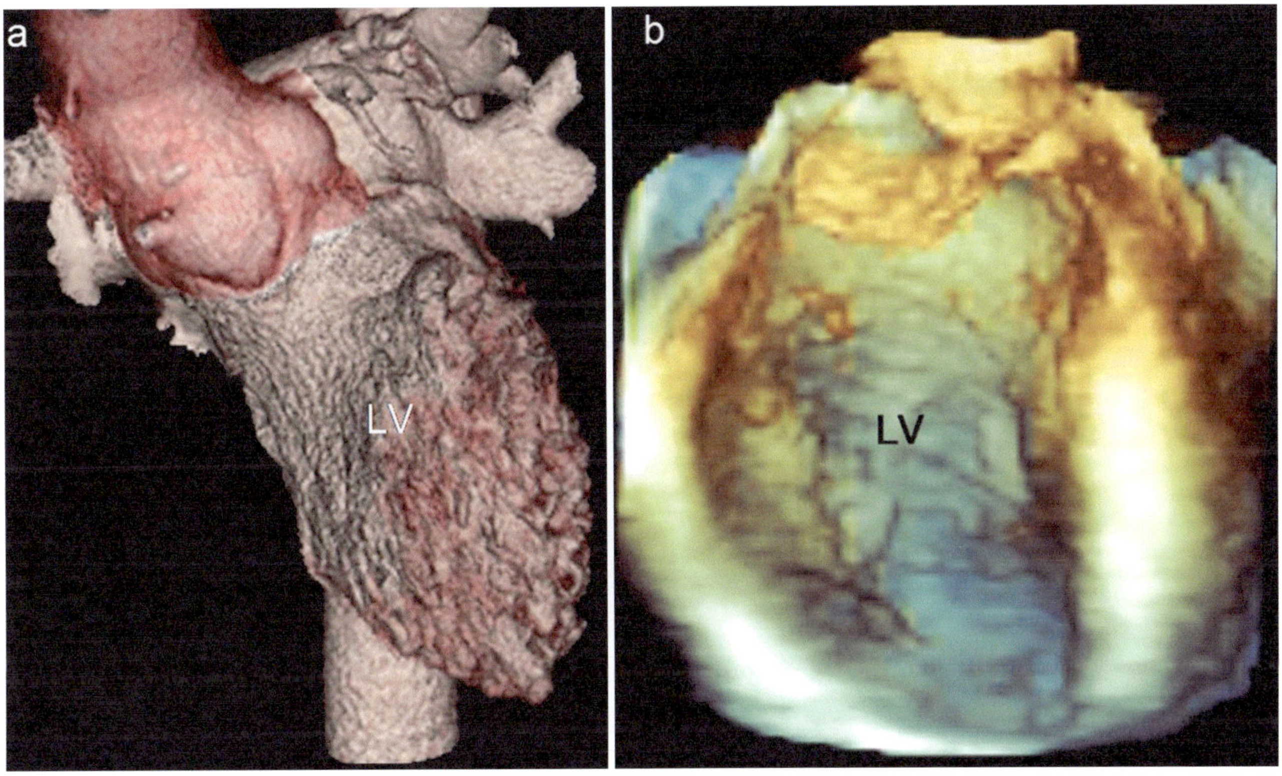

Fig. 1.53 Two different modalities to represent the shape of the left ventricle (*LV*). (**a**) By the MSCT volume data set, an "electronic" cast of the LV is obtained by making completely transparent the myocardium and increasing the opacity of the intracavitary contrast. (**b**) RT 3D TEE after having cut the lateral wall of the LV. Both techniques show the *elliptic.-shaped left* ventricular cavity

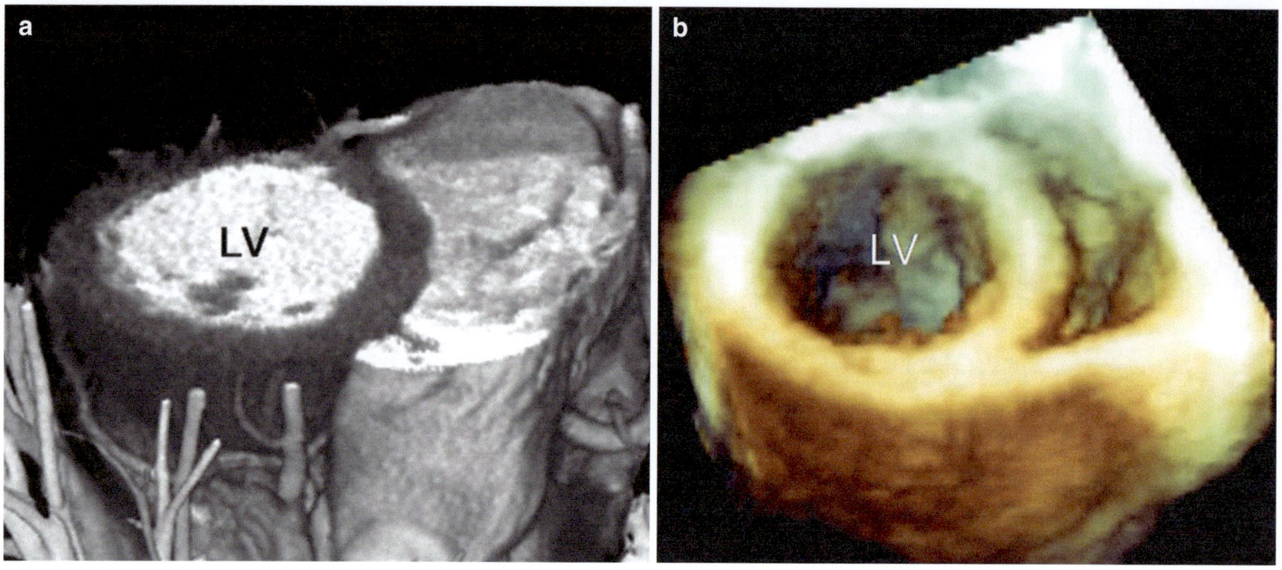

Fig. 1.54 (**a**) MSCT 3D volume rendering and (**b**) RT 3D TEE in short axis cross sections both showing the roughly circular geometry of the *left ventricle*

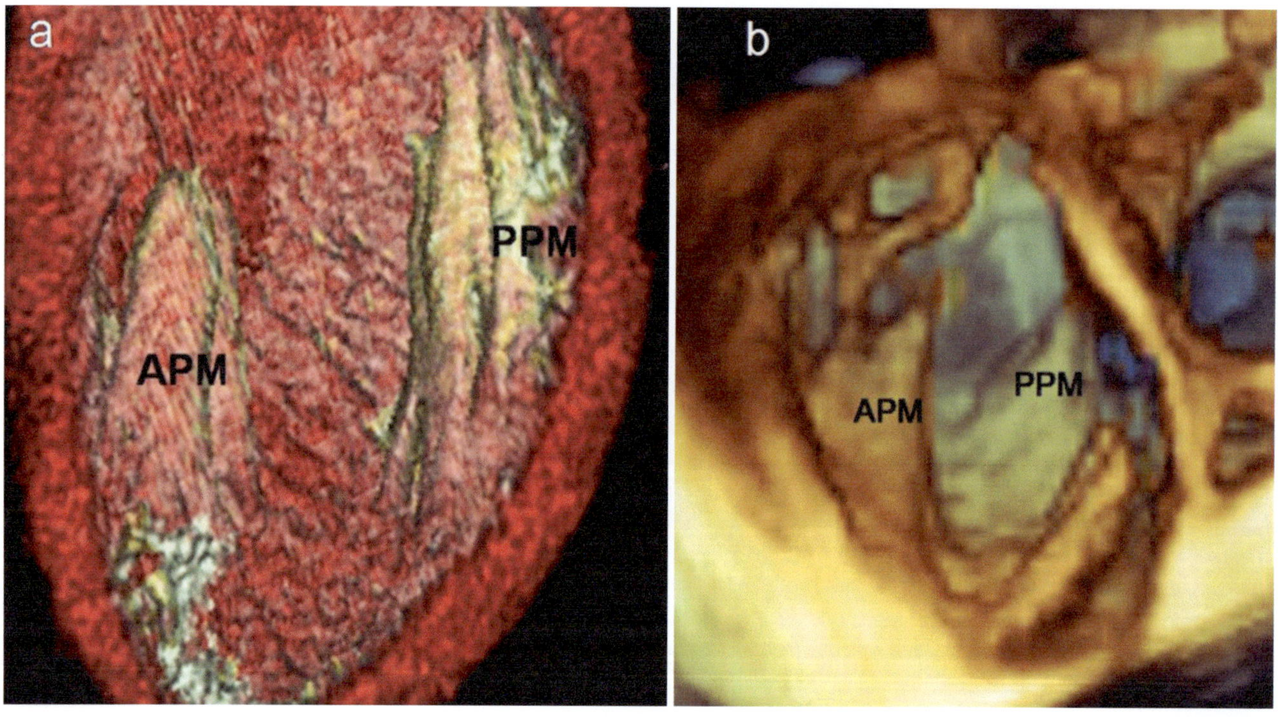

Fig. 1.55 (**a**) MSCT 3D volume rendering and (**b**) RT 3D TEE both showing the anterolateral (*AMP*) and posteromedial (*PPM*) groups of papillary muscles

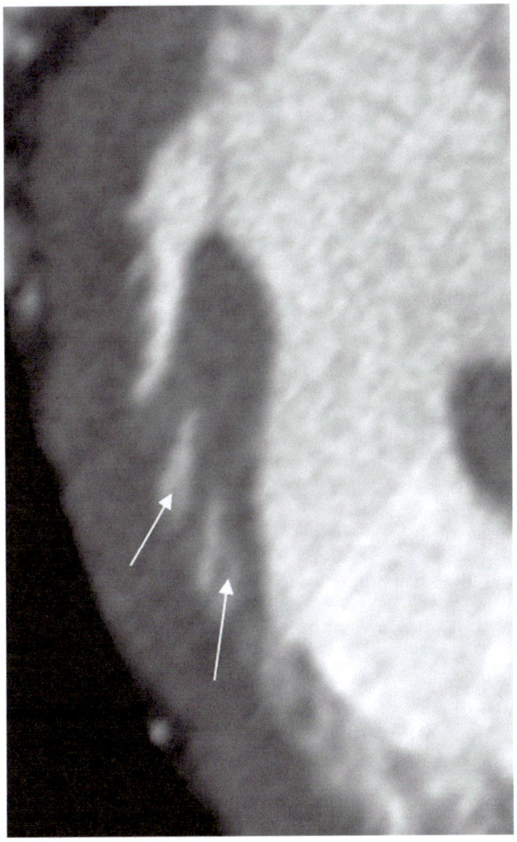

Fig. 1.56 CT slice image showing the base of papillary muscle joining with the network of trabeculations (*arrow*) lining the cavity rather than directly to the solid portion of the wall

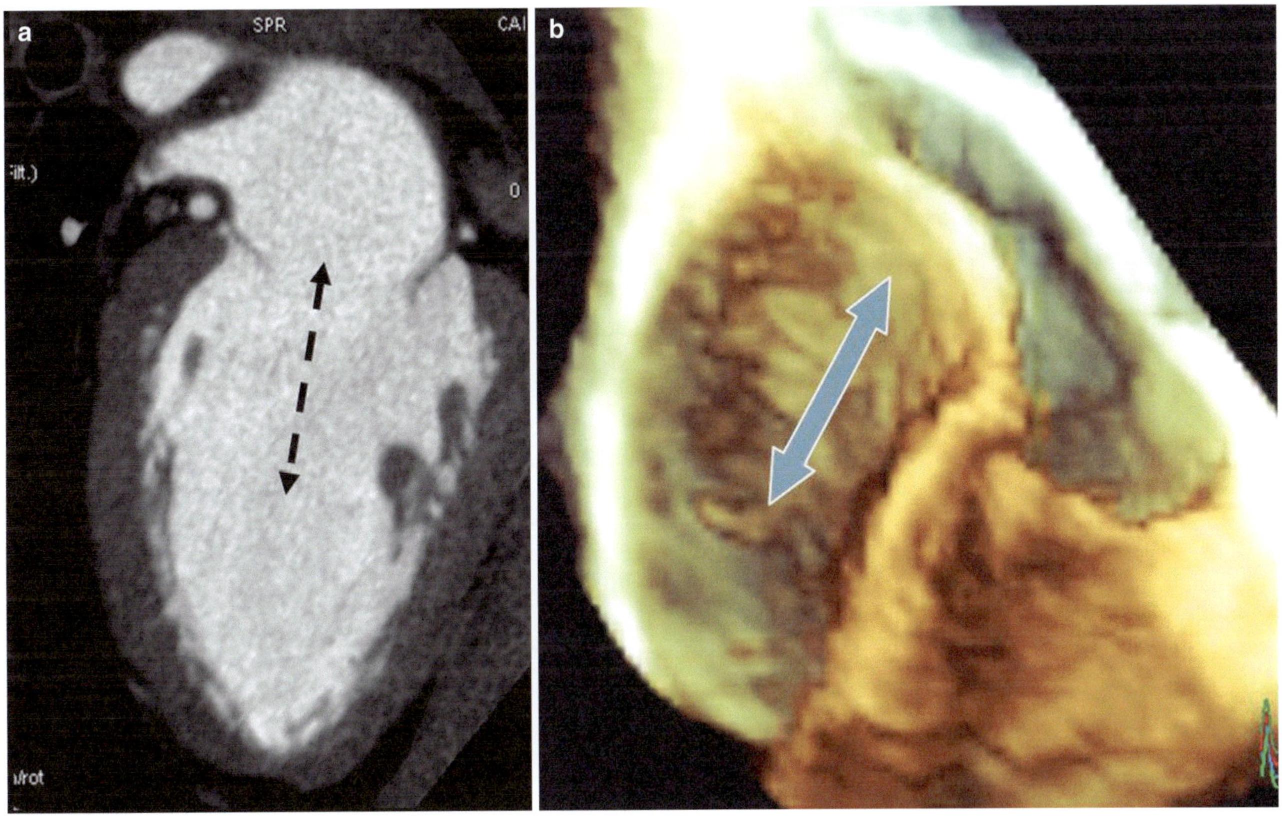

Fig. 1.57 (**a**) CT slice image and (**b**) RT 3D TEE showing the extent of the inlet portion of the left ventricle (*double-headed arrows*)

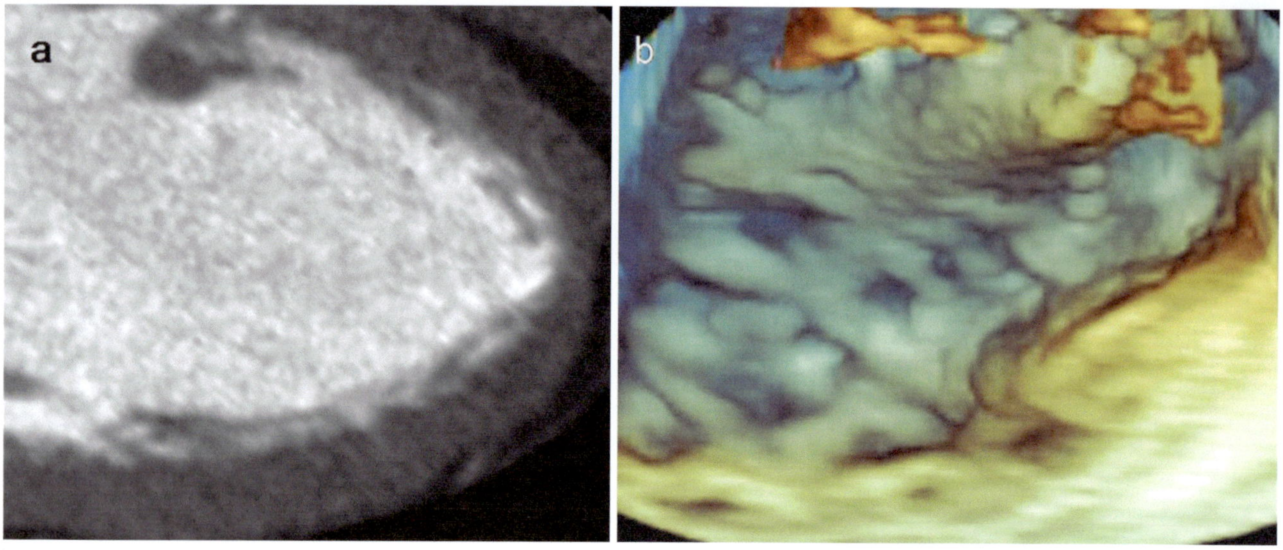

Fig. 1.58 (**a**) CT slice image and (**b**) RT 3D TEE showing the fine trabeculation of the apical portion of the *left ventricle*

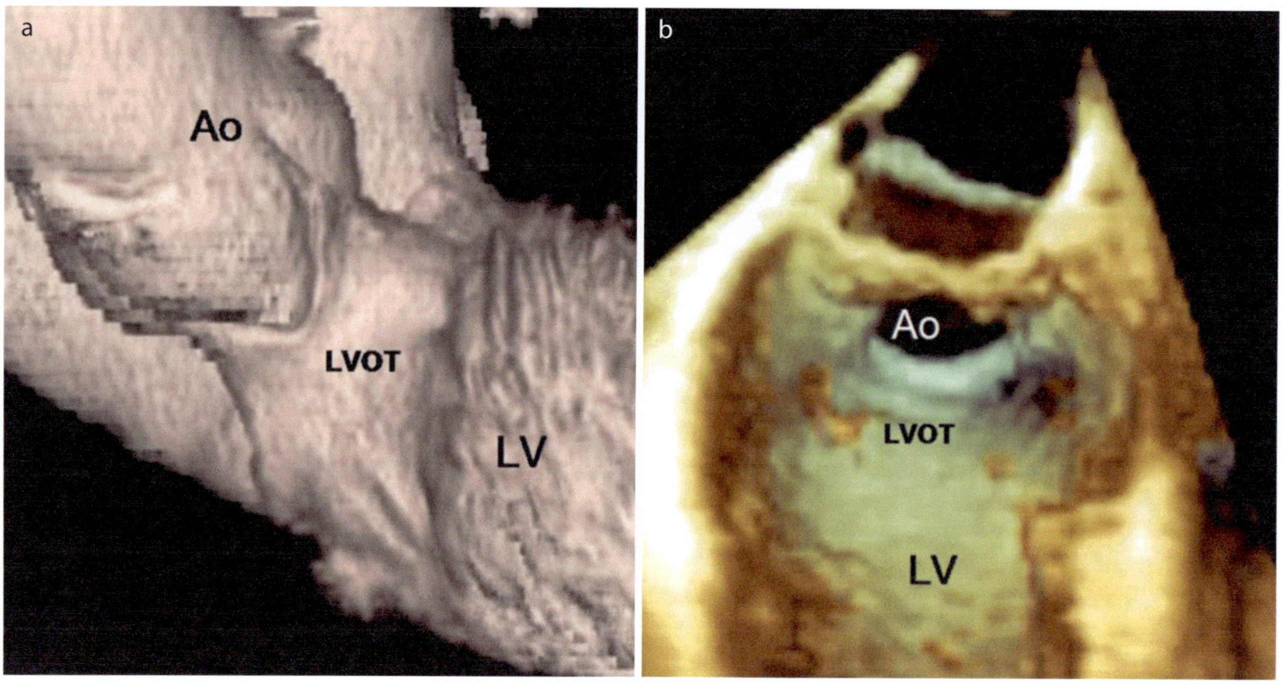

Fig. 1.59 (**a**) Electronic cast by MSCT and (**b**) RT 3D TEE showing the left ventricular outflow tract (*LVOT*). *LV* left ventricle, *Ao* Aorta

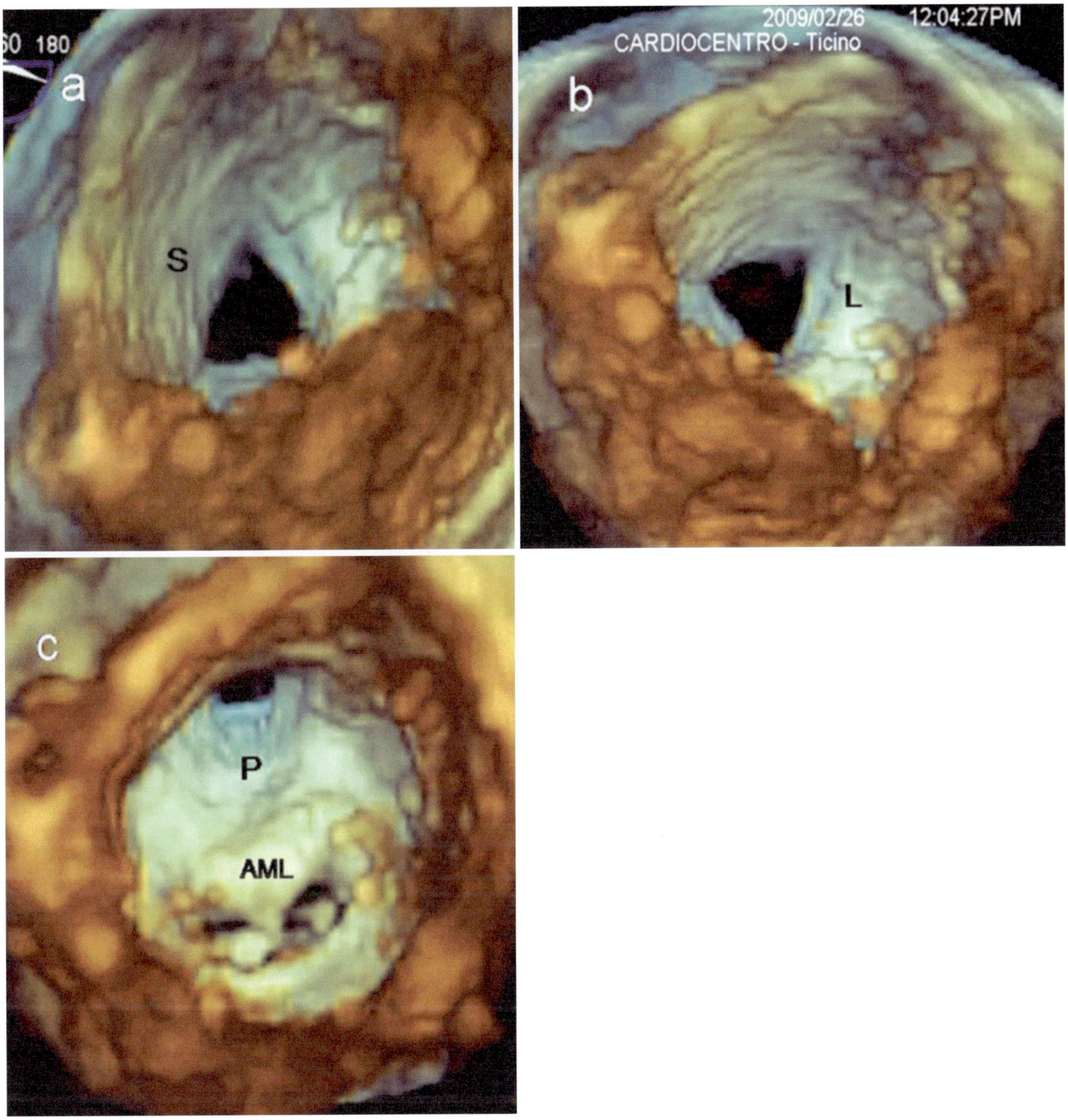

Fig. 1.60 RT 3D TEE images showing (**a**) the septal, (**b**) the anterolateral, and (**c**) the posterior quadrant of the left infundibulum

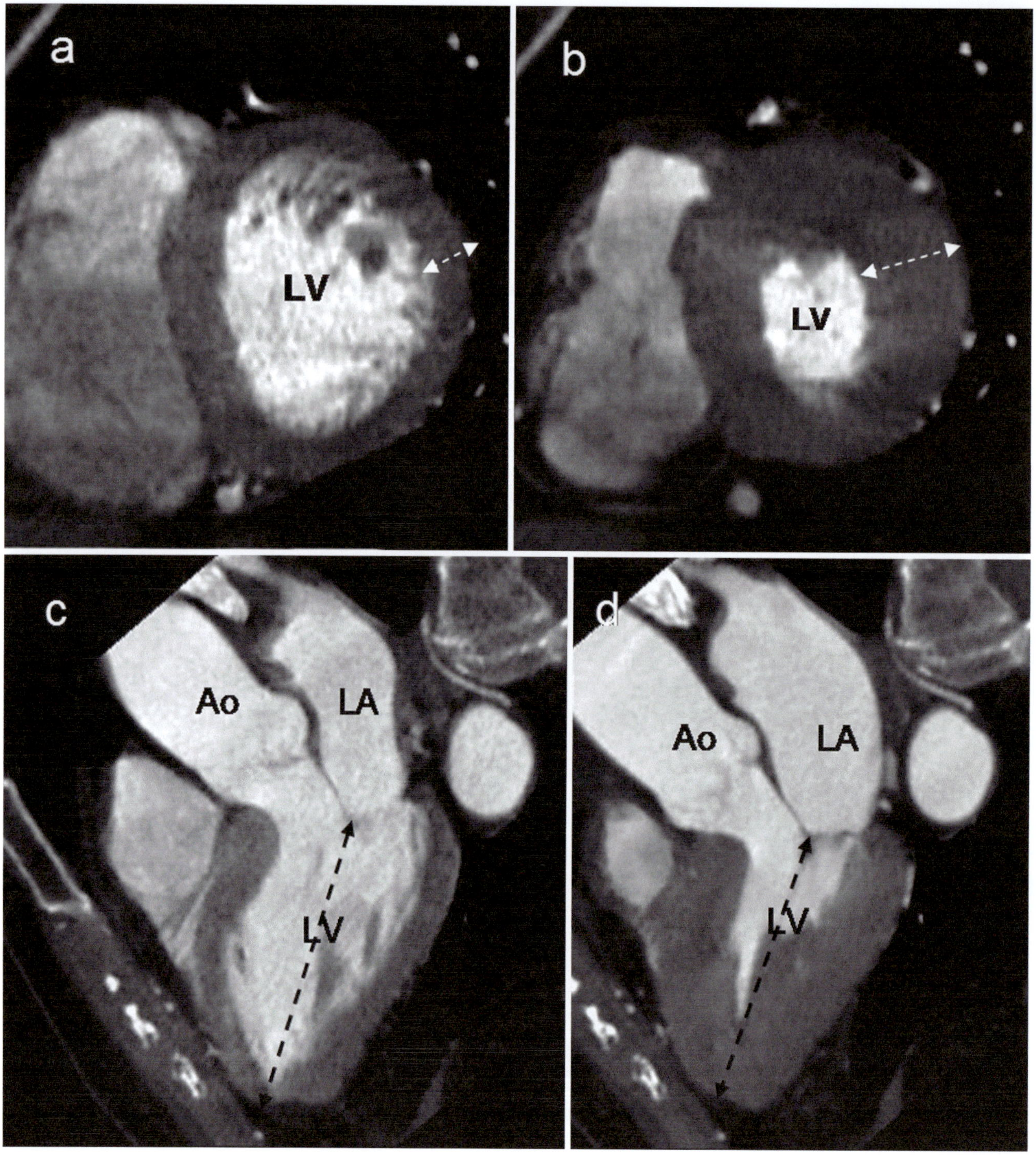

Fig. 1.61 Slice images by CT showing short axis view in diastole (**a**) and systole (**b**) and long axis view in diastole (**c**) and systole (**d**). During systole, the left ventrcicular myocardium thickens radially (*white arrows*) and shortens longitudinally (*black arrows*)

References

1. Cosio FG, Anderson RH, Kuck KH, et al. Living anatomy of the atrioventricular junction. A guide to electrophysiologic mapping: a Consensus Statement from the Cardiac Nomenclature Study Group, Working Group of Arrhythmias, European Society of Cardiology, and the Task force in Cardiac Nomenclature from NASPE. *Circulation*. 1999;100:e31-e37.
2. Ho SY, Anderson RH, Sanchez-Quintana D. Atrial structure and fibres: morphologic bases of atrial conduction. *Cardiovasc Res*. 2002;54:325-336.
3. Loukas M, Tubbs RS, Tongson JM, et al. The clinical anatomy of crista terminalis, pectinate muscles and the teniae sagettalis. *Ann Anat*. 2008;190:81-87.
4. Kalman JM, Olgin JE, Karch MR, et al. Cristal tachycardias: origin of right atrial tachycardias from crista terminalis identified by intracardiac echocardiography. *J Am Coll Cardiol*. 1998;31:451-459.
5. Cosio FG, Lopez-Gil M, Goicolea A, Arribas F. Electrophysiologic studies in atrial flutter. *Clin Cardiol*. 1992;15(9):667-673.
6. Tai CT, Huang JL, Lee PC, Ding YA, Chang MS, Chen SA. High-resolution mapping around the crista terminalis during typical atrial flutter: new insights into mechanisms. *J Cardiovasc Electrophysiol*. 2004;15(4):406-414.
7. Cabrera JA, Sanchez-Quintana D, Ho SY, Medina A, Anderson RH. The architecture of the atrial musculature between the orifice of the inferior caval vein and the tricuspid valve: the anatomy of the isthmus. *J Cardiovasc Electrophysiol*. 1998;9(11):1186-1195.
8. Da Costa A, Romeyer-Bouchard C, Dauphinot V, et al. Cavotricuspid isthmus angiography predicts atrial flutter ablation efficacy in 281 patients randomized between 8 mm. and externally irrigated-tip catheter. *Eur Heart J*. 2006;27:1833-1840.
9. Knecht S, Castro Rodriguez J, Verbeet T, et al. Multidetector 16-slice CT scan evaluation of cavotricuspid isthmus anatomy before radiofrequency ablation. *J Interv Card Electrophysiol*. 2007;20:29-35.
10. Saremi F, Pourzand L, Krishnan S, et al. Right atrial cavotricuspid isthmus: anatomic characterization with multi-detector row CT. *Radiology*. 2008;247:658-668.
11. Lim KY, Murray C, Liu H, Weerasooriya R. Pre-ablation magnetic resonance imaging of the cavotricuspid isthmus. *Europace*. 2007;9:149-153.
12. Kirchhof P, Ozgun M, Zellerhoff S, et al. Diastolic isthmus length and 'vertical' isthmus angulation identify patients with difficult catheter ablation of typical atrial flutter: a pre-procedural MRI study. *Europace*. 2009;11:42-47.
13. Kalman JM, Olgin JE, Saxon LA, Fisher WG, Lee RJ, Lesh MD. Activation and entrainment mapping defines the tricuspid annulus as the anterior barrier in typical atrial flutter. *Circulation*. 1996;94(3):398-406.
14. Lo LW, Tai CT, Lin Y-J, et al. Characteristics of the cavotricuspid isthmus in predicting recurrent conduction in the long-term follow-up. *J Cardiovasc Electrophysiol*. 2009;20:39-43.
15. Asirvatham SJ. Correlative anatomy and electrophysiology for the interventional electrophysiologist: right atrial flutter. *J Cardiovasc Electrophysiol*. 2009;20:113-122.
16. Chauvin M, Shah DC, Haissaguerre M, Marcellin L, Brechenmacher C. The anatomic basis of connections between the coronary sinus musculature and the left atrium in humans. *Circulation*. 2000;101(6):647-652.
17. Martin RE, Ellenbogen KA, Lau YR, et al. Phased-array intracardiac echocardiography during pulmonary vein isolation and linear ablation for atrial fibrillation. *J Cardiovasc Electrophysiol*. 2002;13(9):873-879.
18. Binder TM, Rosenhek R, Frank H, Gwechenberger M, Maurer G, Baumgartner H. Congenital malformations of the right atrium and the coronary sinus: an analysis based on 103 cases reported in the literature and two additional cases. *Chest*. 2000;117(6):1740-1748.
19. Karaca M, Bilge O, Dinckal MH, Ucerler H. The anatomic barriers in the coronary sinus: implications for clinical procedures. *J Interv Card Electrophysiol*. 2005;14(2):89-94.
20. Sorgente A, Truong QA, Conca C, et al. Influence of left atrial and ventricular volumes on the relation between mitral valve annulus and coronary sinus. *Am J Cardiol*. 2008;102:890-896.
21. Takahashi Y, Jais P, Hocini M, et al. Acute occlusion of the left circumflex coronary artery during mitral isthmus linear ablation. *J Cardiovasc Electrophysiol*. 2005;16(10):1104-1107.
22. Alonso C, Leclercq C, Revaul d'Allonnes F, et al. Six years experience of transvenous left ventricular lead implantation for permanent biventricular pacing in patients with advanced heart failure: technical aspects. *Heart*. 2001;86:405-410.
23. Matsumoto Y, Krishnan S, Fowl SJ, et al. The detection of phrenic nerves and their relation to cardiac anatomy using 64-slice multi-detector computed tomography. *Am J Cardiol*. 2007;100(1):133-137.
24. Hegen PT, Shollz DG, Edwards WD. Incidence and size of patent foramen ovale during the first 10 decades of life: an autopsy study of 965 normal hearts. *Mayo Clin Proc*. 1984;59:17-20.
25. Ho SY, Sanchez-Quintana D, Cabrera JA, Anderson RH. Anatomy of left atrium: implications for radiofrequency ablation of atrial fibrillation. *J Cardiovasc Electrophysiol*. 1999;10:1525-1533.
26. Becker AE. Left atrial isthmus: anatomic aspects relevant for linear catheter ablation procedures in humans. *J Cardiovasc Electrophysiol*. 2004;15:809-812.
27. Ernst G, Stollberger C, Abzieher F, et al. Morphology of the left atrial appendage. *Anat Rec*. 1995;242:553-561.
28. Pappone C, Oral H, Santinelli V, et al. Atrio-esophageal fistula as a complication of percutaneous transcatheter ablation of atrial fibrillation. *Circulation*. 2004;109:2724-2726.
29. Cury RC, Abbara S, Schmidt S, et al. Relationship of the esophagus and aorta to the left atrium and pulmonary veins: implication for catheter ablation of atrial fibrillation. *Heart Rhythm*. 2005;2:1317-1323.
30. Nathan H, Eliakim M. The junction between the left atrium and the pulmonary veins. An anatomic study of human hearts. *Circulation*. 1966;34(3):412-422.
31. Ho SY. Pulmonary vein ablation in atrial fibrillation: does anatomy matter? *J Cardiovasc Electrophysiol*. 2003;14:156-157.
32. Haissaguerre M, Shah DC, Jais P, et al. Electrophysiological breakthroughs from the left atrium to the pulmonary veins. *Circulation*. 2000;102(20):2463-2465.
33. Wazni OM, Tsao HM, Chen SA, et al. Cardiovascular imaging in the management of atrial fibrillation. *J Am Coll Cardiol*. 2006;48:2077-2084.
34. Kato R, Lickfett L, Meininger G, et al. Pulmonary vein anatomy in patients undergoing catheter ablation of atrial fibrillation: lessons learned by use of magnetic resonance imaging. *Circulation*. 2003;107(15):2004-2010.
35. Cronin P, Sneider MB, Kazerroni EA, et al. MDCT of the left atrium and pulmonary veins in planning radiofrequency ablation for atrial fibrillation: a how-to guide. *AJR Am J Roentgenol*. 2004;183:767-778.
36. Mansour M, Holmvang G, Sosnovik D, et al. Assessment of pulmonary vein anatomic variability by magnetic resonance imaging: implications for catheter ablation techniques for atrial fibrillation. *J Cardiovasc Electrophysiol*. 2004;15(4):387-393.
37. Cabrera JA, Ho SY, Climent V, et al. The architecture of the left lateral atrial wall: a particular anatomic region with implications for ablation of atrial fibrillation. *Eur Heart J*. 2008;29:356-362.
38. Ho SY, Nihoyannopoulos P. Anatomy echocardiography and normal right ventricular dimensions. *Heart*. 2006;92(suppl I):i2-i13.
39. Ho SY. Anatomic insights for catheter ablation of ventricular tachycardia. *Heart Rhythm*. 2009;6(suppl 8):S77-S80.

40. Kosinski A, Nowinski J, Kozlowski D, et al. The crista supraventricularus in the human heart and its role in the morphogenesis of the septomarginal trabecula. *Ann Anat*. 2007;189:447-456.
41. Marcus FI, Fontaine GH, Guiraudon G, et al. Right ventricular dysplasia: a report of 24 adult cases. *Circulation*. 1982;65(2):384-398.
42. Sorrell VL, Kumar S, Kalra N. Cardiac imaging in right ventricular cardiomyopathy/dysplasia – how does cardiac imaging assist in understanding the morphologic, functional, and electrical changes of the heart in this disease? *J Electrocardiol*. 2009;42(2):e1-10.
43. Bellenger NG, Francis JM, Davies CL, Coats AJ, Pennell DJ. Establishment and performance of a magnetic resonance cardiac function clinic. *J Cardiovasc Magn Reson*. 2000;2(1):15-22.
44. Corrado D, Basso C, Leoni L, et al. Three-dimensional electroanatomical voltage mapping and histologic evaluation of myocardial substrate in right ventricular outflow tract tachycardia. *J Am Coll Cardiol*. 2008;51(7):731-739.
45. Ho SY. Anatomy and myoarchitecture of the left ventricular wall in normal and in disease. *Eur J Echocardiogr*. 2009;10(8):113-117.

Magnetic Resonance Imaging: Description of Technology and Protocols

2

Gaston R. Vergara and Nassir F. Marrouche

Abstract

Since its introduction in the late 1970s, catheter-based radiofrequency ablation has evolved from a primitive and experimental procedure to the mainstay for arrhythmia management it is today. Initial intracardiac catheter navigation was fluoroscopy based, and therefore subject to x-ray limitations and side effects. However, accurate catheter location within the cardiac chambers has required electrophysiologic confirmation of catheter positioning. This led to the development of conventional cardiac mapping techniques. The limitations of fluoroscopy and conventional mapping techniques led to the development of electro-anatomical mapping systems (EAM), in which information regarding catheter position in a 3D space is combined with electrophysiological information in real time to provide an accurate localization of the catheter tip while, at the same time, data regarding electrophysiological properties of the underlying myocardial substrate. Eventually, the mechanisms of more complex arrhythmias, such as atrial fibrillation and scar-based monomorphic ventricular tachycardia, started to be elucidated. This was followed by more difficult ablation procedures that required more accurate mapping systems able to provide real-time information. The introduction of EAM combined with Cardiac Computerized Tomography (CCT), cardiac Magnetic Resonance Imaging (cMRI), and real-time intracardiac echocardiography (ICE) allows for more precise mapping with significant improvement in cure rates for ablation procedures. However, most of these techniques are essentially x-ray based and expose the patient and the operator to the noxious effects of ionizing radiation.

Keywords

Catheter-based radiofrequency ablation • Electro-anatomical mapping systems • Cardiac Computerized Tomography • Cardiac Magnetic Resonance Imaging • Intracardiac echocardiography

Since its introduction[1] in the late 1970s, catheter-based radiofrequency ablation has evolved from a primitive and experimental procedure to the mainstay for arrhythmia management it is today. Initial intracardiac catheter navigation was fluoroscopy based, and therefore subject to x-ray limitations and side effects. However, accurate catheter location within the cardiac chambers has required electrophysiologic confirmation of catheter positioning. This led to the development of conventional cardiac mapping techniques. Initial and current conventional electrophysiologic mapping techniques rely on astute observations and maneuvers to uncover the arrhythmia anatomic substrate and pathophysiologic mechanisms. However, as progress was made in the understanding of the mechanisms underlying arrhythmias, the limitations of fluoroscopy and conventional mapping techniques became apparent.

N.F. Marrouche (✉)
Division of Cardiology, Comprehensive Arrhythmia Research & Management Center, University of Utah Health Sciences Center, Salt Lake City, UT, USA
e-mail: nassir.marrouche@hsc.utah.edu

A. Auricchio et al. (eds.), *Cardiac Imaging in Electrophysiology*,
DOI 10.1007/978-1-84882-486-7_2, © Springer-Verlag London Limited 2012

This led to the development of electro-anatomical mapping systems (EAM), in which information regarding catheter position in a 3D space is combined with electrophysiological information in real time to provide an accurate localization of the catheter tip while, at the same time, data regarding electrophysiological properties of the underlying myocardial substrate.

Eventually, the mechanisms of more complex arrhythmias, such as atrial fibrillation and scar-based monomorphic ventricular tachycardia, were slowly being elucidated. This was followed by more difficult ablation procedures which required more accurate mapping systems able to provide real-time information.

The introduction of EAM combined with Cardiac Computerized Tomography (CCT), cardiac Magnetic Resonance Imaging (cMRI) and real-time intracardiac echocardiography (ICE) allows for more precise mapping with significant improvement in cure rates for ablation procedures. However, most of these techniques are essentially x-ray based and expose the patient and the operator to the noxious effects of ionizing radiation.

2.1 MRI for Arrhythmic Substrate Evaluation: Tissue Characterization and Anatomic Considerations

2.1.1 Atrial Fibrillation Ablation

2.1.1.1 Anatomical Considerations

Atrial fibrillation (AF) is the most common sustained cardiac arrhythmia, affecting more than two million people in the United States,[2] with an incidence rate of 0.4%[3] of the general population. Electrical pulmonary vein isolation (PVI) using radiofrequency (RF) ablation is effective in symptomatic, drug-refractory AF. Still, reported success rates of the procedure vary significantly with reported AF recurrences ranging from 25% to 60%.

Ever since it was first published in 1998 by Haissaguerre et al., pulmonary vein (PV) triggers have been recognized as the most common source of paroxysmal atrial fibrillation; electrical isolation of the PV has remained the cornerstone of atrial fibrillation ablation.[4] Most ablation techniques include, in one way or the other, a group of lesions distributed in a circular fashion to electrically isolate the PV so that it becomes of utmost importance then to clearly define the left atrial (LA) and PV anatomy prior to any ablation.

PV anatomy is variable in the general population, and this is more significant in the AF patient population. Kato et al.[5] observed up to 38% anatomical variants in patients with AF, these patients typically had larger PV diameter than controls. Wazni et al.[3] confirmed the presence of a right middle PV in 18–29% of patients undergoing evaluation for AF ablation, and this structure has been described as a focus for AF initiation. The importance of a clear understanding of the patient's anatomy is of paramount importance when planning an ablation procedure. cMRI can very clearly demonstrate the presence, location, and anatomical variants of PV's prior to ablation; allowing for procedural planning.

2.1.1.2 Integration Between Left Atrium cMRI and Non-fluoroscopy Based Mapping Systems

Integration of LA cMRI images with a non-fluoroscopy-based mapping system is a crucial step in AF ablation, since it allows for precise catheter monitoring in a real-time three-dimensional manner during ablation. Integration typically consists in fusing two images: CCT or cMRI with an electro-anatomical map (EAM) or shell of the LA. This process usually consists of three steps: (1) image acquisition, (2) segmentation, and (3) registration. Accuracy of integration is crucial for safe catheter navigation and positioning; however, pitfalls related to integration of CCT/cMRI with EAM systems could occur due to registration errors and changes in the LA volume, size, and shape between the time of image acquisition and integration with the EAM system.

2.1.1.3 Tissue Characterization, Staging of Atrial Fibrillation, and Prediction of AF Ablation Success

Late gadolinium enhancement-MRI (LGE-MRI) of the LA has been used as a marker for LA fibrosis and structural remodeling. Oakes et al.[6] have shown that the amount of LGE in the LA is a powerful predictor of AF ablation outcome. The rate of AF recurrence post-ablation was directly related to the degree of LA LGE pre-ablation.[6] The amount of LGE of the LA as a marker of scar formation post-AF ablation has also been directly correlated with ablation success in a pilot study.[7]

The use of LGE-MRI pre-ablation for risk stratification and ablation success prediction has allowed for the development of a personalized management approach to atrial fibrillation. Upon initial clinical evaluation and after determining the AF burden, a cardiac MRI was acquired. The following image acquisition parameters are used.

2.1.1.4 MRI Image Acquisition

Pre-ablation cardiac MRI is obtained either on a 1.5 T Avanto or on a 3.0 T Veerio scanners (Siemens Medical Solutions, Erlangen, Germany) using a TIM phased-array receiver coil. The scan is acquired 15 min after 0.1 mmol/kg Multihance (Bracco Diagnostic Inc., Princeton, NJ) contrast agent injection, using a 3D inversion recovery, respiration-navigated, ECG-gated, and gradient-echo pulse sequence. Typical acquisition parameters were free-breathing using navigator gating, a transverse imaging volume with voxel size = 1.25 × 1.25 × 2.5 mm, and GRAPPA with $R=2$ and 46 reference

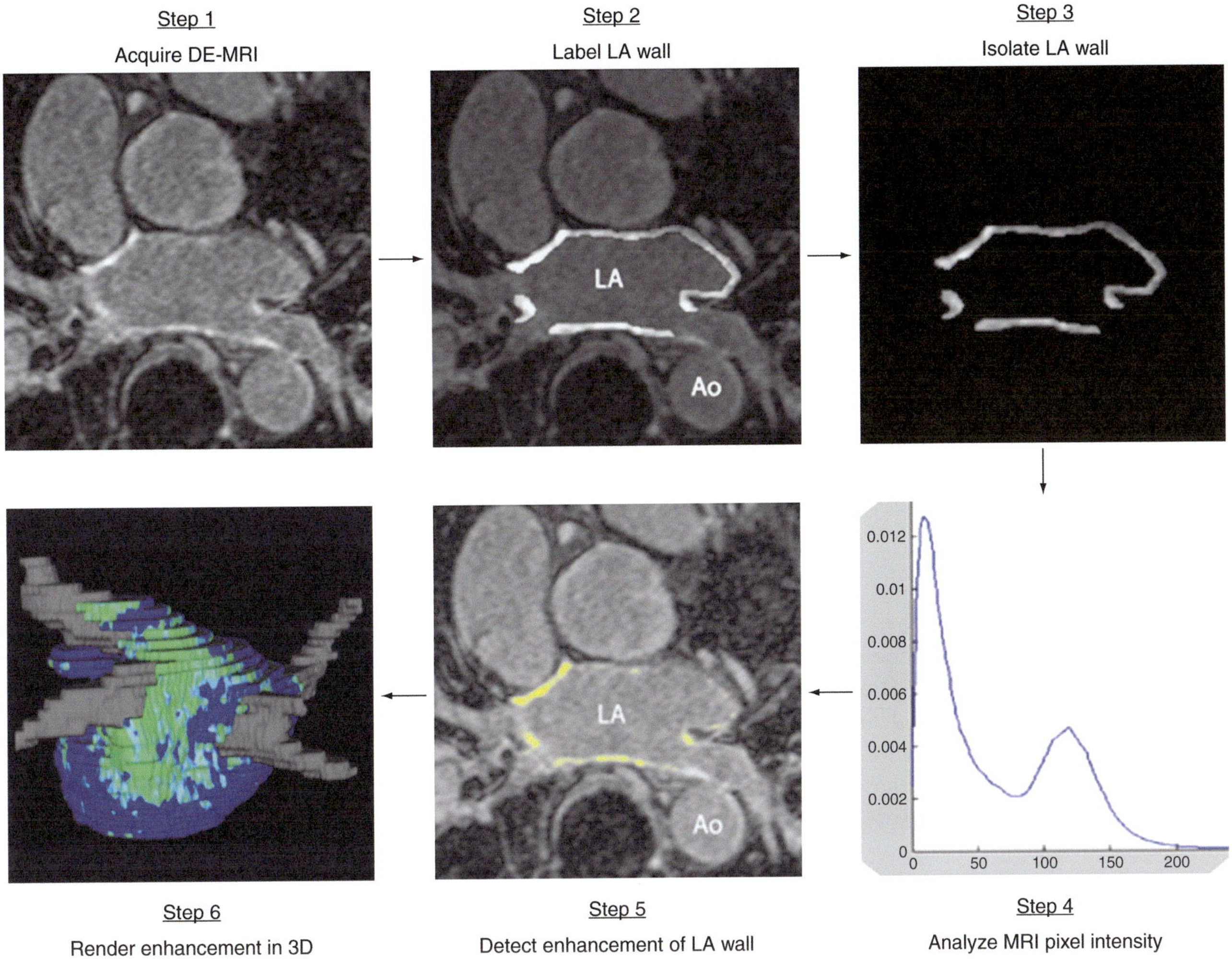

Fig. 2.1 LGE-MRI quantification of pre-ablation fibrosis/structural remodeling and postablation scarring. After LGE-MR images are obtained (*1*), the endocardial and epicardial borders are manually contoured and isolated (*2, 3*), and the extent of LGE is then quantified using a pixel intensity distribution (*4*), qualitative confirmation is then performed, a color lookup table mask is then applied to differentiate enhanced and non-enhanced tissue (*5*), and finally a 3D rendering of the LA is generated allowing for better visualization and spatial localization of the late gadolinium enhancement (*6*)

lines. ECG gating is used to acquire a small subset of phase encoding views during the diastolic phase of the LA cardiac cycle. The time interval between the R-peak of the ECG and the start of data acquisition was defined using the cine images of the LA. Fat saturation is used. The TE of the scan (2.3 ms) is chosen such that fat and water are out of phase and the signal intensity of partial volume fat-tissue voxels was reduced allowing improved delineation of the LA wall boundary. The T1 value for the LGE-MRI scan is identified using a scout scan. Typical scan time for the LGE-MRI study is 5–10 min.

2.1.1.5 LGE-MRI Quantification of Pre-ablation Fibrosis/Structural Remodeling and Post Ablation Scarring

After image acquisition, the epicardial and endocardial LA borders are manually contoured using the CoreView image display and analysis software. The relative extent of pre-ablation enhancement and post-ablation scar is then quantified within the LA wall with a threshold-based algorithm utilizing pixel intensities from normal based on a bimodal distribution (Fig. 2.1). Since post-ablation scar pixel intensity is significantly higher than pre-ablation delayed enhancement, a different threshold is used for analysis and imaging of scar.

2.1.1.6 Staging AF Using MRI

Supported by outcome data we have established at the University of Utah, a clinical staging system composed of four stages based on the amount of pre-ablation delayed enhancement (fibrosis) as a percentage of the volume of the left atrial wall.[8] This clinical staging system includes four stages: Utah I≤5% enhancement, Utah II>5–20%, Utah III>20–35%, and Utah IV>35%. When performing a

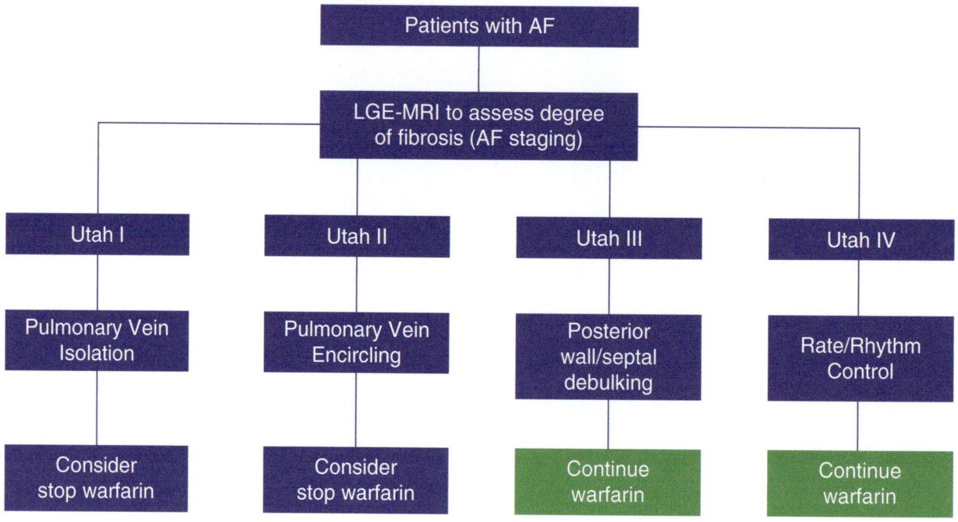

Fig. 2.2 University of Utah proposed LGE-MRI-based management algorithm for patients with AF

multivariate analysis, it was found that the number of PV isolated in patients with Utah stage II and the total amount of scar in those with Utah stage III were predictors of success. Patients with minimal pre-ablation fibrosis, Utah stage I, did well regardless of the number of PV isolated or the total amount of scar, whereas those with advanced atrial remodeling as assessed by LGE-MRI, Utah stage IV, did poorly regardless.[8]

Moreover, in a multivariate regression model, LGE-MRI evaluation of the left atrial substrate was shown to improve the predictive value of the $CHADS_2$ score, allowing defining patients at higher risk of stroke despite having a low or moderate $CHADS_2$ score.[9] Patients with a previous stroke had a significantly higher percentage of LA fibrosis compared to those without ($24.4\% \pm 12.4$ vs. $16.1\% \pm 9.8$, $p = <0.001$). There was also a significant difference in the rate of thromboembolism between patients with Utah stage I and those with stage IV. Also it was found that patients with higher risk for stroke ($CHADS_2$ score ≥ 2) had higher amounts of LA fibrosis. Using univariate and multivariate regression analysis, LGE-MRI quantified left atrial structural remodeling was independently associated with stroke.[9] Based on this staging system, a comprehensive cMRI-based AF management algorithm (Fig. 2.2) has been developed, which helps in triaging patients to AF ablation, as well as planning a corresponding ablation strategy and future anticoagulation strategy.

2.1.2 Safety

Control of collateral damage is critical during AF ablation. The LA is anatomically related with several vital structures; the pulmonary artery runs along the LA dome, the ascending aorta relates with the LA anterior wall and dome, the descending aorta with the posterior wall, the phrenic nerve is anterior to the right pulmonary veins, and the esophagus runs behind the posterior wall and the left inferior PV. Understanding of these relationships and monitoring of these anatomical structures during ablation is of paramount importance to avoid disastrous complications. LGE of the esophagus has been used to monitor for post-ablation injury.[10] In one report, Badger et al.[11] studied 41 patients' LGE-MRI pre-AF ablation, 24 h post-AF ablation, and 3 months after the ablation. Five patients demonstrated esophageal enhancement 24 h post-ablation and esophageal injury confirmed by esophagogastroduodenoscopy (EGD). EGD and cMRI were repeated a week later and confirmed resolution of esophageal LGE and endoscopic resolution of these lesions as well. Follow-up cMRI at 3 months post ablation demonstrated no LGE on the esophageal wall (Fig. 2.3).

2.2 Ventricular Tachycardia Ablation

Arrhythmia substrate evaluation is critical for ventricular tachycardia (VT) evaluation and ablation strategy planning. cMRI has the capacity to assess not only ventricular systolic function but also, and simultaneously, to provide insights into the myocardial underlying pathology.

2.2.1 Scar-Based Monomorphic Ventricular Tachycardia: Ischemic VT

VT associated with myocardial scars, either ischemic (Fig. 2.4a–c), due to sarcoidosis, or cardiomyopathy, is

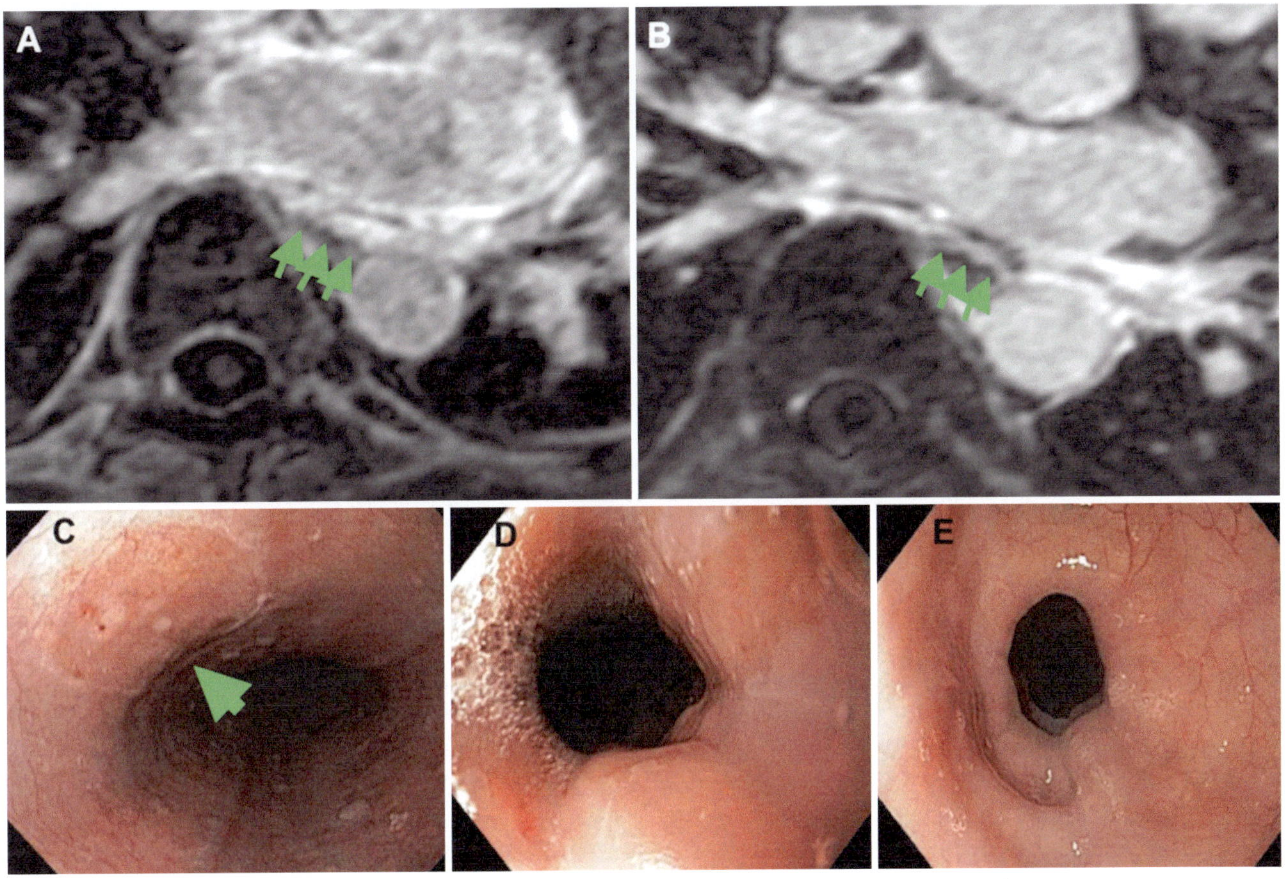

Fig. 2.3 *LGE-MRI of the esophagus and EGD.* (**a**) LGE-MRI demonstrates enhancement of the anterior esophageal wall (*arrows*) which correlates with a lesion (*green arrow*) found on EGD (**c**). (**b**) A week later, there has been resolution of late gadolinium enhancement on MRI (*arrows*) and resolution of the lesion on EGD (**d** and **e**)

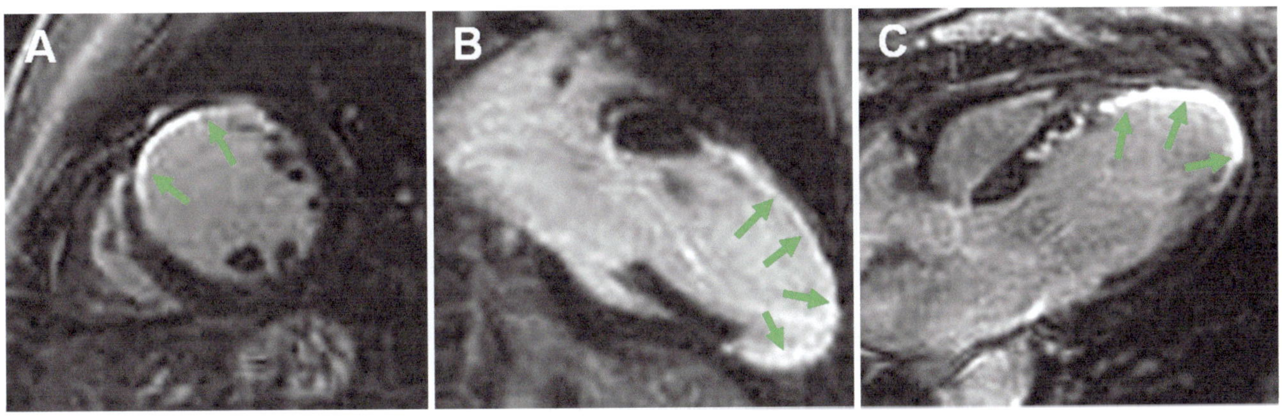

Fig. 2.4 Characteristic cMRI of patients with ischemic scar. Cardiac LGE-MRI (late gadolinium enhancement phase sensitive inversion recovery (PSIR) sequence) of patient with ischemic cardiomyopathy and scar-based monomorphic ventricular tachycardia (**a**: short axis view, **b**: two chamber view, and **c**: long axis view) demonstrating a scar (*green arrowheads*) in the distal antero-septal segments of the LV (bright area)

typically a monomorphic re-entrant arrhythmia dependent upon the presence of a conduction isthmus. This isthmus could be inside the scar, around the scar, or around a fixed anatomical structure (i.e., cardiac valves). These arrhythmias are usually not well tolerated hemodynamically.

Different strategies for the mapping of these VTs include scar/substrate assessment with electro-anatomical mapping system, pace mapping, and evaluation of diastolic potentials. These different strategies, however, are time consuming, adding length and risk to these procedures.

LGE-MRI provides a reliable assessment of myocardial scar, particularly in ischemic substrates. Bello et al. found, in 18 patients, a correlation between infarct surface area and infarct mass as defined by LGE-MRI and VT inducibility on EPS.[12] On another larger study, Schmidt et al. demonstrated the association between "scar border zone," a distinct zone than dense scar based on pixel intensity on LGE-MRI, and inducibility in EPS.[13] In this study, the amount of scar border zone was a good predictor of inducibility, whereas the total amount of dense scar was not. Information about VT substrate has been used, albeit experimentally, to predict the VT circuit. Ashikaga et al. correlated, in an animal model, surface ventricular mapping with ex-vivo high-resolution cMRI and found correlation between exit sites and conduction isthmus with isles of viable myocardium within the scar.[14]

2.2.2 Arrhythmogenic Right Ventricular Dysplasia/Cardiomyopathy

Arrhythmogenic right ventricular dysplasia/cardiomyopathy (ARVD/C) is cardiomyopathy which affects mainly the right ventricle (RV). It is characterized by fatty/fibro-fatty replacement and myocyte loss, ventricular aneurysms, ventricular arrhythmias, and right ventricular failure. There is mounting evidence that the underlying etiology of ARVD/C is desmosomal dysfunction.[15] Its prevalence is estimated to be around 1:5,000 in the United States, and accounts for 5% of all sudden cardiac death in patients younger than 35 years old in the United States. Its diagnosis is based on a set of major and minor criteria established by the Task Force of Cardiomyopathy.[16] They include evaluation for structural and electrophysiological abnormalities, as well as elements from the patient history.

Cardiac MRI is a very useful noninvasive tool for the evaluation of ARVD/C since it can define the presence of myocardial fat infiltration, observed in T1-weighted sequences,[15] and it can also allow for evaluation of the structure of the RV and quantification of its function.

2.2.3 Ventricular Tachycardia in Structurally Normal Ventricles (Idiopathic Ventricular Tachycardia)

Approximately 10% of all ventricular tachycardias occur in ventricles that are structurally normal.[17] The presence of subclinical structural abnormalities is not always evident in the echocardiogram and/or coronary angiogram which are usually normal. MRI in these cases may assist in the differential diagnosis and point toward a different etiology.

2.3 Radiofrequency Ablation Lesion Characterization

Characterization of the myocardial changes following RF ablation is of importance since it would allow for validation of therapy delivered and ultimately for ablation endpoints.

2.3.1 Acute Wall Edema Post Ablation

Acute edema, enhancement on T2w images performed immediately after AF ablation, correlates significantly with low voltage areas (defined as <0.05 mV) mapped using the CARTO system. However, the area enhanced with T2w imaging is much larger than the area covered by LGE on MRI acutely post-AF ablation.[18] Acute post-ablation edema is seen not only in regions directly subjected to RF energy but also in distant regions (Fig. 2.5) and it does not predict final scar formation defined by LGE-MRI at 3 months.[18] A LGE-MRI at 3 months after AF ablation shows loss of enhancement on T2w images consistent with edema resolution in areas free of scar. Edema seen acutely in regions other than in ablated areas suggests a mechanism other than direct radiofrequency thermal lesion as its cause.

Finally, the presence of edema in regions away from areas that result in scar formation, as well as its association with low voltage on electro-anatomical mapping may explain, at least partially, the presence of acute PV disconnection, and late reconnection with edema resolution, or ventricular myocardial recovery following VT ablation.[18]

2.3.2 Late Gadolinium-Enhanced Defined Scar and Non-reflow Phenomenon

Heterogeneity in the LA wall is seen on acute post-ablation LGE-MRI scans with portions showing very little or no enhancement at all even in areas that received direct RF energy (Fig. 2.6). In a porcine model of ablation, these areas correlated well with lesion formation, particularly with areas with the highest amount of injury. Within minutes there is resolution of these areas of non-enhancement and they manifest all the features of ablated/scarred areas. These areas of no-enhancement are believed to correspond to areas of no-reflow, phenomenon similar to that seen in ventricles in the immediate post-MI period.[18]

2.3.3 Late Imaging and Recurrences

The amount of scar and the number of circumferentially scarred PVA on LGE-MRI is associated with better outcomes

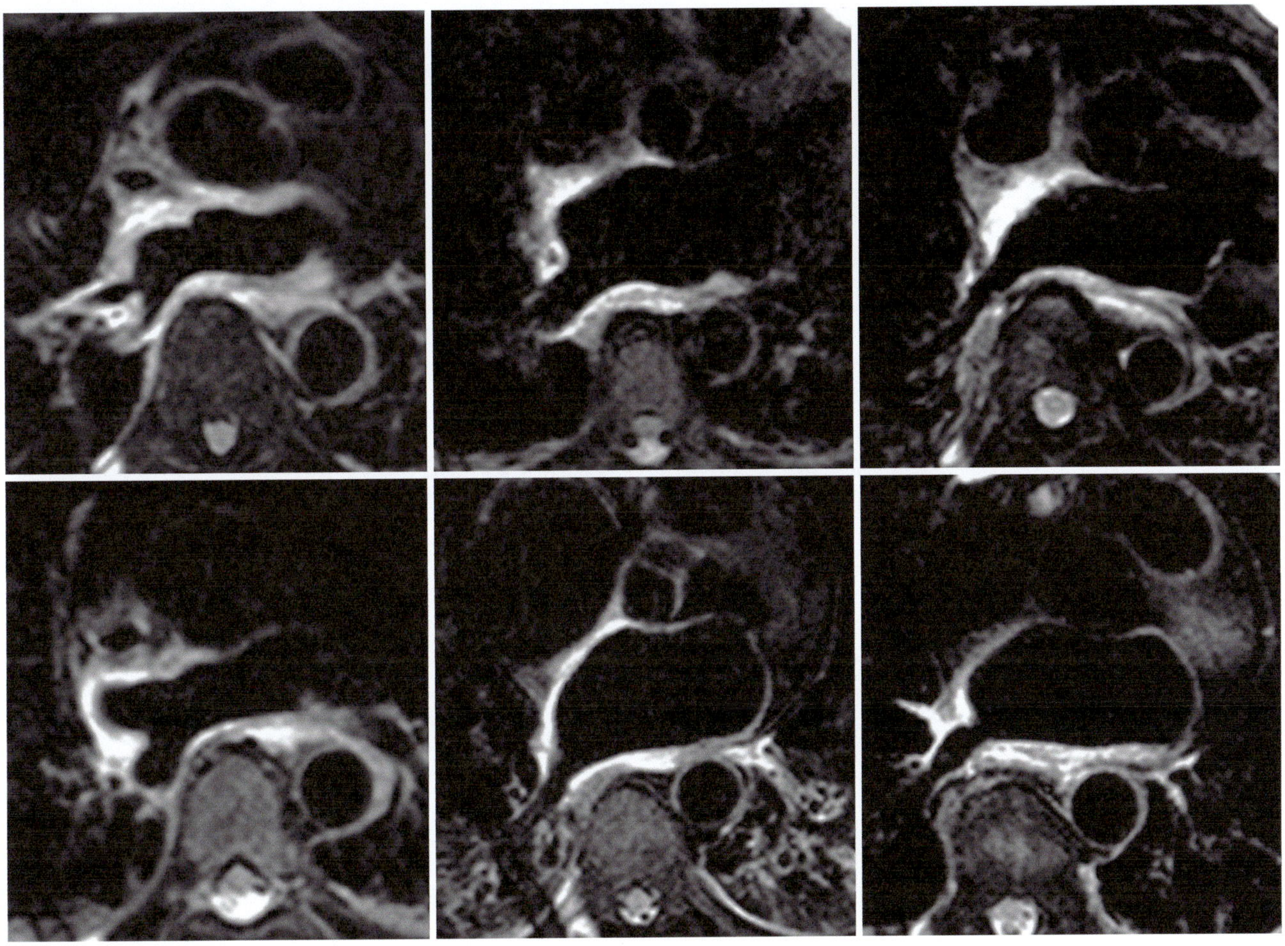

Fig. 2.5 *Post-ablation edema extends beyond ablated regions.* Cardiac MRI from six patients post-AF ablation demonstrates edema (bright T2w signal) extending not only in regions where RF energy was delivered (posterior wall and PV antrum) but also remote LA regions (anterior wall/dome and lateral wall)

for AF ablation, confirming earlier studies that total LA ablation scar burden is associated with AF termination.[19] However, complete PVA isolation is difficult to achieve and complicated by the fact that certain changes seen acutely are reversible over a 3-month period.

Acutely post-ablation voltage and LGE-MRI defined scar do not have a good correlation. However, acute LGE-MRI areas correlate well with areas of low voltage at 3 months. These areas of acute LGE-MRI likely represent areas with irreversible damage from RF ablation whereas the larger area of low voltage during the acute post-ablation period likely represents a combination of tissue edema, other reversible changes, and areas that will scar completely.

LGE-MRI can also accurately identify the location of breaks in ablation lesion sets, and its correlation with conduction recovery, which may explain post-ablation AF recurrences.[19] Badger et al.[20] demonstrated that AF recurrences following ablation are associated with significant gaps between lesions, and that these gaps correlated well with recovery of local EGMs or PV electrical conduction. This has allowed to better plan and tailor re-do procedures for patients with PV tachycardias, atrial/flutters, or atrial fibrillation.

2.4 The Future: Real-Time-MRI

The assessment of lesion formation during electrophysiologic procedures has always been a challenge; cMRI allows for visualization of location and extent of RF ablation lesion, scar formation in the myocardium, and potentially real-time assessment of lesion formation. Real-time MRI (RT-MRI)–based imaging and ablation system has the potential advantage of tissue lesion visualization during RF delivery, which could be used as an ablation end point. Electrophysiology RT-MRI-guided procedures have been carried out by a few laboratories.

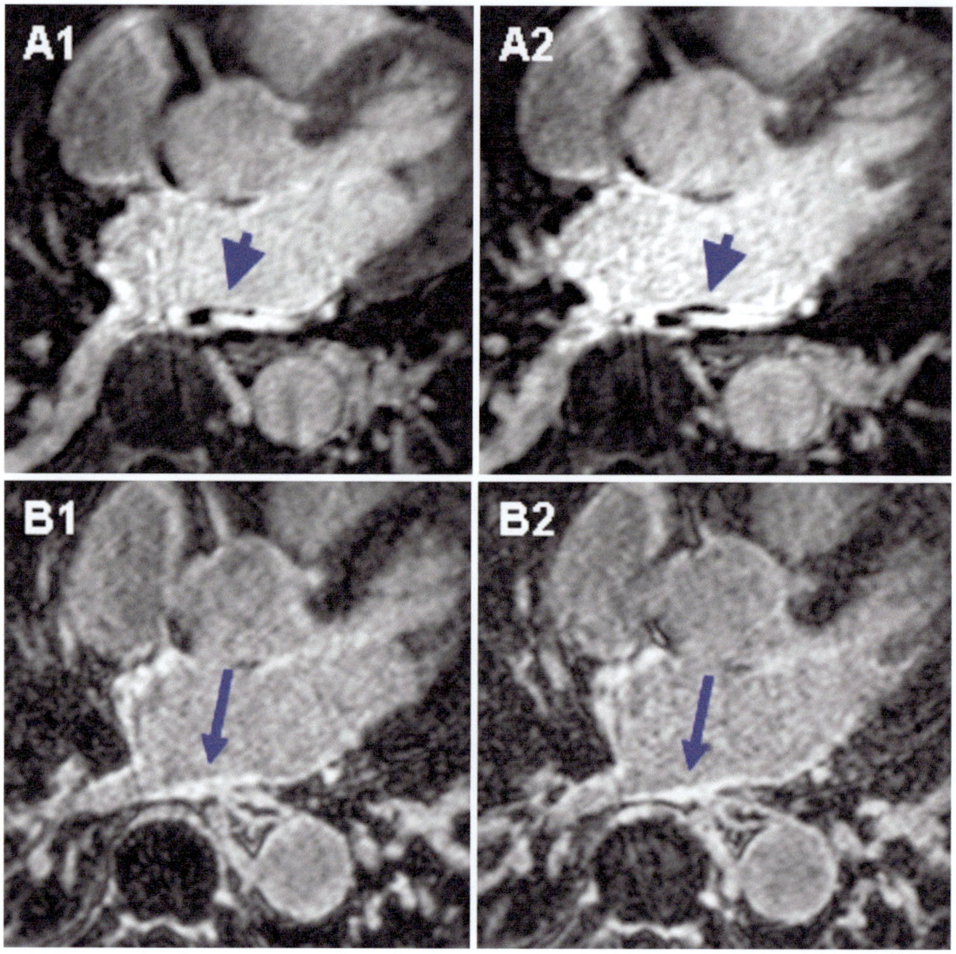

Fig. 2.6 *Non-reflow phenomenon on LGE-MRI.* (**a1, a2**) Cardiac LGE-MRI of two patients immediately following AF ablation demonstrates areas of LGE mixed with areas of no enhancement in the posterior wall (*blue arrowheads*). These same patients underwent LGE-MRI at 3 months post ablation. (**b1**, **b2**)The above areas correlated well with scar formation in the posterior wall (*blue arrows*)

MRI-guided ablation in the atrium has recently been reported by Schmidt et al.[21] and by Hoffmann et al.[22] In one of these studies, MRI angiography of the atrium was acquired, the atrium surface was segmented, and real-time catheter navigation was then carried out using this 3D reconstruction; however, no images were acquired during ablation[21]; rather, immediately postablation lesion formation was confirmed by LGE imaging. In the other study,[22] the catheters were navigated using RT-MRI sequences; however, there was no immediate tissue visualization during RF delivery and lesion formation, although there was T2w evaluation of the ablation site just before and after the ablation of the cavo-tricuspid isthmus. These two studies were done in 1.5-T MRI.

At our EP-MRI suite, we could demonstrate the feasibility to safely navigate and pace and record intracardiac EGMs in the atrial chambers under 3-T real-time MRI guidance.[23]

We used a novel 3-T real-time (RT) MRI based porcine RF ablation model with visualization of lesion formation in the atrium during RF energy delivery (Fig. 2.7).

In this model, RF energy was delivered under cMRI visualization at 3-T using custom RT-MRI software. A novel MRI-compatible mapping and ablation catheter was also used. Under RT-MRI, this catheter was guided and positioned within either the left or right atrium. Unipolar and bipolar electrograms were recorded. The catheter tip-tissue interface was then visualized with a T1w FLASH (T1-weighted fast low angle shot) sequence. RF energy was then delivered in a power-controlled fashion, and myocardial changes and lesion formation were visualized with a T2w HASTE (half Fourier with single shot turbo spin echo) sequence during the ablation. The presence of a lesion was confirmed by LGE- MRI and macroscopic tissue examination.

According to these studies, MRI-compatible catheters can be navigated and RF energy safely delivered under 1.5- and 3 T RT MRI guidance. It was also feasible to record EGMs in the atrium and ventricle during real-time image acquisition. Real-time visualization of lesion as it forms during

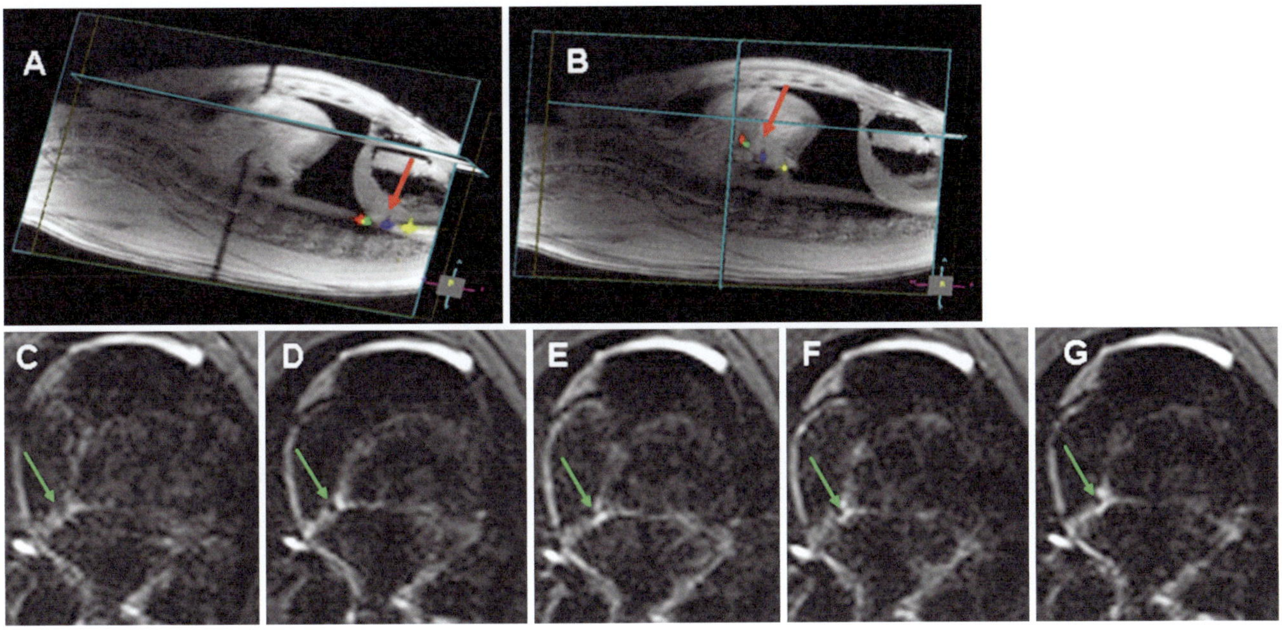

Fig. 2.7 *Real-time MRI ablation and lesion visualization at 3-T.* (**a** and **b**) MRI-compatible RF catheter guided under RT-MRI from the IVC into RA, the signal from the tracking elements is displayed and color coded (*red*: distal and *yellow*: proximal) to allow the operator catheter visualization. (**c–g**) Real-time 20 W power lesion can be seen (T2w HASTE) as it is being formed from catheter touch down (**c**) to 45 (**g**) (*green arrows*)

delivery of RF energy was possible and demonstrated using T2-w HASTE imaging under 3-T.

Finally, catheter visualization and myocardial tissue imaging under RT-MRI during RF energy could help improve ablation procedure outcomes by immediate assessment of ablation endpoints in the myocardium.

References

1. Mitsui T, Ijima H, Okamura K, Hori M. Transvenous electrocautery of the atrioventricular connection guided by the His electrogram. *Jpn Circ J*. 1978;42(3):313-318.
2. Feinberg WM, Blackshear JL, Laupacis A, Kronmal R, Hart RG. Prevalence, age distribution, and gender of patients with atrial fibrillation. Analysis and implications. *Arch Intern Med*. 1995;155: 469-473.
3. Wazni OM, Tsao HM, Chen SA, et al. Cardiovascular imaging in the management of atrial fibrillation. *J Am Coll Cardiol*. 2006;48(10): 2077-2084. Epub 2006 Nov 1.
4. Haïssaguerre M, Jaïs P, Shah DC, et al. Spontaneous initiation of atrial fibrillation by ectopic beats originating in the pulmonary veins. *N Engl J Med*. 1998;339(10):659-666.
5. Kato R, Lickfett L, Meininger G, et al. Pulmonary vein anatomy in patients undergoing catheter ablation of atrial fibrillation: lessons learned by use of magnetic resonance imaging. *Circulation*. 2003; 107(15):2004-2010.
6. Oakes RS, Badger TJ, Kholmovski EG, et al. Detection and quantification of left atrial structural remodeling with delayed-enhancement magnetic resonance imaging in patients with atrial fibrillation. *Circulation*. 2009;119(13):1758-1767.
7. Peters DC, Wylie JV, Hauser TH, et al. Recurrence of atrial fibrillation correlates with the extent of post-procedural late gadolinium enhancement: a pilot study. *JACC Cardiovasc Imaging*. 2009;2(3): 308-316.
8. Akoum N, Daccarett M, McGann C, et al. Atrial fibrosis helps select the appropriate patient and strategy in catheter ablation of atrial fibrillation: a DE-MRI guided approach. *J Cardiovasc Electrophysiol*. 2011;22(1):16-22.
9. Daccarett M, Badger TJ, Akoum N, et al. Association of left atrial fibrosis detected by delayed-enhancement magnetic resonance imaging and the risk of stroke in patients with atrial fibrillation. *J Am Coll Cardiol*. 2011;57(7):831-838.
10. Meng J, Peters DC, Hsing JM, et al. Late gadolinium enhancement of the esophagus is common on cardiac MR several months after pulmonary vein isolation: preliminary observations. *Pacing Clin Electrophysiol*. 2010;33(6):661-666.
11. Badger TJ, Adjei-Poku YA, Burgon NS, et al. Initial experience of assessing esophageal tissue injury and recovery using delayed-enhancement MRI after atrial fibrillation ablation. *Circ Arrhythm Electrophysiol*. 2009;2(6):620-625.
12. Bello D, Fieno DS, Kim RJ, et al. Infarct morphology identifies patients with substrate for sustained ventricular tachycardia. *J Am Coll Cardiol*. 2005;45(7):1104-1108.
13. Schmidt A, Azevedo CF, Cheng A, et al. Infarct tissue heterogeneity by magnetic resonance imaging identifies enhanced cardiac arrhythmia susceptibility in patients with left ventricular dysfunction. *Circulation*. 2007;115(15):2006-2014.

14. Ashikaga H, Sasano T, Dong J, et al. Magnetic resonance-based anatomical analysis of scar-related ventricular tachycardia: implications for catheter ablation. *Circ Res*. 2007;101(9):939-947.
15. Jain A, Tandri H, Calkins H, Bluemke DA. Role of cardiovascular magnetic resonance imaging in arrhythmogenic right ventricular dysplasia. *J Cardiovasc Magn Reson*. 2008;10(1):32.
16. McKenna WJ, Thiene G, Nava A, et al. Diagnosis of arrhythmogenic right ventricular dysplasia/cardiomyopathy. Task Force of the Working Group Myocardial and Pericardial Disease of the European Society of Cardiology and of the Scientific Council on Cardiomyopathies of the International Society and Federation of Cardiology. *Br Heart J*. 1994;71(3):215-218.
17. Klein LS, Shih HT, Hackett FK, Zipes DP, Miles WM. Radiofrequency catheter ablation of ventricular tachycardia in patients without structural heart disease. *Circulation*. 1992;85(5):1666-1674.
18. Vergara GR, Marrouche NF. Tailored management of atrial fibrillation using a LGE-MRI based model: from the clinic to the electrophysiology laboratory. *J Cardiovasc Electrophysiol*. 2011;22(4):481-487.
19. McGann CJ, Kholmovski EG, Oakes RS, et al. New magnetic resonance imaging-based method for defining the extent of left atrial wall injury after the ablation of atrial fibrillation. *J Am Coll Cardiol*. 2008;52(15):1263-1271.
20. Badger TJ, Daccarett M, Akoum NW, et al. Evaluation of left atrial lesions after initial and repeat atrial fibrillation ablation: lessons learned from delayed-enhancement MRI in repeat ablation procedures. *Circ Arrhythm Electrophysiol*. 2010;3(3):249-259.
21. Schmidt EJ, Mallozzi RP, Thiagalingam A, et al. Electroanatomic mapping and radiofrequency ablation of porcine left atria and atrioventricular nodes using magnetic resonance catheter tracking. *Circ Arrhythm Electrophysiol*. 2009;2(6):695-704.
22. Hoffmann BA, Koops A, Rostock T, et al. Interactive real-time mapping and catheter ablation of the cavotricuspid isthmus guided by magnetic resonance imaging in a porcine model. *Eur Heart J*. 2010;31(4):450-456. Epub 2009 Nov 5.
23. Vergara GR, Vijayakumar S, Kholmovski EG, et al. Real time MRI guided radiofrequency atrial ablation and visualization of lesion formation at 3-Tesla. *Heart Rhythm*. 2011;8(2):295-303.

Cardiac Computed Tomography: Description of Technology and Protocols

3

Quynh A. Truong, Subodh B. Joshi, and Udo Hoffmann

Abstract

A typical CT scanner has several main components: (1) the X-ray tube that emits X-ray beam, (2) the detector array located at the opposite side that collects the X-ray beam, (3) the gantry that holds and rotates the X-ray tube and detector array, and (4) the table where the patient lies on and moves through the rotating gantry. The human body lies in the center of the scanner within the field of view of the X-ray source, which emits a fan-like beam of X-rays to the detector array. The basic principle of cardiac CT is the use of a two-dimensional (2D) X-ray beam that passes through the body in continuous circular rotations to generate multiple 2D X-ray images to create a three-dimensional (3D) image of the heart. Although the gantry rotates a full 360°, only approximately 180° of data is needed to reconstruct one image: so-called half-scan reconstruction. Depending on the tissue content, there is attenuation of the X-ray beam as the body absorbs it. The photons that pass through the body are then collected by the detector array. Data acquired onto the detector array are digitalized as gray-scale pixels using "filtered back projection" technique, a method similar to that used in nuclear cardiology, to produce a cross-sectional image slice made up of a pixel matrix in the X–Y plane (typically 512 × 512 pixels). Multiple 2D cross-sectional image slices are added in series in the Z-direction (or longitudinal direction) to become a 3D reconstructed image.

Keywords

Cardiac computed tomography • CT technology • CT protocols for cardiology • CT data acquisition • Cardiac CT in electrophysiology

3.1 Description of CT Technology

3.1.1 History and Basic Principle of Computed Tomography

Sir Godfrey Hounsfield, an electrical engineer, constructed the first computed tomography (CT) scanner to image the brain in 1971. The brain CT scanner became commercially available 1 year later in 1972. He subsequently developed the first whole body CT scanner in 1974 and together with Allan Cormack, a physicist, who developed the signal-processing algorithms enabling CT image reconstructions, won the Nobel Prize for Physiology and Medicine in 1979 for the development of X-ray computed tomography.[1]

A typical CT scanner has several main components: (1) the X-ray tube that emits X-ray beam, (2) the detector array located at the opposite side that collects the X-ray beam, (3) the gantry that holds and rotates the X-ray tube and detector array, and (4) the table where the patient lies on and moves through the rotating gantry (Fig. 3.1a). The human body lies in the center of the scanner within the field of view of the

U. Hoffmann (✉)
Department of Radiology, Massachusetts General Hospital and Harvard Medical School, Boston, MA, USA
e-mail: uhoffmann@partners.org

A. Auricchio et al. (eds.), *Cardiac Imaging in Electrophysiology*,
DOI 10.1007/978-1-84882-486-7_3, © Springer-Verlag London Limited 2012

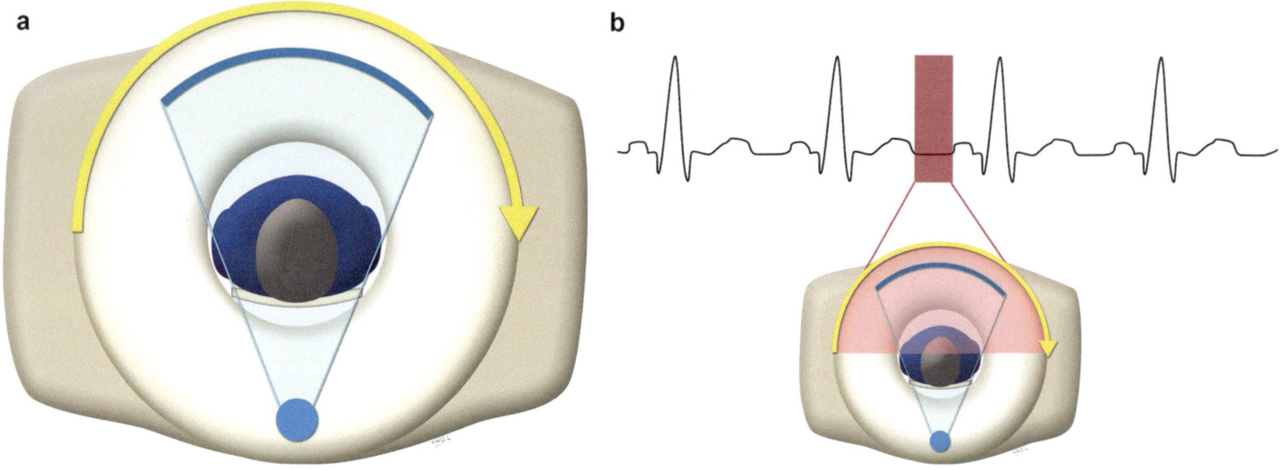

Fig. 3.1 (**a**) Single-source computed tomography with a gantry that houses the X-ray tube and detector array. The patient lies in the center on a table that moves through the rotating gantry. (**b**) With half-scan reconstruction, although the gantry rotates in circular 360° rotations continuously, data from 180° is needed to reconstruct one image

X-ray source, which emits a fan-like beam of X-rays to the detector array. The basic principle of cardiac CT is the use of a two-dimensional (2D) X-ray beam that passes through the body in continuous circular rotations to generate multiple 2D X-ray images to create a three-dimensional (3D) image of the heart. Although the gantry rotates a full 360°, only approximately 180° of data is needed to reconstruct one image: so-called half-scan reconstruction (Fig. 3.1b). Depending on the tissue content, there is attenuation of the X-ray beam as the body absorbs it. The photons that pass through the body are then collected by the detector array. Data acquired onto the detector array are digitalized as gray-scale pixels using "filtered back projection" technique, a method similar to that used in nuclear cardiology, to produce a cross-sectional image slice made up of a pixel matrix in the X–Y plane (typically 512 × 512 pixels). Multiple 2D cross-sectional image slices are added in series in the Z-direction (or longitudinal direction) to become a 3D reconstructed image. There are newer reconstruction algorithms being developed, such as iterative reconstruction, to improve image quality, limit noise, and reduce artifacts but the filter back projection method remains ubiquitous in most CT scanners.

The individual pixel has a value that is the equivalent of the linear transformation of the X-ray attenuation coefficient as referenced to the value of water and is called "Hounsfield units" or HU (Table 3.1). Pixels brighter or whiter than water will have a more positive HU, while those darker or blacker than water will have a negative HU, allowing for differentiation of iodinated contrast from other internal bodily structures. With proper timing of contrast injection, the appropriate cardiac chambers and vasculature can be opacified and isolated from the non-opacified structures.

With a 64-slice multi-detector CT (MDCT) system, the 64-slice refers to the maximum number of slices that could be acquired by that CT scanner in one rotation. However,

Table 3.1 Typical Hounsfield Units (HU) of common body substances

Substances	HU values
Air	−1,000
Lung	−800
Fat	−120 to −50
Water	0
Muscle	+40 to +80
Non-contrast-enhanced blood	+30 to +50
Iodinated contrast-enhanced blood	+130 to +500
Calcification	+130 to +600
Bone	+400 to +1,500
Metal	+1,000 or more

not all 64-slice MDCT scanners have the same number of detector rows or detector slice width. A detector array contains a number of detector rows, each with a particular slice width. The detectors can be either fixed array detectors or adaptive array detectors, dependent on the manufacturer. The fixed array detector design consists of detector rows with equal sizes (or slice width) to each other in the longitudinal direction, while the adaptive array detector design consists of detector rows with different sizes (or slice width). The focal spot (or X-ray source) emits a fan-shaped beam of X-rays, which is then narrowed by the collimator before the X-ray beams are transmitted to the patient and determines the collimated slice width at the scanner's isocenter (the point at the center of the gantry rotation). The CT operator can then pre-define the various slice collimation settings, which combines the signals of the detector rows and the desired collimated slice width, to yield the actual volume covered in one rotation. For example, with a 64-row fixed detector array and collimated slice width of 0.5–0.625 mm (dependent on manufacturer), 64-slices of data would provide 3.2–4 cm of volume coverage. The average volume needed to cover an entire heart in the Z-direction is approximately

12–16 cm; thus, obtaining full coverage of the entire heart would require summation of data over several heartbeats. With certain adaptive array detector 64-slice CT scanners, a "z-flying focal spot" or double z-sampling technique is used with periodic motion of the focal spot in the Z-direction and 40-row detector design, of which 32 rows had 0.6 mm detector collimation at isocenter. Two sequential 32-slices of data with 0.3 mm oversampling of data at isocenter resulted in 64 overlapping 0.6 mm slices per rotation, and a sampling scheme corresponding to that of a 64 × 0.3 mm detector with 64 slices in one rotation.[2] Regardless of the type of 64-slice MDCT scanner, depending on the user-defined slice collimation (typically 0.3–0.625 mm), 2–4 cm of heart volume could be covered in one heartbeat, and the entire heart could be covered in 4–8 heartbeats with one breath-hold of less than 15 s as the patient moves through the scanner. How fast a patient moves through the scanner is dependent upon the pitch of the scanner. The pitch is defined as the table movement per rotation/single slice collimation. If the pitch is set too fast, then there may be gaps in the data being collected. To avoid this problem, most modern cardiac CT scanners have an automated pitch adaptation based on the patient's heart rate, so that there is slight overlap of the data collected to ensure no gaps in image data collection and to improve longitudinal resolution.

3.1.2 Spatial and Temporal Resolution

Since the advent of cardiac CT, significant improvement in both spatial and temporal resolution has enabled its widespread applicability into the realm of electrophysiology pre- and post-procedurally. It is worth defining and distinguishing the difference between spatial and temporal resolution in order to comprehend the technological aspects of cardiac CT.

Spatial resolution is how close two objects can be and still be discernible from each other. As the spatial resolution improves, the CT scanner is able to detect smaller size objects. The spatial resolution of CT is comprised of the X–Y pixel size in the imaging plane (or the in-plane resolution of the axial slice) as well as the Z-plane (or longitudinal plane), creating a "voxel," which is the 3D volume counterpart for a pixel. Factors that can affect spatial resolution include slice collimation width, power of the X-ray source (both the tube current and voltage), and reconstruction algorithm kernels. Because the thickness of the fan beam (or slice collimation) determines the slice thickness and spatial resolution in the Z-axis as the patient moves through the CT scanner, modern multi-detector CT scanners of 64-slice or above now have isotropic or near-isotropic spatial resolution of 0.4–0.5 mm. This value, while still inferior, is near comparable to that of cardiac angiography, which has 2D spatial resolution of 0.2 mm, and is far superior to that of echocardiography and nuclear cardiology. While the in-plane resolution and even the Z-plane resolution of cardiac magnetic resonance imaging (CMR) could potentially be comparable to that of cardiac CT, this would be at the cost of prolonged imaging acquisition time. Presently, CT remains superior to CMR in spatial resolution, particularly with respect to the Z-axis.

Temporal resolution is how fast an image could be acquired such that motion does not create image blurring, similar to the shutter speed of a camera. The temporal resolution is most dependent on the speed of the gantry rotation. This speed is how fast the gantry is able to make one circular 360° rotation. Typical modern multi-detector CT scanners have gantry rotation speed of 330–500 ms. Since image acquisition in a single-source CT scanner requires just a 180° rotation for a complete cross-sectional image, the temporal resolution is thus half of the gantry speed rotation (i.e., 165–250 ms, dependent on the make and model of the CT scanners). As the temporal resolution improves, the CT scanner is able to acquire motion-free images at higher heart rate with fewer requirements for strict control of heart rate to <65 beats/min (bpm) for excellent image quality, thus sometimes negating the need for beta-blockers. The faster the temporal resolution of the CT, the faster the pitch of the table allowing for the patient to be expedited through the CT scanner and subjected to less radiation exposure.

3.1.3 Types of Cardiac CT Scanners

CT technology has advanced rapidly over the past decade, particularly since the introduction of the multi-detector CT (MDCT) in the late 1990s. In 1998, the four-slice MDCT became commercially available, allowing for simultaneous acquisition of four slices of images and with a gantry rotation of 500 ms, enabled full volumetric coverage of the heart over multiple heartbeats but with a single breath-hold of approximately 30 s. These innovative concepts produced the earliest mechanical cardiac CT images.[3,4] The detector width plays a special role in this respect. With four-slice detectors, more rotations of the X-ray source and thus more heartbeats are needed to get full coverage of the entire heart, leading to longer breath-hold requirement. CT technology continued to improve with wider detector array that could cover the range of 16-slices and then 64-slices in one rotation. When the 64-slice MDCT scanners became available in 2004, the rise in use of cardiac CT was rampant. Simultaneous acquisition of 64-slices with faster temporal resolution (165–250 ms) due to faster gantry rotation (330–500 ms) enabled full heart coverage with fewer heartbeats and in a single breath-hold of approximately 10–15 s. Shorter scan duration also had the added benefit of less radiation exposure. The main caveat with the 16- and 64-slice MDCT scanner was that the baseline heart rate ideally should be <65 bpm and if possible <60 bpm, to obtain adequate motion-free cardiac images,

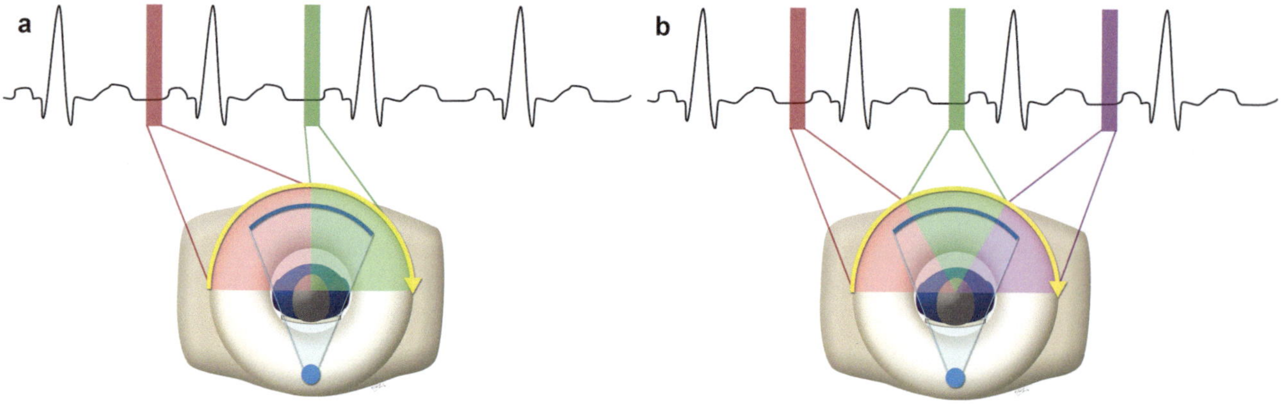

Fig. 3.2 (**a**) Bi-segment reconstruction algorithm combines data from 90° rotation of two consecutive heartbeats to achieve better temporal resolution. (**b**) Multi-segment reconstruction algorithm combines data from more than two consecutive heartbeats to further improve temporal resolution

thus, requiring the use of either oral or intravenous beta-blockers. There were certain "sweet spot" higher heart rates with the 64-slice MDCT that if regular, a special "bi-segment" reconstruction algorithm could be employed to improve the temporal resolution (Fig. 3.2a). This special sequence algorithm combines the data from two adjacent heartbeats; so instead of the required 180° rotation, it combines data from 90° rotation of the two heartbeats to achieve a temporal resolution of 83 ms, which is a quarter of the gantry rotation speed. However, this temporal resolution is only possible at when the heart rate is regular and without variability. For example, with a 64-slice MDCT scanner with a gantry rotation of 330 ms, the "sweet spot" heart rates are 66, 81, or 104 bpm.[5] Some CT vendors allow for "multi-segment" reconstruction by combining data from more than two consecutive heartbeats to further improve the temporal resolution (Fig. 3.2b).

In contrast to the single-source 64-slice CT scanners which have one X-ray tube source and one detector array located on opposite ends making cross-sectional images for each 180° rotation for a temporal resolution of 165–250 ms, newer dual-source CT scanners available since 2005 have two X-ray tube sources and detector arrays, aligned 90° apart from each other, with a gantry rotation of 330 ms (Fig. 3.3). With two X-ray tubes and detector arrays positioned at right angles to each other, instead of a full 180° rotation required from image acquisition, only 90° rotation is needed and the temporal resolution is decreased to 83 ms. Using a "bi-segment" reconstruction algorithm, the temporal resolution can be decreased even further to up to 42 ms with a mean temporal resolution of 60 ms.[5] This algorithm has great potential in certain electrophysiology research studies where improved temporal resolution is integral, such as the assessment for dyssynchrony; however, this is at the cost of slightly higher radiation dose since certain dose-saving algorithms cannot be employed under this current algorithm.

Newer MDCT scanners with wider detectors, such as 256-slice and 320-slice, have been available since 2007. While the spatial and temporal resolution of these scanners are comparable to that of 64-slice MDCT, the main benefit of these newer scanners is that the larger detector width and increase in number of slices allow for coverage of the entire heart in one heartbeat, with 12 cm of coverage for the 256-slice MDCT and 16 cm of coverage for the 320-slice MDCT. This has great potential for radiation dose reduction and may negate the issue of misregistration or slab artifacts, although beta-blockers are still needed to achieve low heart rate for motion-free images because of the temporal resolution of the CT scanner. Most recently, the dual-source CT "Flash" which has two X-ray sources and two 128-slice detectors promises to improve the temporal resolution further to 75 ms using single-segment reconstruction. The wider detector array will allow for full cardiac volume coverage in one heartbeat with an acquisition time of less than 1 s, typically 250 ms for 14 cm of coverage, which will significantly reduce radiation dose.

3.1.4 Types of CT Scan Acquisition

A *non-gated* CT can be acquired without electrocardiographic (ECG) synchronization. However, this is at the cost of severe motion artifact, particularly when imaging the rapidly moving heart. A non-gated CT scan is rarely used in cardiac imaging except as part of the test bolus scan when the structure of interest is the ascending aorta or for a pulmonary vein scan if the patient is in atrial fibrillation, where ECG-gating is not beneficial. When a patient is in sinus rhythm, *ECG-gated* CT is typically performed since synchronization of the images to the cardiac cycle allows for near motion-free images. Depending on the detector width, summation of the images from multiple consecutive heartbeats would yield a

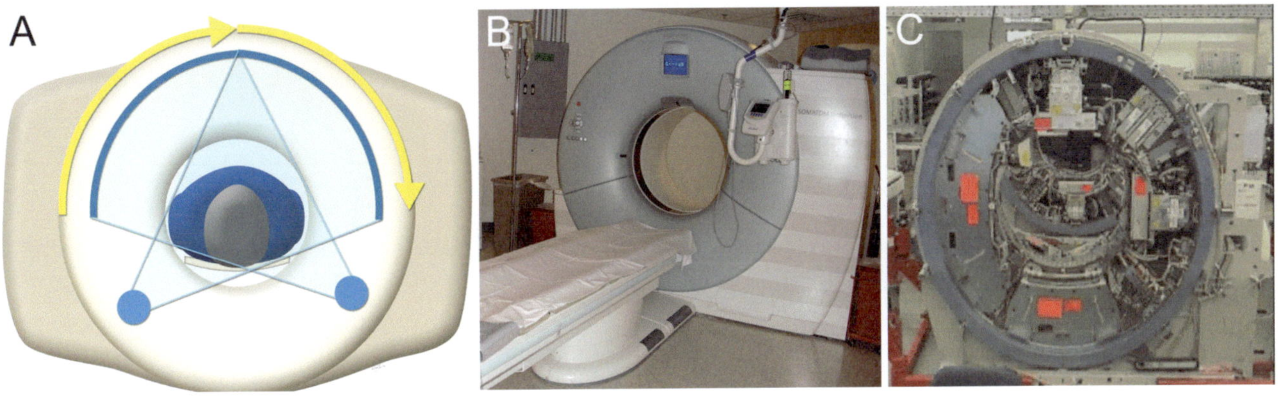

Fig. 3.3 (**a**) Dual-source computed tomography with a gantry that houses two X-ray tubes and detector arrays that are aligned 90° from each other. (**b**) External view of a dual-source computed tomography scanner in a typical CT scanner room. (**c**) Internal view of a dual-source computed tomography scanner [Fig. 3c – Courtesy of Stephan Achenbach and Christianne Leidecker (Siemens Medical Solutions USA)]

3D image of the heart. The ECG-gating can be either prospectively triggered or retrospectively gated (Fig. 3.4).[6]

In *prospectively ECG-triggered scanning* (Fig. 3.4c), the images are acquired sequentially at a user-defined set duration after the R-wave of the QRS complex for each heartbeat. This is also referred to as *"sequential"* or *"step-and-shoot"* scanning. Most modern software algorithms have a built-in arrhythmia detection mechanism, which will ignore and not trigger image acquisition after an ectopic beat. The primary benefit of prospective-triggered scans is the reduction in radiation dose, since the X-ray tube is only "on" during that one phase of the cardiac cycle when the images are being acquired. The disadvantage of a prospective-triggered scan is that functional assessment, such as left ventricular ejection fraction, which requires multiple phases throughout the cardiac cycle, is not possible because only one phase was acquired and available for analysis. In addition, image quality may be compromised with misregistration and slab artifacts or breath-hold may be prolonged if ectopic beats, arrhythmias, or heart rate variability occurs during the CT acquisition. Thus, the heart rate should ideally be regular and slow (<65 beats/min) with a prospectively triggered scan.

In *retrospectively ECG-gated scans* (Fig. 3.4a), the images are acquired continuously with the X-ray source "on" spinning in a spiral fashion as the CT table moves the patient in the Z-direction through the CT scanner. This is also called *"helical"* or *"spiral"* scanning. After the scan is completed, images can then be reconstructed throughout the entire cardiac cycle, allowing for the evaluation of functional assessment and ECG-editing if there is motion artifact or misregistration from an ectopic beat or arrhythmia. The trade-off is, of course, the increase in radiation dose since the X-ray tube is "on" continuously throughout several cardiac cycles. However, several dose-saving algorithms are available that can minimize the radiation dose a patient receives. In ECG-correlated tube current modulation, the X-ray tube current is decreased to 20% or even as low as 4% during the phases of the cardiac cycle not of interest (Fig. 3.4b).

3.1.5 Utilization of Cardiac CT in Electrophysiology

Cardiac CT has becoming an important part of electrophysiology for procedure planning, to improve procedural efficiency, and to monitor for complications. With respect to cardiac CT imaging in electrophysiology, the current appropriateness criteria from the major societies recommend use of cardiac CT for two purposes: (1) evaluation of pulmonary vein anatomy prior to invasive radiofrequency ablation for atrial fibrillation and (2) noninvasive coronary vein mapping prior to placement of biventricular pacemaker.[7] While there will be brief mention of the emerging role of cardiac CT for other potential electrophysiology indications, these remain investigational and should not be considered part of standard of care at the present time.

In atrial fibrillation management with the rise of pulmonary vein ablation, cardiac CT can define the pulmonary vein anatomy pre-ablation, facilitate co-registration with electroanatomic mapping intra-procedurally, and diagnose potential complications such as pulmonary vein stenosis. CT may have an emerging role in the pre-procedural planning and post-procedural assessment of left atrial appendage occluder devices (i.e., WATCHMAN), which are used to prevent thrombus formation and reduce the risk of stroke.[8,9] In addition, there is great potential for CT as a single imaging modality for cardiac resynchronization therapy (CRT) in defining coronary venous anatomy, evaluation for intraventricular dyssynchrony, and possibly myocardial scar assessment.[10,11]

In cardiac imaging of electrophysiology, every noninvasive imaging modality has its pros and cons. There is overlap

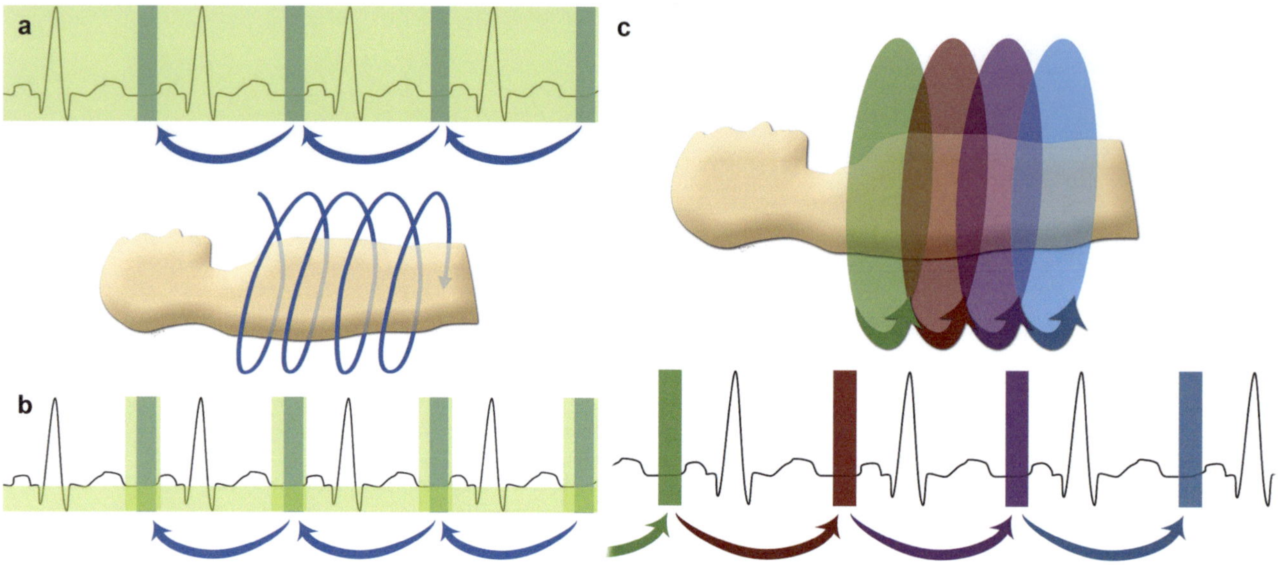

Fig. 3.4 (**a**) In retrospectively ECG-gated scan, the X-ray tube current (*in yellow*) is on maximum capacity and images are acquired continuously in a "*spiral*" or "*helical*" mode. Image reconstruction is performed afterward. (**b**) ECG-tube current modulation is a commonly used dose-saving algorithm employed with retrospectively ECG-gated scan. The X-ray tube current (*in yellow*) is decreased to 20% or as low as 4% during the phases of the cardiac cycle not of interest to reduce radiation exposure to the patient. (**c**) Prospectively ECG-triggered scan is another dose-saving algorithm mode. This "*sequential*" or "*step-and-shoot*" scanning method is triggered at a user-defined set duration after the QRS complex. The X-ray tube current is on only during image acquisition. Benefit of this mode is low radiation dose, but disadvantage is that functional datasets are not available for assessment

and sometimes exchangeability with cardiac magnetic resonance imaging (CMR) and echocardiography. Several factors should be taken into consideration with regard to which imaging modality should be performed. First, the expertise of the cardiac imager and what imaging modality is offered at the local institution is integral. Depending on the patient's tolerance on laying flat and issues of claustrophobia, cardiac CT could be preferable given the rapid scan acquisition time (typically <15 min for the entire scan) as compared to CMR (which are typically >45 min). Cardiac CT can be performed in patients with metal devices as opposed to CMR; however, this may be less preferable when repeated studies are needed given the issue of cumulative radiation dose, particularly in young patients. Renal impairment affects both cardiac CT and CMR. While contrast-induced nephropathy for CT has been a known entity, a new disease arose with the use of gadolinium-based contrast agents, called nephrogenic systemic fibrosis, which was first observed in 1997 and described in 2000. This irreversible skin complication occurred in renal impaired patients, particularly end-stage renal disease on dialysis.[12] This led many institutions to adopt the policy of low-dose gadolinium in patients with mild-to-moderate renal impairment (estimated glomerular filtration rate [eGFR] of 30–60 mL/min/1.73 m^2) and for severe renal disease (eGFR < 30 mL/min/1.73 m^2) including hemodialysis to perform the CMR without gadolinium or substitute with another alternative imaging modality. For this reason, for dialysis patients, cardiac CT would be the preferred modality over CMR since the iodinated contrast can be removed from the circulation by hemodialysis.

An added benefit of cardiac CT over other noninvasive imaging modalities is that simultaneous assessment for coronary artery disease is possible and easy to obtain with minimal modification of the scanning protocol. Noncardiac incidental findings, such as pulmonary disease, which are within the field of view should also be interpreted and could have the downside of repeated imaging tests that results in invasive procedures but may also lead to the early detection of cancer.[13]

For pulmonary vein ablations, both cardiac CT and CMR can be used for pulmonary venous and left atrial anatomy assessment, co-registration with electro-anatomical mapping intra-procedurally, and for post-procedural concern for pulmonary vein stenosis.[14-19] A potential advantage of cardiac CT is that left atrial appendage thrombus or slow flow can be identified as a filling defect on contrast-enhanced cardiac CT, which either persists (thrombus) or fills in (slow flow) with a 30-s to 1-min delayed non-contrast prospectively triggered CT scan.[20] However, at this time, confirmation should still be sought with transesophageal echocardiography. Contrast-enhanced CMR lacks the diagnostic accuracy for the detection of thrombi in the left atrial appendage,[21] though it has the benefit of repeated exams without the concern of cumulative radiation dose, particularly in younger patients. A recent feasibility study using CT image integration with remote catheter navigation system appears exciting.[22]

In the CRT patient population, 2D echocardiography is the most widely used for dyssynchrony assessment, but has been limited by issues of reliability and inability to predict CRT response.[23] Three-dimensional echocardiography[24-26] faces many of the same challenges intrinsic to 2D echocardiography with regard to operator dependency, limited acquisition windows, varying angle planes, and poor spatial resolution. While CMR[27-31] has great potential for dyssynchrony assessment, its general applicability may be limited by long scan time (typically >45 min), complex protocols (requires highly trained cardiac imager), and metal incompatibility. Since many heart failure patients with severe systolic dysfunction already have implantable defibrillators or biventricular devices, upgrades from implantable defibrillators or lead revisions for those with preexisting CRT devices precludes the use of CMR. In this population, cardiac CT can provide detailed coronary venous anatomy[15,32-37] due to its superior spatial resolution. A proof of principle study using cardiac CT with changes in radial wall thickness assessment for dyssynchrony has excellent reproducibility, though needs to be validated prospectively using better temporal resolution CT scanners with correlation to CRT response.[11] CT has the potential to be used a single imaging modality for assessment of coronary veins, dyssynchrony, and myocardial scar, though the latter two are still at the investigational stage.[10]

3.1.6 Risks of Cardiac CT

3.1.6.1 Radiation

CT scanning results in a measurable radiation exposure. This issue of radiation and risk of cancer from CT is of recent concern.[38-41] The median effective dose of cardiac CT exams in 2007 from 50 international centers in 1,965 patients was a dose-length product (DLP) of 885 mGy × cm, which corresponds to an estimated radiation dose of 12 mSv.[41] This amount of radiation is still within the acceptable annual radiation exposure limit. To put the radiation dose of a cardiac CT scan[42] in a wider perspective, Table 3.2 lists the estimated radiation dose from various sources.

Table 3.2 Estimated radiation dose

	Estimated radiation dose
Natural background radiation	3.6 mSv/year
Allowed exposure medical radiation workers	50 mSv/year
Diagnostic coronary angiogram	3–10 mSv
Nuclear stress test (rest and stress)	14–22 mSv
Elective/emergency PCI	9–29 mSv
64-slice cardiac CTA	9–12 mSv

Radiation dose is cumulative and the lifetime risk of malignancy is highly dependent on age, with female gender more vulnerable due to the proximity of the breasts.[38] Furthermore, the estimated risks of fatal malignancy or death resulting from radiation exposure or lifetime odds of dying per 1,000 individuals from a 10 mSv cardiac CT angiography is 0.5, while that from a motor vehicle accident is 11.9, pedestrian accident is 1.6, and being struck by lightning is 0.013.[43] Although repeated cardiac CT scanning should be avoided when possible, this needs to be weighed with respect to the risk-benefit profile of the patient. In young patients, especially women, CMR would be preferable over CT for pre-imaging of pulmonary vein ablation given the possibility of repeated scans should suspicion of pulmonary vein stenosis or need for re-do ablations be required. However, other factors need to be considered such as cardiac imager expertise, institutional capability, scanner availability, patient claustrophobia, and ability to perform adequate breath-hold instructions, end-stage renal disease contraindication, and metal incompatibility issues. For the elderly patients and CRT population, the decade or more required for malignancy to develop makes radiation less of an issue. Particularly, in the high-risk group of refractory New York Heart Association (NYHA) Class III–IV heart failure patients with 50% 5-year mortality,[44] judicious use of CT likely has a favorable benefit-risk profile when being used to help guide appropriate usage of CRT, a life-saving treatment.

Regardless, if cardiac CT should be performed, dose-saving algorithms are highly recommended to reduce the radiation exposure to the patient. There are several ways to reduce the radiation dose. The most commonly used dose-saving algorithm in clinical practice was ECG-correlated tube current modulation resulted in 25% reduction of dose estimates. Less frequently used algorithms of lower 100 kV tube current instead of the typical 120 kV tube current resulted in 46% reduction and sequential or prospectively triggered scan resulted in a 78% reduction. The automatic exposure control resulted in no change in radiation dose. Other independent factors attributable to higher radiation dose were patient weight, absence of stable sinus rhythm, scan length, or range.[41] Thus, to adhere to the ALARA (As Low As Reasonably Achievable) principle, ECG-tube current modulation, lower tube voltage of 100 kVp and tube current in thinner patients, prospectively triggered scans if functional assessment is not needed, and minimizing the scan range should be employed when possible to reduce radiation dose.[38,41,43] In addition, future dose reductions may be achieved by the addition of appropriate organ improved detector efficiency, advanced organ shields, advanced filters and post-processing algorithms. However, radiation dose reduction at the expense of diagnostic image quality should be avoided. Hopefully, as newer CT technology continues to be developed where sub-mSv whole heart CT scans can be performed in one heartbeat, many of these dose-saving algorithms may become obsolete in the future.

3.1.6.2 Contrast Nephropathy

Contrast-induced nephropathy (CIN) can develop in patients with already limited renal function, and while its definition is highly variable, typically occurs 48 h after contrast administration, it persists for 2–5 days, and usually it is reversible by 7–10 days.[45] In patients with normal renal function (eGFR >60 mL/min/1.73 m^2), this event is unlikely and occurs in less than 2% of patients.[46] Oral hydration is recommended the day prior to the CT scan to reduce the risk of CIN. For subjects with diabetes on metformin (glucophage), they will be require to stop metformin (glucophage) for at least 48 h after the administration of contrast due to the possibility of developing lactic acidosis. For patients with eGFR 30–60 mL/min/1.73 m^2, pre-hydration with oral and intravenous fluid should be considered as prophylactic measure, if they are not heart failure patients. Optional *N*-acetylcysteine (Mucomyst) 600 mg orally twice a day one day prior and after could be considered; however, there is no convincing data on the efficacy of this therapy. For patients with eGFR <30 mL/min/1.73 m^2 not on hemodialysis, contrast-enhanced CT should be avoided if possible, with the risk-benefit profile individualized for every patient and consultation with a nephrologist should the CT scan be deemed necessary. For patients on hemodialysis, contrast-enhanced CT could be performed with dialysis scheduled within 1–3 days.

3.1.6.3 Extravasation

Even when checked for proper positioning, extravasation can occur when contrast material leaks out of the intravenous (IV) line and collects under the skin of the arm. If available, extravasation electrode detection patch should be used to alarm the CT technologist of possible extravasation. CT departments should have extravasation protocols in place and treatment would depend on the amount of contrast extravasated as well as whether there is tissue compromise. Typically, observation is all that is needed when there is no evidence of tissue compromise. In rare severe cases, if compartment syndrome is suspected, then vascular surgery consultation will be need. Rarely an infection may develop.

3.1.6.4 Allergy

The administration of iodinated contrast can result in allergic reactions. Allergic reactions such as skin reactions and in rare occasions anaphylaxis have been reported. Most of them are very mild (itching, rash) with only 1:1,000,000 fatal cases reported. In the unlikely case of a serious allergic reaction, intravenous epinephrine, antihistamines, and corticosteroids should be available on-site as well as immediate referral to the emergency department. If a patient reports a prior history of mild contrast reaction, such as hives or less, then premedication with steroid and anti-histamine should be given prior to the contrast administration. A typical premedication dose regimen would be Prednisone 50 mg orally 13, 7, and 1 h before the scan and diphenhydramine (Benadryl) 50 mg orally 1 h prior to receiving intravenous contrast. The association between shellfish allergy and iodinated contrast is weak and should not preclude the use of contrast; however, a detailed inquiry on the severity of the shellfish allergy should be taken with consideration for premedicating if the reaction to shellfish is severe.

3.2 Overview of Basic CT Protocols

The technique for performing cardiac CT for EP applications is similar to that used for coronary artery imaging with some notable exceptions. Nitroglycerin is not necessary and should not be given for either pulmonary vein or coronary vein protocols, unless the coronary arteries are to be evaluated. Giving nitroglycerin to a coronary vein study may make interpretation more difficult, since the coronary arteries will be dilated and overlap even more so with the coronary veins. In addition, for both pulmonary and coronary vein imaging, reconstructions during systole will bring out the maximum size of the left atrium and cardiac veins. Furthermore, acquiring the CT scans in the caudo-cranial direction may be helpful to avoid the contrast interference in the superior vena cava for the pulmonary veins and since the coronary sinus is an inferiorly positioned cardiac structure. The breathing instructions, need for heart rate control, duration of contrast administration, and timing of scan acquisition may vary slightly depending on the indication.

3.2.1 Patient Preparation

Proper patient preparation is key to obtaining an excellent image quality CT scan. Ensure that at minimum, a working 18-gauge or larger IV is in place, preferably in the right antecubital position. The IV line should be checked and any presence of air bubbles should be removed. This is particularly important if the patient has a patent foramen ovale to avoid the risk of air embolism to the left-sided circulatory system.

Specifics on whether a non-gated versus ECG-gated CT scan has to be performed will be discussed in the later section of this chapter. If an ECG-gated CT scan is to be performed, then four ECG leads should be attached to non-metallic ECG electrodes on the thorax. Two ECG electrodes should be near each shoulder-clavicular region and the other two electrodes on each side of the thorax at the level of the diaphragm. The optimal position would avoid having the lead wires over the chest and the heart, since this may cause artifacts from the wires.

Breath-holding instruction is crucial to minimize respiratory motion artifacts, which cannot be fixed by ECG-editing or other post-processing steps. It is important to practice the breath-hold instruction several times with the patient, paying close attention to the patient's diaphragm and abdomen for any signs of air escaping. Simultaneous evaluation of the ECG monitor for heart rate changes, variability, and ectopic beats during breath-hold is also helpful. The breath-hold can be at inspiration or expiration, depending on the protocol.

Lastly, forewarn the patients regarding how many CT image acquisitions (i.e., breath-hold) sequences they will have to undergo. Let them know which scans require contrast and that they may feel a warm, flushed, hot sensation going up their arm from where the IV is located to all over their body. They may feel like they have to urinate and then have a mild shiver afterward. If a test bolus scan was performed, let them know that whatever they felt with the test bolus scan will be greatly intensified with the subsequent CT contrast-enhanced scan. Reassure them that it is not an allergic reaction but just the contrast transiting through their body. A well-informed patient will be a happy patient. All the meanwhile, watch for signs of an allergic reaction.

3.2.2 CT Data Acquisition

There are many parameters that can be modified for acquiring images for a diagnostic CT study. The below is an example but is not meant to be exclusive as alterations to sample protocols are variable and may differ by institutions and CT scanners. The basic CT scan consists of 2–4 steps, depending on the protocol.

The first image acquisition is the scout image or topogram (Fig. 3.5a). The heart position is localized in a projectional topographic scan of the chest. This low-dose projectional image of the chest is similar to an anterior-posterior chest radiograph and ensures that we are imaging the chest and not some other body part. From the topogram, we can set up our next image acquisition by localizing the imaging slice for the test bolus or bolus tracking and set up the scan coverage for the actual CT angiography or venography.

To determine the optimal time to image the contrast-enhanced heart, either a test bolus scan or bolus tracking technique can be used. We prefer the test bolus scan because it gives the patient an extra "practice" with their breath-hold, allows us to see how the contrast affects the heart rate (which may require us to give more beta blocker), and also reassures us that the IV is working properly with the high-power contrast injector.

With a test bolus scan (Fig. 3.5b), a small volume of iodinated contrast is injected (typically 10–20 mL of contrast followed by 40 mL of saline chase) at a rate of 5–6 mL/s. Repeated non-gated low-dose images at one slice level are obtained, with each image being 2 s apart. The typical position is 1 cm below the level of the carina bifurcation for the ascending aorta. At this level, we should see the ascending aorta as a circular structure anterior to the right pulmonary artery. The contrast should appear first in the pulmonary artery then ascending aorta, and image acquisition can stop once peak opacification of the ascending aorta has occurred. A region of interest in the ascending aorta is placed to determine the image number where peak opacification has occurred. The contrast agent transit time is then calculated as the summation of this peak image opacification time plus the delay time set on the CT scanner for the breathing instruction. While the ascending aorta is used for most standard arterial phase imaging, modification with the region of interest placed at the descending aorta can also be used.

For arterial phase imaging, this contrast agent transit time will be the delay time set for the CT angiography. The amount of volume used for the CT angiography is calculated from the scan range time plus an extra couple of seconds multiplied by the same flow rate used in the test bolus of 5 or 6 mL/s. The addition of the extra 2 s will ensure that there will be enough contrast covering the desired scan range even with heart rate variability.

Alternatively, an automated bolus tracking scan could be used and monitors a region of interest in the ascending aorta in real time.[47] Once a pre-defined attenuation value of about 110–150 HU is reached, the breath-hold instruction is given (~4–6 s), and the CT scanner automatically initiates the CT angiography image acquisition with contrast enhancement increased to the desired level of enhancement. Using this technique, the CT technologist needs to pay close attention and manually override and initiate image acquisition should the CT scanner not automatically start acquiring images.

For the pulmonary vein protocol, an additional non-contrast prospectively ECG-triggered low-dose CT scan can be performed after the contrast-enhanced CT to confirm left atrial thrombus. The decision to use either inspiratory or expiratory breath-hold, test bolus versus bolus tracking, prospectively triggered or retrospectively gated CT scanning, and use of dose-saving algorithms needs to be made prior to the CT scan acquisition.

3.2.3 Image Reconstructions and Post-processing Techniques

In cardiac CT scans, knowledge on the options for image reconstruction and post-processing techniques is integral for interpretation of the CT scan, particularly with the retrospectively ECG-gated CT scans. While the coronary arteries are typically reconstructed in diastole when there is minimal cardiac motion, for both pulmonary vein and coronary vein assessments, image reconstruction is better during systole when the left atrium and the cardiac veins are most plethoric.[35] Consideration for reconstruction using absolute time intervals after the onset of the QRS (for example, at 250 ms) instead of the typical percentage phase reconstruction (for example, at 35%) may provide

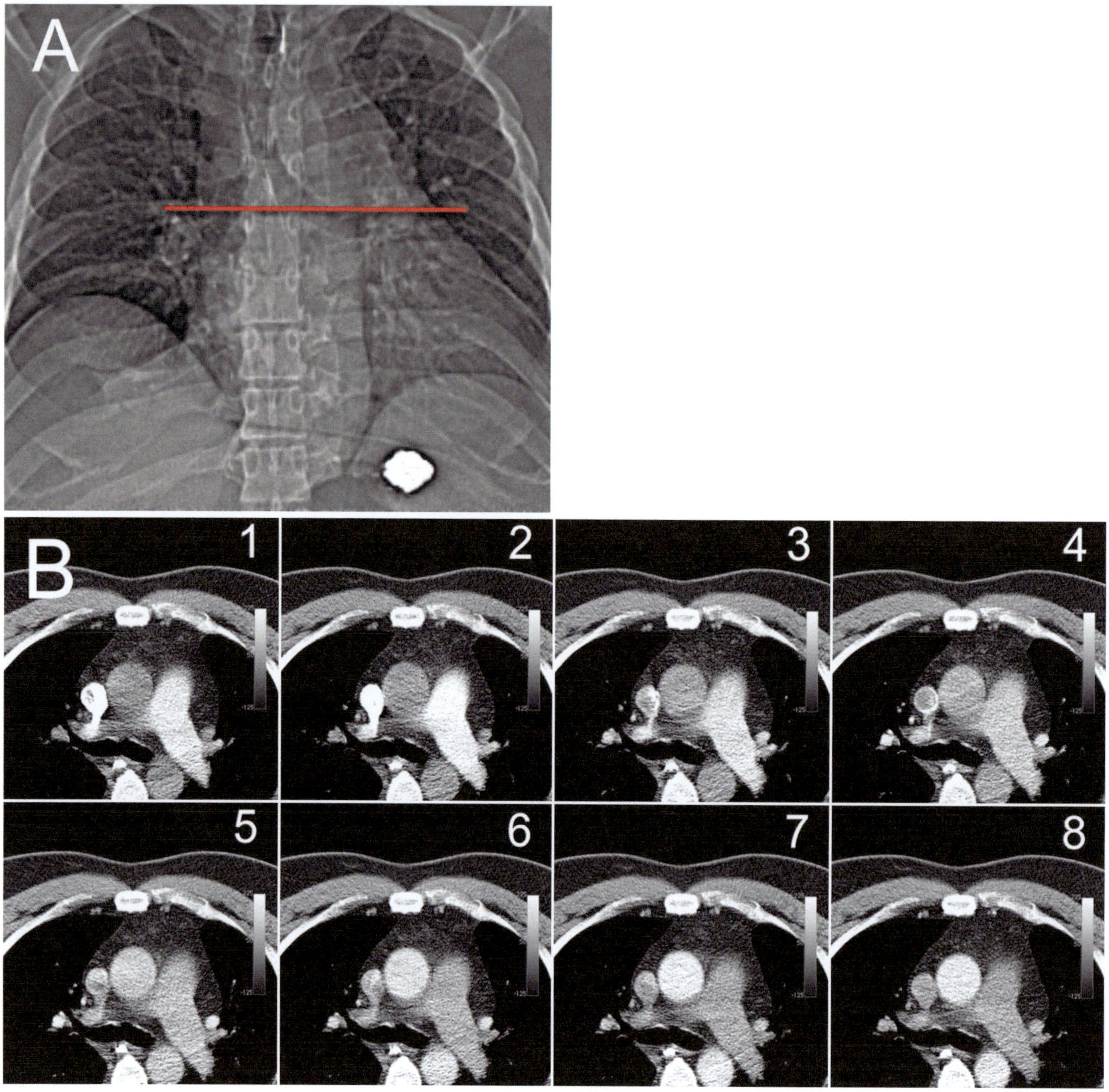

Fig. 3.5 (**a**) Scout localization with a "*topogram*" is the first image acquisition of a CT scan. The red line is positioned at the image slice location for the test bolus or bolus tracking. At this slice level (approximately 1 cm below bifurcation of the carina), the ascending aorta and pulmonary artery should be in the field of view. Note the bright irregularly shaped circular density is the oral barium in the stomach. (**b**) The test bolus scan is acquisition of repeated sequential images at the user-defined slice level each obtained 2 s apart. The ascending aorta is the circular structure anterior to the right pulmonary artery. Contrast is seen first opacifying the pulmonary artery with peak opacification of the ascending aorta at image 7. The contrast agent transit time can then be optimally calculated from this test bolus scan

less motion artifacts and result in better image quality in patients with atrial fibrillation or frequent ectopy, since the duration of diastole may be more variable than systole. While the details of image reconstructions and techniques are beyond the scope of this chapter, a brief description on the various types of image reconstructions is worth mentioning.

Trans-axial images are the source data that are the basic images to any cardiac CT scan. These images are independent of whether a scan is non-gated, prospectively ECG-triggered, or retrospectively ECG-gated. The axial images have typically an in-plane resolution of 512 × 512 pixels and do not require special cardiac CT software for viewing. They can be visualized by scrolling up and down the images in the Z-direction.

Multi-planar reformations (MPR) are images that are reconstructed from the original trans-axial images and can be viewed in 3D planes with all voxels on that plane visualized in the planar image. Standard MPR images are oriented in the axial, sagittal, and coronal planes but can be manipulated by the user and viewed in any obliquity, including the double-oblique view to get the true short-axis of a lumen.

In maximum intensity projections (MIP), the voxels with the maximum HUs are accentuated in the 2D parallel plane of the X-ray, thus producing better contrast differentiation. The MIP images can be reconstructed at different slab or slice thickness and at any orientation plane similar to the MPR planes. These images are frequently used to view the longitudinal plane of a contrast-enhanced structure.

In volume-rendering techniques (VRT), advanced 3D post-processing software algorithm is applied to the 3D stack of data and each voxel is assigned a color based on a voxel-intensity histogram. Most cardiac CT software packages have automated VRT available that can segment out the contrast-filled structure and remove high-attenuation structures, such as bone, from the image to provide a visually pleasing color 3D-volumetric image. For both pulmonary vein and coronary venous scans, VRT images are frequently made due to its ease of interpretation of anatomy for the electrophysiologist. An advanced form of VRT commonly used in pulmonary vein studies is the "virtual" endoluminal view or "fly-through" view.

Multi-phase reconstructions (also commonly abbreviated as MPR) can be performed with retrospective ECG-gating CT scans for functional analysis. This is not to be confused with multi-planar reformations. In multi-phase reconstructions, images are reconstructed at fixed intervals of the cardiac cycle, allowing for 4D-imaging (with time being the fourth dimension). With the 64-slice MDCT scanners, images can be reconstructed at every 10% increment of the cardiac cycle or 10 phases. While technically, reconstructing at 5% increments of the cardiac cycle or 20 phases with the 64-slice CT scanner would appear to provide better temporal resolution, however, this would be artificial and is limited by the temporal resolution of the CT scanner itself. The use of 5% intervals for 20 phases is not limited with the faster temporal resolution CT scanners, such as the dual-source CT, and could be utilized to achieve better visualization of cardiac function. In multi-phase reconstructions, any of the 3D reconstruction algorithms such as MPR, MIP, and VRT can be applied for 4D-viewing.

3.3 Description of Electrophysiology-Specific CT Protocols

This section will focus on the specific CT protocols for the two purposes deemed "appropriate" by the major societies for use of cardiac CT: pulmonary vein and coronary vein anatomy.[7] While there are numerous variations for these protocols due to variable patient profile (i.e., whether the patient is in atrial fibrillation or sinus rhythm, whether the heart rate is fast or slow) and different CT scanners (i.e., 64-slice CT or dual-source CT), the below protocols can provide an overview on parameters that could then be modified and optimized for the individual patient.

3.3.1 Pulmonary Veins

Whether a pulmonary vein study is performed for pre-atrial fibrillation ablation or for evaluation of post-procedural complication, such as pulmonary vein stenosis, the protocol is similar. If a patient is in atrial fibrillation for a pulmonary vein study, then a non-gated cardiac CT should be considered. Since the pulmonary veins and left atrial anatomy are large objects, detailed sub-millimeter spatial resolution is not a requirement. Alternatively, a retrospectively ECG-gated CT could be performed with reconstruction performed during systole although beta-blockers should be given to slow the heart rate to as close as 60 bpm with the 64-slice MDCT scanner and less than 100 bpm with the dual-source CT scanner to achieve the best temporal resolution the CT scanner could provide.

Because the contrast enhancement of the left atrium and pulmonary veins occurs slightly earlier than the opacification of the ascending aorta, arterial phase imaging similar to that of a coronary artery study could be used. Nitroglycerin is not necessary unless the coronary arteries are also in question. Beta-blockers can be given if necessary to optimize the heart rate. For pre-ablation patients, because of the potential risk of atrio-esophageal fistula during the ablation procedure, a teaspoon of oral barium can be given prior to the CT scan to coat the esophagus to localize its relationship to the left atrium and pulmonary veins.[48-51] Oral barium may cause streak artifact in the CT dataset and is optional, depending on the referring electrophysiologist's preference. Oral barium is not necessary if the CT exam is to evaluate for pulmonary vein stenosis. To optimize co-registration with the electro-anatomical mapping where the ablation procedure is performed with free-breathing, CT image acquisition should be obtained with an expiration breath-hold (Fig. 3.6a).

After the topogram, either a test bolus or bolus tracking technique can be utilized. Scan coverage should include the aortic arch vessels to the level of the bottom of the diaphragm. Caudo-cranial acquisition would minimize the interference of contrast flux in the superior vena cava. During the CT contrast scan acquisition, inspection of the left atrial appendage for a potential filling defect is recommended with immediate reconstruction of one dataset for closer evaluation. If a filling defect is suspected, then an additional non-contrast-enhanced delayed CT scan should be performed within 30 s to 1 min after the original contrast CT scan to differentiate between a left atrial thrombus (which will remain as a persistent filling defect) versus slow flow (which will "fill in" with resolution of the filling defect). To minimize the patient's radiation dose, the delayed scan should be prospectively triggered and image acquisition set at the level of the aortic arch to the middle of the ventricle just to target the left atrial appendage region (Fig. 3.7).

Image reconstruction is typically performed during systole when the left atrium is largest using either fixed distance after the QRS duration (i.e., 250 ms) or a percentage phase

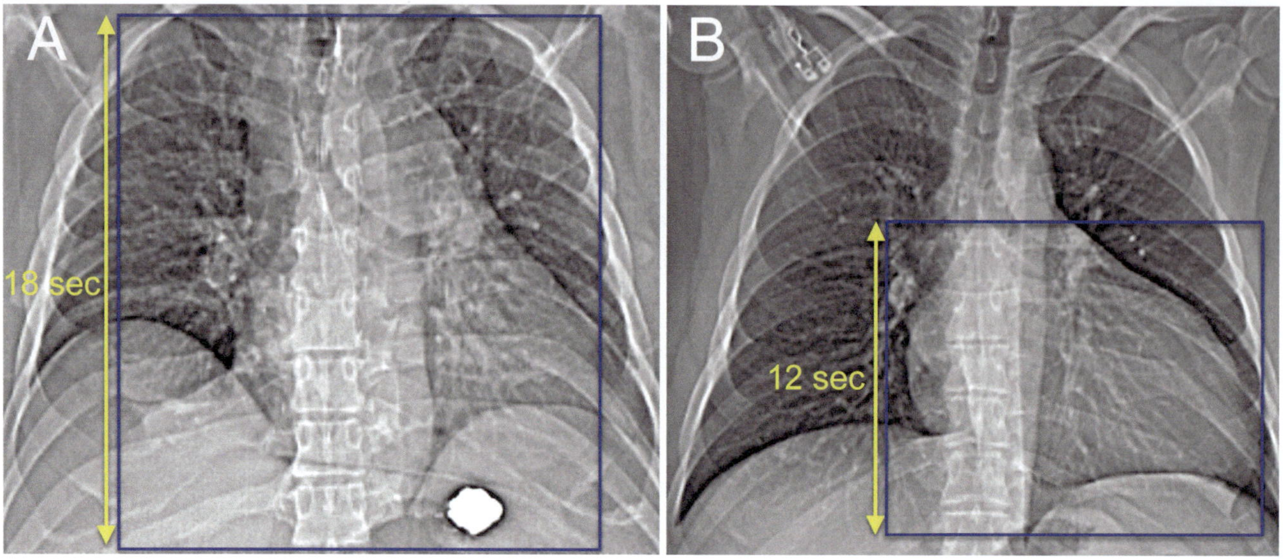

Fig. 3.6 (**a**) Topogram of a pulmonary vein study is notable for an expiratory breath-hold. The scan coverage for the contrast-enhanced CT is highlighted in the *blue box* and should range from the aortic arch vessels to the diaphragm. The scan coverage time can be multiplied by the rate of contrast infusion to obtain the estimate total contrast volume needed for the scan. (**b**) Topogram of a coronary vein study is notable for an inspiratory breath-hold. Note the scan coverage for the CT venography is similar to that of a CT angiography and less than a pulmonary vein study since coverage up to the arch vessel is not necessary for coronary venous anatomy

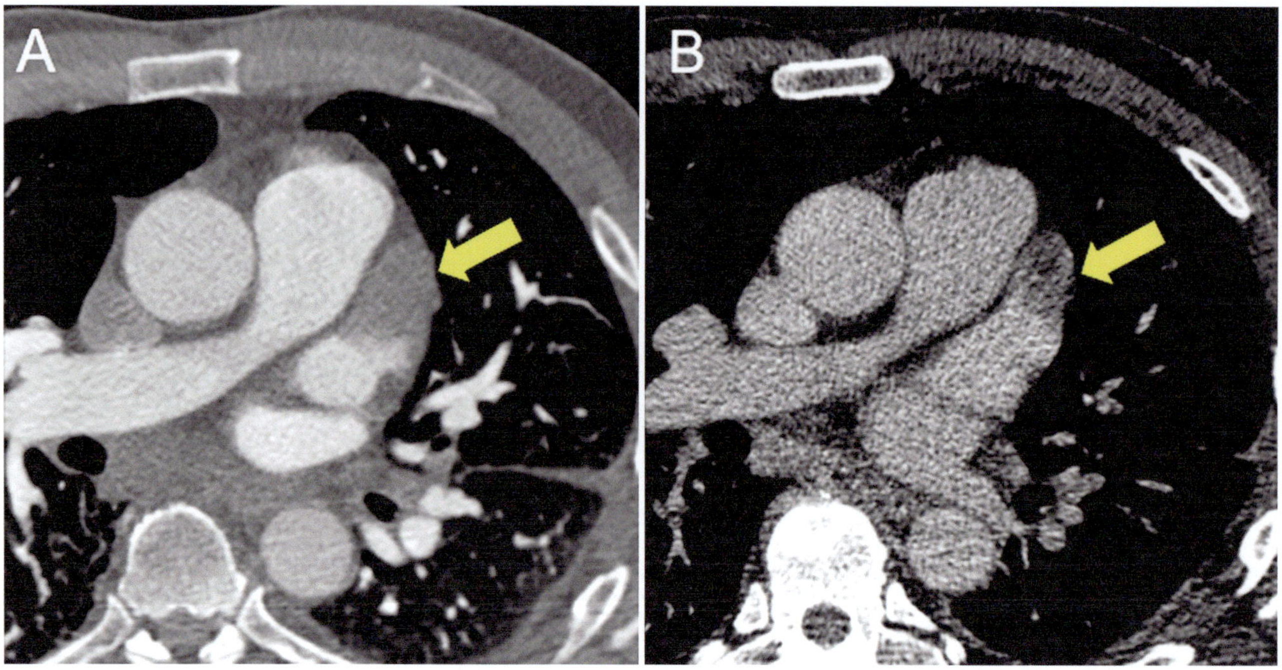

Fig. 3.7 (**a**) Trans-axial CT image of a patient undergoing a pulmonary vein study during arterial phase shows a filling defect within the left atrial appendage. (**b**) A non-contrast delayed scan performed within 1 min after contrast administration shows a persistent filling defect, thus highly suggestive of a left atrial appendage thrombus instead of slow flow

reconstruction (i.e., 35%). For measuring luminal diameters, double obliques of the MPR could provide true short-axis diameters. For anatomical description, evaluation of the pulmonary veins from the axial images and the longitudinal plane with MIP reconstruction provides good anatomical detail of the variants of pulmonary veins, the number of aortic arch vessel, and location of the esophagus in relation to the left atrium. Volume-rendered techniques are additionally helpful for the electrophysiologist. The 3D-VRT can isolate just the left atrium with pulmonary veins and aorta, while the "fly-through" or endoluminal view can provide intra-luminal images of each of the pulmonary veins

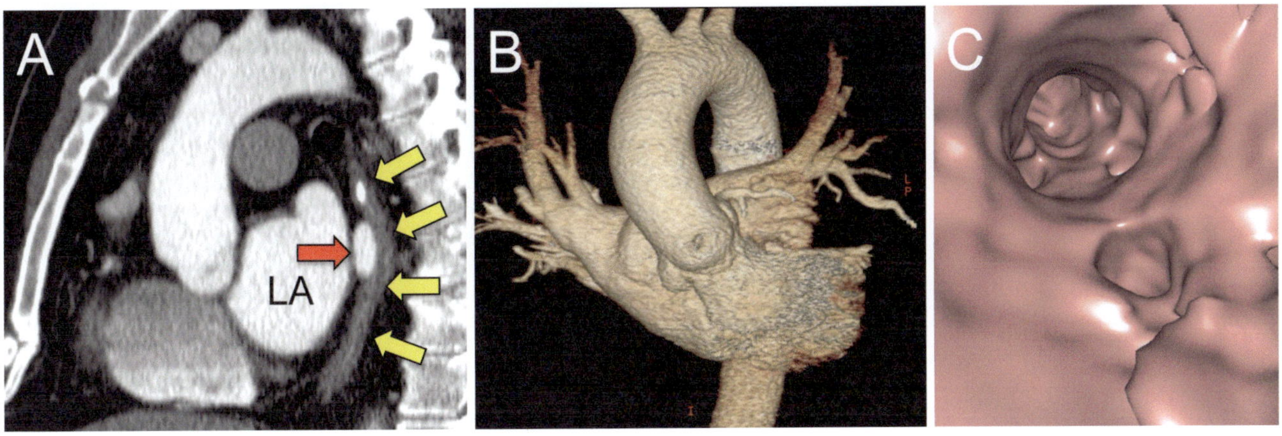

Fig. 3.8 (**a**) Maximum intensity projection (MIP) CT image with 2-mm slice thickness in an oblique sagittal view shows barium within the esophagus and the location of the esophagus (*yellow arrows*) with respect to the left atrium and left inferior pulmonary vein (*red arrow*). (**b**) Three-dimensional volume-rendered (VRT) CT image of the left atrium, pulmonary veins, and three-vessel aortic arch. (**c**) Virtual endoluminal or "*fly-through*" view of the left atrium depicts the intraluminal ostium of the right superior pulmonary vein

(Fig. 3.8). The data from the contrast-enhanced CT can then be sent to the electrophysiologist for "merging" and co-registration for electro-anatomical mapping using specialized software.

3.3.2 Coronary Veins

The coronary sinus and venous tree with its tributaries can be well visualized and can be defined pre-procedurally by MDCT.[15,32-36,52] If the patient is in atrial fibrillation for a coronary vein study, this is not an ideal situation; however, an ECG-gated CT could be performed with the dual-source CT scanner and beta-blockers should be given to decrease the heart rate less than 100 bpm and as close to 60 bpm as possible. We do not recommend scanning atrial fibrillation patients with the 64-slice MDCT scanner due to its limited temporal resolution; however, if deemed necessary, then decreasing the heart rate below 60 bpm would allow for the best chance of reconstructing interpretable images.

The CT imaging protocol is nearly similar to that used for a standard coronary CT angiography with several important exceptions. For coronary venous imaging, inspiratory breath-hold is similar to that from a coronary artery study (Fig. 3.6b). Nitroglycerin is not needed as dilation of the coronary arteries may interfere with the evaluation of the cardiac veins, which course in similar directions as the coronary arteries. Depending on the type of CT scanner, the use of beta-blockers to achieve optimal heart rate may be considered. Caudo-cranial image acquisition is also preferred given that the coronary sinus is inferiorly positioned in the heart, although there does not appear to be much difference in our experience when cranio-caudal acquisition is performed.

After the scout image, a test bolus protocol is preferred over bolus tracking to calculate the contrast agent transit time. For patients with severe left ventricular dysfunction, inform them that the test bolus scan is the longest of the image acquisitions and may take up to 30 s for the breath-hold. This is due to the poor cardiac output and prolonged transit time from the antecubital vein to the ascending or descending aorta. However, the subsequent CT venography breath-hold should be less than 15 s and is based on the scan range volume coverage of the heart. Scan coverage should be from the carina bifurcation to the diaphragm, similar to that of a coronary artery scan. For optimal enhancement of the coronary venous system, the most notable distinction is that venous phase imaging is desired. An empiric 10–15 s delay should be added to the contrast agent transit time so that image acquisition will be performed during the venous and not arterial phase. We reserve an 15 s delay for patients with severely impaired left ventricular function. Whatever additional delay time used for venous phase imaging should be added to the volume of contrast, although this additional volume could be given at a slower rate (i.e., 2–3 mL/s) to minimize the contrast load. In our experience, the typical delay time to set on the CT scanner (using either the Siemens 64-slice CT or dual-source CT scanners) for CT venography is anticipated to be 40 s or longer for the CRT population, though this time may be variable depending on vendors.

With image reconstruction, quick review of all the phases obtained from the multi-phase reconstructions would give guidance to which phase provides the best image. Typically, a systolic percentage phase of either 35% or 45% yields good visualization of the cardiac venous system. However, occasionally, a small cardiac vein may appear at a diastolic phase; thus, close review of each of the phases is recommended. Evaluation of the coronary sinus, the presence of a Thesbesian valve,[53] and the cardiac veins that supply the posterolateral wall[54-56] of the heart are most relevant to the electrophysiologist. 3D-VRT images should be reconstructed as

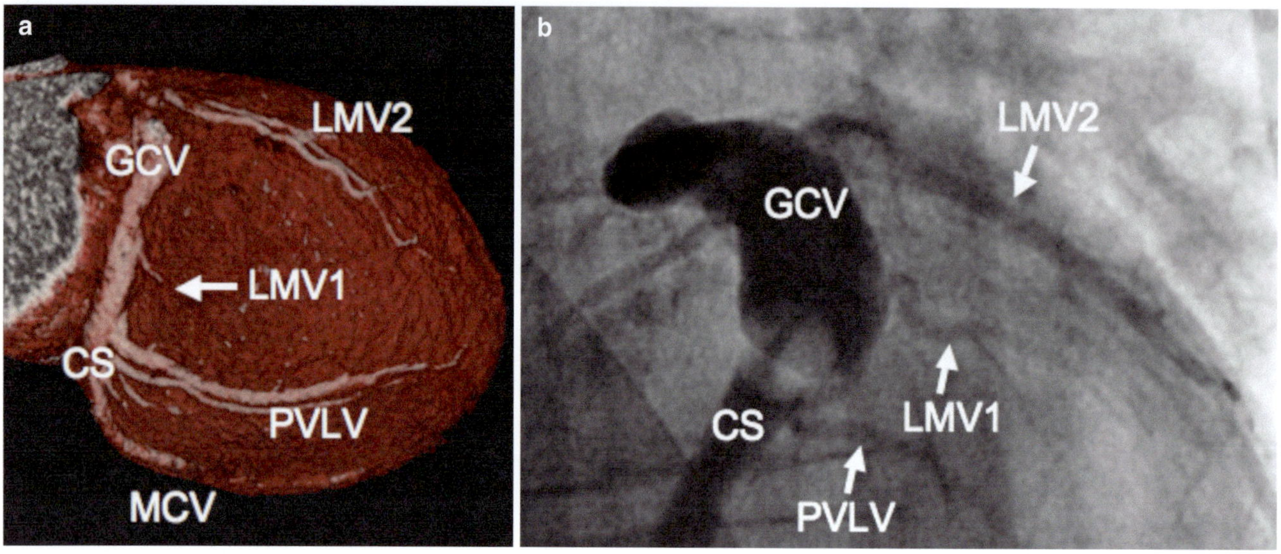

Fig. 3.9 (**a**) Volume-rendered (VRT) CT image of the coronary venous system using a dedicated coronary venous protocol. (**b**) Corresponding invasive coronary venography with occlusive balloon-tipped catheter inflated between the posterior vein of the left ventricle (*PVLV*) and the first left marginal vein (*LMV1*) *CS* coronary sinus, *GCV* great cardiac vein, *LMV2* second left marginal vein, *MCV* middle cardiac vein

they provide images of the heart and cardiac veins that the electrophysiologists can easily identify and correlate to the invasive coronary venography (Fig. 3.9).

Acknowledgment We would like to thank Susanne Loomis in the Massachusetts General Hospital Radiology Educational Media Services for providing us with the illustrations used in this chapter.

References

1. Oransky I. Sir Godfrey N. Hounsfield. *Lancet*. 2004;364:1032.
2. Flohr T, Stierstorfer K, Raupach R, Ulzheimer S, Bruder H. Performance evaluation of a 64-slice CT system with z-flying focal spot. *Rofo*. 2004;176:1803-1810.
3. Achenbach S, Giesler T, Ropers D, et al. Detection of coronary artery stenoses by contrast-enhanced, retrospectively electrocardiographically-gated, multislice spiral computed tomography. *Circulation*. 2001;103:2535-2538.
4. Nieman K, Oudkerk M, Rensing BJ, et al. Coronary angiography with multi-slice computed tomography. *Lancet*. 2001;357:599-603.
5. Flohr TG, McCollough CH, Bruder H, et al. First performance evaluation of a dual-source CT (DSCT) system. *Eur Radiol*. 2006; 16:256-268.
6. Moloo J, Shapiro MD, Abbara S. Cardiac computed tomography: technique and optimization of protocols. *Semin Roentgenol*. 2008; 43:90-99.
7. Hendel RC, Patel MR, Kramer CM, et al. ACCF/ACR/SCCT/SCMR/ASNC/NASCI/SCAI/SIR 2006 appropriateness criteria for cardiac computed tomography and cardiac magnetic resonance imaging: a report of the American College of Cardiology Foundation Quality Strategic Directions Committee Appropriateness Criteria Working Group, American College of Radiology, Society of Cardiovascular Computed Tomography, Society for Cardiovascular Magnetic Resonance, American Society of Nuclear Cardiology, North American Society for Cardiac Imaging, Society for Cardiovascular Angiography and Interventions, and Society of Interventional Radiology. *J Am Coll Cardiol*. 2006;48:1475-1497.
8. Sick PB, Schuler G, Hauptmann KE, et al. Initial worldwide experience with the WATCHMAN left atrial appendage system for stroke prevention in atrial fibrillation. *J Am Coll Cardiol*. 2007;49: 1490-1495.
9. Heist EK, Refaat M, Danik SB, Holmvang G, Ruskin JN, Mansour M. Analysis of the left atrial appendage by magnetic resonance angiography in patients with atrial fibrillation. *Heart Rhythm*. 2006; 3:1313-1318.
10. Truong QA, Hoffmann U, Singh JP. Potential uses of computed tomography for management of heart failure patients with dyssynchrony. *Crit Pathw Cardiol*. 2008;7:185-190.
11. Truong QA, Singh JP, Cannon CP, et al. Quantitative analysis of intraventricular dyssynchrony using wall thickness by multidetector computed tomography. *JACC Cardiovasc Imaging*. 2008;1: 772-781.
12. Kuo PH, Kanal E, Abu-Alfa AK, Cowper SE. Gadolinium-based MR contrast agents and nephrogenic systemic fibrosis. *Radiology*. 2007;242:647-649.
13. Lehman SJ, Abbara S, Cury RC, et al. Significance of cardiac computed tomography incidental findings in acute chest pain. *Am J Med*. 2009;122:543-549.
14. Malchano ZJ, Neuzil P, Cury RC, et al. Integration of cardiac CT/MR imaging with three-dimensional electroanatomical mapping to guide catheter manipulation in the left atrium: implications for catheter ablation of atrial fibrillation. *J Cardiovasc Electrophysiol*. 2006;17:1221-1229.
15. Tops LF, Bax JJ, Zeppenfeld K, et al. Fusion of multislice computed tomography imaging with three-dimensional electroanatomic mapping to guide radiofrequency catheter ablation procedures. *Heart Rhythm*. 2005;2:1076-1081.
16. Patel AM, Heist EK, Chevalier J, et al. Effect of presenting rhythm on image integration to direct catheter ablation of atrial fibrillation. *J Interv Card Electrophysiol*. 2008;22:205-210.
17. Kistler PM, Rajappan K, Jahngir M, et al. The impact of CT image integration into an electroanatomic mapping system on clinical outcomes of catheter ablation of atrial fibrillation. *J Cardiovasc Electrophysiol*. 2006;17:1093-1101.
18. Kistler PM, Rajappan K, Harris S, et al. The impact of image integration on catheter ablation of atrial fibrillation using electroanatomic mapping: a prospective randomized study. *Eur Heart J*. 2008;29:3029-3036.

19. Saad EB, Marrouche NF, Saad CP, et al. Pulmonary vein stenosis after catheter ablation of atrial fibrillation: emergence of a new clinical syndrome. *Ann Intern Med.* 2003;138:634-638.
20. Hur J, Kim YJ, Lee HJ, et al. Left atrial appendage thrombi in stroke patients: detection with two-phase cardiac CT angiography versus transesophageal echocardiography. *Radiology.* 2009;251:683-690.
21. Mohrs OK, Nowak B, Petersen SE, et al. Thrombus detection in the left atrial appendage using contrast-enhanced MRI: a pilot study. *AJR Am J Roentgenol.* 2006;186:198-205.
22. Reddy VY, Neuzil P, Malchano ZJ, et al. View-synchronized robotic image-guided therapy for atrial fibrillation ablation: experimental validation and clinical feasibility. *Circulation.* 2007;115:2705-2714.
23. Chung ES, Leon AR, Tavazzi L, et al. Results of the predictors of response to CRT (PROSPECT) trial. *Circulation.* 2008;117: 2608-2616.
24. Krenning BJ, Szili-Torok T, Voormolen MM, et al. Guiding and optimization of resynchronization therapy with dynamic three-dimensional echocardiography and segmental volume–time curves: a feasibility study. *Eur J Heart Fail.* 2004;6:619-625.
25. Szili-Torok T, Krenning BJ, Voormolen MM, Roelandt JR. Dynamic three-dimensional echocardiography combined with semi-automated border detection offers advantages for assessment of resynchronization therapy. *Cardiovasc Ultrasound.* 2003;1:14.
26. Kapetanakis S, Kearney MT, Siva A, Gall N, Cooklin M, Monaghan MJ. Real-time three-dimensional echocardiography: a novel technique to quantify global left ventricular mechanical dyssynchrony. *Circulation.* 2005;112:992-1000.
27. England B, Lee A, Tran T, et al. Magnetic resonance criteria for future trials of cardiac resynchronization therapy. *J Cardiovasc Magn Reson.* 2005;7:827-834.
28. Chalil S, Stegemann B, Muhyaldeen S, et al. Intraventricular dyssynchrony predicts mortality and morbidity after cardiac resynchronization therapy: a study using cardiovascular magnetic resonance tissue synchronization imaging. *J Am Coll Cardiol.* 2007;50:243-252.
29. Tecelao SR, Zwanenburg JJ, Kuijer JP, et al. Quantitative comparison of 2D and 3D circumferential strain using MRI tagging in normal and LBBB hearts. *Magn Reson Med.* 2007;57:485-493.
30. van der Land V, Germans T, van Dijk J, et al. The effect of left bundle branch block on left ventricular remodeling, dyssynchrony and deformation of the mitral valve apparatus: an observational cardiovascular magnetic resonance imaging study. *Int J Cardiovasc Imaging.* 2007;23:529-536.
31. Westenberg JJ, Lamb HJ, van der Geest RJ, et al. Assessment of left ventricular dyssynchrony in patients with conduction delay and idiopathic dilated cardiomyopathy: head-to-head comparison between tissue doppler imaging and velocity-encoded magnetic resonance imaging. *J Am Coll Cardiol.* 2006;47:2042-2048.
32. Auricchio A, Sorgente A, Singh JP, et al. Role of multislice computed tomography for preprocedural evaluation before revision of a chronically implanted transvenous left ventricular lead. *Am J Cardiol.* 2007;100:1566-1570.
33. Abbara S, Cury RC, Nieman K, et al. Noninvasive evaluation of cardiac veins with 16-MDCT angiography. *AJR Am J Roentgenol.* 2005;185:1001 1006.
34. Van de Veire NR, Schuijf JD, De Sutter J, et al. Non-invasive visualization of the cardiac venous system in coronary artery disease patients using 64-slice computed tomography. *J Am Coll Cardiol.* 2006;48:1832-1838.
35. Tada H, Kurosaki K, Naito S, et al. Three-dimensional visualization of the coronary venous system using multidetector row computed tomography. *Circ J.* 2005;69:165-170.
36. Muhlenbruch G, Koos R, Wildberger JE, Gunther RW, Mahnken AH. Imaging of the cardiac venous system: comparison of MDCT and conventional angiography. *AJR Am J Roentgenol.* 2005;185: 1252-1257.
37. Van de Veire NR, Marsan NA, Schuijf JD, et al. Noninvasive imaging of cardiac venous anatomy with 64-slice multi-slice computed tomography and noninvasive assessment of left ventricular dyssynchrony by 3-dimensional tissue synchronization imaging in patients with heart failure scheduled for cardiac resynchronization therapy. *Am J Cardiol.* 2008;101:1023-1029.
38. Einstein AJ, Henzlova MJ, Rajagopalan S. Estimating risk of cancer associated with radiation exposure from 64-slice computed tomography coronary angiography. *JAMA.* 2007;298:317-323.
39. Brenner DJ, Hall EJ. Computed tomography–an increasing source of radiation exposure. *N Engl J Med.* 2007;357:2277-2284.
40. Einstein AJ, Moser KW, Thompson RC, Cerqueira MD, Henzlova MJ. Radiation dose to patients from cardiac diagnostic imaging. *Circulation.* 2007;116:1290-1305.
41. Hausleiter J, Meyer T, Hermann F, et al. Estimated radiation dose associated with cardiac CT angiography. *JAMA.* 2009;301: 500-507.
42. Siemens A. *Application Guide: SOMATOM Sensation Cardiac 64.* Forchheim: Siemens AG Medical Solutions; 2004.
43. Gerber TC, Carr JJ, Arai AE, et al. Ionizing radiation in cardiac imaging: a science advisory from the American heart association committee on cardiac imaging of the council on clinical cardiology and committee on cardiovascular imaging and intervention of the council on cardiovascular radiology and intervention. *Circulation.* 2009;119:1056-1065.
44. Rosamond W, Flegal K, Friday G, et al. Heart disease and stroke statistics–2007 update: a report from the American Heart Association Statistics Committee and Stroke Statistics Subcommittee. *Circulation.* 2007;115:e69-e171.
45. Gleeson TG, Bulugahapitiya S. Contrast-induced nephropathy. *AJR Am J Roentgenol.* 2004;183:1673-1689.
46. Rao QA, Newhouse JH. Risk of nephropathy after intravenous administration of contrast material: a critical literature analysis. *Radiology.* 2006;239:392-397.
47. Cademartiri F, Nieman K, van der Lugt A, et al. Intravenous contrast material administration at 16-detector row helical CT coronary angiography: test bolus versus bolus-tracking technique. *Radiology.* 2004;233:817-823.
48. Tsao HM, Wu MH, Higa S, et al. Anatomic relationship of the esophagus and left atrium: implication for catheter ablation of atrial fibrillation. *Chest.* 2005;128:2581-2587.
49. Cury RC, Abbara S, Schmidt S, et al. Relationship of the esophagus and aorta to the left atrium and pulmonary veins: implications for catheter ablation of atrial fibrillation. *Heart Rhythm.* 2005;2: 1317-1323.
50. Cummings JE, Schweikert RA, Saliba WI, et al. Brief communication: atrial-esophageal fistulas after radiofrequency ablation. *Ann Intern Med.* 2006;144:572-574.
51. Pappone C, Oral H, Santinelli V, et al. Atrio-esophageal fistula as a complication of percutaneous transcatheter ablation of atrial fibrillation. *Circulation.* 2004;109:2724-2726.
52. Singh JP, Houser S, Heist EK, Ruskin JN. The coronary venous anatomy: a segmental approach to aid cardiac resynchronization therapy. *J Am Coll Cardiol.* 2005;46:68-74.
53. Christiaens L, Ardilouze P, Ragot S, Mergy J, Allal J. Prospective evaluation of the anatomy of the coronary venous system using multidetector row computed tomography. *Int J Cardiol.* 2008;126: 204-208.
54. Chalil S, Stegemann B, Muhyaldeen SA, et al. Effect of posterolateral left ventricular scar on mortality and morbidity following cardiac resynchronization therapy. *Pacing Clin Electrophysiol.* 2007;30: 1201-1209.
55. Bleeker GB, Kaandorp TA, Lamb HJ, et al. Effect of posterolateral scar tissue on clinical and echocardiographic improvement after cardiac resynchronization therapy. *Circulation.* 2006;113: 969-976.
56. Helm RH, Byrne M, Helm PA, et al. Three-dimensional mapping of optimal left ventricular pacing site for cardiac resynchronization. *Circulation.* 2007;115:953-961.

4 Nuclear-Based Imaging: Description of Technology and Protocols

Juhani Knuuti

Abstract

In general, nuclear cardiology techniques are considered robust and accurate for clinical imaging of heart disease. Thus, they will likely continue to play a key role in the assessment of myocardial perfusion, function, viability, and neural function. The technology is rapidly developing toward smaller and faster devices with improved sensitivity and resolution dedicated to cardiac applications. The development of novel tracers will further expand the clinical applications. Especially the new molecular imaging techniques can enable more personalized decision making. This chapter reviews the basic aspects of nuclear imaging techniques and imaging protocols.

Keywords

Nuclear-based Imaging • Nuclear Imaging technology • Radionuclide imaging of the heart • Single photon emission computed tomography • SPECT imaging of the heart • PET imaging of the heart

Radionuclide imaging of the heart is an established technique for the clinical diagnostic and prognostic workup of cardiac diseases. For the detection of coronary artery disease (CAD), myocardial perfusion single photon emission computed tomography (SPECT) has been widely used and its usefulness is supported by a very large body of evidence.[1] Positron emission tomography (PET) has also been available for decades but has long been considered mainly a research tool because of limited availability. However, during recent decade, cardiac PET is now increasingly used in clinical cardiology.[2]

In general, nuclear cardiology techniques are considered robust and accurate for clinical imaging of heart disease. Thus, they will likely continue to play a key role in the assessment of myocardial perfusion, function, viability, and neural function. The technology is rapidly developing toward smaller and faster devices with improved sensitivity and resolution dedicated to cardiac applications. The development of novel tracers will further expand the clinical applications.[3] Especially the new molecular imaging techniques can enable more personalized decision making. This chapter will review the basic aspects of nuclear imaging techniques and imaging protocols.

4.1 Imaging Technology

4.1.1 SPECT

Myocardial SPECT imaging is typically performed using a multidetector gamma camera system, which rotates around the chest to obtain tomographic images of single emitted photons. The patient is typically positioned supine on the table.[4] Recently also small footprint dedicated cardiac scanners are commercially available and in those devices the patient is in upright position. The most recent high sensitive devices allow very rapid 2–4 min image acquisition of the heart.

The general-purpose design has been replaced by some manufacturers with systems with multiple detectors focused on the heart yielding 5–10 times the sensitivity of conventional SPECT. Some novel designs also use novel detectors

J. Knuuti
Turku PET Centre, Turku University Hospital, Turku, Finland
e-mail: juhani.knuuti@utu.fi

A. Auricchio et al. (eds.), *Cardiac Imaging in Electrophysiology*,
DOI 10.1007/978-1-84882-486-7_4, © Springer-Verlag London Limited 2012

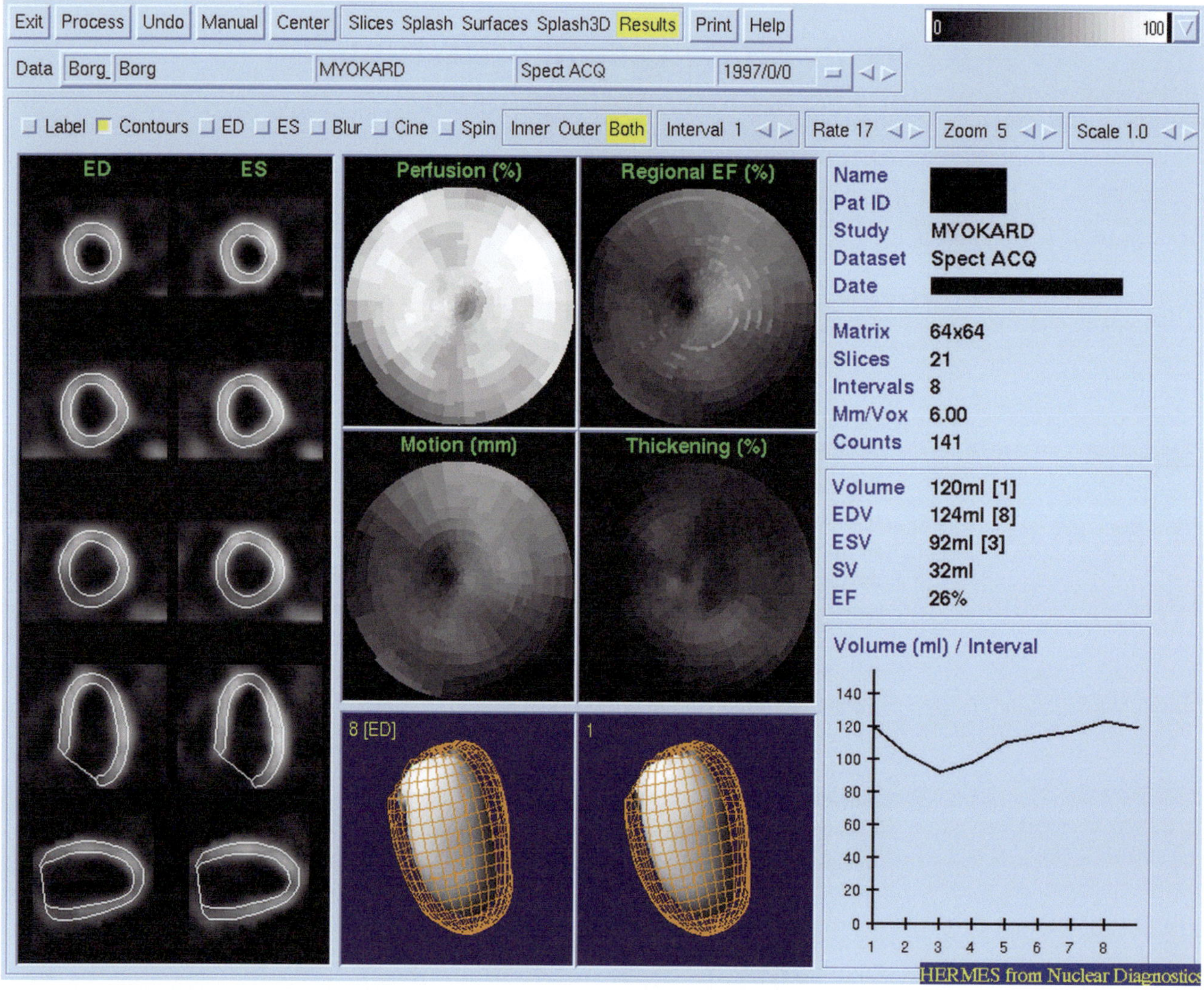

Fig. 4.1 An example of standard automatic SPECT image analysis report. End-systolic (*ES*) and end-diastolic (*ED*) short axis slices with automatic contour detection are displayed (*left panel*). Polar plots of perfusion, EF, motion and thickening are displayed (*middle panel*). Finally numerical data from the analysis with LV volume curve are displayed (*right panel*)

with improved energy resolution. Thus, SPECT cardiac studies with these novel techniques are faster and more accurate, while radiotracer dose and thus radiation exposure can be reduced.[5]

After acquisition, images are reconstructed and filtered using dedicated reconstruction algorithms. The resulting tomographic data sets are reoriented along the left ventricular (LV) short and long axes to facilitate review of myocardial tracer distribution and comparison of rest and stress studies (Fig. 4.1). Commercial software tools have been developed which employ automatic contour detection algorithms, and they create polar maps from the tomographic images.[6-9] These polar maps are a two-dimensional display of the three-dimensional (3D) tracer distribution throughout the myocardium which allow comparison of patient data with normal databases and also quantitative analysis of defects.

The acquisition of SPECT using electrocardiographic (ECG) gating has become a standard procedure especially with perfusion imaging. The gating allows quantitative measurement of global and regional LV function (ejection fraction, EF and volumes).[10] Gating also improves the diagnostic accuracy of perfusion imaging since the problems with attenuation artifacts may be solved by gated images.[11] For ECG-gating, the patient should have a fairly regular heart rhythm. The cardiac cycle is usually divided into 8, and sometimes into 12 or 16 time bins. For reproducible functional analysis, softwares are available which semi-automatically generate 3D myocardial contours throughout the cardiac cycle. Volumetric data from the contours can then be used for 3D display and calculation of quantitative global parameters.

The nuclear imaging is subject to certain artifacts which need to be controlled. Attenuation of the radiation in the body can introduce artifacts in the images. Furthermore, the radiation scatter degrades the image quality. The amount of attenuation in a clinical study depends, among others, on the shape and the thickness of the body. Thus, correction for attenuation requires exact knowledge of the characteristics of each patient. Nowadays the attenuation correction is usually generated individually by transmission imaging, increasingly using the X-ray CT in hybrid systems. It has been documented that attenuation correction improves image quality and image interpretation[12-14] if scatter correction is also performed.

The data analysis is based on a systematic visual review of raw data and reconstructed images on a computer screen.[15] The raw projection data are reviewed to identify motion artifacts and assess tracer distribution in organs other than the heart. Then, reoriented tomographic images are reviewed without and with corrections. Gated images are visualized and reviewed. Thereafter, software-derived quantitative data are reviewed and used to strengthen the previous impressions of images. Finally, the image analysis is integrated with clinical data and reported in a standardized format.

4.1.2 PET

PET differs in many aspects with SPECT. The geometry of scanner and the detection principle are different. The correction of attenuation and scatter is routinely applied to get images of quantitative tracer distribution and systems are fast enough to allow dynamic imaging protocols. These characteristics allow quantification of perfusion and molecular processes in absolute terms. In addition, positron-emitting radioisotopes have much shorter half-lives which increase the flexibility of imaging protocols. On the other hand, the availability of these short half-life tracers is currently the limiting factor of PET since they require production site close to the imaging site.[2]

To allow absolute quantification, a series of 3D volumes are acquired over time to create the time–activity curves and investigate the kinetics of tracer uptake and release from different tissues. Compared with radionuclides emitting single gamma-ray photons, the positron emission leads to emission of pair of 511 keV annihilation photons which gives PET imaging higher detection efficiency, better uniformity of spatial resolution, and easier correction for attenuation (scattering) of photons in the tissue. Also the attenuation correction property is different from SPECT and more robust.[2]

To utilize the full potential of PET several corrections need to be performed. The geometry of a PET system introduces variation in the detection sensitivity and this is corrected by performing normalization procedures. The detector dead time is the period when a detector is unable to record an event.[16,17] This might be because the electronics are busy or when more than one photon strikes a detector within its resolving time. Traditionally, an external positron-emitting source has been used to measure the attenuating factor before the administration of radioactivity,[18-20] but currently most of the PET scanners are hybrid systems (PET-CT) and a CT scan has substituted the conventional PET transmission scan.[21] The correction methods for scatter and random coincidence events are very important and are routinely employed but the methods vary between the vendors. Also the good temporal resolution allows effective motion correction using list mode acquisition. ECG-gating is a standard procedure for use with PET, and studies have also been carried out on respiratory gating, making feasible the monitoring of and the correction for both cardiac and respiratory movements of the patient.[22-24]

Despite the developments in imaging technology, basic physics determine that the spatial resolution is limited. This leads to so-called partial volume effect, which actually is effective in all imaging techniques. In addition, the signal in a particular region will be spilled over to the surrounding tissues. Combination of transmission and emission data has been used to correct for this effect in the myocardium.[25]

4.2 Current Imaging Procedures

4.2.1 SPECT

SPECT is widely used for the clinical workup of suspected or known CAD and its diagnostic and prognostic values are supported by a large body of evidence.[1] Most importantly, SPECT derived information is utilized for clinical decision making. Assessment of myocardial perfusion and function are the two major clinical applications.[15,26] Furthermore, imaging of cardiac neural function has recently gained increasing interest.[27]

4.2.1.1 Myocardial Perfusion

For perfusion imaging, several tracers are (^{201}Tl, ^{99m}Tc-sestamibi, and -tetrofosmin) commercially available. The first-pass extraction of ^{201}Tl by the myocardium is high (88%) and the uptake proportionally increases with perfusion over a relatively large range.[28] ^{201}Tl redistributes over several hours, thus allowing delayed images to be acquired that are independent of perfusion and reflect viability. ^{201}Tl has been used clinically for almost three decades, but does have certain limitations such as the long half-life associated with high radiation burden, low photon energy resulting frequently in attenuation artifacts, and low injected activity contributing to a low signal-to-noise ratio. After an administration of ^{201}Tl at stress, SPECT images are obtained early after injection. The redistribution images are acquired 2–4 h later which reflect the resting perfusion and viability.[15] The stress and redistribution images are compared for the relative regional

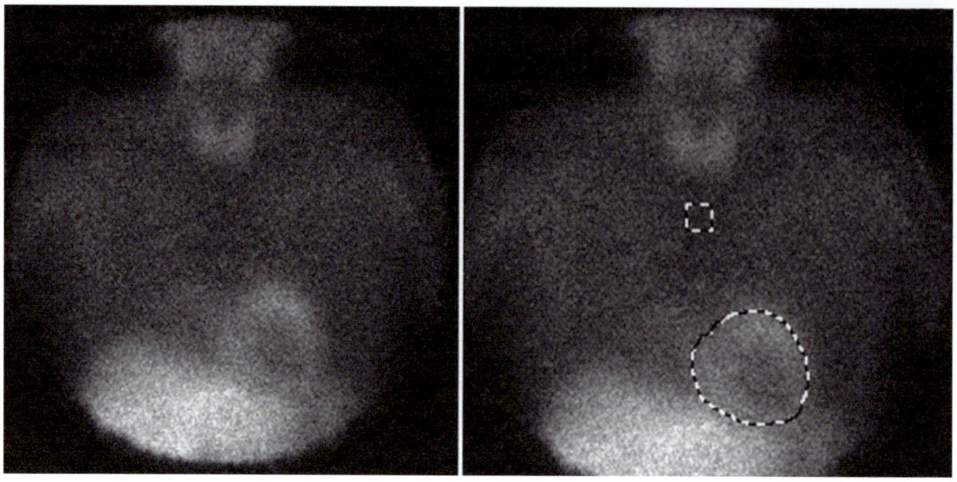

Fig. 4.2 An example of planar 123I-MIBG images and the positioning of the regions-of-interests (ROIs). The most simple analysis is based on the ratio of the whole heart ROI to mediastinal ROI (H/M ratio)

stress-induced ischemia and fixed defects are considered to represent myocardial necrosis. However, the redistribution may be incomplete at 4 h and a second injection of ^{201}Tl or delayed imaging can be performed for a more accurate assessment of myocardial viability.[29]

The higher energy of ^{99m}Tc generally leads to better quality images as compared with ^{201}Tl. Moreover, the short half-life permits much higher activities to be administered, giving better counting statistics and results in lower radioactivity doses. However, tracer kinetic properties are somewhat inferior when compared with ^{201}Tl.[30] The uptake of ^{99m}Tc-labeled tracers is less avid and defects may be less profound. The retention is based on intact mitochondria indicating that viable myocytes are needed for the uptake.[31,32] The tracer molecules stay within the myocytes and do not redistribute so that two injections are necessary to obtain stress and rest images. There are some differences between the commercially available ^{99m}Tc-labeled tracers. Hepatic clearance of tetrofosmin is slightly more rapid than in the case of sestamibi.[33,34] Two-day, same-day stress–rest or same-day rest–stress imaging protocols have been established.[35,36] The image acquisition typically begins 30–60 min after injection to allow for hepatobiliary clearance. Longer delays are required for resting images and with vasodilator stressors. For the assessment of myocardial viability, resting injections can be given following nitrate administration.[37] Also a dual-isotope imaging has been used. This consists of rest ^{201}Tl injection, followed immediately by stress and a ^{99m}Tc-compound injection.[38]

4.2.1.2 Ventricular Function

In addition to give additional prognostic information, the assessment of LV function and volumes also improves the diagnostic and prognostic accuracy of cardiac SPECT.[11,39,40] Additionally, assessment of the function of the right ventricle (RV) is recognized to be important in some diseases such as arrhythmogenic RV and pulmonary hypertension. Finally, determination with equilibrium radionuclide ventriculography of LV-EF is recognized as one of the methods-of-choice for monitoring the cardiotoxicity of cytotoxic anticancer drugs.[41] Functional radionuclide cardiac studies include several techniques, and ECG-gating is a key point in these methods. Equilibrium radionuclide ventriculography is performed after ^{99m}Tc labeling of red blood cells.[26] It provides high-quality planar images and may even be performed as a SPECT study for accurate separation of RV and LV, and for accurate assessment of regional wall motion.[42]

4.2.1.3 Innervation Imaging

The sympathetic nervous system plays an important role in cardiovascular physiology. Both SPECT with ^{123}I-MIBG and PET can be used to visualize the sympathetic innervation of the heart and the abnormalities in innervation caused by, for example, ischemia, heart failure, and arrhythmogenic disorders.[27] Furthermore, cardiac neuronal imaging allows early detection of autonomic neuropathy in diabetes mellitus.

Most experience has been obtained with SPECT and ^{123}I-MIBG. ^{123}I-MIBG is a false neurotransmitter, an analog of norepinephrine, which uses similar uptake mechanisms in the presynaptic nerve terminals as norepinephrine. The tracer is primarily transported into the presynaptic nerve terminals. In the nerve terminal, no degradation of ^{123}I-MIBG takes places resulting in an accumulation of ^{123}I-MIBG with high signal intensity.

Early planar and SPECT imaging is performed at 10–20 min after MIBG administration, whereas delayed planar and SPECT imaging is performed 3–4 h after tracer injection. From planar images, global cardiac MIBG uptake can be assessed visually or semi-quantitatively using early and late heart-to-mediastinum (H/M) ratio (Fig. 4.2). To calculate H/M ratio, regions-of-interest are manually drawn over the heart and upper mediastinum and the mean of myocardial counts per pixel is divided by the mean of mediastinal counts per pixel. Another planar-based parameter is the cardiac washout rate which indicates the rate by which MIBG is released from the myocardium between early and delayed imaging. To calculate cardiac washout rate, late

H/M ratio is subtracted from the early H/M ratio and divided by the early H/M ratio.

Assessment of sympathetic nerve activity in patients with heart failure has been shown to provide important prognostic information, and cardiac neuronal imaging can potentially identify patients who are at increased risk of sudden death and those patients with ICD discharge[43,44] or to predict inducibility of ventricular arrhythmias on electrophysiological testing.[45]

The available studies have shown that cardiac innervation imaging holds great potential for risk stratification and prognostification of heart failure patients. Recent studies suggest that particularly MIBG imaging may play a major role in identifying patients with LV dysfunction at elevated risk of heart failure death or arrhythmic death.[27,43-45] The results of these initial studies are promising, and more studies will help to determine the precise role of innervation imaging in risk stratification of heart failure patients.

4.2.2 PET

While SPECT has been the mostly used nuclear imaging technique, PET offers broader insights into cardiac physiology and pathophysiology. The current clinical applications of PET imaging in cardiology can be divided into three main categories: studies of regional myocardial blood flow, metabolism, and molecular function.

4.2.2.1 Myocardial Perfusion

Mainly three tracers are used for the perfusion imaging using PET: ^{15}O-labeled water ($H_2{}^{15}O$)[46,47] ^{13}N-labeled ammonia ($^{13}NH_3$),[48-51] and the potassium analog ^{82}Rubidium (^{82}Rb).[52] The kinetic models[46,47,49,51,53-56] have been developed and validated for quantification of perfusion using these tracers. The extraction of ^{82}Rb is lower and more dependent on the flow than the other two tracers,[56] but it is generator produced and, thus, more widely available. All tracers have short physical half-lives (2–10 min) which allow repeated perfusion measurements in the same session.[57]

PET has been used to detect impairments of MBF in asymptomatic subjects with various cardiovascular risk factors[58,59] and in diabetic patients without symptoms of cardiac disease.[60] In patients with CAD, the measurement of absolute perfusion or perfusion reserve is useful for the assessment of the functional significance of coronary stenoses.[61] Quantification of perfusion is particularly effective in those circumstances where the perfusion is diffusely (and not only regionally) blunted, e.g., in patients with hypertrophic or dilated cardiomyopathy and patients with balanced coronary artery disease.[61]

4.2.2.2 Myocardial Metabolism

PET imaging allows imaging myocardial substrate and oxidative metabolism. Several different tracers have been used such as ^{18}FDG for glucose metabolism,[61] ^{11}C-palmitate and ^{18}F-FTHA for fatty acid metabolism,[62] and ^{11}C-labeled acetate[63] for myocardial oxygen consumption.[64,65] Various kinetic models have been developed to quantify these processes. Most of these above-mentioned methods have been used in research protocols, only ^{18}FDG has been widely used in clinical cardiology, mainly for the detection of viability and recently also for inflammatory cardiac diseases such as sarcoidosis.

4.2.2.3 Imaging of Molecular Function

The attachment of a radioisotope to a biomolecule to follow its distribution throughout the body is suitable for noninvasive detection of biologic processes at the level of tissue and cells.[3,66-68] Different tracers have been used to study the presynaptic sympathetic terminals: ^{18}F-labeled fluorometaraminol,[69,70] ^{11}C-labeled hydroxyephedrine,[71] and ^{11}C-labeled epinephrine[72] which compete with endogenous noradrenaline for the transport into the presynaptic nerve terminals. Also beta-blocker drugs have been labeled with ^{11}C to act as radioligands for the study of postsynaptic beta-adrenoceptors.[73] In addition to studies of the sympathetic nervous system, the density and affinity of myocardial muscarinic receptors can be evaluated noninvasively with ^{11}C-MQNB (methylquinuclidinyl benzilate).[74,75] In patients with congestive heart failure, mean receptor concentration was significantly higher compared with normal subjects[74] suggesting that congestive heart failure is associated with an upregulation of myocardial muscarinic receptors paralleling the down-regulation of beta-adrenoceptors.

Myocardial metabolism and sympathetic function have emerged as the first applications of clinical molecular cardiac radionuclide imaging. While those are increasingly entering the clinical stage, a broad spectrum of other tracers for specific biologic targets in the cardiovascular system is being evaluated on the preclinical level. These targets include the renin–angiotensin system, integrins, matrix metalloproteinases, cell death, reporter genes, and transplanted stem cells. The goal is the visualization of key mechanisms involved in subjects of ongoing basic cardiovascular science, including early disease development and novel therapeutic interventions. It is expected that molecular imaging, early disease detection, and molecular therapy will progress hand-in-hand from the preclinical to the clinical level in the future.

4.3 Hybrid Imaging

Hybrid scanners combining PET or SPECT with high-resolution multidetector CT are becoming the standard for almost all commercially available systems. Hybrid scanners offer the ability to assess the anatomy of the heart and coronary arteries, and the functional evaluation either at stress (for

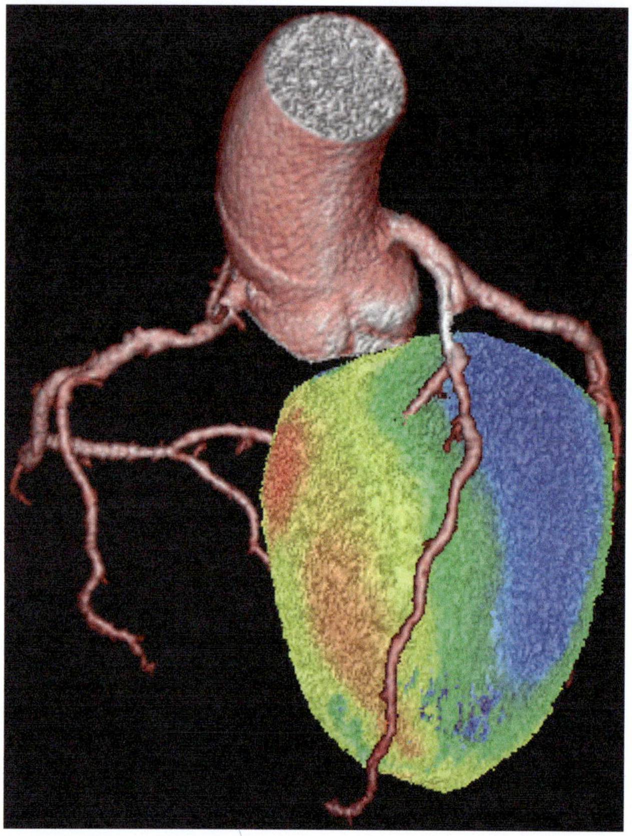

Fig. 4.3 An example of the hybrid PET/CT imaging. CT coronary angiography is overlaid with absolute PET perfusion images creating 3D rendered parametric images. Large perfusion defect is located in the area supplied by occluded diagonal branch while left anterior descending artery was patent

assessment of induced ischemia) (Fig. 4.3), or at rest (for viability) in association with the left ventricular systolic function. Therefore, combining functional information from PET or SPECT is appealing.[76,77]

References

1. Marcassa C, Bax JJ, Bengel F, et al. Clinical value, cost-effectiveness, and safety of myocardial perfusion scintigraphy: a position statement. *Eur Heart J*. 2008;29:557-563.
2. Le Guludec D, Lautamaki R, Knuuti J, Bax JJ, Bengel FM. Present and future of clinical cardiovascular PET imaging in Europe-a position statement by the European council of nuclear cardiology (ECNC). *Eur J Nucl Med Mol Imaging*. 2008;35(9):1709-1724.
3. Higuchi T, Bengel FM. Cardiovascular nuclear imaging: from perfusion to molecular function: non-invasive imaging. *Heart*. 2008;94:809-816.
4. Segall GM, Davis MJ. Prone versus supine thallium myocardial SPECT: a method to decrease artifactual inferior wall defects. *J Nucl Med*. 1989;30:548-555.
5. Patton JA, Slomka PJ, Germano G, Berman DS. Recent technologic advances in nuclear cardiology. *J Nucl Cardiol*. 2007;14:501-513.
6. Ficaro EP, Lee BC, Kritzman JN, Corbett JR. Corridor4DM: the Michigan method for quantitative nuclear cardiology. *J Nucl Cardiol*. 2007;14:455-465.
7. Klein JL, Garcia EV, DePuey EG, et al. Reversibility bull's-eye: a new polar bull's-eye map to quantify reversibility of stress-induced SPECT thallium-201 myocardial perfusion defects. *J Nucl Med*. 1990;31:1240-1246.
8. Nekolla SG, Miethaner C, Nguyen N, Ziegler SI, Schwaiger M. Reproducibility of polar map generation and assessment of defect severity and extent assessment in myocardial perfusion imaging using positron emission tomography. *Eur J Nucl Med*. 1998;25:1313-1321.
9. Slomka PJ, Nishina H, Berman DS, et al. Automated quantification of myocardial perfusion SPECT using simplified normal limits. *J Nucl Cardiol*. 2005;12:66-77.
10. Germano G, Kiat H, Kavanagh PB, et al. Automatic quantification of ejection fraction from gated myocardial perfusion SPECT. *J Nucl Med*. 1995;36:2138-2147.
11. Choi JY, Lee KH, Kim SJ, et al. Gating provides improved accuracy for differentiating artifacts from true lesions in equivocal fixed defects on technetium 99m tetrofosmin perfusion SPECT. *J Nucl Cardiol*. 1998;5:395-401.
12. Bateman TM, Cullom SJ. Attenuation correction single-photon emission computed tomography myocardial perfusion imaging. *Semin Nucl Med*. 2005;35:37-51.
13. Hendel RC, Berman DS, Cullom SJ, et al. Multicenter clinical trial to evaluate the efficacy of correction for photon attenuation and scatter in SPECT myocardial perfusion imaging. *Circulation*. 1999;99:2742-2749.
14. Tonge CM, Manoharan M, Lawson RS, Shields RA, Prescott MC. Attenuation correction of myocardial SPECT studies using low resolution computed tomography images. *Nucl Med Commun*. 2005;26:231-237.
15. Hesse B, Tagil K, Cuocolo A, et al. EANM/ESC procedural guidelines for myocardial perfusion imaging in nuclear cardiology. *Eur J Nucl Med Mol Imaging*. 2005;32:855-897.
16. Cranley K, Millar R, Bell T. Correction for deadtime losses in a gamma camera data analysis system. *Eur J Nucl Med*. 1980;5:377-382.
17. Daube-Witherspoon ME, Carson RE. Unified deadtime correction model for PET. *IEEE Trans Med Imaging*. 1991;10:267-275.
18. deKemp RA, Nahmias C. Attenuation correction in PET using single photon transmission measurement. *Med Phys*. 1994;21:771-778.
19. Karp JS, Muehllehner G, Qu H, Yan XH. Singles transmission in volume-imaging PET with a 137Cs source. *Phys Med Biol*. 1995;40:929-944.
20. Yu SK, Nahmias C. Single-photon transmission measurements in positron tomography using 137Cs. *Phys Med Biol*. 1995;40:1255-1266.
21. Burger C, Goerres G, Schoenes S, Buck A, Lonn AH, Von Schulthess GK. PET attenuation coefficients from CT images: experimental evaluation of the transformation of CT into PET 511-keV attenuation coefficients. *Eur J Nucl Med Mol Imaging*. 2002;29:922-927.
22. Dawood M, Kösters T, Fieseler M, et al. Motion correction in respiratory gated cardiac PET/CT using multi-scale optical flow. *Med Image Comput Comput Assist Interv*. 2008;12:155-162.
23. Lamare F, Teras M, Kokki T, et al. Correction of respiratory motion in dual gated cardiac imaging. *J Nucl Med*. 2008;49(suppl 1):389.
24. Teräs M, Kokki T, Durand-Schaefer N, et al. Dual-gated cardiac PET-clinical feasibility study. *Eur J Nucl Med Mol Imaging*. 2010;37(3):505-516.
25. Boussion N, Hatt M, Lamare F, Rest CC, Visvikis D. Contrast enhancement in emission tomography by way of synergistic PET/CT image combination. *Comput Methods Programs Biomed*. 2008;90:191-201.
26. Hesse B, Lindhardt TB, Acampa W, et al. EANM/ESC guidelines for radionuclide imaging of cardiac function. *Eur J Nucl Med Mol Imaging*. 2008;35:851-885.

27. Henneman MM, Bengel FM, van der Wall EE, Knuuti J, Bax JJ. Cardiac neuronal imaging: application in the evaluation of cardiac disease. *J Nucl Cardiol.* 2008;15(3):442-455.
28. Grunwald AM, Watson DD, Holzgrefe HH Jr, Irving JF, Beller GA. Myocardial thallium-201 kinetics in normal and ischemic myocardium. *Circulation.* 1981;64:610-618.
29. Dilsizian V, Smeltzer WR, Freedman NM, Dextras R, Bonow RO. Thallium reinjection after stress-redistribution imaging. Does 24-hour delayed imaging after reinjection enhance detection of viable myocardium? *Circulation.* 1991;83:1247-1255.
30. Takahashi N, Reinhardt CP, Marcel R, Leppo JA. Myocardial uptake of 99mTc-tetrofosmin, sestamibi, and 201Tl in a model of acute coronary reperfusion. *Circulation.* 1996;94:2605-2613.
31. Beanlands RS, Dawood F, Wen WH, et al. Are the kinetics of technetium-99m methoxyisobutyl isonitrile affected by cell metabolism and viability? *Circulation.* 1990;82:1802-1814.
32. Meerdink DJ, Leppo JA. Comparison of hypoxia and ouabain effects on the myocardial uptake kinetics of technetium-99m hexakis 2-methoxyisobutyl isonitrile and thallium-201. *J Nucl Med.* 1989;30:1500-1506.
33. Jain D, Wackers FJ, Mattera J, McMahon M, Sinusas AJ, Zaret BL. Biokinetics of technetium-99m-tetrofosmin: myocardial perfusion imaging agent: implications for a one-day imaging protocol. *J Nucl Med.* 1993;34:1254-1259.
34. Munch G, Neverve J, Matsunari I, Schroter G, Schwaiger M. Myocardial technetium-99m-tetrofosmin and technetium-99m sestamibi kinetics in normal subjects and patients with coronary artery disease. *J Nucl Med.* 1997;38:428-432.
35. Berman DS, Kiat HS, Van Train KF, Germano G, Maddahi J, Friedman JD. Myocardial perfusion imaging with technetium-99m-sestamibi: comparative analysis of available imaging protocols. *J Nucl Med.* 1994;35:681-688.
36. Heo J, Kegel J, Iskandrian AS, Cave V, Iskandrian BB. Comparison of same-day protocols using technetium-99m-sestamibi myocardial imaging. *J Nucl Med.* 1992;33:186-191.
37. Sciagra R, Bisi G, Santoro GM, Rossi V, Fazzini PF. Nitrate versus rest myocardial scintigraphy with technetium 99m-sestamibi: relationship of tracer uptake to regional left ventricular function and its significance in the detection of viable hibernating myocardium. *Am J Card Imaging.* 1995;9:157-166.
38. Berman DS, Kiat H, Van Train K, Friedman JD, Wang FP, Germano G. Dual-isotope myocardial perfusion SPECT with rest thallium-201 and stress Tc-99m sestamibi. *Cardiol Clin.* 1994;12:261-270.
39. Sharir T, Germano G, Kang X, et al. Prediction of myocardial infarction versus cardiac death by gated myocardial perfusion SPECT: risk stratification by the amount of stress-induced ischemia and the poststress ejection fraction. *J Nucl Med.* 2001;42:831-837.
40. Germano G, Berman DS. On the accuracy and reproducibility of quantitative gated myocardial perfusion SPECT. *J Nucl Med.* 1999;40:810-813.
41. Klocke FJ, Baird MG, Lorell BH, et al. ACC/AHA/ASNC guidelines for the clinical use of cardiac radionuclide imaging-executive summary: a report of the American college of cardiology/ American heart association task force on practice guidelines (ACC/AHA/ ASNC committee to revise the 1995 guidelines for the clinical use of cardiac radionuclide imaging). *Circulation.* 2003;108:1404-1418.
42. Daou D, Van Kriekinge SD, Coaguila C, et al. Automatic quantification of right ventricular function with gated blood pool SPECT. *J Nucl Cardiol.* 2004;11:293-304.
43. Arora R, Ferrick KJ, Nakata T, et al. I-123 MIBG imaging and heart rate variability analysis to predict the need for an implantable cardioverter defibrillator. *J Nucl Cardiol.* 2003;10(2):121-131.
44. Nagahara D, Nakata T, Hashimoto A, et al. Predicting the need for an implantable cardioverter defibrillator using cardiac metaiodobenzylguanidine activity together with plasma natriuretic peptide concentration or left ventricular function. *J Nucl Med.* 2008;49(2):225-233.
45. Bax JJ, Kraft O, Buxton AE, et al. 123-I-MIBG scintigraphy to predict inducibility of ventricular arrhythmias on cardiac electrophysiology testing. *Circ Cardiovasc Imaging.* 2008;1(2):131-140.
46. Araujo LI, Lammertsma AA, Rhodes CG, et al. Noninvasive quantification of regional myocardial blood flow in coronary artery disease with oxygen-15-labeled carbon dioxide inhalation and positron emission tomography. *Circulation.* 1991;83:875-885.
47. Bergmann SR, Herrero P, Markham J, Weinheimer CJ, Walsh MN. Noninvasive quantitation of myocardial blood flow in human subjects with oxygen-15-labeled water and positron emission tomography. *J Am Coll Cardiol.* 1989;14:639-652.
48. Schelbert HR, Phelps ME, Hoffman EJ, Huang SC, Selin CE, Kuhl DE. Regional myocardial perfusion assessed with N-13 labeled ammonia and positron emission computerized axial tomography. *Am J Cardiol.* 1979;43:209-218.
49. Huang SC, Williams BA, Krivokapich J, Araujo L, Phelps ME, Schelbert HR. Rabbit myocardial 82Rb kinetics and a compartmental model for blood flow estimation. *Am J Physiol.* 1989;256:H1156-H1164.
50. Bellina CR, Parodi O, Camici P, et al. Simultaneous in vitro and in vivo validation of nitrogen-13-ammonia for the assessment of regional myocardial blood flow. *J Nucl Med.* 1990;31:1335-1343.
51. Hutchins GD, Schwaiger M, Rosenspire KC, Krivokapich J, Schelbert H, Kuhl DE. Noninvasive quantification of regional blood flow in the human heart using N-13 ammonia and dynamic positron emission tomographic imaging. *J Am Coll Cardiol.* 1990;15:1032-1042.
52. Herrero P, Markham J, Shelton ME, Weinheimer CJ, Bergmann SR. Noninvasive quantification of regional myocardial perfusion with rubidium-82 and positron emission tomography. Exploration of a mathematical model. *Circulation.* 1990;82:1377-1386.
53. Anagnostopoulos C, Almonacid A, El Fakhri G, et al. Quantitative relationship between coronary vasodilator reserve assessed by 82Rb PET imaging and coronary artery stenosis severity. *Eur J Nucl Med Mol Imaging.* 2008;35:1593-1601.
54. Lortie M, Beanlands RS, Yoshinaga K, Klein R, Dasilva JN, DeKemp RA. Quantification of myocardial blood flow with 82Rb dynamic PET imaging. *Eur J Nucl Med Mol Imaging.* 2007;34:1765-1774.
55. Shah A, Schelbert HR, Schwaiger M, et al. Measurement of regional myocardial blood flow with N-13 ammonia and positron-emission tomography in intact dogs. *J Am Coll Cardiol.* 1985;5:92-100.
56. Araujo L, Schelbert HR. Dynamic positron emission tomography in ischaemic heart disease. *Am J Card Imaging.* 1984;1:117-124.
57. Kaufmann PA, Gnecchi-Ruscone T, Yap JT, Rimoldi O, Camici PG. Assessment of the reproducibility of baseline and hyperemic myocardial blood flow measurements with 15O-labeled water and PET. *J Nucl Med.* 1999;40:1848-1856.
58. Kaufmann PA, Gnecchi-Ruscone T, di Terlizzi M, Schafers KP, Luscher TF, Camici PG. Coronary heart disease in smokers: vitamin C restores coronary microcirculatory function. *Circulation.* 2000;102:1233-1238.
59. Pitkänen OP, Nuutila P, Raitakari OT, et al. Coronary flow reserve in young men with familial combined hyperlipidemia. *Circulation.* 1999;99(13):1678-1684.
60. Pitkänen OP, Nuutila P, Raitakari OT, et al. Coronary flow reserve is reduced in young men with IDDM. *Diabetes.* 1998;47(2):248-254.
61. Knuuti J, Kajander S, Mäki M, Ukkonen H. Quantification of myocardial blood flow will reform the detection of CAD. *J Nucl Cardiol.* 2009;16(4):497-506.
62. Schelbert HR, Henze E, Schon HR. C-11 palmitic acid for the non-invasive evaluation of regional myocardial fatty acid metabolism

with positron computed tomography. IV. In vivo demonstration of impaired fatty acid oxidation in acute myocardial ischaemia. *Am Heart J*. 1983;106:736-750.
63. Buxton DB, Schwaiger M, Nguyen A, Phelps M, Schelbert HR. Radiolabeled acetate as a tracer of myocardial tricarboxylic acid cycle flux. *Circ Res*. 1988;63:628-634.
64. Armbrecht JJ, Buxton DB, Brunken R, Phelps M, Schelbert HR. Regional myocardial oxygen consumption determined noninvasively in humans with [1–1 1 C] acetate and dynamic positron tomography. *Circulation*. 1989;80:863-872.
65. Walsh MN, Geltman EM, Brown MA. Noninvasive estimation of regional myocardial oxygen consumption by positron emission tomography with carbon-11 acetate in patients with myocardial infarction. *J Nucl Med*. 1989;30:1798-1808.
66. Knuuti J, Bengel FM. Positron emission tomography and molecular imaging. *Heart*. 2008;94:360-367.
67. Schwaiger M, Bengel FM. From thallium scan to molecular imaging. *Mol Imaging Biol*. 2002;4:387-398.
68. Wu JC, Bengel FM, Gambhir SS. Cardiovascular molecular imaging. *Radiology*. 2007;244:337-355.
69. Goldstein DS, Chang PC, Eisenhofer G, et al. Positron emission tomographic imaging of cardiac sympathetic innervation and function. *Circulation*. 1990;81:1606-1621.
70. Wieland DM, Rosenspire KC, Hutchins GD, et al. Neuronal mapping of the heart with 6-[18F] fluorometaraminol. *J Med Chem*. 1990;33:956-964.
71. Schwaiger M, Kalff V, Rosenspire K, et al. Noninvasive evaluation of sympathetic nervous system in human heart by positron emission tomography. [see comment]. *Circulation*. 1990;82:457-464.
72. Munch G, Nguyen NTB, Nekolla SG, et al. Evaluation of sympathetic nerve terminals with [11 C]epinephrine and [11 C] hydroxyephedrine and positron emission tomography. *Circulation*. 2000;101:516-523.
73. Syrota A. Positron emission tomography: evaluation of cardiac receptors. In: Marcus ML, Schelbert HR, Skorton DJ, Wolf GL, eds. *Cardiac Imaging: A Companion to Braunwald's Heart Disease*. Philadelphia: W B Saunders; 1991:1256-1270.
74. Le Guludec D, Cohen-Solal A, Delforge J, Delahaye N, Syrota A, Merlet P. Increased myocardial muscarinic receptor density in idiopathic dilated cardiomyopathy: an in vivo PET study. *Circulation*. 1997;96:3416-3422.
75. Le Guludec D, Delforge J, Syrota A, et al. In vivo quantification of myocardial muscarinic receptors in heart transplant patients. *Circulation*. 1994;90:172-178.
76. Di Carli MF, Hachamovitch R. New technology for noninvasive evaluation of coronary artery disease. *Circulation*. 2007;115:1464-1480.
77. Bax JJ, Beanlands RS, Klocke FJ, et al. Diagnostic and clinical perspectives of fusion imaging in cardiology: is the total greater than the sum of its parts? *Heart*. 2007;93(1):16-22.

Echocardiography: Technology and Protocols

5

José Zamorano and José Alberto de Agustín

Abstract

Imaging techniques for the study of heart disease have experienced a great development in recent decades and are essential in current clinical practice. The main value of imaging techniques in Electrophysiology is to determine the underlying heart disease, which helps to define the patient management and prognosis. Noninvasive imaging of cardiac electrophysiology is still a major goal despite all recent technical innovations. However, electrophysiological procedures require an intimate knowledge of heart anatomy and classical imaging techniques may be unable to visualize structures involved in arrhythmia mechanisms. Novel methods, such as intracardiac echocardiography and three-dimensional echocardiography, have enabled a more accurate imaging during electrophysiology procedures. This may reduce ionizing radiation exposure and shorten procedure time. This chapter describes the echocardiographic technology and protocols currently used for the study of most common heart diseases involving arrhythmias.

Keywords

Electrophysiology • Echocardiography • Protocols • Atrial fibrillation and echocardiography • Arrhythmogenic right ventricular dysplasia • Hypertrophic cardiomyopathy and echocardiography • Non-compacted cardiomyopathy and echocardiography

The main value of imaging techniques in Electrophysiology is to determine the underlying heart disease, which helps to define the patient management and prognosis. Identification of patients at significant risk of arrhythmia and sudden cardiac death is one of the major cardiology challenges. Noninvasive imaging of cardiac electrophysiology is still a major goal despite all recent technical innovations. However, classical imaging techniques may be unable to visualize structures involved in arrhythmia mechanisms and therapy.

Electrophysiological procedures are increasingly used to diagnose and treat ventricular or supraventricular tachycardias. These procedures require a sound knowledge of heart anatomy. Until recently, fluoroscopy was the most important guidance for advancing and positioning catheters. Technological advances in electroanatomic mapping have reduced the use of fluoroscopy, and have increased accuracy in electrophysiological procedures. However, these techniques have several limitations. Geometries generated by the magnetic field-based mapping system CARTO are

J. Zamorano (✉)
Unidad de Imagen Cardiovascular,
Hospital Clínico San Carlos, Madrid, Spain
e-mail: jzamorano.hcsc@salud.madrid.org

A. Auricchio et al. (eds.), *Cardiac Imaging in Electrophysiology*,
DOI 10.1007/978-1-84882-486-7_5, © Springer-Verlag London Limited 2012

limited in resolution by the number of points acquired. Furthermore these techniques do not provide real-time visualization of cardiac structure and function, but static geometric maps only.

Novel methods, such as intracardiac echocardiography (ICE) and three-dimensional echocardiography, have enabled a more accurate imaging during electrophysiology procedures. ICE has improved our abilities in electrophysiological procedures, thus serving as an adjunct to fluoroscopy by identifying critical anatomical landmarks and enabling precise navigation of the ablation catheter within the heart. Other potential applications of ICE include confirmation of lesion formation, immediate identification of complications, and assistance in transseptal puncture.

There is also a great interest in using transesophageal echocardiography (TEE) for guidance of electrophysiological procedures. This may reduce ionizing radiation exposure and shorten procedure time guiding mapping and ablation catheters to pulmonary vein ostia. However, difficulties are often encountered in accurately localizing catheter tips, particularly when they curve out of plane. Three-dimensional transesophageal echocardiography (3DTEE) has also been used for electrophysiological procedures. Volume rendering and multiple display slices provide detailed three-dimensional visualization of anatomic landmarks and catheter placement and orientation. This method can be applied toward guidance of cardiac ablation and also for guidance of the pacing leads placement for cardiac resynchronization therapy. This chapter describes the echocardiographic technology and protocols currently used for the study of most common heart diseases involving arrhythmias.

5.1 Echocardiography in Atrial Fibrillation

Atrial fibrillation (AF) is the most common arrhythmia in clinical practice. Therefore, it is essentially an accurate echocardiographic assessment to guide the patient management.

5.1.1 Transthoracic Echocardiography

Transthoracic echocardiography is recommended for all patients with AF.[1] TTE can detect the underlying heart disease and the risk of complications. In the past, rheumatic heart disease was the more frequent etiology for AF development. Currently it has increased the incidence of AF associated to cardiomyopathies, ischemic heart disease, and hypertensive heart disease. Hypertension is the most prevalent cause of AF. The left ventricular (LV) wall thickness, left atrial (LA) size, and LV systolic and diastolic function should be evaluated in all patients with FA.

LA anteroposterior diameter can be measured from parasternal long-axis view, with M-mode or two-dimensional echocardiography. Recurrence after cardioversion is more frequent when LA anteroposterior diameter is >45–50 mm, while the success rate is very low in patients with severe LA enlargement (anteroposterior diameter >60 mm).[2] However, this value provides an inaccurate measure of LA size, because it does not take into account the other LA dimensions. LA volume has been associated with cardiovascular risk burden and long-term prognosis,[3] and is recommended in current clinical practice. LA volume can be obtained from the apical two and four chambers view, measured in the frame immediately before mitral valve opening. However, this is valid for patients in sinus rhythm but not in patients with chronic AF.[4] It is also possible to assess auricular stunning after cardioversion by strain and strain rate analyses. Patients with higher atrial strain and strain rate appear to have a greater likelihood of staying in sinus rhythm. Chronic anticoagulation may be considered in those with lower atrial strain and strain rate measurements.[5]

5.1.1.1 Transesophageal Echocardiography

Transesophageal approach allows an accurate assessment of posterior cardiac structures including the LA, interauricular septum, and pulmonary veins. TEE also offers a more accurate assessment of valvular function, particularly in prosthetic valves. Currently TEE is also the imaging technique of choice to examine the LA appendage and in the detection of thrombus (Fig. 5.1). The LA appendage has a complex morphology that is difficult to study by magnetic resonance or computed tomography. The modern multiplanar probes in an expert's hands detect thrombus with a high sensitivity and specificity (95–100%).

TEE can guide electrical cardioversion in AF of longer than 48 hours' duration. This option has advantages as opposed to the classic management with oral anticoagulation for 4 weeks before cardioversion. Cardioversion guided by TEE reduces embolic events recognizing atrial thrombus, and shortens the time of anticoagulation, thus reducing the bleeding complications. The Acute study[6] compared both strategies. After 8 weeks of follow-up, a lower incidence of bleeding in the group guided by TEE was found, without differences in embolic events or mortality rate. TEE predictors of thromboembolism in AF are listed in Table 5.1. TEE should also be performed to confirm thrombus resolution after 7–8 weeks of oral anticoagulation.

Sometimes it is difficult to distinguish left appendage thrombus from pectineus muscle. The pectineus muscles have more echogenicity than thrombus wherein many are often aligned in parallel, and have synchronous movement with left appendage. Lower echogenicity and an erratic motion are characteristics of thrombus. TEE also detects the presence of LA spontaneous contrast described as smoke-like

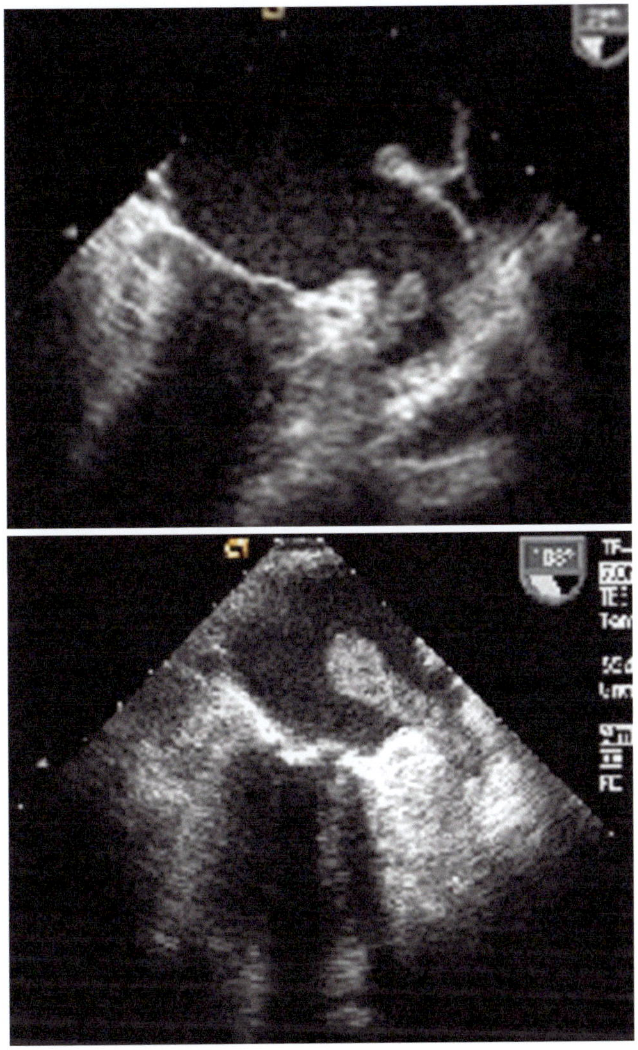

Fig. 5.1 Transesophageal echocardiography assessment of left atrial appendage thrombus

Table 5.1 Echocardiographic predictors of thromboembolism in atrial fibrillation

– Rheumatic valvular disease
– Left ventricular systolic dysfunction
– Left atrial enlargement
– Presence of thrombus
– Left appendage flow rates <20 cm/s
– Aortic plaques

discrete reflectances in atrial cavity. It is a marker of blood stasis and is associated with a further risk of thromboembolic events.[7-9] TEE allows identifying other sources of embolization, as complex atheroma of the ascending aorta and arch.[10]

The left appendage mechanical function can be assessed by pulsed wave Doppler with a sample volume placed 1 cm into the left appendage. Flow rate <20 cm/s is associated with the presence of spontaneous contrast and thrombus formation,[11,12] while flow rate >40 cm/s increases the probability of remaining in sinus rhythm 1 year after cardioversion (Fig. 5.2). Another way to assess the contractile function of the left appendage is the use of tissue Doppler imaging (TDI). Patients in sinus rhythm who develop embolic stroke have higher left appendage velocities with similar flow rates. Thus, TDI in the left appendage complements pulsed wave Doppler to assess the risk of embolism.[13]

An important concept is LA "stunning" after electrical cardioversion. It can be assessed by A wave obtained from transmitral pulsed wave Doppler analysis. LA stunning increases the risk of thromboembolic events. The severity and duration of atrial stunning appears to reflect the duration of antecedent AF. Excluding patients with rheumatic valvular disease, the majority of thrombi are in the left appendage. Thus, its elimination would obviate anticoagulation therapy in patients not likely to receive it. The intervention can be surgical or percutaneous, inserting an oclusor device. The guide from the percutaneous approach can be done with TEE.[14] In this way, echocardiography has emerged as an essential technique for the insertion and monitoring of the device.

TEE has been used as a guide for AF ablation in combination with conventional fluoroscopy, enabling the identification and cannulation of pulmonary veins, and obviating the use of angiography.[15] TEE is a useful tool for assessing the interauricular septum. Thickening of interauricular septum is associated with the onset of AF.[16] TEE also allows a more accurate evaluation of valvar lesions, particularly prosthetic dysfunction.

Other utility of TEE in AF is assisting transesophageal cardioversion. Preliminary data suggest it is well tolerated, requires less energy, and allows for hemodynamic monitoring during and immediately after the cardioversion.[17] Clinical trials are needed to elucidate the role of this strategy.

5.1.1.2 Real-Time Three-Dimensional Transesophageal Echocardiography

Studies have shown the usefulness of traditional two-dimensional TEE for guiding catheter-based cardiac procedures. However, difficulties are often encountered in accurately localizing catheter tips, particularly when they curve out of plane. Real time 3DTEE provides catheter and pacing lead visualization simultaneously with functional volumetric cardiac imaging. This technology has been also used to image the cardiac anatomy in animal models, showing the feasibility of real-time 3DTEE for guidance in electrophysiological procedures.[18] 3DTEE provides better visualization of catheter placement and orientation without repositioning of interventional catheters. Multiple simultaneous viewing planes and the addition of real-time three-dimensional color flow Doppler enable confirmation of anatomy with minimal adjustment of the esophageal

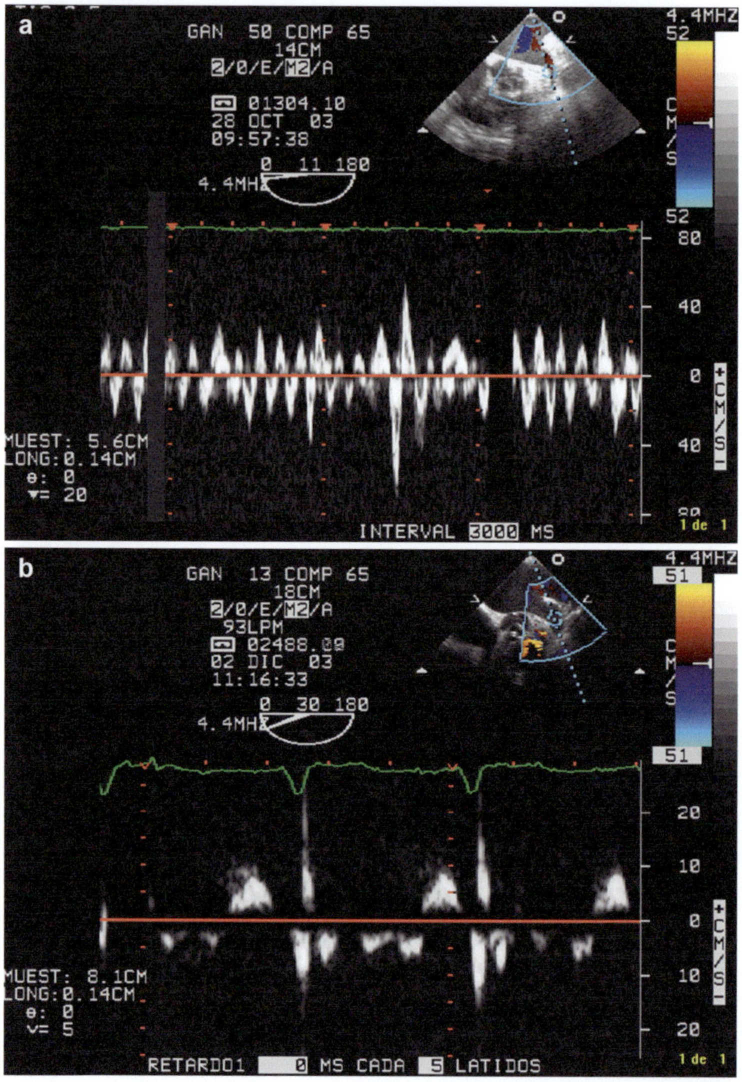

Fig. 5.2 Left appendage mechanical function assessment by pulsed wave Doppler. Patients with lower flow rate (**b**) have increased probability of thrombus formation

probe. This technology could be integrated with current electroanatomic mapping systems for combined electrical information and true real-time visualization. In addition electroanatomic mapping systems do not provide visualization of other anatomic reference landmarks outside of the heart, such as the esophagus. Thus, 3DTEE enables monitoring of any possible esophageal injury.

5.1.1.3 Intracardiac Echocardiography

Intracardiac echocardiography improves identification of anatomical landmarks and enables precise navigation of the ablation catheter within the heart. Six to twelve megahertz transducers assembled in catheters of 6–10 Frenchs are introduced through a femoral access. There are two different ICE technologies available. The first one uses a 9 MHz single element transducer mounted at the tip of an 8F catheter. It provides cross-sectional images in a 360° radial plane (similar to the used in intracoronary ultrasound). Pulling the catheter within the heart, three-dimensional reconstruction of the anatomy can be obtained. The other technology uses phased-array ultrasound-tipped catheter that consists of a 64-element transducer. The high-resolution, multiple frequency transducer (5–10 MHz) is incorporated into a 10F steerable catheter (four directions) and provides 90° sector images with depth control (similar to transesophageal echocardiography). The catheter allows for the whole spectrum of Doppler imaging capabilities and provides information in real time during the ablation procedures. These capabilities make use of the second system preferable in interventional electrophysiology.

ICE plays an important role in imaging pulmonary venous ostia, assisting accurate placement of ablation catheter, monitoring of lesion morphology and detecting procedural complications.[19] It is also a guide for the transseptal puncture procedure. It has been found a good correlation between ICE and computed tomography or magnetic resonance in the assessment of location and diameters of the pulmonary veins ostium assisting in selection the appropriate catheter.[20]

A common antrum on the left side is detected in approximately 80% of cases.[21,22] Right pulmonary veins tend to enter the atrium separately. However, supernumerary veins are more often present on the right side. These anatomical variants can be detected with a high degree of precision with ICE.

Transseptal catheterization has become a very important procedure in electrophysiology in order to obtain access into the LA for catheter ablation of atrial fibrillation. Traditionally, it has relied on fluoroscopic guidance but it requires great operator experience and is not free of complications, especially in patients with structural heart disease. Potential life-threatening complications of transseptal puncture include aortic puncture, pericardial puncture or tamponade, systemic embolism, and perforation of the inferior cava vein.[23-25] ICE could practically avoid these serious complications.

ICE enables to define the characteristics of interauricular septum (elastic, lipomatous hypertrophic, aneurysmatic, or double layer septum) and guide transseptal puncture by direct imaging of the needle tip within the fossa ovalis region.[26] A cross-sectional view of the fossa ovalis is best obtained with the ICE catheter placed near the septum. For catheter ablation of AF, the preferable puncture site is posterior and inferior. ICE displays tenting of the septum with the needle tip and advancement of the assembly into the LA.

Confirmation of the morphological lesions produced during ablation is another utility of ICE through monitoring of microbubble formation.[27] It corresponds with tissue overheating and it allows prevention of pop formation.[28] This may avoid unnecessary applications. Radiofrequency induces lesion morphologic changes including wall thickening, echodensity, and/or crater formation. Extensive ablation within the LA may increase risk of thromboembolic complications.

One of the most important roles of ICE imaging is early diagnosis of procedural complications during complex procedures. Potential acute complications are cardiac tamponade, atrioesophageal fistula, and thrombus formation. All of them can be identified early by ICE.[29] Other important complication is pulmonary vein stenosis after radiofrequency catheter ablation of atrial fibrillation. An incidence up to 3% has been reported.[30] The visualization of acute changes requires a closed monitoring of these patients. The flow changes in the pulmonary veins after application of radiofrequency have been studied.[31] Minor increases in pulmonary flow velocities are relatively common after the ablation. Values <150 cm/s are well tolerated and seem to return to baseline within 3 months after the procedure. Significant stenoses are characterized by higher peak flow velocities.

The usefulness of ICE is not restricted to AF. It is also applied in ablation of idiopathic ventricular tachycardias, inappropriate sinus tachycardia, and paroxysmal supraventricular tachycardia associated with congenital heart disease.[32] ICE can guide mapping and ablation of idiopathic ventricular tachycardias. ICE facilitates delineation of the myocardial substrate such as a postinfarction scar. In some cases, catheter ablation of the cavotricuspid isthmus may be challenging by the presence of an abnormal anatomy of the same one. ICE can visualize anatomy of the isthmus and anomalous diverticula or a prominent Eustachian ridge.[33] Catheter ablation of supraventricular tachycardias after correction of complex congenital heart disease is one of the most promising utilities of ICE. ICE provides an excellent guidance for navigation of the ablation catheter. In summary, ICE results in reduction of fluoroscopy time and maximizes safety and efficacy of complex ablation procedures.

Table 5.2 Diagnostic criteria for arrhythmogenic right ventricular dysplasia

Mayor criteria	• Severe dilatation of the right ventricle and reduced right ventricular ejection fraction, without left ventricular involvement • Localized right ventricular aneurysms • Severe segmental dilatation of the right ventricle
Minor criteria	• Mild dilatation of the right ventricle or reduced right ventricular ejection fraction • Mild segmental dilatation of the right ventricle • Regional right ventricular hypokinesia

5.2 Echocardiography in Arrhythmogenic Right Ventricular Dysplasia

Arrhythmogenic right ventricular dysplasia (ARVD) is a genetic cardiomyopathy, characterized by fatty or fibrofatty infiltration of the right ventricular (RV) myocardium that leads to RV enlargement, RV dysfunction, and ventricular arrhythmias with a left bundle branch block. Its diagnosis is challenging because sudden death is often the first manifestation of the disease. Echocardiography is considered the reference standard for the diagnosis of ARVD. The echocardiographic criteria for diagnosis are listed in Table 5.2.[34] However, echocardiography has been shown to be relatively insensitive and nonspecific in the detection of structural or functional abnormalities of the RV myocardium. The RV is a complex structure wrapped around the LV and is not completely visualized from any single plane. Therefore, different projections are needed for a proper study. Parasternal short- and long-axis view, apical four chambers, and subcostal view are needed for the RV assessment (Fig. 5.3). The free wall is usually <5 mm and should be measured is in subcostal view at the R wave peak of the QRS complex, at the level of tricuspid subvalvular apparatus.[35] It is essential to avoid the measurement of epicardial fat and prominent trabeculae. The RV dimensions are obtained from the apical four-chamber view and the normal values are shown in Table 5.3. Although the finding of normal RV does not exclude the diagnosis of

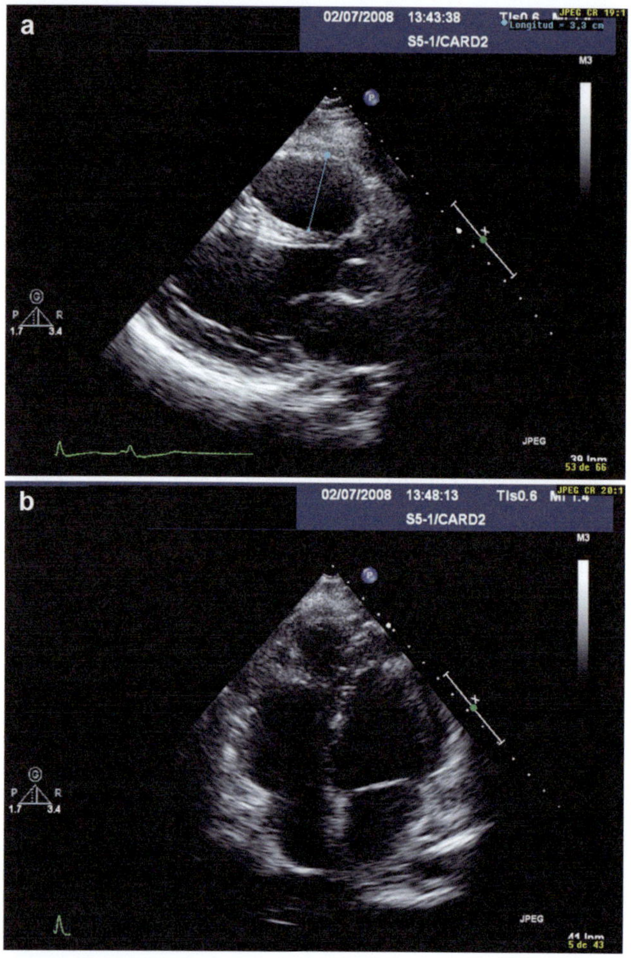

Fig. 5.3 Echocardiographic assessment in a patient with arrhythmogenic right ventricular dysplasia. Parasternal long-axis view (**a**) and apical four chambers view (**b**)

Table 5.3 Normal right ventricular dimensions (mm)

	Normal	Mild dilation	Moderate dilation	Severe dilation
Tricuspid annulus diameter	20–28	29–33	34–38	>39
Midventricular diameter	27–33	34–37	38–41	>42
Longitudinal diameter	71–79	80–85	86–91	>92

ARVD. The segment most frequently affected is the RV outflow tract. A RV outflow tract long-axis diastolic dimension >30 mm is required for the diagnosis of ARVC. It should be obtained from parasternal long-axis view.[36]

RV systolic function is usually estimated qualitatively. Quantitative assessment can be obtained measuring tricuspid annular plane systolic excursion (TAPSE). Normal values are >15 mm. The TAPSE has a good correlation with the RV ejection fraction.[37] TDI is also useful in the evaluation of RV systolic function (Fig. 5.4). Tricuspid annulus S wave is >16 cm/s in healthy subjects. Patients with ARVD have decreased TAPSE and Tricuspid annulus S wave by TDI. The volumetric and ejection fraction assessment remains problematic given the complex geometry of the RV. Mild disease is characterized by normal RV dimensions and localized hypokinetic or akinetic regions; moderate disease is characterized by mild RV enlargement and localized akinetic and/or dyskinetic areas; and severe disease is characterized by an enlarged RV and widespread areas of akinesis and/or dyskinesis. The LV may be affected in advanced cases. The presence of LV systolic dysfunction or congestive heart failure carries a poor long-term prognosis.

5.3 Echocardiography in Hypertrophic Cardiomyopathy

Hypertrophic cardiomyopathy (HCM) is a genetic heart disease characterized by the presence of idiopathic myocardial hypertrophy not explained by another cardiac or systemic disease, such as systemic hypertension or aortic stenosis. Patients with HCM have a considerably increased LV wall thickness with a small non-compliant but apparently well-contracting LV. Color Doppler and 2-dimensional transthoracic echocardiography is the standard imaging approach for the diagnosis of HCM. These modalities can demonstrate cardiac morphology and function, the extent of systolic anterior movement of the mitral valve, the degree of mitral regurgitation, and the severity of the LV outflow tract gradient. The most common diagnostic criterion of HCM is LV wall thickness ≥15 mm. The degree of thickening shows a direct relationship with the risk of sudden death.[38]

Early M-mode echocardiographic studies defined the characteristic features as asymmetrical hypertrophy of the ventricular septum, with or without systolic anterior motion of the mitral valve, and premature closure of the aortic valve. A disadvantage of M-mode is that only a small section of the LV can be examined with a single ultrasound beam, usually passing through the anterior septum and posterior walls. Two-dimensional echocardiography allows a more complete anatomical description of HCM (Fig. 5.5). The degree and distribution of LV hypertrophy are variable, and include septal hypertrophy with or without obstruction to LV outflow, concentric hypertrophy, apical hypertrophy, hypertrophy of the LV free wall, and RV hypertrophy. However, there is no evidence that this distribution has a relationship to prognosis.

Echocardiographic assessment of HCM requires studying the LV from several projections, including parasternal long-axis, serial short-axis views, and imaging from the apical and subcostal windows. The parasternal long-axis view is of pivotal importance. It is important that the beam transects the LV perpendicularly. Oblique images may lead to the overestimation of wall thickness and cavity dimensions. Parasternal long-axis view examines the profile of the ventricular septum and LV outflow tract. From this view, the relations between

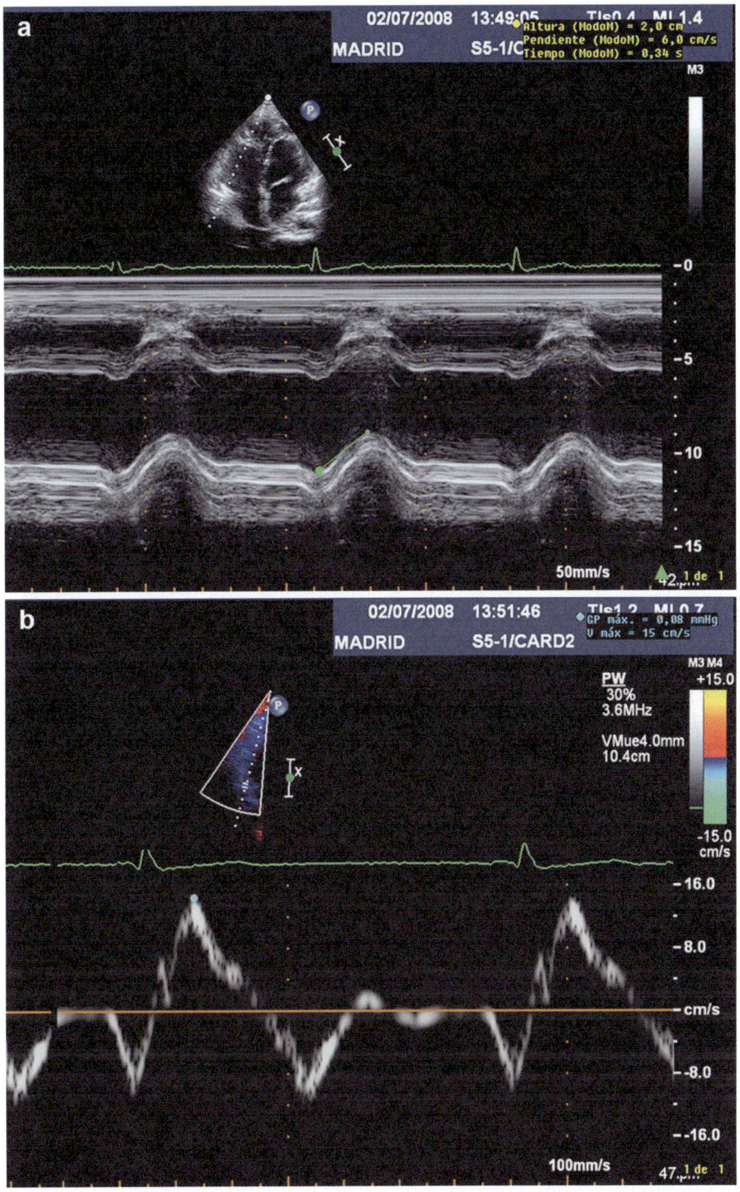

Fig. 5.4 Right ventricular systolic function obtained measuring tricuspid annular plane systolic excursion (**a**), and by tissue Doppler imaging (**b**)

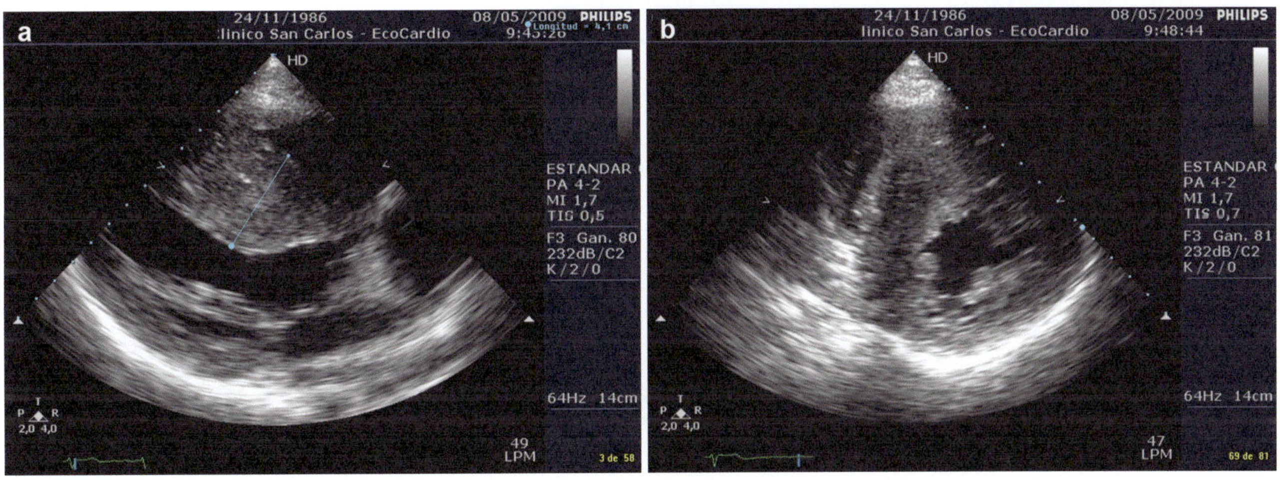

Fig. 5.5 Parasternal long-axis (**a**) and short-axis (**b**) view in a patient with hypertrophic cardiomyopathy. The ultrasonic beam should transect the left ventricle perpendicularly. Oblique images may overestimate the wall thickness

Table 5.4 Echocardiographic diagnostic criteria for hypertrophic cardiomyopathy

– Concentric or asymmetric left ventricular hypertrophy
– Enlargement of left atrium
– Non-dilated ventricles
– Absence of other causes of hypertrophy (aortic stenosis, coarctation, hypertension, renal failure, or amyloidosis)
– Normal left ventricular systolic function

the basal septum, mitral valve, and aortic valve during the cardiac cycle can be appreciated. Echocardiographic diagnostic criteria for HCM are listed in Table 5.4.

Doppler echocardiography allows determining the presence or absence of a dynamic LV outflow tract obstruction. The transducer position is modified in order to obtain the maximum velocity signal from the LV outflow tract. Particular care should be taken to separate the LV outflow tract signals from those due to mitral regurgitation. A resting pressure gradient in LV outflow tract is present in 25% of patients. In many other patients, a gradient is present only during physical or pharmacologic maneuvers. Evaluation of LV obstruction can also be seen on the M-mode echocardiogram. Aortic valve motion displays early-systolic closure and a "peak and dome" configuration of aortic pressure and velocity, which corresponds to transient mid systolic obstruction and a reduction in stroke volume.

The differential diagnosis between HCM and the so-called "heart of athlete" is crucial. LV wall thicknesses of 13–16 mm have been identified in a minority of elite rowers and cyclists. Various criteria for making this distinction have been described including[39]:

- The presence of systolic anterior motion of the anterior mitral valve suggests MCH.
- End-systolic diameter ≥55 mm is relatively common in trained athletes, but infrequent in MCH. Other useful parameter is the ratio of end-diastolic interventricular septum and end-diastolic LV volume. A value >0.26 provides a good sensitivity and specificity for diagnosis HCM.
- Most cases of HCM have alterations in ventricular filling, assessed by transmitral pulsed wave Doppler flow and TDI.
- Female athletes rarely have a wall thickness >11 mm. Therefore, values of 13–15 mm are more suggestive of HCM.
- After the training, a wall thinning of 2–5 mm occurs in about 3 months.

Other important issues in the echocardiographic assessment of HCM are:

- Diastolic function assessment: Diastolic dysfunction can be detected by Doppler echocardiography including mitral valve inflow, pulmonary vein, and tissue Doppler parameters. Early diastolic (Ea) velocity is reduced in patients with HCM. The ratio of mitral E to annular Ea (the E/E' ratio) allows us to estimate LV filling pressures.
- Asynchrony assessment. An intraventricular delay >45 ms identifies a subgroup of patients with ventricular tachycardia during Holter monitoring with a sensitivity of 90.9% and specificity of 95.8%,[40] and is the most powerful predictor of sudden death during the 4-year follow-up.[41]
- Myocardial deformation. Longitudinal, circumferential, and radial strains are decreased in patients with HCM. Myocardial deformation can be assessed by TDI or speckle-tracking technology.

In most cases, the diagnosis can be conveniently performed with echocardiography. However, this technique presents inherent difficulties: This method needs adequate acoustic window and sometimes images are inevitably obliques. Echocardiographic assessment is also difficult in those cases with involvement of anterolateral free wall as a result of the poor spatial resolution that prevents the recognition of the epicardial edge. In this regard, cardiac magnetic resonance (RMC) offers the ability to acquire tomographic cuts in any direction.

5.4 Echocardiography in Not-Compacted Cardiomyopathy

LV non-compaction, also called LV hypertrabeculation or spongy myocardium, is a rare disorder due to an interruption of the embryonic myocardial compaction process, occurring at 5–8 weeks gestation. This process is performed from the base toward the apex, epicardium toward endocardium and from the septum toward the lateral wall. This explains the characteristic distribution of the non-compaction mainly affecting the apex.[42] The clinical course is characterized by a significant morbidity caused by heart failure requiring occasionally transplants, malignant ventricular arrhythmias, sudden death, and embolic episodes. An early diagnosis is therefore necessary.

The diagnosis of Not-compacted cardiomyopathy is usually established by echocardiography. However, prominent LV trabeculations can be found in healthy people (70% of autopsied healthy hearts have some degree of compaction),[43] and can be observed in hypertrophic hearts secondary to dilated, valvular, or hypertensive cardiomyopathy. Thus, the differentiation between variants and not-compacted cardiomyopathy may occasionally be challenging. The main discriminating feature is the markedly thickened LV wall consisting of two layers: a thin compacted epicardial layer and a markedly thickened endocardial layer with numerous prominent trabeculations and deep recesses which are filled with blood from the ventricular cavity, without evidence of communication to the epicardial coronary artery system. A ratio of non-compacted to compacted myocardium >2:1 at end-systole in the parasternal short-axis view is characteristic.[44] Color Doppler shows blood flow within the deep intertrabecular recesses (Fig. 5.6).

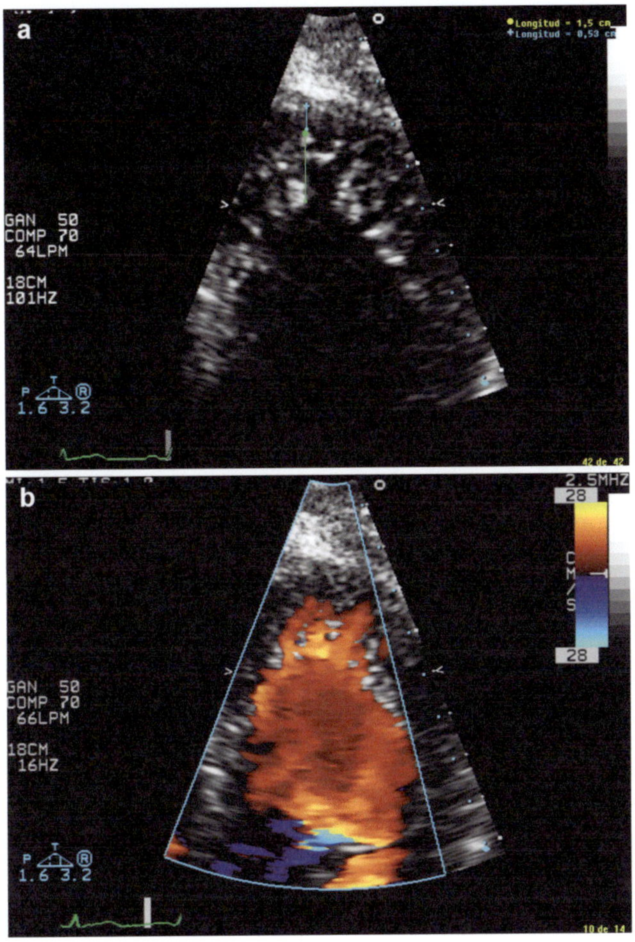

Fig. 5.6 Apical view in a patient affected by not-compacted cardiomyopathy. Color Doppler shows endocardial non-compacted zone filled with blood from the ventricular cavity

Non-compacted myocardium is predominantly found in the apical and mid-ventricular areas of both the inferior wall and the lateral wall. The affected segments are often hypokinetic in symptomatic patients or in those patients with impaired LV systolic function. The RV apex in healthy people is often intensely trabecular, which makes difficult to distinguish normal and pathologic patterns. However, prominent trabeculae and hypokinesis of RV wall accompanied by LV non-compaction permits diagnosis of RV involvement. Other findings that may be seen on echocardiography include reduced global LV systolic function, LV thrombi, diastolic dysfunction, and abnormal papillary muscle structure. In the case of poor image quality, contrast echocardiography clearly demarcating the endocardial borders may be helpful and may facilitate the diagnosis. Non-compacted myocardium is occasionally seen accompanying other congenital cardiac disorders such as Ebstein's anomaly, bicuspid aortic valve, aorta-to-left ventricular tunnel, congenitally corrected transposition, and isomerism of the LA appendage. Non-compacted myocardium has also been noted in patients with cardiomyopathies due to neuromuscular disorders.[45]

5.5 Conclusion

Echocardiography has an important role in the assessment of cardiac structure and function, risk stratification, and increasingly in guiding the management of cardiac arrhythmias. Technical progress continues to open new horizons in interventional electrophysiology. Expansion of catheter ablation procedures has led to a change of the paradigm from an electrophysiologically guided procedure to a procedure guided anatomically. Accurate identification of anatomical landmarks and catheters becomes mandatory. The development of intracardiac echocardiography enables real-time guidance of electrophysiological procedures. By using intravascular ultrasound imaging systems in the cardiac chambers, direct endocardial visualization can be provided. The widespread use of this technologies results in the reduction of fluoroscopy time and maximizes safety and efficacy of complex ablation procedures.

References

1. Fuster V, Rydén LE, Asinger RW, et al. ACC/AHA/ESC guidelines for the management of patients with atrial fibrillation: executive summary. A report of the American College of Cardiology/American Heart Association task force on practice guidelines and the European Society of Cardiology committee for practice guidelines and policy conferences (committee to develop guidelines for the management of patients with atrial fibrillation) developed in collaboration with the North American Society of Pacing and Electrophysiology. *Circulation*. 2001;104:2118-2150.
2. Duytschaever M, Haerynck F, Tavernier R, Jordaens L. Factors influencing long term persistence of sinus rhythm after a first electrical cardioversion for atrial fibrillation. *Pacing Clin Electrophysiol*. 1998;21:284-287.
3. Kizer JR, Bella JN, Palmieri V, et al. Left atrial diameter as an independent predictor of first clinical cardiovascular events in middle-aged and elderly adults: the strong heart study. *Am Heart J*. 2006; 151:412-418.
4. Tsang TS, Abhayaratna WP, Barnes ME, et al. Prediction of cardiovascular outcomes with left atrial size: is volume superior to area or diameter? *J Am Coll Cardiol*. 2006;47:1018-1023.
5. Di Salvo G, Caso P, Lo Piccolo R, et al. Atrial myocardial deformation properties predict maintenance of sinus rhythm after external cardioversion of recent-onset lone atrial fibrillation: a color Doppler myocardial imaging and transthoracic and transesophageal echocardiographic study. *Circulation*. 2005;112:387-395.
6. Klein AL, Grimm RA, Murray RD, et al. Use of transesophageal echocardiography to guide cardioversion in patients with atrial fibrillation. *N Engl J Med*. 2001;344:1411-1420.
7. SPAF III Investigators. Transesophageal echocardiographic correlates of thromboembolism in high-risk patients with nonvalvular atrial fibrillation. The Stroke Prevention in Atrial Fibrillation Investigators Committee on Echocardiography. *Ann Intern Med*. 1998;128:639-647.
8. Fukuchi M, Watanabe J, Kumagai K, et al. Increased von Willebrand factor in the endocardium as a local predisposing factor for thrombogenesis in overloaded human atrial appendage. *J Am Coll Cardiol*. 2001;37:1436-1442.

9. Bernhardt P, Schmidt H, Hammerstingl C, Lüderitz B, Omran H. Patients with atrial fibrillation and dense spontaneous echo contrast at high risk a prospective and serial follow-up over 12 months with transesophageal echocardiography and cerebral magnetic resonance imaging. *J Am Coll Cardiol*. 2005;45:1807-1812.
10. Krahn AD, Manfreda J, Tate RB, Mathewson FA, Cuddy TE. The natural history of atrial fibrillation: incidence, risk factors, and prognosis in the Manitoba Follow-Up study. *Am J Med*. 1995;98: 476-484.
11. Troughton RW, Asher CR, Klein AL. The role of echocardiography in atrial fibrillation and cardioversion. *Heart*. 2003;89:1447-1454.
12. [No authors listed]. Risk factors for stroke and efficacy of antithrombotic therapy in atrial fibrillation. Analysis of pooled data from five randomized controlled trials. *Arch Intern Med*. 1994;154:1449-1457.
13. Parvathaneni L, Mahenthiran J, Jacob S, et al. Comparison of tissue Doppler dynamics to Doppler flow in evaluating left atrial appendage function by transesophageal echocardiography. *Am J Cardiol*. 2005;95:1011-1014.
14. Sievert H, Lesh MD, Trepels T, et al. Percutaneous left atrial appendage transcatheter occlusion to prevent stroke in high-risk patients with atrial fibrillation: early clinical experience. *Circulation*. 2002;105:1887-1889.
15. Kinnaird TD, Uzun O, Munt BI, Thompson CR, Yeung-Lai-Wah JA. Transesophageal echocardiography to guide pulmonary vein mapping and ablation for atrial fibrillation. *J Am Soc Echocardiogr*. 2004;17:769-774.
16. López-Candales A, Grewal H, Katz W. The importance of increased interatrial septal thickness in patients with atrial fibrillation: a transesophageal echocardiographic study. *Echocardiography*. 2005;22:408-414.
17. Scholten MF, Thornton AS, Jordaens LJ, Roelandt JR, Kerber RE, Kronzon I. Usefulness of transesophageal echocardiography using a combined probe when converting atrial fibrillation to sinus rhythm. *Am J Cardiol*. 2004;94:470-473.
18. Pua EC, Idriss SF, Wolf PD, Smith SW. Real-time three-dimensional transesophageal echocardiography for guiding interventional electrophysiology: feasibility study. *Ultrason Imaging*. 2007;29:182-194.
19. Ren J, Marchlinsky F. Intracardiac echocardiographic imaging in radiofrequency catheter ablation for atrial fibrillation. In: Ren J, Marchlinsky F, Callans D, eds. *Practical Intracardiac Echocardiography in Electrophysiology*. 1st ed. Malden: Blackwell; 2006: 88-116.
20. Wood MA, Wittkamp M, Henry D, et al. A comparison of pulmonary vein ostial anatomy by computerized tomography, echocardiography, and venography in patients with atrial fibrillation having radiofrequency catheter ablation. *Am J Cardiol*. 2004;93:49-53.
21. Jongbloed MR, Bax JJ, Lamb HJ, et al. Multislice computed tomography versus intracardiac echocardiography to evaluate the pulmonary veins before radiofrequency catheter ablation of atrial fibrillation: a head-to-head comparison. *J Am Coll Cardiol*. 2005;45: 343-350.
22. Mlcochová H, Tintera J, Porod V, Peichl P, Cihák R, Kautzner J. Magnetic resonance angiography of pulmonary veins: implications for catheter ablation of atrial fibrillation. *Pacing Clin Electrophysiol*. 2005;28:1073-1080.
23. De Ponti R, Zardini M, Storti C, Longobardi M, Salerno-Uriarte JA. Trans-septal catheterization for radiofrequency catheter ablation of cardiac arrhythmias: results and safety of a simplified method. *Eur Heart J*. 1998;19:943-950.
24. Lindeneg O, Hansen AT. Complication in transseptal left heart catheterization. *Acta Med Scand*. 1996;180:395-399.
25. Roelke M, Smith AJ, Palacios IF. The technique and safety of transseptal left heart catheterization: the Massachusetts general hospital experience with 1279 procedures. *Cathet Cardiovasc Diagn*. 1994;32:332-339.
26. Hung JS, Fu M, Yeh KH, Chua S, Wu JJ, Chen YC. Usefulness of intracardiac echocardiography in transseptal puncture during percutaneous transvenous mitral commis. *Am J Cardiol*. 1993;72:853-854.
27. Marrouche NF, Martin DO, Wazni O, et al. Phased-array intracardiac echocardiography monitoring during pulmonary vein isolation in patients with atrial fibrillation: impact on outcome and complications. *Circulation*. 2003;107:2710-2716.
28. Wood MA, Shaffer KM, Ellenbogen AL, Ownby ED. Microbubbles during radiofrequency catheter ablation: composition and formation. *Heart Rhythm*. 2005;2:397-403.
29. Ren JF, Marchlinski FE, Callans DJ. Left atrial thrombus associated with ablation for atrial fibrillation: identification with intracardiac echocardiography. *J Am Coll Cardiol*. 2004;43:1861-1867.
30. Saad E, Cole C, Marrouche N, et al. Use of intracardiac echocardiography for prediction of chronic pulmonary vein stenosis after ablation of atrial fibrillation. *J Cardiovasc Electrophysiol*. 2002;13: 986-989.
31. Ren J, Marchlinski F, Callans D, Zado E. Intracardiac Doppler echocardiographic quantification of pulmonary vein flow velocity. *J Cardiovasc Electrophysiol*. 2002;13:1076-1081.
32. Vaseghi M, Cesario D, Majan A, et al. Catheter ablation of right ventricular outflow tract tachycardia: value of defining coronary anatomy. *J Cardiovasc Electrophysiol*. 2006;17:632-637.
33. Morton JB, Sanders P, Davidson NC, Sparks PB, Vohra JK, Kalman JM. Phased-array intracardiac echocardiography for defining cavotricuspid isthmus anatomy during radiofrequency ablation of typical atrial flutter. *J Cardiovasc Electrophysiol*. 2003;14: 591-597.
34. McKenna WJ, Thiene G, Nava A, et al. Diagnosis of arrhythmogenic right ventricular dysplasia/cardiomyopathy. Task Force of the Working Group Myocardial and Pericardial Disease of the European Society of Cardiology and of the Scientific Council on Cardiomyopathies of the International Society and Federation of Cardiology. *Br Heart J*. 1994;71:215-218.
35. Foale R, Nihoyannopoulos P, McKenna W, et al. Echocardiographic measurement of the normal adult right ventricle. *Br Heart J*. 1986;56:33-44.
36. Yoerger DM, Marcus F, Sherrill D, et al. Echocardiographic findings in patients meeting task force criteria for arrhythmogenic right ventricular dysplasia: new insights from the multidisciplinary study of right ventricular dysplasia. *J Am Coll Cardiol*. 2005;45:860-865.
37. Kaul S, Tei C, Hopkins JM, Shah PM. Assessment of right ventricular function using two-dimensional echocardiography. *Am Heart J*. 1984;107:526-531.
38. Spirito P, Bellone P, Harris KM, Bernabo P, Bruzzi P, Maron BJ. Magnitude of left ventricular hypertrophy predicts the risk of sudden death in hypertrophic cardiomyopathy. *N Engl J Med*. 2000;342:1778-1785.
39. Maron BJ. Distinguishing hypertrophic cardiomyopathy from athlete's heart: a clinical problem of increasing magnitude and significance. *Heart*. 2005;91:1380-1382.
40. D'Andrea A, Caso P, Severino S, et al. Association between intraventricular myocardial systolic dyssynchrony and ventricular arrhythmias in patients with hypertrophic cardiomyopathy. *Echocardiography*. 2005;22:571-578.
41. D'Andrea A, Caso P, Severino S, et al. Prognostic value of intra-left ventricular electromechanical asynchrony in patients with hypertrophic cardiomyopathy. *Eur Heart J*. 2006;27:1311-1318.
42. Petersen SE, Selvanayagam JB, Wiesmann F, et al. Left ventricular non-compaction: insights from cardiovascular magnetic resonance imaging. *J Am Coll Cardiol*. 2005;46:101-105.
43. Boyd MT, Seward JB, Tajik AJ, Edwards WD. Frequency and location of prominent left ventricular trabeculations at autopsy in 474 normal human hearts: implications for evaluation of mural thrombi by two-dimensional echocardiography. *J Am Coll Cardiol*. 1987;9:323-326.

44. Jenni R, Oechslin E, Schneider J, Attenhofer Jost C, Kaufmann PA. Echocardiographic and pathoanatomical characteristics of isolated left ventricular non-compaction: a step towards classification as a distinct cardiomyopathy. *Heart*. 2001;86:666-671.

45. Stollberger C, Finsterer J, Blazek G. Left ventricular hypertrabeculation/noncompaction and association with additional cardiac abnormalities and neuromuscular disorders. *Am J Cardiol*. 2002;90:899-902.

Part II

Catheter Ablation

Anatomy Assessment for Atrial Arrhythmias

6

Sheldon M. Singh and Andre d'Avila

Abstract

Catheter ablation of atrial arrhythmias has increased dramatically over the past decade, mainly due to an increase in the number of ablation procedures performed for atrial fibrillation, the most common ablation procedure performed worldwide. Given the complexity of the left and right atria as well as the intricacy of ablation procedures for these arrhythmias, knowledge of atrial anatomy and its variants is vital when planning a safe and effective ablation strategy. In this chapter, we will review the role and importance of anatomy assessment including the use of cardiac imaging prior to atrial ablation procedures.

Keywords

Cardiac anatomy by imaging • Cardiac imaging for atrial arrhythmias • Atrial arrhythmia and imaging • Left atrium anatomy • Right atrium anatomy • Inter-atrial septum

Catheter ablation of atrial arrhythmias has increased dramatically over the past decade, mainly due to an increase in the number of ablation procedures performed for atrial fibrillation, the most common ablation procedure performed worldwide.[1] Given the complexity of the left and right atria as well as the intricacy of ablation procedures for these arrhythmias, knowledge of atrial anatomy and its variants is vital when planning a safe and effective ablation strategy. In this chapter, we will review the role and importance of anatomy assessment including the use of cardiac imaging prior to atrial ablation procedures.

A. d'Avila (✉)
Helmsley Electrophysiology Center,
Mount Sinai School of Medicine, New York, NY, USA
e-mail: andre.d'avila@mountsinai.org

6.1 Right Atrium

While right atrial ablation is occasionally performed during AF ablation procedures, ablation of primary right-sided atrial arrhythmias including atrial flutter and atrial tachycardia is quite common. Although considered "simple" ablation procedures by many, a thorough understanding of the right atrial anatomy is essential especially when troubleshooting the difficult case.

6.1.1 Cavo-Tricuspid Isthmus-Dependent Atrial Flutter

"Typical" or "type 1" atrial flutter describes a macro-reentrant right-sided tachycardia involving the cavo-tricuspid isthmus, an area of tissue situated over the lower right atrium between the inferior vena cava and the tricuspid isthmus (Fig. 6.1). Placement of radiofrequency energy within this region to create bidirectional conduction block is frequently performed with a high rate of success to cure patients of typical atrial flutter.[2] While difficulties rarely arise, they will be encountered by electrophysiologists performing a large

A. Auricchio et al. (eds.), *Cardiac Imaging in Electrophysiology*,
DOI 10.1007/978-1-84882-486-7_6, © Springer-Verlag London Limited 2012

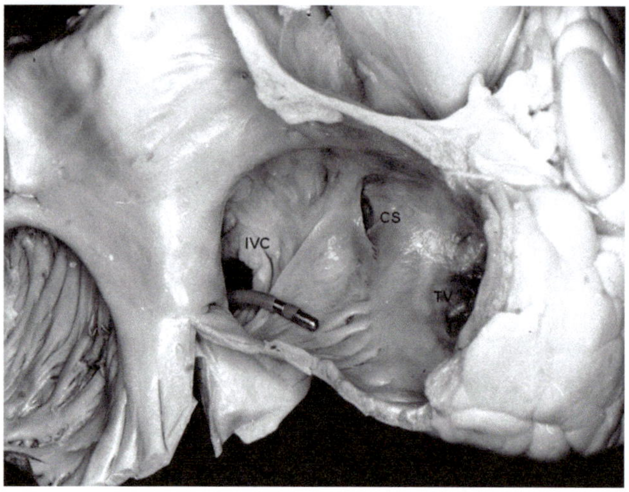

Fig. 6.1 A view of the cavo-tricuspid isthmus. An ablation catheter can be seen on the cavo-tricuspid isthmus. *IVC* inferior vena cava, *TV* tricuspid valve, CS coronary sinus (Courtesy of Eduardo Sosa)

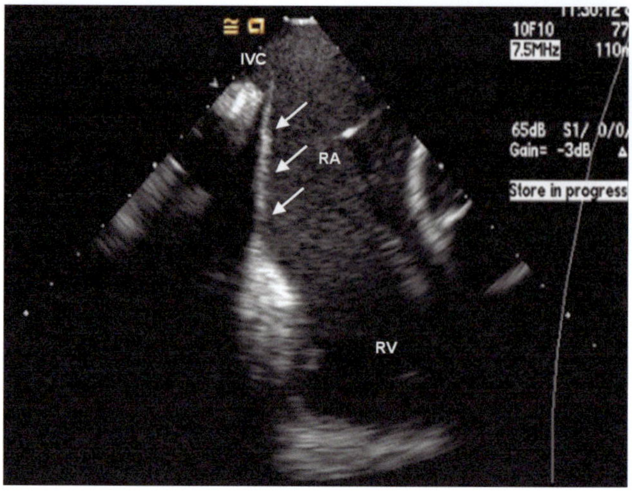

Fig. 6.2 Intracardiac echocardiography imaging from the right atrium demonstrating a flat cavo-tricuspid isthmus. The isthmus is denoted by the *arrows*. *IVC* inferior vena cava, *RA* right atrium, *RV* right ventricle

number of these procedures. Knowledge of the size, geometry, and composition (i.e., myocardial versus fibrofatty tissue) of the cavo-tricuspid isthmus region will allow one to troubleshoot these difficult cases.

The fibers of the cavo-tricuspid isthmus originate septally in the region of the coronary sinus and extend inferolaterally. The optimal site for ablation is the central portion of the cavo-tricuspid isthmus, at approximately 6 o'clock in the left anterior oblique view with fluoroscopy as the length of the isthmus in this area is shortest (19 ± 4 mm), thinnest (2.3 ± 1.3 mm), comprised of fibrofatty tissue, and thus less likely to resist radiofrequency ablation.[3] In fact, the presence of fibrofatty tissue suggests that the length of the isthmus requiring ablation may in fact be shorter than the actual distance between the tricuspid annulus and the inferior vena cava.[4] Ablation in this region is also advantageous as more septal ablation has an associated risk of AV block due to collateral ablation of AV nodal extensions and/or the AV nodal artery which lies in close proximity to the septal isthmus in approximately 10% of patients.[3,4] Anseleme and colleagues highlighted this risk of AV block when ablating the septal isthmus[5]; in their study, 5 of 36 patients developed AV block (4 transient and 1 permanent) during ablation of the septal isthmus whereas no patient developed AV block during ablation of the central isthmus.

Discontinuous lesion placement or the inability to deliver power while ablating along the cavo-tricuspid isthmus can contribute to procedural failure. Both situations occur due to variations in normal anatomy since the cavo-tricuspid isthmus is not simply a flat surface. Sub Eustachian pouches and ridges may both result in suboptimal catheter positioning and ineffective ablation.[6] Pouch-like recesses may complicate cavo-tricuspid isthmus ablation. These recesses, which occur between the tricuspid annulus and the Eustachian ridge, have varying degrees of depression.[6,7] Although the myocardium tends to be thinner, decreased blood flow within the pouch frequently results in suboptimal power delivery. Moreover, if not recognized, there will be gaps in the ablation lesion set if the catheter is simply pulled back in a planar fashion as it will not be in contact with the isthmus. While recognizing the presence of a pouch may be difficult with fluoroscopy alone due to the limited soft tissue visualization, abnormal catheter tip motion may alert the operator regarding the presence of a pouch. Another clue is difficulty with coronary sinus cannulation as the presence of a Thebesian valve, which guards the os of the coronary sinus, is frequently associated with sub-Eustachian pouches.[7] Imaging tools such as a pre-acquired CT scan, intra-procedural use of right atrial angiography, or intracardiac echocardiography (Fig. 6.2) can allow the operator to obtain a better appreciation of the topography of the isthmus and thus plan the ablation strategy. Moreover, the use of phased array intracardiac echocardiography can allow the operator to visualize catheter tip contact during the procedure to ensure adequate contact when ablating in the pouch.[8]

Visualization of a pouch may also alter the operator's approach to catheter positioning or ablation strategy. In our experience, improved catheter contact often occurs with the catheter looped in the region of the pouch with progressive relaxation of the curve allowing for ablation within the pouch. A prominent Eustachian ridge may be present in these patients (Fig. 6.3); extensive ablation of this structure is often not necessary as it is fibrous in nature.[6] However, its presence may impact catheter contact with the isthmus. In these situations, a long vascular sheath may be of assistance during the procedure.

While ablation of the lateral isthmus can avoid the risk of AV block as well as difficulties associated with the presence

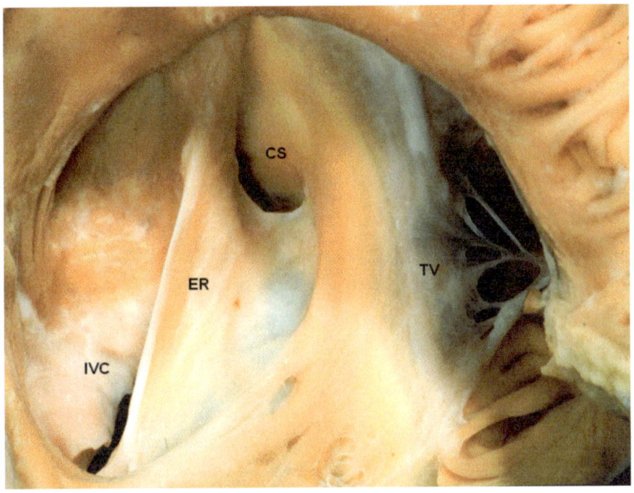

Fig. 6.3 A prominent Eustachian ridge is noted on this specimen. Pouches often co-exist with a prominent Eustachian ridge. *ER* Eustachian ridge, *TV* tricuspid valve, *IVC* inferior vena cava, *CS* coronary sinus (Courtesy of Anton Becker)

of a pouch, ablation in this region may be complicated by the presence of prominent pectinate muscle bundles which fan out across the isthmus.[9] These muscle bundles may complicate ablation for multiple reasons including a lack of catheter stability due to the presence of a "bumpy" surface, the inability to deliver power when the catheter is wedged between muscle bundles, and the inability to obtain transmural lesions secondary to the thick myocardium associated with these muscle bundles. As with the presence of pouches, recognition of pectinate muscle bundles may also be difficult. Electrogram amplitude is also very helpful and correlates the anatomic and electrophysiologic features in the region – in general, electrogram voltage >1.5 mV usually corresponds to areas with myocardium >5 mm.[10] In these situations, more central or septal ablation may be considered to avoid the pectinate muscles. Alternatively, the use of an irrigated tip catheter to enhance power delivery may also be considered in these regions with thick atrial myocardium.[11]

6.1.2 Right Atrial Tachycardia

While sustained atrial tachycardia is relatively rare, it does account for 5–20% of patients undergoing an electrophysiology study.[12,13] Although the mechanism of atrial tachycardia can be heterogeneous, the focus of this arrhythmia is not randomly distributed throughout the atrium. Kalman demonstrated that approximately two-thirds of right atrial tachycardias in patients with structurally normal hearts originated in the cristae terminalis, the junction between the pectinate (anterior) right atrium and smooth-walled (posterior) venous component.[14] Moreover, a gradation in frequency from high to low cristae was also noted. Knowledge of this anatomic predilection may guide mapping and ablation in the electrophysiology laboratory.

While intracardiac echocardiography may be helpful to demonstrate the location of the catheter in relation to the *cristae terminalis,* in general, specialized anatomic assessment is rarely required prior to atrial tachycardia ablation. In fact, electroanatomic mapping systems to catalog the activation sequence of the atrial activation may be more helpful to ensure that other less common locations such as the coronary sinus os (7% of ATs), tricuspid annulus (13% of AT), parahisian region, right atrial appendage, or even the left atrium are not the sites of origin of the tachycardia.

As a close relationship exists between the right phrenic nerve and the superior vena cava superiorly and lateral border of the inferior vena cava and right atrium inferiorly, caution must be exercised when performing ablation in this region to minimize phrenic nerve injury. High-output pacing to demonstrate the presence or absence of phrenic nerve capture should be performed in these regions to ensure one does not deliver energy directly on the phrenic nerve. Clues to early phrenic nerve injury during ablation include a reduction in diaphragmatic excursion during energy application, or patient coughing or hiccupping during energy delivery. Recently it has been shown that the right phrenic nerve can accurately be identified on multidetector computed tomography either directly[15] or indirectly by identifying the phrenic artery which runs with it.[16] This, especially when integrated into an electroanatomic mapping systems, may allow operators to alter an ablation strategy should there be a high risk of injuring the phrenic nerve.

6.2 Interatrial Septum

An appreciation of the anatomy of the interatrial septum and its relationship to adjacent structures is crucial for interventional electrophysiologists managing left-sided atrial arrhythmias as access to the left atrium is often achieved via transseptal punctures, which can only safely be undertaken if one has an intimate knowledge of the interatrial septum. It must be appreciated that the true interatrial septum is a structure which, when removed from the heart, does not allow one to exit the heart.[17] Thus, the true interatrial septum is limited to the floor of the fossa ovalis, the anterior-inferior rim of the fossa, and the flap itself. The region between the SVC and the superior border of the fossa, although septal in position, is not part of the true septum as it is an infolding of the left and right atrial walls with adipose tissue between. Performing a transseptal puncture in this region will result in the transseptal sheath exiting the heart.

Pre-procedural imaging such as 2D echocardiography or CT/MRI imaging may identify the presence of an aneurismal interatrial septum, delineate the relationship between the left

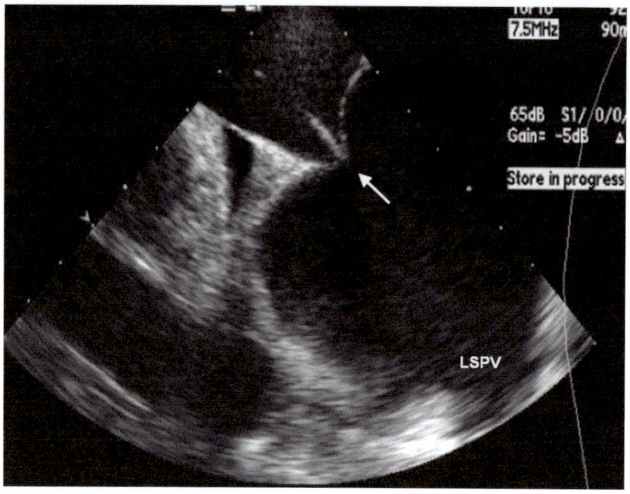

Fig. 6.4 An intracardiac echocardiography image of a transseptal system (sheath and needle) positioned at the fossa ovalis during a transseptal puncture. The image is obtained with the intracardiac echocardiography imaging probe placed in the right atrium. Tenting of the interatrial septum is noted. In this image, the transseptal apparatus is aligned with the left superior pulmonary vein (*LSPV*) in the imaging view. This provides an optimal trajectory to puncture the septum during atrial fibrillation ablation procedures

atrium and aorta, and provide an assessment of the left atrial size. However, these modalities do not provide real-time assessment of these anatomic variants and complications which may arise during transseptal catheterization. While fluoroscopy provides real-time imaging and adequate information for transseptal puncture in many cases, it lacks sufficient soft tissue visualization and does not allow one to truly appreciate the interatrial septum and adjacent structures compared to intracardiac echocardiography which provides operators with a real-time appreciation of the interatrial septum during transseptal catheterization (Fig. 6.4) and has the advantage of not requiring a separate operator, sedation, or obstructing fluoroscopy as occurs with the use of intra-procedural transesophageal echocardiography. Knowledge of the orientation of the transseptal sheath in relation to structures such as the aorta (anteriorly) and posterior wall can be obtained with intracardiac echocardiography, and may allow the operator to readjust the transseptal system to obtain an optimal trajectory.

A novel fiber-optic catheter (Iris, Voyage Medical, Campbell, CA) is currently under investigation, which provides real-time direct visualization of the interatrial septum and allows one to identify the fossa ovalis, thereby ensuring an appropriate site for transseptal puncture is selected (Fig. 6.5).[18]

6.3 Left Atrium

Over the last decade, left atrial ablation is increasingly being performed for management of atrial fibrillation. Given the extensive ablation necessary for managing this arrhythmia, a thorough understanding of the left atrium and all its components (vestibule, venous connections, and appendage), as well as adjacent structures (such as the aorta, esophagus, and phrenic nerve) is necessary to safely and effectively perform these procedures (Fig. 6.6). In addition to the complexity associated with normal left atrial-pulmonary vein anatomy, approximately 40% of individuals undergoing atrial fibrillation ablation procedures have variant anatomy.[19] This finding underscores the importance of pre-procedural imaging in order to assess the left atrial anatomy and hence guide an ablation strategy.

6.3.1 Left Atrial Appendage

Prior to proceeding with an ablation procedure for atrial fibrillation or flutter, one must rule out the presence of a left atrial appendage thrombus in patients without adequate anticoagulation. Currently transesophageal echocardiography is considered the gold standard for detection of left atrial appendage thrombus. In many cases, intracardiac echocardiography can confirm the presence of a left atrial appendage thrombus; however, this modality remains to be rigorously validated against the gold standard transesophageal echocardiography. To date, a small study comparing intracardiac echocardiography to transesophageal echocardiography demonstrated that while intracardiac echocardiography could visualize the interatrial septum and LA body for the presence of LA thrombus with similar concordance rates to transesophageal echocardiography, intracardiac echocardiography fell short in identifying left atrial appendage thrombus with a more moderate positive predictive value compared to transesophageal echocardiography.[20] Thus, intracardiac echocardiography alone may be insufficient to evaluate the left atrial appendage. Of note, imaging the left atrium has also become progressively important for the implantation of left atrial appendage occluder devices (e.g., Watchman device).

Multidetector CT scanning has also been evaluated as a tool for detecting left atrial appendage thrombus. Similar to intracardiac echocardiography, the positive predictive value with an abnormal MDCT scan is quite low and, in the largest comparative series to date, reported at only 23% whereas the negative predictive value of this tool was reported to be 100%.[21] It was suggested that in patients with a CHADS2 score <1, a negative MDCT scan could effectively rule out the presence of a LAA thrombus and avoid the need for additional testing with a transesophageal echocardiogram. However, due to the low positive predictive value, transesophageal echocardiography should be performed to definitively rule in the presence of a left atrial appendage thrombus when any abnormality on MDCT is noted.[21,22]

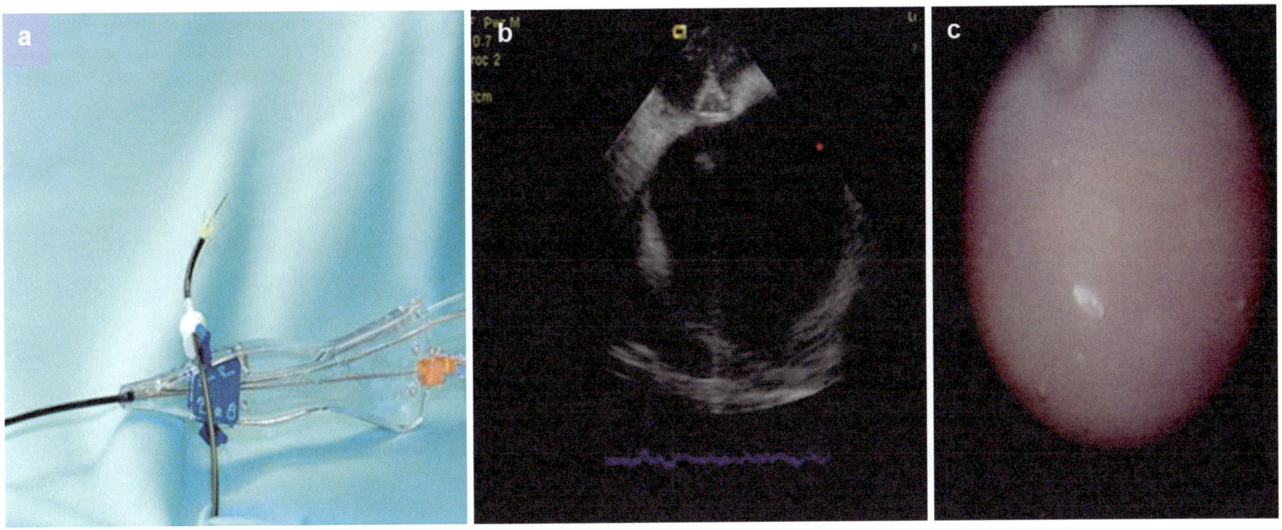

Fig. 6.5 The IRIS catheter (**a**) consists of a 2Fr fiber-optic endoscope which allows one to directly visualize the septum. The distal aspect of the catheter has a hood which excludes blood from the field of view, thereby allowing one to obtain an unobstructed view of the septum. The catheter system is placed along the septum. (**b**) Intracardiac echocardiography image from the right atrium demonstrating placement of the IRIS catheter at the site of the fossa ovalis. (**c**) The corresponding endoscopic image of the fossa ovalis obtained with the fiber-optic endoscopy located within the IRIS catheter

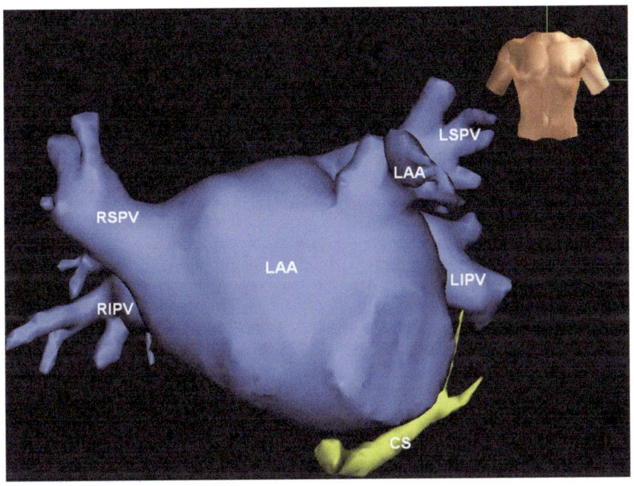

Fig. 6.6 The left atrial geometry (*purple*) in this figure is a 3D reconstruction of the left atrium using a multi-electrode catheter and the NavX system. The relationship between the left atrial body (LA), the appendage (*LAA*), and the pulmonary veins (*LSPV* left superior, *LIPV* left inferior, *RSPV* right superior, *RIPV* right inferior) is demonstrated in this figure. The location of the coronary sinus is also noted (in *yellow*)

Multidetector CT scanning can identify the presence of unusual left atrial appendage insertions into the left atrium.[23] Although rare, such variant anatomy may compromise procedural safety if not recognized.

6.3.2 Left Atrial Ablation for Atrial Fibrillation

Current strategies for atrial fibrillation ablation have evolved from focal ablation targeting arrhythmogenic foci to proximal ablation of the pulmonary venous antral regions with circumferential pulmonary vein isolation. As such, this approach requires operators to understand the relationship between the pulmonary veins and the left atrial body (Fig. 6.6). This relationship is often very complex with significant inter- and intra-patient variability in the location, ostial diameter, and branching patterns of these veins. Challenges to ablation include ensuring that one is not within the ostium of the pulmonary veins in order to avoid pulmonary vein stenosis, ensuring adequate catheter stability while delivering radiofrequency energy, and delivering the appropriate power to balance adequate lesion formation with the risk of perforation given the variability of the left atrial wall thickness. Knowledge of any variation in anatomy is vital when performing these procedures. As such, pre-procedural and intra-procedural imaging can provide important details to guide left atrial ablation.

6.3.2.1 Pulmonary Veins

Safe ablation within the antral region of the left atrium requires the operator to be able to identify the left atria-pulmonary vein junction. Additionally, ablation within the ridges between pulmonary veins or between the left pulmonary veins and the left atrial appendage (Figs. 6.7–6.9) requires an appreciation of the size and location of inter-vein ridges as inadvertent application of radiofrequency energy within a pulmonary vein may result in pulmonary vein stenosis or, in the case of the left superior pulmonary vein–left atrial appendage ridge, ablation within the left atrial appendage may result in perforation. An understanding of each patient's left atrial anatomy is vital as approximately 40% of patients have some variation in normal

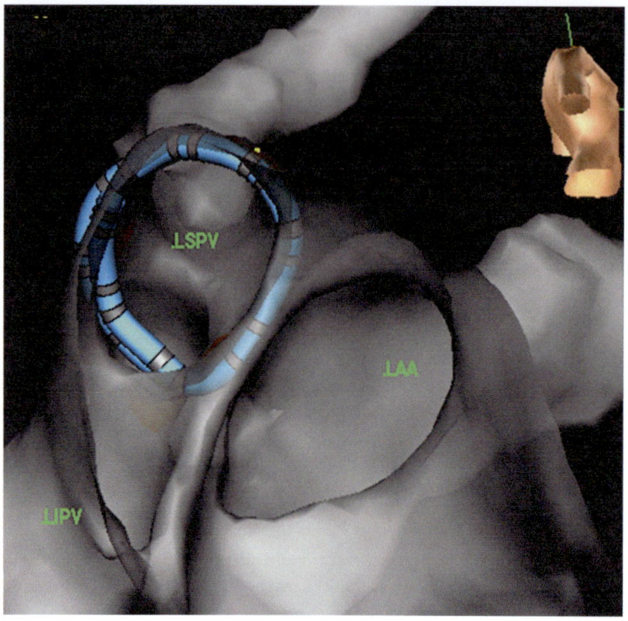

Fig. 6.7 This is an endoscopic view of the left-sided pulmonary veins (superior (*LSPV*) and inferior (*LIPV*)) and left atrial appendage (*LAA*) using the NavX system. A 3D reconstruction of the left atrium, pulmonary veins, and left atrial appendage is constructed with a multi-electrode catheter and the NavX system. The multi-electrode Lasso catheter is seen in the left superior pulmonary vein (*LSPV*). The ridge between the LSPV and LAA is visualized with this endoscopic view. The ridge width and length is variable among patients

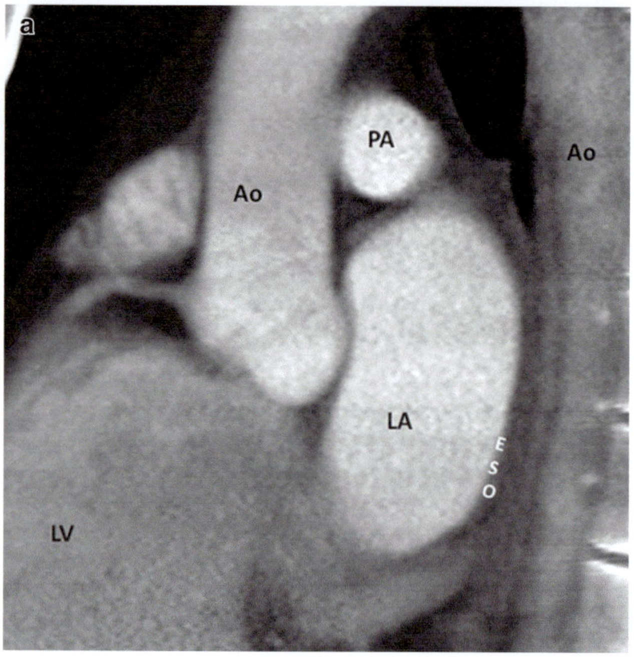

Fig. 6.8 The relationship between the left atrium, esophagus, and aorta is demonstrated above. The esophagus frequently lies in close proximity to the posterior left atrium and maybe vulnerable to injury with posterior left atrial ablation. The position varies among patients (*left versus right*) and can change within a patient during an ablation procedure. The location of the ascending and descending aorta is illustrated in this image. (**a**) Location of these structures on CT scan. (**b**) Left-sided esophagus. (**c**) Right-sided esophagus in a separate patient. *Ao* aorta, *PA* pulmonary artery, *LA* left atrium, *DA* descending aorta, *ESO* esophagus, *LV* left ventricle

pulmonary vein anatomy.[19] Marom demonstrated 3–5 right-sided pulmonary veins in 28% of patients, a single right-sided pulmonary vein in 2% of patients, and a single left-sided pulmonary vein in 14% of patients.[24] Ho also demonstrated variability in the mean diameter of the pulmonary vein os, which ranged from 8 to 21 mm.[25] Mansour characterized the variability in the width of the ridge between the left pulmonary vein and left atrial appendage and the ridge between the right pulmonary veins (Figs. 6.7 and 6.8). Ninety two percent of patients had a left pulmonary vein-left atrial appendage ridge <5 mm and the average diameter of the ridge between the right-sided pulmonary veins was approximately 3 mm.[26]

Positioning an ablation catheter in these small regions can be a challenge. Fluoroscopy does not provide sufficient additional information on catheter placement. Electrogram analysis may also not provide any incremental information especially as most patients are in atrial fibrillation at the time of the procedure with significant variability in the amplitude of the electrogram signals. In these situations, pre-procedural imaging with CT or MRI may be of help to better delineate this complex anatomy. Additionally, integration of pre acquired images with electroanatomic maps may provide detailed information to guide the operator during ablation along these narrow ridges.

Contrast venography also provides real-time imaging of the pulmonary veins; however, it may not be as accurate as CT or MRI imaging due to overestimation or underestimation of the PV ostial diameter when obtained in a single view.[27] Intracardiac echocardiography can provide detailed anatomic information of the left atrial ridges and has the advantage of providing real-time imaging, thereby allowing the operator to visualize and hence facilitate catheter placement in the antrum and along these narrow ridges. However, like contrast venography, this mode of imaging is also only 2-dimensional. Recently it has become possible to use intracardiac echocardiography to obtain a real-time 3-dimensional reconstruction of the LA including the antrum and pulmonary vein junction.[28] This advancement may allow a better appreciation of the 3-dimensional anatomy of the LA that one may obtain with CT and MRI with the added benefit of real-time imaging which avoids the need to account for changes in pre-load or rhythm at the time the pre-acquired CT or MRI was obtained. Additionally, it is well known that

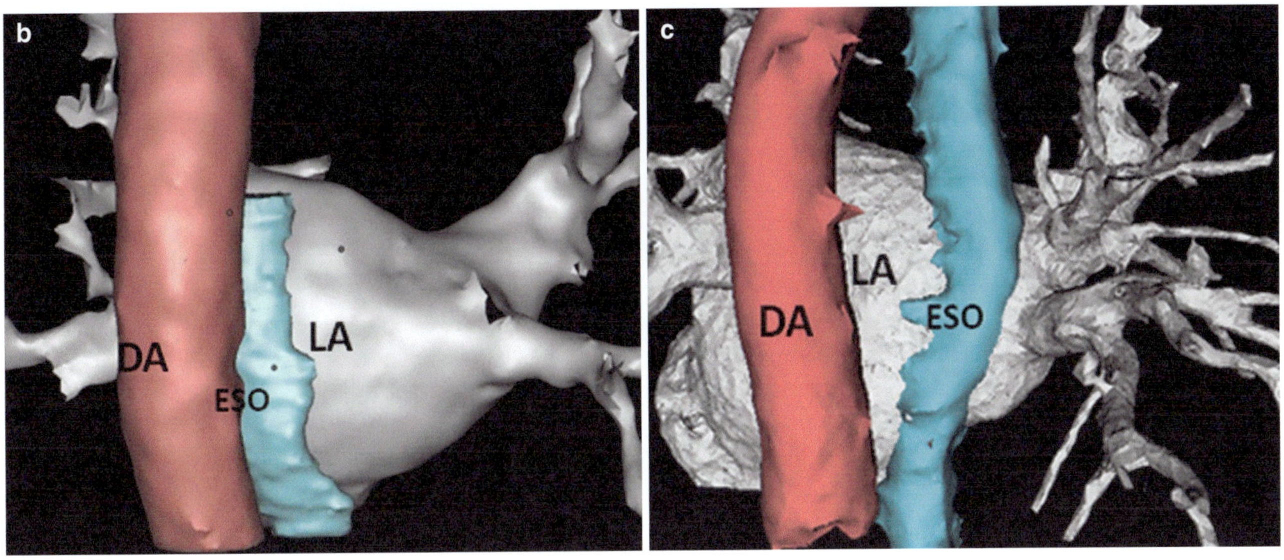

Fig. 6.8 (continued)

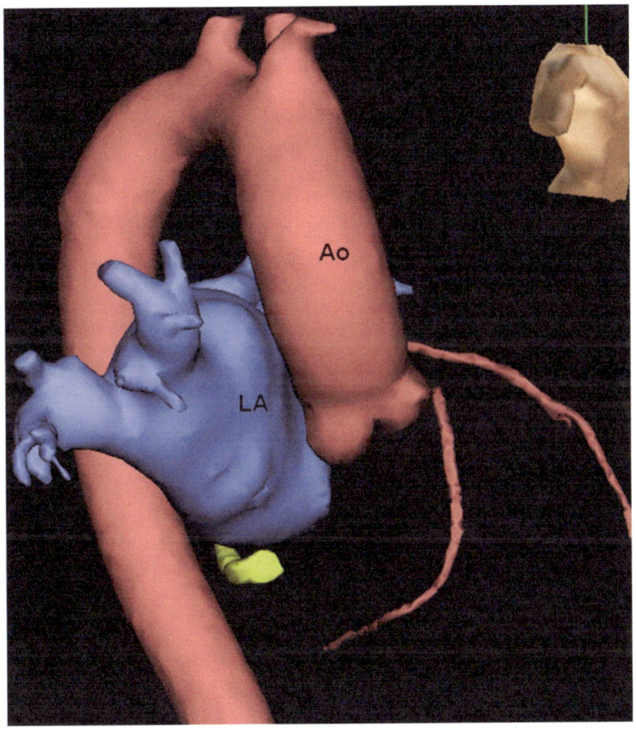

Fig. 6.9 The anterior left atrium is compressed by the ascending aorta in a patient without aortic dilation

pulmonary veins are not static in nature. MRI imaging has demonstrated variation in size (possibly due to pulmonary vein contractility during the cardiac cycle)[29] and orientation with respiration.[30] Real-time imaging with ICE will allow one to account for these factors during ablation procedures.

6.3.2.2 Left Atrial Wall Thickness

Knowledge of the lack of uniformity in the left atrial wall thickness is also very important during left atrial ablation procedures as a lower intensity or duration of radiofrequency energy may be required in thinner areas of the left atrium where less atrial myocardial tissue is present. Inappropriate delivery of radiofrequency energy may cause collateral damage to adjacent structures or result in left atrial perforation. Hall and colleagues assessed the left atrial wall thickness in 34 consecutive human heart specimens during routine postmortem evaluation.[31] Significant differences existed between different anatomic areas within the left atrium. The roof and posterior wall of the LA were thinnest while the anterior wall and septum thickest. The mean LA wall thickness was 1.06 ± 0.49 mm at the roof, 1.40 ± 0.46 mm in the posterior wall, 1.86 ± 0.59 mm in the anterior wall, 2.20 ± 0.82 mm in the septum, and 1.60 ± 0.48 mm in the mitral isthmus region. Additionally, LA walls were thinner at the roof and septum in women compared to men, despite accounting for variability in heart weight. Platonov examined the posterior wall LA thickness in 298 consecutive patients and demonstrated a gradation in posterior wall thickness from the superior to inferior posterior wall with the inferior posterior wall being thicker than the superior aspect of the posterior wall. Moreover, patients with atrial fibrillation had even thinner LA posterior wall thickness.[32] One should be cognizant of these anatomic findings during ablation especially when ablating in the roof and posterior wall of the left atrium, especially in patients with chronic atrial fibrillation who likely have dilated thinner walled atria and women in order to avoid complications.

6.3.2.3 Pouches and Septal Ridges

Other anatomic variations include the presence of left atrial pouches or septal ridges.[33,34] Pouches, mainly located at the roof, may impact ablation due to inadequate catheter tissue contact if not recognized with resulting gaps in the ablation lesion set. Wongchareon performed CT scans in 49 patients with drug-refractory atrial fibrillation and 47 controls. Roof pouches were demonstrated in approximately 15% of patients with the diameter of the pouch varying from 4.4 to 13 mm and depth from 2.9 to 8.7 mm.[33] Detection of this variant prior to the procedure may improve upon the safety and efficacy of the procedure. Ridges, possible remnants of a septal raphe, have been described to occur in the septal aspect and anterior wall of the LA. The significance of these ridges is unclear. However, their presence may impact upon the contiguity of ablation lesions placed in this region. Wongchareon also demonstrated the presence of septal and anterior ridges in 32% of individuals with atrial fibrillation and 23% of controls.[33] Pre-procedural imaging with CT or MRI scans may provide a road map to the operator and demonstrate the presence of these anatomic variants. Moreover, image integration may also be useful in these situations as it may explain unexpected catheter motion noted on fluoroscopy and guide ablation lesion placement.[34]

6.3.2.4 Adjacent Structures: The Esophagus

Knowledge of structures in adjacent to the left atrium is also of vital importance during left atrial ablation to minimize the risk of collateral damage. One such structure is the esophagus. Minimizing ablation in close proximity to the esophagus may minimize thermal injury and possibly the risk of the fatal complication of atrio-esophageal fistula.[35] CT scan imaging has suggested that the esophagus is in close relationship to the posterior wall of the left atrium. Fatty tissue is present between both structures, but measures approximately 0.9 ± 0.2 mm.[36] While this tissue may provide a degree of thermal insulation and mitigate esophageal injury, it must be noted that this tissue is discontinuous in 98% of patients, and often absent at the midposterior wall.[37] Thus, one needs to ensure that methods to minimize esophageal injury are employed.

Pre-procedural imaging of the esophagus is easily performed with CT or MRI imaging and can delineate its course along the posterior left atrium (Fig. 6.8). While the identification of the location of the esophagus on pre-procedural imaging is helpful, one must keep in mind that the esophagus is a dynamic structure whose position may change from the time that pre-procedural imaging was obtained and even during the procedure. Nolker performed a 3D reconstruction of the left atrium and esophagus with a cardiac C-arm in patients undergoing atrial fibrillation ablation procedures. There was a poor correlation ($r = 0.53$) between the esophageal location documented on the 3D reconstruction the morning of the procedure and that obtained on CT scanning 24 h prior.[37] Moreover, the maximum movement of the esophagus was almost 30% of the distance between the superior pulmonary veins. Thus, in addition to pre-procedural imaging, intraprocedural imaging and assessment of the esophageal location seems prudent. This may be achieved with esophageal temperature monitoring which in addition to providing an estimate of the esophageal temperature also provides the esophageal location on fluoroscopy.[38] Barium swallow may also provide information on the fluoroscopic location of the esophagus but carries a risk of aspiration. The use of intracardiac echocardiography may be very helpful as it provides real-time imaging of the esophagus and can locate the esophagus in relation to the posterior wall, thereby alerting operators when ablating over this region.[39]

6.3.2.5 Adjacent Structures: The Aorta

Knowledge of the relationship between the aorta and left atrium is also of importance as atrial distortion may occur with and without a dilated aorta even in the absence of musculoskeletal deformities (Fig. 6.9). This relationship can best be appreciated with pre-procedural imaging such as CT or MRI scanning. Specifically, compression leading to pulmonary vein narrowing (usually the left inferior pulmonary vein in patients with a left-sided aortic arch) can occur due to compression from the descending aorta.[40] In these cases, care must be taken when performing circumferential ablation to avoid further stenosis of these vessels. Distortion of the interatrial septum due to compression from the ascending aorta may also occur. In this situation, intracardiac echocardiography is very helpful to minimize complications such as inadvertent puncture of the aorta or posterior wall.[40]

6.3.2.6 Adjacent Structures: The Phrenic Nerve

As discussed earlier, the right phrenic nerve runs in close proximity to the lateral surface of the superior vena cava and right atrium. Given the close proximity of the right superior pulmonary vein to these structures, it is no surprise that injury to the right phrenic nerve has been reported during ablation of the right superior pulmonary vein[41] likely due to direct heat transfer from ablation in this region. As it is often impractical to pace multiple sites near the right superior pulmonary vein to ensure absence of diaphragmatic capture, one should monitor diaphragmatic motion on fluoroscopy during ablation. Of course, this may not be helpful should the procedure be performed under general anesthesia with paralysis. A more proximal (i.e., antral) ablation strategy may mitigate this risk due to increasing distance from the phrenic nerve with antral pulmonary vein ablation. Identification of the course of the right phrenic nerve and/or the adjacent phrenic artery on pre-procedural imaging and integration of this in electroanatomic mapping systems has been reported and may also be helpful to minimize phrenic nerve injury during ablation.[15,16]

6.3.3 Post-Procedural Assessment

While there are no established guidelines on left atrial assessment post-atrial fibrillation ablation, many centers perform repeat imaging (CT or MRI) at 3–12 months to screen for pulmonary vein stenosis which cannot be predicted by initial pulmonary vein size or duration of radiofrequency ablation, but rather by catheter position within the pulmonary vein during the ablation procedure. Post-procedural imaging with CT scan is also useful for identifying additional complications such as pulmonary vein thrombosis or atrio-esophageal fistula, early findings of the latter include the presence of mediastinitis centered around the posterior left atria-esophageal region (i.e., changes within the mediastinal fat, presence of fluid collections, or gas between the esophagus and posterior wall).[42]

Post-procedural imaging with delayed enhancement MRI can also demonstrate the degree of atrial scar post ablation and has the potential to improve our understanding of the ablation process and possibly predict recurrence.[43] Moreover, identifying areas without scar from prior ablation procedures may represent areas of arrhythmia breakthrough allowing an operator to focus additional ablation at these sites, thereby reducing mapping time and improving the efficacy of the procedures.[44]

6.4 Special Situations: Congenital Heart Disease

Atrial arrhythmias are being recognized as an increasingly important cause of morbidity in patients with either surgically treated congenital heart disease. While multiple etiologies of these arrhythmias exist, they usually are secondary to a re-entrant circuit around a fixed barrier such as scar or a suture line. Overall the rate of successful ablation for these arrhythmias is between 55% and 90%[45]; part of this reduced success is related to the presence of multiple re-entrant circuit and the complex anatomic alterations which may exist (either due to the native congenital condition or the corrective reconstructive surgery). As fluoroscopic imaging has limited ability to assess the complex anatomic changes in these patients, additional modalities are necessary to provide the operator with an assessment of each patient's unique anatomy including scars which may act as anatomic obstacles. Pre-procedural imaging with CT or MRI is quite helpful with this assessment. Additionally, electroanatomic mapping may also identify scar and provide complementary physiological information in addition to the anatomy.

6.5 Conclusion

Pre-procedural and intra-procedural imaging has become indispensible for complex atrial ablation procedures. Consequently, it has become imperative for interventional electrophysiologists to be not only well versed with the atrial anatomy, but also with the consistently evolving imaging modalities.

References

1. Calkins H. Prevention of esophageal injury during catheter ablation of atrial fibrillation: Is intracardiac echocardiography the answer? *Heart Rhythm*. 2006;3:1162-1163.
2. Perez FJ, Schubert CM, Parvez B, Pathak V, Ellenbogen KA, Wood MA. Long term outcomes after catheter ablation of cavo-tricuspid isthmus dependent atrial flutter: a meta-analysis. *Circ Arrhythm Electrophysiol*. 2009; Aug 2(4);393-401
3. Cabrera JA, Sanchez-Quintana D, Farre J, Rubio JM, Ho SY. The inferior right atrial isthmus: further architectural insights for current and coming ablation technologies. *J Cardiovasc Electrophysiol*. 2005;16:402-408.
4. Nako M, Saoudi N. More on isthmus anatomy for safety and efficacy. *J Cardiovasc Electrophysiol*. 2005;16:409-410.
5. Anseleme F, Klug D, Scanu P, et al. Randomized comparison of two targets in typical atrial flutter. *Am J Cardiol*. 2000;85: 1302-1307.
6. Asirvathnam SJ. Correlative anatomy and electrophysiology for the interventional electrophysiologist: right atrial flutter. *J Cardiovasc Electrophysiol*. 2009;20:113-122.
7. Asirvatham S, Friedman P, Packer D, Edwards W. Prevalence of a right atrial pouch and extension of pectinate muscles across the tricuspid valve-IVC isthmus. *Circulation*. 2001;104:409.
8. Morton JB, Sanders P, Davidson NC, Sparks PB, Vohra JK, Kalman JK. Phased array intracardiac echocardiography for defining cavotricuspid anatomy during radiofrequency ablation of typical flutter. *J Cardiovasc Electrophysiol*. 2003;14:591-597.
9. Becker R, Bauer A, Metz S, et al. Intercaval block in normal canine hearts: role of the terminal crest. *Circulation*. 2001;103: 2521-2526.
10. Okishige K, Kawabata M, Yamashiro K, et al. Clinical study regarding the anatomical structures of the right atrial isthmus using intra-cardiac echocardiography. Implication for catheter ablation of common atrial flutter. *J Interv Card Electrophysiol*. 2005;12:9-12.
11. DaCosta A, Jamon Y, Romeyer-Bouchard C, Thevenin J, Messier M, Issaz K. Catheter selection for ablation of cavotricuspid isthmus for treatment of typical atrial flutter. *J Interv Card Electrophysiol*. 2006;17:93-101.
12. Rodriguez LM, de Chillou C, Schlapfer J, et al. Age at onset and gender of patients with different types of supraventricular tachycardias. *Am J Cardiol*. 1992;70:1213-1215.
13. Porter MJ, Morton JB, Denman R, et al. Influence of age and gender on the mechanism of supraventricular tachycardia. *Heart Rhythm*. 2004;4:393-396.
14. Kalman JM, Olgin JE, Karch MR, Hamden M, Lee RJ, Lesh MD. "Cristal tachycardias": origin of right atrial tachycardias from the crista terminalis identified by intracardiac echocardiography. *J Am Coll Cardiol*. 1998;31:451-459.
15. Yamashiro K, Ehara M, Ozawa T, Satoh K, Suzuki T. Accuracy of right phrenic nerve image using 64 slice multi-detector computed

tomography in patients with atrial fibrillation. *J Am Coll Cardiol.* 2009;53(suppl):A104.

16. Horton R, Sanchez J, Hepler E, et al. Locating the right phrenic nerve by imaging the right pericardiophrenic artery with computerized tomography. *J Am Coll Cardiol.* 2009;53(suppl):A107.
17. Anderson RH, Brown NA. Anatomy of the heart revisited. *Anat Rec.* 1996;246:1-7.
18. Thiagalingam A, D'avila A, Foley L, et al. Full-color direct visualization of the atrial septum to guide transseptal puncture. *J Cardiovasc Electrophysiol.* 2008;19:1310-1315.
19. Kato R, Lickfett L, Meininger GR, et al. Pulmonary vein anatomy in patients undergoing catheter ablation of atrial fibrillation. *Circulation.* 2003;107:2004-2010.
20. Saksena S, Sra JS, Jordaens L. Intracardiac Echocardiography-Guided Cardioversion Helps Interventional Procedures (ICE-CHIP) trial. Heart Rhythm Society 2007 Scientific Sessions; Late Breaking Clinical Trials II, May 11, 2007; Denver.
21. Martinez MW, Kirsch J, Williamson EE, et al. Utility of nongated multidetector computed tomography for detection of left atrial thrombus in patients undergoing catheter ablation of atrial fibrillation. *JACC Cardiovasc Imaging.* 2009;2:69-76.
22. Garcia MJ. Detection of left atrial appendage thrombus by cardiac computed tomography. *JACC Cardiovasc Imaging.* 2009;2:77-79.
23. Iuliano A, Stabile G, De Simone A. Unusual insertion and course of the left atrial appendage: how imaging can guide atrial fibrillation ablation. *Europace.* 2007;9:643-644.
24. Marom EM, Herndon JE, Kim YH, McAdans HP. Variations in pulmonary venous drainage to the left atrium: implications for radiofrequency ablation. *Radiology.* 2004;230:824-829.
25. Ho SY, Cabrera JA, Tran VH, Farre J, Anderson RH, Sanchez-Quintana D. Architecture of the pulmonary veins: relevance to radiofrequency ablation. *Heart.* 2001;86:265-270.
26. Mansour M, Refaat M, Heist EK, et al. Three-dimensional anatomy of the left atrium by magnetic resonance angiography: implications for catheter ablation of atrial fibrillation. *J Cardiovasc Electrophysiol.* 2006;17:719-723.
27. McGavigan AD, Kalmann JM. Atrial anatomy and imaging for atrial fibrillation ablation. *J Cardiovasc Electrophysiol.* 2006;17:S8-S15.
28. Singh SM, Heist EK, Donaldson D, et al. Image integration using intracardiac ultrasound to guide catheter ablation of atrial fibrillation. *Heart Rhythm.* 2008;5:1548-1555.
29. Thiagalingam A, Reddy VY, Cury RC, et al. Pulmonary vein contraction: characterization of dynamic changes in pulmonary vein morphology using multiphase multislice computed tomography scanning. *Heart Rhythm.* 2008;5:1645-1650.
30. Noseworthy PA, Malchano ZJ, Ahmed J, Homvang G, Ruskin JN, Reddy VY. The impact of respiration on left atrial and pulmonary venous anatomy: implications for image-guided intervention. *Heart Rhythm.* 2005;2:1173-1178.
31. Hall B, Jeevanantham V, Simon R, Filippone J, Vorobiof G, Daubert J. Variation in left atrial transmural wall thickness at sites commonly targeted for ablation of atrial fibrillation. *J Interv Card Electrophysiol.* 2006;17:127-132.
32. Platonov PG, Ivanov V, Ho SY, Mitrofanov L. Left atrial posterior wall thickness in patients with and without atrial fibrillation: data from 298 consecutive autopsies. *J Cardiovasc Electrophysiol.* 2008;19:689-692.
33. Weerasooriya R, Murray C. Left atrial roof pouch. *Europace.* 2007;9:1141.
34. Wongcharoen W, Tsao H, Wu M, et al. Morphologic characterization of the left atrial appendage, roof and septum: implications for the ablation of atrial fibrillation. *J Cardiovasc Electrophysiol.* 2006;17:951-956.
35. Cummings JE, Schwelkert RA, Salida W, et al. Brief communication: atrial-esophageal fistulas after radiofrequency ablation. *Ann Intern Med.* 2006;144:572-574.
36. Lemola K, Sneider M, Desjardins B, et al. Computed tomographic analysis of the anatomy of the left atrium and the esophagus: implications for left atrial catheter ablation. *Circulation.* 2004;110:3655-3660.
37. Nolker G, Gutleben KJ, Marschang H, et al. Three-dimensional left atrial and esophagus reconstruction using cardiac C-arm computed tomography with image integration into fluoroscopic views for ablation of atrial fibrillation: accuracy of a novel modality in comparison with multislice computed tomography. *Heart Rhythm.* 2008;5:1651-1657.
38. Singh SM, Doshi SK, Brugge WR, et al. Esophageal injury and temperature monitoring during atrial fibrillation ablation. *Circ Arrhythm Electrophysiol.* 2008;1:162-168.
39. Helms A, West JJ, Patel A, et al. Real-time rotational ICE imaging of the relationship of the ablation catheter tip and the esophagus during atrial fibrillation ablation. *J Cardiovasc Electrophysiol.* 2009;20:130-137.
40. Ho I, Heist EK, Aryana A, et al. Compression of the left atrium by the thoracic aorta in patients undergoing pulmonary vein isolation procedure for atrial fibrillation. *J Interv Card Electrophysiol.* 2007;19:29-36.
41. Sacher F, Jais P, Stephenson K, et al. Phrenic nerve injury after catheter ablation of atrial fibrillation. *Indian Pacing Electrophysiol J.* 2007;1:1-6.
42. Stojanovska J, Cronin P. Computed tomography imaging of the left atrium and pulmonary veins for radiofrequency ablation of atrial fibrillation. *Semin Roentgenol.* 2008;43:154-166.
43. Peters DC, Wylie JV, Hauser TH, et al. Recurrence of atrial fibrillation correlates with the extent of post-procedural late gadolinium enhancement: a pilot study. *JACC Cardiovasc Imaging.* 2009;2:317-318.
44. Reddy VY, Schmidt EJ, Holmvang G, Fung M. Arrhythmia recurrence after atrial fibrillation ablation: Can magnetic resonance imaging identify gaps in atrial ablation lines? *J Cardiovasc Electrophysiol.* 2008;19:434-437.
45. Walsh EP. Arrhythmia in patients with congenital heart disease. *Card Electrophysiol Rev.* 2002;6:422-430.

Anatomical Assessment for Catheter Ablation of Ventricular Tachycardia

7

Paolo Della Bella, Giuseppe Maccabelli, and Corrado Carbucicchio

Abstract

The anatomical assessment required in patients undergoing catheter ablation (CA) for ventricular tachycardia (VT) specifically aims at study and knowledge of the main structural characteristics of the left/right ventricular chambers, which are expressions of the underlying cardiac disease: these characteristics are strictly related to the arrhythmia and may also strongly condition the approach for CA. The anatomical assessment comprises the evaluation of all vascular structures that may be involved in the catheterization procedure, as well as of the pericardium when a direct percutaneous epicardial approach is undertaken. An accurate diagnostic evaluation must precede each electrophysiological procedure and aims at a complete clinical assessment. The point of our analysis is to finalize our diagnostics tools to an effective strategy for ablation and avoid complications. This is of particular relevance today, as CA is used for VT treatment in a wide population of patients with advanced cardiac disease and frequent comorbidities, who may undergo complex procedures of mapping and ablation and require multiple catheterizations.

Keywords

Catheter ablation • Ventricular tachycardia • Anatomy of ventricular tachycardia • Arrhythmia anatomy • Echocardiography for catheter ablation

The anatomical assessment required in patients undergoing catheter ablation (CA) for ventricular tachycardia (VT) specifically aims at the study and knowledge of the main structural characteristics of the left/right ventricular chambers, which are expressions of the underlying cardiac disease: These characteristics are strictly related to the arrhythmia and may also strongly condition the approach to CA. The anatomical assessment comprises the evaluation of all vascular structures that may be involved in the catheterization procedure, as well as the pericardium when a direct percutaneous epicardial approach is undertaken.

An accurate diagnostic evaluation must precede each electrophysiological procedure and aims at a complete clinical assessment. The result of our analysis is to finalize our diagnostic tools to an effective strategy for ablation and to avoid complications. This is of particular relevance today, as CA is used for VT treatment in a wide population of patients with advanced cardiac diseases and frequent comorbidities, who may undergo complex procedures of mapping and ablation and require multiple catheterizations.

7.1 Ordinary and Extraordinary Diagnostic Assessment

The ordinary preliminary evaluation for patients with structural heart disease consists of a complete noninvasive diagnostic assessment, mainly achieved by chest x-ray and echocardiography. *Chest x-ray examination* focuses on the

P.D. Bella (✉)
Arrhythmia Unit, San Raffaele Hospital, Via Olgettina 60, 20132 Milan, Italy
e-mail: dellabella.paolo@hsr.it

A. Auricchio et al. (eds.), *Cardiac Imaging in Electrophysiology*,
DOI 10.1007/978-1-84882-486-7_7,

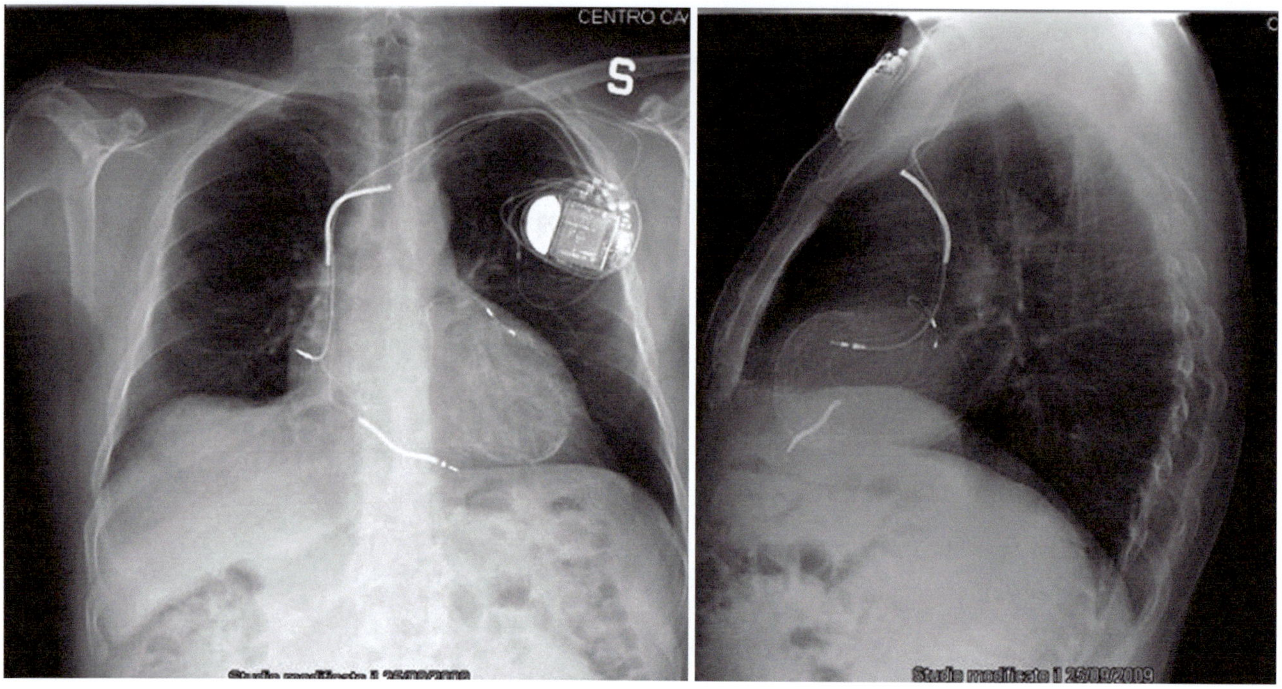

Fig. 7.1 Chest x-ray shows a big calcified thrombus within a large anterior left ventricular aneurism

global characteristics of the heart and its anatomical relationships. Its role is well defined with regard to the preliminary diagnostic assessment of each patient; however, its sensitivity for the anatomical characterization of the underlying disease is poor, and specific findings are limited to few cases (Fig. 7.1). On the other side, additional information is provided by the examination of the lung fields mainly for evaluation of pulmonary congestion and redistribution of blood flow.

Transthoracic echocardiography is more relevant in defining the morphology of both ventricles, focusing on chamber dimensions and wall characteristics. Baseline assessment includes the ordinary evaluation of the valvular apparatus to exclude any significant disease; more specifically, the detection of moderate-to-severe aortic and mitral valve disease may preclude the feasibility of any retrograde transaortic or anterograde transseptal approach, respectively.

Important specific information relates to each underlying cardiomyopathy. Area(s) of akinesia/dyskinesia, or areas of segmental contraction abnormalities are the rule in patients with post-myocardial infarction cardiomyopathy, but are frequently detected also in patients with arrhythmogenic right ventricular dysplasia or post-myocarditis cardiomyopathy. Diseased areas must be accurately defined; the comparison with the ECG recording during sinus rhythm and with the ECG pattern of activation during VT may contribute to support the site of origin of the arrhythmia. When a ventricular aneurism is recognized, the size and the borders of the lesion must be accurately defined, as well as the thickness of the corresponding part of the ventricular wall. The presence of a thrombus may be related to any underlying area of akinesia or dyskinesia (Fig. 7.2), but can also be the consequence of a diffused ventricular hypokinesia. Its characteristics must be evaluated and the detection of a soft or mobile thrombus must be considered as an additional risk for thromboembolism during endocardial mapping and ablation maneuvers; in selected cases, the procedure should be modified and undertaken after an adequate period of anticoagulation for safety reasons. Moreover, the intraventricular thrombus may preclude contact with the endocardial surface of the ventricle, thus limiting the efficacy of radiofrequency delivery on the site of interest. In these cases, the opportunity to undertake only an epicardial mapping must be considered as extraordinary.

Echocardiography is equally a first-line tool for the diagnosis of cardiac tumors; although uncommon, the detection of any tumoral lesion is of great significance, as it may clarify the cause of the arrhythmia and justify a surgical intervention (Figs. 7.2b and 7.2c).

Transesophageal echocardiography is not routinely required for the preliminary assessment of patients undergoing CA of VT. It is however indicated in the majority of suitable patients with chronic atrial fibrillation to exclude atrial thrombosis, because of the risk of thromboembolic accidents

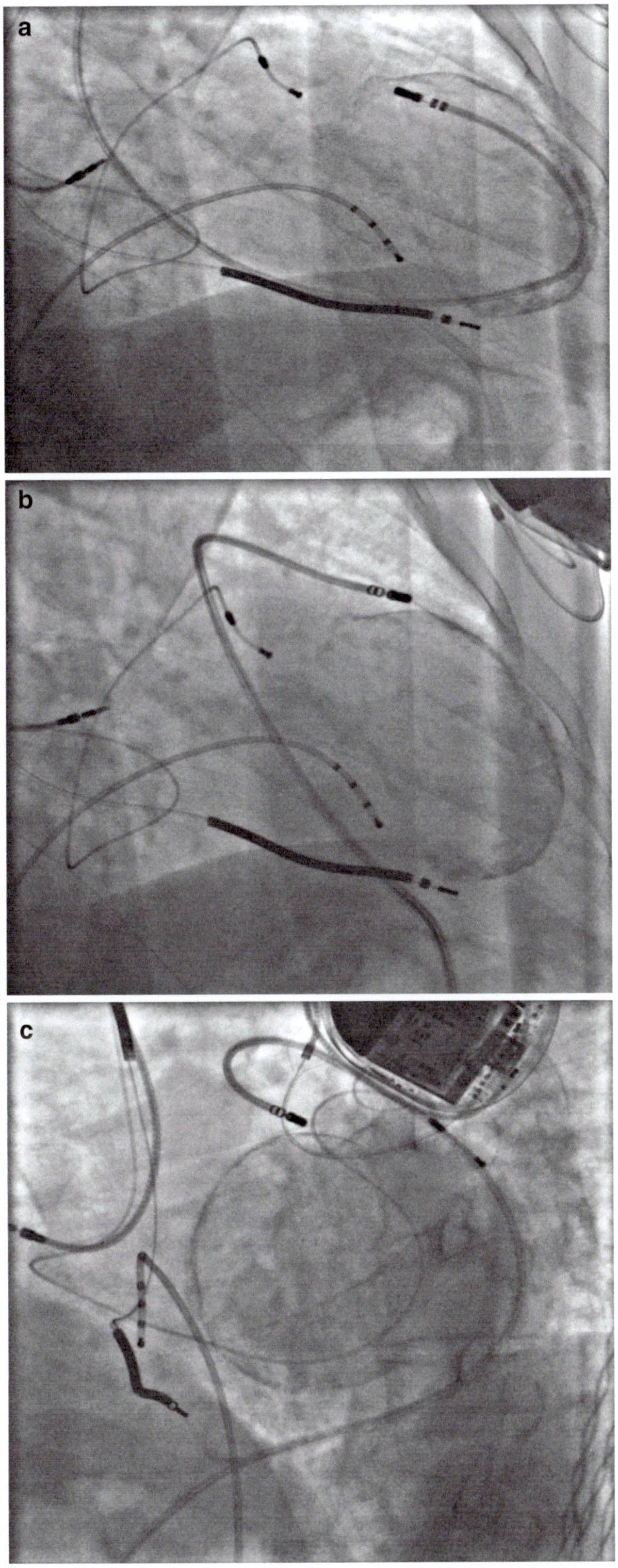

Fig. 7.2 The contour of the thrombus is clearly defined in different LAO and RAO radioscopic views obtained during the procedure. Efficacious ablation was achieved by delivering radiofrequency current at the superior border of the aneurism from the endocardium (**a**) and from the epicardium (**b, c**), where a discrete area of slow conduction involved in the VT reentry was identified

subsequent to sinus rhythm restoration after the delivery of an external DC shock. This may be considered an unlikely event, however, in patients with chronic atrial fibrillation, under continuous anticoagulation treatment, and in those who have already experienced multiple internal ICD interventions – either successful or unsuccessful in the termination of atrial fibrillation.

Evaluation of coronary artery circulation in patients with post-infarction cardiomyopathy and, more specifically, of any surgical graft is generally required to guarantee the safety of the procedure and to exclude ischemia as the cause of the arrhythmia: this is however related to an old post-MI scar in the majority of patients. The coronary assessment is ordinarily achieved by coronary angiography, during the same catheterization procedure, or, in selected cases, by performing a high-definition TC scan before. The need for a coronary angiography must be considered also in those patients with dilated cardiomyopathy of uncertain etiology. In the case of epicardial CA, the coronary angiography is required to clarify the relationship with the main epicardial arterial vessels before delivering radiofrequency energy, as discussed below; moreover, CA targeting arrhythmia foci localized at the sinus of Valsalva may require the angiographic visualization of the ascending aorta to better characterize the contour of the aortic bulb and to avoid intracoronary radiofrequency delivery.

The *ventricular angiographic evaluation*, usually during the CA procedure itself, is required only in selected cases. Left ventricular angiography has a poor diagnostic role and therefore it is rarely required in patients with dilated cardiomyopathy. On the contrary, the right ventricular angiogram may be of diagnostic value in patients with arrhythmogenic right ventricular dysplasia/cardiomyopathy and, more specifically, may help in the radioscopic identification of areas of akinesia or dyskinesia.

An anatomical assessment of *the arterial and venous system* is preliminarily achieved by the ordinary physical examination; the main point is to exclude any significant disease of the abdominal aorta and the major peripheral arterial vessels. A chest X-ray may contribute to presume the presence of any significant atherosclerotic disease in the thoracoabdominal district, mainly by the detection of calcified lesions. More specific information is required (1) when a significant thoracoabdominal aortic disease has been previously documented and (2) in case of any surgical or interventional vascular correction (that regarding the abdominal aorta, the Carrefour, or the iliac-femoral artery in the majority of cases). This is usually achieved by Doppler examination, by angiography or TC-angiography, based on the kind of lesion and on the vascular district of interest. The vascular situation, once clarified, may condition the access to the left ventricular chamber (that might be limited to a transseptal catheterization) or preclude the possibility for a multiple left ventricular catheterization, when an advanced mapping technique is required.

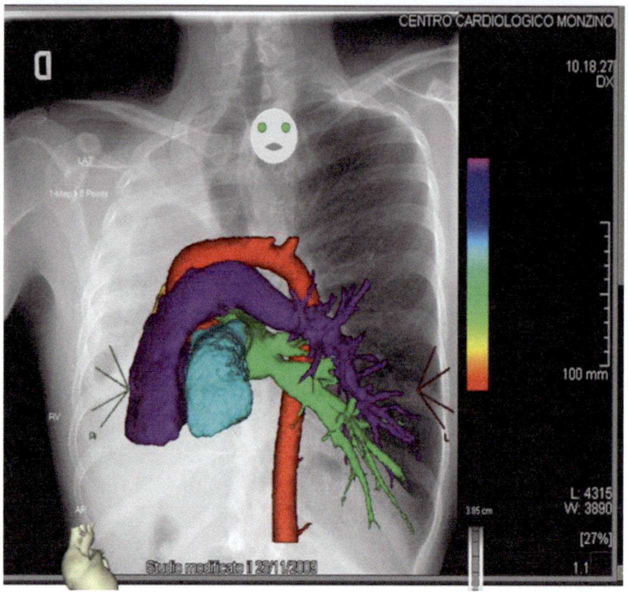

Fig. 7.3 The case of a patient with right-positioned heart (right heart displacement) secondary to right pulmonary agenesia is shown. The TC scan reconstruction is superimposed on a posteroanterior chest x-ray view. *Light blue* represents the left ventricle; *purple* the right ventricle giving rise to the left pulmonary artery; *red* the ascending aorta; *green* the left atrium with the left pulmonary veins. The right atrium is covered by the right ventricle; right pulmonary veins are absent

In conclusion, preliminary diagnostic evaluation aiming at the anatomical assessment plays a main role to define the patient's characteristics with regard to the underlying cardiac disease, to confirm safety and feasibility, and to direct to the preferential approach and mapping mode; the information acquired may provide the reason for more specific diagnostic tools.

The case of a patient with right-positioned heart (right heart displacement) secondary to right pulmonary agenesia is shown in Fig. 7.3. The TC scan reconstruction is superimposed on a posteroanterior chest x-ray view. Light blue represents the left ventricle; purple the right ventricle giving rise to the left pulmonary artery; red the ascending aorta; and green the left atrium with the left pulmonary veins. The right atrium is covered by the right ventricle; right pulmonary veins are absent.

Another case of a patient with ventricular tachycardia arising from a left ventricle aneurism with endocardial calcification is shown in Fig. 7.4. X-ray analysis permits clear identification of the endocardial calcification (Fig. 7.4). The endocardial ablation was unable to ablate the arrhythmia circuit and epicardial approach was not possible for pericardial adherence due to previous cardiac surgery. Figure 7.6 shows removal of the endocardial calcification during surgical procedure. For this reason, the patient was submitted to a surgical ablation performed with cryo energy after the removal of the endocardial calcification (Fig. 7.5). Figure 7.6 shows the absence of the apical calcification after the surgical procedure.

7.2 Anatomical Insights for Left and Right Ventricular Mapping and Ablation

Left and right ventricles significantly differ with regard to their anatomical and functional features. Morphologically, the left ventricle is shaped like an egg and presents a circular section in its midportion.[1,2] Its wall is three to four times as thick as that of the right ventricle and accounts for about 75% of the whole cardiac mass. The interventricular septum, represented by the anterior aspect of the left chamber, is generally considered as part of it, but also contains longitudinal fibers belonging to the right ventricle. The right ventricle is commonly described as having two sections, the inflow portion – containing the tricuspid valve and its chordae – and the outflow, i.e., the conus or infundibulum, which is distinctly derived from the *bulbus cordis*. These two parts are defined by specific anatomical structures as the *moderator band*, the *crista supraventricularis*, and the muscular region between the tricuspid valve and the outflow tract (septal band and parietal band). The right ventricle has a thinner wall (3–5 mm at the ventricular body, 1–2 mm at the junction), but its trabeculae are more pronounced and irregular when compared with the left ventricle.

Of note, the line of each pulmonary valve leaflet crosses the ventriculoarterial junction, so that small amounts of myocardial tissue are comprised in the three pulmonary sinuses, and are responsible for so-called supravalvular locations of ventricular arrhythmias. In a similar way, the semilunar hinge lines of the aortic leaflets enclose a segment of ventricular myocardium in the nadirs of the right and of the left coronary aortic sinus.

Of specific interest, among the left ventricular endocardial structures, the *chordae tendineae,* which tether both mitralic cusps to the anterior and posterior papillary muscles, may limit the access to the subvalvular region. Myocardial tissue also projects within the ventricular cavity, resulting in specific endocavitary electrically active structures such as the papillary muscle, the moderator band, and the false tendon: In selected cases, advanced methods of mapping may prove that the automatic or reentry mechanism responsible for the arrhythmia correlates with these anatomical structures. Moreover, they may limit or complicate catheter manipulation during VT mapping and, for these reasons, additional value is assigned to intracardiac echocardiography: the three-dimensional visualization that can be achieved may prevent the damage of the valvular

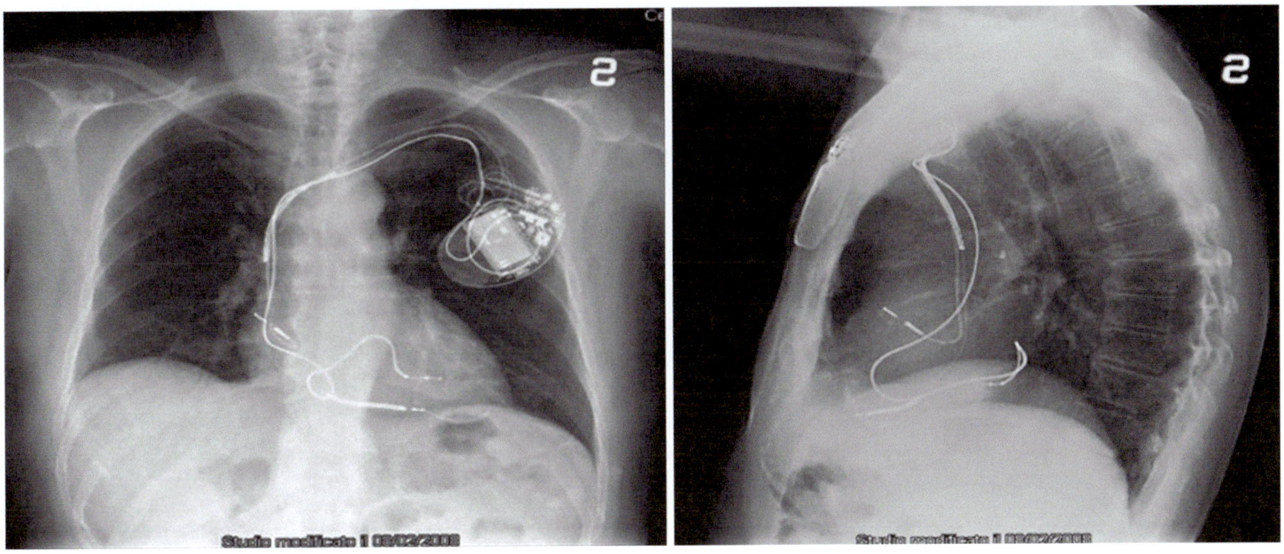

Fig. 7.4 Another case of a patient with ventricular tachycardia arising from a left ventricle aneurism with endocardial calcification is shown. X-ray analysis permit clear identification of the calcification. The endocardial ablation was unable to ablate the arrhythmia circuit and epicardial approach wasn't possible for pericardial adherence due to previous cardiac surgery

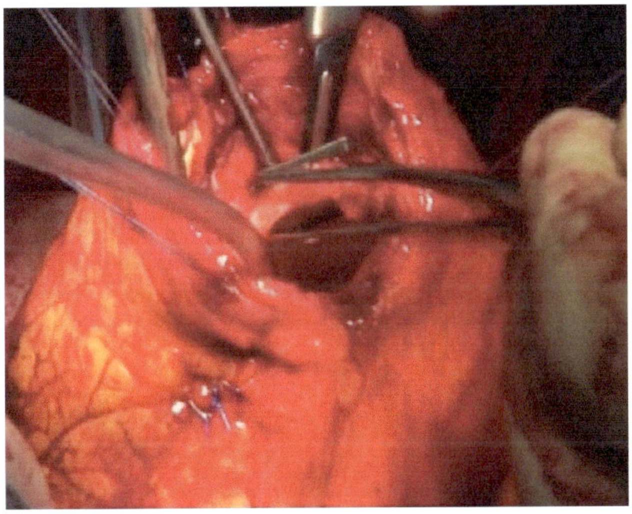

Fig. 7.5 Removal of the endocardial calcification is shown during the surgical procedure

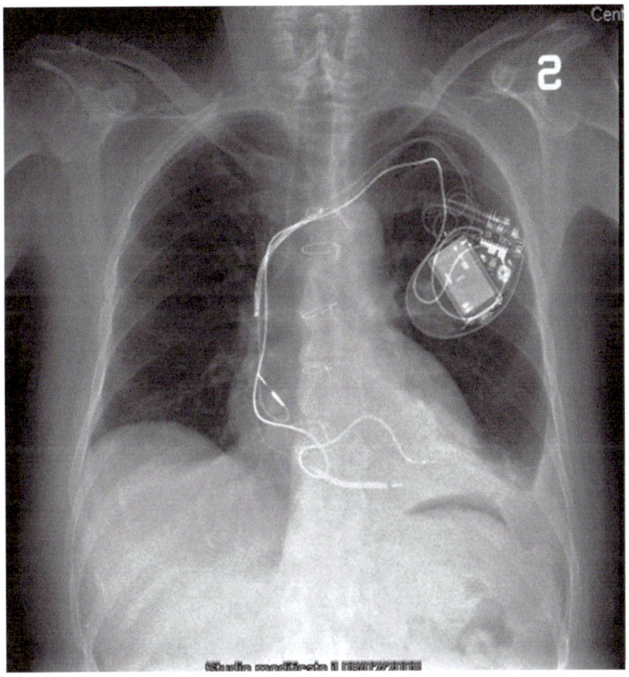

Fig. 7.6 X-ray analysis of the patient shown in Fig. 7.5 shows the absence of the apical calcification after the surgical procedure

structures and may facilitate contact optimization during radiofrequency delivery.

Effectiveness of mapping for ventricular arrhythmias is moreover conditioned by the adequacy of the contact of the tip of the catheter with the endocardial surface: This results from the precise definition of the internal contour of the chamber, from the identification of specific anatomical or pathological structures, but also benefits from the possibility to properly advance and stabilize the tip of the catheter at the site of interest. For this reason, steerable sheets are commonly used to support the mapping catheter during right and left ventricular procedures; in the latter case, the possibility to enter the left ventricle with an appropriate sheet is limited to the transseptal approach only.

7.3 Advanced Tools for the Improvement of Anatomy Understanding

Although in a normal anatomical condition the navigation inside the left ventricle may be workable without significant problems, sometimes in specific pathological situations the precise definition of the structures may become pivotal. Recently, technological improvements have made a lot of tools available, such as the electroanatomical mapping system that can be used in the setting of ventricular ablation. With this system and the nonfluoroscopic navigation of the catheters in a three-dimensional space and the opportunity to analyze in a 3D mode the electrical characteristics of the scars, the integration with the images of the heart obtained with CT or MRI scan is now available. This technique is currently extensively used in the atrial procedure where it has achieved reasonable accuracy and usability, providing important information about the anatomy of the pulmonary vein, the ridge, and the auricula that may help guide precise localization of the RF lesions. The same technique can be used for the left ventricle even if in this case currently pre-procedural anatomic images have some disadvantages due to the improper registration technique that does not allow the detailed reconstruction of the anatomical structure present in the inner part of the left ventricle.

Nevertheless, image integration may be particularly useful in cases of left ventricle mapping where the achievement of some part of the chamber may be really difficult. In this case, special attention must be paid to choose the appropriate curve of the mapping catheter and a dual approach (retrograde transaortic and transseptal) must be used. Also with these concerns, the inability of the system to give precise information about the contact of the catheter tip with the endocardial surface may, in some cases, contribute to the creation of a false surface that, usually for the poor contact, has the electrical characteristics of a scar. The availability of a 3D shape of the left ventricle is of the utmost importance to avoid these mistakes that can deeply alter the result of the procedures.

Another important tool available with the electroanatomical mapping system is the intracardiac echo (CARTO SOUND) that may be particularly useful in special anatomical conditions. This system uses a specific phased-array intracardiac ultrasound catheter equipped with an electroanatomic sensor positioned at the tip that provides the orientation and the location of each registered image used for creating a 3-D shape. This imaging gives detailed anatomical information about normal structure like papillary muscle, valve, and endocardial surface and about pathological structures like aneurysm border or intraventricular thrombus that can be integrated into each segmented volume. Moreover, the system can provide important information about the contact of the catheter tip with the endocardial surface.

7.4 Anatomy for Epicardial Ablation

In the last years, epicardial ablation has undertaken an important role in the treatment of ventricular tachycardias related to some pathological disease. This particular approach requires the knowledge of several anatomical issues necessary to perform safely the pericardial access and the ablation procedure. In this procedure, two pivotal problems may be recognized (1): the pericardial puncture, and (2) problems related to the pericardial ablation linked to the anatomical structures present on pericardial surface.

7.4.1 Pericardial Access

The technique, originally described by Sosa et al.,[3] was initially used for the epicardial ablation in patients with Chagas disease. Briefly the subxiphoidal approach is carried out with a Tuohy needle. The pericardial puncture is performed just below the xiphoidal apophysis toward the left scapula or the left shoulder with a 45° angle from the skin surface and from the middle line of the sternum. The angle with the skin surface may be adjusted according to the wall that one wants to access: lower than 45° for the anterior wall, 45° or more for the inferior wall. These rules are helpful for the vast majority of patients, but it must be remembered that in some cases the different dimensions and positions of the heart inside the thorax require necessary but some modification of the suggested technique. Before the procedure it is important to evaluate the chest x-ray to understand the position of the heart inside the thorax and particularly the relationship with the sternum that may be very different, particularly in obese or long-limbed patients where particularly in the latter, the heart may be slightly distant from the xiphoidal process. In these patients, to reduce the needle's route, the puncture may be carried out in anti-Trendelenburg position or, if the patient is not in deep sedation or in general anesthesia, during forced inhaling. In any case to better understand the anatomy, a catheter placed in the right ventricular apex and in the coronary sinus may be useful reference to guide the procedure. The puncture must be carried out verifying the position of the needle in AP, LAO, and RAO projection to reach the middle portion of the right ventricle, where based on the coronary angiography no major coronary arteries are found. Starting from the xiphoidal process, the needle comes across subcutaneous and muscle tissue and reaches the pericardium without encountering important vascular structures. The puncture of the diaphragmatic muscle must be avoided because of risk of abdominal bleeding, in case of injury of a small arterial vessel, that may require abdominal surgery to resolve. This complication must always be suspected in case of slow and progressive reduction of the arterial

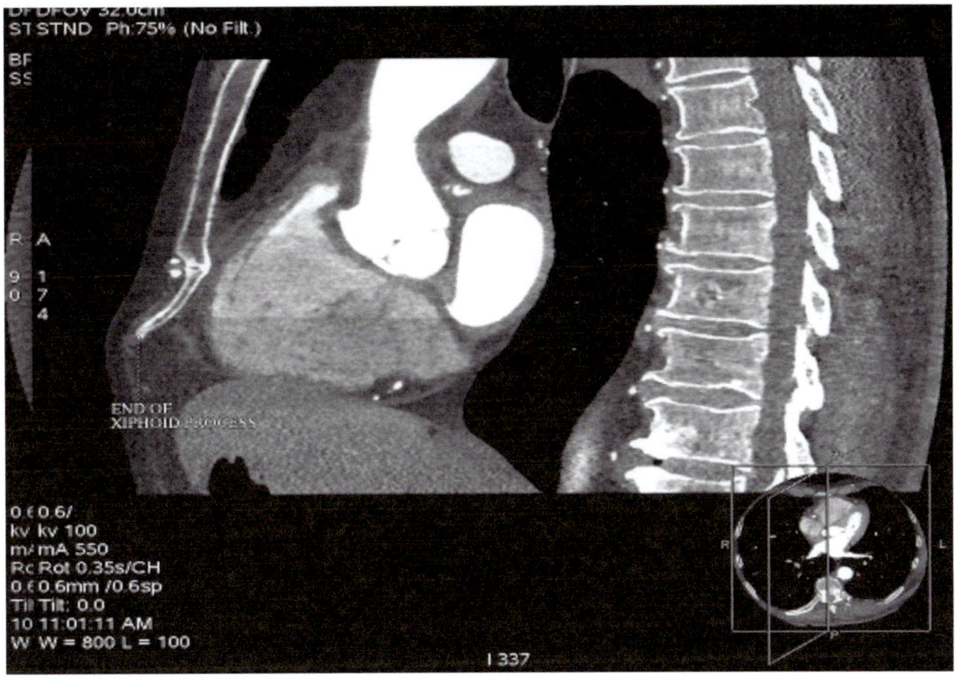

Fig. 7.7 The xiphoidal apophysis is much higher as regards the inferior border of the heart and in this case, with the usual slope of the needle, the inferior wall of the ventricle is unapproachable

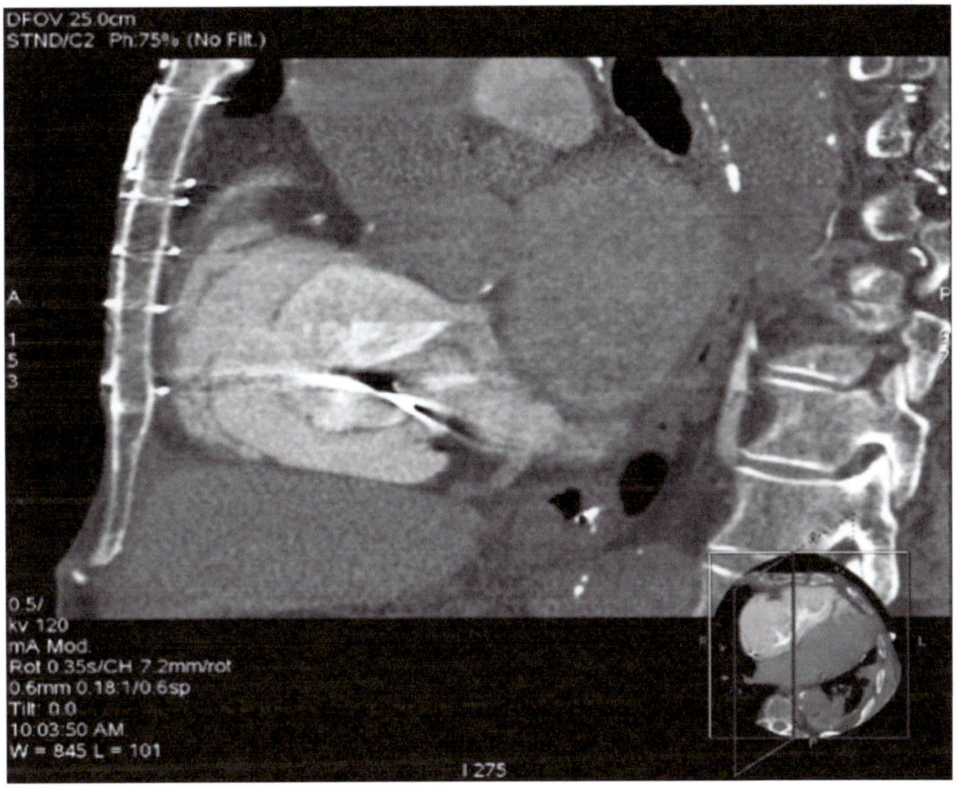

Fig. 7.8 A case is shown where, with standard routes, the puncture of the liver is unavoidable

pressure during the procedure without evidence of pericardial bleeding. In our experience, these rules are sufficient to perform the pericardial puncture minimizing the risk of complication.

Figures 7.7 and 7.8 show two borderline cases. In the first, the xiphoidal apophysis is much higher as regards the inferior border of the heart, and in this case, with the usual slope of the needle, the inferior wall of the ventricle is unapproachable. The second shows a case where with standard routes, the puncture of the liver is unavoidable.

In any case and particularly in patients with chronic heart decompensation, a careful clinical evaluation and, if neces-

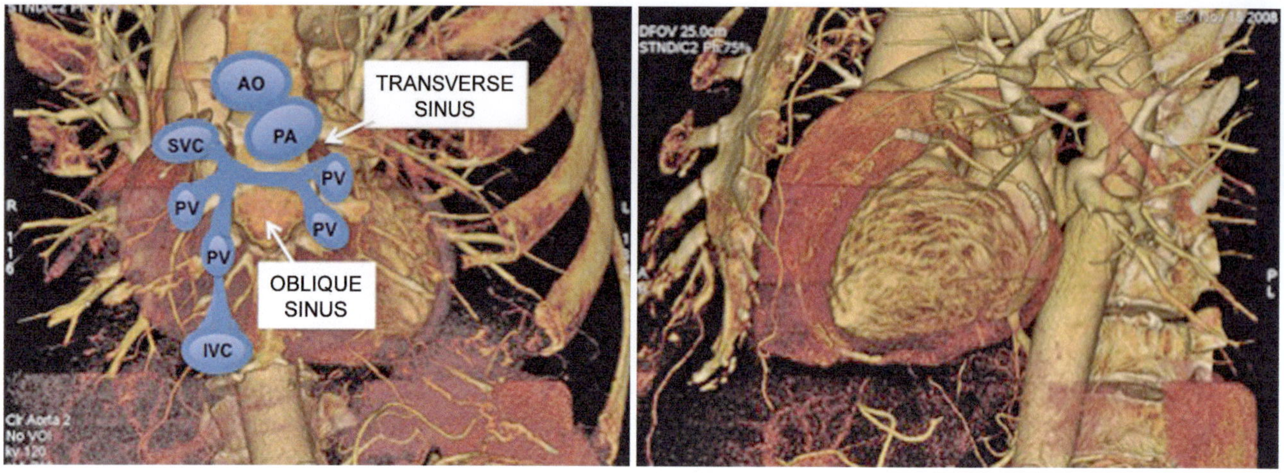

Fig. 7.9 The serous layer splits itself from the fibrous one near the aorta and the pulmonary vessels to cover a little portion of the vessels and the heart upon which it forms the epicardium. In the pericardial space, it can be recognized as two channels: one in the anterosuperior position is named transverse sinus and one, the oblique sinus, posteriorly, is located under the left atrium

sary, an echo or CT scan evaluation must be performed for a complete understanding of the anatomical correlation.

Special attention in the pre-procedural evaluation must be reserved for the patients in whom a current or old inflammatory pericardial process may be suspected. In this patient, it is mandatory to perform a CT scan and, in the absence of implantable devices, an MRI evaluation. Pericardial calcifications can also be revealed by the chest X-ray analysis in about 50% of patients with constrictive pericarditis[4] and are well detected by the CT scan. This finding together with the evidence of pericardial thickening (above 6 mm), and pericardial effusion may identify a pathological process of the pericardium that, especially if chronic, because of local adherence, may render the access to the pericardial space or the navigation inside really problematic if not impossible.

Another very rare event that can be investigated is the tumors more frequently secondary to the pericardium and the absence of the pericardium. The latter condition, which can also be related to a cardiac trauma or to previous cardiac surgery, may be congenital, showing a wide variety of anatomical spectrum ranging from little defect to a total absence of the pericardium. The most common form is the partial absence usually of the left side of the heart. Being asymptomatic, this condition becomes evident only after a failed attempt to access the pericardial space and can be evaluated with CT and MRI scans that highlight the absence of the pericardium and a lot of secondary signs such as a posterior and leftward shift of the heart, a prominent left atrial appendage, and a separation of the aorta and the main pulmonary artery by the lung.[5]

7.4.2 Pericardial Ablation

The normal pericardium is composed of two layers: the outside fibrous layer and the serous inner layer. The two layers are tightly coupled to form the parietal pericardium that coats the entire heart. It can be highlighted with CT and MRI scanning. With MRI it appears as a structure of a few millimeters of thickness with low-intensity signal interposed between the pericardial and mediastinal fat characterized by high-intensity signal. The serous layer splits itself from the fibrous one near the aorta and the pulmonary vessels to cover a small portion of the vessels and the heart upon which it forms the epicardium. In the pericardial space, it can be recognized as two channels (Fig. 7.9): one in the anterosuperior position is named transverse sinus and one, the oblique sinus, posteriorly, is located under the left atrium. These structures do not involve the pericardial surface of the ventricles, which is free of anatomical obstacles and therefore can be navigated without particular problems.

The first problem encountered during the navigation in the pericardial space is the presence of the coronary arteries. Usually, they are covered or surrounded by the epicardial fat and, with the conventional electroanatomical bipolar mapping, are unnoticeable from the normal pericardial muscle. Before RF ablation it is important to recognize them because RF delivery above the artery may cause a lot of problems. It was observed[6] that when RF energy is delivered close to the coronary arteries, only minor changes may occur such as replacement of the media with extracellular matrix. But when the energy is delivered above the artery, severe intimal hyperplasia with intravascular thrombosis

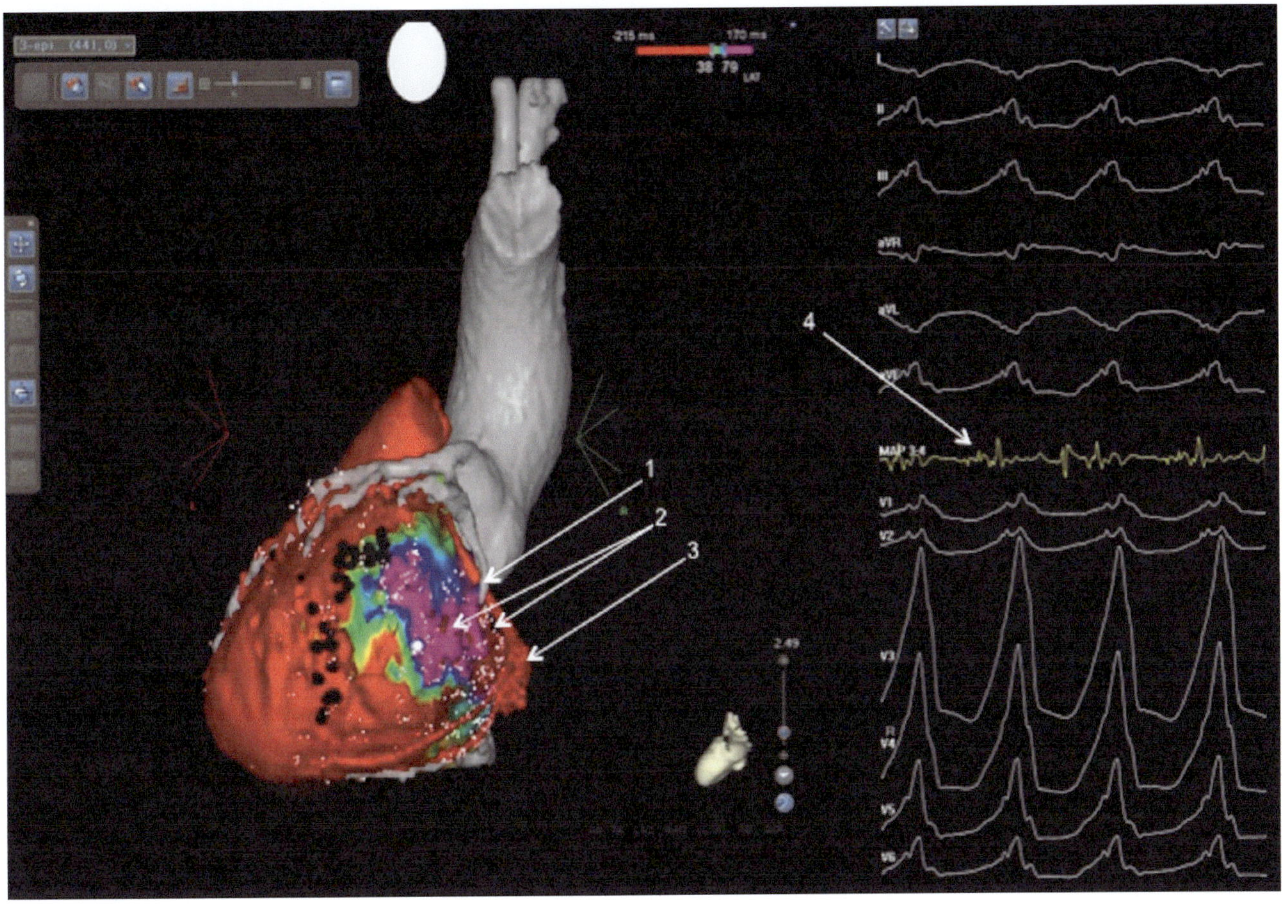

Fig. 7.10 The pericardial surface with a late potentials map is shown. Purple area represents the late potential zone. (1) coronary artery, (2) ablation points, (3) venous system. During ventricular tachycardia, with the catheter tip positioned in this area, diastolic potentials were registered (4)

occurs. Although the damage is inversely related to the dimension of the coronary arteries because greater arteries are protected by the blood flow,[7] careful attention must be paid, before RF delivery, to verify the position of the coronary arteries. Multiple projections must be observed to identify the exact distance of the coronary artery from the ablating catheter, and even though Thyer et al.[8] have suggested the use of intracoronary chilled saline irrigation to protect the endothelium from the heat-induced damage of the RF energy, at present, the only reasonable solution is to maintain a distance of at least 5 mm.

Another possible way to recognize the course of the coronary arteries is the use of imaging obtained with CT scan and MRI. At present, the quality and resolution of images obtained with these diagnostic tools are good, and one can follow the entire course of the arteries over the pericardial surface up to the thinnest branch, but the processing necessary to use these with the Carto system irreparably causes a loss of resolution that makes it hard to visualize the arteries. Nevertheless, if the images obtained with the protocol used for coronary arteries are of good quality, sometimes with manual processing, which can require sometimes require a long time, it is possible to reconstruct over the pericardial surface a part of the course of the arteries. In this case, it is important to segment a little portion of the aorta (in detail, the ascending aorta) because the left coronary ostium can be used as a landmark for the visual alignment. Once the merge process is completed, considering the accuracy of the spatial localization of the catheter tip, the ablation can be carried out without the need for coronary angiography (Fig. 7.10).

Another anatomical issue that can be encountered with the epicardial ablation is the presence of the phrenic nerve. The right phrenic nerve is located along the right anterolateral wall of the superior vena cava, after that it heads slightly posterior where it reaches the cavoatrial junction and then passes near the right superior pulmonary vein close to the junction with the left atrium and right atrium. The terminal intra-thoracic part of the nerve passes near the right atrium. For this reason, it is not an obstacle for epicardial ventricular ablation. The condition for the left nerve is different, however. An anatomical and histological study published by Sanchez Quintana et al.[9] has shown that the course of the left phrenic nerve in 79% of cases passes along the obtuse cardiac margin near the left obtuse marginal artery and vein; in the remaining cases, it can run more anterior over the sternocostal surface, over the main stem of the left coronary artery or the anterior descending artery. Because of this variability, it always important, before RF delivery, to test if the ablating catheter is located near the phrenic nerve with high-intensity pacing (20 mA). A good method is to identify the course of the nerve and mark it with a black dot over the pericardial map.

The solution proposed by Matsuo et al.[10] to avoid damage of the phrenic nerve when the ablation point is placed near it is to inject 100 mL of air via the intrapericardial sheet to create a space between the nerve and the ablation catheter. Di Biase et al.[11] in a recent paper suggest the creation of "hydropneumopericardium" because the combination of fluid and air seems more effective in creating space between the nerve and the pericardial surface with lower hemodynamic impact.

The last important anatomical issue that must be well known in pericardial ablation is the presence of epicardial fat that may be liable for the inefficacy of the procedure. Epicardial fat is present in variable amounts over the epicardial surface of the heart, usually around the coronary arteries, the left ventricular apex, the right free wall, the two appendages, and the atrioventricular and interventricular grooves. The amount of the epicardial fat that seems correlated with the myocardial ventricular mass[12,13] and with the subcutaneous fat[14] is significantly increased in patients with ischemic cardiomyopathy and correlated with the cardiomyopathy staging.[15] Moreover, thickness of epicardial fat is greater in patients with ischemic cardiomyopathy than in those without (4.0 vs. 1.5 mm, $p>0.001$).[16] Transthoracic echocardiography as well as CT and MRI scan have been extensively used for its quantification.

The presence of fat interposed between the tip of the catheter and the epicardium may prevent lesion creation. Hong et al.[17] demonstrated that the epicardial fat can limit the lesion formation and the transmurality using an energy source that utilizes conductive heating like the RF. While standard RF ablation is unable to produce lesion over fat, cooled-tip RF ablation may produce lesion of 4.1 ± 2 mm in the presence of 2.6 ± 1.2 mm of fat, but in the presence of a fat layer of 3.5 mm or more, it is unable to produce epicardial lesions.[18] At present, a method is not available that permits the exact recognition of the presence of fat during the ablation procedure.

References

1. Ho SY. Anatomic insights for catheter ablation of ventricular tachycardia. *Heart Rhythm*. 2009;6(8 suppl):S77-S80.
2. Abouezzeddine O, Suleiman M, Buescher T. Relevance of endocavitary structures in ablation procedures for ventricular tachycardia. *J Cardiovasc Electrophysiol*. 2010;21(3):245-254.
3. Sosa E, Scanavacca M, D'Avila A. A new technique to perform epicardial mapping in the electrophysiology laboratory. *J Cardiovasc Electrophysiol*. 1996;7:531-536.
4. Stephen WM. Imaging pericardial disease. *Radiol Clin North Am*. 1989;27:1113.
5. Costa Lopez I, Bhalla S. Computed tomography and magnetic resonance imaging of the pericardium. *Semin Roentgenol*. 2008;43(3): 234-245.
6. Sosa E, Scanavacca M, d'Avila A. Transthoracic epicardial catheter ablation to treat recurrent ventricular tachycardia. *Curr Cardiol Rep*. 2001;3:451-458.
7. d'Avila A, Gutierrez P, Scanavacca M. Effects of radiofrequency pulses delivered in the vicinity of the coronary arteries: implications for nonsurgical transthoracic epicardial catheter ablation to treat ventricular tachycardia. *Pacing Clin Electrophysiol*. 2002;25: 1488-1495.
8. Thyer IA, Kovoor P, Barry MA. Protection of the coronary arteries during epicardial radiofrequency ablation with intracoronary chile saline irrigation: assessment in an in vitro model. *J Cardiovasc Electrophysiol*. 2006;17(5):544-549.
9. Sánchez-Quintana D, Cabrera JA, Climent V. How close are the phrenic nerves to cardiac structures? Implications for cardiac interventionalists. *J Cardiovasc Electrophysiol*. 2005;16(3): 309-313.
10. Matsuo S, Jaïs P, Knecht S. Images in cardiovascular medicine. Novel technique to prevent left phrenic nerve injury during epicardial catheter ablation. *Circulation*. 2008;117(22):e47.
11. Di Biase L, Burkhardt JD, Pelargonio G. Prevention of phrenic nerve injury during epicardial ablation: comparison of methods for separating the phrenic nerve from the epicardial surface. *Heart Rhythm*. 2009;6:957-961.
12. Corradi D, Maestri R, Callegari S, et al. The ventricular epicardial fat is related to the myocardial mass in normal, schemic and hypertrophic hearts. *Cardiovasc Pathol*. 2004;13:313-316.
13. Iacobellis G, Ribaudo MC, Zappaterreno A, et al. Relation between epicardial adipose tissue and left ventricular mass. *Am J Cardiol*. 2004;94:1084-1087.
14. Schejbal V. Epicardial fat on the right ventricle-morphology, morphometry and functional significance. *Pneumologie*. 1989;43: 490-499.
15. Silaghi A, Piercecchi-Marti MD, Grino M. Epicardial adipose tissue extent: relationship with age, body fat distribution, and coronaropathy. *Obesity*. 2008;16(11):2424-2430.
16. Ahn SG, Lim HS, Joe DY. Relationship of epicardial adipose tissue by echocardiography to coronary artery disease. *Heart*. 2008; 94(3):e7.
17. Hong KN, Russo MJ, Liberman EA. Effect of epicardial fat on ablation performance: a three-energy source comparison. *J Card Surg*. 2007;22(6):521-524.
18. d'Avila A, Houghtaling C, Gutierrez P. Catheter ablation of ventricular epicardial tissue: a comparison of standard and cooled-tip radiofrequency energy. *Circulation*. 2004;109(19):2363-2369.

Intra-Procedural Techniques: Fluoroscopy and Angiography

8

Matthew Wright, Sébastien Knecht, and Pierre Jaïs

Abstract

X-ray fluoroscopic imaging of the heart is the standard to which all other imaging modalities are currently compared to, and no electrophysiology laboratory is designed without an X-ray system. Even though magnetic resonance (MR) imaging is being used by more laboratories for pre-procedural imaging and is being developed for use as a real-time imaging modality to allow interventional procedures to be carried out within the MR environment, all such systems in use today incorporate an X-ray system (XMR). The pre-eminence of fluoroscopy above imaging modalities is due to a number of factors. Fluoroscopic systems are easy to use, provide real-time information, catheters are easily visualized, electrical interference with recording systems is minimal and virtually no time is required to set up the system.

Keywords

Fluoroscopy • Rotational angiography • Radiation • Angiography • Cardiac fluoroscopy • Cardiac angiography

X-ray fluoroscopic imaging of the heart is the standard to which all other imaging modalities are currently compared to, and no electrophysiology laboratory is designed without an X-ray system. Even though magnetic resonance (MR) imaging is being used by more laboratories for pre-procedural imaging and is being developed for use as a real-time imaging modality to allow interventional procedures to be carried out within the MR environment, all such systems in use today incorporate an X-ray system (XMR). The pre-eminence of fluoroscopy above imaging modalities is due to a number of factors. Fluoroscopic systems are easy to use, provide real-time information, catheters are easily visualized, electrical interference with recording systems is minimal and virtually no time is required to set up the system.

However, there are a number of drawbacks to the sole use of X-ray fluoroscopy in the electrophysiology (EP) laboratory. The images obtained are only two-dimensional representations of three-dimensional structures and visualization of the cardiac chambers is indirect as only the cardiac silhouette can be seen. Additionally, tissue contact is only inferred when catheter movement becomes restricted and not before, the effects of ablation on the tissue are not seen, and both the patient and operator are exposed to radiation. These limitations have led to the development of non-fluoroscopic three-dimensional mapping systems, intracardiac echocardiography,[1] rotational angiography,[2-4] and remote catheter manipulation using either robotic[5] or magnetic guidance.[6] All of these technologies are still used, however, in combination with X-ray fluoroscopy. Using these systems in combination has a number of positive benefits, radiation to both the patient and operator are reduced, in keeping with the ALARA principle[7] (as low as reasonably achievable), three-dimensional representation of individual cardiac chambers can be seen, and an approximation of tissue contact can be made. Although the use of lead aprons is

P. Jaïs (✉)
Department of Cardiology,
Hôpital Cardiologique du Haut- Lévêque,
CHU Bordeaux, Pessac,, France
e-mail: pierre.jais@chu-bordeaux.fr

A. Auricchio et al. (eds.), *Cardiac Imaging in Electrophysiology*,
DOI 10.1007/978-1-84882-486-7_8, © Springer-Verlag London Limited 2012

associated with serious long-term problems for operators,[8,9] the use of radiation protection cabins or remote catheter manipulation the risk to the operator of serious orthopedic complications is minimized.[10,11]

Knowledge of cardiac radiographic anatomy is therefore imperative for the practicing electrophysiologist, in order to facilitate catheter manipulation and to avoid complications, and is covered in earlier chapters. Our current practice utilizes fluoroscopy as our primary imaging modality, and this chapter will reflect how we use fluoroscopy to help guide our procedures.

8.1 Safety Considerations

The American College of Cardiology consensus document on radiation safety has a number of recommendations to minimize the risk from radiation of interventional electrophysiological procedures to both the patient and the operator while maximizing image quality.[7] The recommendations regarding reducing radiation exposure can be categorized into equipment factors, operator-dependent functions, laboratory maintenance, operator shielding, monitoring, and training. The following suggestions to minimize radiation exposure while maintaining high-quality images have been made by Farre et al.[12]:

1. Keep fluoroscopy time as low as reasonably achievable.
2. Use x-ray beam systems entering the posterior and not the anterior side of the patient, thus attenuating radiation to thyroid, breast, and eye tissues of the patient.
3. Limit the size of the explored field with collimation.
4. Use the largest possible field of the image intensifier since magnification increases the dose.
5. Use pulsed fluoroscopy with <12.5 pulses/second instead of continuous fluoroscopy.
6. Use digital fluorography at <12.5 images/second rather than 35-mm filming to store positions of catheters or angiographic information.
7. Use all possible protections like a leaded acrylic shield between patient and operator, leaded aprons, neck collars, glasses, and filtration of the primary x-ray beam.
8. Maintain all the personnel as far as possible from the x-ray beam.
9. Manipulate catheters as little as possible from a subclavian or jugular approach that result in more scattered radiation for the exploring physician than a femoral approach.
10. Manipulate catheters as much as possible in the right anterior oblique (RAO) projection since the left anterior oblique (LAO) projection is the worst in terms of secondary radiation for the exploring physician while the posteroanterior (PA) or frontal projection is intermediate.

In our practice we find that pulsed fluoroscopy at 3.75 frames/second provides adequate information to aid ablation, but for transseptal punctures we use 15 frames/second. Additionally by using a radiation protection cabin, only the hands are exposed to scattered radiation, which is minimized by the use of the leaded acrylic shield.[10] The use of the radiation protection cabin has the additional benefit of freeing the operator from wearing protective lead aprons, reducing the risk of back pain and consequent days off work.[9] When protective aprons need to be worn, for example percutaneous pericardial puncture, a thorough knowledge of how to prevent spinal injuries when using protective aprons is required.[13,14]

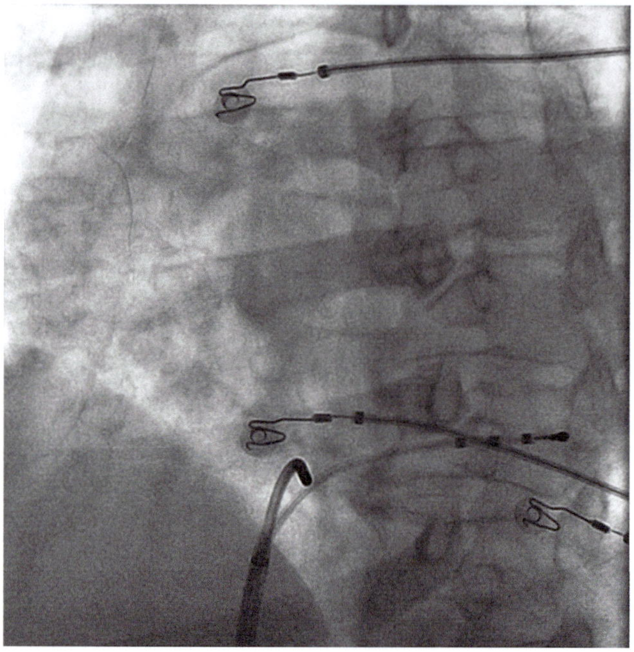

Fig. 8.1 Ablation of cavotricuspid dependent atrial flutter. The isthmus is at 6 o'clock in the LAO projection, which is often used for ablation along the isthmus

8.2 Fluoroscopic Views

In order to better appreciate catheter position inside the heart volume using fluoroscopy, different projection angles are utilized. The attitudinally correct nomenclature of the cardiac chambers that has been endorsed by the European Society of Cardiology and HRS (previously the North American Society for Pacing and Electrophysiology) helps the operator manipulate catheters within the cardiac chambers in a logical manner.[15] In electrophysiology, the most useful views are PA, LAO, and RAO. The lateral projection is useful during transseptal puncture and for percutaneous pericardial access.[16]

In the PA projection, the superior vena cava (SVC) and the right atrium form the right cardiac border and the aortic arch, the left main pulmonary artery, the left atrial appendage and the left ventricle form the left cardiac border, from cranial to caudal respectively.

The RAO projection is useful to assess anterior and posterior positions as well as superior and inferior. In LAO, the superior, inferior, lateral, and medial locations can be visualized. The LAO projection is often used when mapping around the mitral or tricuspid annulus typically (Fig. 8.1) when localizing accessory pathways, or when attempting to engage the coronary sinus.

8.3 Transseptal Puncture

Transseptal puncture has become a standard technique in the EP laboratory with the increase in patients attending for an atrial fibrillation (AF) ablation,[17] and ablation within the left atrium. Originally described by Ross[18] with subsequent modifications[19] the technique and equipment has changed little since then. It is important to remember that the technique was originally developed to measure left heart pressures in patients with predominantly mitral valve disease and subsequently large left atria, and that the catheters were designed for this application.

The true interatrial septum, whereby crossing it does not result in exiting from the heart, is the area of the fossa ovalis and its immediate inferior muscular rim.[20,21] In most patients this is covered by a membranous flap; however, approximately 25% of the population has a probe patent foramen ovale, which is located in the anterosuperior part of the foramen ovale.[20,22] This anterosuperior location makes manipulation of catheters within the left atrium more challenging, particularly when trying to isolate the pulmonary veins; however, similar clinical results are obtained when compared to access via a transseptal puncture.[22]

In order to perform a safe transseptal puncture, the boundaries of the foramen ovale need to be visualized (Fig. 8.2). There are several different ways of doing this. Anterior to the fossa is the aortic bulge, and to avoid puncture of the aorta, this can be marked either by a pigtail catheter placed in the aortic root or by noting where a His bundle recording can be made, which overlies the most inferior aspect of the noncoronary cusp of the aorta. Having a catheter placed within the coronary sinus can mark the posterior border of the left atrium, and defines the intrathoracic rotation of the heart.

In the PA projection, the level of the puncture can be assessed by comparison of the left main bronchus, which defines the roof of the left atrium to where a His recording is made, and the ostium of the coronary sinus. In the RAO projection, the fossa is posterior and either at the same level or superior to the site where the His signal can be recorded.[12] In the LAO projection, the site of puncture is superior to the coronary sinus, and the plane of the transseptal assembly should be in the same plane as a catheter inserted into the coronary sinus. A left lateral projection can be helpful, and in this view the transseptal assembly is almost perpendicular to the screen, with less than 10° posterior rotation, with the posterior margin of the left atrium marked by the coronary sinus catheter.

To perform a transseptal puncture, a 0.035 or 0.032[2] guidewire is introduced into the SVC via an eight French hemostatic sheath, placed in the right femoral vein. This sheath is then exchanged for the transseptal assembly that consists of a long sheath and a slightly longer dilator that has been flushed with heparinized saline. It is important that the operator is familiar with the transseptal sheath and dilator as they come in a variety of lengths and angulations to help with different anatomies, with the standard lengths being 63 and 67 cm respectively, and a needle length of 71 cm.

Continuous pressure monitoring is mandatory via the sidearm of the long sheath, which is continually flushed with heparinized saline. Once the transseptal sheath and dilator are in the SVC, the guidewire is withdrawn and the Brokenborough needle is introduced into the dilator and passed until it is just proximal to the very tip of the dilator. At this point the sheath and the needle are aligned so that they are facing posteromedially at between 4 and 5 o'clock. In either a PA projection or the LAO projection, the transseptal assembly is then pulled down into the fossa. There are two characteristic movements of the assembly, the initial drop and rightward movement indicates the drop from the SVC into the right atrium, followed by another jump and rightward movement into the fossa ovalis. At this point the limbus of the fossa is engaged by a slight forward movement of the transseptal assembly.

The correct positioning can be confirmed in the RAO projection where the transseptal assembly should be posterior to the area of the His bundle recording and, if used, a pigtail catheter in the aortic root. Confirmation is finally performed with injection of radio-opaque contrast. When at the fossa, the contrast stains with a characteristic round shape, like a pea, with a well-defined border. If the needle is oriented to the muscular septum or the posterior wall, the contrast will spread diffusely in the myocardium with irregular and hazy borders. Keeping the transseptal assembly steady the needle is then advanced and tenting of interatrial septum can be visualized just prior to crossing the septum. Continuous pressure monitoring will change from a damped trace on the septum to a left atrial trace when the septum is successfully crossed. The needle is then aspirated and oxygenated blood should be withdrawn. Contrast is then injected and when the left atrium is partially opacified, attention should be paid to its contours, notably the roof and the lateral wall, to avoid trauma when advancing the transseptal sheath. If contrast delineates the cardiac silhouette, the needle is in the pericardial space and should be withdrawn, whereas if contrast flows superiorly in a pulsatile manner, aortic puncture should be suspected. Once satisfied that the needle has entered the left atrium, the sheath assembly is advanced through a short distance so that the tip of the dilator enters the left atrium. Although at this stage the sheath can be pushed into the left atrium, there is a risk that the dilator could perforate the lateral wall of the left atrium. It is for this reason that we prefer to introduce the guidewire into the left atrium via the dilator. Here the guidewire can be positioned in the left superior pulmonary vein and the sheath assembly passed into the left atrium. This also allows the transseptal hole to be dilated, by passing the sheath and dilator across several times, with rotation anterior and posterior, until no resistance is felt when the lip between the dilator and the sheath crosses the septum. This is only necessary if a further catheter is to be introduced into the left atrium without a second transseptal puncture, or

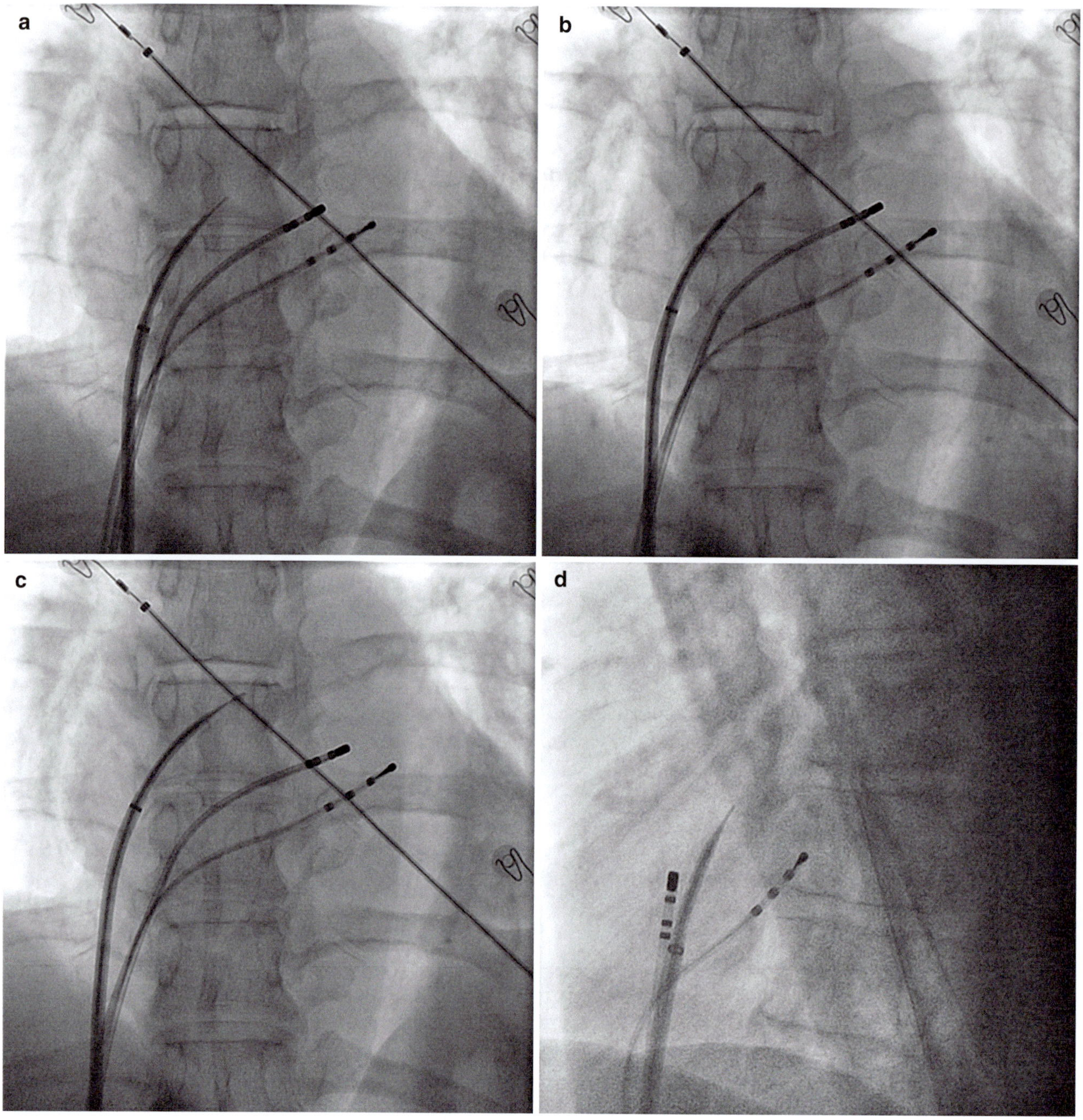

Fig. 8.2 Transseptal puncture. Performing a transseptal puncture is now routine in the EP lab, given the increase in AF ablation. (**a**) The ablation catheter is positioned at the His position, for orientation of the aortic root, which is behind the catheter. The transseptal apparatus is pulled down from the SVC assessing for a double jump, and engaging in the fossa ovalis. (**b**) Well-demarcated contrast staining of the fossa clearly demonstrates that the needle is in the correct position. Diffuse staining would indicate that the needle is not in the fossa, and the needle should be repositioned. (**c**) In this example, even with a lot of force applied, the transseptal needle does not cross the septum. This is often the case when, as in this example, the patient has had previous AF ablations. (**d**) The left lateral view confirms that the transseptal apparatus is posterior to the His and anterior to the posterior margin of the heart as assessed by the coronary sinus catheter. (**e**) RF energy was applied and the transseptal puncture is successful. Contrast swirls around the left atrium. (**f**) The transseptal hole is dilated multiple times with the sheath and dilator, until no further resistance is felt. The guidewire is then introduced into the left superior pulmonary vein. (**g**, **h**, **i**) The ablation catheter can then be passed through the transseptal puncture. (**j**, **k**, **l**, **m**) The transseptal sheath and dilator is then moved into the left atrium, removing the dilator and guidewire prior to the circular mapping catheter being moved into the left atrium via the sheath

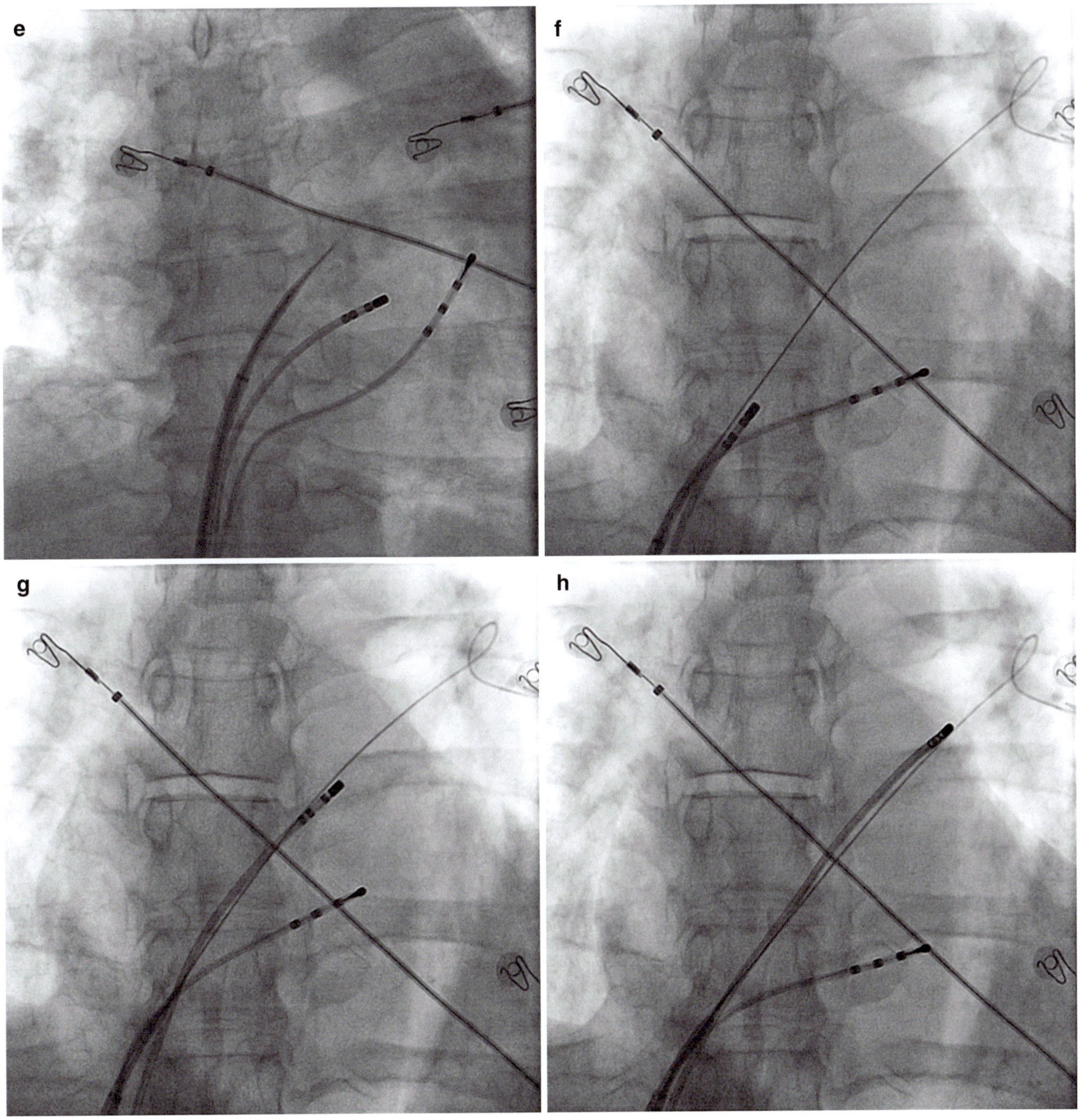

Fig. 8.2 (continued)

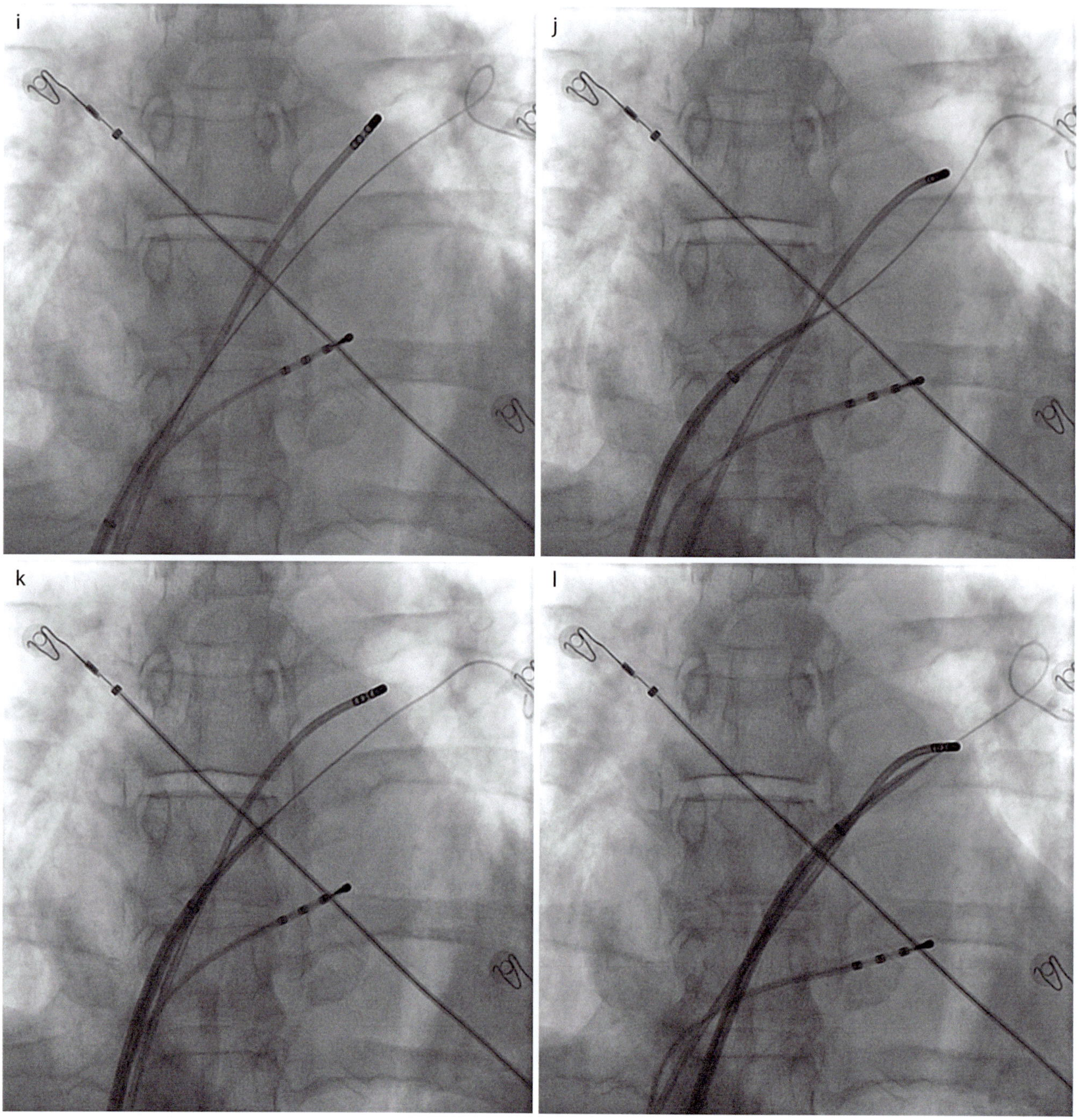

Fig. 8.2 (continued)

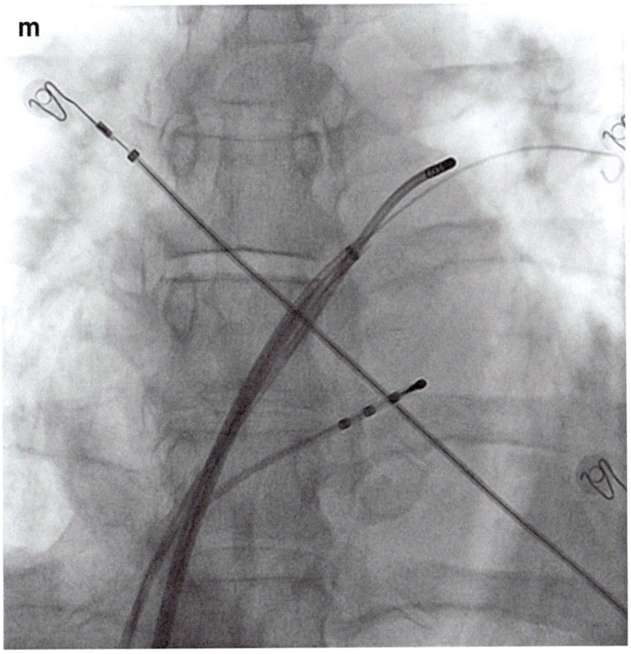

Fig. 8.2 (continued)

if the sheath does not move freely, hindering catheter manipulation. If a catheter is to be introduced into the left atrium in addition to the sheath, the transseptal assembly is withdrawn from the left atrium into the right atrium, while leaving the guidewire in the left superior pulmonary vein, and the catheter is then advanced into the left atrium. The transseptal assembly is then advanced into the left atrium, and finally once the sheath is across, the dilator is withdrawn. Once the dilator has been withdrawn, the sheath is aspirated to ensure no air or thrombus enters into the left atrium and flushed with heparinized saline, and systematic heparin administered.

If transseptal puncture is difficult or prolonged, it is important to aspirate and flush the sheath as thrombus may have formed. The two main problems that are encountered are thickened tough septae that are difficult to puncture, or distorted anatomy, either due to cardiac malformation or severe skeletal malformations.

Although intracardiac echocardiography or transesophageal echocardiography can be used to aid with transseptal punctures[1,23-27] we do not consider these tools to be mandatory for standard transseptal catheterization; however, they do allow the operator to directly visualize the transseptal needle and the atrial septum and fossa ovalis, and are useful when there is distorted anatomy. For cases of toughened septae, we use the ablation catheter to assist with the puncture.[28] Once satisfied that the transseptal assembly is located at the fossa, the needle is pushed out of the dilator, with slight pressure to get tenting of the septum. Using the ablation catheter, unipolar radiofrequency energy of 30 W is transmitted from the ablation catheter to transseptal needle by simply touching the distal electrode to the proximal end of the needle. Energy is then concentrated at the tip of the needle, with the long plastic sheath isolating the body of the needle from other anatomical structures. It is important that when the septum is crossed (normally within 2 s) the energy is stopped rapidly.

8.4 AF Ablation

Pulmonary vein isolation, whether ostial or antral, is the cornerstone in treatment for paroxysmal AF, and is the starting point in the stepwise ablation of persistent AF. X-ray fluoroscopy allows accurate assessment of the ostia of the pulmonary veins to aid ablation. Selective pulmonary venous angiography gives detailed information that can then be recalled throughout the case. A bolus injection of contrast into the left atrium, with or without manipulations to minimize atrial emptying (adenosine, rapid ventricular pacing) will gives additional information on the walls of the left atrium. Alternatively, manipulation of the catheter within the left atrium and pulmonary veins can give similar information.

Patients with persistent AF require a more extensive ablation, often necessitating ablation at the base of the left atrial appendage (LAA) (Fig. 8.3), the inferior left atrium (Fig. 8.4), the septum and within the coronary sinus, as well as linear lesions, for instance the left atrial roof line or the mitral isthmus line. Again, with knowledge of the anatomy, assessment of catheter movement checked with fluoroscopy, and electrogram recognition, these structures can be accurately defined.

8.5 Pulmonary Vein Anatomy

An understanding of pulmonary vein anatomy is crucial both to understand the limitations of X-ray fluoroscopy and to perform pulmonary vein isolation. The majority of patients have four pulmonary veins, two superior and two inferior with independent ostia.[29] A number of variations have been described in the literature.[29-33] In one series, a common ostium for the left pulmonary veins was described in 83% of patients; however, a common ostium for the right pulmonary veins was less frequent, occurring in 40% of patients.[34] Other reports confirm that a common ostium for the superior and inferior veins is more frequent on the left than the right but the overall frequency is much less at 15% and 10% of patients respectively.[31] One of the reasons for the difference in the reported frequency of common pulmonary vein ostia are the different imaging modalities used Magnetic resonance angiography (MRA[32]), rotational angiography,[35-37] Computed tomography (CT) angiography,[34,38] Echocardiography[39] and

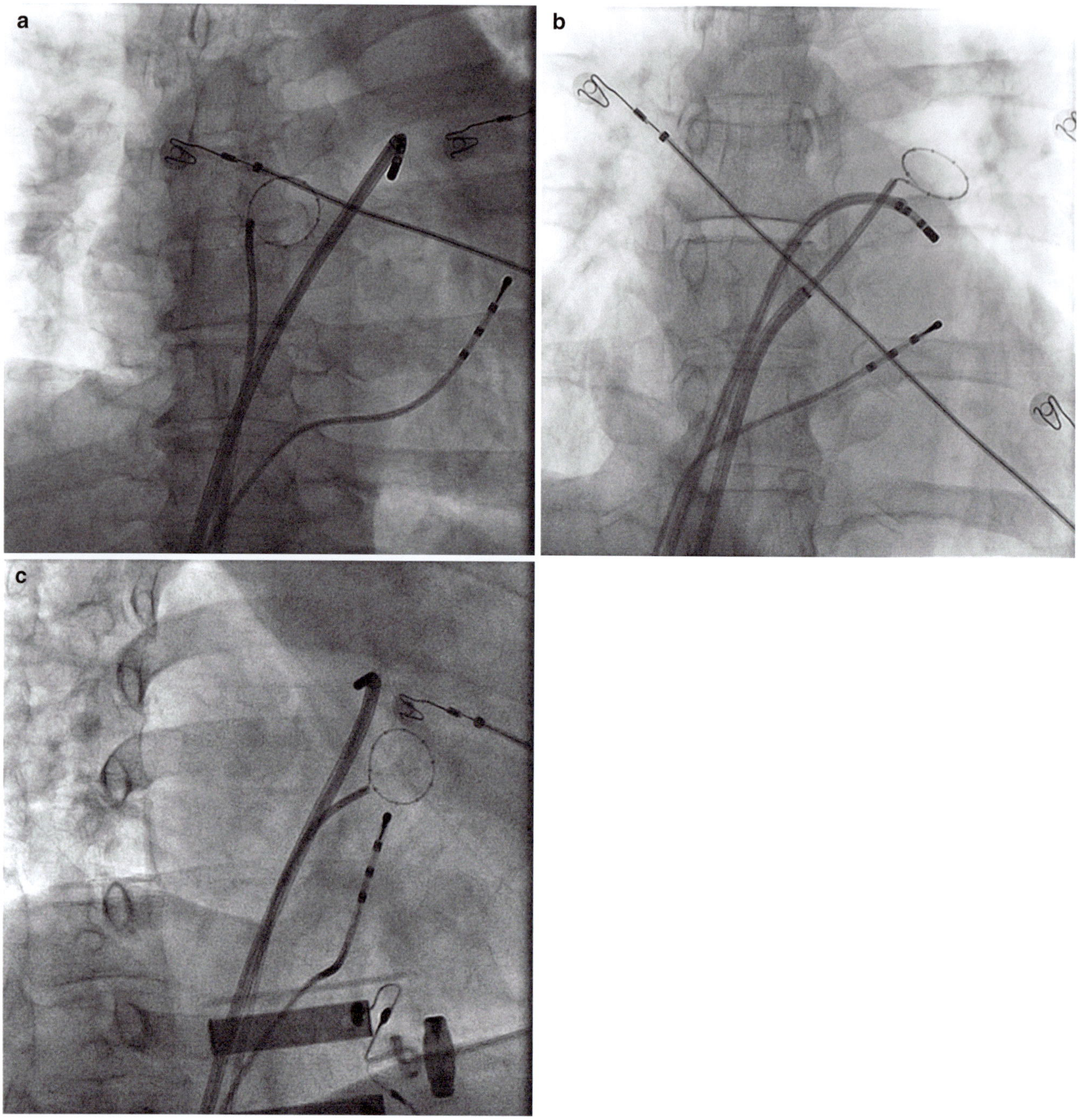

Fig. 8.3 Ablation of the left atrial appendage. Ablation at the base, and sometime within the left atrial appendage is often necessary in the ablation of persistent AF. (**a** and **b**) Ap view of catheter and sheath positions used for steady contact at the base of the appendage. (**c**) RAO view

post mortem specimens[31]), and the definition used for the border between the left atrial wall and the pulmonary vein.

The other common variant described in the literature is a separate origin for the right middle pulmonary vein. The majority of the population will have the middle pulmonary vein drain into the right superior pulmonary vein. A separate origin has been described in 17–23% and insertion into the right inferior pulmonary vein the least frequent, 3–8%, in surgical series.[40,41] More recently, a separate origin for the middle pulmonary vein has been described in 22% in one study, using multi-detector row CT,[34] but only 10% in another study using MRA,[32] and with none being seen in a postmortem study utilizing 20 hearts.[31] Additional left-sided pulmonary veins are far less frequent, but are described,[34,42] as has been a variant in which the pulmonary vein enters into the roof of the left atrium.[37,43] All of these variations in anatomy can be identified using left atrial angiography, with or without techniques to slow atrial emptying and therefore increase the time the contrast spends in the left atrium and pulmonary veins.

The diameter of the pulmonary veins is also important both for risk of pulmonary vein stenosis, which is greater in smaller veins,[32] and sizing of catheters. A study compared pulmonary vein diameters in patients with AF as measured by intracardiac echocardiography (ICE) and multislice CT.[39] The majority of patients were in PAF and there was a statistically significant difference in the measured size of the pulmonary veins between the two imaging modalities. The diameter of the veins tended to be undersized when using ICE compared to CT. The mean diameters of the pulmonary veins in patients vary between 11 and 19 mm for left-sided veins and 15 and 19 mm for right-sided veins. The average

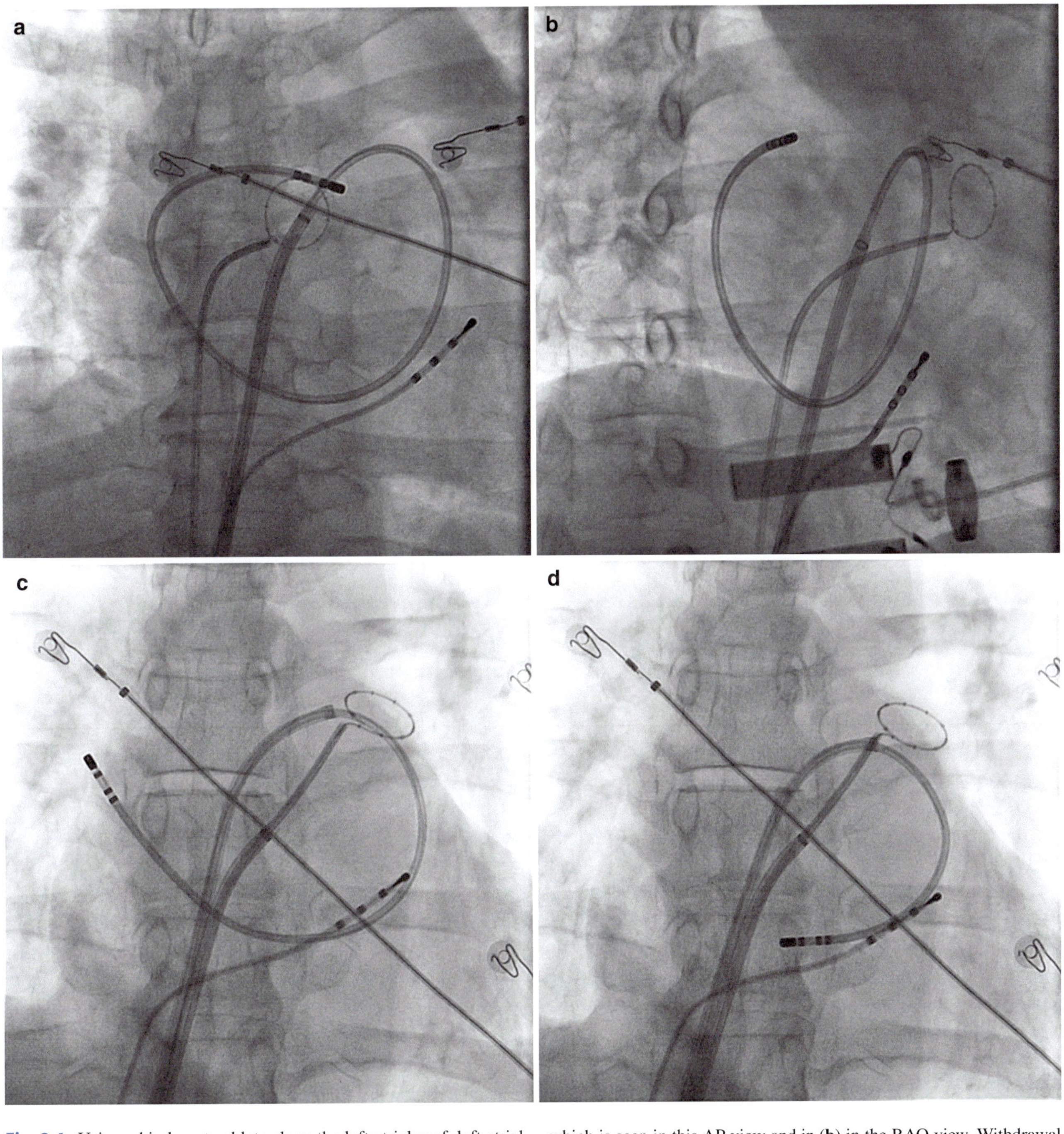

Fig. 8.4 Using a big loop to ablate along the left atrial roof, left atrial septum, and the inferior left atrium. (**a**) A large loop can be made with the ablation catheter by pushing the transseptal sheath anterior to the left atrial appendage and then flexing the ablation catheter. This will then track round the left atrium to the left superior pulmonary vein, which is seen in this AP view and in (**b**) in the RAO view. Withdrawal of the catheter allows ablation along the left atrial roof, anterior to the right-sided pulmonary veins (**c** and **d**), and then along the inferior left atrium (**e** and **f**). This can be extended to the lateral mitral isthmus

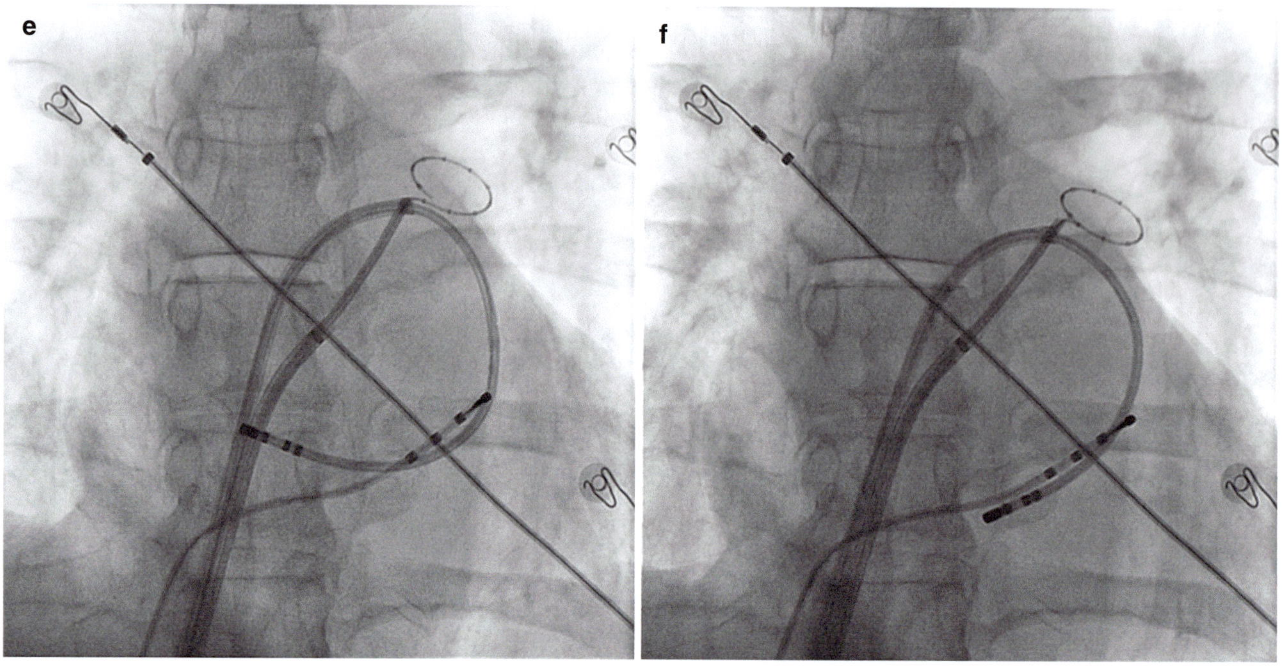

Fig. 8.4 (continued)

diameter for common ostia is between 18 and 27 mm for common left-sided ostia and 26–34 mm for right-sided common ostia. These results encompass a number of imaging modalities and patient populations.[29-33,39,42,44]

Although size is important, it is less crucial today as a consensus favored atrial lesions (antral ablation) as compared to the "ostial" initial strategy. A number of new ablation catheters are balloon based and are circular in the axial plane. However, several studies have shown that the pulmonary veins themselves are not circular.[32,34,39] All studies that have reported upon the geometry of the pulmonary veins have shown that the left-sided veins are oval, with a greater superior-inferior diameter compared to anterior-posterior diameter. The right-sided veins tend to be more circular, but they still tend to have a larger superior-inferior diameter. In a study utilizing MRA, the ratio of the superior-inferior diameter to the anterior-posterior diameter was 1.5 for the left-sided pulmonary veins and 1.2 for the right-sided veins.[32] When using projection imaging, such as fluoroscopy, the oval nature of the left-sided pulmonary veins is not as easily appreciated as when using CT or MRA reconstructions. This means that pulmonary vein stenosis could be missed if fluoroscopy is the only imaging modality is used.

8.6 Technique

Electrical isolation of all pulmonary veins is the endpoint of ablation in patients with paroxysmal AF, and this objective measure of pulmonary vein disconnection is both easy to confirm and results in maintenance of sinus rhythm in between 60% and 85% of patients.[45,46]

The technique of pulmonary vein isolation is similar for paroxysmal AF and long-lasting persistent AF. Pulmonary vein isolation is currently performed as the initial ablation step in all patients with long-lasting persistent in our laboratory.[46,47] Although isolation restricted to veins identified as "arrhythmogenic" has been advocated by others,[48] uniform isolation of all pulmonary veins identified is routinely performed and confirmed using a circumferential mapping catheter positioned at the venous ostium (Fig. 8.5). Veins are isolated individually or as ipsilateral pairs in accordance with the venous anatomy and operator's preference. Confirmed isolation is critical as the electrically inactive tissue enclosed within the ablation lesions serves as an anchor for the roof and mitral isthmus lines that may be necessary should AF persist.

As previously mentioned the absolute pulmonary vein/left atrial junction is difficult to define, by any imaging modality. However, given the risk of pulmonary vein stenosis, isolation is preferentially performed within the left atrium. Recently the antral area of the pulmonary veins has been demonstrated to be an area particularly arrhythmogenic, and for this reason, as well as minimizing the risk of pulmonary vein stenosis, isolation is preferentially performed about 1 cm from the ostium of the pulmonary vein.

One trick to simply locate the inferior portion of the pulmonary vein ostia is to advance the catheter into the pulmonary vein with downward deflection of the tip and then dragging back while fluoroscopically monitoring the drop off the ostial edge of the catheter. Pulmonary vein isolation is guided by the use of a circumferential mapping catheter. However, antral ablation means that at least for the right veins, a full circle of lesions have to be delivered. As a consequence, the first step is anatomical. Then, ablation preferentially targets regions of earliest pulmonary vein activation when it is asynchronous or is performed by ablating around the pulmonary vein circumference during sinus rhythm or ongoing AF. The extent of circumference ablated is variable between pulmonary veins. However, because ablation is performed proximally and during AF, more circumferential ablation often is required to achieve pulmonary vein isolation. Ablation is started randomly at either the right or left pulmonary veins and is performed individu-

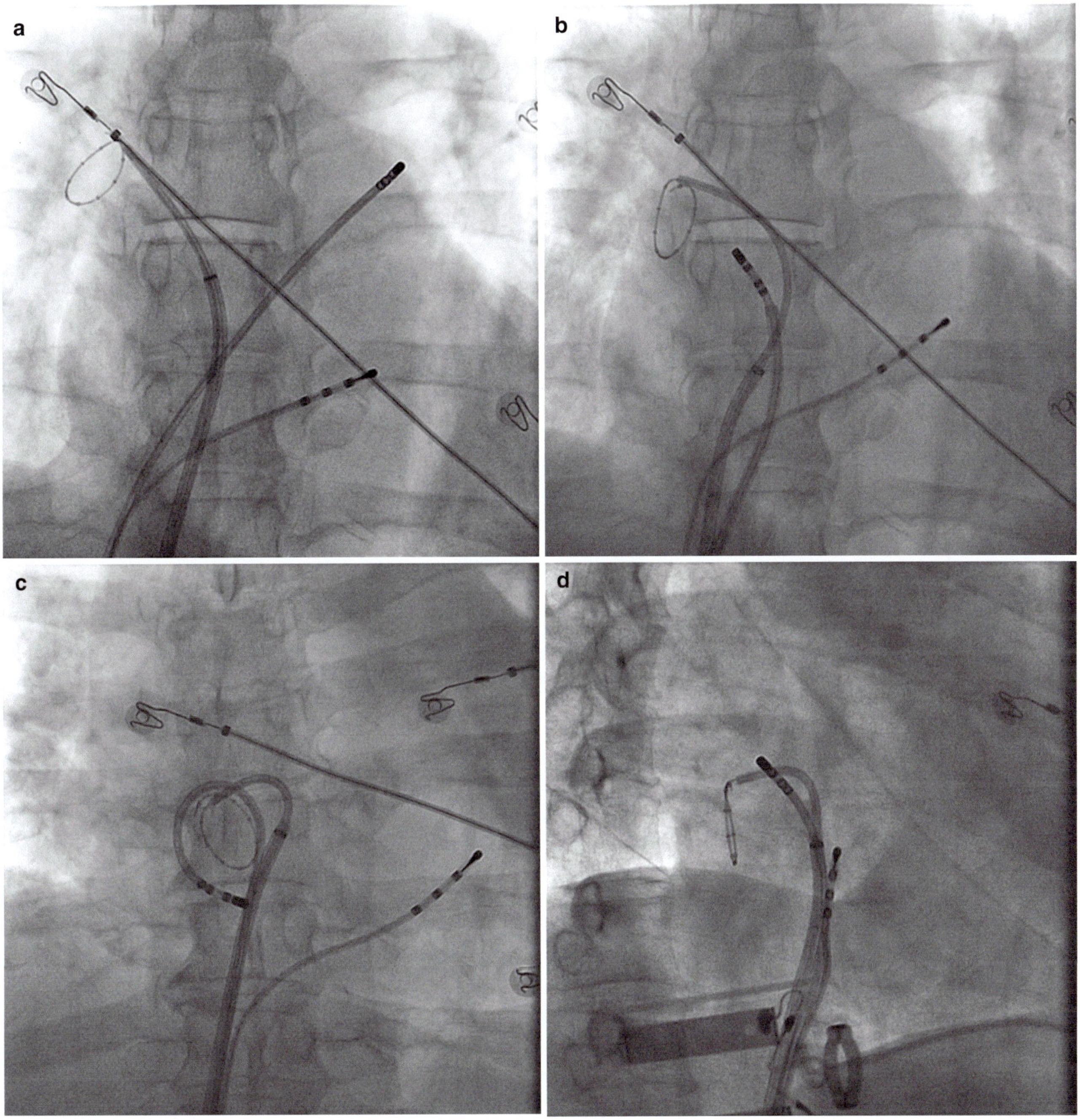

Fig. 8.5 Pulmonary vein ablation unequivocal pulmonary vein isolation is a pre-requisite for AF ablation. (**a**) The lasso catheter is in the RSPV, and the ablation catheter in the LSPV in this AP projection. (**b**) The ablation catheter is manipulated to the anterior aspect of the RSPV in this AP view. (**c**) The lasso catheter has been moved into the right inferior pulmonary vein to ease identification of LA-PV connections. (**d**) The right inferior pulmonary vein in the RAO view. (**e** and **f**) The left superior and inferior pulmonary veins viewed in the AP projection, with the ablation catheter anterior and posterior respectively

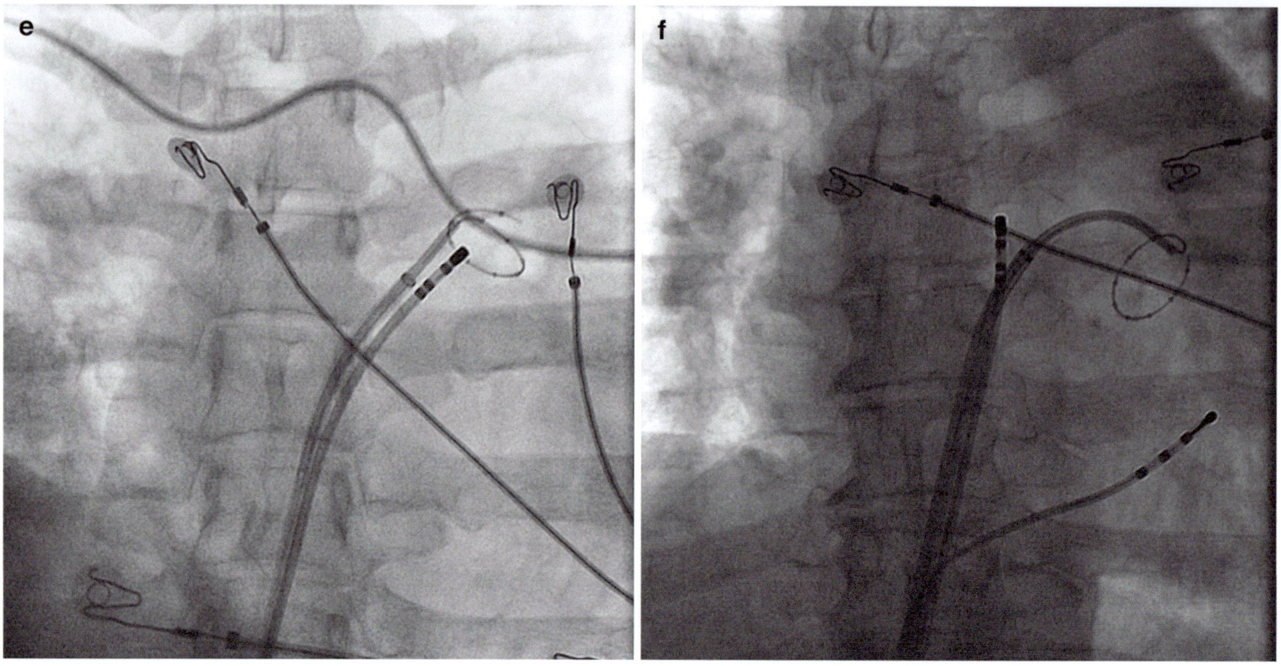

Fig. 8.5 (continued)

ally or en bloc. Ablation is started at the posterior wall (usually fluoroscopically facing the border of the spine in the anteroposterior projection) and continued around the venous perimeter. The posterior wall of the right pulmonary vein is reached by counterclockwise torque of the catheter and the left pulmonary vein by clockwise torque. When ablation is required at the anterior portions of the left superior and inferior pulmonary veins, it is often necessary to enter the first few millimeters of the veins. This is due to the ridge that separates the veins from the left anterior appendage is of a variable thickness and adequate tissue contact is often not possible at this site.

8.6.1 Left Atrial Linear Lesions

One of the key techniques to master for successful ablation of persistent AF is how to perform left atrial linear lesions. The two lines that have been demonstrated to be of use are the left atrial roof line and the lateral mitral isthmus line; please quote as well Meleze for the roof and me for the MIG.[49, 74, 75]

8.6.2 Roof Line

The roof line refers to a contiguous line of ablation lesions joining the right and left superior pulmonary veins (Fig. 8.6). The ablation catheter is introduced into the left atrium via a long sheath to achieve stability and allow orientation of the catheter tip toward the roof of the left atrium. Ablation starts at the encircling lesion at the left superior pulmonary vein, and the sheath and catheter assembly are then rotated clockwise posteriorly and dragged toward the right superior pulmonary vein. To achieve catheter stability along the roof of the left atrium, the catheter is directed toward the left superior pulmonary vein and the sheath rotated to face the right pulmonary veins and vice versa. If this catheter and sheath position fails, there are two alternative methods that can be used. A large loop can be made with the catheter around the lateral, inferior, and septal walls to arrive at the roof and then the left superior pulmonary vein, ablation can then be commenced while dragging the catheter back from the left to the right superior pulmonary vein ostia. Regardless of the technique utilized, ablation is preferably performed cranially rather than posteriorly to minimize the risk of esophageal fistula. RF energy (25–30 W) is delivered for 20–120 s at each point until the local potential is eliminated or there are double potentials. The electrophysiological end point of ablation was by demonstration of a complete line of block joining the two superior pulmonary veins.

Evaluation of complete linear block is performed after the restoration of sinus rhythm to allow pacing of the anterior left atrium adjacent to the line. Anterior left atrium pacing could be achieved by pacing with the proximal poles of a decapolar catheter while its distal end seats in the left atrium appendage. Alternatively, pacing can be performed by advancing the catheter to the anterior aspect of the coronary sinus. Complete linear block was defined by the following criteria: (1) Demonstration by point-by-point

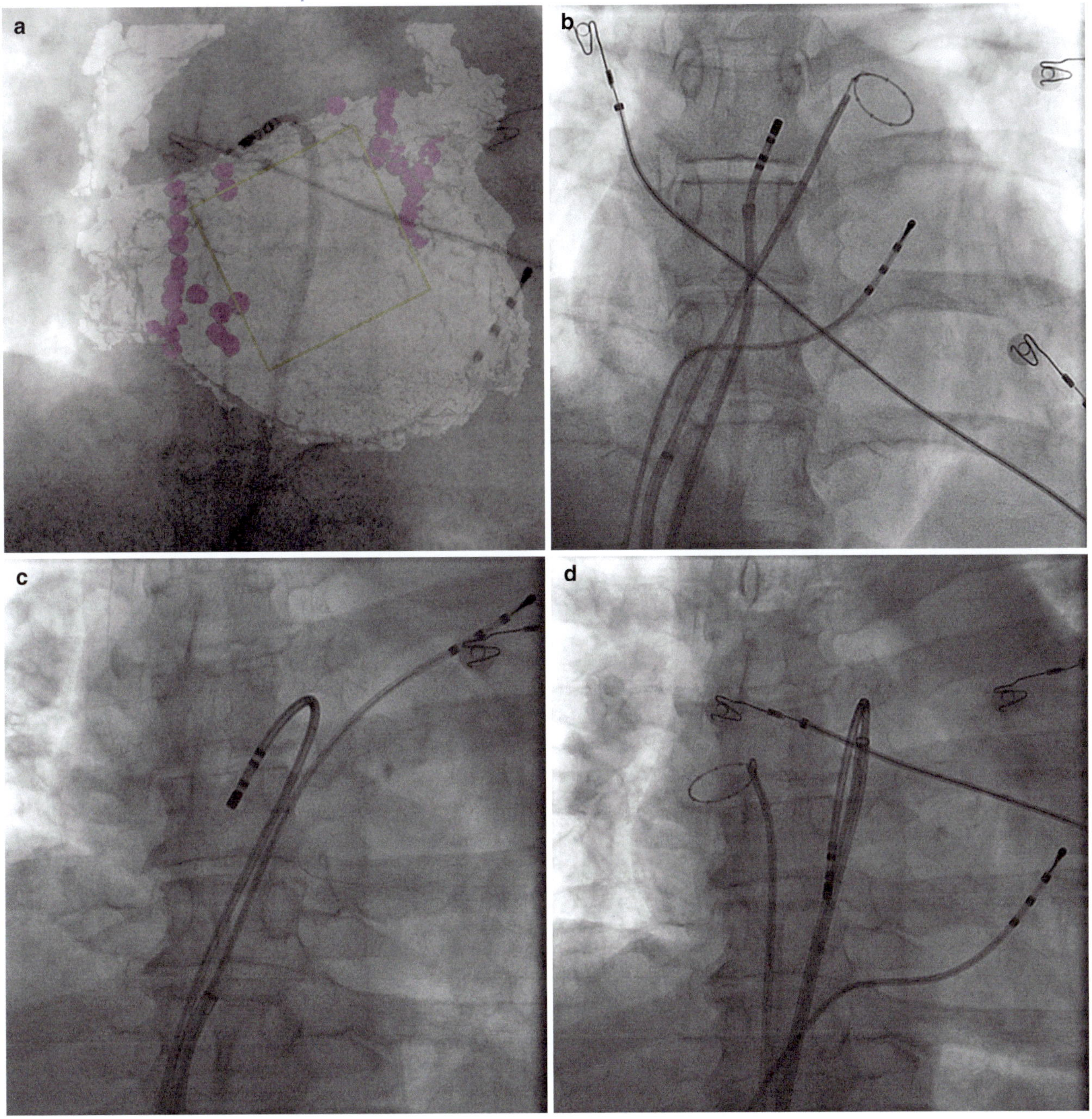

Fig. 8.6 Left atrial roof line. Left atrial linear lesions are normally required for a successful outcome in patients with persistent atrial fibrillation. (**a**) Demonstrates a horizontal orientation of the ablation catheter along the left atrial roof. This is a very safe orientation for ablation, with a low risk of steam pops. In (**b**) a more perpendicular orientation is used. (**c** and **d**) Demonstrate the catheter positions used to check for block of the roof line by pacing anterior to the line from the quadripolar catheter in the left atrial appendage and assessing conduction along the posterior wall. In cases of block of the roof line, the wavefront of activation is ascending up the posterior wall (from low to high, **d** and **c**, respectively), whereas in cases where the line is not complete, activation can go via the left atrial roof and thus descend down the posterior wall

mapping of an online corridor of double potentials along the entire length of the roof during pacing of the anterior left atrium; and (2) demonstration of an activation detour circumventing the right and left pulmonary veins to activate caudocranially the posterior wall with no conduction through the left atrial roof.

8.6.3 Mitral Line

A mitral line is performed to disrupt a circuit around the mitral annulus or in case of persisting AF as a last resort. As such several lines can be performed, for example an anterior line from the anterior mitral annulus to a complete roof line or the isolated right superior pulmonary vein or a septal mitral isthmus line from the septal mitral annulus to the isolated right inferior pulmonary vein. However, the most commonly performed line is the lateral mitral isthmus line, from the lateral mitral annulus to the isolated left inferior pulmonary vein. For this line the coronary sinus catheter is positioned to bracket the potential linear lesion between its proximal and distal bipoles. The ablation catheter is then curved between 90° and 180° and introduced via the long sheath to the ventricular edge of the lateral mitral annulus, with an atrioventricular electrogram ratio of between 1:1 and 2:1. Ablation is then commenced and the sheath and catheter are rotated clockwise to extend the lesion posteriorly, ending at the left inferior pulmonary vein ostium. Ablation energy (35 W) is delivered for up to 120 s at each site. The lesion is monitored by observing the conduction delay between the local electrogram during pacing relative to the coronary sinus bipole immediately septal to the lesion. If the initial attempt failed to produce complete block, ablation is performed in a more lateral position, at the base of the appendage. Persisting epicardial conduction is suspected when the linear lesion resulted in an endocardial conduction delay recorded on the ablation catheter but not on the adjacent distal bipole of the coronary sinus catheter (lateral of the line). In such cases, ablation needs to be performed within the coronary sinus, which is approximately 70% of the time. Ablation within the coronary sinus is performed with a flow rate of between 17 and 60 mL/min (maximal flow at the distal coronary sinus), a target temperature of 50°C, and power of 20–30 W.

To assess bidirectional block of the mitral isthmus, differential pacing is performed. The ablation catheter is placed just lateral to the line. Using the coronary sinus catheter, stimulation starts using the bipole just septal the linear lesion, then the pacing site is changed to the next proximal bipole of the coronary sinus catheter without moving any of the catheters. Stimulus-to-atrial electrogram timing on either the ablation catheter or a bipole on the coronary sinus catheter that is lateral to the line is then measured to the same point on the matching electrogram component before and after changing the pacing site. With complete block, the stimulus-to-electrogram timing is shorter after shifting the pacing site from the distal to the proximal bipole. Pacing lateral to the line through the ablation catheter demonstrates a proximal-to-distal activation sequence along the coronary sinus septal of the line, thus confirming bidirectional conduction block. In addition, widely separated local double potentials along the length of the ablation line during coronary sinus pacing septal to the line can be mapped.

8.7 Assessment of Complications

The risks of a complication associated with catheter ablation of cardiac arrhythmias are small. However, when complications do occur, prompt recognition and action can be life saving. X-ray fluoroscopy can help with the early recognition of certain complications of catheter ablation, and since this is the primary imaging modality that is used in the catheter, the specific signs that can be detected need to be known.

8.7.1 Cardiac Tamponade

During transseptal catheterization and during atrial fibrillation ablation, several maneuvers carry the risk for cardiac wall injury and subsequent pericardial effusion and tamponade. One of the earliest signs of tamponade, even before a significant drop in blood pressure occurs, is decreased excursion of the cardiac silhouette,[50] best seen at the left superior cardiac border in an AP projection. As fluid accumulates in the pericardial space, it is no longer possible to see the myocardial contraction on fluoroscopy. Prompt recognition of this sign can lead to early intervention.

8.7.2 Phrenic Nerve Injury

Injury to the right phrenic nerve can occur while ablating near the SVC, on the lateral wall of the right atrium and inside the right superior pulmonary vein. The left phrenic nerve can be injured by ablation deep inside the left atrial appendage. Phrenic nerve lesion results in ipsilateral hemidiaphragmatic paralysis, which is reversible if recognized early, but can be devastating if it goes unnoticed and irreversible damage is done.[51,52] Direct visualization of diaphragmatic excursion during inspiration confirms integrity of the phrenic nerve, and should be monitored during ablation of the aforementioned regions. Prior to ablation, pacing from the ablation catheter at maximal output should be performed to check for the absence of phrenic capture. During radiofrequency delivery, coughing or absence of normal diaphragmatic movement should prompt immediate cessation of RF delivery. A novel method to avoid left phrenic nerve injury is to inject air into the pericardial space to lift the pericardium and the overlying phrenic nerve away from the heart.[53]

8.7.3 Pulmonary Vein Stenosis

Pulmonary vein stenosis is a known complication of pulmonary veins isolation, occurring in around 1–3% of cases,[54] but is more common when ablation is performed inside the vein. In redo procedures, selective angiography of the pulmonary veins can detect asymptomatic stenosis (Fig. 8.7) and avoid further injury to the vein. However, if PV stenosis is suspected on the basis of clinical symptoms (exertional dyspnea, hemoptysis), computed tomography or magnetic resonance angiography should be performed.

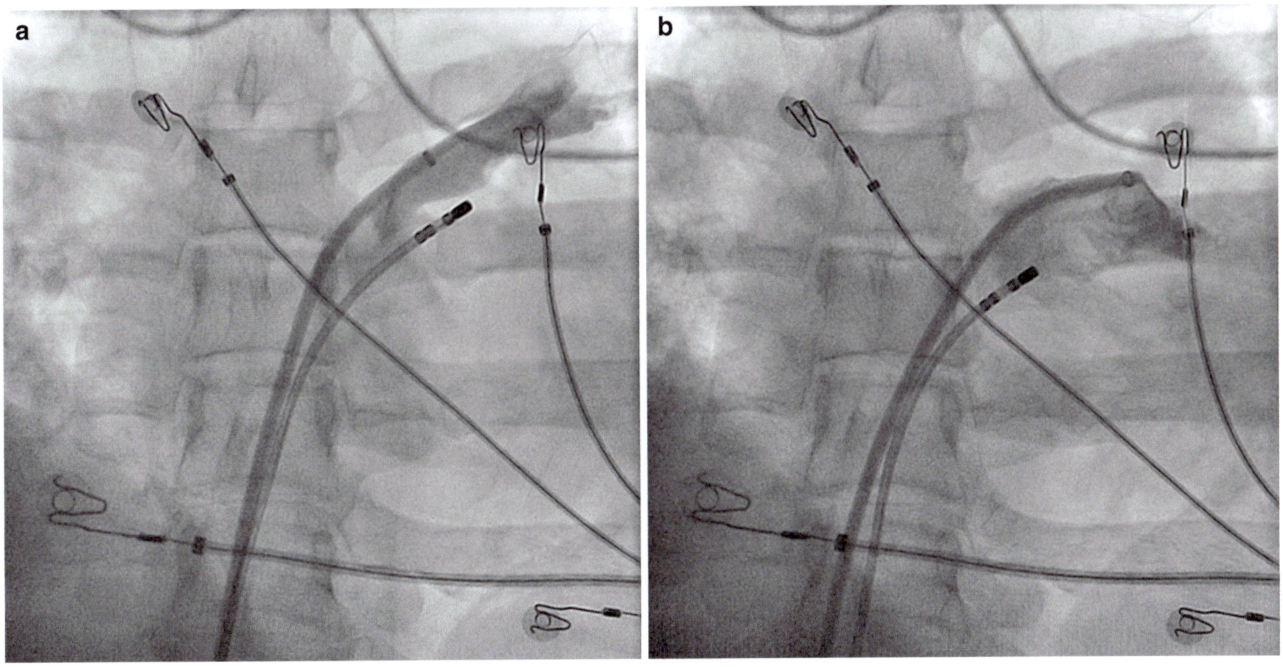

Fig. 8.7 Pulmonary vein stenosis. Pulmonary vein stenosis is a serious complication of AF ablation, but is now rare as ablation is targeted at the ostium and the antrum of the pulmonary vein. Pulmonary venography clearly demonstrates PV stenosis in the left superior pulmonary vein (**a**) and in another patient in the left inferior pulmonary vein (**b**)

8.7.4 Esophageal Fistula

Esophageal fistula is one of the most feared complications of atrial fibrillation catheter ablation.[55-57] In order to avoid esophageal injury, care should be taken when ablating in the posterior left atrium. A number of strategies have been suggested to try and avoid this complication. The most effective of these we believe is to limit the power delivered and duration of individual ablation lesions when on the posterior wall. Direct visualization of the esophagus is useful, and can be done with barium paste, or with overlay of a CT scan or rotational angiography onto the live fluoroscopy screen, but a limitation is that the esophagus is a mobile structure, and its position can change during the ablation.[2,58-61] In addition, there are some data suggesting that conscious sedation (instead of general anesthesia)[62] and proton pump inhibitors after the ablation are able to reduce the incidence of esophageal lesions.[58]

8.8 Advanced Fluoroscopic Imaging Solutions

While simple X-ray fluoroscopy allows visualization of the cardiac silhouette, ablation, and mapping catheters, the operator has to have a high level of experience to understand the anatomy of the various cardiac chambers, as they are not directly visualized. This has led to different imaging modalities being used to help the operator. Pre-procedural CT of the left atrium and pulmonary veins results in a high-quality image that can be used at the time of the ablation,[60] either by overlaying the 3-dimensional shell on the live fluoroscopy screen or by integrating the shell into an electroanatomic mapping system (CARTO Merge or NavX Fusion). Although this technique results in a high-quality 3D representation of the left atrium, and other cardiac chambers, registration of the 3D shell is sometimes difficult due to changes that have occurred between the scan taking place and the ablation procedure. For example, the patient's rhythm may have changed from sinus to AF, the fluid status of the patient is likely to be different, and perhaps most importantly, the patient's position on the operating table is different to when having a CT scan.

To overcome these difficulties, and to reduce the radiation dose to the patient, rotational angiography (3D ATG) has been developed (Fig. 8.8).[2-4,37,63] This utilizes the fluoroscopy system that is used in the EP lab routinely. In this system after the left atrium has been isocentered, contrast is injected into the heart, either from the inferior vena cava/right atrial junction, pulmonary vein, or directly into the left atrium, and the C-arm then rotates around the patient in a 240° arc over 4 s. This creates a data set that can be read by the EP system as a CT scan. Automatic segmentation of the left atrium is performed, and this is checked and manually and corrected as necessary. Following this the 3D shell is automatically overlaid on the real-time fluoroscopy. A benefit of this system

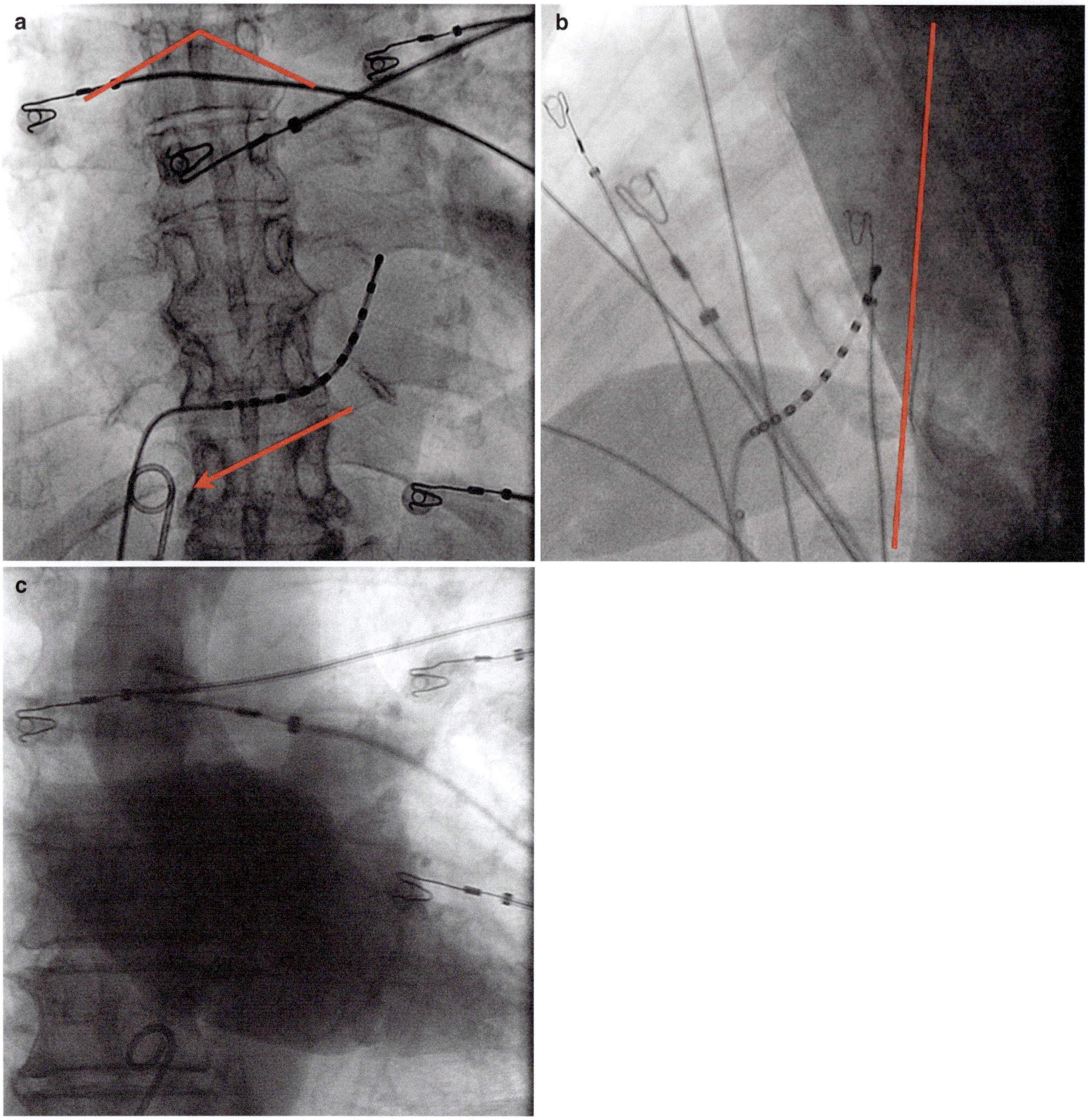

Fig. 8.8 Angiorotation. (**a**) The left atrium is isocentered. In the AP plane, the carina is centered (*see red line*) and the coronary sinus catheter can be used as the inferior border of the left atrium. The pigtail catheter is at the junction of the right atrium and the IVC the left atrium (*arrow*). (**b**) In the left lateral view, the posterior margin of the heart can be identified with the coronary sinus catheter, with the ablation catheter anterior on the His and the transseptal apparatus in the fossa ovalis. The anterior margin of the spine is kept in view (*line*). (**c**) A still from the angiorotation. All catheters have been removed to reduce artifact. (**d**) Segmentation. Using the EP navigator, the segmentation of the left atrium is checked, and corrected as necessary. (**e**) The left atrial shell is overlaid on real-time fluoroscopy. In this example, the cutting plane is used to demonstrate the posterior wall of the left atrium. The guidewire has been placed in the left superior pulmonary vein. (**f**) Registration can be checked by assessing catheter positions, or by pulmonary venography in orthogonal planes. As the image has been taken just prior to ablation, with no change in fluid balance or attitudinal changes (compared to CT angiography), registration errors are minimal, and manual adjustment is only necessary if the patient moves. (**g**) In this LAO view, the ostium of the right inferior pulmonary vein is clearly seen. As the C-arm rotates, as does the overlaid anatomy. (**h**) Ablation lesions are marked on the left atrial shell. (**i**) The anterior surface of the left atrium is easily visualized. (**j**) Ablation directed to the inferior left atrium targeting complex electrograms can be marked in different colors

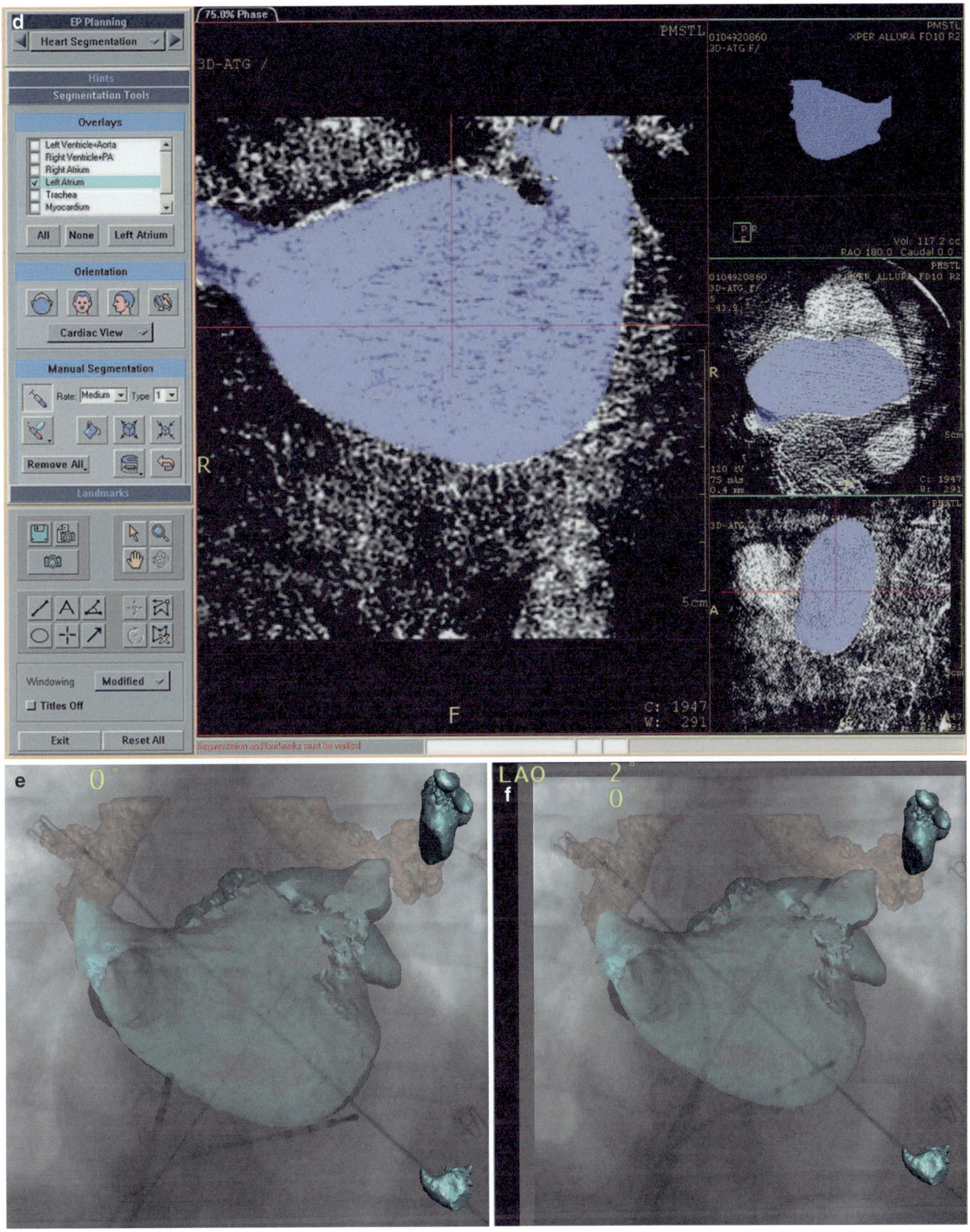

Fig. 8.8 (continued)

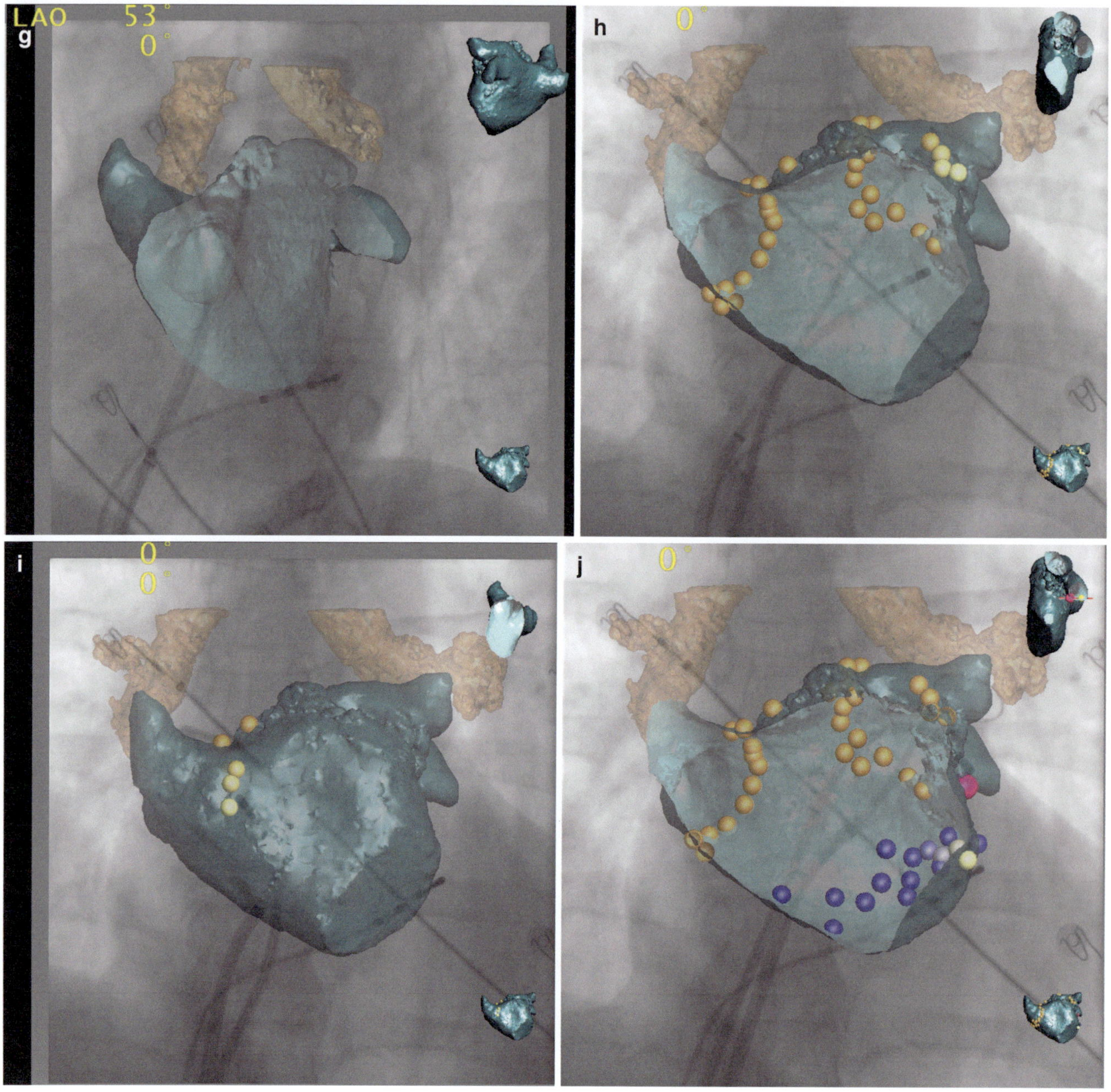

Fig. 8.8 (continued)

is that registration is not required, as long as the patient has not moved, given the scan has been performed using the same equipment. Registration accuracy can be rapidly checked, normally by placing a catheter within the superior pulmonary veins and looking for the drop off between the pulmonary vein ostium and the left atrial body. Following this step the ablation can then be performed with the operator being able to see exactly where critical structures, such as the ostia of the pulmonary veins, the ridge between the left atrial appendage and the left superior vein, and any anomalous pulmonary veins, are (see movie in Chap. 8 folder on Springer Extras). The direct visualization of these structures obviously allows for accurate catheter placement, and helps avoid potentially dangerous areas, such as the posterior wall with the underlying esophagus (which can be visualized and segmented with the 3D rotation), and helps to accurately determine the level of isolation of the pulmonary veins (antral or ostial).

A recent two center study comparing CARTO to 3D ATG with over 90 patients recruited has demonstrated equivalent results using both systems in terms of acute and short-term procedural success, with equivalent fluoroscopy times whether using the fluoroscopy-based system or the CARTO system.[64] Although these images, whether CT, 3D ATG or electroanatomic systems, intrinsically make the operator feel more comfortable when ablating in the left atrium, whether this translates into clinical benefit is less certain. In a retrospective study utilizing CARTO, Merge suggested that there was a clinical benefit to using the system.[65] However, in a later study by the same group no clinical benefit was seen, leading to the conclusion that a successful procedure is guided by electrical isolation of pulmonary veins rather than the technology used to achieve it.[66] A more recent randomized study, in which 290 patients were recruited and randomized between ablation using CARTO Merge and conventional ablation, demonstrated a benefit in outcomes in favor of using image integration.[67]

While this could be seen as disappointing in terms of the benefit of imaging systems, there are two important caveats. Firstly, a learning curve was seen when using the CT overlay in the former study,[65] and secondly, these studies are typically performed by experienced operators. As was discussed earlier there is a large population of patients who simply do not have access to a trained electrophysiologist who is capable of performing these difficult procedures, and while technological advances are not a surrogate for training in the basic principles of cardiac electrophysiology, the accurate anatomical representation of the cardiac chambers should help less experienced operators. Additionally most complications occur when operators are learning a new technique, and as AF ablation will be increasingly performed throughout the world, these imaging techniques may help avoid unnecessary complications due to a lack of experience.

8.9 Integration with EP Recording Systems

Although rotational angiography has clear advantages over CT-based systems, one of the limitations is that it only provides anatomical information. Electrophysiology is the integration of discrete anatomical structures with discrete electrical properties.

We are currently working with BARD and Philips on a software prototype whereby the electrical data from the EP recording system is displayed in a color-coded format on the 3D overlay (Fig. 8.9). This potentially allows rapid recognition of the direction of the wavefront for regular atrial tachycardias, allows for color-coded entrainment mapping, as has shown to be of benefit in a recent study,[68] and easily demonstrates presence or absence of conduction block across linear lesions. Additionally the prototype software allows for dominant frequency mapping in atrial fibrillation, and wavelet analysis to try and localize areas that are critical to the AF process.

8.9.1 Epicardial Ablation

There is increasing interest in percutaneous epicardial ablation, predominantly for the treatment of ventricular tachycardia[69,70] but also for atrial arrhythmias.[71] The technique of subxiphoid percutaneous pericardial access for catheter mapping was first described by Sosa[72] in three patients with Chagas disease, and was a modification of the original technique of Marfan.[73] A variety of fluoroscopic views can be used to help guide the puncture, but we have found that the left lateral fluoroscopic view to be most helpful.[16] Briefly, with the patient lying flat on the table, a Tuohy needle is aimed superiorly and medially, entering the skin about 2 cm inferior to the xiphoid process. We do believe that by keeping the needle as superficial as possible, immediately beneath the sternum, we reduce the risk for abdominal organs injury. This is associated with an anterior stick of the pericardial space and differs from the original description that was posterior. Using the left lateral fluoroscopic view, when the needle is thought to be close to the pericardium, a small amount of contrast is injected (Fig. 8.10). This allows visualization of tenting of the pericardium when the needle is pushed gently forward before entering the pericardial space. Once the pericardial space has been entered, further contrast is injected which should distribute around the silhouette of the heart. A guidewire is then introduced that should track around the heart crossing both the left and right ventricles. Once satisfied that the needle or wire has not entered into the heart, a hemostatic sheath and dilator can be advanced over the wire.

8.10 Future Improvements

Although X-ray fluoroscopy remains the standard imaging modality in EP, there are a number of limitations that have already been highlighted. Future developments of X-ray fluoroscopy systems need to enable the operator to perform procedures in a more efficient manner while decreasing the total radiation dose that the patient receives. Rotational angiography has already been shown to give additional information to the operator, but this is still a technology in progress. Future developments will include respiratory motion compensation, and the ability to perform ECG-gated scans, and hopefully imaging of the ventricles. If multiple imaging modalities are combined, it may be that physiological data could be displayed on an anatomical shell overlaid on live fluoroscopy. For example visualization of myocardial scar from cardiac MR in patients with scar-related ventricular tachycardia. However, such technologies need to have proven benefit compared to currently used imaging modalities.

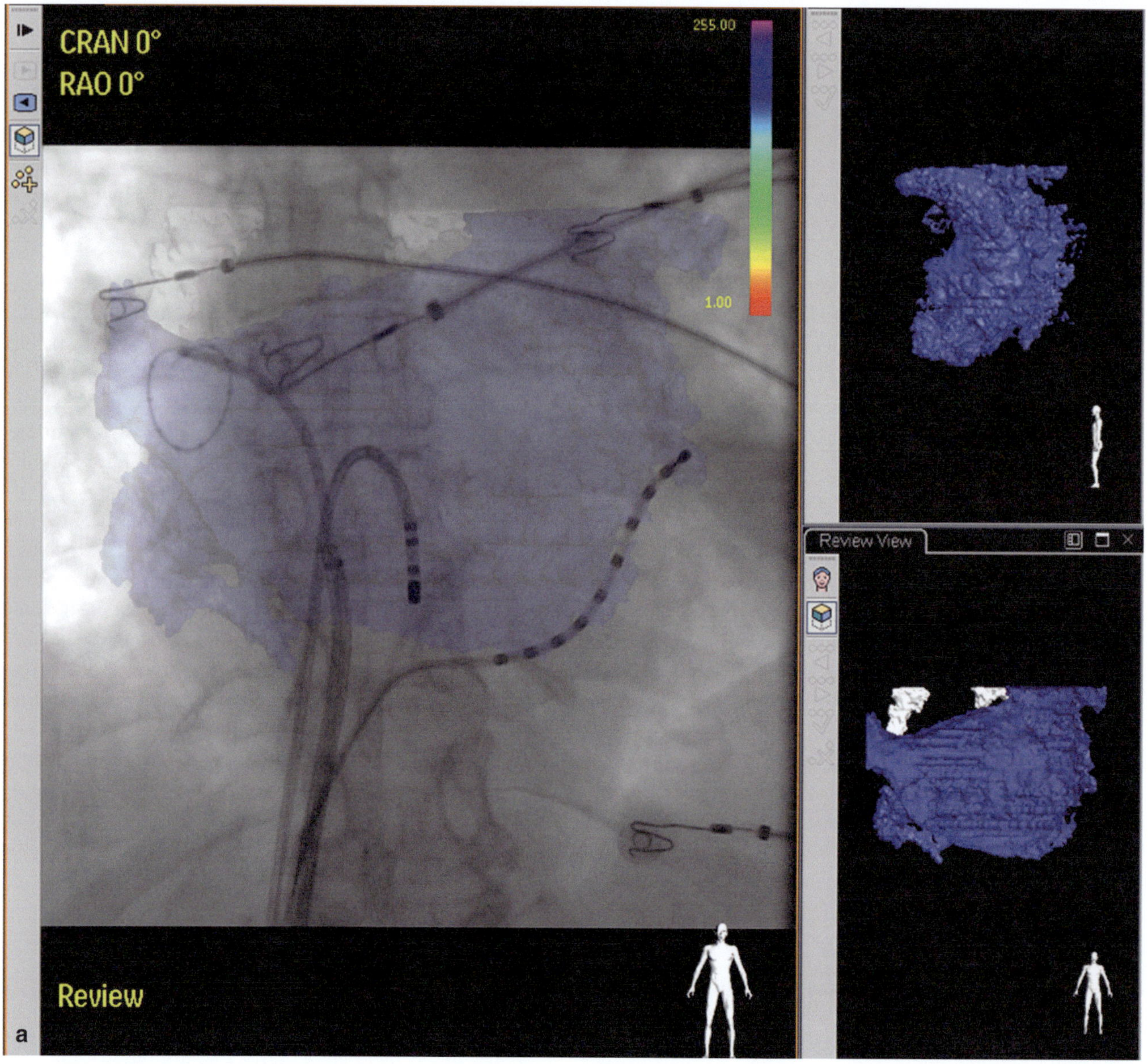

Fig. 8.9 ElectroNav. Integration of the angiorotation and intracardiac electrograms is being developed by Bard EP and Philips. (**a**) A point is taken on the overlaid anatomy, in this case the posterior left atrium. (**b**) When the point is taken, a "line of sight" is displayed, as the point can lie anywhere along this (i.e., anterior and posterior in the example). (**c**) Another view will confirm the position, here in LAO. (**d**) For illustrative purposes only, the same point in the RAO projection. (**e**) Once sufficient points have been taken, an understandable activation map is seen, here activation and entrainment mapping, and finally ablation confirmed a localized reentrant atrial tachycardia on the anterior left atrium, in a patient with previous ablations for persistent atrial fibrillation

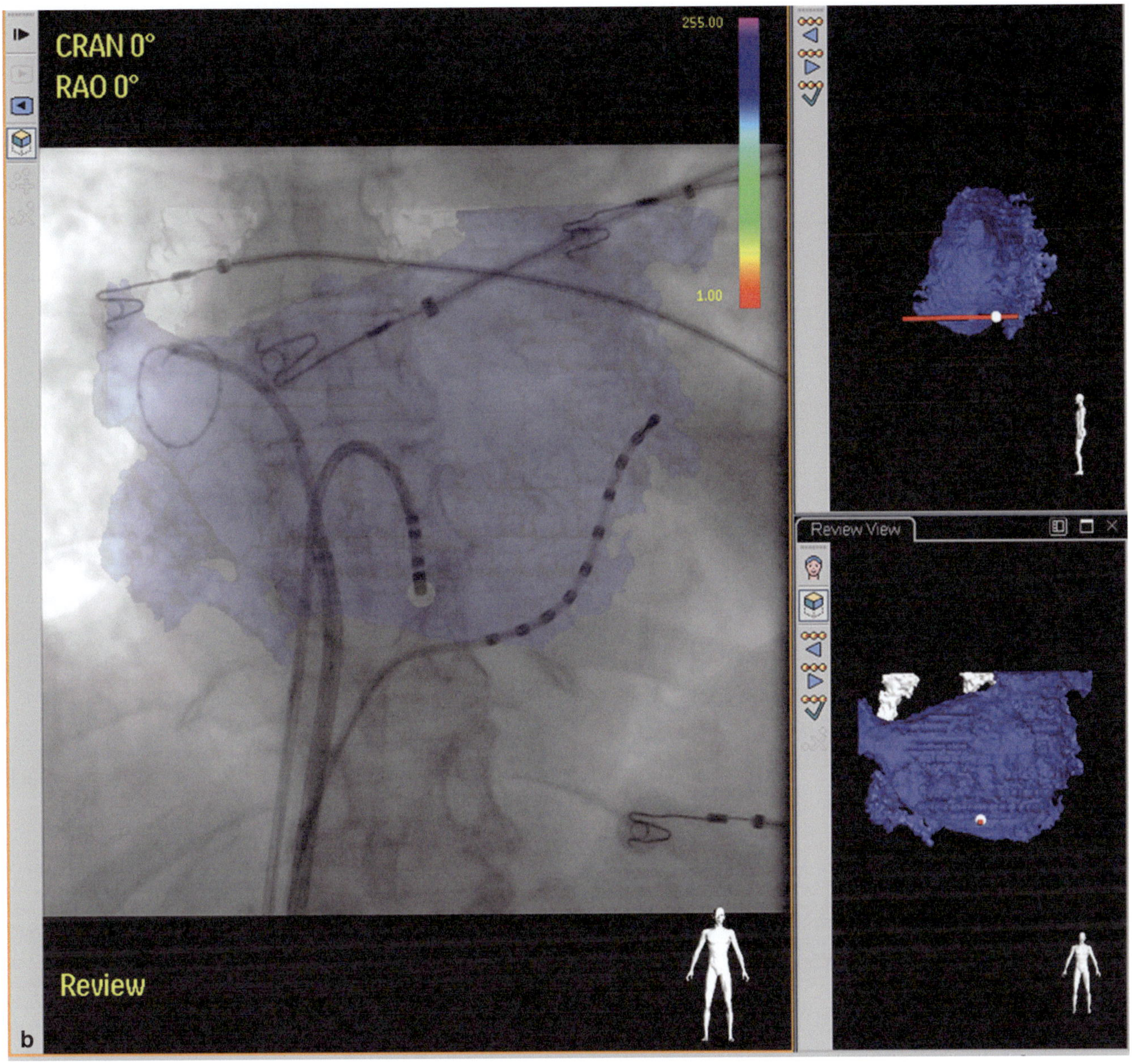

Fig. 8.9 (continued)

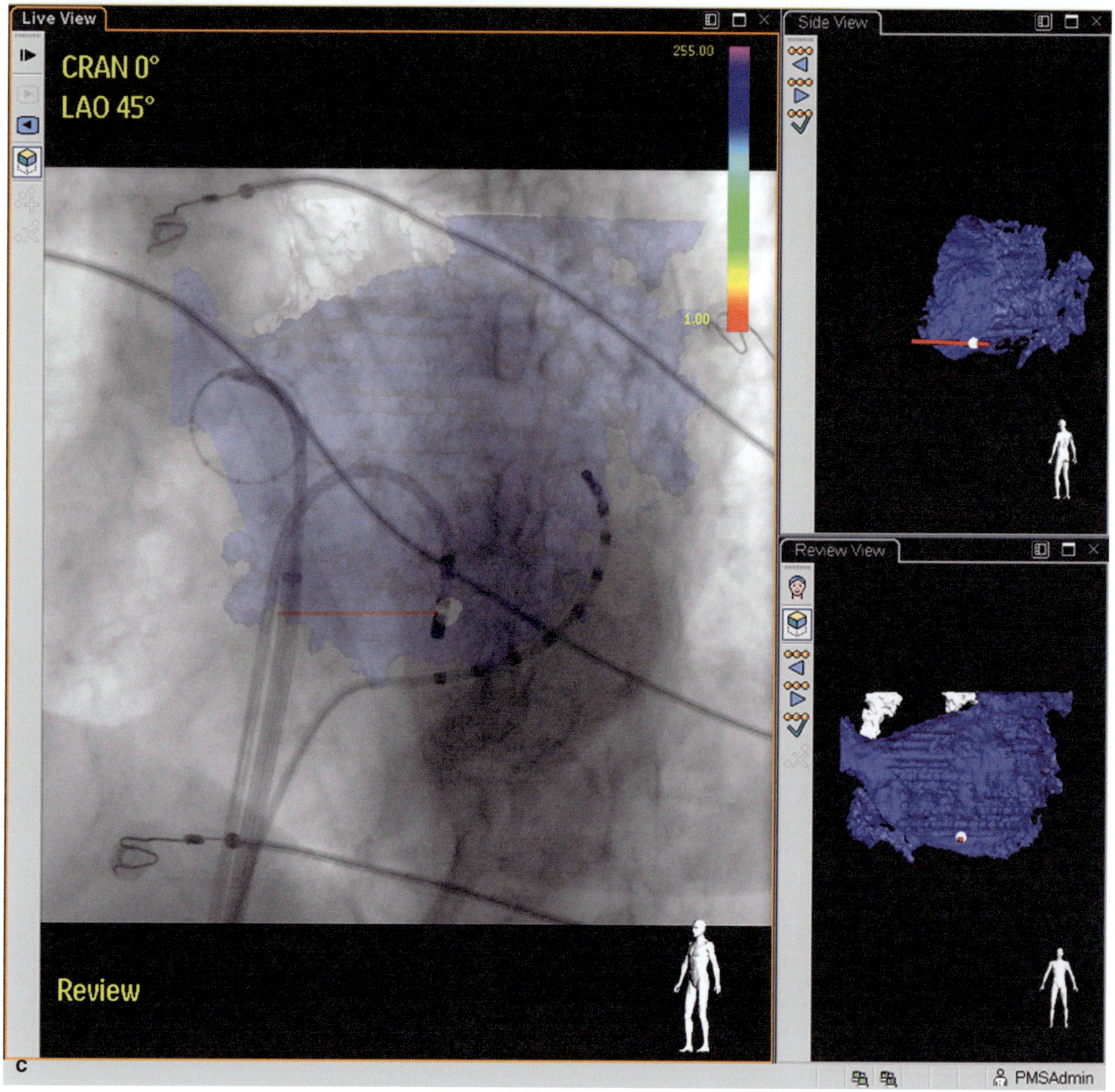

Fig. 8.9 (continued)

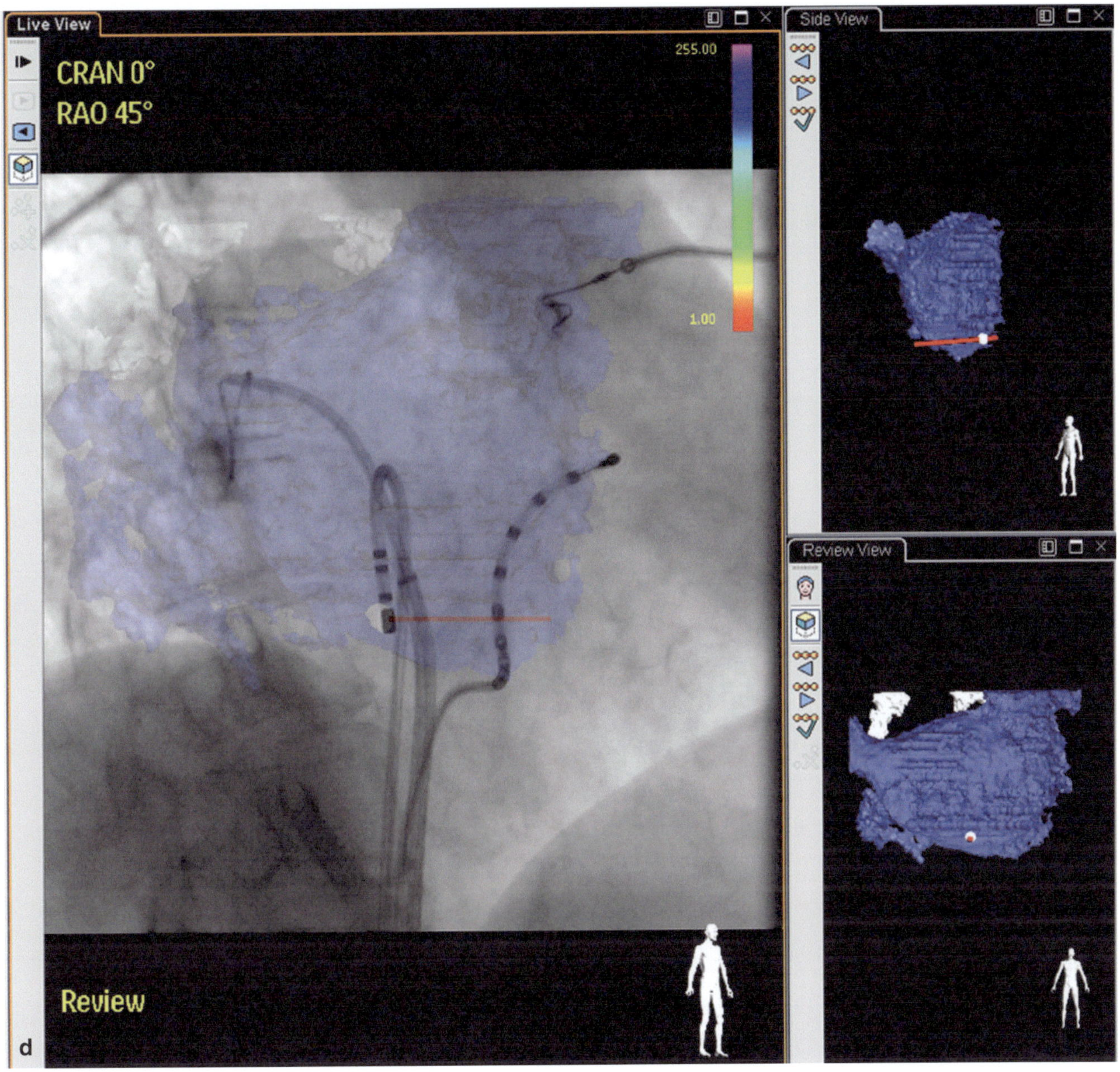

Fig. 8.9 (continued)

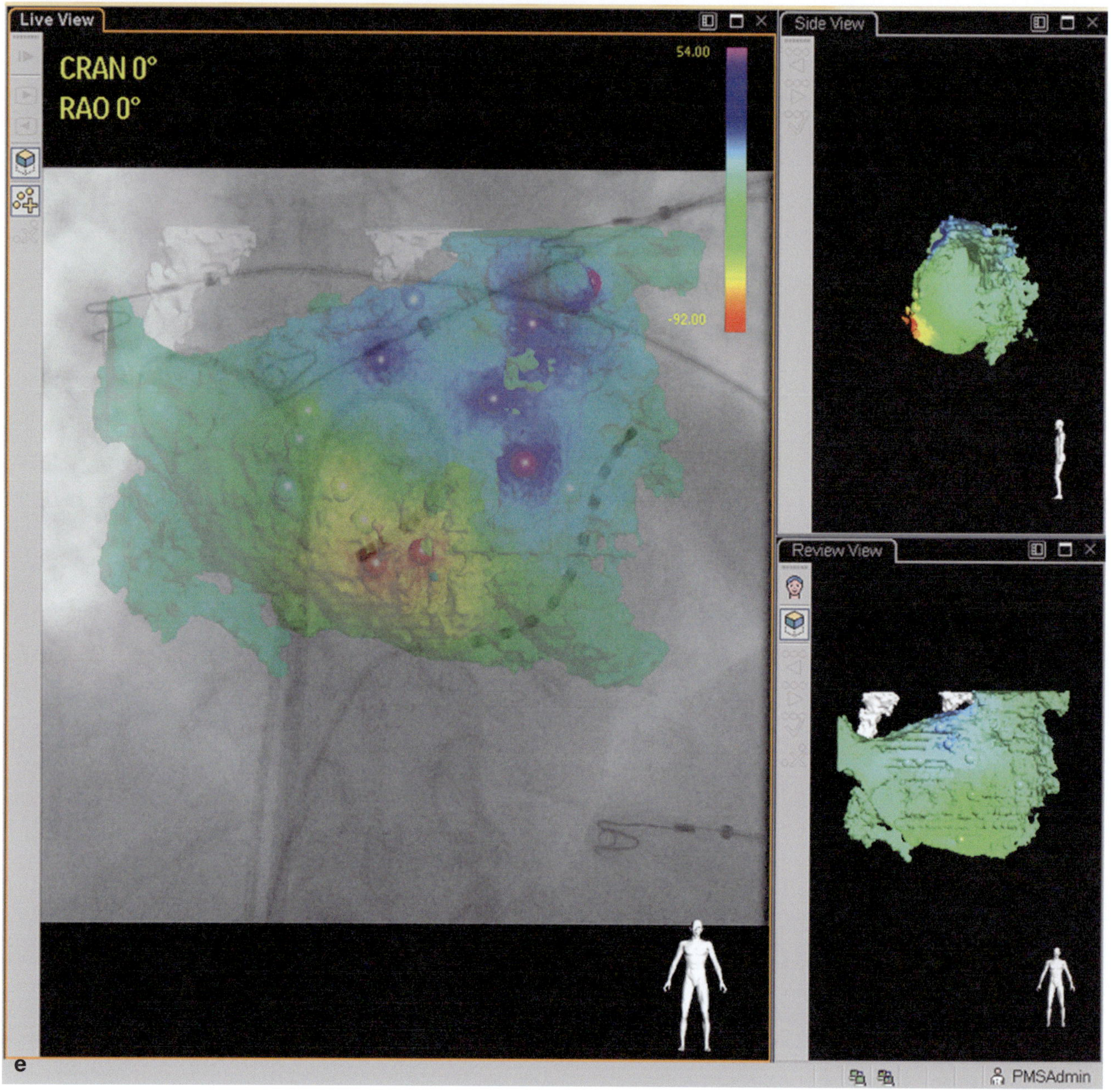

Fig. 8.9 (continued)

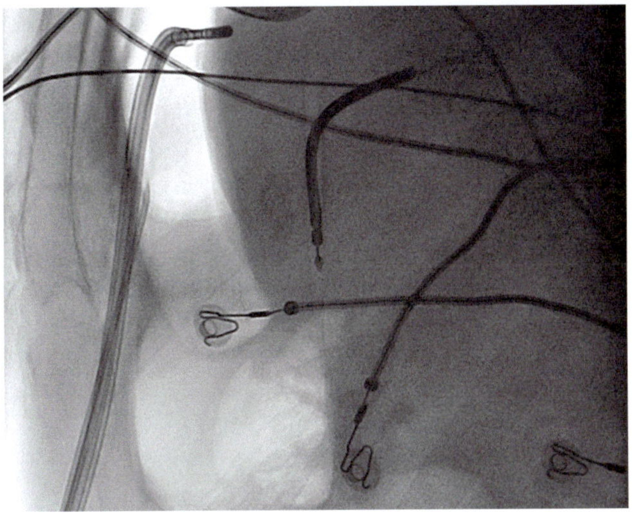

Fig. 8.10 Epicardial access. This panel shows a left lateral view for accessing the epicardium, an increasingly used method for ablation of VT. In this case, air has been introduced after the first pericardial puncture to facilitate a second pericardial puncture

Acknowledgments Matthew Wright acknowledges financial support from the Department of Health via the National Institute for Health Research (NIHR) comprehensive Biomedical Research Centre award to Guy's and St Thomas' NHS Foundation Trust in partnership with King's College London and King's College Hospital NHS Foundation Trust.

References

1. Daoud EG, Kalbfleisch SJ, Hummel JD. Intracardiac echocardiography to guide transseptal left heart catheterization for radiofrequency catheter ablation. *J Cardiovasc Electrophysiol*. 1999;10:358-363.
2. Orlov MV, Hoffmeister P, Chaudhry GM, et al. Three-dimensional rotational angiography of the left atrium and esophagus – a virtual computed tomography scan in the electrophysiology lab? *Heart Rhythm*. 2007;4:37-43.
3. Nolker G, Gutleben KJ, Marschang H, et al. Three-dimensional left atrial and esophagus reconstruction using cardiac C-arm computed tomography with image integration into fluoroscopic views for ablation of atrial fibrillation: accuracy of a novel modality in comparison with multislice computed tomography. *Heart Rhythm*. 2008; 5:1651-1657.
4. Kriatselis C, Tang M, Nedios S, et al. Intraprocedural reconstruction of the left atrium and pulmonary veins as a single navigation tool for ablation of atrial fibrillation: a feasibility, efficacy, and safety study. *Heart Rhythm*. 2009;6(6):733 741.
5. Saliba W, Cummings JE, Oh S, et al. Novel robotic catheter remote control system: feasibility and safety of transseptal puncture and endocardial catheter navigation. *J Cardiovasc Electrophysiol*. 2006;17:1102-1105.
6. Pappone C, Vicedomini G, Manguso F, et al. Robotic magnetic navigation for atrial fibrillation ablation. *J Am Coll Cardiol*. 2006; 47:1390-1400.
7. Limacher MC, Douglas PS, Germano G, et al. ACC expert consensus document. Radiation safety in the practice of cardiology. American College of Cardiology. *J Am Coll Cardiol*. 1998;31: 892-913.
8. Lickfett L, Mahesh M, Vasamreddy C, et al. Radiation exposure during catheter ablation of atrial fibrillation. *Circulation*. 2004; 110:3003-3010.
9. Ross AM, Segal J, Borenstein D, et al. Prevalence of spinal disc disease among interventional cardiologists. *Am J Cardiol*. 1997; 79:68-70.
10. Dragusin O, Weerasooriya R, Jais P, et al. Evaluation of a radiation protection cabin for invasive electrophysiological procedures. *Eur Heart J*. 2007;28:183-189.
11. Klein LW, Miller DL, Balter S, et al. Occupational health hazards in the interventional laboratory: time for a safer environment. *Heart Rhythm*. 2009;6:439-444.
12. Farre J, Anderson RH, Cabrera JA, et al. Fluoroscopic cardiac anatomy for catheter ablation of tachycardia. *Pacing Clin Electrophysiol*. 2002;25:76-94.
13. Fadl YY, Ellenbogen KA, Grubb RLJ, et al. A review of spinal injuries in the invasive cardiologist: part 1. Biomechanics and pain generation. *Pacing Clin Electrophysiol*. 2007;30:1012-1019.
14. Fadl YY, Ellenbogen KA, Grubb RLJ, et al. A review of spinal injuries in the invasive cardiologist II: prevention and treatment. *Pacing Clin Electrophysiol*. 2007;30:1149-1157.
15. Cosio FG, Anderson RH, Becker A, et al. Living anatomy of the atrioventricular junctions. A guide to electrophysiological mapping. A consensus statement from the Cardiac Nomenclature Study Group, Working Group of Arrhythmias, European Society of Cardiology, and the Task Force on Cardiac Nomenclature from NASPE. North American Society of Pacing and Electrophysiology. *Eur Heart J*. 1999;20:1068-1075.
16. Weerasooriya R, Jais P, Sacher F, et al. Utility of the lateral fluoroscopic view for subxiphoid pericardial access. *Circ Arrhythm Electrophysiol*. 2009;2:e15-e17.
17. Earley MJ. How to perform a transseptal puncture. *Heart*. 2009;95:85-92.
18. Ross JJ, Braunwald E, Morrow AG. Transseptal left heart catheterization: a new diagnostic method. *Prog Cardiovasc Dis*. 1960;2: 315-318.
19. Brockenbrough EC, Braunwald E, Ross JJ. Transseptal left heart catheterization. A review of 450 studies and description of an improved technique. *Circulation*. 1962;25:15-21.
20. Ho SY, Sanchez-Quintana D, Cabrera JA, et al. Anatomy of the left atrium: implications for radiofrequency ablation of atrial fibrillation. *J Cardiovasc Electrophysiol*. 1999;10:1525-1533.
21. Anderson RH, Webb S, Brown NA. Clinical anatomy of the atrial septum with reference to its developmental components. *Clin Anat*. 1999;12:362-374.
22. Knecht S, Wright M, Lellouche N, et al. Impact of a patent foramen ovale on paroxysmal atrial fibrillation ablation. *J Cardiovasc Electrophysiol*. 2008;19(12):1236-1241.
23. Chierchia GB, Capulzini L, de Asmundis C, et al. First experience with real-time three-dimensional transesophageal echocardiography-guided transseptal in patients undergoing atrial fibrillation ablation. *Europace*. 2008;10:1325-1328.
24. Chu E, Kalman JM, Kwasman MA, et al. Intracardiac echocardiography during radiofrequency catheter ablation of cardiac arrhythmias in humans. *J Am Coll Cardiol*. 1994;24:1351-1357.
25. Epstein LM, Smith T, TenHoff H. Nonfluoroscopic transseptal catheterization: safety and efficacy of intracardiac echocardiographic guidance. *J Cardiovasc Electrophysiol*. 1998;9:625-630.
26. Helms A, West JJ, Patel A, et al. Real-time rotational ICE imaging of the relationship of the ablation catheter tip and the esophagus during atrial fibrillation ablation. *J Cardiovasc Electrophysiol*. 2009;20:130-137.
27. Szili-Torok T, Kimman G, Theuns D, et al. Transseptal left heart catheterisation guided by intracardiac echocardiography. *Heart*. 2001;86:E11.

28. Knecht S, Jais P, Nault I, et al. Radiofrequency puncture of the fossa ovalis for resistant transseptal access. *Circ Arrhythm Electrophysiol.* 2008;1:169-174.
29. Marom EM, Herndon JE, Kim YH, et al. Variations in pulmonary venous drainage to the left atrium: implications for radiofrequency ablation. *Radiology.* 2004;230:824-829.
30. Lin WS, Prakash VS, Tai CT, et al. Pulmonary vein morphology in patients with paroxysmal atrial fibrillation initiated by ectopic beats originating from the pulmonary veins: implications for catheter ablation. *Circulation.* 2000;101:1274-1281.
31. Ho SY, Cabrera JA, Tran VH, et al. Architecture of the pulmonary veins: relevance to radiofrequency ablation. *Heart.* 2001;86:265-270.
32. Wittkampf FH, Vonken EJ, Derksen R, et al. Pulmonary vein ostium geometry: analysis by magnetic resonance angiography. *Circulation.* 2003;107:21-23.
33. Hassink RJ, Aretz HT, Ruskin J, et al. Morphology of atrial myocardium in human pulmonary veins: a postmortem analysis in patients with and without atrial fibrillation. *J Am Coll Cardiol.* 2003;42:1108-1114.
34. Jongbloed MR, Dirksen MS, Bax JJ, et al. Atrial fibrillation: multidetector row CT of pulmonary vein anatomy prior to radiofrequency catheter ablation–initial experience. *Radiology.* 2005;234:702-709.
35. Meyer C, Manzke R, Peters J, et al. Automatic intra-operative generation of geometric left atrium/pulmonary vein models from rotational X-ray angiography. *Med Image Comput Comput Assist Interv.* 2008;11:61-69.
36. Thiagalingam A, Manzke R, D'Avila A, et al. Intraprocedural volume imaging of the left atrium and pulmonary veins with rotational X-ray angiography: implications for catheter ablation of atrial fibrillation. *J Cardiovasc Electrophysiol.* 2008;19:293-300.
37. Kriatselis C, Tang M, Roser M, et al. A new approach for contrast-enhanced X-ray imaging of the left atrium and pulmonary veins for atrial fibrillation ablation: rotational angiography during adenosine-induced asystole. *Europace.* 2009;11:35-41.
38. Monteiro MM, Saraiva C, Castelo Branco J, et al. Characterization of pulmonary vein morphology using multi-detector row CT study prior to radiofrequency ablation for atrial fibrillation. *Rev Port Cardiol.* 2009;28:545-559.
39. Jongbloed MR, Bax JJ, Lamb HJ, et al. Multislice computed tomography versus intracardiac echocardiography to evaluate the pulmonary veins before radiofrequency catheter ablation of atrial fibrillation: a head-to-head comparison. *J Am Coll Cardiol.* 2005;45:343-350.
40. Tsao HM, Wu MH, Yu WC, et al. Role of right middle pulmonary vein in patients with paroxysmal atrial fibrillation. *J Cardiovasc Electrophysiol.* 2001;12:1353-1357.
41. Yazar F, Ozdogmus O, Tuccar E, et al. Drainage patterns of middle lobe vein of right lung: an anatomical study. *Eur J Cardiothorac Surg.* 2002;22:717-720.
42. Ghaye B, Szapiro D, Dacher JN, et al. Percutaneous ablation for atrial fibrillation: the role of cross-sectional imaging. *Radiographics.* 2003;23:S19-S33. discussion S48-50.
43. Lickfett L, Kato R, Tandri H, et al. Characterization of a new pulmonary vein variant using magnetic resonance angiography: incidence, imaging, and interventional implications of the "right top pulmonary vein". *J Cardiovasc Electrophysiol.* 2004;15:538-543.
44. Dill T, Neumann T, Ekinci O, et al. Pulmonary vein diameter reduction after radiofrequency catheter ablation for paroxysmal atrial fibrillation evaluated by contrast-enhanced three-dimensional magnetic resonance imaging. *Circulation.* 2003;107:845-850.
45. Mainigi SK, Sauer WH, Cooper JM, et al. Incidence and predictors of very late recurrence of atrial fibrillation after ablation. *J Cardiovasc Electrophysiol.* 2007;18:69-74.
46. Hocini M, Sanders P, Jais P, et al. Techniques for curative treatment of atrial fibrillation. *J Cardiovasc Electrophysiol.* 2004;15:1467-1471.
47. Sanders P, Hocini M, Jais P, et al. Pulmonary vein isolation for atrial fibrillation. In: Huang SK, Wood MA, eds. *Catheter ablation of cardiac arrhythmias.* Philadelphia: Saunders Elsevier; 2006:269-287.
48. Oral H, Chugh A, Good E, et al. A tailored approach to catheter ablation of paroxysmal atrial fibrillation. *Circulation.* 2006;113:1824-1831.
49. Knecht S, Hocini M, Wright M, et al. Left atrial linear lesions are required for successful treatment of persistent atrial fibrillation. *Eur Heart J.* 2008;29(19):2359-2366.
50. Nanthakumar K, Kay GN, Plumb VJ, et al. Decrease in fluoroscopic cardiac silhouette excursion precedes hemodynamic compromise in intraprocedural tamponade. *Heart Rhythm.* 2005;2:1224-1230.
51. Sacher F, Monahan KH, Thomas SP, et al. Phrenic nerve injury after atrial fibrillation catheter ablation: characterization and outcome in a multicenter study. *J Am Coll Cardiol.* 2006;47:2498-2503.
52. Sanchez-Quintana D, Cabrera JA, Climent V, et al. How close are the phrenic nerves to cardiac structures? Implications for cardiac interventionalists. *J Cardiovasc Electrophysiol.* 2005;16:309-313.
53. Matsuo S, Jais P, Knecht S, et al. Images in cardiovascular medicine. Novel technique to prevent left phrenic nerve injury during epicardial catheter ablation. *Circulation.* 2008;117:e471.
54. Cappato R, Calkins H, Chen SA, et al. Prevalence and causes of fatal outcome in catheter ablation of atrial fibrillation. *J Am Coll Cardiol.* 2009;53:1798-1803.
55. Sonmez B, Demirsoy E, Yagan N, et al. A fatal complication due to radiofrequency ablation for atrial fibrillation: atrio-esophageal fistula. *Ann Thorac Surg.* 2003;76:281-283.
56. Scanavacca MI, D'avila A, Parga J, et al. Left atrial-esophageal fistula following radiofrequency catheter ablation of atrial fibrillation. *J Cardiovasc Electrophysiol.* 2004;15:960-962.
57. Pappone C, Oral H, Santinelli V, et al. Atrio-esophageal fistula as a complication of percutaneous transcatheter ablation of atrial fibrillation. *Circulation.* 2004;109:2724-2726.
58. Cummings JE, Schweikert RA, Saliba WI, et al. Assessment of temperature, proximity, and course of the esophagus during radiofrequency ablation within the left atrium. *Circulation.* 2005;112:459-464.
59. Wang SL, Ooi CG, Siu CW, et al. Endocardial visualization of esophageal-left atrial anatomic relationship by three-dimensional multidetector computed tomography "navigator imaging". *Pacing Clin Electrophysiol.* 2006;29:502-508.
60. Knecht S, Skali H, O'Neill MD, et al. Computed tomography-fluoroscopy overlay evaluation during catheter ablation of left atrial arrhythmia. *Europace.* 2008;10:931-938.
61. Martinek M, Bencsik G, Aichinger J, et al. Esophageal damage during radiofrequency ablation of atrial fibrillation: impact of energy settings, lesion sets, and esophageal visualization. *J Cardiovasc Electrophysiol.* 2009;20:726-733.
62. Di Biase L, Saenz LC, Burkhardt DJ, et al. Esophageal capsule endoscopy after radiofrequency catheter ablation for atrial fibrillation: documented higher risk of luminal esophageal damage with general anesthesia as compared with conscious sedation. *Circ Arrhythm Electrophysiol.* 2009;2:108-112.
63. Gerds-Li JH, Tang M, Kriatselis C, et al. Rapid ventricular pacing to optimize rotational angiography in atrial fibrillation ablation. *J Interv Card Electrophysiol.* 2009;26(2):101-107.
64. Knecht S, Akrivakis ST, Wright M, et al. Randomized prospective evaluation of 3D rotational atriography versus CARTO to guide AF ablation. *Heart Rhythm.* 2009;6:S396-S397.
65. Kistler PM, Rajappan K, Jahngir M, et al. The impact of CT image integration into an electroanatomic mapping system on clinical outcomes of catheter ablation of atrial fibrillation. *J Cardiovasc Electrophysiol.* 2006;17:1093-1101.

66. Kistler PM, Rajappan K, Harris S, et al. The impact of image integration on catheter ablation of atrial fibrillation using electroanatomic mapping: a prospective randomized study. *Eur Heart J*. 2008;29: 3029-3036.
67. Della Bella P, Fassini G, Cireddu M, et al. Image integration-guided catheter ablation of atrial fibrillation: a prospective randomized study. *J Cardiovasc Electrophysiol*. 2009;20:258-265.
68. Esato M, Hindricks G, Sommer P, et al. Color-coded three-dimensional entrainment mapping for analysis and treatment of atrial macroreentrant tachycardia. *Heart Rhythm*. 2009;6:349-358.
69. Tedrow U, Stevenson WG. Strategies for epicardial mapping and ablation of ventricular tachycardia. *J Cardiovasc Electrophysiol*. 2009;20:710-713.
70. Soejima K, Stevenson WG, Sapp JL, et al. Endocardial and epicardial radiofrequency ablation of ventricular tachycardia associated with dilated cardiomyopathy: the importance of low-voltage scars. *J Am Coll Cardiol*. 2004;43:1834-1842.
71. Phillips KP, Natale A, Sterba R, et al. Percutaneous pericardial instrumentation for catheter ablation of focal atrial tachycardias arising from the left atrial appendage. *J Cardiovasc Electrophysiol*. 2008;19:430-433.
72. Sosa E, Scanavacca M, D'Avila A, et al. A new technique to perform epicardial mapping in the electrophysiology laboratory. *J Cardiovasc Electrophysiol*. 1996;7:531-536.
73. Marfan AB. Ponction du pericarde par l'epigastre. *Ann Med Chir Inf*. 1911;15:529-531.
74. Jais P, Hocini M, Hsu LF et al. Technique and results of linear ablation at the mitral isthmus. *Circulation* 2004; 110:2996–3002.
75. Hocini M, Jais P, Sanders P et al. Techniques, evaluation, and consequences of linear block at the left atrial roof in paroxysmal atrial fibrillation: a prospective randomized study. *Circulation* 2005; 112:3688–3696.

Intraprocedural Techniques: Transesophageal (2D/3D) and Intracardiac Echocardiography

9

Josef Kautzner and Petr Peichl

Abstract

Expansion of indications to catheter ablation has led to a change of the paradigm from electrophysiologically guided procedure to a procedure guided predominantly anatomically. Accurate online identification of cardiac structures and catheters becomes mandatory and ICE is at present the only imaging modality that can accomplish easily this task. It is a valuable tool for guiding procedures such as transseptal puncture or for monitoring complex ablation procedures like ablation of atrial fibrillation or ventricular tachycardias. It also allows prevention and/or early detection of intraprocedural complications. The widespread use of ICE is likely to result in reduction of fluoroscopy time and to maximize safety and efficacy of complex ablation procedures.

Keywords

Transesophageal echocardiography • Intracardiac echocardiography– 3D • echocardiography • Catheter ablation

Progress in the field of catheter ablation over the past decade has resulted in a paradigm shift in interventional electrophysiology. While in conventional ablation procedures (e.g., ablation of accessory pathways or AV nodal reentry) the definition of the target site is determined primarily by electrograms and fluoroscopy is sufficient for spatial navigation of ablation catheters, the situation is different in catheter ablation of complex arrhythmias (e.g., atrial fibrillation, arrhythmias after previous correction for congenital heart disease, ventricular tachycardias in structural heart disease, etc.). Such cases require modified approach to identify the ablation targets. These are not anymore defined by electrophysiological recordings only (like zone of slow conduction or earliest activation) but by a complex interplay between electrograms and anatomical structures. Moreover, variable individual anatomy emphasizes the need for novel imaging techniques. So far, echocardiography is the only available clinical tool that allows direct online visualization of both cardiac anatomy and catheters during the electrophysiology procedures. This chapter will review clinical utility of transesophageal (TEE) and intracardiac (ICE) echocardiography for intraprocedural navigation during complex ablation procedures.

J. Kautzner (✉)
Department of Cardiology, Institute for Clinical and Experimental Medicine, Prague, Czech Republic
e-mail: josef.kautzner@medicon.cz

9.1 Echocardiographic Imaging Tools

Some benefits of intraprocedural echocardiography for electrophysiologic interventions was shown nearly two decades ago. The first reports used TEE for navigation during ablation of ventricular tachycardias[1] and/or accessory pathways.[2] However, the widespread use of TEE for lengthy electrophysiologic procedures remains limited by the necessity for deep sedation or general anesthesia.

Miniaturization of the ultrasound technology has allowed development of truly intracardiac catheters that can be introduced into cardiac chambers.[3,4] ICE has significant advantage against TEE since it does not require prolonged esophageal intubation and is not associated with the risk of

A. Auricchio et al. (eds.), *Cardiac Imaging in Electrophysiology*,
DOI 10.1007/978-1-84882-486-7_9, © Springer-Verlag London Limited 2012

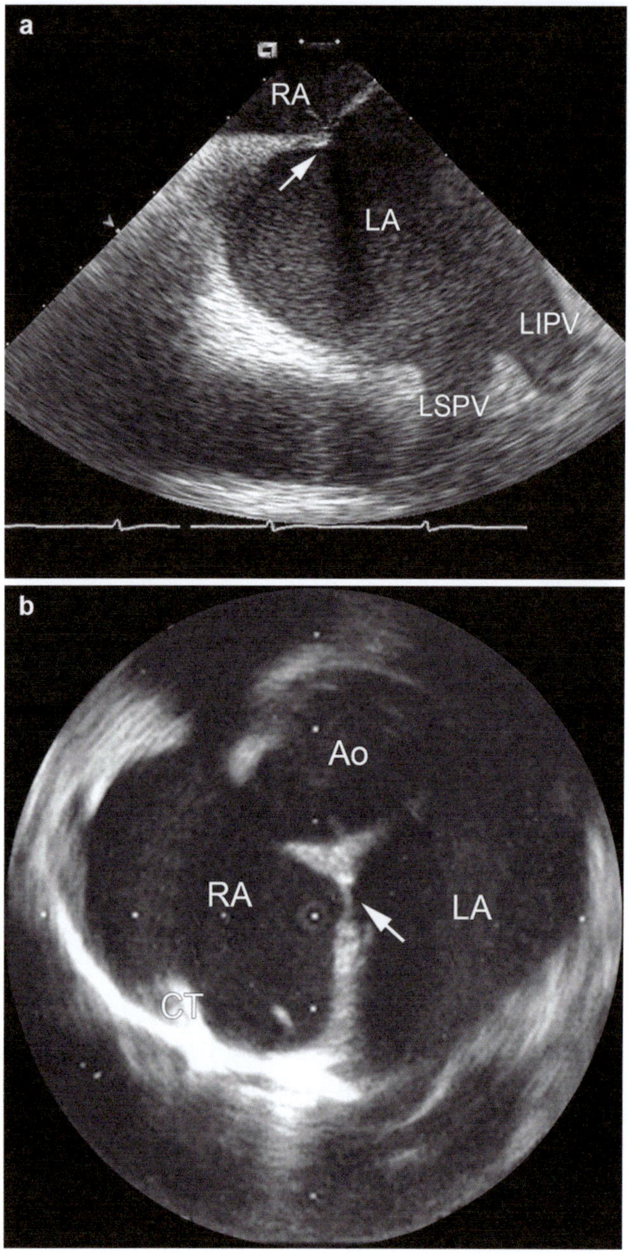

Fig. 9.1 Two types of intracardiac ultrasound systems. (**a**) Image of interatrial septum obtained by the phase array ultrasound probe (AcuNav, Acuson). (**b**) Image in a 360° radial plane perpendicular to the axis of the ultrasound catheter (Clearview, Cardiovascular Imaging Systems) positioned in the right atrium. The *arrow* marks the fossa ovalis. *Ao* aorta, *CT* crista terminalis, *LA* left atrium, *LIPV* left inferior pulmonary vein, *LSPV* left superior pulmonary vein, *RA* right atrium

esophageal damage or discomfort for the patient. In addition, it could be routinely performed by the electrophysiologist alone without the need for other personnel for image acquisition and interpretation.

At present, two different ICE technologies are available (Fig. 9.1). One is mechanical ultrasound transducer-tipped catheter (Clearview, Cardiovascular Imaging Systems, Inc, Freemont, California, USA) connected to the ultrasound machine (Boston Scientific Corporation, San Jose, California, USA). For intracardiac use, the catheter uses a 9 MHz single element transducer mounted at the tip of an 8F catheter. A piezoelectric crystal rotates at 1,800 rpm providing cross-sectional images in a 360° radial plane with depth of visualized field up to 5 cm. The other technology uses phased-array ultrasound tipped catheter (AcuNav Diagnostic Ultrasound Catheter, Acuson – Siemens, Mountain View, California, USA or ViewFlex Plus, St. Jude Medical, St. Paul Minnesota, USA). High-resolution, multiple frequency transducer (4.5–10 MHz) is incorporated in an 8F or 10F steerable catheter (2–4 directions) and provides 90° sector images with depth of visualization reaching up to 21 cm. The phased-array catheters allow for the whole spectrum of Doppler imaging capabilities. Sector imaging, similar to TEE, flexibility in changing frequency and full Doppler applications make the use of this system preferable. The following text will focus mainly on the utility of phased-array ICE.

9.2 Intracardiac Imaging Technique by ICE

Understanding the relation between the ultrasound beam fan of the ICE catheter (90°) and surrounding cardiac structures is critical for proper catheter manipulation. The phased-array ICE catheter resembles the longitudinal transesophageal probe and 2D sector scanning shows a cross-sectional view oriented from the tip to the shaft of the catheter (Fig. 9.2; see Chap. 9 movies on Springer Extras). The left/right orientation marker on the image indicates position of the catheter's shaft. When it is set to the left of the image and the catheter is introduced via the inferior vena cava to the right atrium, the cranio-caudal axis projects from the right to the left (cranial is on the right side of the image fan). It is at operator's discretion where to put the marker. We prefer to set it to the left and all accompanied images were obtained in this fashion.

The operator may adjust orientation of imaging plane by a simple catheter advancement or withdrawal, by gentle catheter rotation in a clockwise or counterclockwise direction, or by tip deflection in two to four directions. For beginners, it is useful to review CT images of the heart in a transversal plane to understand which structures can be imaged from the right atrium and/or right ventricle. Positioning the ICE catheter in the right atrium, just above the orifice of the inferior vena cava with leftward and anterior orientation of the transducer face, provides the so-called home-position or the reference position of the catheter (Fig. 9.2a). To obtain this position requires no adjustment in tip orientation of the catheter. The resulting image shows adjacent right atrium with the cavotricuspid isthmus, the tricuspid valve, and the right ventricle. Counterclockwise

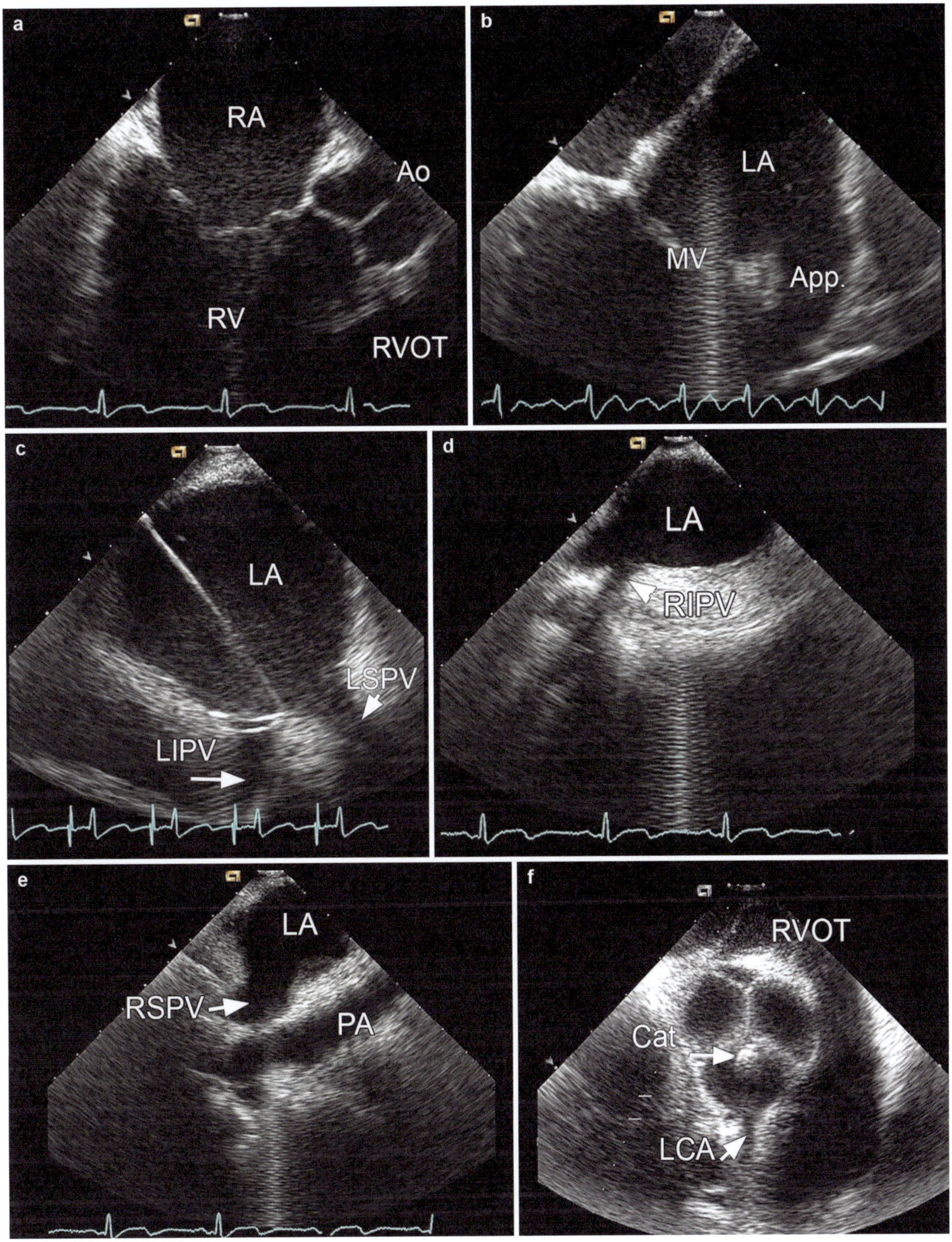

Fig. 9.2 (**a**) Cardiac chambers and structures as depicted by ICE: (**a**) "Home position" obtained by introduction of ICE probe into the right atrium (*RA*) that shows the tricuspid annulus, the right ventricle (*RV*), aorta (*Ao*) and right ventricular outflow tract(*RVOT*), (**b**) the left atrium (*LA*), the left appendage (*App*) and the mitral annulus (*MV*), (**c**) the left superior pulmonary vein (*LSPV*) and inferior (*LIPV*) with lasso catheter in the ostium, (**d**) the right inferior pulmonary vein (*RIPV*), (**e**) the right superior pulmonary vein (*RSPV*) and pulmonary artery (*PA*), (**f**) the aorta in short axis (*Ao*) with the mapping catheter (*Cat*), the ostium of left coronary artery and (*LCA*) and the right ventricular outflow tract (*RVOT*)

rotation opens the view of the lateral wall of the right atrium and provides images of the crista terminalis and the right appendage. Gentle clockwise rotation from the "home-position" allows for imaging the aortic root in the long axis and the right ventricle including its outflow tract. More clockwise rotation shows the mitral valve and the left appendage (Fig. 9.2b). Continuing clockwise rotation opens the view of the fossa ovalis (proximally to the probe) and the left pulmonary veins (on the opposite side of the image fan) with the inferior vein oriented to the left of the image and the superior on the opposite side (Fig. 9.2c). Further rotation shows typically the left atrial cavity with an esophagus coursing behind the posterior wall. Additional clockwise rotation brings more or less cross-sectional view of the right pulmonary veins (depending on the position of the heart). In order to get more longitudinal views, the catheter tip has usually to be deflected slightly posteriorly. It shows first the right inferior vein and then, with more rotation (and sometimes even some lateral deflection – this is feasible only with Acuson catheter), the right superior vein with the adjacent right pulmonary artery (Fig. 9.2d, e). When the transducer is placed in the mid-right atrium and deflected to the right, a short axis view of the aortic root with aortic valve cusps is visualized (Fig. 9.2f).

For ventricular imaging, it may be preferable to deflect the tip of the catheter into the right ventricle. It allows imaging of the right ventricle or the left ventricle and both outflow tracts. This position is also useful for evaluation of pericardial effusion around the both ventricles. Sometimes it might be useful to introduce the ICE catheter into the proximal coronary sinus. It provides long axis views of the left ventricle and could also display left atrial appendage. In patients with atrial septal defect or foramen ovale patens, the ICE catheter can be introduced freely into the left atrium. Crossing the fossa ovalis with the ICE catheter using the long sheath may help in the case of poor acoustic window from the right atrium in patients without natural communication on atrial level. From the body of the left atrium, both the pulmonary veins and/or the left appendage could be better displayed.

Table 9.1 Clinical applications of ICE in electrophysiology

• Assistance in transseptal puncture
• Evaluation of intracardiac thrombus
• Assessment of catheter contact and positioning
• Monitoring of radiofrequency current delivery and/or lesion formation
• Assessment of pericardial effusion
• Guidance of complex ablation procedures – Atrial fibrillation – Arrhythmias after repair of complex congenital heart disease – Difficult cases of atrial flutter – Idiopathic ventricular tachycardias originating from aortic or pulmonary cusps – Ventricular tachycardias in structural heart disease
• Other applications

9.3 Clinical Utility of ICE in Electrophysiology

The clinical application of ICE in electrophysiological interventional procedures is summarized in Table 9.1.

9.3.1 Transseptal Puncture Guidance

Transseptal catheterization has become very important procedure in electrophysiology in order to get access into the left atrium or the left ventricle. Traditional technique is based on fluoroscopic guidance in which the important anatomical structures are not displayed directly and the needle with a sheath were guided entirely by their movement during pull-back from the superior vena cava and by their position within cardiac silhouette relative to catheter in coronary sinus and/or aortic root.[5,6] More recently, some modifications using nitinol thin wire for safe fossa ovalis puncture have been described.[7] However, there is a lack of data on clinical performance of these techniques in large series of patients.

Many authors use TEE to guide transseptal puncture.[8] It allows direct imaging of the needle tip within the fossa ovalis region and enables safe puncture. Even easier guidance of the transseptal puncture can be obtained by means of ICE.[9] A cross-sectional view of the fossa ovalis is best acquired with the ICE catheter placed near the septum. Using this guidance, the optimum puncture site can be subselected according to the clinical need. For accessory pathway ablation or access to the left ventricle, it is preferable to perform puncture more anteriorly. On the contrary, for catheter ablation around pulmonary veins, the preferable puncture site is more posterior and inferior. Both TEE and ICE display clearly tenting of the septum (Fig. 9.3a) with the needle tip/dilator and advancement of the assembly into the left atrium. The main advantage of ultrasound guidance is safe navigation in case of anatomical abnormalities such as lipomatous hypertrophy of the septum, atrial septal aneurysm, or double layered fossa ovalis (Fig. 9.3b, c). ICE can also help to re-navigate into the site of former puncture in cases when catheter slips back to the right atrium during manipulation. The catheter tip can be displayed and color Doppler flow imaging can easily detect initial puncture site.

Even though the complications of traditionally guided transseptal puncture are infrequent in highly experienced centers, they occasionally occur.[5,6] They include aortic puncture, pericardial puncture or tamponade, systemic embolism, and perforation of the vena cava inferior. It is believed that ICE guidance would practically avoid serious complications of the transseptal puncture. In a large series of 1,692 transseptal punctures guided by ICE in our facility in the past 4 years, no serious complication was observed and efficacy reached 100% (J. Kautzner, Unpublished data).

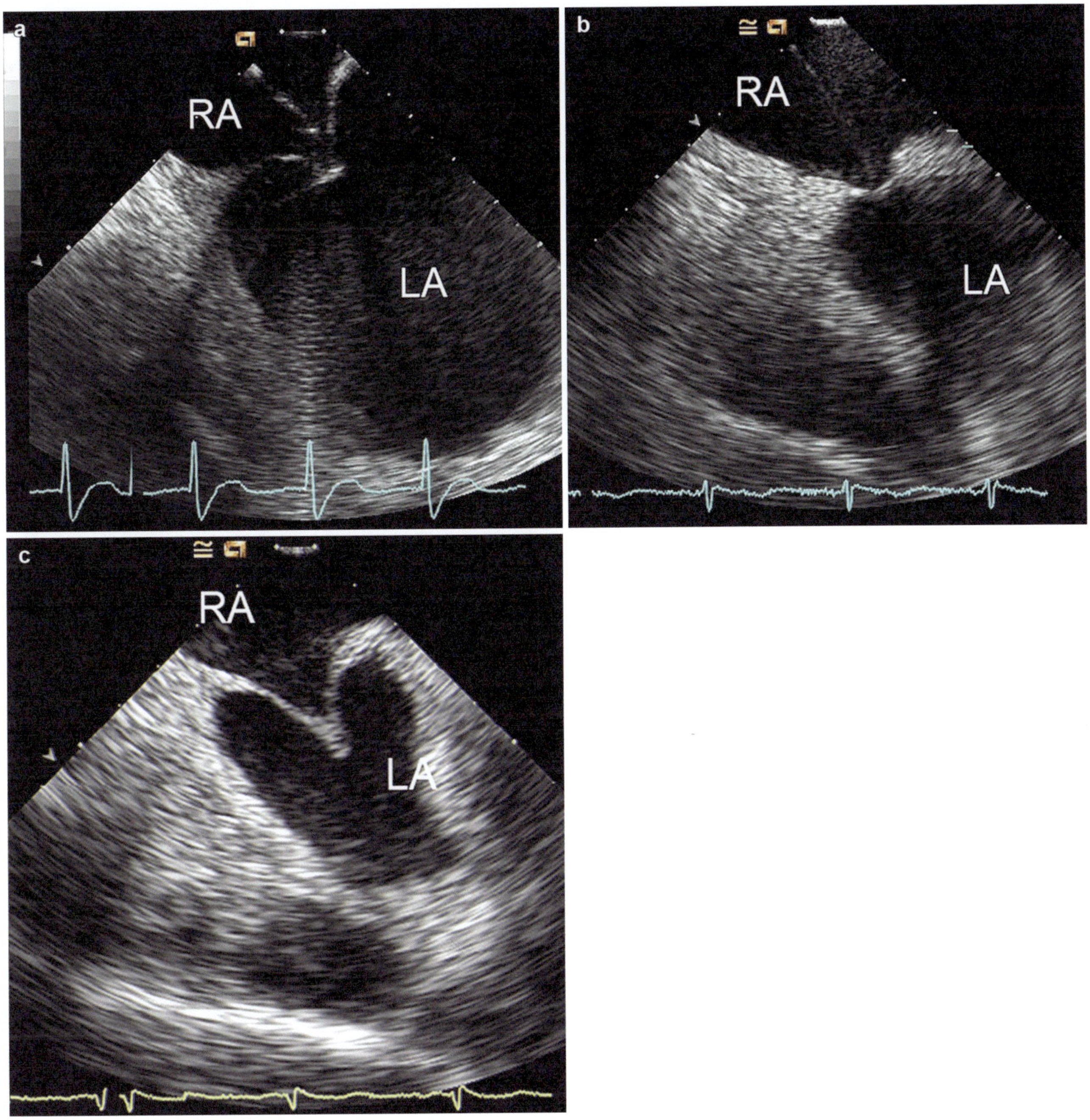

Fig. 9.3 Anatomical variants of the interatrial septum during transseptal puncture: (**a**) tenting of the interatrial septum before the puncture; (**b**) relatively small fossa ovalis in patient with lipomatous hypertrophy of interatrial septum; (**c**) aneurysm of interatrial septum extended by needle during transseptal puncture. *LA* left atrium, *RA* right atrium

9.3.2 Catheter Ablation of Atrial Fibrillation

The pioneering work of the Bordeaux group has ushered a new era in non-pharmacological treatment of atrial fibrillation.[10] Since their original description of triggering foci in the pulmonary veins and subsequent concept of electrical isolation of all pulmonary veins, several ablation strategies have been developed.[11-13] Pulmonary venous ostia are the key structure for most of these. However, lessons have been learned from anatomical and imaging studies that individual anatomy of the pulmonary veins and their junctions with the left atrium (the so-called antra) varies tremendously from one individual to another. Improper understanding of anatomy may result in serious complications such as pulmonary venous stenosis or closure. Extensive ablation within the left atrium may increase risk of thromboembolic complications

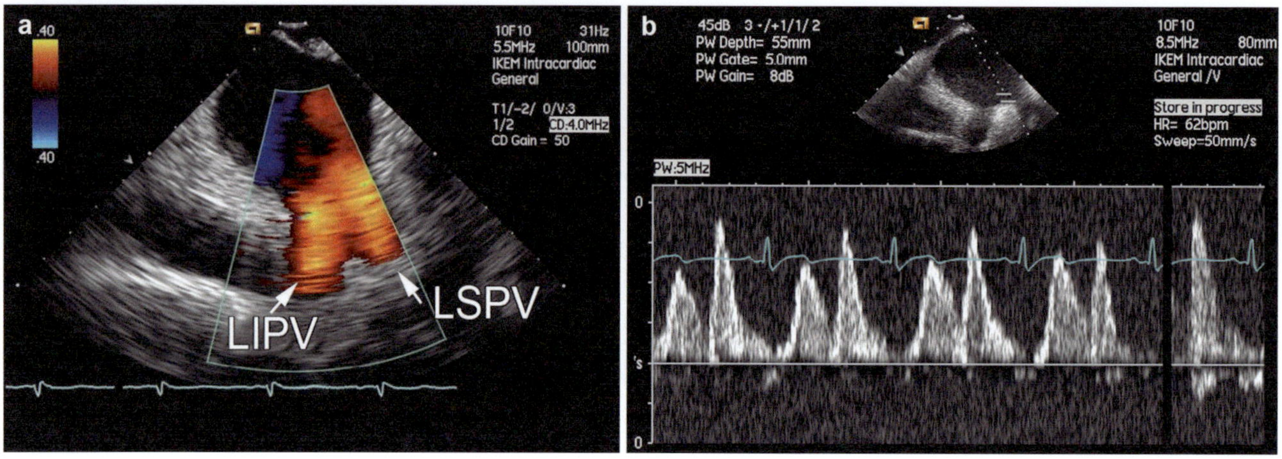

Fig. 9.4 Examples of Doppler evaluation of pulmonary veins. (**a**) Color Doppler depiction of flow in the left superior (*LSPV*) and inferior (*LIPV*) pulmonary vein. (**b**) Flow in the *LSPV* obtained by pulsed Doppler maging

and/or cardiac tamponade. Proximity of the esophagus to the posterior wall of the left atrium creates potential for another serious complication – atrioesophageal fistula. In this respect, ICE imaging plays an important role in online imaging of pulmonary venous ostia and/or esophagus, assisting accurate placement of both mapping and ablation catheter, monitoring of lesion morphology and flow changes in pulmonary veins (Fig. 9.4), and detecting procedural complications.

Imaging of pulmonary veins. Data from 3-D imaging studies have clearly demonstrated that the prevailing pattern of pulmonary venous anatomy is a common antrum on the left side with short or long common portion between the veins and the left appendage. This pattern rather than two separate ostia can be detected in approximately 80% of cases.[14,15] Anteriorly, the antrum is separated from the appendage by a carina of a various thickness. On the contrary, posterior transition into the left atrium is more gradual and often reaches to the middle of the posterior wall. Right pulmonary veins are more rounded in their shape and tend to enter the atrium separately. Especially right superior vein is practically always of rounded, funnel-shaped appearance with gradual transition into the left atrium. Supranumerary veins are more often present on right side (in about 20–20% cases) with diameter that could be less than 10 mm. All these anatomical variants can be detected with a high degree of precision with ICE[14] and its use alleviates the need for preprocedural CT or MR imaging.

Imaging of the esophagus. In order to avoid risk of atrioesophageal fistula, many authors suggested different ways of delineation of the course of esophagus (CARTO, probe, barium, CT, etc.). Some of these techniques rely on preprocedural display of the esophagus. However, spontaneous movement of the esophagus during the procedure was observed in some patients. In this respect, ICE provides excellent online imaging of the esophagus and its course (Fig. 9.5).

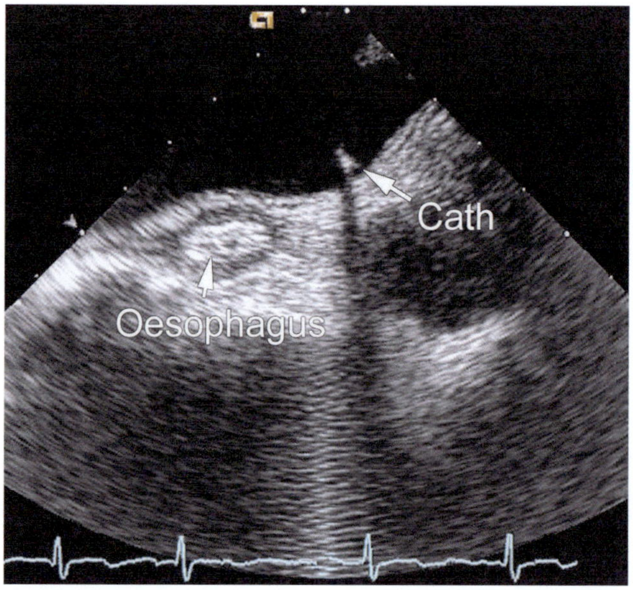

Fig. 9.5 The esophagus in short axis behind the left atrial wall. Note its proximity to the left atrial posterior wall. Arrow depicts the tip of the ablation catheter

Guiding mapping and ablation catheter. ICE can be used for accurate measurement of pulmonary venous diameters and thus, assist in the selection of appropriate size of circular mapping catheter and/or balloon ablation tools. The above-described anatomical pattern of pulmonary veins can make positioning of circular catheter challenging. Especially in relatively flat left veins, the circular mapping catheter tends to distort its shape and could align against posterior wall or slip far into the vein. ICE provides instant imaging of both mapping and ablation catheter and prevents ablation deep inside the pulmonary vein. When using balloon ablation tools, ICE allows online assessment of the position of any balloon catheter within the pulmonary venous ostium and

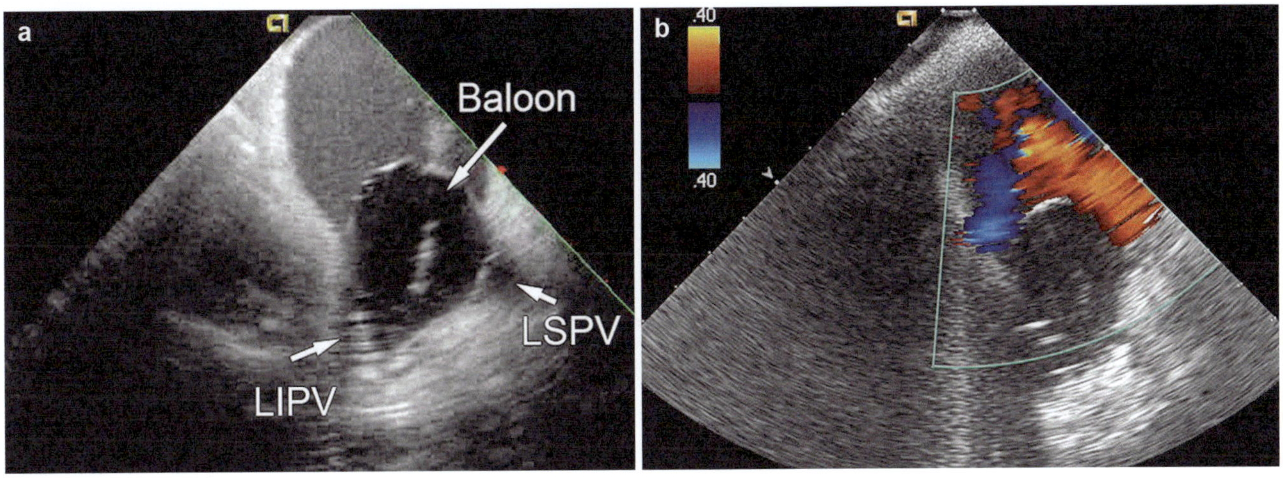

Fig. 9.6 (**a**) Endoscopic laser balloon ablation system (Cardiofocus, Marlborough, MA) with a compliant balloon wedged in the common left pulmonary venous ostium and occluding both left superior (*LSPV*) and inferior (*LIPV*) pulmonary veins. (**b**) Laser balloon positioned in left inferior pulmonary vein of a patient with more separate pulmonary venous ostia with color Doppler depicting flow only from the left superior vein

evaluation of the residual blood flow around the circumference of the balloon (Fig. 9.6).

Reduction of fluoroscopy time. The utility of ICE for ablation catheter navigation can be demonstrated by the fact that it enables to perform catheter ablation of atrial fibrillation without fluoroscopy. Ferguson et al.[16] in 21 patients undergoing catheter ablation of atrial fibrillation used ICE and electroanatomic mapping for procedure guidance and in 19/21 patients no fluoroscopy was used and the staff did not wear any protective lead. This technique may be of particular benefit to the obese patients, children, and pregnant women who are at the highest risk from x-ray exposure.

Monitoring of radiofrequency current delivery. Radiofrequency current delivery at left atrial wall induces lesion morphologic changes including increased wall thickening, echodensity, and/or crater formation. Using an 8-mm tip electrode, ICE can be used to monitor effectively heating of the tissue though monitoring of microbubble formation.[11] The experience of the Cleveland group revealed that this approach significantly improved the success rate and decreased complications of a lesion formation. Subsequent experimental work confirmed that the appearance of microbubbles in ICE imaging corresponds with tissue overheating and allows prevention of pop formation.[17] Based on our empirical experience, even in cool tip catheter with open irrigation, one can prevent tissue overheating by stopping radiofrequency energy delivery when sudden increase in number of microbubbles is visible. ICE can be also used to visualize tissue edema and/or crater after the steam pop occurrence.

Prevention of thromboembolism. Intraprocedural ICE can minimize risk of thromboembolism during ablation of atrial fibrillation (Fig. 9.7). The Pennsylvania group demonstrated

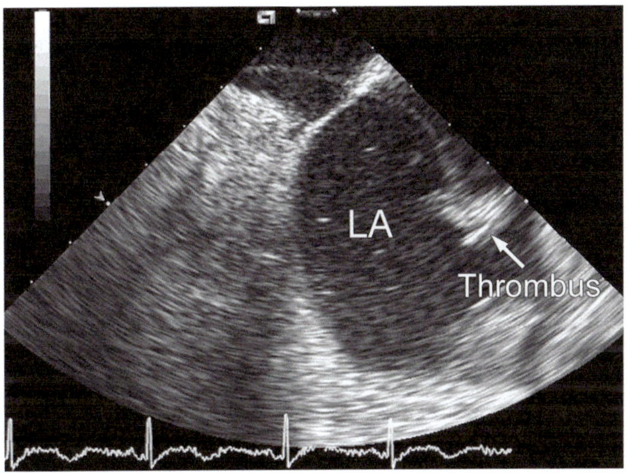

Fig. 9.7 Thrombus on the transseptal sheath in the left atrium (*LA*) that was detected by ICE just after transseptal puncture

for the first time that maintaining average level of anticoagulation with ACT around 250 s does not provide complete protection against thrombus formation in the left atrium. They revealed approximately 10% rate of thrombus occurrence and ICE allowed safe removal of the catheters with thrombus. Spontaneous echocontrast was identified as the only independent risk predictor of subsequent thrombus formation. Therefore, ICE may help to identify both high-risk subjects and display thrombus whenever it develops. The use of ICE for transseptal puncture guidance allows heparin administration even before the crossing of the fossa ovalis.[18] We typically administer 100 IU/kg body weight of heparin before the first transseptal puncture with adjustment of the activated clotting time around 350 s.

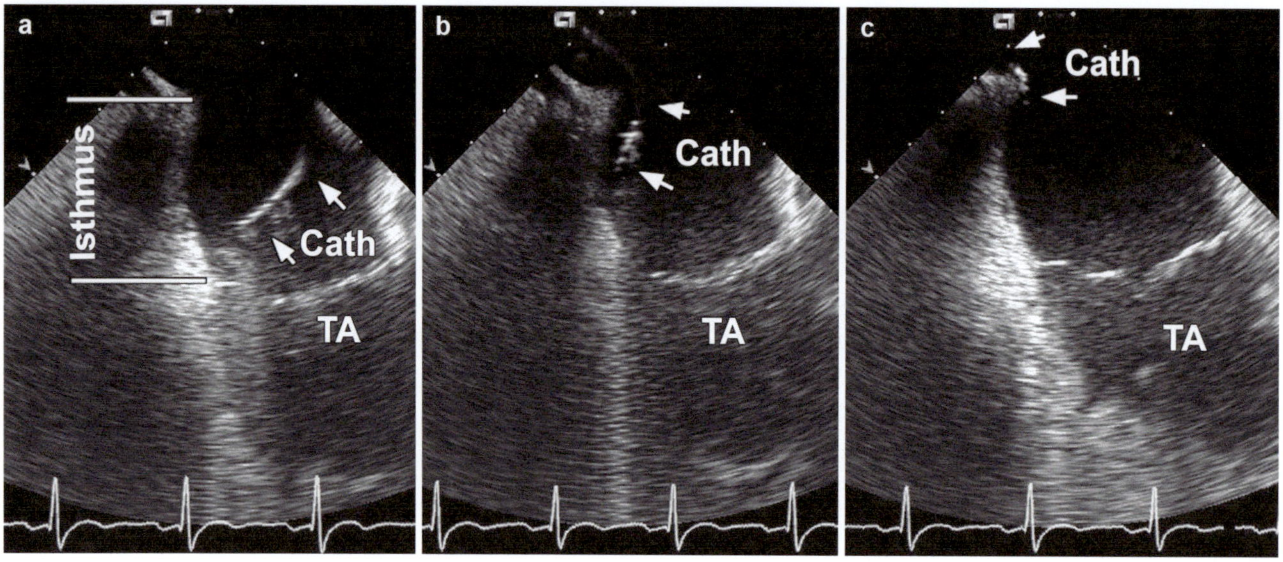

Fig. 9.8 Monitoring of the catheter–tissue contact during the ablation of the cavotricuspid isthmus. (**a**–**c**) Depict the position of the ablation catheter (*Cath*) along the isthmus between the tricuspid valve (*TA*) and the inferior vena cava

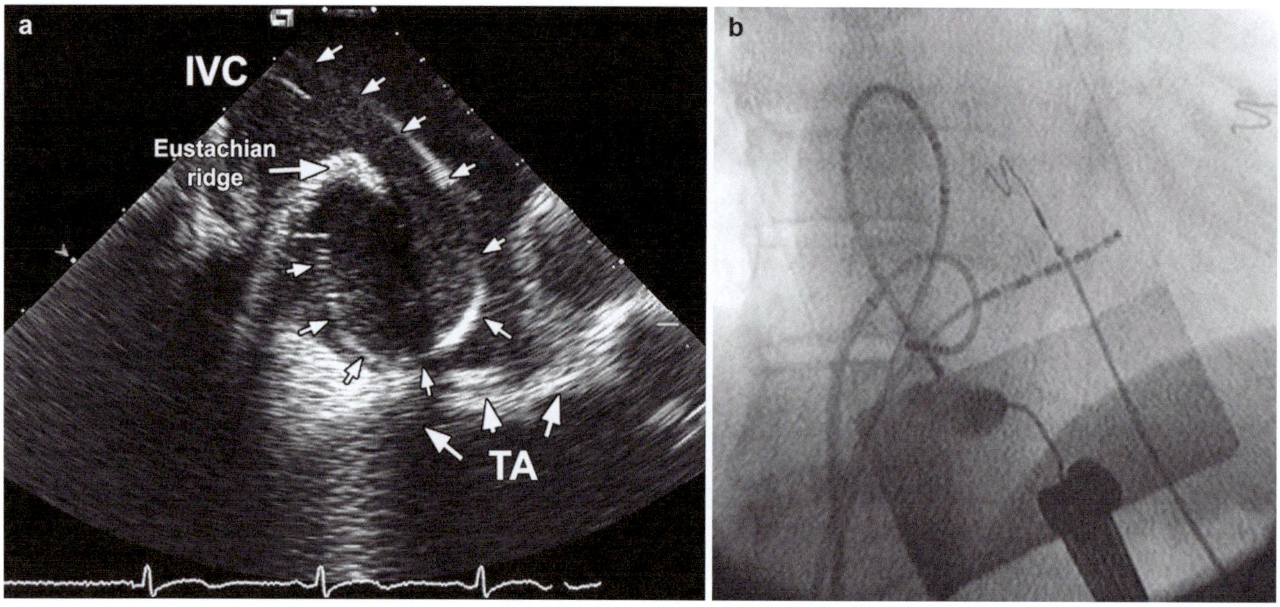

Fig. 9.9 A variant of cavotricuspid isthmus with prominent Eustachian ridge. (**a**) In order to reach the myocardium below the ridge with the ablation catheter, it is necessary to loop its distal end and bring it down to the isthmus (*small arrows*). (**b**) Corresponding fluoroscopic image of the ablation catheter in *PA* view. *IVC* – inferior vena cava, *TA* – tricuspid anulus

9.3.3 Catheter Ablation of Difficult Cases of Atrial Flutter

Despite the fact that catheter ablation of the cavotricuspid isthmus can be considered as a standard ablation technique that does not require special mapping or imaging techniques, some cases may be challenging. It appears that the main reason for such difficulties is abnormal anatomy of the isthmus. ICE can be used to visualize anatomy (Fig. 9.8) of the isthmus and anomalous diverticula or prominent Eustachian ridge (Fig. 9.9).[19,20] In some cases, the cavotricuspid could be very mobile with significant folding of the tissue during cardiac cycle (Fig. 9.10). Such hypermobility of the isthmus usually results in difficulties in obtaining stable catheter contact with the tissue. By enabling depiction of both the isthmus morphology and catheter–tissue contact, the ICE may be of a significant value in patients in whom conventional approach failed.

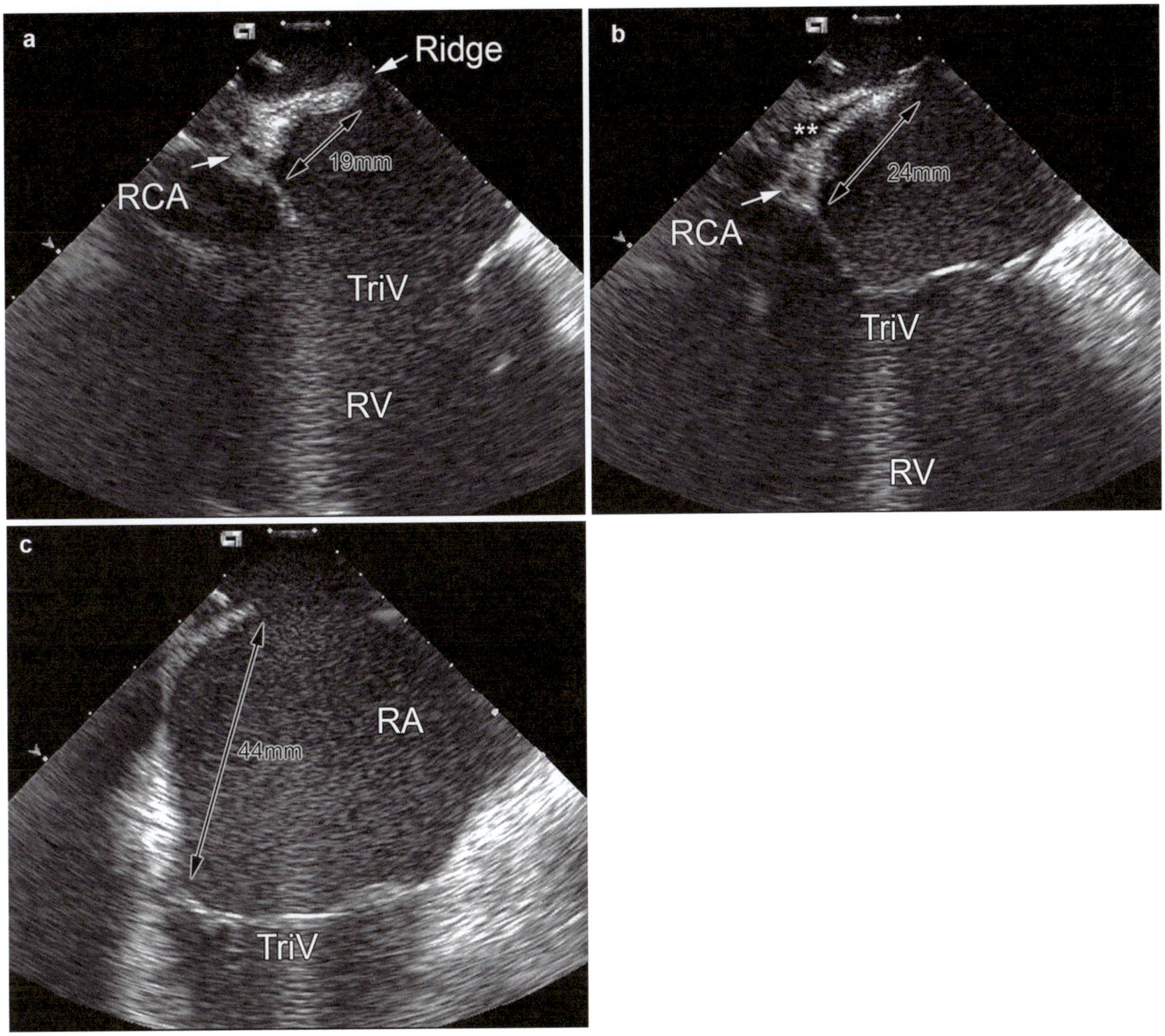

Fig. 9.10 Hypermobile cavotricuspid isthmus. (**a–c**) Depict the change of the size and shape of the isthmus during the cardiac cycle. Note the close relationship of the right coronary artery that runs through the isthmus. *RA* right atrium, *RCA* right coronary artery, *RV* right ventricle, *TriV* tricuspid valve

9.3.4 Catheter Ablation of Arrhythmias in Patients with Congenital Heart Disease

Catheter ablation of postincisional tachycardias in patients after correction of complex congenital heart disease (e.g., after Senning or Mustard procedure) is one of the most challenging procedures in electrophysiology. In these patients, ICE provides an excellent guidance to safe puncture via the baffle and enables monitoring of contact of the ablation catheter[21] (Fig. 9.11). In patients with Wolff–Parkinson–White syndrome and Ebstein anomaly, the ICE can be used to guide the catheter to the boundary of the atrialized ventricle and "true" right ventricular myocardium.[22]

9.3.5 Catheter Ablation of Ventricular Tachycardia

ICE appears to be one of the most suitable imaging techniques to guide mapping and ablation of idiopathic ventricular tachycardias from aortic or pulmonary cusps.[23] Positioning of the ablation catheter within a specific cusp and its relation to commissure or to the coronary artery ostium can be guided by ICE (Fig. 9.12), alleviating the need for concomitant cusp and/or coronary angiography.

ICE with the transducer placed in the right ventricle can provide imaging of the left ventricular structures and, thus, facilitate delineation of myocardial substrate such as

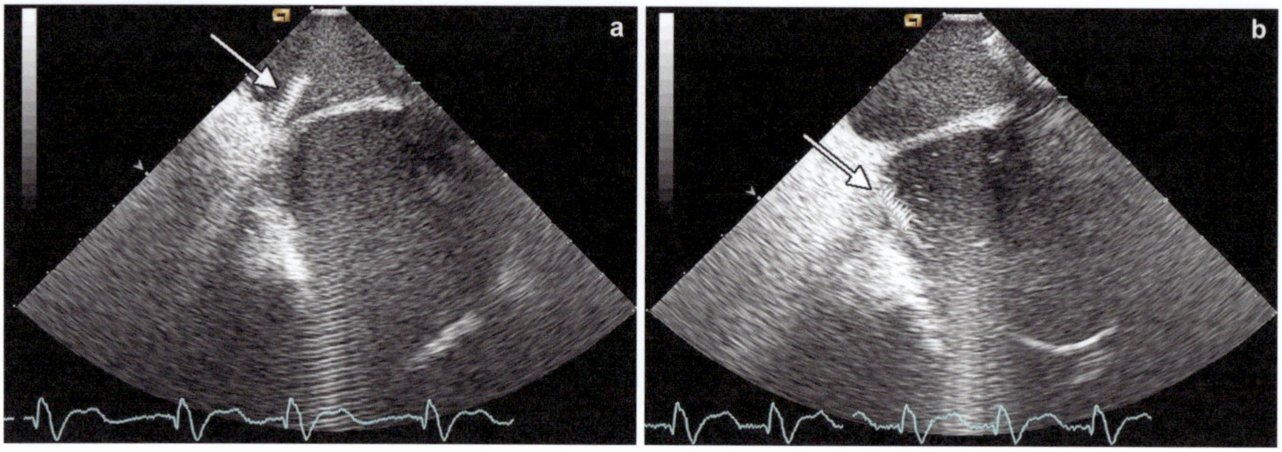

Fig. 9.11 ICE image of cavotricuspid isthmus in patient after Mustard correction of d-transposition of the great arteries. After surgery, the isthmus is divided into two portions by an interatrial baffle. One is accessible from systemic venous atrium, the other only from the atrium of the pulmonary veins. Both panels show position of the ablation catheter on the cavotricuspid isthmus from both sides of the intra-atrial baffle. (**a**) In the systemic venous atrium. (**b**) In the pulmonary venous atrium

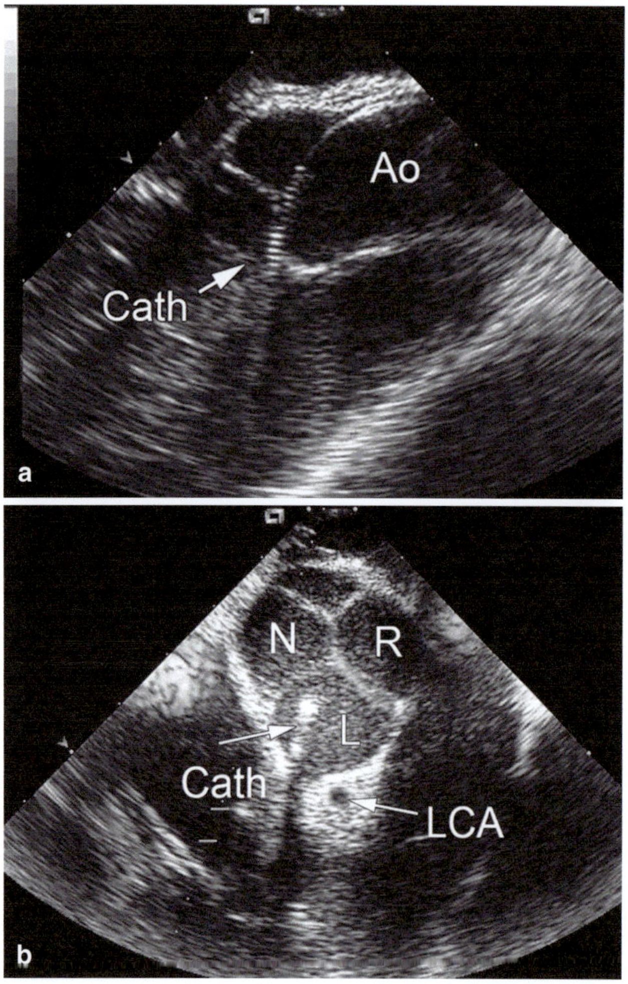

Fig. 9.12 Visualization of the mapping catheter in patient with idiopathic ventricular tachycardia originating from the aortic cusp. The aorta can be visualized in long axis (**a**) and short axis (**b**). *Cath* catheter, *N* noncoronary, *R* right and *L* left aortic cusp, *LCA* left coronary artery

postinfarction scar with akinesis or dyskinesis. It can display ablation catheter tip contact or guide ablation at specific structures such as papillary muscle[24,25] (Fig. 9.13a, b) or false tendon.[26] In addition, ICE can monitor lesion formation and prevent tissue overheating.

Our experience also suggests that ICE can be very useful to guide ablation of ventricular tachycardia originating from the right ventricle in arrhythmogenic right ventricular cardiomyopathy. It can assess precise catheter positioning and contact along the tricuspid annulus and within right ventricular fissures (Fig. 9.13c).

9.3.6 Monitoring to Detect Complications

One of the important roles of ICE imaging in electrophysiology is early diagnosis and prevention of procedural complications during complex procedures. Some of these applications such as prevention of thromboembolism or esophageal injury have already been discussed above. This section will summarize other aspects.

Damage to cardiac structures due to inadvertent manipulation of the catheter may cause damage to adjacent structures such as appendage, mitral or tricuspid valve, and atrial or ventricular wall. As mentioned above, radiofrequency energy application may cause perforation due to tissue superheating and "crater" formation.[27] Monitoring of the position of the ablation catheter in relationship to specific anatomical structures prevents inadvertent delivery of radiofrequency current in high-risk regions such as proximal part of the pulmonary vein. Recent experience shows that navigation by ICE can practically eliminate pulmonary venous stenosis.[11] In addition, ICE allows for measurement of the diameter of the

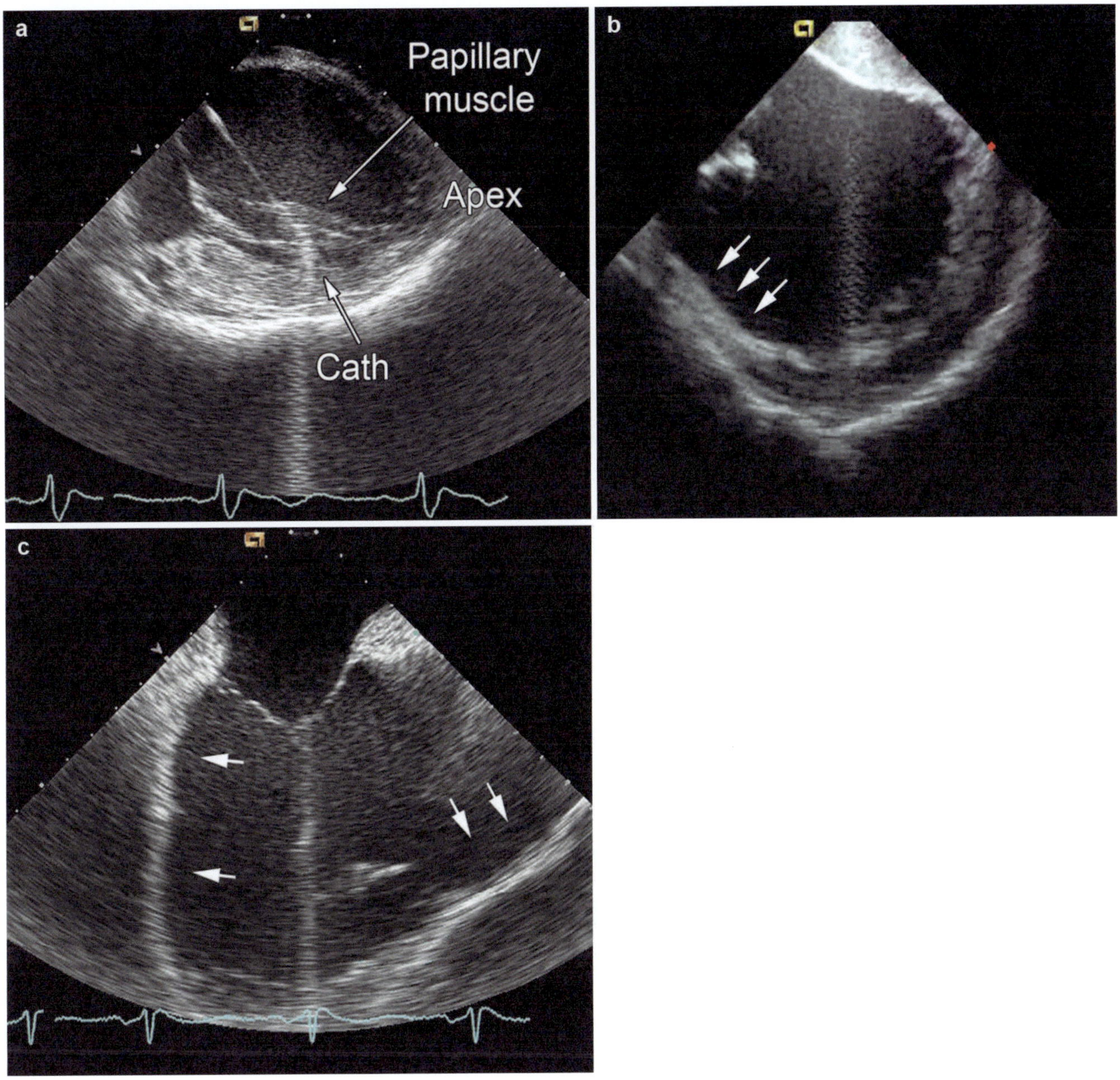

Fig. 9.13 (**a**) Position of the catheter within the left ventricle in a patient with ventricular tachycardia originating from the anterolateral papillary muscle. (**b**) ICE image of the left ventricle in a patient with ventricular tachycardias after myocardial infarction. *Arrows* show high echogenicity on the lateral left ventricular wall, indicating local scarring. (**c**) Aneurysmatic enlargement (*arrows*) of the right ventricle and outflow tract in a patient with arrhythmogenic right ventricular cardiomyopathy

pulmonary veins and safe placement of balloon or multi electrode catheters. Doppler modality could be used anytime during the procedure to check pulmonary venous velocities.[28] Experience shows that pulmonary vein ostial peak velocity increasing to more than 150 cm/s with Doppler pattern of turbulent flow indicates a moderate narrowing effect. Significant stenoses were characterized by higher peak flow velocities.

Pericardial effusion is one of the common complications associated with catheter manipulation and radiofrequency current delivery within the left atrium (Fig. 9.14). In such cases, ICE can provide early diagnosis and, thus, prevent tamponade due to early reversal of anticoagulation.[29] In case of tamponade, ICE allows online monitoring of the patient's status.

Tissue edema occurs after multiple application of radiofrequency energy targeted into small area (Fig. 9.15). This increases the tissue thickness, which results in the diminishing of amplitude of local electrograms making location of gap within the ablation line more difficult to trace. Furthermore, creation of the transmural lesion within edematous tissue is more challenging.

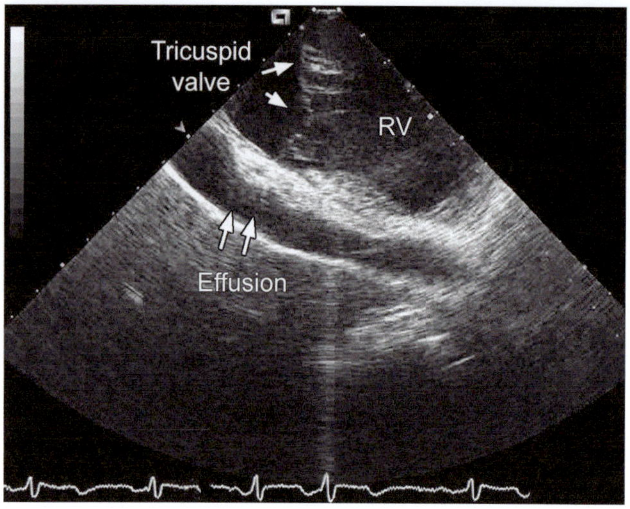

Fig. 9.14 Pericardial effusion below the right ventricle

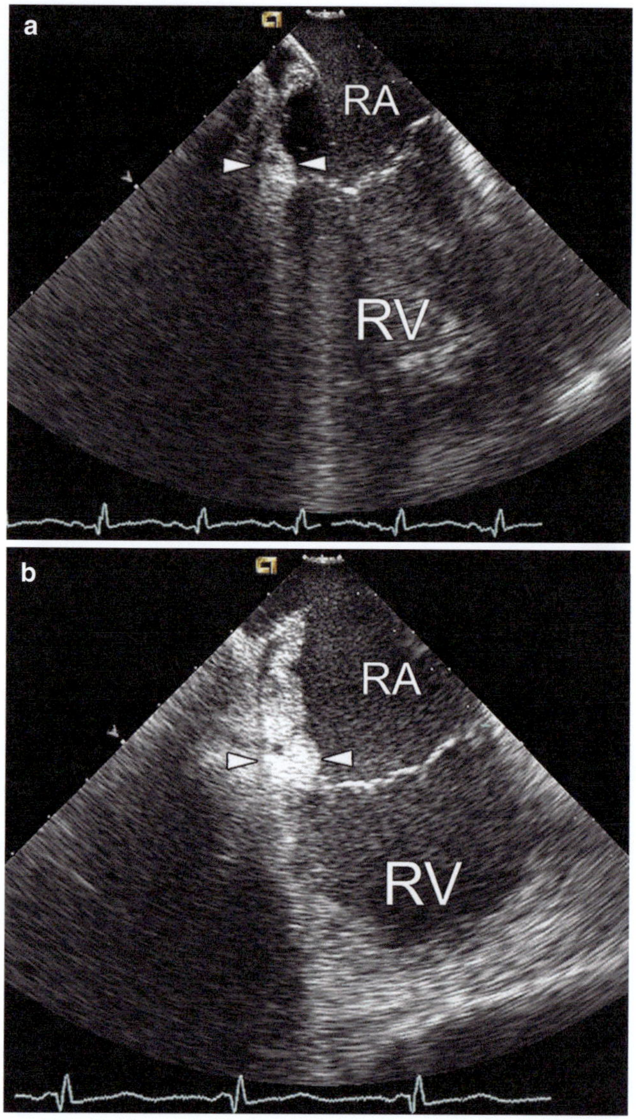

Fig. 9.15 Monitoring the application of radiofrequency energy. (**a**) The cavotricuspid isthmus before the ablation. (**b**) After multiple applications of the radiofrequency energy on the isthmus, the tissue is edematous and shows higher echogenicity. *RA* right atrium, *RV* right ventricle

9.4 3D Echocardiography for Catheter Ablation

2D ICE imaging has several limitations. For example, curved catheter shaft often cannot be fully visualized with single plane image, thus necessitating acquisition of multiple imaging planes and image adjustments. This requires specific skills and could be time consuming and, thus, distracting during an invasive procedure. Therefore, work is in the progress to provide 3D visualization. 3D imaging should enable visualization of the entire shaft of intracardiac catheters, along with a clear depiction of their positions in relation to other cardiac structures.[30] Additionally, certain anatomic regions can be depicted in an "en face" view (Fig. 9.16) and, thus, demonstrate their complex anatomy. Finally, reliable 3D imaging of specific cardiac chambers may obviate the need for virtual electroanatomic mapping and make any mapping or ablation procedure more precise and faster. The availability of 3D imaging would also open the possibility for creation of 4D datasets (3D+time).

9.4.1 3D Transesofageal/Transthoracic Echocardiography

Over the past several years, improvements in transducer technologies have allowed the development of a full matrix array transducer which can be used to acquire pyramidal-shaped ultrasound data sets. These data sets can be processed both online and offline to allow the display of cardiac structures and catheters as they move in time and space. Recently, a miniaturized matrix probe has been coupled with a transesophageal probe. This allows the acquisition of high-quality real-time 3D images.[31] The potential applications of the 3D echocardiography include wide spectrum of cardiac interventions including occlusion of atrial/ventricular septal defect, patent foramen ovale; percutaneous mitral valve valvuloplasty and repair; left atrial appendage obliteration; percutaneous closures of prosthetic valve dehiscence, catheter ablation of atrial fibrillation, etc. However, the use of 3D transesophageal echocardiography for electrophysiology procedures is limited by patient's discomfort and necessity of sedation or general anesthesia. Whether 3D imaging may affect performance of procedures and translate into shortened procedure times, decreased radiation and contrast exposure and safer procedures is yet to be determined.

Matrix array transthoracic 3D echocardiography could be used in electrophysiology to identify the site of origin of ventricular ectopic activity, especially in anatomically complex region of ventricular outflow tracts.[32] Using tissue tracking imaging, the site of arrhythmia origin could be defined as the site at which the earliest color-coded signal detected in the myocardium during ventricular premature contractions. This may help to target the region of interest directly and possibly simplify the procedure.

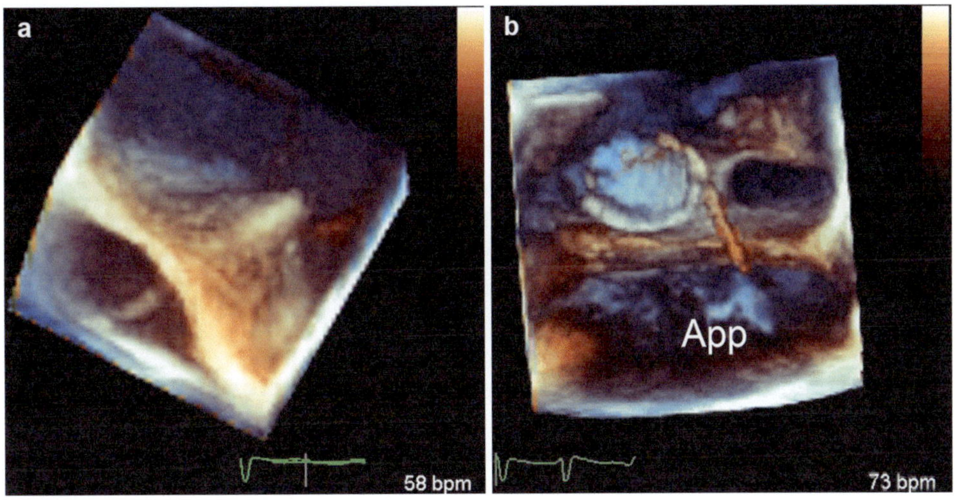

Fig. 9.16 Examples of 3D transesophageal echocardiography. (**a**) Tenting of the fossa ovalis during transseptal puncture. (**b**) 3D reconstruction of the ridge between the left veins and the appendage (*App*). The circular mapping catheter is positioned in the ostium of the left inferior pulmonary vein

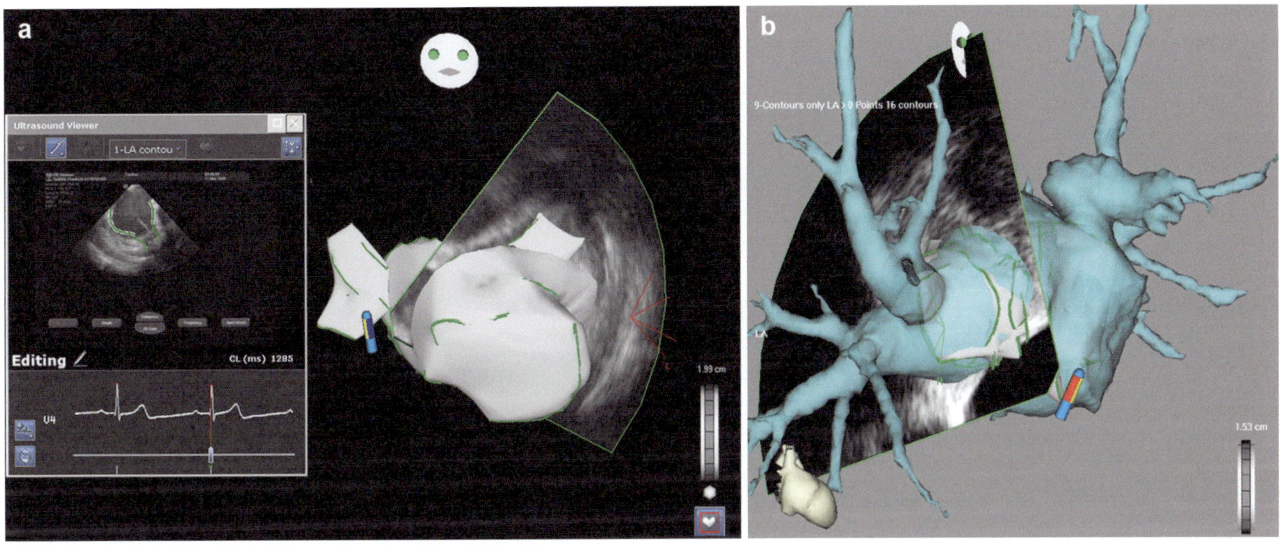

Fig. 9.17 (**a**) An example of automatic contour tracing on CartoSound image of the left atrium (*green line*) and the resulting 3D reconstruction. (**b**) Shows reasonable fit of the CartoSound ICE image obtained from the right atrium with 3D image of the left atrium and pulmonary veins obtained from CT angiography before the procedure

9.4.2 3D Intracardiac Echocardiography

ICE catheter allowing direct 3D real-time imaging is not available yet. Currently, three strategies of 3D reconstruction using ICE are being pursued. One 3D technique emerges from a marriage of the phased-array ICE catheter (AcuNav Diagnostic Ultrasound Catheter, Acuson - Siemens, Mountain View, California, USA) with a special sensor of electromagnetic field that is used in catheters for electroanatomical mapping (CARTO, Biosense Webster, Inc, Diamond Bar, California, USA). The resulting new image integration software module is called CartoSound™. It enables to trace semiautomatically contours of the chamber of interest in different planes (Fig. 9.17a, b), and subsequent addition of these contour points into 3D electroanatomical map.[33] In this way, 3D electroanatomical map of the left atrium or the left ventricle could be constructed from a series of images obtained with ICE catheter within the right atrium or right ventricle. Clinical data suggests feasibility of such an approach and high accuracy in creation of 3D map of a relevant cardiac chamber both for mapping within atria[34] and ventricles.[35] Tracing orifices of different veins manually in different echocardiographic views allows either to let these areas labeled by several contours or enables creation of separate small 3D maps of each vein as a rough estimate of the real shape. This suboptimal reproduction of true anatomy can be offset by merging of 3D echo image with CT or MR angiography reconstruction (Fig. 9.17). Due to accurate integration of both images, only CT or MR anatomical shell could be used for catheter navigation.

For 3D reconstruction of images from mechanical ultrasound (Clearview, Cardiovascular Imaging Systems, Inc, Freemont, California, USA), a special pullback device has been developed.[36] The device is controlled by 3D workstation and uses a stepping motor to move the ICE catheter linearly in cranio-caudal direction. The station also receives both respiration and ECG signal. Cardiac cycles are recorded when they fall into a preset limit around the mean RR and respiration interval. Pulling the catheter within the heart, 3D reconstruction of the anatomy can be obtained. Clear details of intracardiac structures such as fossa ovalis, coronary sinus, or triangle of Koch could be displayed. However, the depth of imaging remains limited.

Finally, the third approach consists of 3D reconstruction of data obtained during rotational scanning with conventional phased-array ICE catheter. In this method, a motor rotates the transducer array around an axis corresponding to the shaft of the catheter. This allows 3D imaging of selected sector images. Knackstedt et al.[37] published their first experience with semiautomated 3D ICE in a cohort of 5 pigs. The ICE catheter was introduced via a straightened sheath into the right atrium and then rotated automatically around the longitudinal axis from 90° to 360° in 2–5° steps using a custom-made stepper motor. The echocardiographic 2D images triggered by respiration and ECG were digitized and a 3D reconstruction was performed by a prototype software (TomTec Imaging, Unterschlessheim, Germany). After experimental validation, the system was tested in 6 patients during an electrophysiologic study. 3D acquisition and partial volume reconstruction of the left atrium and ventricles was obtained within 3–5 min. This was achieved from a single spot within the right atrium, precluding the need for catheter position readjustments. The main advantage of the system was that it allowed precise assessment of pulmonary venous anatomy and diameter. Further development of the system is expected to result in a clinically applicable 3D imaging platform for guiding of complex ablation procedures.

Ongoing research focuses also on visualization of 3D echocardiographic data in the 3D space (virtual reality). Dynamic holographic imaging of the 3D echocardiographic data appears to be feasible and may help in providing a preview of cardiac anatomy for complex ablation procedures. Such imaging may be also used in the future for sophisticated simulator of complex ablation procedures.[38]

9.5 Conclusions

Expansion of indications to catheter ablation has led to a change of the paradigm from electrophysiologically guided procedure to a procedure guided predominantly anatomically. Accurate online identification of cardiac structures and catheters becomes mandatory and ICE is at present the only imaging modality that can accomplish easily this task. It is a valuable tool for guiding procedures such as transseptal puncture or for monitoring complex ablation procedures like ablation of atrial fibrillation or ventricular tachycardias. It also allows prevention and/or early detection of intraprocedural complications. The widespread use of ICE is likely to result in reduction of fluoroscopy time and to maximize safety and efficacy of complex ablation procedures.

References

1. Saxon LA, Stevenson WG, Fonarow GC, et al. Transesophageal echocardiography during radiofrequency catheter ablation of ventricular tachycardia. *Am J Cardiol*. 1993;72(9):658-661.
2. Goldman AP, Irwin JM, Glover MU, et al. Transesophageal echocardiography to improve positioning of radiofrequency ablation catheters in left-sided Wolff-Parkinson-white syndrome. *Pacing Clin Electrophysiol*. 1991;14(8):1245-1250.
3. Ren JF, Schwartzman D, Callans DJ, et al. Intracardiac echocardiography (9 MHz) in humans: methods, imaging views and clinical utility. *Ultrasound Med Biol*. 1999;25(7):1077-1086.
4. Ren JF, Weiss JP. Imaging equipment and right heart catheterization technique. In: Ren JF, Marchlinski FE, Callans DJ, Schwartzman D, eds. *Practical Intracardiac Echocardiography in Electrophysiology*. Malden: Blackwell Publishing Inc; 2006:5-17.
5. De PR, Zardini M, Storti C, et al. Trans-septal catheterization for radiofrequency catheter ablation of cardiac arrhythmias. Results and safety of a simplified method. *Eur Heart J*. 1998;19(6): 943-950.
6. Roelke M, Smith AJ, Palacios IF. The technique and safety of transseptal left heart catheterization: the Massachusetts general hospital experience with 1,279 procedures. *Cathet Cardiovasc Diagn*. 1994;32(4):332-339.
7. Wieczorek M, Hoeltgen R, Akin E, et al. Use of a novel needle wire in patients undergoing transseptal puncture associated with severe septal tenting. *J Interv Card Electrophysiol*. 2010;27(1):9-13.
8. Hahn K, Gal R, Sarnoski J, et al. Transesophageal echocardiographically guided atrial transseptal catheterization in patients with normal-sized atria: incidence of complications. *Clin Cardiol*. 1995;18(4):217-220.
9. Hung JS, Fu M, Yeh KH, et al. Usefulness of intracardiac echocardiography in transseptal puncture during percutaneous transvenous mitral commissurotomy. *Am J Cardiol*. 1993;72(11):853-854.
10. Haissaguerre M, Jais P, Shah DC, et al. Spontaneous initiation of atrial fibrillation by ectopic beats originating in the pulmonary veins. *N Engl J Med*. 1998;339(10):659-666.
11. Marrouche NF, Martin DO, Wazni O, et al. Phased-array intracardiac echocardiography monitoring during pulmonary vein isolation in patients with atrial fibrillation: impact on outcome and complications. *Circulation*. 2003;107(21):2710-2716.
12. Nademanee K, McKenzie J, Kosar E, et al. A new approach for catheter ablation of atrial fibrillation: mapping of the electrophysiologic substrate. *J Am Coll Cardiol*. 2004;43(11):2044-2053.
13. Pappone C, Rosanio S, Oreto G, et al. Circumferential radiofrequency ablation of pulmonary vein Ostia: a new anatomic approach for curing atrial fibrillation. *Circulation*. 2000;102(21):2619-2628.
14. Jongbloed MR, Bax JJ, Lamb HJ, et al. Multislice computed tomography versus intracardiac echocardiography to evaluate the pulmonary veins before radiofrequency catheter ablation of atrial fibrillation: a head-to-head comparison. *J Am Coll Cardiol*. 2005; 45(3):343-350.

15. Mlcochova H, Tintera J, Porod V, et al. Magnetic resonance angiography of pulmonary veins: implications for catheter ablation of atrial fibrillation. *Pacing Clin Electrophysiol*. 2005;28(10):1073-1080.
16. Ferguson JD, Helms A, Mangrum JM, et al. Catheter ablation of atrial fibrillation without fluoroscopy using intracardiac echocardiography and electroanatomic mapping. *Circ Arrhythm Electrophysiol*. 2009;2(6):611-619.
17. Martin RE, Ellenbogen KA, Lau YR, et al. Phased-array intracardiac echocardiography during pulmonary vein isolation and linear ablation for atrial fibrillation. *J Cardiovasc Electrophysiol*. 2002;13(9):873-879.
18. Wazni OM, Rossillo A, Marrouche NF, et al. Embolic events and char formation during pulmonary vein isolation in patients with atrial fibrillation: impact of different anticoagulation regimens and importance of intracardiac echo imaging. *J Cardiovasc Electrophysiol*. 2005;16(6):576-581.
19. Morton JB, Sanders P, Davidson NC, et al. Phased-array intracardiac echocardiography for defining cavotricuspid isthmus anatomy during radiofrequency ablation of typical atrial flutter. *J Cardiovasc Electrophysiol*. 2003;14(6):591-597.
20. Peichl P, Čihák R, Kautzner J. Morphology of cavotricuspid isthmus defined by intracardiac echocardiography. *Interv Akut Kardiol*. 2006;6:203-206.
21. Peichl P, Kautzner J, Gebauer R. Ablation of atrial tachycardias after correction of complex congenital heart diseases: utility of intracardiac echocardiography. *Europace*. 2009;11(1):48-53.
22. Traykov VB, Pap R, Bencsik G, et al. Ventricular location of a part of the right atrial isthmus after tricuspid valve replacement for Ebstein's anomaly: a challenge for atrial flutter ablation. *J Interv Card Electrophysiol*. 2009;25(3):199-201.
23. Lamberti F, Calo' L, Pandozi C, et al. Radiofrequency catheter ablation of idiopathic left ventricular outflow tract tachycardia: utility of intracardiac echocardiography. *J Cardiovasc Electrophysiol*. 2001;12(5):529-535.
24. Seiler J, Lee JC, Roberts-Thomson KC, et al. Intracardiac echocardiography guided catheter ablation of incessant ventricular tachycardia from the posterior papillary muscle causing tachycardia-mediated cardiomyopathy. *Heart Rhythm*. 2009;6(3):389-392.
25. Yamada T, McElderry HT, Doppalapudi H, et al. Real-time integration of intracardiac echocardiography and electroanatomic mapping in PVCs arising from the LV anterior papillary muscle. *Pacing Clin Electrophysiol*. 2009;32(9):1240-1243.
26. Jongbloed MR, Bax JJ, van der Burg AE, et al. Radiofrequency catheter ablation of ventricular tachycardia guided by intracardiac echocardiography. *Eur J Echocardiogr*. 2004;5(1):34-40.
27. Takagi M, Toda I, Yoshikawa J. The "pop phenomenon" detected by phased array intracardiac echocardiography. *Heart*. 2003;89(7):706.
28. Ren JF, Marchlinski FE, Callans DJ, et al. Intracardiac Doppler echocardiographic quantification of pulmonary vein flow velocity: an effective technique for monitoring pulmonary vein Ostia narrowing during focal atrial fibrillation ablation. *J Cardiovasc Electrophysiol*. 2002;13(11):1076-1081.
29. Ren JF, Marchlinski FE. Early detection of iatrogenic pericardial effusion: importance of intracardiac echocardiography. *JACC Cardiovasc Interv*. 2010;3(1):127.
30. Lange A, Palka P, Burstow DJ, et al. Three-dimensional echocardiography: historical development and current applications. *J Am Soc Echocardiogr*. 2001;14(5):403-412.
31. Perk G, Lang RM, Garcia-Fernandez MA, et al. Use of real time three-dimensional transesophageal echocardiography in intracardiac catheter based interventions. *J Am Soc Echocardiogr*. 2009;22(8):865-882.
32. Tada H, Toide H, Naito S, et al. Tissue tracking imaging as a new modality for identifying the origin of idiopathic ventricular arrhythmias. *Am J Cardiol*. 2005;95(5):660-664.
33. Schwartzman D, Zhong H. On the use of CartoSound for left atrial navigation. *J Cardiovasc Electrophysiol*. 2009;21(6):656-664.
34. Singh SM, Heist EK, Donaldson DM, et al. Image integration using intracardiac ultrasound to guide catheter ablation of atrial fibrillation. *Heart Rhythm*. 2008;5(11):1548-1555.
35. Khaykin Y, Skanes A, Whaley B, et al. Real-time integration of 2D intracardiac echocardiography and 3D electroanatomical mapping to guide ventricular tachycardia ablation. *Heart Rhythm*. 2008;5(10):1396-1402.
36. Szili-Torok T, McFadden EP, Jordaens LJ, et al. Visualization of elusive structures using intracardiac echocardiography: insights from electrophysiology. *Cardiovasc Ultrasound*. 2004;2:6.
37. Knackstedt C, Franke A, Mischke K, et al. Semi-automated 3-dimensional intracardiac echocardiography: development and initial clinical experience of a new system to guide ablation procedures. *Heart Rhythm*. 2006;3(12):1453-1459.
38. van den Bosch AE, Koning AH, Meijboom FJ, et al. Dynamic 3D echocardiography in virtual reality. *Cardiovasc Ultrasound*. 2005;3:37.

10 Intraprocedural Techniques: Angioscopy and Optical Imaging

Paul J. Wang, Amin Al-Ahmad, Henry H. Hsia, Paul C. Zei, Mintu Turakhia, Marco Perez, and Christian S. Eversull

Abstract

The treatment of electrical disorders of the heart has evolved enormously over the past two decades. New technologies have been developed to treat the range of bradyarrhythmias, supraventricular arrhythmias, ventricular arrhythmias, and heart failure. These treatments largely have been developed in two general categories: implantable heart rhythm devices and intracardiac procedures such as catheter ablation. Imaging has played a major role in these procedures, and new technologies exhibit considerable promise in aiding the advancement of these procedures. New imaging technologies may overcome previously insurmountable challenges such as directly assessing lesion formation and guiding catheters and leads into specific anatomic structures such as the coronary sinus.

Keywords

Angioscopy and optical imaging • Optical imaging and angioscopy • Electrical disorders of the heart • Catheter ablation

The treatment of electrical disorders of the heart has evolved enormously over the past two decades. New technologies have been developed to treat the range of bradyarrhythmias, supraventricular arrhythmias, ventricular arrhythmias, and heart failure. These treatments largely have been developed in two general categories: implantable heart rhythm devices and intracardiac procedures such as catheter ablation. Imaging has played a major role in these procedures and new technologies exhibit considerable promise in aiding the advancement of these procedures. New imaging technologies may overcome previously insurmountable challenges such as directly assessing lesion formation and guiding catheters and leads into specific anatomic structures such as the coronary sinus.

P.J. Wang (✉)
Cardiac Arrhythmia Service and Cardiac Electrophysiology Laboratory, Cardiovascular Medicine, Stanford University, Stanford, CA, USA
e-mail: paul.j.wang@stanford.edu

10.1 Conventional Imaging of Intracardiac Structures

Imaging of critical structures in the heart is essential to successful heart rhythm procedures. Fluoroscopy, while the primary imaging modality, has significant limitations since it does not permit clear delineation of heart structures from the circulating blood pool. Ultrasound-based techniques such as transesophageal echocardiographic imaging have provided very accurate real-time assessment of cardiac structures including the heart chambers, heart valves, great vessels, and surrounding anatomy. Intracardiac echocardiography has permitted detailed imaging of ablation catheters and cardiac structures from within the heart. New modalities such as real-time rotational CT which creates a complete three-dimensional cardiac reconstruction have further expanded the ability to examine the heart chambers within the electrophysiology laboratory. In this chapter, we will review the technological advances that have enabled direct imaging of the endocardial surface of the heart during interventional and diagnostic procedures.

A. Auricchio et al. (eds.), *Cardiac Imaging in Electrophysiology*,
DOI 10.1007/978-1-84882-486-7_10,

10.2 Direct Optical Visualization of Heart Structures

10.2.1 Cardioscopy

Sakibara[1] first reported directly imaging the inside of the heart during surgery using a cardioscope in 1939, thus demonstrating the feasibility of this novel form of optical visualization. Silander laid the foundation for less invasive direct visualization in the early 1960s by introducing a cardioscope in the heart without a thoracotomy.[2]

10.2.2 Intracardiac Direct Visualization

Direct visualization within the heart is limited predominantly by blood filling its chambers. In the visible spectrum, imaging is limited by the light absorption and scattering properties of blood which prevent effective transmission of light for visualization. In order to circumvent this limitation, a primary strategy has been to use an optically transparent fluid-filled balloon to displace blood from the path between an imaging element and the wall of the heart. Because of the large volume of blood and high flow rate within the heart, it is not feasible to displace the entire blood volume from the heart chambers in the clinical setting.

One of the first descriptions of angioscopy in large vessels and the heart was by Gamble and Innis in 1967.[3] They equipped a flexible 12 French cardioscope with a latex balloon and filled it with 4–7 mL of air. Glass fibers were embedded in the balloon to maintain the balloon's shape and facilitate visualization through its distal wall, which was transparent. A bundle of 70,000 fibers was used to transmit the image. The image was recorded on 8-mm black-and-white motion picture film. In the initial experiments with a rigid scope, access was obtained from the jugular vein or carotid artery. Using a flexible scope, access from the femoral arteries and veins was also feasible. Pictures of the aortic valve were obtained and the scope was passed into the left ventricle easily. The color of the endocardium could be visualized. When the scope was advanced into the right side of the heart, the right-sided structures such as the atrial septum, fossa ovalis, orifice of the coronary sinus, and tricuspid valve could all be visualized.

Following the report of Gamble and Innis, there were remarkable advances in fiber-optic imaging technology. Tools such as gastrointestinal endoscopes led to this advance, becoming technologically extremely sophisticated. Working in the cardiovascular system, Shure et al. in 1981 and 1984 demonstrated that fiber-optic angioscopy could be used to image pulmonary emboli and pulmonary arterial obstruction, respectively.[4,5] Miniaturization of this technology led to fiber-optic angioscopy of small blood vessels including the coronary arteries. The use of a miniature scope to visualize structures within the human coronary arteries was revolutionary. The scope consisted of a fiber-optic element terminating at the end of a steerable catheter that could be inserted via a guiding sheath into the coronary vasculature. In small vessels, angioscopy is achieved by using a saline flushing mechanism to push the blood out of the visual field.[6,7] The coronary angioscope helped to confirm the appearance of thrombus on underlying coronary artery plaques as the primary finding in acute myocardial infarction and unstable angina.

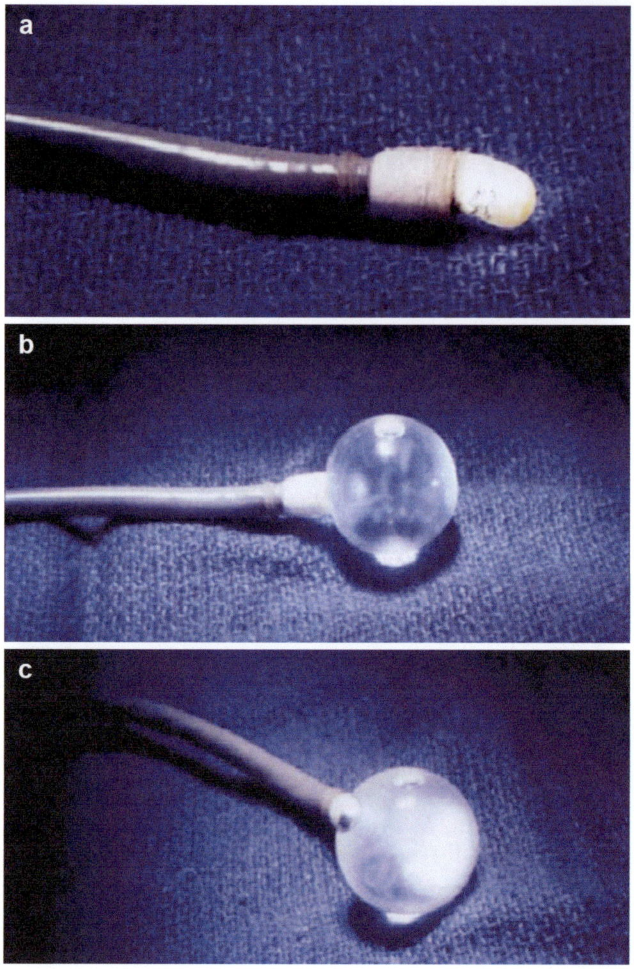

Fig. 10.1 (**a**) The tube-balloon system with the balloon deflated. The Olympus fiber-optic scope is inserted inside the tube-balloon system. (**b**) The balloon is inflated with normal saline and illuminated. (**c**) Frontal view of **b** (From Kuo and Koch[11], with permission)

Uchida et al. demonstrated that a fiber-optic angioscopy could be used to image the heart chambers, heart valves, and great vessels.[8-10] Uchida et al. also demonstrated that fiber-optic angioscopy could be used to guide left ventricular myocardial biopsy.[9]

In 1994, Kuo et al. demonstrated the ability to visualize ablative lesions within the canine heart using a saline-filled balloon.[11] Kuo et al. used a 3.3-mm fiber-optic endoscope placed inside a polyurethane tube attached to a latex balloon (Fig. 10.1). Using fluoroscopic guidance, the scope assembly was advanced into the right atrium. The latex balloon

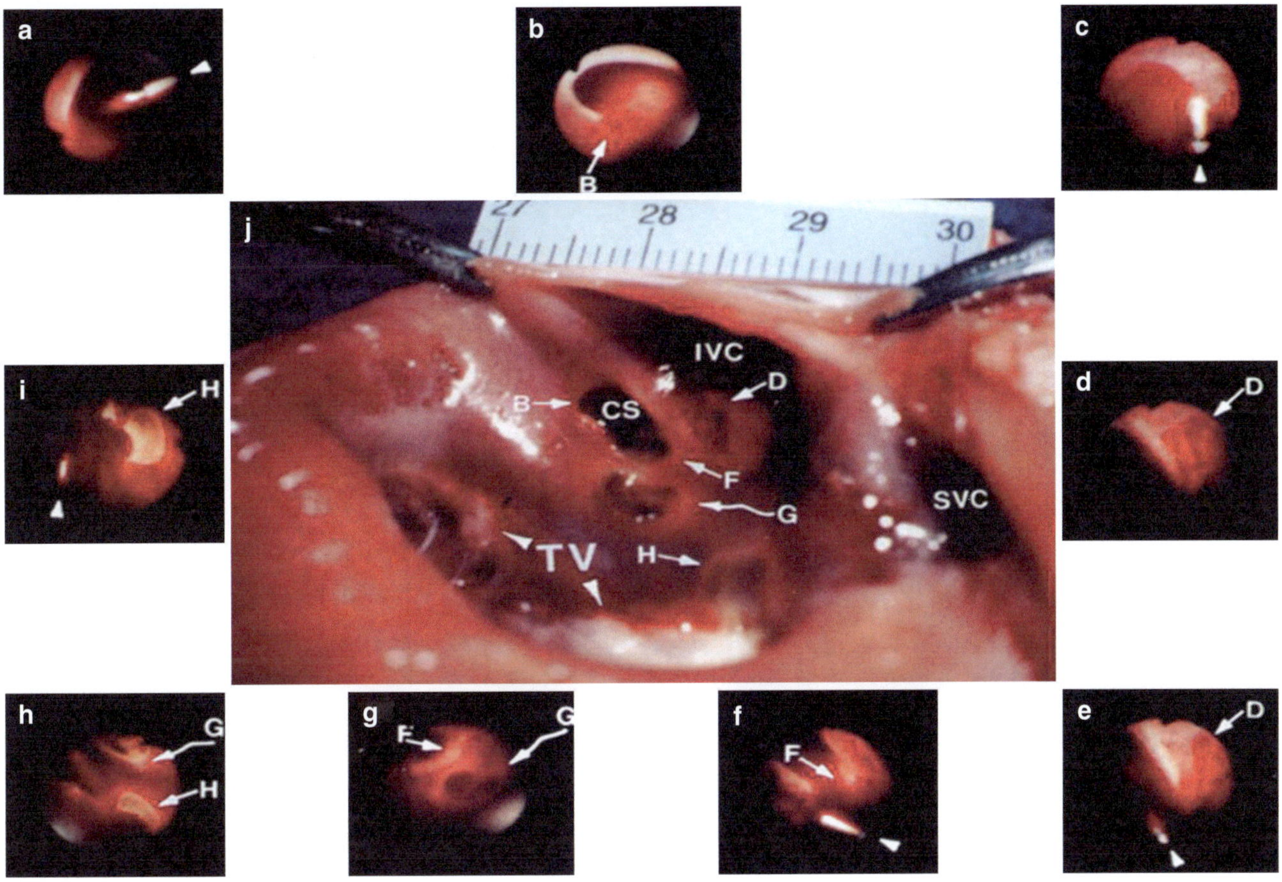

Fig. 10.2 Visualization of endocardial structures. (**a**, **b**) Coronary sinus ostium. (**a**) Dark, unsaturated blood. (**b**) Strand-like structure in orifice. (**c**) Purkinje network. (**d**) Tricuspid leaflet at end-systole. (**e**) Right atrial pectinate muscle. (**f**) Heart worm-like structure near superior vena cava. (**g**) Postmortem appearance (From Kuo and Koch[11], with permission) (h) and (**i**) lesion in A-V node region (**j**) appearance and lesion location

was then inflated with 5–10 mL of normal saline. Normal saline was found to be superior to air because the image it generated was clearer. The endoscope was used to image the right atrium, superior vena cava, tricuspid valve, and right ventricle. Further, in order to guide advancement of an electrode catheter into the coronary sinus, the endoscope was positioned so that the coronary sinus ostium could be visualized and the electrode catheter was then advanced into the coronary sinus. Radiofrequency ablation could also be visualized using the endoscope, including both the tip of the radiofrequency ablation catheter and the ablation lesion. Kuo et al. demonstrated that inflation of the balloon up to 25 mL in the small canine right atrium decreased the mean cardiac output from 4.2 ± 1.6 to 3.6 ± 1.3 L/min, but did not cause significant hemodynamic compromise. Kuo et al. described the anatomy of intracardiac structures in great detail (Fig. 10.2) including the semi-lunar structure of the Thebesian valve of the coronary sinus and the appearance of the floor of the coronary sinus ostium. They described the difference in color of the dark unsaturated blood draining from the coronary sinus. The investigators could even identify small venules draining directly into the right atrium. They could also define changes in the contour of the atrial endocardium, identifying the trabeculated and smooth portions accurately. The interatrial septum and the fossa ovalis could be accurately identified as could the trileaflet structure of the tricuspid valve. An area of yellowish appearance was identified as the location of the central fibrous body and the atrioventricular node. Using both videotape and photography, Kuo et al. recorded the intracardiac structures described. Motion of the cardiac valves could be appreciated using videotape recording. The endocardial trabeculation of the right ventricle and the related Purkinje network could be observed. Radiofrequency ablative lesions were delivered in order to later correlate the anatomic position with the position of the endoscopic visualization. The investigators noted that sometimes the volume of the balloon interfered with the ability to move the electrode catheter. On the other hand, the balloon catheter could also be used to stabilize the ablation catheter against the wall of the heart. On several occasions the balloon was perforated during ablation, emphasizing the need to utilize heat-resistant materials if ablation is used during imaging.

Fujimura et al.[12] used a 36-mm endoscope with a latex balloon at its distal end to image catheter ablation. The endoscope was inserted via femoral vein cutdown and advanced into the right atrium without the need for fluoroscopy. During

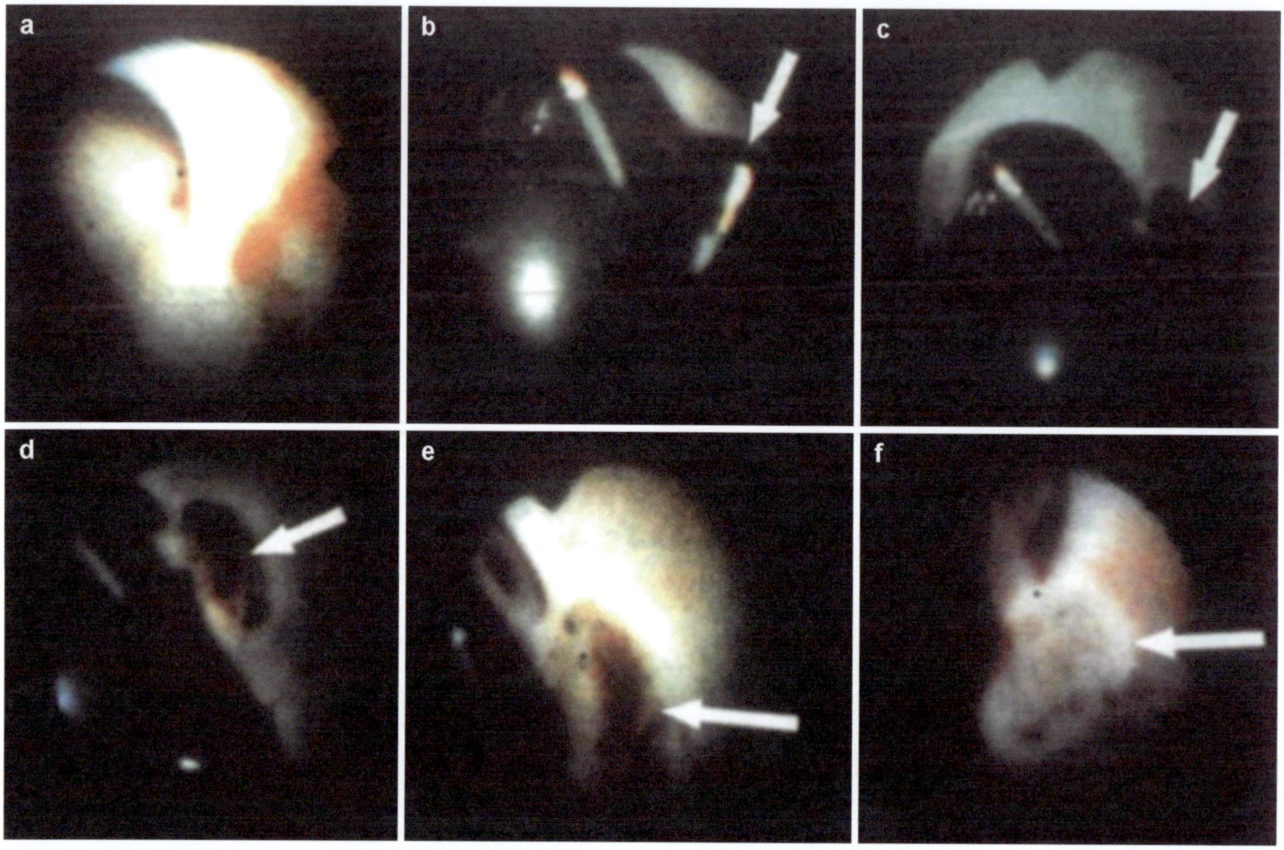

Fig. 10.3 Visualization of radiofrequency lesions. (**a**) Pre-ablation normal tissue. (**b**) Image during ablation with 20 W for 30 s. No change in tip of catheter seen. (**c**) Lesion immediately after energy delivery. (**d**) Close-up view of lesion in **c**. Lesion is covered by black coagulum. (**e**) Same lesion 5 min later with less coagulum and yellowish appearance. (**f**) Same lesion 30 min later with coagulum washed away and well-demarcated yellowish lesion (From Fujimura et al.[12], with permission)

radiofrequency ablation with a steerable ablation catheter, the endoscope visualized the ablation process and lesion formation. The investigators noted that frequently ablation lesions were covered with a black coagulum when initially visualized using the endoscope. After 5 min, however, the coagulum was no longer present and the ablation lesion appeared yellowish or reddish in color (Fig. 10.3). They noted that there was no significant bubble formation during creation of the ablation lesions. Lesions observed ranged in size from 38 to 314 mm^3. As in the report by Kuo et al., one balloon was destroyed during radiofrequency energy delivery.

Although most of the groundbreaking work was done in the mid-1990s, there was very little work in the field until after 2000. One of the largest breakthroughs in intracardiac imaging was the development of a 7F fiber-optic deflectable visualization catheter (previously made by Acumen Medical, Inc., Sunnyvale, CA) (Fig. 10.4).[13] The catheter has a distal imaging lens positioned within a 15-mm saline-filled balloon. The image is carried via a coherent fiber bundle (6,000–7,000 pixels) to an imaging sensor located in the device handle outside the body. A visible light source within the device handle delivers the light via a fiber bundle for illumination. Real-time video output was achieved via a cable to a standard video monitor.

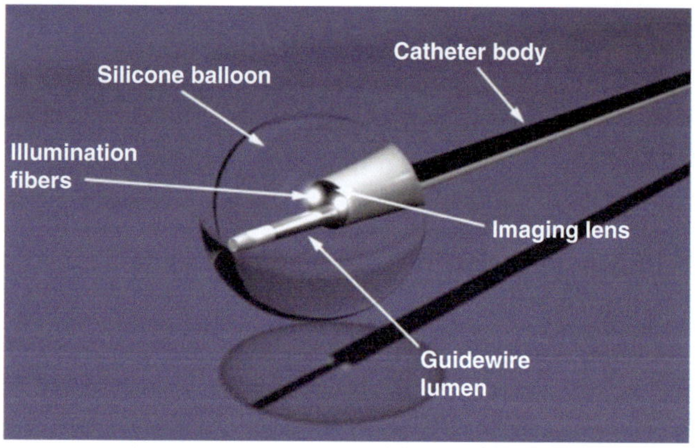

Fig. 10.4 Saline-filled balloon steerable endoscope. A compliant saline-filled balloon surrounds the imaging element. Imaging is achieved using a distal imaging lens and illumination fiber optics. A 0.032″ lumen extends from the catheter (From Eversull et al.[13], with permission)

In a study by Irani et al.[14] radiofrequency ablation lesions were made in vitro in excised porcine hearts. Gaps between lesions were purposefully created to determine whether gaps in ablation lines could be accurately identified. In total, there were 230 ablation lesions with 94 gaps. Figure 10.5[14] shows a comparison of images from the visualization catheter with photographs of the gross specimen. Lesions varied in size from approximately 3 to 4 mm in diameter. The mean gap distance was 2.6 mm ± 1.7 mm with a median gap distance of 2.4 mm. Inter-lesion gaps were identified by direct visualization with greater than 98% accuracy. In addition, Eversull et al. demonstrated in vivo that it is possible to observe radiofrequency ablation in real time using this visualization device and to identify the ablation lesion after creation in vivo (Fig. 10.6).[13] In this study, there was no damage to the visualization balloon or imaging system from the radiofrequency ablation. Eversull et al. demonstrated that in vivo direct visualization is capable of providing remarkable detail in imaging structures such as the right atrial appendage (Fig. 10.7).[13] At present there are no commercially available visualization balloon devices.

10.2.3 Visualization for Placement of Pacing Leads for Cardiac Resynchronization

The visualization device described by Eversull et al. was developed initially for the purpose of guiding the placement of left ventricular pacing leads via the coronary sinus. Anh et al.[15] were able to visualize the coronary sinus ostium in 98 of 100 consecutive patients undergoing left ventricular lead placement. In these 98 patients, 53 patients had an identified Thebesian valve. Anh et al.[15] described in humans the extremely variable anatomy of the coronary sinus valve using this technique. The high resolution of imaging real time provided the ability to define anatomic variations such as cardiac venous branches and Thebesian valves (Fig. 10.8). In a separate study, Ahn et al.[16] described the ability of this visualization catheter to facilitate coronary sinus cannulation. In 58 consecutive patients, visualization of the coronary sinus was possible in all patients. The mean time from insertion of the visualization catheter to visualization of the coronary sinus ostium was 6 ± 5 min. The overall success rate for implantation of the left ventricular lead was 55 of 58 patients (94.8%). The mean time from coronary sinus visualization to cannulation of the coronary sinus was 2 ± 4 min. However, there is no comparison non-visualization group in this study. In four cases, the visualization catheter could not be advanced into the coronary sinus, but in 3 out of 4 of these cases, left ventricular lead placement was still successful by other means.[15,16]

10.2.4 Non-balloon Visualization Systems

Although most of the approaches for intracardiac visualization have included a balloon to exclude the blood pool, a novel design for direct visualization involves a hood-like structure that permits a low volume flow of saline to push away the blood in order to image endocardial structures. In a study by Thiagalingam et al.[17] this technique is used to visualize the atrial septum for the purposes of guiding transseptal puncture. These investigators used a 12 French IRIS visualization catheter (Voyage Medical, Inc., Redwood City, CA) with a 12-mm collapsible hood that has a 3-mm eccentric opening and eight 1-mm openings around the perimeter of the hood (Fig. 10.9). The IRIS catheter is flushed continuously at the rate of 33 mL per min. Once the atrial septum was visualized, the catheter was positioned at the fossa ovalis and a needle and subsequently a guidewire were advanced across the atrial septum. The guidewire was then withdrawn and the IRIS catheter was again positioned at the atrium septum (Fig. 10.9). The site of the previous puncture was visualized and a second puncture was performed (Fig. 10.10). In this study of 6 female pigs, the fossa ovalis surface area was measured at 82.0 ± 32.3 mm^2. There were 12 punctures successfully performed in the 6 animals, requiring 6.8 ± 3.6 min and 299 ± 94 mL of fluid. The site of each puncture was clearly visible. This study demonstrates the feasibility of using direct visualization to facilitate transseptal puncture.

10.2.5 Infrared Imaging of Intracardiac Structures

Although blood must be excluded from the visual field to directly visualize the endocardial surface using visible light, the absorption spectrum of blood exhibits local minima which permit visualization of structure through blood using the infrared range of the electromagnetic spectrum. A novel infrared endoscope was developed by CardioOptics, Inc. (Boulder, CO) to enable visualization of intracardiac structures. Nazarian et al.[18] demonstrated the ability of this flexible, steerable infrared endoscope to visualize the coronary sinus and its branches to guide placement of left ventricular pacing leads for cardiac resynchronization.

10.2.6 Combination of Balloon Ablation and Visualization

With the development of balloon ablation systems to isolate the pulmonary veins for the treatment of atrial fibrillation, a new strategy for combining ablation and visualization has become possible. A laser ablation device developed by CardioFocus, Inc. (Marlborough, MA) uses a fluid-filled

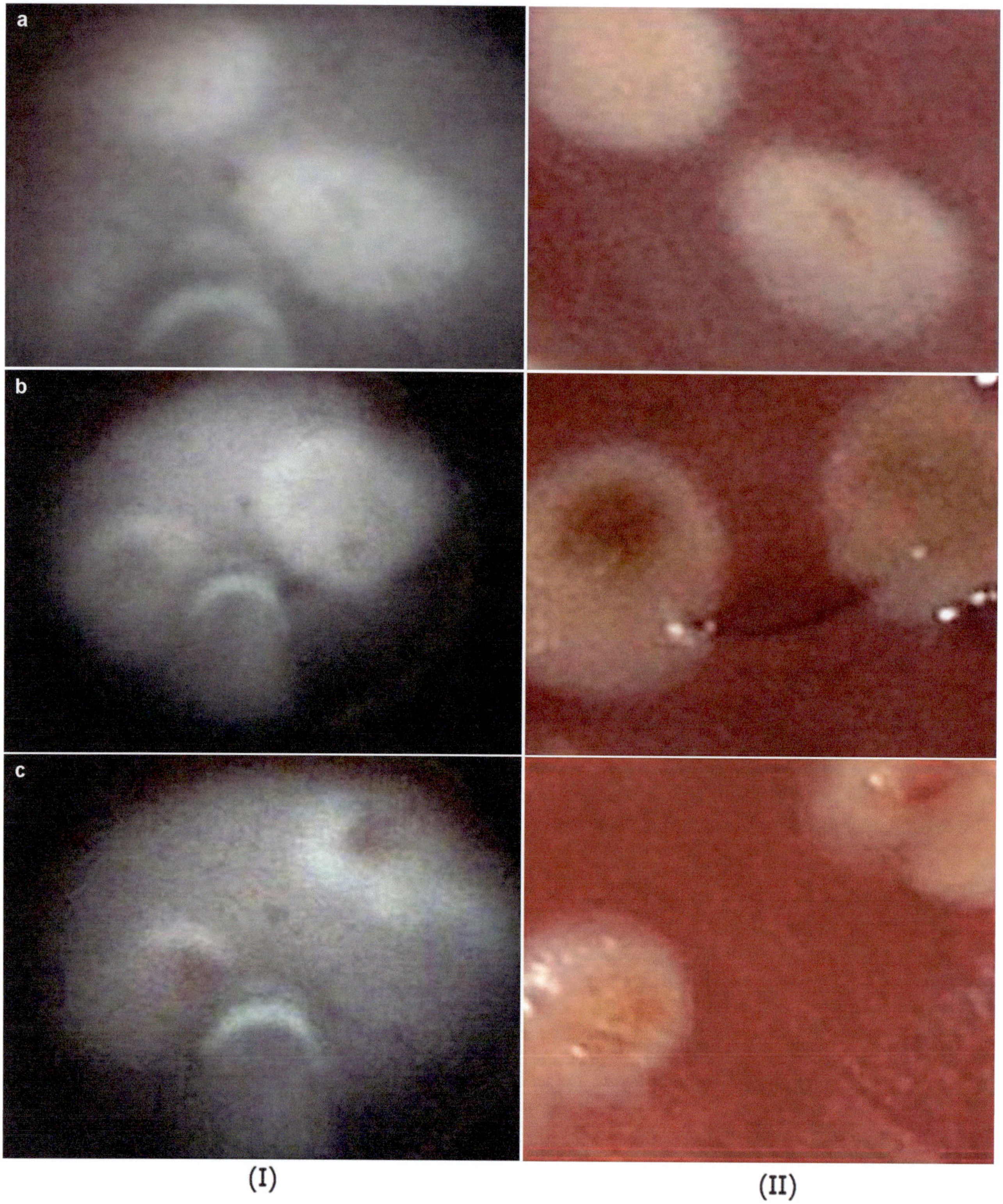

Fig. 10.5 In vitro direct visualization of an ablation lesions using saline-filled balloon steerable endoscope. In the *left column* are the endoscopic images and in the *right column* are the corresponding photographs of the gross specimen. Gaps in the ablation lines were purposely made. Three gaps between two ablation lesions are shown here, measuring 0.5 mm (**a**), 1.6 mm (**b**), and 3.3 mm (**c**) (From Irani et al.[14], with permission)

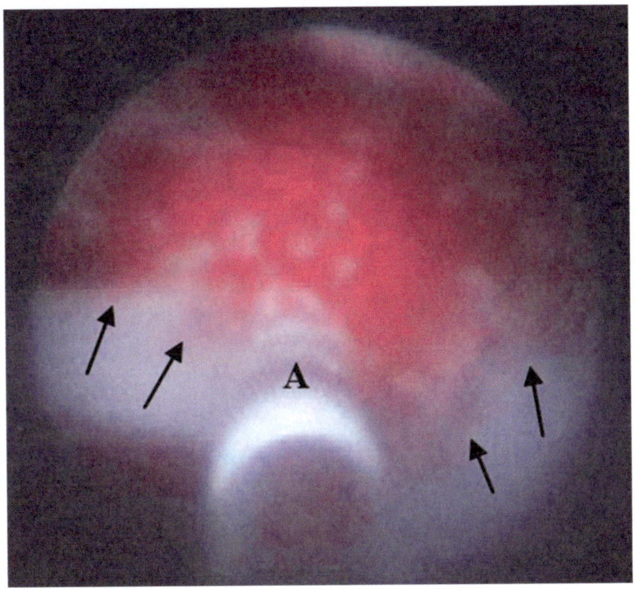

Fig. 10.6 In vivo direct visualization of an ablation lesion in the right atrium using saline-filled balloon steerable endoscope. A conductive element was passed through the central lumen to the distal tip (*A*) of the direct visualization catheter and radiofrequency energy was delivered via this element. A confluence of previously created lesions appears in the *upper half* of the field of view with pristine endocardial tissue seen in the *lower half* of the field of view with *arrows* denoting the lesion borders (From Eversull et al.[13], with permission)

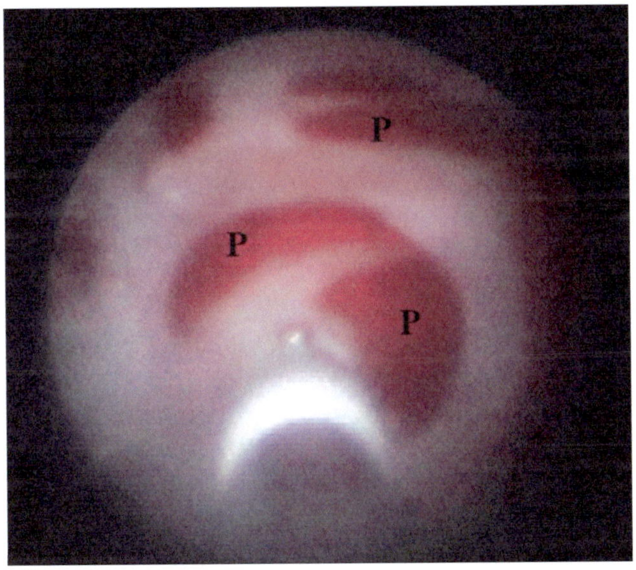

Fig. 10.7 In vivo direct visualization of the right atrial appendage using saline-filled balloon steerable endoscope. The blood-filled pockets (*P*) accentuate the atrial appendage (From Eversull et al.[13], with permission)

balloon to create an optically transparent path to cardiac tissue. The device utilizes an aiming beam comprised of visible light in order to guide positioning of a therapeutic laser beam used to create ablation lesions. By positioning an endoscope within the laser balloon, visualization of lesion formation during laser energy delivery is possible. Themistoclakis et al.[19] used a 12 French ablation catheter to deliver continuous-wave laser light energy projected as a ring through a 25-mm balloon. In this study of five patients, 16 of 18 pulmonary veins could be accessed using this device. Based on the endoscopic appearance, complete circumferential contact was confirmed in 15 out of 16 of these pulmonary veins. Circumferential contact was also determined using intracardiac echocardiography and color Doppler flow. Two gaps in contact were seen by the fiber-optic endoscope. The endoscope was not able to separate the left atrium–pulmonary vein junction from the distal portion of the vein. The ability to visualize contact and to adjust the location of the laser energy delivery may be critical in achieving circumferential lesion formation, avoiding gaps in ablation, and achieving optimal and long-lasting ablative lesions.

Schmidt et al.[20] used a second-generation endoscopic laser ablation system which utilizes a compliant balloon to perform ablation in 30 patients with paroxysmal atrial fibrillation. The balloon is capable of ablating pulmonary veins as small as 9 mm and as large as an average of 32 mm. Each individual ablation lesion covers 30° of arc and is overlapped by 30–50% to create a contiguous lesion. The ablation lines were created as a figure-of-eight at the ipsilateral pulmonary vein ostia. The ablation system was used to treat a total of 116 PVs (four patients with LCPV) with the endpoint of pulmonary vein isolation being achieved in 114/116 (98%) of pulmonary veins. In 12/60 pulmonary veins (20%) with an individual isolation attempt pulmonary vein isolation was not achieved after a single ablation circle. Following mapping with a spiral catheter to determine the conduction gap, all pulmonary veins were successfully isolated with a median of six laser applications.

In a separate study, Dukkapati[21] performed ablation using the laser ablation system in 101 pulmonary veins in 27 patients. Complete occlusion and positioning of the balloon is demonstrated by contrast injection (Fig. 10.11). The endoscopic system permits visualization of the pulmonary vein anatomy, the position of the aiming beam, and ablation lesion that is formed (Fig. 10.11). Using the endoscopic visual guidance, 84.2% (85/101) PVs were isolated with the first attempt for visually guided circumferential placement of contiguous point-by-point lesions. Using the same spiral catheter mapping technique as Schmidt et al.,[19] 100% of the remaining pulmonary veins were isolated, resulting in an average of 1.3 attempts to isolate each vein.

10.2.7 Future Visualization Ablation Systems

Future visualization systems will utilize technology that will optimize the ability to see ablation lesions in the human myocardium. In addition, they will combine visualization

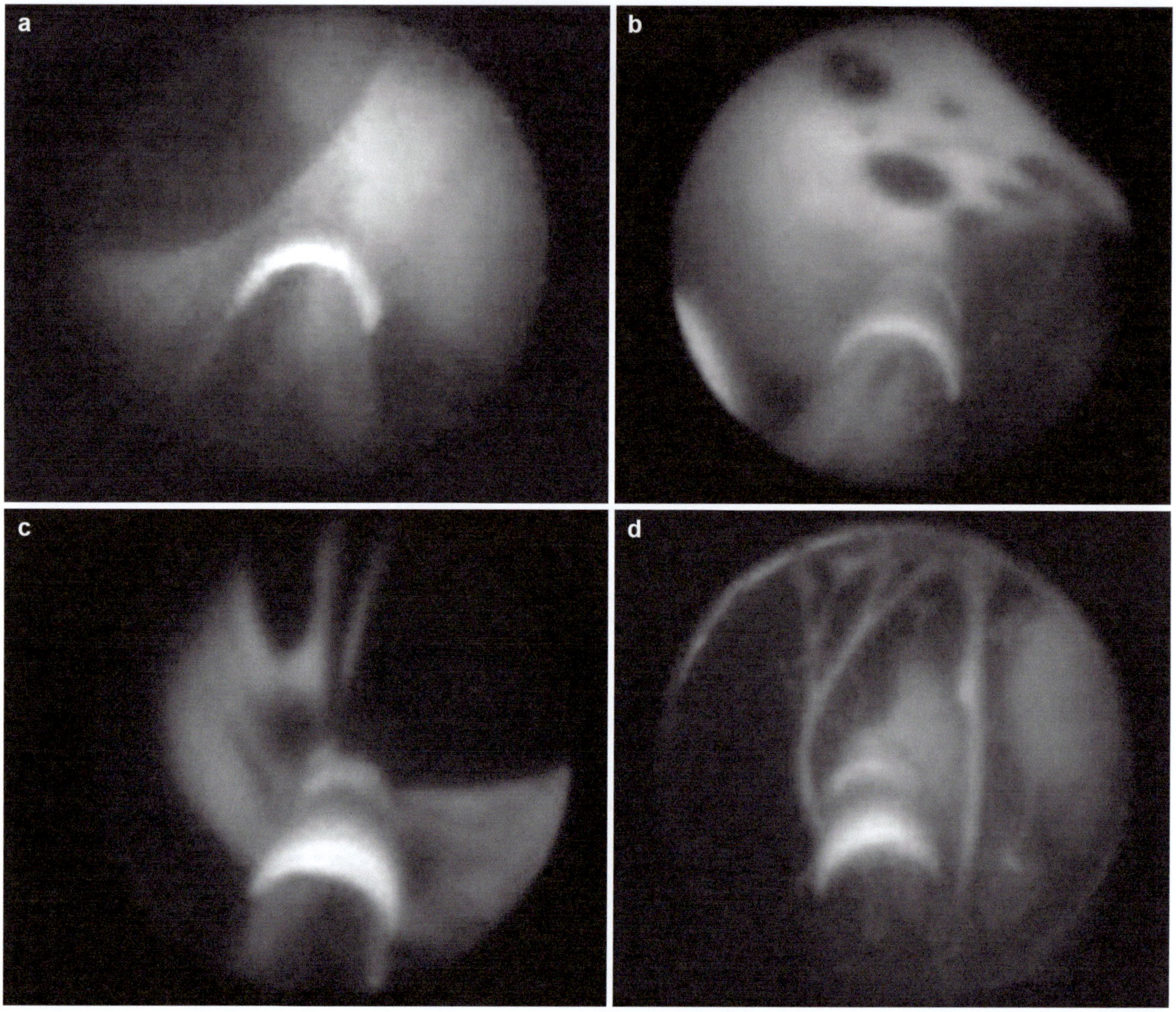

Fig. 10.8 Images seen from the fiber-optic saline-filled balloon catheter. (**a**) Simple semi-lunar valve. (**b**) Fenestrated valve. (**c**) Valve and strands and bands within the ostium. (**d**) Strands and bands within the ostium (From Irani et al.[14], with permission)

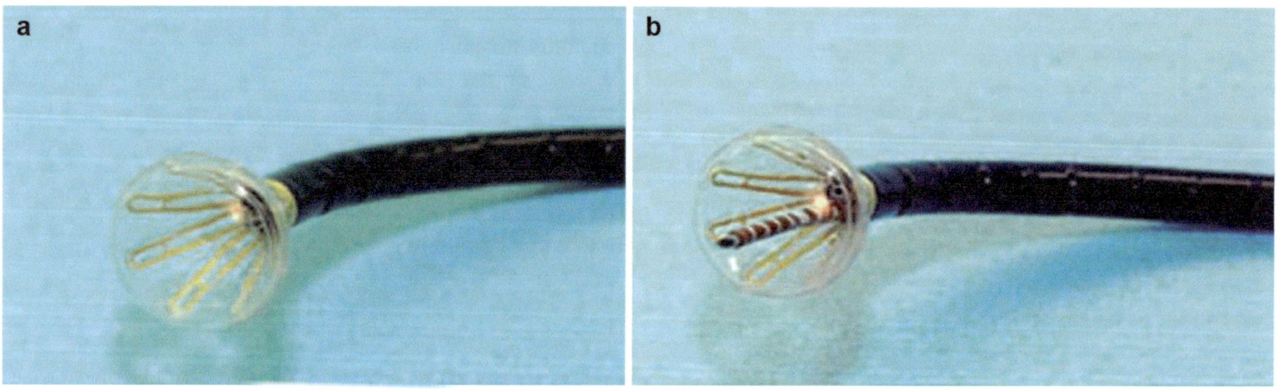

Fig. 10.9 The IRIS visualization catheter has a hood that is kept open by gold-plated nitinol struts (**a**). The needle sheath (*brown and white striped*) and needle were extended (**b**). The catheter has been flexed (**c**). The fiberscope has 10,000 pixel imaging (From Thiagalingam et al.[17], with permission)

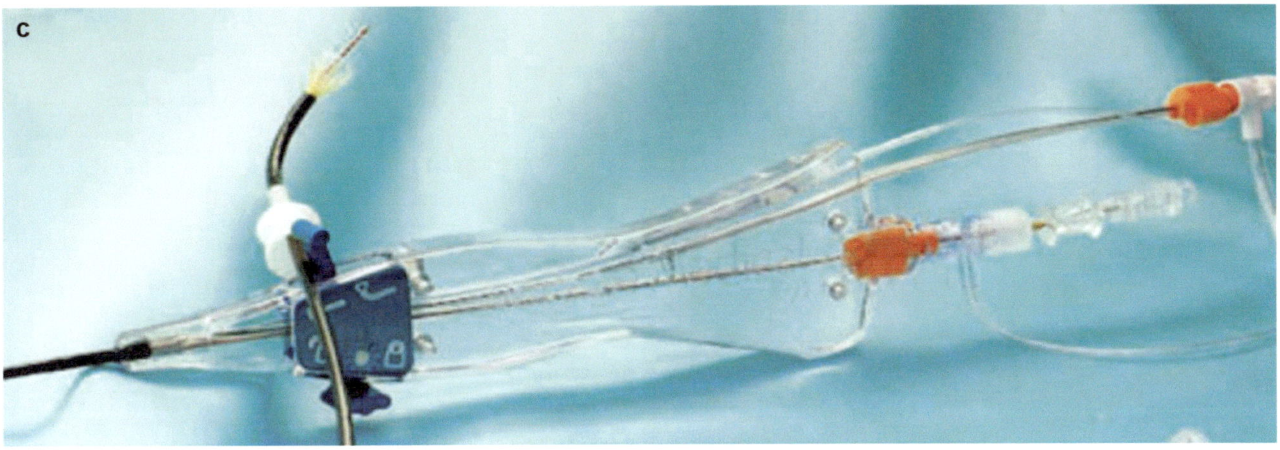

Fig. 10.9 (continued)

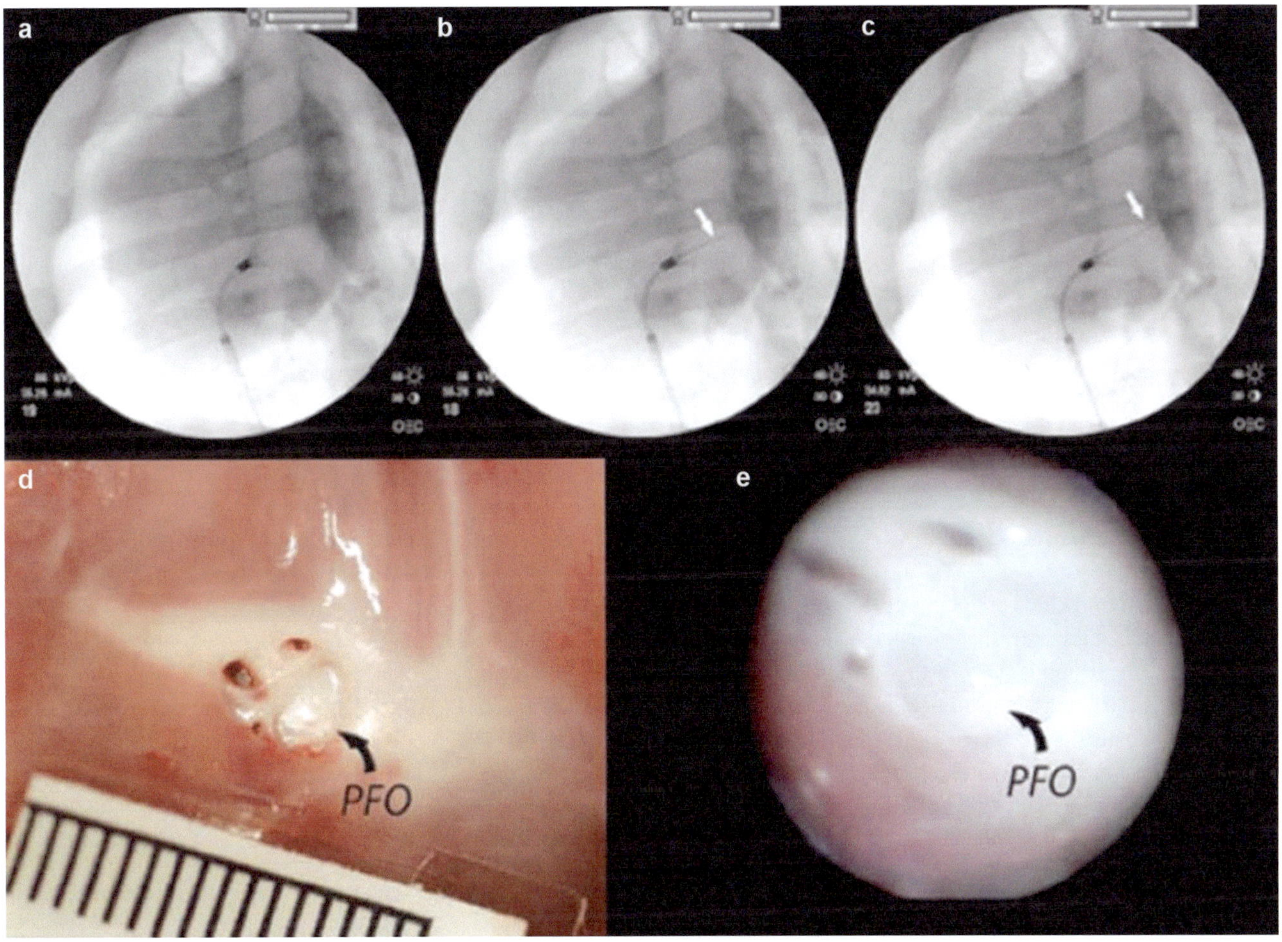

Fig. 10.10 The IRIS catheter is positioned at the fossa ovalis (**a**). The transseptal needle is advanced (**b**). The needle is withdrawn and the wire is advanced into the left atrium (**c**). A postmortem specimen (**d**) and the image from the IRIS catheter (**e**) are shown. There are three atrial septal punctures proximal to the patent foramen ovale (From Thiagalingam et al.[17], with permission)

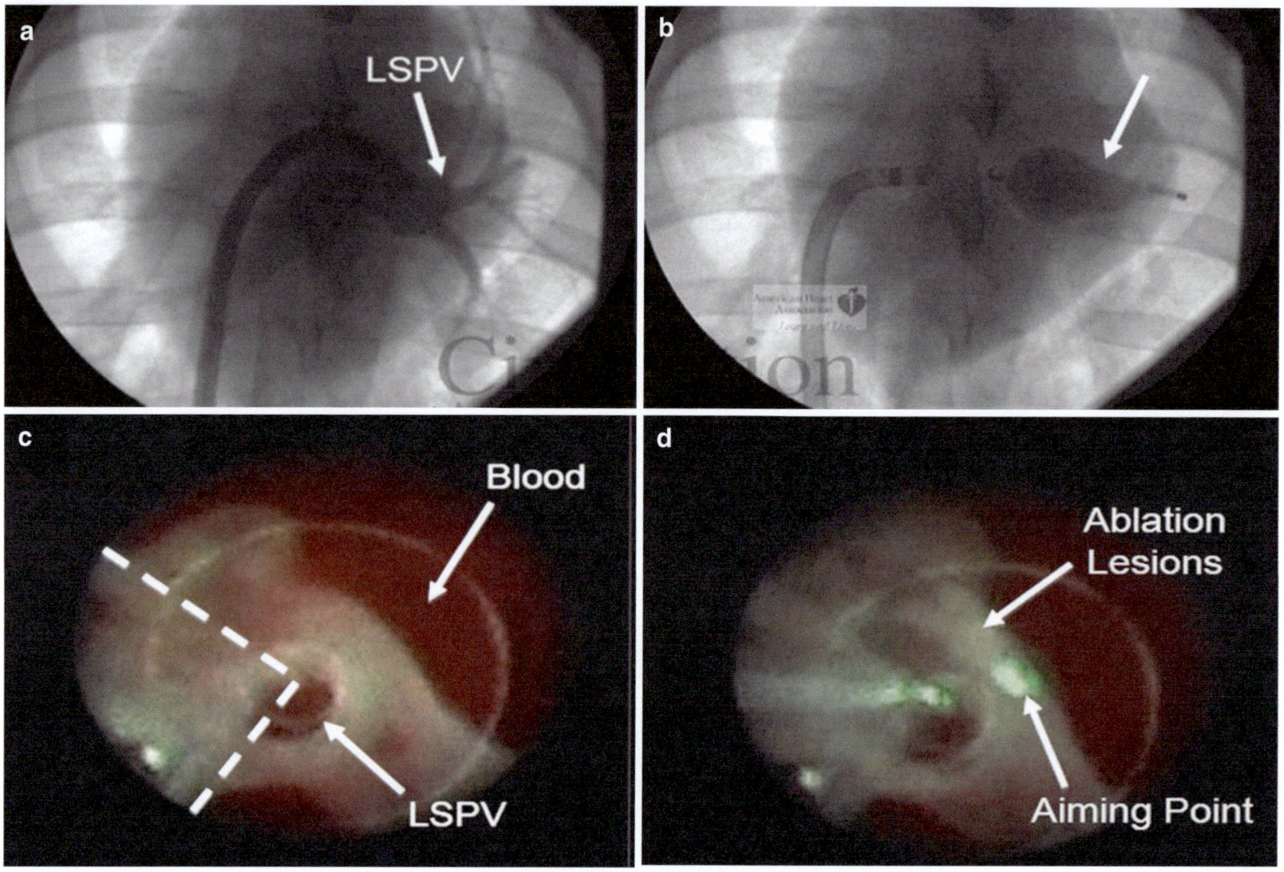

Fig. 10.11 The positioning of the Cardiofocus endoscopic light ring balloon catheter demonstrating complete circumferential occlusion (**a**, **b**), with endoscopic view of aiming point and ablation lesion (**c**, **d**) (From Schmidt et al.[20], with permission)

with other ablation energies such as radiofrequency, ultrasound, and cryoablation. Future visualization systems will also be designed to facilitate lead placement.

10.3 Conclusion

Although the foundations of direct intracardiac imaging were laid several decades ago, the pace of technology development and clinical application in this area have accelerated during the last 10 years. The clinical feasibility, and in some cases clinical utility, of this imaging modality have been demonstrated for device implantation and catheter-based ablation. Despite the advances in this technology, there remain a number of areas where further development might be anticipated. It is likely that as image sensors become smaller and less expensive, the sensors positioned at the distal catheter tip will eliminate the need to use coherent fiber bundles for image transmission. The resulting increase in imaging resolution will yield more anatomical detail and facilitate larger fields of view. Improved blood displacement elements and the combination of visible light imaging with infrared imaging may provide for greater depth of imaging field within the heart. Electromagnetic energy in and around the visible spectrum may be used to selectively enhance or differentiate the appearance of the specific features of interest, such as newly created ablation lesions. Overall, the continued use of direct visualization in the electrophysiology laboratory has the potential to improve navigation within the heart and improve procedural outcomes.

References

1. Sakikibara T. Cardiosocope. *Nippon Geka Gakkai Zasshi (J Japanese Surg Soc)*. 1939;40:905-911.
2. Silander T. Cardioscopy without thoracotomy. *Acta Chir Scand*. 1964;127:67-84.
3. Gamble WI, Innis RE. Experimental intracardiac visualization. *N Engl J Med*. 1967;276:1397-1403.
4. Shure D, Moser KM, Harrell JH, Hartman MT. Identification of pulmonary emboli in the dog: comparison of angioscopy and perfusion scanning. *Circulation*. 1981;64:618-621.
5. Shure D, Gregoratos G, Moser K. Fiberoptic angioscopy: role in the diagnosis of chronic pulmonary arterial obstruction. *Ann Intern Med*. 1985;103:844-850.
6. Sherman CT, Litvack F, Grundfest W, et al. Coronary angioscopy in patients with unstable angina. *N Engl J Med*. 1986;315(15): 913-919.

7. Lee G, Garcia JM, Corso PJ, et al. Correlation of coronary angioscopy to angiographic findings in coronary artery disease. *Am J Cardiol*. 1986;58(3):238-241.
8. Uchida Y, Tomaru T, Nakamura F, Sonoki H, Sugimoto T. Fiberoptic angioscopy of cardiac chambers, valves, and great vessels using a guiding balloon catheter in dogs. *Am Heart J*. 1988;115:1297-1302.
9. Uchida Y, Nakamura F, Tsukamoto M, You S, Kido H, Sugimoto T. Percutaneous ventricular endomyocardial biopsy with angioscopic guidance. *Am Heart J*. 1989;118:1039-1041.
10. Uchida Y, Tomaru T, Nakamura F, Oshima T, Fujimori Y, Hirose J. Percutaneous fiberoptic cardioscopy of the left ventricle. *Jpn Heart J*. 1991;32:455-471.
11. Kuo CS, Koch CA. In vivo angioscopic visualization of right heart structure in dogs by means of a balloon-tipped fiberoptic endoscope: potential role in percutaneous ablative procedures. *Am Heart J*. 1994;127:187-197.
12. Fujimura O, Lawton MA, Koch CA. Direct in vivo visualization of right cardiac anatomy by fibreoptic endoscopy: observation of radiofrequency-induced acute lesions around the ostium of the coronary sinus. *Eur Heart J*. 1994;15:534-540.
13. Eversull CS, Lin B, Irani A, et al. Direct visualization of cardiac radiofrequency ablation lesions. *J Cardiovasc Transl Res*. 2009;2:198-201.
14. Irani A, Lin B, Eversull C, et al. Visualizing ablation gaps in vitro using a deflectable fiber optic endocardial visualization catheter. *J Interv Card Electrophysiol*. 2009;25:107-110.
15. Anh DJ, Eversull CS, Chen HA, et al. Characterization of human coronary sinus valves by direct visualization during biventricular pacemaker implantation. *Pacing Clin Electrophysiol*. 2008;31:78-82.
16. Anh DJ, Chen HA, Eversull CS, et al. Early human experience with use of a deflectable fiberoptic endocardial visualization catheter to facilitate coronary sinus cannulation. *Heart Rhythm*. 2006;3:875-878.
17. Thiagalingam A, D'Avila A, Foley L, et al. Importance of catheter contact force during irrigated radiofrequency ablation: evaluation in a porcine ex vivo model using a force-sensing catheter. *J Cardiovasc Electrophysiol*. 2010;21(7):806-811.
18. Nazarian S, Knight BP, Dickfeld TL. Direct visualization of coronary sinus ostium and branches with a flexible steerable fiberoptic infrared endoscope. *Heart Rhythm*. 2005;2:844-848.
19. Themistoclakis S, Wazni OM, Saliba W. Endoscopic fiberoptic assessment of balloon occlusion of the pulmonary vein ostium in humans: comparison with phased-array intracardiac echocardiography. *Heart Rhythm*. 2006;3:44-49.
20. Schmidt B, Metzner A, Chun KRJ, et al. Feasibility of circumferential pulmonary vein isolation using a novel endoscopic ablation system. *Circ Arrhythm Electrophysiol*. 2010;3(5):481-488. Epub 2010 Jul 24.
21. Dukkipati S, Neuzil P, Skoda J, et al. Persistent pulmonary vein isolation visual balloon-guided point-by-point ablation: reliable, reproducible, and persistent pulmonary vein isolation. *Circ Arrhythm Electrophysiol*. 2010;3(3):266-273. Epub 2010 May 26.

Contact and Noncontact Electroanatomical Mapping

11

Narayanan Namboodiri, Anthony G. Brooks, and Prashanthan Sanders

Abstract

Cardiac arrhythmias are traditionally mapped by assessing the timing of electrical activation at various locations in the heart (endocardial, epicardial, or both) under fluoroscopic guidance. These observed timing data are interpreted by the physicians, and the potential sites of ablation are targeted. However, the importance of anatomic structures in the initiation and maintenance of cardiac arrhythmia is increasingly being recognized, mandating a better understanding and definition of the patient's cardiac anatomy for successful electrophysiological procedures. There has been a significant evolution of these electroanatomic systems to considerably enhance the electrophysiologists' understanding of real-time position management and mapping, thus to enable more precise catheter navigation in different chambers of the heart. Once the image of the cardiac chamber of interest is created, the ablation catheter is moved to the appropriate target sites. Contact mapping technique involves the direct recording of electrical activity with electrodes positioned in selected places within the endocardial surface of the heart or the epicardium. In contrast, noncontact mapping does not require direct contact with the endocardium; the electrograms and the activation sequence are computed based on far-field interpolated voltage differences generated by myocardial depolarization. Image integration of the generated image with an anatomical image of computerized tomography (CT) or magnetic resonance imaging (MRI), although not traditionally included in the arsenal of electrophysiology, could be seen as useful adjuncts for precise and reproducible catheter guidance.

Keywords

Electroantomical mapping • Contact mapping • Noncontact mapping systems • Cardiac mapping • Catheter ablation guidance

Cardiac arrhythmias are traditionally mapped by assessing the timing of electrical activation at various locations in the heart (endocardial, epicardial, or both) under fluoroscopic guidance. These observed timing data are interpreted by the physicians, and the potential sites of ablation are targeted. However, the importance of anatomic structures in the initiation and maintenance of cardiac arrhythmia is increasingly being recognized, mandating a better understanding and definition of the patient's cardiac anatomy for successful electrophysiological procedures.

Appreciation of the three-dimensional cardiac chamber anatomy has been substantially improved by the computer-generated 3D-representation of data using electroanatomic systems. These can objectively reinforce a physician's subjective interpretations. Even though conventional electrophysiological mapping is adequate for the majority of cardiac arrhythmias, advanced mapping systems may facilitate the mapping of complex arrhythmias and reduce fluoroscopic

P. Sanders (✉)
Department of Cardiac Electrophysiology, Centre for Heart Rhythm Disorders, University of Adelaide and Royal Adelaide Hospital, Adelaide, SA, Australia
e-mail: prash.sanders@adelaide.edu.au

A. Auricchio et al. (eds.), *Cardiac Imaging in Electrophysiology*,
DOI 10.1007/978-1-84882-486-7_11, © Springer-Verlag London Limited 2012

exposure while potentially reducing the procedure duration and risk. Electroanatomic 3D mapping techniques, both contact and noncontact, utilize specialized catheters and/or patches to determine position within the heart. This positional information is combined with electrical timing data to provide a map of the cardiac chamber of interest in these computer-based mapping systems. These new mapping systems can function in a complementary role to the conventional mapping technique, or can be used independently.

There has been a significant evolution of electroanatomic systems to considerably enhance the electrophysiologists' understanding of real-time position management and mapping to enable more precise catheter navigation in different chambers of the heart. Once the image of the cardiac chamber of interest is created, the ablation catheter is moved to the appropriate target sites. Contact mapping technique involves the direct recording of electrical activity with electrodes positioned in selected places within the endocardial surface of the heart or the epicardium. In contrast, noncontact mapping does not require direct contact with the endocardium; the electrograms and the activation sequence are computed based on far-field interpolated voltage differences generated by myocardial depolarization. Image integration of the generated image with an anatomical image of computerized tomography (CT) or magnetic resonance imaging (MRI), although not traditionally included in the arsenal of electrophysiology, could be seen as useful adjuncts for precise and reproducible catheter guidance.

11.1 Advantages of 3-D Mapping over Conventional Mapping System

Although successful for many applications, the conventional endocardial catheter mapping has certain inherent limitations. Fluoroscopy cannot reproduce the three-dimensional heart structure using orthogonal fluoroscopic projections. Sequential catheter mapping to determine the activation patterns is cumbersome, relationships between points are not inherently obvious, and exposure to ionizing radiation is often high. This is compounded further by the lack of accurate reproducibility of intracardiac catheter localization by x-ray projection alone.

In contrast, electroanatomic mapping systems help recreate cardiac anatomy, evaluate electrical activation during arrhythmias, allow real-time catheter localization, and guide catheter placement for delivery of radiofrequency current. A system which plots the position of and activation time at a roving mapping catheter assists in identifying sites of early activation for focal arrhythmias, and, combined with entrainment mapping, appears to be useful in identifying critical isthmuses in complex reentry circuits, such as those resulting from atrial scars and incisions late after repair of congenital heart disease. The 3D constructs of electrogram voltage may also help define areas of electrical scarring and infarction. Voltage, local activation timing, and complex fractionated maps are helpful tools displaying electrical signals of the heart superimposed on the 3D heart chamber to demonstrate the complex relation between the anatomical and functional barriers in the complex arrhythmias. In fact, atrial fibrillation, atrial flutter with complex reentrant circuits, scar-related ventricular tachycardia, and postoperative arrhythmias which are not routinely considered for ablation using conventional mapping systems can be treated effectively with 3D assistance.

While the greatest benefit of electroanatomic systems have been seen with mapping and ablation of complex arrhythmias, it is well documented that these systems can significantly reduce fluoroscopic exposure to the operator and patient.[1,2] Theoretically, by the availability of better anatomical representation, these systems should facilitate the reduction in complications; however, this has not been reported to date. Importantly, it has been shown, particularly in AF ablation, that the use of these systems may produce better clinical outcomes.[3]

11.2 Contact Electroanatomic Mapping Systems

11.2.1 Overview

At present, four real-time catheter localization systems (EnSite NavX, CARTO XP, Real-time Position Management system and Electro View) utilizing different technologies are available for clinical use; however, CARTO and EnSite NavX systems have market dominance.

The EnSite NavX® system (Endocardial Solutions, St. Jude Medical, Inc., St. Paul, MN, USA) is capable of displaying 3D positions of multiple catheters based on impedance measurement derived from the voltage gradient that appears across the tissue when a current is applied through a pair of surface electrodes.[4] For 3D navigation, six surface patch electrodes are placed in three orthogonal pairs: anterior to posterior, left to right lateral, and superior (neck) to inferior (left leg) with the heart at the intersection of these axes. A low-level, 5.6 kHz current is alternately delivered through each pair of these electrodes to form a 3D transthoracic electrical field. The absolute range of voltage along each axis (X, Y and Z) varies from each other, influenced by the volume and type of tissue subtended between each pair of surface electrodes. Each level of impedance, as calculated by dividing the voltage gradient by the known applied current along each axis, corresponds to a specific anatomic location within the thorax. As catheters are maneuvered within the heart, each catheter electrode senses the corresponding level of impedance, derived from the measured voltage. Timed

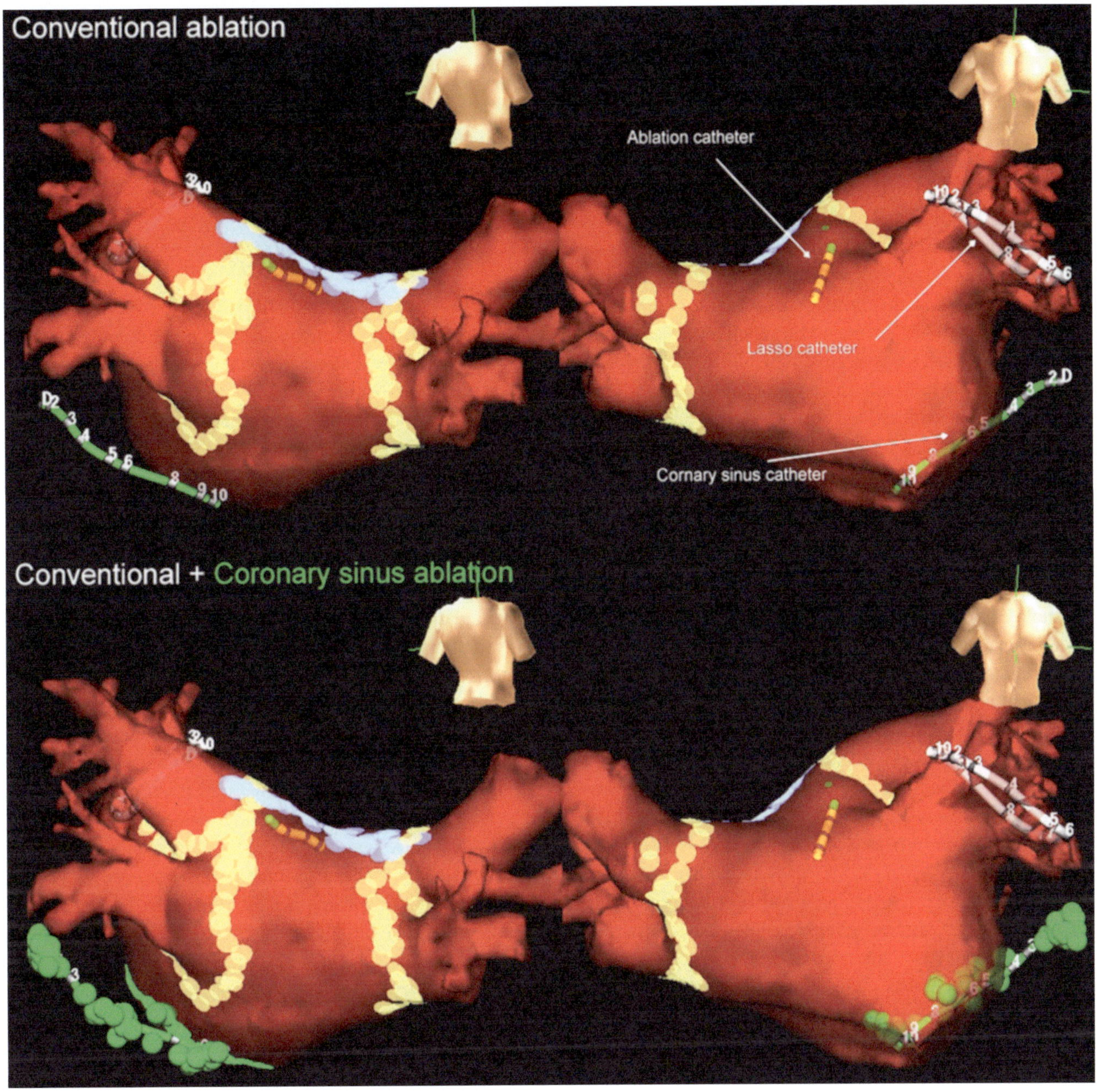

Fig. 11.1 The linear ablation points – pulmonary vein isolation (*yellow in color*), roofline (*blue*) and coronary sinus (*green*) – are marked on the chamber geometry created by EnSite NavX system and fused with the CT image (NavX fusion) in a patient with permanent atrial fibrillation. The intracardiac catheters – circular multipolar (lasso), quadripolar ablation, and decapolar coronary sinus catheters – are seen in real time within the geometry

with the current delivery, NavX calculates the X-Y-Z positional coordinates at each catheter electrode to graphically locate the catheters in real time (Fig. 11.1). The distance of the intracardiac catheters from each skin patch, and ultimately, their location in space, can be triangulated with either external or intracardiac positional references. Up until its recent release (Version 7) the nonlinearity of the impedance field over the heart has limited the anatomical relevance of NavX geometries (i.e., the similarity between the NavX geometry and the reconstructed CT scan), even though they have been shown to provide the same level of navigational accuracy.[5] However, in the latest iteration of NavX, "field scaling" adjusts for impedance nonlinearity against a known electrode spacing constant such that now the NavX geometry is physiologically real. In addition, it can be fused to CT to locate multiple catheters within an anatomically correct structure. The most recent version of EnSite NavX is EnSite Velocity which uses the same technology in a smaller, slimline hardware platform with a significantly updated user interface to expedite 3D mapping.

The CARTO mapping system (Biosense, Diamond Bar, CA, USA) utilizes low-level magnetic fields (5×10^{-6} to 5×10^{-5} T) which are generated by coils within a locator beneath the patient[6-9] to locate a magnetic-tipped catheter in 3D space. A tripod beneath the patient emits three electromagnetic waves at unique frequencies. The location sensors embedded proximal to the tip of a specialized mapping catheter detect the strengths of these magnetic fields and allow computation of catheter position in 3D space as well as catheter orientation (pitch, roll, and yaw). The distance between the sensor and coil determines the strength of each coil's magnetic field. Hence, by integrating each coil's field strength and converting this measurement into a distance, the location of the sensor at catheter tip can be triangulated, when the catheter tip is considered against an external reference patch electrode. In contrast to the NavX system, a point-by-point 3D map is built such that it contains both electrical and anatomical information. CARTO-3 is a recent upgrade of the Biosense mapping platform that provides, for the first time on the CARTO system, the ability to visualize multiple catheters simultaneously. Using Active Current Localization (ACL) technology, which is similar in principle to the NavX impedance-based catheter localization, non-magnetic catheters can be visualized in the CARTO 3D environment. Suffering the same nonlinearity issues as EnSite NavX, the localization of catheters using ACL is adjusted against a "known" location of a magnetic CARTO catheter (i.e., NAVI-STAR), such that the navigational accuracy of non-magnetic catheters improves when they are visualized near a magnetic catheter.

The Real-time Position Management system (Boston Scientific, Natick, MA, USA) uses ultrasound ranging to localize reference and mapping/ablation catheter positions.[10] This ultrasound ranging technique involves a transmitter/receiver unit and two intracardiac reference catheters with ultrasound transducers. The ultrasound-transmitting and -receiving device emits continuous ultrasound energy at 558.5 kHz, which is received by the transducers housed within the reference and ablation catheters. The distance between transducers on the reference catheters is calculated from the velocity of sound transmission in the heart and by measuring the delay between departure and reception of the ultrasound pulse. Once the locations and orientations of reference catheters are established, triangulation algorithms define the position and curve of a mapping catheter. The geometry is generated from inside out, with sequential acquisition of points.

The fourth and more recent system is a simplification of the above concepts. The Electro View is a 3D mapping extension of LabSystem Pro acquisition system from Bard Electrophysiology. In this system, the data points are positioned by the operator on a preexisting geometry of the heart derived from CT or MRI, according to catheter position on fluoroscopic image and the activation map is made (Fig. 11.2). The advantage of this system is that it requires no proprietary equipment other than a conventional Bard Electrophysiology system. The limitation lies in the dependence of localization of catheter position on 3D geometry and hence its possible misalignment. This system is currently undergoing trials for combination with fluoroscopic systems to facilitate navigation.

11.2.2 Principle of Operation

11.2.2.1 Fundamental Principles

The basic principle of any electroanatomic mapping system combines three essential steps to diagnose arrhythmia: (1) outline 3D chamber anatomy; (2) record signal morphology, timing, and voltage; and (3) interpret results to delineate the arrhythmia mechanism, in isolation or combined with other techniques like entrainment mapping.

11.2.2.2 Defining the Chamber Anatomy

Depending on the mapping system, chamber anatomy can be reconstructed by sequentially moving the catheter in the endocardial surface, wherein the location of the roving catheter tip is recorded along with the local electrogram (CARTO) or a more complex geometry is created first and then point-by-point data are projected onto the geometry (NavX). Using either mapping system, the resolution of the map to diagnose arrhythmia depends on the number of data points taken. Higher point numbers reduce anatomical interpolation and in the case of CARTO allow for a finer anatomical resolution.

The NavX system can "project" data onto a more complex previously created geometry; however, the diagnostic accuracy of timing maps is still related to the number of points taken. The creation of geometry using this system is best performed using a multipolar catheter as the system allows the simultaneous and automatic collection of geometry points. These then need to be edited offline to exclude erroneous points due to pressure effects on the surface. The use of a multipolar catheter facilitates the collection of a dense geometric data set with greater number of system references to correct changes in impedance (a process called field scaling) and thereby provide an accurate anatomical image. The benefit of this system is the rapid nature of the geometry creation. The latest iteration of this system (EnSite-NavX velocity) allows the creation of the geometry simultaneously with acquisition of electrophysiological information. The CARTO system to date has collected the geometry and electrophysiological information at the same time. However, the process requires the physical acquisition of a point to acquire either information; therefore, the creation of the geometry can be a longer process. The latest iteration of this system (CARTO-3) is adapted to also acquire geometry using a multipolar catheter using additional impedance mapping techniques as described above.

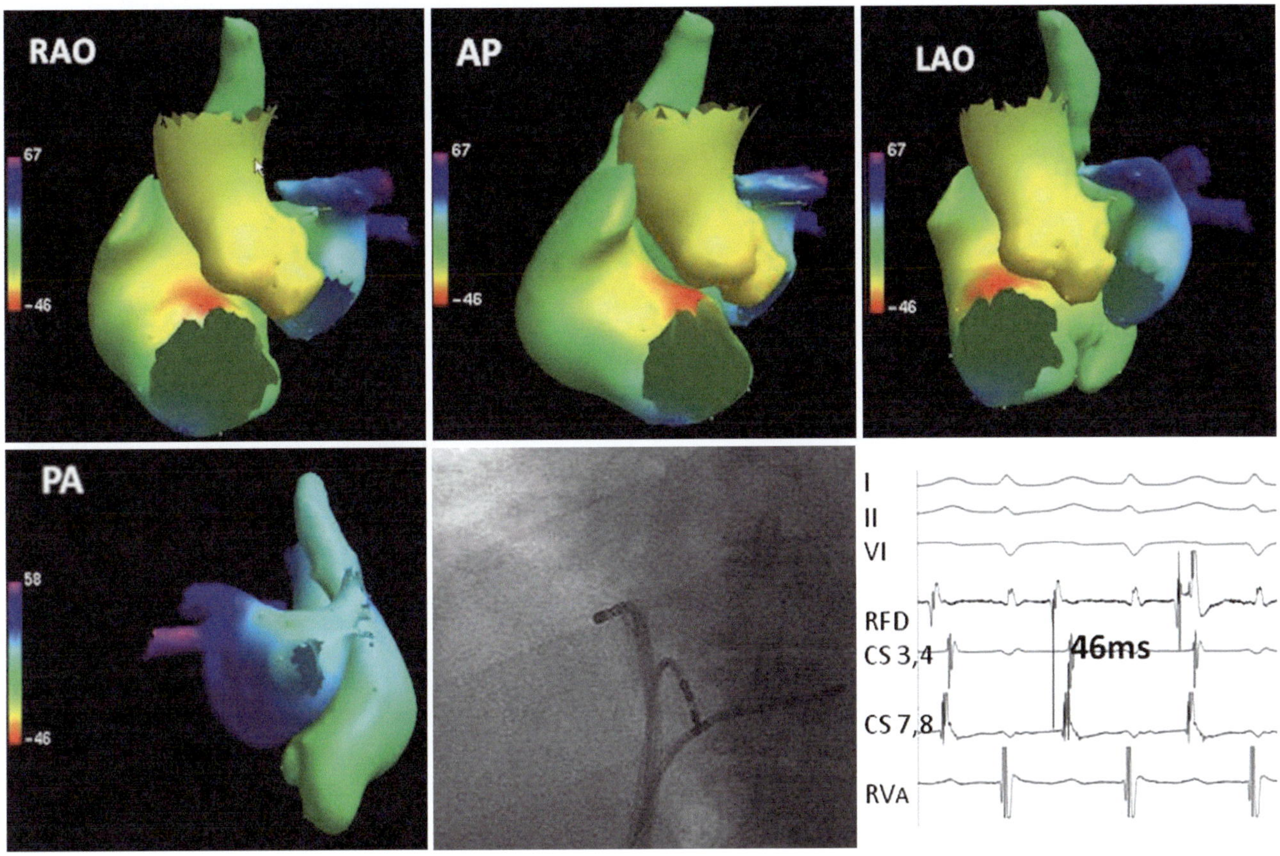

Fig. 11.2 Electro view mapping of right atrium in a patient with atrial tachycardia. The mapping panels show the color-coded activation maps in *RAO*, *AP*, *LAO*, and *PA* views of respectively. The fluoroscopic view shows the intracardiac position of the catheters, with the ablation catheter at the site of successful ablation, in LAO view. The tachycardia was successfully ablated at the anterosuperior tricuspid annuls, where the intracardiac electrogram preceded the onset of surface p (which was simultaneous to the atrial electrogram in CS 7–8) by 46 ms

Both systems have limitations when mapping complex and adjoining structures, especially in the areas like the junction between chambers or connections of the chamber with structures like valves, veins, or arteries. One way to minimize this "interpolation obliteration" of those junctions with smoothing over the angles is to treat the adjoining structures as separate volumes or maps, which some systems readily allow. It is a good practice to identify the anatomic structures such as the vena cava, atrioventricular annuli, and pulmonary veins and display them in the initial phase of geometry creation itself. Validating the identity of these anatomical landmarks with fluoroscopy or intracardiac potentials in the initial phase of chamber geometry creation can serve as a skeleton for further mapping and improve its overall diagnostic sensitivity. Care should be taken in distinguishing internal locations from the endocardial surface during this phase, especially in systems such as CARTO, which simultaneously collects geometric and electrical data. The final 3D geometry can be displayed in conventional projections, freely rotated about its axes or orientated in endoscopic views to enable effective catheter localization and arrhythmia diagnosis (Figs. 11.1 and 11.3).

To overcome the confounding effects of patient movement, mapping systems use anatomical references, which in the NavX system can be an intracardiac or extracardiac electrode or using CARTO, an external reference patch. Intra- and extracardiac reference electrodes/patches have their benefits and limitations. The extracardiac reference patch in the CARTO system provides a fixed point of interest; however, upon patient movement, this patch moves with relation to the under-bed locator and as such catheter localization becomes inaccurate within the "old" 3D map. In other words, the catheter is being accurately located within a different 3D space, which is not compatible with the previous map. In this case, the patient can be moved back to the original position or another electroanatomical map maybe created. Rather than an under-bed fixed locator, the NavX system utilizes "on-patient" patches and usually an intracardiac reference electrode so that this mapping system can be more stable in the setting of a "mobile" patient. In an intracardiac position, a reference electrode is also more confluent with respiratory movement. If the intracardiac electrode moves or is pulled out of position, new geometric and electrical data still need to be collected. However, in the NavX system, if the internal

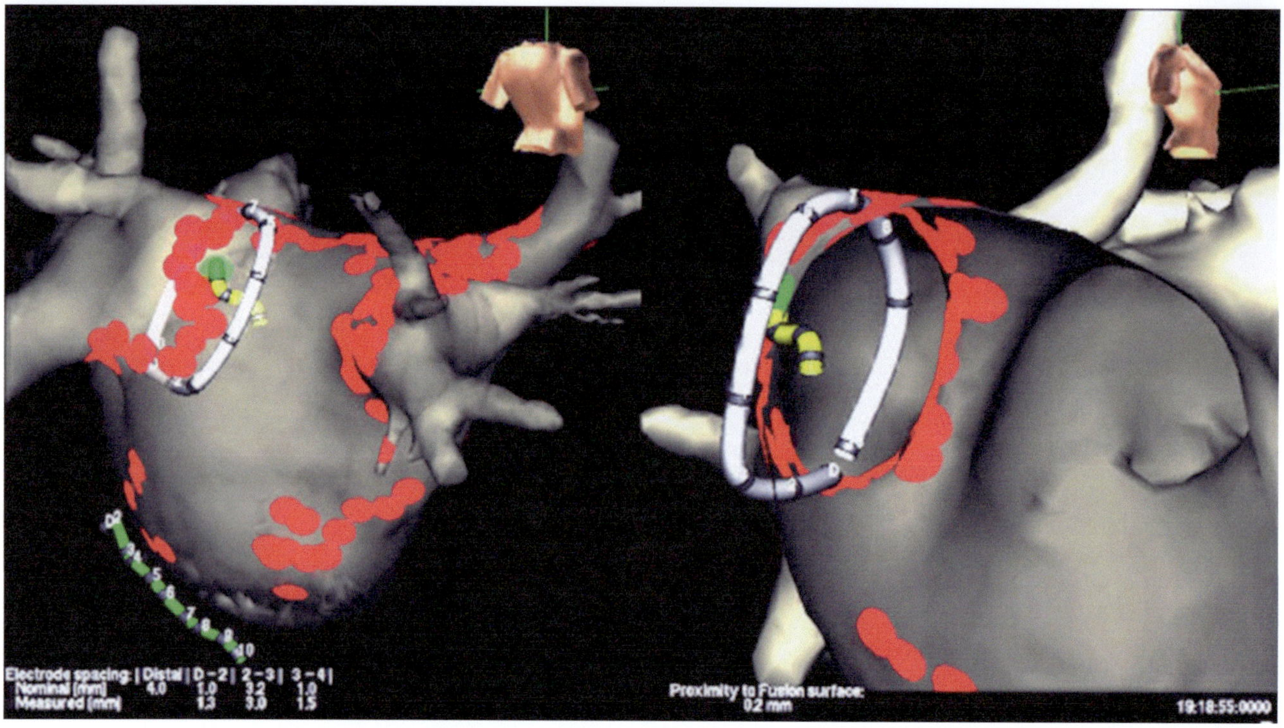

Fig. 11.3 NavX map of the left atrium and accompanying pulmonary veins as established during atrial fibrillation ablation. The endoscopic view, shown on right, shows the encircling decapolar catheter and the ablation catheter at the antrum of the left pulmonary vein. The defining bridge between the posteriorly placed left pulmonary vein and anterior left atrial appendage is well visualized

reference is moved or becomes unstable, an external reference can be selected, albeit it is more sensitive to respiration artifact given its location on the patient's abdominal wall.

11.2.2.3 Recording Electrical Information

Local activation timing requires an anatomically stable and electrically well-defined reference electrogram, to which the activation time of a roving catheter can be compared. Any surface electrocardiographic lead or intracardiac electrode (bipolar or unipolar) may serve as the reference electrogram. Often, an electrogram in the coronary sinus is chosen as a reference because of its anatomical stability especially for maps of supraventricular tachycardia. A surface lead often serves as the electrical reference for studies of ventricular tachycardia. The reference electrode should remain stable during the entire study; so confirmation of its position fluoroscopically in relation to other landmarks is essential to ensure its temporal stability. In addition, reproducibility in automatic sensing of the reference is essential for map accuracy. Furthermore, the malsensing of a ventricular electrogram for an atrial electrogram can confound the mapping, especially when a coronary sinus electrode is taken as the reference. Automated sensing of mapping and reference electrograms is usually accomplished by detecting peak amplitude or maximal slope. Conventionally, maximum amplitude is taken for bipolar electrograms and maximum slope for unipolar electrograms. The steepest negative intrinsicoid deflection in a unipolar electrogram is known to correlate with maximal Na+ conductance[11] and, thus, has the advantage of providing precise measurement of local activation times. But poor signal-to-noise ratio, especially in areas of scar or low voltage potentials, can compromise the reproducible interpretation of this fiducial point.

While mapping tachycardia, apart from the electrical reference, a window for assigning activation times on the mapping catheter is required. This is the time interval relative to the fiducial point, during which the local activation time is determined in the mapping (i.e., roving) channel. Only those activation times falling within this window will be acquired. These activation timings are labeled "early" or "late" relative to the electrical reference within this window. The total length of the window of interest should always be less than the tachycardia cycle length to avoid two activations in the same point falling during this window period. The boundaries of this window relative to the reference electrogram and its range should be set based on the presumed activation range of the tachycardia being mapped. For example, it is important to select a broader range for window, nearly approximating the tachycardia cycle length, while mapping macroreentrant tachycardias. However, while doing so, it is important to remember that the designation of activation time as early or late in a chamber is arbitrary, as some point of the chamber is activated at any given point during macroreentrant atrial flutters. Hence, a change in window of interest in these cases would only result in a phase-shift of the activation pattern, still maintaining the same activation sequence within the circuit.

11.2.2.4 Displaying Electrical Information

Electrical information obtained during the signal recording can be presented in the form of activation timing, voltage amplitude, or other user-defined measurements. The values obtained are color coded and assigned to the positional location of the roving catheter. These values can be used to generate activation, isochronal, propagation, voltage, or signal complexity maps depending upon the nature of arrhythmias. Activation maps display color-coded local activation time superimposed on the chamber geometry, and are useful for mapping both focal and macroreentrant tachycardias (Fig. 11.4) and to demonstrate "activation detour" following linear ablation (Fig. 11.5). The propagation maps show the dynamic propagation of active wave front across the endocardial chambers. As per conventional color-coding for activation and propagation mapping, red indicates the early activation sites, blue and purple late activated sites, and yellow and green the intermediate local activation sites. Voltage maps are created by the system based on the peak-to-peak amplitude of the local electrograms. The value is color coded with purple and red representing areas of highest and lowest amplitude respectively (Fig. 11.6). Conventionally, bipolar voltage amplitude <1.5 mV is considered abnormal in the ventricles, and an electrically "dense scar" is inferred in regions demonstrating voltages <0.5 mV.[12] However, in the atria, scar is defined as an area with bipolar voltage less than 0.05 mV.[13-17] This approach is highly useful to identify the areas of electrical scarring and defining the potential channels within it. However, one should remember that these areas with low voltage ("electrical scarring") do not necessarily represent areas of pathological scar. Signal complexity maps, another way of depicting the electrical information, are often generated based on user-defined algorithms (Fig. 11.7).

The accuracy of electroanatomic maps largely depends on their spatial resolution and temporal stability. High-density maps are often required near the area of interest to avoid erroneous interpolation between distant neighboring points and thus misinterpretation of the map. Similarly, identification and annotating the presence and location of double potentials, fractionated electrograms and pathway, diastolic or Purkinje potentials on the anatomic shell may superiorly define the tachycardia circuit and may even provide additional diagnostic information for targeted ablation.

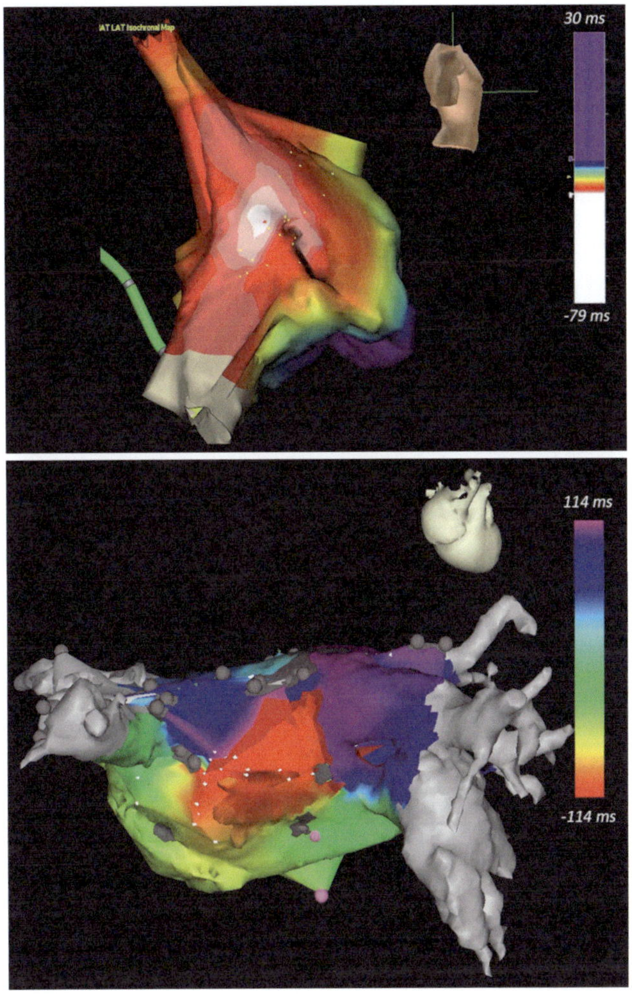

Fig. 11.4 Isochronal mapping of focal and macroreentrant tachycardias. The activation time-scale is color coded with the *white* followed by *red color* denoting the earliest sites of activation and the *purple* the last activated site. (**a**) The centrifugal activation from the mid-crista of right atrium in a patient with atrial tachycardia of focal mechanism. Only 109 ms could be mapped in the right atrium while the tachycardia cycle length was 298 ms, suggesting centrifugal activation. (**b**) Mapping of left atrial flutter in a patient with previous history of ablation for atrial fibrillation. The tachycardia cycle length was 242 ms. 228 ms (>90%) of the flutter cycle length could be mapped in the left atrium. The *dark red band* at the roof depicts the "early meets late" activation pattern. The wave front propagation pattern, which is perpendicular to the isochrones, suggests "roof dependent flutter" in this case, possibly related to the recovery of conduction across the roof line. The areas colored *grey* in the roof represent the areas with scar related to the previous ablation

11.2.3 Optimal Clinical Application

11.2.3.1 Interpretation of Maps

The primary aim of the electroanatomic maps is to assist in defining the tachycardia mechanism and thereby its treatment. A focal tachycardia is characterized by a localized area of early activation where the electrical signals usually precede the surface P or QRS by 30–70 ms. From this point of early activation, the impulse spread in a centrifugal fashion, which can be depicted by the color-coded activation or propagation maps.[18] In a focal mechanism, the points of the earliest and the latest points of electrical activation are anatomically well separated in the chamber being mapped, unless a unidirectional conduction pattern results close to the focus of arrhythmia due to fixed or functional conduction

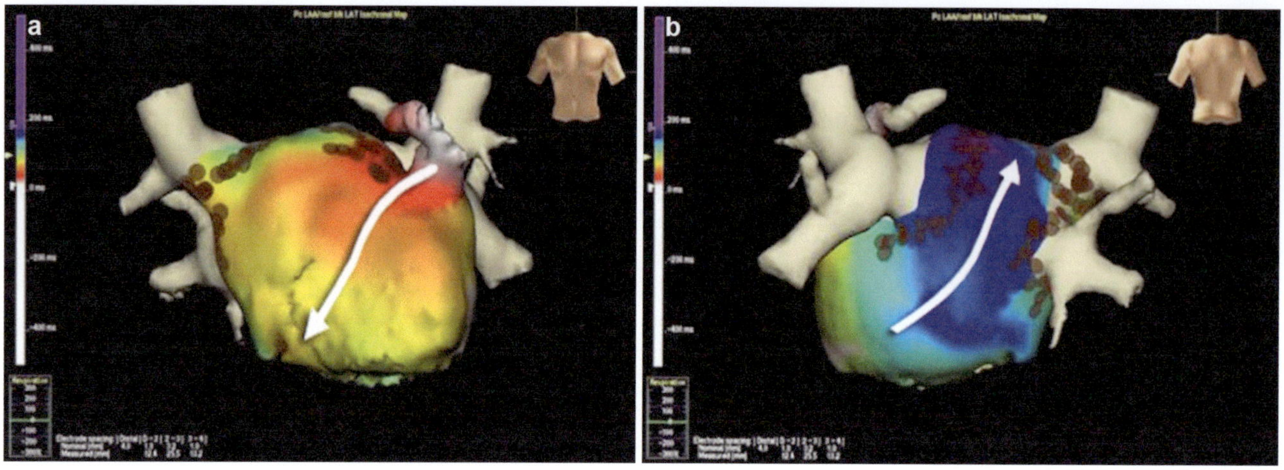

Fig. 11.5 Isochronal mapping of left atrium on pacing from left atrial appendage demonstrating complete conduction block across the roof-line. The roofline block results in "activation detour" of the wave front craniocaudally in the anterior wall (**a**, *anteroposterior view*) and caudocranially in the posterior wall (**b**, *posteroanterior view*). The *brown circles* indicate the points ablated

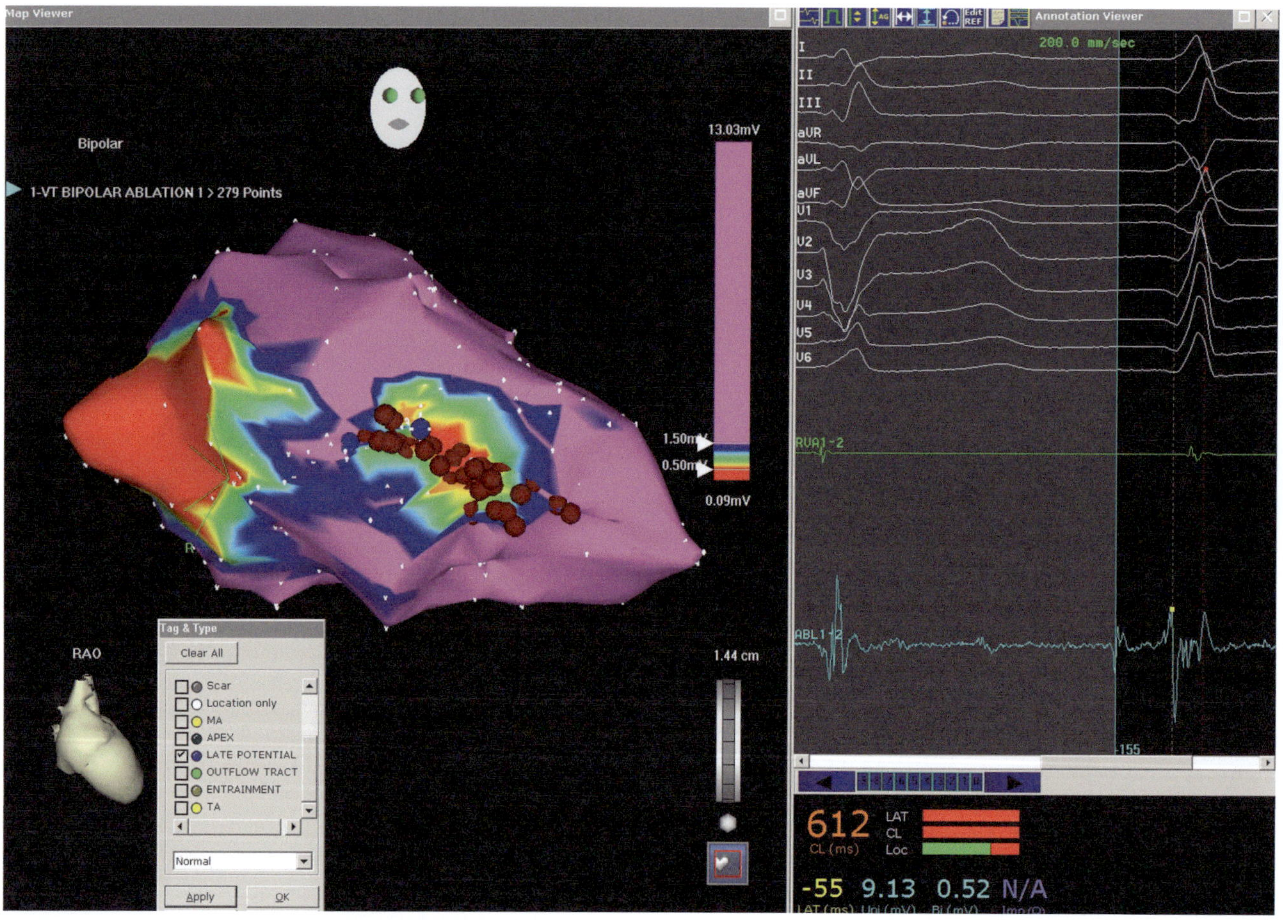

Fig. 11.6 Voltage map of left ventricle in RAO view in an elderly male with previous anterior wall myocardial infarction, left bundle branch block, and recurrent ventricular tachycardia. Local bipolar electrogram voltage of ≤0.5 mV has been arbitrarily selected as the threshold of dense scar (*red in color*), ≥1.5 mV as normal (*purple in color*) and 0.5–1.5 mV the borderline zone. The electrogram window shows a sinus beat followed by the initial beat of ventricular tachycardia (VT). The local electrogram from the area marked with *blue circle* at the scar border zone shows mid-diastolic and presystolic potential in the diastolic interval preceding the initiation of VT. Linear ablation as shown by *brown circles* in the figure made the tachycardia non-inducible

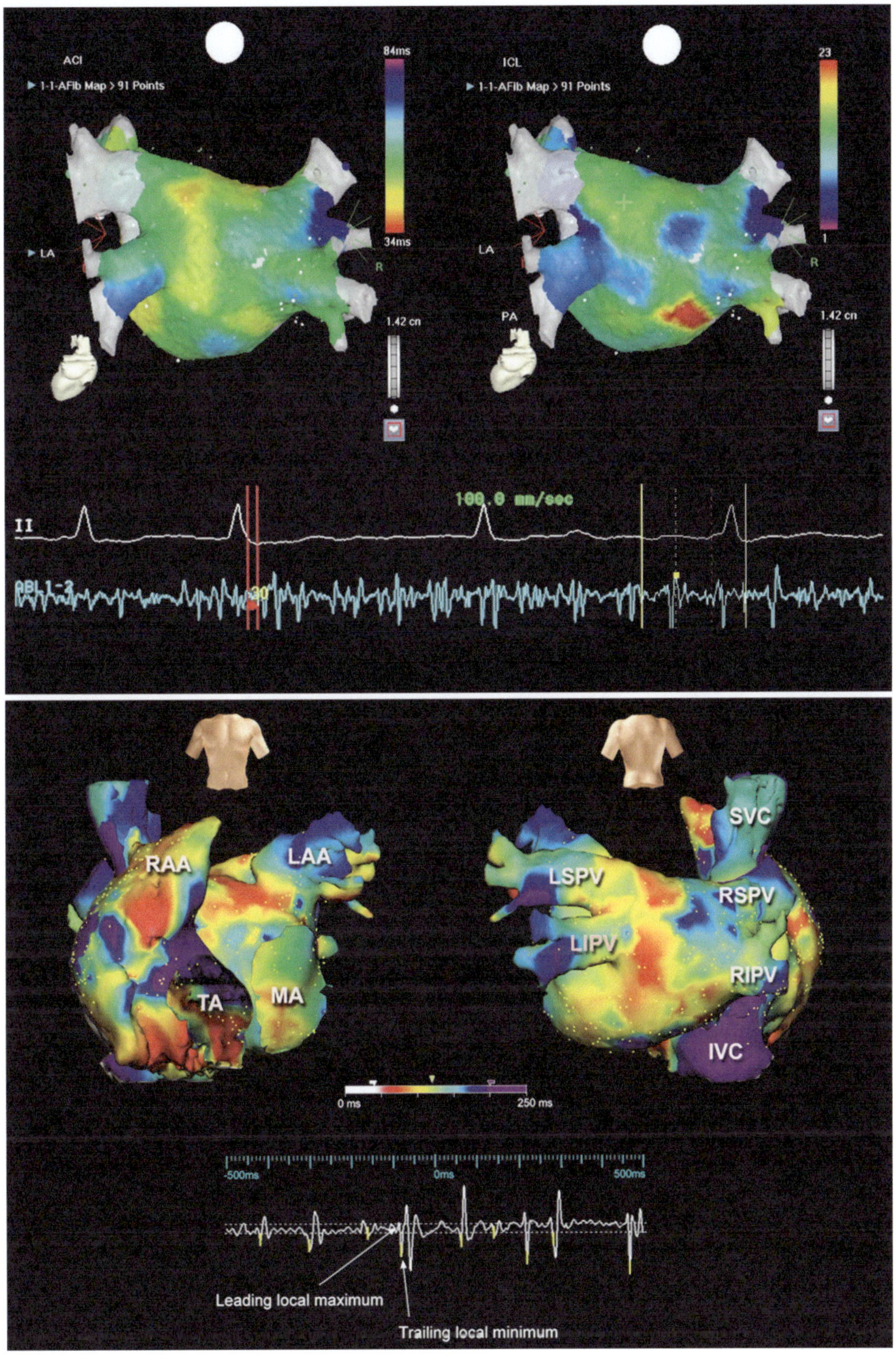

Fig. 11.7 Integrated fractionated electrogram color maps (*top panel*: CARTO; *bottom panel* NavX), with representative bipolar electrograms (CARTO – 2.5 s sample; NavX – 1 s sample) from each system. The CARTO maps demonstrate both Average Confidence Interval (*left-sided map*) and Interval Confidence Level (*right-sided map*) quantification of the fibrillatory electrograms. The NavX maps demonstrate high-density biatrial electrogram maps using complex fractionated electrogram mean (CFE-mean) to assess relative signal complexity. *Yellow tick marks* on the sample electrogram (1 s) represent detection of deflection events. *IVC* inferior vena cava, *LAA* left atrial appendage, *LIPV* left inferior pulmonary vein, *LSPV* left superior pulmonary vein, *MA* mitral annulus, *RAA* right atrial appendage, *RIPV* right inferior pulmonary vein, *RSPV* right superior pulmonary vein, *SVC* superior vena cava, *TA* tricuspid annulus

block. Although regions of conduction block and slow conduction are not unusual bystanders in a focal mechanism, they are not critical for maintenance of the tachycardia circuit. The total chamber activation time in these tachycardias represents the conduction time from the earliest to the last point of activation time within the chamber, and is expected to be substantially less than the tachycardia cycle length.

In contrast, macroreentrant tachyarrhythmias would show adjacent areas of early and late activation, separated by areas of intermediate values of activation. More than 90% of the tachycardia cycle length can be mapped in macroreentrant tachycardias. Another useful feature of the mapping software helps to detect "early meets late activation" pattern (Fig. 11.4a). By coding these adjacent regions a specific color, sequential activation of the late to early sites can be depicted, as would be expected in a reentrant circuit. This algorithm avoids interpolation of activation times between the early and late points and avoids the false impression of a focal mechanism to the tachycardia. The direction of propagation of the reentrant wave front can be interpreted based on the isochronal maps, which runs perpendicular to the isochronal lines. However, the radial activation of the bystander sites can confound interpretation of isochronal maps in many cases. It is advisable to combine entrainment mapping with electroanatomic mapping in these particular situations.

Increasingly, the role of localized reentry in the genesis of arrhythmias is recognized, particularly occurring in patients with previous ablation. Unfortunately, due to the annotation used by all of these mapping systems unless the electrograms are carefully scrutinized, these arrhythmias may be missed as the color interpolation summates the region. Thus, it is important to combine the findings of conventional mapping and entrainment techniques with that obtained from the 3D mapping systems for expeditious and accurate mapping of arrhythmias.

11.2.3.2 Caveats: The Possible Sources of Error

Although electroanatomic mapping systems have largely contributed to better understanding of the electrophysiologic substrate predisposing to various clinical arrhythmias[14-16,19-22] and thereby improvements in the approach to complex arrhythmias,[12,13,23-34] the correct interpretation of the electrical information obtained during mapping is still underpinned by basic electrophysiology principles. The following sources of error need to be considered during this process.

1. *Identification of the chamber(s) harboring arrhythmia.* Arrhythmias arising and limited to one of the chambers can be commonly mistaken for that arising from the other. This is especially true for focal arrhythmias arising close to septum. Furthermore, if mapping is limited to the right atrium, left atrial tachycardias may be mistaken for focal arrhythmias originating in the region of dominant interatrial conduction like Bachmann bundle, the interatrial septum, or the coronary sinus. Similarly, ventricular tachycardias arising from left ventricular outflow including aortic sinuses may be mistaken for arrhythmias arising from right ventricular outflow region if only right ventricle is mapped (Fig. 11.8). Broader areas of early activation along the intervening septum and the absence of significant prematurity of the local activation time in the mapped areas in relation to the P wave or QRS can serve as a clue in recognizing this septal breakthrough. In addition, the utility of entrainment mapping from the septal regions in the identification of these septal breakthroughs during macroreentrant tachycardias cannot be overemphasized.
2. *Inappropriate activation window.* Focal arrhythmias with adjacent areas of conduction blocks or slow passive activation of the chamber with prolonged chamber activation time are not unusual. If the activation window selected approximates the tachycardia cycle length, this could result in ambiguous map especially in tachyarrhythmias with cycle length variations. This inappropriate window selection may generate a spurious "early meets late activation" pattern mimicking a macroreentrant mechanism. Additional measures like identification of the functional and anatomical barriers contributing to this delay and use of entrainment mapping can be useful in these settings.
3. *Low-resolution maps.* A poor-resolution map can lead on to incorrect interpretation about components of the arrhythmia circuit. Interpolation between anatomically distant activation times may hide the true mechanism of arrhythmia. Radial activation from a macroreentrant circuit can also give intermediate activation times in the surrounding region not directly involved in the circuit. The direction of wave front propagation can be correctly interpreted in these cases by the "early meets late" algorithm and entrainment mapping. Similarly, dense mapping is often required in the scar region in cases of scar-related atrial or ventricular complex reentrant arrhythmias to identify the low voltage signals, which may serve as potential targets of ablation in the diastolic isthmus.[13]
4. *Areas of anatomical barriers and fixed conduction block.* Many areas in the heart like crista terminalis, scars, venoatrial junctions, valves, or even infarct-related scars can act as areas of conduction block. The interpolation of activation through these areas of conduction block can give an appearance of focal arrhythmia with centrifugal spread of activation, masking a macroreentrant mechanism. Dense mapping in these areas and looking for the presence of double potentials can often help in discrimination.
5. *Complex reentrant circuits.* The components of circuits in reentrant arrhythmias such as double loop reentrant tachycardias may remain undetected unless suspected and specifically looked for. A low-resolution map can particularly contribute to this problem. Dense mapping, noting the markers of local conduction abnormalities like the fraction-

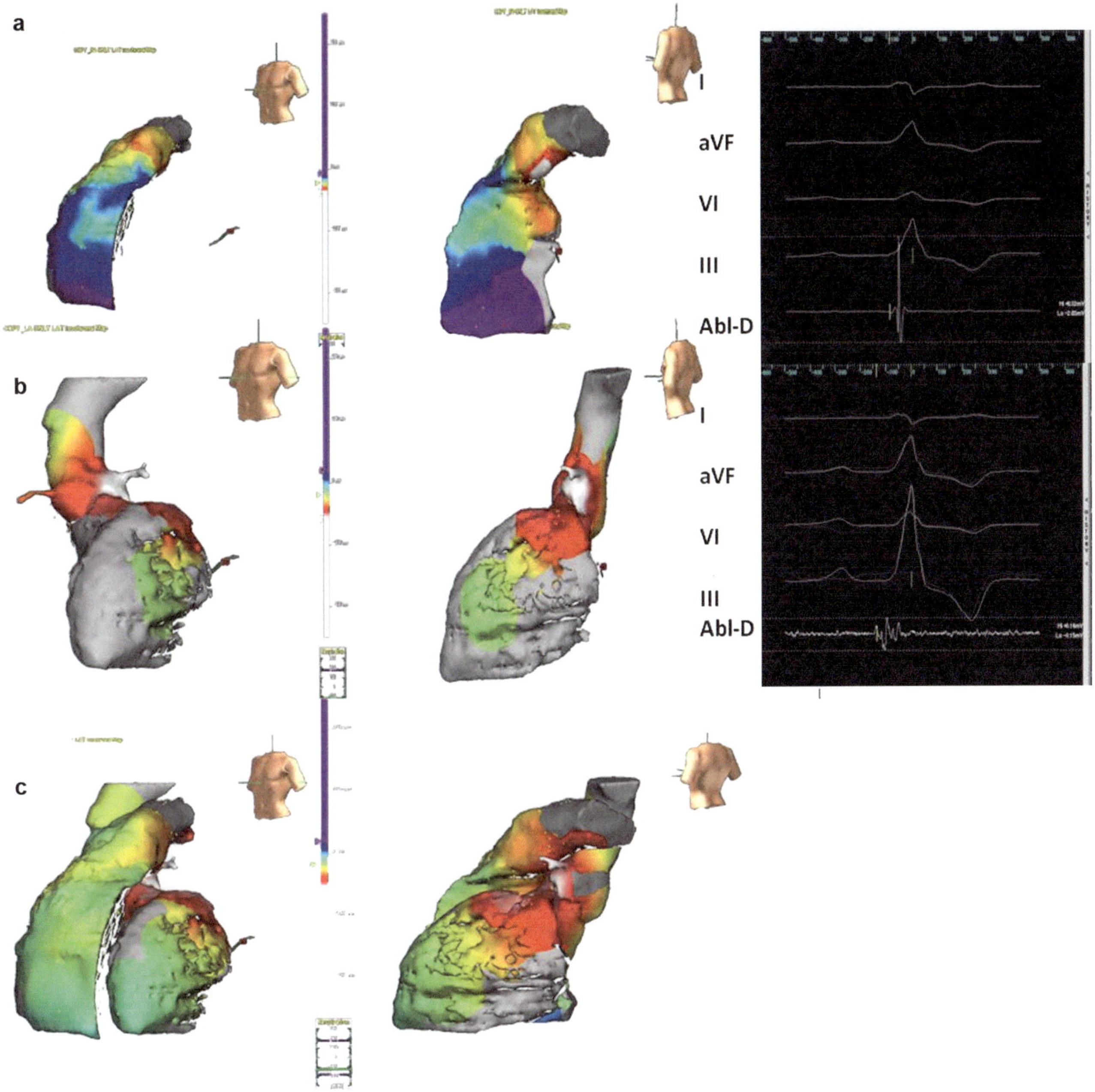

Fig. 11.8 Activation mapping of outflow tract ventricular ectopics using EnSite NavX system. Initial mapping of the right ventricle showed earliest activation at upper septal region in this case (**a**). However, the endocardial signals preceded the surface QRS onset only by 24 ms. Further mapping on the left ventricular outflow region (**b**) showed centrifugal activation from a location immediately beneath left main artery origin within the left coronary cusp with the local electrogram preceding the surface QRS by 45 ms. The combined view of maps from both ventricular outflow tracts is shown in (**c**)

ated electrogram, double potentials or local conduction delay, and entrainment mapping can help to identify and localize the reentrant circuits in these cases. Similarly, in scar-related ventricular tachycardias, maps solely based on the activation timing and voltage of the signals should be interpreted with caution. As in any reentrant circuits, many innocent bystander loops can have similar activation timings without necessarily being part of the circuit. Recognition of diastolic isthmus in these complex reentrant circuits may demand supplementary entrainment maneuvers.

6. *Fractionated electrograms*. Assigning an activation time for highly fractionated and wide potentials requires a cautious approach. In bipolar maps, larger amplitude signals usually represent the local activation time; however, misannotation of these points may lead to incorrect map interpretation (Fig. 11.9).

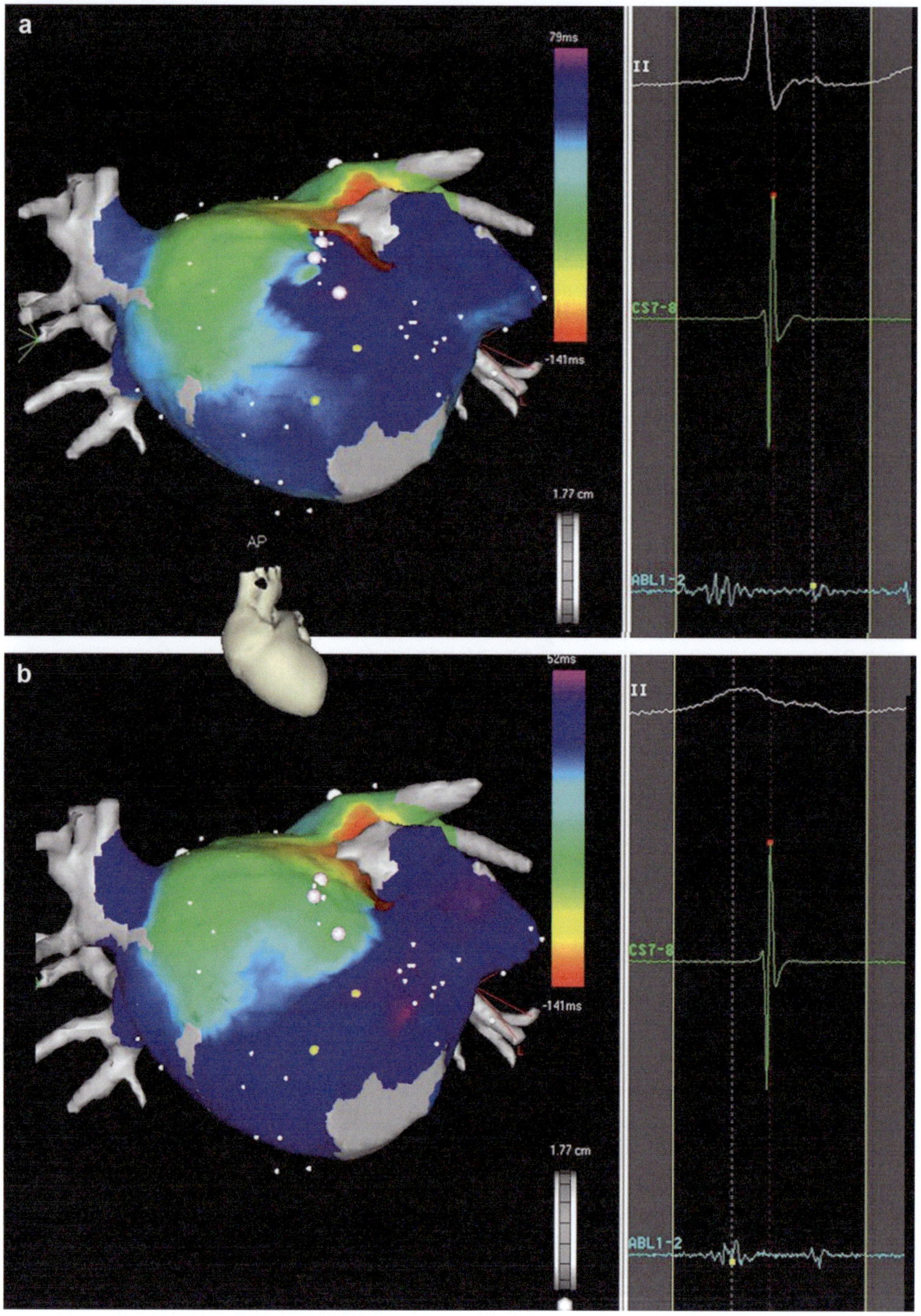

Fig. 11.9 Left atrial isochronal mapping during atrial flutter with the reentry limited to the roof of left atrium near the base of left atrial appendage. The areas with fractionated electrograms (*white circles*) are seen in anterosuperior wall of left atrium (**a**). The coronary sinus electrogram is chosen as the reference, and the sensing window is –150 and +150 ms relative to the reference. The tachycardia cycle length is 320 ms, and the total activation time mapped in the small area was 193 ms. The second potential in the fractionated electrogram with smaller amplitude has been annotated in (**a**) as the local potential giving "early meets late appearance" in the anterior wall. However, the entrainment response suggested this region "out of the circuit." The correct annotation of the larger one among the fractionated electrogram as the local potential suggested this area of double potentials was simply an "innocent bystander" with radial activation from the reentrant circuit (**b**). The response to entrainment and subsequently to ablation confirmed this interpretation

7. *Movement artifacts*. Movement artifacts occurring after the geometry has been acquired can distort the anatomical correlation. Cardiac and respiratory movements may decrease the accuracy of catheter localization; however, this movement can be filtered (NavX) or catheter location can be updated only on specific time within the cardiac cycle (CARTO). If the mapping has been performed during various activation sequences such as sinus rhythm, arrhythmia, or cardiac pacing, apparent shifts in the catheter localization might occur because of change in chamber geometry and gating. An intracardiac spatial reference will largely reduce the shifts due to the respiratory excursions. Selection of points during the same phase of respiratory cycle can also theoretically reduce this problem, though difficult to perform practically.

11.3 Noncontact Mapping

In cases of nonsustained, hemodynamically poorly tolerated tachycardias or multiple tachycardias, sequential acquisition of data points over several cardiac cycles becomes virtually impossible. This would limit the number of points acquired and the resolution of the images created. This inherent limitation of contact mapping systems is overcome using noncontact mapping systems that allow simultaneous acquisition of virtual electrogram data from a whole chamber over a single beat. Both contact multielectrode basket catheters and noncontact multielectrode array catheters are capable of doing this. However, suboptimal resolution due to wide spacing and poor apposition to the endocardial surface of the chamber limits the clinical utility of maps using the former technology.[35-38]

11.3.1 Multi-Electrode Array (MEA): Principle of Operation

The commercially available noncontact mapping system uses a MEA catheter (EnSite, Endocardial solutions Inc., St. Paul, MN, USA) to map entire cardiac cycles in real time without requiring sequential point-to-point acquisition.[39-41] The system contains a catheter-based, noncontact MEA (9F, 64 electrodes) mounted on a 7.5 mL ellipsoid balloon, a reference patch electrode, amplifiers, filters, and a computer workstation. The balloon is surrounded by 64 electrically insulated wires with each wire having a break in its insulation, allowing it to function as a unipolar electrode. The reference for this unipolar recording is the ring electrode, located 16 cm proximal to the MEA.

The MEA placed in the chamber cavity records intracavitary far-field potentials that are sampled at 1.2 kHz and digitally filtered at 0.1–300 Hz. However, these cavitary potentials detected by the array are not easily interpretable in their raw form because of the low amplitude and frequency. These signals are analyzed and mathematically transformed at the EnSite workstation based on the inverse solution to Laplace's equation to calculate how a signal detected by the MEA would appear on an endocardial surface that has previously been traced by a contact catheter located in 3D space using the MEA.[42] The MEA is capable of localizing any conventional roving electrode catheter by emitting a low current "locator signal" (5.68 Hz) from the roving catheter tip and receiving by two ring electrodes located proximally and distally on the MEA. Using this method, the system calculates the signal angles and updates the positional location of the catheters 400 times per second. The system can reconstruct and display 3,360 virtual unipolar electrograms simultaneously and can generate isovoltage maps together with virtual electrograms from any point on traced endocardial surface. While the accuracy of this system has been validated repeatedly,[41,43-47] it is limited by the distance from the MEA to the surface being examined, with distances >4 cm conferring a reduced accuracy.[39,41] This limitation makes this technology of limited use in the very large chamber.

11.3.2 Recording of Electrical Signals

The MEA catheter is advanced through a special 9Fr sheath into the chamber of interest and deployed by injection of a 50/50 mixture of contrast media and saline. An additional guide wire or the pigtail configuration at the tip is used to anchor the distal catheter. This is crucial as displacement of MEA will invalidate the electrical and anatomic information acquired during the study. As the accuracy of isopotential mapping is also directly related to the distance to endocardial surface, the array must be positioned as close as possible to the area of interest within the chamber. To assist this stable positioning, the system continuously displays the distance between the mapping catheter and the center of the balloon.

A color-coded isopotential map representing the spatial changes of depolarization as activation wave fronts can be created on the virtual endocardial surface (Fig. 11.10) and unipolar electrograms can be selected from the surface of interest to verify the slope of depolarization (–dV/dT), far-field signals, and fractionated/double potentials. In the review mode, the color-coded isopotential map can be tracked back in time to demonstrate the earliest endocardial activation. As the low-frequency noise, far-field signals, and repolarization waves in the chamber being mapped are magnified into larger signals, these can interfere with the identification of true local unipolar activation signals of clinical interest. A larger high-pass filtering (HPF) setting would help to filter out the large repolarization waves but may

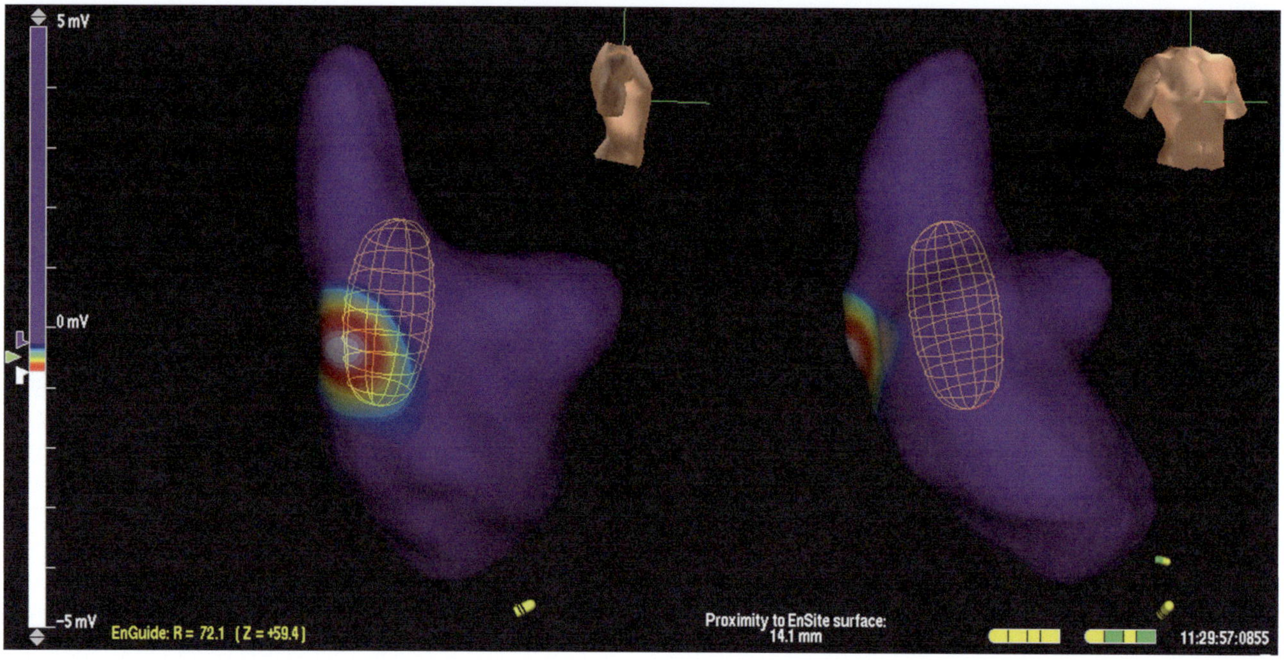

Fig. 11.10 Isopotential maps of the right atrium in a patient with atrial tachycardia originating from the mid-crista region, shown in right lateral and RAO views. Color scale has been set so that *white* indicates most negative potential and *blue* indicates least negative potential. The activation wave front shows centrifugal propagation from the mid-crista region

attenuate the critical information in the signals of lower frequency and amplitude. Ideally, the signal sampling processing should have filter settings that are sensitive and selective to enable proper delineation of early low amplitude signal components and suppress repolarization-related far-field signal elements as much as possible. Low HPF setting enables an improved display of early and slow rising activation – the signals of interest in isopotential mapping – and a better display of the wave fronts, whereas higher values of HPF reduce repolarization-related signal components and noise. Slow and fast activation wave fronts are displayed including potential repolarization components at HPF of 0.5–2.0 Hz. Raising the HPF setting to about 4–12 Hz, the wave front is sharpened and the unwanted signal components become more suppressed. Still higher values like a HPF of 16 Hz and more display mainly wave front changes only. The HPF is generally adjusted between 1 and 32 Hz. Though the different HPF levels influence the spatial analysis of noncontact mapping, the effect on the final ablation result is probably not marked.

11.4 Caveats and Limitations

Despite the novel method and its application in complex arrhythmias,[43-47] this mapping system is not without pitfalls. Accuracy of the electrogram reconstruction and subsequent localization is best obtained at endocardial sites less than 4 cm away from the balloon surface. This is a limitation in patients with dilated ventricular chambers and may necessitate balloon repositioning during the study. The balloon catheter cannot be moved after completion of geometry creation as it would change the activation localization and result in distortion of isopotential maps. The bulkiness of the inflated balloon dimensions (4.6 × 1.8 cm^2) can restrict the mapping catheter manipulations especially in the atrial chambers. Because of this high profile, strict anticoagulation is to be ensured during the study. As the system faithfully records the earliest activation in the mapped chamber as the earliest activation within the heart ("chamber-centric map'), synchronized mapping of more than one chamber may be demanded in some particular situations like arrhythmias arising from septal structures or macroreentrant circuits involving more than a chamber. Moreover, the acquired chamber geometry is often more distorted as compared to contact mapping systems, requiring multiple set points to clearly establish complicated structures such as the left atrial appendage or pulmonary veins. Low-voltage-generating acquisition sites may also be missed in the isopotential mapping.

11.5 Integrated Anatomy-Based Mapping

Despite the marked technological improvements in the application of these electroanatomic mapping systems, the geometry or electroanatomic maps created by them are only an approximation to the true anatomy of the cardiac chambers,

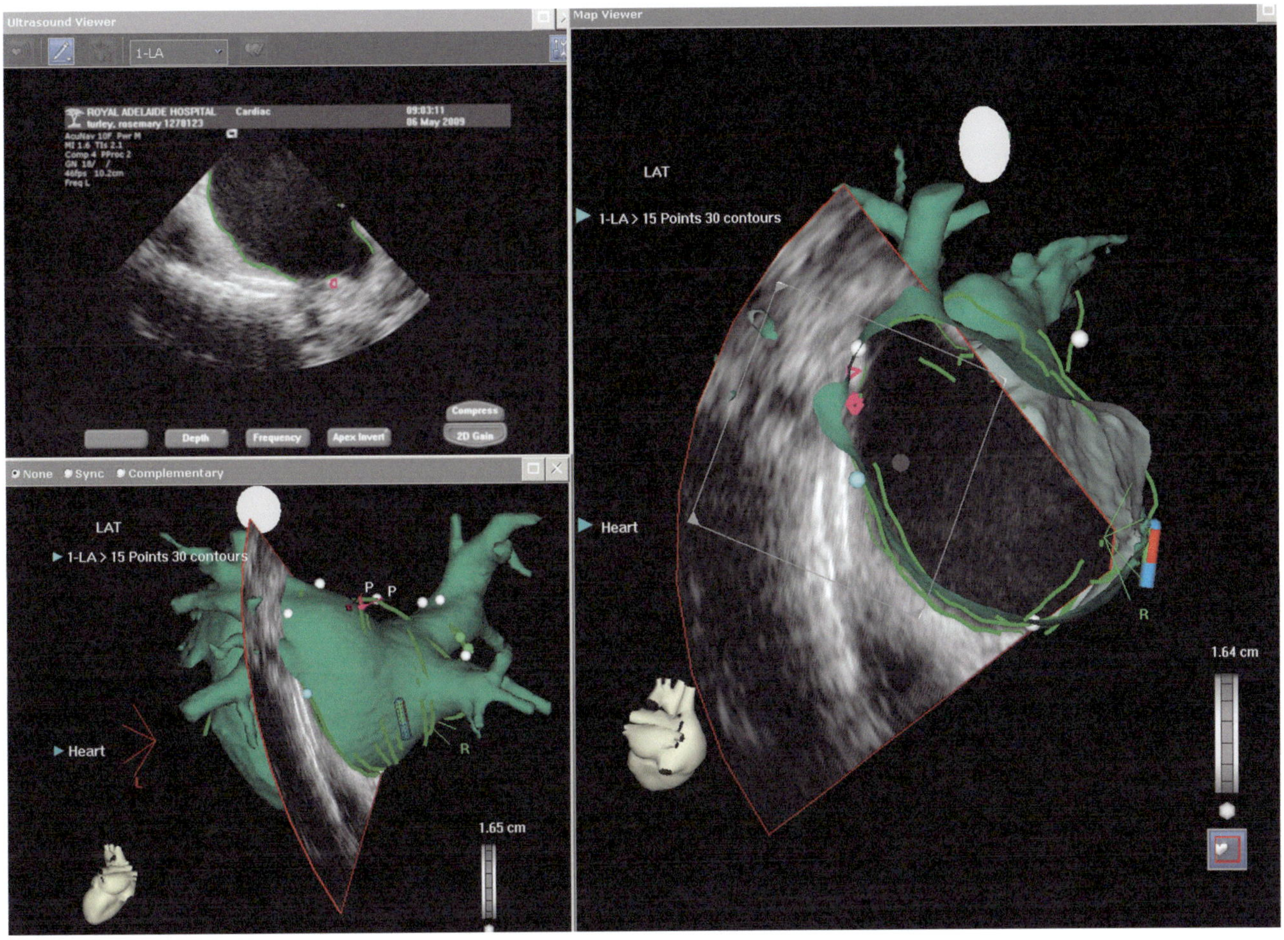

Fig. 11.11 A screen capture from CARTO-sound anatomical map development showing a 2D echocardiography slice taken from the right atrium, highlighting the carina of the left pulmonary veins (*pink dot*). The Sound-Star catheter tip in the right atrium, its orientation and slice angle are seen clearly in the right-sided image. Outlines demonstrate the traced edges of sequentially collected echocardiographic anatomical information that is required for interpolation of 3D geometry. White geometry points are collected with a conventional magnetic-tipped catheter and can be added to the anatomical map where required

and are heavily operator dependent. Furthermore, important but subtle anatomic variations with possible implications, such as the precise location of the pulmonary vein–left atrial junction, size and location of the left atrial appendage, and presence and location of ridges and pouches, are not well defined by them.[48-55] On the other hand, highly detailed information on the complex anatomical structures can be obtained by complementary imaging modalities like CT and MRI[51-55] (Fig. 11.11). Combining the electroanatomical maps with anatomical imaging can facilitate the catheter navigation and may increase the efficacy and safety of ablative procedure. The two available modalities allowing integration of the anatomical images with electroanatomic maps are the CARTO-Merge™ (Biosense Webster, Diamond Bar, CA, USA) and the NavX Fusion™ (St Jude Medical, St Paul, MN, USA) systems. Both of them use fundamentally different methodologies and algorithms for mapping and image integration, and have been used as an aid in catheter ablation procedures for atrial fibrillation.

In CARTO-Merge, the image integration involves different steps like image acquisition, "segmentation" of the acquired image (dividing the image into different regions to select the structures of interest), and "registration' (alignment of the electroanatomical map and anatomical images).[56,57] The main processes of registration used are "landmark registration" and "surface registration." "Landmark registration" involves operator assignment of a "landmark' to a particular anatomical structure (e.g., the veno-atrial junction or the left atrial appendage) on both the electroanatomic map and the CT or MRI image. These landmark pairs are used to align the two images. By using a minimum of three landmark pairs, appropriate registration along three orthogonal axes can be acquired. The anterior left atrial points should be avoided during the merge due to inherent mobility of these structures compared to posterior atrial structures. "Surface registration" involves the alignment of the whole electroanatomic map and the CT or MRI image based on a specific algorithm which allows the best

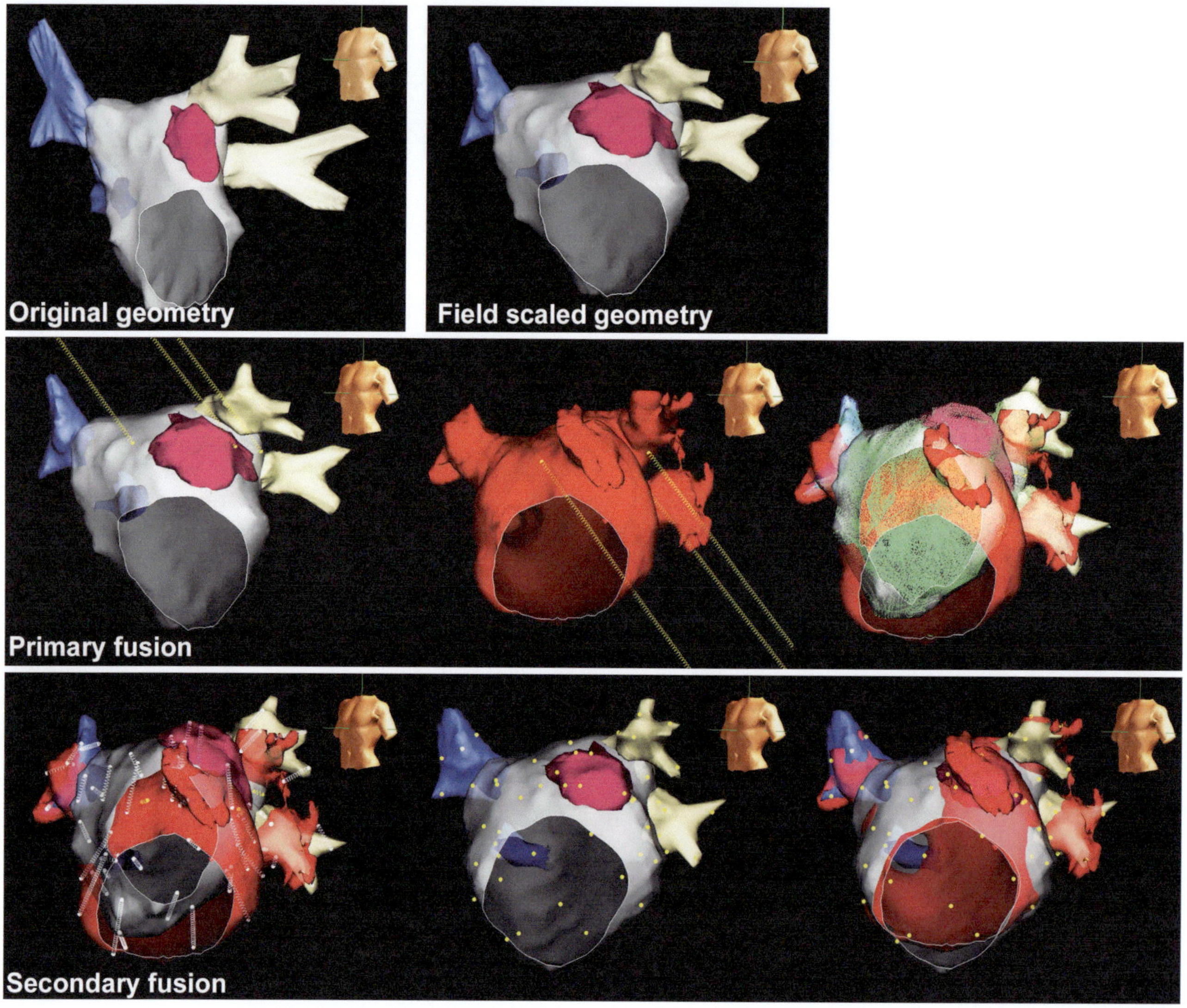

Fig. 11.12 Screenshot series showing progressive improvement in geometry similarity to computed tomography for original geometry, field scaled and primary fused geometry, and secondary fused geometry. Note significant dragging together of the surfaces during the secondary fusion stage

fit between the two structures by minimizing the distance between all mapping points and the CT or MRI image. Compared to the CARTO-Merge system, NavX fusion has been validated in fewer clinical trials.[5] The image integration process in NavX fusion consists of several steps, including "field scaling" of the reconstructed geometry, fusion of the structures using fiducial markers (landmarks), and optimization of the integration by adjusting ("molding and bending") the reconstructed geometry (Fig. 11.12). Both image integration systems, contemporarily in clinical use, use a geometry reconstructed from a set of endocardial data points collected during electrophysiological mapping. A new method that allows the creation and analysis of the maps based on the geometry derived from CT, not on the shape reconstructed from limited number of recordings of mapping catheter positions, has been described recently.[58]

With the better definition of cardiac anatomy offered by image integration, more accurate catheter navigation and lesion placement would logically be expected. This is expected to decrease the procedural and fluoroscopy time and may improve the outcome of the procedure.[59] A better arrhythmia-free outcome in patients who underwent AF ablation guided by image integration compared to electroanatomic or noncontact mapping has been reported. However, information on clinical outcomes is limited, and many theoretical benefits remain unproven at this stage.

Many factors could possibly interfere with the accuracy of the integration of the electroanatomic images with CT or MRI images.[60] The spatial resolution of CT is better than MRI and may therefore result in improved registration accuracy. The discrepancies in the phase of respiratory cycle during the image acquisition and the catheter ablation, and the time delay between the time of imaging and the ablation may adversely affect the process. In addition, the changes in left atrial volume between the procedures may influence the image integration accuracy. Lastly, the changes in heart rhythm may result in transient changes in left atrial volume and anatomy, and subsequently may affect the image integration process.

Intracardiac ultrasound, another imaging modality, is highly useful to provide focused real-time images of the endocardial surfaces critical for positioning of catheters, establishment of catheter/tip tissue contact, and for monitoring energy delivery in the beating heart.[61,62] The real-time monitoring with intracardiac echocardiography is expected to take care of the drawbacks of integration with anatomic images like some of the issues created by changing chamber volumes or arrhythmias. The value of 3D imaging on a real-time basis may be considerable in such a setting, although this is yet to be proven. Intracardiac ultrasound as an adjuvant in ablative procedures has a theoretical advantage over the other two previously described image integration methods in that it is online real time and it does not consider heart as a rigid body over which maps are displayed. As a consequence, the major limitation of the approach to merge CT or MRI images with the electroanatomic maps previously described (differences created during the time interval between acquisition and intervention, as well as changes in volume, rhythm, cardiac cycle, or respiratory cycle) cannot influence this modality.

11.6 The Selection of Mapping Systems

The CARTO, NavX, and RPM systems can be used effectively for mapping sustained, stable arrhythmias. However, nonsustained arrhythmias and infrequent atrial or ventricular premature beats are better mapped by the noncontact mapping system using the MEA catheter. The noncontact mapping array works well in these settings, although the maps can be filter-frequency dependent. The noncontact mapping with the MEA catheter provides a rapid assessment of the activation pattern during unstable VTs, obviating the need for long periods of tachycardia required by contact mapping systems. However, the precision of mapping can be limited in the setting of various cardiomyopathies with ventricular enlargement. In these settings, dynamic substrate modification or scar mapping along with pace mapping and entrainment mapping could act as useful alternatives to noncontact mapping in infarct-related ventricular tachycardia. CARTO and NavX perform well in this regard. Although these systems are comparable in the majority of clinical situations, the choice of mapping system depends ultimately on the experience of the operator and the laboratory. The geometry creation and the integration of maps are acceptably straightforward. In contrast, the noncontact mapping system requires some extra-effort in the creation of a user-friendly working geometry. Obviously, each of these systems is in the development stage and the various capabilities could change substantially over the next few years.

It is important to remember that the use of these advanced mapping systems, capable of three-dimensional rendering of cardiac chambers and superimposition of electrical information, is not designed to replace conventional mapping techniques but to be used as an adjunctive tool in the analysis and treatment of complex arrhythmias.

References

1. Rotter M, Takahashi Y, Sanders P, et al. Reduction of fluoroscopy exposure and procedure duration during ablation of atrial fibrillation using a novel anatomical navigation system. *Eur Heart J.* 2005; 26:1415-1421.
2. Tondo C, Mantica M, Russo G, et al. A new nonfluoroscopic navigation system to guide pulmonary vein isolation. *Pacing Clin Electrophysiol.* 2005;28(Suppl 1):S102-S105.
3. Della Bella P, Fassini G, Cireddu M, et al. Image integration-guided catheter ablation of atrial fibrillation: a prospective randomized study. *J Cardiovasc Electrophysiol.* 2009;20:258-265.
4. Krum D, Goel A, Hauck J, et al. Catheter location, tracking, cardiac chamber geometry creation, and ablation using cutaneous patches. *J Interv Card Electrophysiol.* 2005;12:17-22.
5. Brooks AG, Wilson L, Kuklik P, et al. Image integration using NavX fusion: initial experience and validation. *Hear Rhythm.* 2008; 5:526-535.
6. Gepstein L, Hayam G, Ben-Haim SA. A novel method for nonfluoroscopic catheter-based electroanatomical mapping of the heart. In vitro and in vivo accuracy results. *Circulation.* 1997;95: 1611-1622.
7. Shpun S, Gepstein L, Hayam G, Ben-Haim S. Guidance of radiofrequency endocardial ablation with real-time three-dimensional magnetic navigation system. *Circulation.* 1997;96:2016-2021.
8. Hayam G, Gepstein L, Ben-Haim S. Accuracy of the in vivo determination of location using a new nonfluoroscopic electroanatomical mapping system [Abstract]. *Pacing Clin Electrophysiol.* 1996;19.
9. Smeets J, Ben-Haim S, Rodriguez L, Timmermans C, Wellens H. New method for nonfluoroscopic endocardial mapping in humans. Accuracy assessment and first clinical results. *Circulation.* 1998; 97:2426-2432.
10. De Groot N, Bootsma M, Van der Velde E, Schalij M. Three-dimensional catheterpositioning during radiofrequency ablation in patients: first application of a real-time position management system. *J Cardiovasc Electrophysiol.* 2000;11:1183-1192.
11. De Bakker JMT, Hauer RNW, Simmers TA. Activation mapping: unipolar versus bipolar recording. In: Zipes DP, Jalife J, eds. *Cardiac Electrophysiology: From Cell to Bedside.* Philadelphia: WB Saunders; 1995:1068-1078.
12. Marchlinski FE, Callans DJ, Gottlieb CD, Zado E. Linear ablation lesions for control of unmappable ventricular tachycardia in patients with ischemic and nonischemic cardiomyopathy. *Circulation.* 2000;101:1288-1296.
13. Nakagawa H, Shah N, Matsudaira K, et al. Characterization of reentrant circuit in macroreentrant right atrial tachycardia after surgical repair of congenital heart disease: isolated channels between scars allow "focal" ablation. *Circulation.* 2001;103:699-709.

14. Sanders P, Morton JB, Davidson NC, et al. Electrical remodeling of the atria in congestive heart failure: electrophysiological and electroanatomic mapping in humans. *Circulation*. 2003;108:1461-1468.
15. Sanders P, Morton JB, Kistler PM, et al. Electrophysiological and electroanatomic characterization of the atria in sinus node disease: evidence of diffuse atrial remodeling. *Circulation*. 2004;109:1514-1522.
16. Kistler PM, Sanders P, Fynn SP, et al. Electrophysiologic and electroanatomic changes in the human atrium associated with age. *J Am Coll Cardiol*. 2004;44:109-116.
17. de Groot NM, Schalij MJ, Zeppenfeld K, Blom NA, Van der Velde ET, Van der Wall EE. Voltage and activation mapping: how the recording technique affects the outcome of catheter ablation procedures in patients with congenital heart disease. *Circulation*. 2003;108:2099-2106.
18. Iwai S, Markowitz SM, Stein KM, et al. Response to adenosine differentiates focal from macroreentrant atrial tachycardia: validation using three dimensional electroanatomic mapping. *Circulation*. 2002;106:2793-2799.
19. Stiles MK, John B, Wong CX, et al. Paroxysmal lone atrial fibrillation is associated with an abnormal atrial substrate: characterizing the "second factor". *J Am Coll Cardiol*. 2009;53:1182-1191.
20. Stiles MK, Brooks AG, Kuklik P, et al. High-density mapping of atrial fibrillation in humans: relationship between high-frequency activation and electrogram fractionation. *J Cardiovasc Electrophysiol*. 2008;19:1245-1253.
21. Stiles MK, Brooks AG, John B, et al. The effect of electrogram duration on quantification of complex fractionated atrial electrograms and dominant frequency. *J Cardiovasc Electrophysiol*. 2008;19:252-258.
22. Kottkamp H, Hugl B, Krauss B, et al. Electromagnetic versus fluoroscopic mapping of the inferior isthmus for ablation of typical atrial flutter: a prospective randomized study. *Circulation*. 2000;102:2082-2086.
23. Earley MJ, Showkathali R, Alzetani M, et al. Radiofrequency ablation of arrhythmias guided by non-fluoroscopic catheter location: a prospective randomized trial. *Eur Heart J*. 2006;27:1223-1229.
24. Jaïs P, Sanders P, Hsu LF, et al. Flutter localized to the anterior left atrium after catheter ablation of atrial fibrillation. *J Cardiovasc Electrophysiol*. 2006;17:279-285.
25. Zeppenfeld K, Schalij MJ, Bartelings MM, et al. Catheter ablation of ventricular tachycardia after repair of congenital heart disease: electroanatomic identification of the critical right ventricular isthmus. *Circulation*. 2007;116:2241-2252.
26. Ernst S, Antz M, Ouyang F, et al. Ostial PV isolation: is there a role for three-dimensional mapping? *Pacing Clin Electrophysiol*. 2003;26:1624-1630.
27. Marchlinski FE, Zado E, Dixit S, et al. Electroanatomic substrate and outcome of catheter ablative therapy for ventricular tachycardia in setting of right ventricular cardiomyopathy. *Circulation*. 2004;110:2293-2298.
28. Szumowski L, Sanders P, Walczak F, et al. Mapping and ablation of polymorphic ventricular tachycardia after myocardial infarction. *J Am Coll Cardiol*. 2004;44:1700-1706.
29. Nademanee K, McKenzie J, Kosar E, et al. A new approach for catheter ablation of atrial fibrillation: mapping of the electrophysiologic substrate. *J Am Coll Cardiol*. 2004;43:2044-2053.
30. Ouyang F, Ernst S, Vogtmann T, et al. Characterization of reentrant circuits in left atrial macroreentrant tachycardia: critical isthmus block can prevent atrial tachycardia recurrence. *Circulation*. 2002;105:1934-1942.
31. Shah D, Sunthorn H, Burri H, et al. Narrow, slow-conducting isthmus dependent left atrial reentry developing after ablation for atrial fibrillation: ECG characterization and elimination by focal RF ablation. *J Cardiovasc Electrophysiol*. 2006;17:508-515.
32. Sanders P, Hocini M, Jais P, et al. Characterization of focal atrial tachycardia using high-density mapping. *J Am Coll Cardiol*. 2005;46:2088-2099.
33. Magnano AR, Argenziano M, Dizon JM, et al. Mechanisms of atrial tachyarrhythmias following surgical atrial fibrillation ablation. *J Cardiovasc Electrophysiol*. 2006;17:366-373.
34. Ernst S, Schluter M, Ouyang F, et al. Modification of the substrate for maintenance of idiopathic human atrial fibrillation: efficacy of radiofrequency ablation using nonfluoroscopic catheter guidance. *Circulation*. 1999;100:2085-2092.
35. Schmitt C, Zrenner B, Schneider M, et al. Experience with a novel multielectrode basket catheter in right atrial tachycardias. *Circulation*. 1999;99:2414-2422.
36. Jenkins KJ, Walsh EP, Colan SD, Van Praagh R, Lock JE, Walsh EP. Multipolar endocardial mapping of the RA during cardiac catheterization: description of a new technique. *J Am Coll Cardiol*. 1993;22:1105-1110.
37. Rodriguez E, Callans D, Kantharia B, Gottlieb C, Marchlinski FE. Basket catheter localization of the origin of atrial tachycardia with atypical morphology after atrial flutter ablation. *Pacing Clin Electrophysiol*. 2000;2:269-272.
38. John B, Stiles MK, Kuklik P, et al. Electrical remodelling of the left and right atria due to rheumatic mitral stenosis. *Eur Heart J*. 2008;29:2234-2243.
39. Khoury D, Berrier K, Badruddin S, Zoghbi WA. Three dimensional electrophysiological imaging of the intact canine left ventricle using a non contact multielectrode cavitary probe: study of sinus, paced and spontaneous premature beats. *Circulation*. 1998;97:399-409.
40. Kadish A, Hauck J, Pederson B, Beatty G, Gornick C. Mapping of atrial activation with a non contact multielectrode catheter in dogs. *Circulation*. 1999;99:1906-1913.
41. Schilling RJ, Peters NS, Davies DW. Simultaneous endocardial mapping in the human left ventricle using a noncontact catheter: comparison of contact and reconstructed electrograms during sinus rhythm. *Circulation*. 1998;98:887-898.
42. Ihara R, Barr RC. Electrocardiographic inverse solution for ectopic origin of excitation in two dimensional propagation model. *Med Biol Eng Comput*. 1994;32:S41-S50.
43. Klemm HU, Ventura R, Steven D, et al. Catheter ablation of multiple ventricular tachycardias after myocardial infarction guided by combined contact and noncontact mapping. *Circulation*. 2007;115:2697-2704.
44. Schilling RJ, Peters NS, Davies DW. Feasibility of a noncontact catheter for endocardial mapping of human ventricular tachycardia. *Circulation*. 1999;99:2543-2552.
45. Voss F, Steen H, Bauer A, Giannitsis E, Katus HA, Becker R. Determination of myocardial infarct size by noncontact mapping. *Heart Rhythm*. 2008;5:308-314.
46. Schilling RJ, Kadish AH, Peters NS, Goldberger J, Davies DW. Endocardial mapping of atrial fibrillation in the human right atrium using a noncontact catheter. *Eur Heart J*. 2000;21:550-564.
47. Ribbing M, Wasmer K, Mönnig G, et al. Endocardial mapping of right ventricular outflow tract tachycardia using noncontact activation mapping. *J Cardiovasc Electrophysiol*. 2003;14:602-608.
48. Wongcharoen W, Tsao HM, Wu MH, et al. Morphologic characteristics of the left atrial appendage, roof, and septum: implications for the ablation of atrial fibrillation. *J Cardiovasc Electrophysiol*. 2006;17:951-956.
49. Noseworthy PA, Malchano ZJ, Ahmed J, Holmvang G, Ruskin JN, Reddy VY. The impact of respiration on left atrial and pulmonary venous anatomy: implications for image-guided intervention. *Heart Rhythm*. 2005;2:1173-1178.

50. Sanders P, Stiles MK, Young GD. Virtual anatomy for atrial fibrillation ablation. *J Cardiovasc Electrophysiol*. 2006;17:349-351.
51. Kato R, Lickfett L, Meininger G, et al. Pulmonary vein anatomy in patients undergoing catheter ablation of atrial fibrillation: lessons learned by use of magnetic resonance imaging. *Circulation*. 2003; 107:2004-2010.
52. Lemola K, Sneider M, Desjardins B, et al. Computed tomographic analysis of the anatomy of the left atrium and the esophagus: implications for left atrial catheter ablation. *Circulation*. 2004;110:3655-3660.
53. Malchano ZJ, Neuzil P, Cury RC, et al. Integration of cardiac CT/MR imaging with three-dimensional electroanatomical mapping to guide catheter manipulation in the left atrium: implications for catheter ablation of atrial fibrillation. *J Cardiovasc Electrophysiol*. 2006;17:1221-1229.
54. Wittkampf FH, Vonken EJ, Derksen R, et al. Pulmonary vein ostium geometry: analysis by magnetic resonance angiography. *Circulation*. 2003;107:21-23.
55. Mansour M, Holmvang G, Sosnovik D, et al. Assessment of pulmonary vein anatomic variability by magnetic resonance imaging: implications for catheter ablation techniques for atrial fibrillation. *J Cardiovasc Electrophysiol*. 2004;15:387-393.
56. Tops LF, Bax JJ, Zeppenfeld K, et al. Fusion of multislice computed tomography imaging with three dimensional electroanatomic mapping to guide radiofrequency catheter ablation procedures. *Heart Rhythm*. 2005;2:1076-1081.
57. Dong J, Calkins H, Solomon SB, et al. Integrated electroanatomic mapping with three-dimensional computed tomographic images for real-time guided ablations. *Circulation*. 2006;113:186-194.
58. Kuklik P, Szumowski L, Zebrowski JJ, Sanders P. Integration of the data from electroanatomical mapping system and CT imaging modality. *Int J Cardiovasc Imaging*. 2009;25:425-432.
59. Kistler PM, Rajappan K, Jahngir M, et al. The impact of CT image integration into an electroanatomic mapping system on clinical outcomes of catheter ablation of atrial fibrillation. *J Cardiovasc Electrophysiol*. 2006;17:1093-1101.
60. Heist EK, Chevalier J, Holmvang G, et al. Factors affecting error in integration of electroanatomic mapping with CT and MR imaging during catheter ablation of atrial fibrillation. *J Interv Card Electrophysiol*. 2006;17:21-27.
61. Zanchetta M, Rigatelli G, Pedon L, et al. Intracardiac echocardiography: gross anatomy and magnetic resonance correlations and validations. *Int J Cardiovasc Imaging*. 2005;21:391-401.
62. Knackstedt C, Franke A, Mischke K, et al. Semi-automated 3-dimensional intracardiac echocardiography: development and initial clinical experience of a new system to guide ablation procedures. *Heart Rhythm*. 2006;3:1453-1459.

12 The Role of Imaging for Catheter Ablation in the Pediatric Population

John Triedman

Abstract

Over the past two decades, catheter ablation has been increasingly widely applied to pediatric patients and patients with major congenital heart defects, often of adult age. As is the case in adult electrophysiology, pediatric ablation is a highly specialized procedure based largely on precise, real-time analysis of anatomical imaging. In general, the imaging problem to be addressed is how best to visualize a catheter or group of catheters within the beating heart. This scope problem is expanded by considering that the specific position of a catheter must be known, not only in terms of absolute coordinates, but more importantly in terms of its anatomical frame of reference, which is constantly in motion. Finally, that position must be known and updated continuously so that the course and trajectory of the visualized catheter can be used to guide the operator as the catheter is precisely navigated from place to place.

Keywords

Catheter ablation imaging • Imaging for catheter ablation • Pediatric catheter ablation • Pediatric imaging for catheter ablation • Anatomical imaging in pediatric patients • Electrophysiology in pediatrics

12.1 Pediatric and Congenital Catheter Ablation

Ablation techniques and the associated imaging and ablative technologies used in the pediatric and congenital heart disease group have largely been translated and adapted from methods used in adult ablative practice. However, significant differences exist in the anatomical and pathophysiological characteristics which describe the pediatric and congenital heart patient, and these in turn have affected the ways in which novel imaging technologies have been used in this group.

For the majority of pediatric ablation cases, the factor that principally distinguishes these patients from adult patients is their age and body size. Most cases are performed in otherwise healthy children with varieties of SVT occurring in anatomically and functionally normal hearts. The primary technical issues relevant to imaging and ablation in these patients are the size of the heart and vasculature, as these may impose limitations on the number and size of catheters which can be inserted. For a smaller number of patients, the primary anatomical issue facing catheter ablation techniques is the presence of congenital heart disease, which is associated with a high incidence of clinically important and often highly atypical atrial and ventricular

J. Triedman
Department of Pediatrics, Harvard Medical School, Boston, MA, USA

Department of Cardiology, Children's Hospital Boston, Boston, MA, USA
e-mail: john.triedman@cardio.chboston.org

A. Auricchio et al. (eds.), *Cardiac Imaging in Electrophysiology*,
DOI 10.1007/978-1-84882-486-7_12, © Springer-Verlag London Limited 2012

tachycardias. The presence of congenital malformation and the surgical procedures typically used to palliate these defects often affect vascular access, intracardiac anatomy, and the pathophysiology that determines the arrhythmia targets themselves.[1]

The specific description of the case distribution varies by pediatric medical center, and the type of population referred for care. At Children's Hospital Boston, the median age of patients undergoing ablative therapy is 11–12 years, with approximately 5% of patient below 18 months of age and 15 kg body weight. About 20% of patients referred for catheter ablation have a concomitant anatomical diagnosis of congenital heart disease, most commonly status post one or more surgical interventions.

12.2 Components of the Ablation Procedure

The technique of catheter ablation can be divided into several functional components for which different types of imaging support might be used. In sequence, catheter ablation may entail: (a) acquisition of vascular access, (b) navigation of peripheral vasculature to the heart, (c) placement of catheters at fixed and predetermined cardiac locations, (d) performance of diagnostic electrophysiology study, (e) navigated mapping of various cardiac landmarks, (f) acquisition of transseptal access, (g) supervised monitoring of catheter position during ablation, and (h) post ablation survey to detect therapeutic effect as well as any adverse or complicating events. This sequence can be conceptually reduced to: "DIAGNOSE – NAVIGATE – ABLATE – EVALUATE." Given the broad range of specific tasks outlined in this sequence, it is not surprising that different imaging modalities might be optimal for different objectives, which can range from broad survey of the thorax, identification of vascular structures, navigation of normal and abnormal anatomy with catheters, and close examination of intracardiac structures.

12.3 Advantages and Limitations of Fluoroscopy in Pediatric Ablation

Images used to support ablation should be a perfectly accurate representation of the heart, easy to generate, fully transparent to the user, imbued with electrophysiological information and anatomical relevance, and immediately available for real-time catheter navigation. An approach to understanding the value and appropriate uses of imaging tools in ablation is to assess the degree to which they meet these criteria.

The original technique used to guide catheter ablation procedure in children and adults has been fluorography. This imaging modality is the legacy technique employed by virtually all interventional catheterization specialists who have trained over the last several decades, there is an enormous installed base of fluoroscopy units in hospitals around the world, and the technology has spawned an entire set of technical professional specialties to support its continuing use. Fluoroscopy has a number of features which render it uniquely valuable for performance of catheter-based procedures. Although it provides a flattened and distorted representation of cardiothoracic anatomy, the use of multiple projections by a skilled operator allows understanding of anatomy and catheter location in remarkable spatial detail. It is a widely available and very mature technology that provides real-time, high-resolution images of the procedural field of interest to those performing cardiac interventions. This feature set allows fluoroscopy to be used both for navigational survey – seeing "the big picture" – as well as precise catheter navigation. Given the materials useful in catheter fabrication, especially for electrophysiology catheters, good imaging contrast allows for precise and easy catheter visualization. Although anatomical interpretation of fluoroscopic images can take many years to master, the initial technical learning curve required to obtain useful images is very short. The development of fluoroscopy-based EP procedures has finally been facilitated by the serendipitous fact that, for many common arrhythmias, important procedural landmarks such as the His bundle and mitral groove may be identified on fluoroscopy by placement of catheters in easily identified, reproducible locations.

There are several drawbacks to fluoroscopy use as well. Fluorography relies on shadowing of structures by transillumination, and interpretation of these images can be quite challenging. Although large field navigation is well enabled, anatomical detail is largely lost, especially within the heart where the relative opacity of blood and tissue to X-ray is the same. Due to beam spreading effects, there are gradients of magnification error within images which are difficult to appreciate. In the absence of biplane fluorography, movements of catheters perpendicular to the plane of view are ambiguous.

Also important in fluoroscopic procedures are the exposure of patient and staff to ionizing radiation. Despite efforts to minimize radiation dosing by improvements in photon detector technology and reduction of pulse rate, large series of catheter ablation procedures in pediatric populations with SVT have reported median fluoroscopy times ranging from ~30 to 45 min.[2,3] Depending on the duration of fluoroscopy, the body size of the patient and the projections used, this can often result in doses totaling hundreds or even thousands of milliGray. This becomes increasingly important for patients undergoing more complex procedures. Potential consequences of radiation exposure include both deterministic

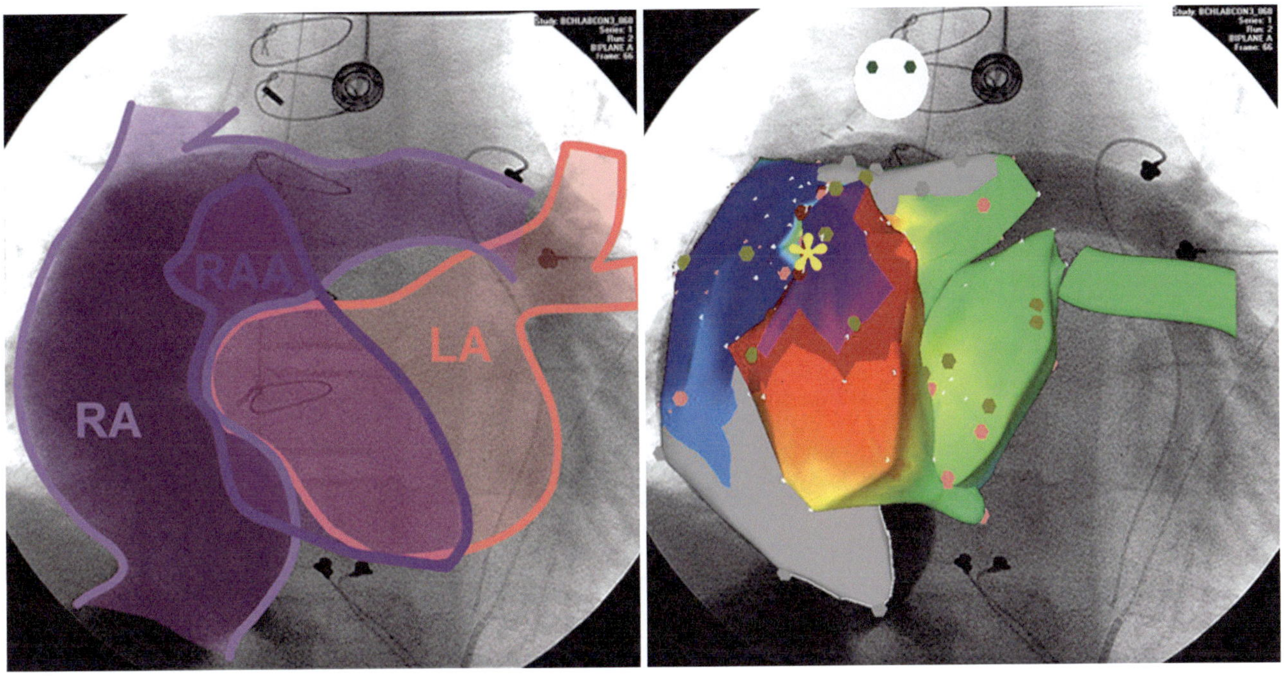

Fig. 12.1 Integration of fluoroscopy and electroanatomical mapping illustrating a complex atrial macroreentry circuit in a patient with a Fontan procedure performed for severe Ebstein's anomaly. *Left panel* – Angiographic anatomy of Fontan pathway (*RA*), created by baffling of IVC to atriopulmonary anastomosis. Pulmonary venous portion of the right atrium (*RAA*) and left atrium (*LA*) are also indicated, determined by additional angiographies. *Right panel* – superposition of three chambers mapped in atrial tachycardia using CARTO on the prior angiographic view. A narrow conducting corridor is shown (*yellow asterisk*) between the RA and RAA, bounded above by the atriopulmonary anastomosis and below by the extent of the surgical right atriotomy used to perform the Fontan

effects (e.g., radiodermatitis at beam entry site) and stochastic effects (e.g., increased late attributable risk of neoplasia); with respect to the latter, no safe lower limit for exposure has been determined.

12.4 Electroanatomical Mapping and Navigation

One important limitation of fluoroscopy is that there is no simple way to organize multiple measurements taken from a roving catheter into an easily visualized and clinically useful model of cardiac electrical activity. Complex atrial and ventricular arrhythmias common in patients with congenital heart defects are not well defined by fluoroscopically identifiable cardiac landmarks. This limits the utility of fluoroscopy for accurate target localization in congenital patients, and is one likely reason that initially reported efficacy of ablation for such arrhythmias is lower than that seen in SVT encountered in anatomically normal hearts.[4,5]

Arrhythmia mapping, as developed in experimental EP and clinical arrhythmia surgery, imposes a spatial frame of reference on the heart to characterize the activation sequence. This power of this methodology, even when reduced to the form of rough, *ad hoc* assignments of electrogram timing to approximate locations on anatomic cartoons, has been amply demonstrated to be of value and has been used to develop important general hypotheses regarding arrhythmia mechanism (e.g., the work by Cosio et al.[6]).

Electromagnetic field–based technologies such as NavX™ (St. Jude), CARTO™ (Biosense Webster), and LocaLisa™ (Medtronic)[7-9] have therefore been widely used to enhance our ability to visualize, map, and navigate the heart in pediatric and congenital practice. These devices each provide an accurate spatial frame of reference in the form of an electromagnetic field that defines a reference grid within the area of interest. Detection of this field either using an unmodified electrode or an electromagnetic sensor allows continuous monitoring of spatial location. By iterative recording of its location, the position of which is bounded by the endocardial surface, this spatial data can be used both to construct three-dimensional images of the heart that closely match chamber anatomy and to navigate catheters in real time through the reference space.

Advanced electroanatomical mapping has allowed increasingly complete characterization of typical and variant anatomical patterns of atrial macroreentrant tachycardias seen in congenital heart patients with common problems such as the Fontan procedure for single ventricle variants (Fig. 12.1), with Mustard or Senning procedures for transposition of the great ventricles (Fig. 12.2), status post repair of

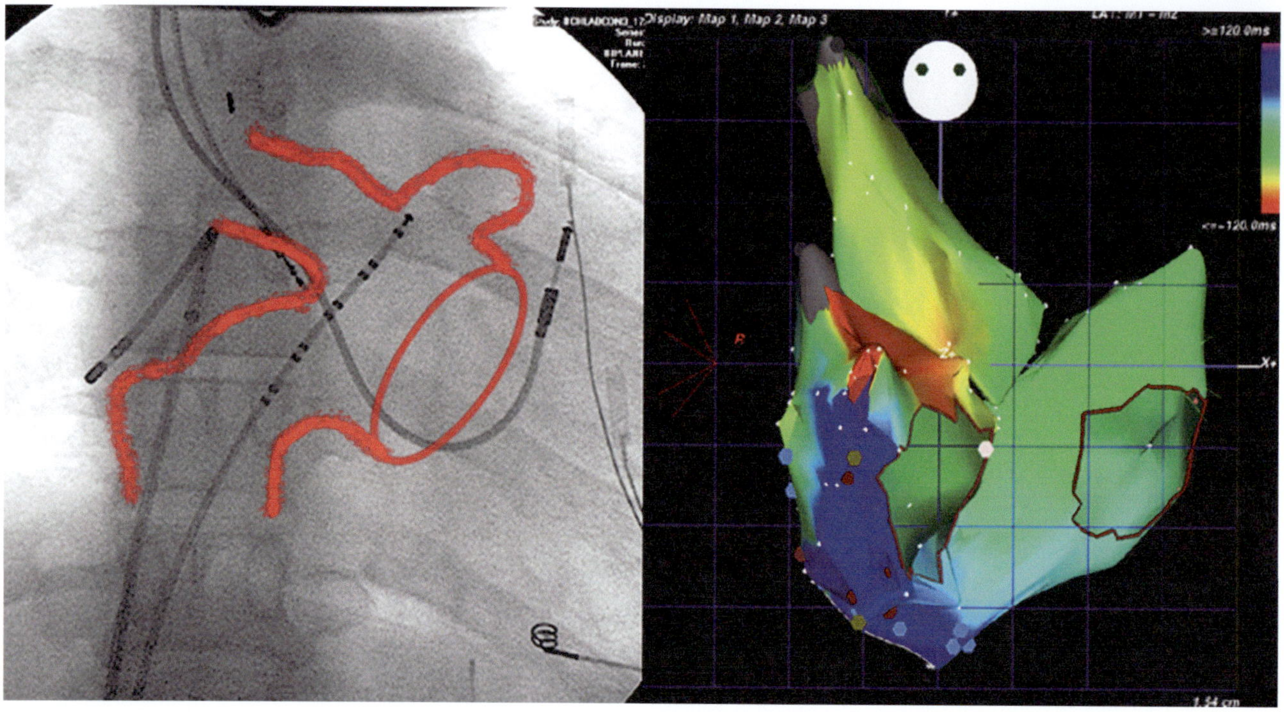

Fig. 12.2 Imaging of atrial macroreentry circuit in a patient who has undergone intra-atrial baffling for palliation of transposition of the great vessels (Mustard procedure). *Left panel* – fluoroscopic catheter view with annotations indicating the location of the "neo-right atrium," consisting of baffled inferior and superior caval pathways and the antrum of the subpulmonary mitral valve. The mapping catheter course is through a transbaffle puncture and in the "neo-left atrium," which redirects blood from the pulmonary veins laterally to the subaortic tricuspid valve. *Right panel* – Multichamber CARTO map of an atrial reentry circuit bounded anteriorly by the tricuspid annulus and posteriorly by a presumed atriotomy scar; the overall pattern is that of a clockwise atrial flutter. The superior caval baffle has been drawn as a separate chamber from the neo-right atrium to help preserve the anatomical accuracy of the map

tetralogy of Fallot and atrial septal defect, and in less common diagnoses in mixed series. The positive effect of the use of such technologies on acute procedure outcomes has been demonstrated.[10] More recently, similar knowledge of common patterns and mapping and ablation methodologies in ventricular tachycardias has also been developed.[11]

A significant limitation of this type of three-dimensional imaging is the discrepancy between spatial and anatomical accuracy. Phasic changes in intrathoracic pressure and volume, heart rate and filling pressures, and the distortion of the endocardium by catheter pressure cause significant variations in chamber size and geometry. These factors can be partially but not fully compensated for by gating acquisition to cardiac and respiratory cycles. They do not negate the value of this type of mapping, but do impose an upper bound on the reliability with which the operator can discriminate two closely spaced anatomical locations.

12.5 Incorporation of Ultrasound in Ablative Imaging

The use of echocardiography is widespread in pediatrics and a cornerstone of diagnostic evaluation in congenital heart disease. It is possibly to identify many of the anatomical structures relevant to ablation procedures using ultrasound, as well as diagnostic and ablation catheters (Fig. 12.3, left panel), using transesophageal and more recently intracardiac echo probes, of both the rotational and phased array variety. Initially adopted as an adjunct imaging modality for support of transseptal puncture, these approaches have been demonstrated to be effective in supporting many aspects of ablation procedures.[12] Importantly, ultrasound modalities at present represent the only widely available imaging technology which can demonstrate the endocardial anatomy in real time, addressing the limitations of electroanatomical mapping mentioned above. Direct contact of ablation catheters with the endocardial surface can be observed (Fig. 12.3, right panel), and recent research has suggested that it may be possible to use ultrasound to monitor the effect of ablation on myocardial tissue.[13]

Each of these characteristics is of high potential value to patients with congenital heart disease, and has been demonstrated in exemplary manner in anecdotal reports and short case series.[14-16] Nonetheless, transesophageal echo and intracardiac ultrasound have not been systematically used to support ablation procedures to date, and no published data documents the specific value of this approach in supporting such procedures. Reasons for this appear to include the unfamiliarity of many electrophysiologists with echocardiographic image

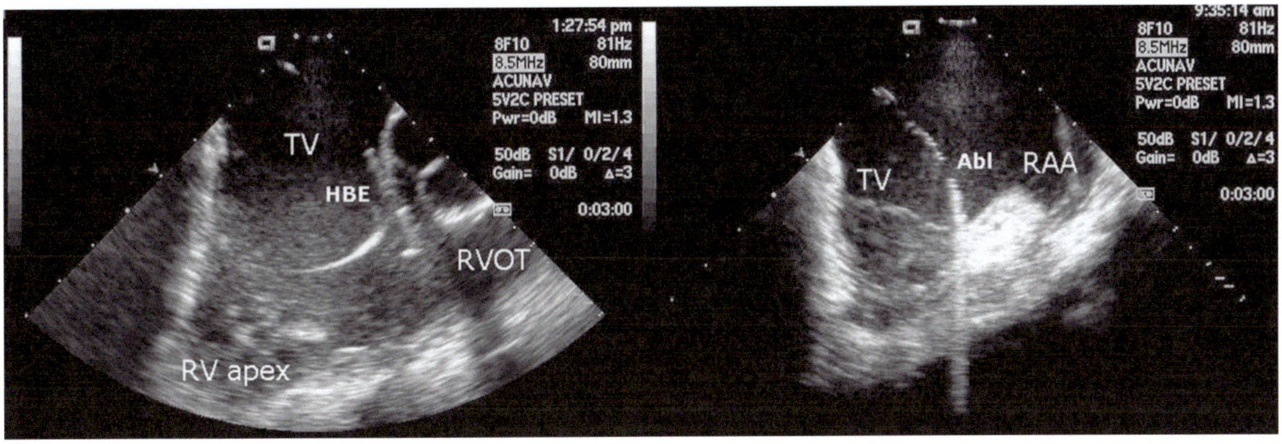

Fig. 12.3 Use of intracardiac echo in support of SVT ablation in children with normal hearts. *Left panel* – standard views facilitate interpretation and can be used to identify proper EP catheter location in relation to relevant anatomical structure. This is a home view for position of *HBE* and *RV* catheters in relation to tricuspid annulus (*TV*), *RV apex*, RV outflow (*RVOT*), and aortic root. *Right panel* – Intracardiac echo is useful in visualization of direct myocardial contact of the ablation catheter (*Abl*). In this view, a right anterolateral pathway is ablated with relations of tricuspid annulus (*TV*), right atrial appendage (*RAA*), and mapping catheter identified

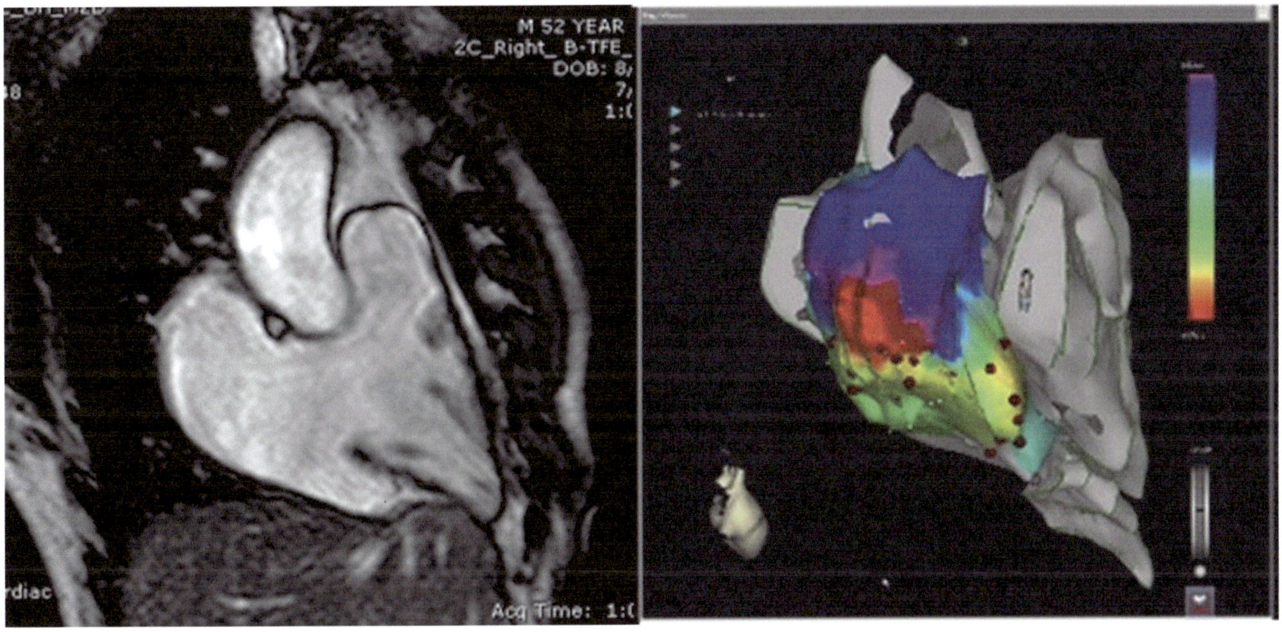

Fig. 12.4 Use of CartoSOUND integration of ICE and electroanatomical navigation to construct chamber volumes for real-time mapping. On the left, an MRI plane of an adult congenital patient with complex Ebstein's anomaly and a large outpouching of the subtricuspid right atrium between the right ventricle and the diaphragm. On the right, a CartoSOUND reconstruction performed during an ablation procedure demonstrating all the relevant anatomic information, as well as the relevant electrophysiologic mapping information

interpretation; difficulty in integration of an unregistered, two-dimensional ultrasound image; and, in the case of transesophageal echo, the need for an additional operator to manipulate the echo probe.

Recently, the problem of integrating an echocardiographic image with more easily recognized three-dimensional maps of the heart has been directly addressed by the introduction of an electroanatomically navigated intracardiac echo probe (CartoSOUND™, Biosense Webster).[17] This device allows the two-dimensional echo sector to be continuously located within the spatial frame of an electroanatomical map, and the endocardial contours detected on echo to be used to create an accurate, interpolated representation of the chamber surface. As seen in Fig. 12.4, this can allow for real-time imaging of quite complex chamber anatomies relevant to ablation in congenital heart patients. In particular, it allows detailed discrimination and examination of anatomical structures relevant to ablation, such as the cavotricuspid isthmus, intra-atrial baffles, and relationships and connections of the great veins, in a direct manner which cannot be achieved using fluoroscopy or electroanatomical mapping.

12.6 Multimodal Image Fusion in Congenital Heart Ablation Procedures

The foundation of ablation procedures in congenital heart patients is an accurate, detailed, and accessible representation of the cardiac anatomy, which can be used to guide mapping and catheter navigation. Another technology which can contribute to this process is the integration of segmented volume images of the heart obtained by CT scan or magnetic resonance imaging with ablation procedures supported by electroanatomical catheter navigation. This approach has been used very effectively to navigate the complex anatomy of the pulmonary veins during atrial fibrillation ablation procedures. Effective use of such images demands that they be properly placed within the electroanatomical reference grid which used to guide catheter movement. Several studies have shown that careful application of techniques to ensure proper coregistration between the volume image and navigational tools can reliably result in accuracy measured in millimeters, which for most ablation applications (e.g., pulmonary venous antral ablation) appears to be sufficient.[18-20]

As is the case with ultrasound, multimodal image fusion has been reported to be of value in congenital heart disease, but not established in most laboratories as a routinely used tool to support these procedures.[21] In addition to the intrinsic limitations related to accuracy mentioned above, application of this technique in congenital patients is often rendered more difficult due to detailed and complex anatomy. This makes accurate segmentation and registration of images simultaneously more difficult and more critically important for effective use during the ablation procedure. It has been our observation as well that this technology, while helpful for understanding of anatomical issues relevant to catheter course and access, often provides less useful information than three-dimensional images constructed in real time using navigated ultrasound technologies as described above.

12.7 Frontiers in Complex Imaging for Pediatric and Congenital Ablation

Technological innovation has been an important driver in the development of sophisticated ablative procedures in pediatric and congenital ablation. Relatively few clinical studies have been performed that demonstrate the specific value of these technologies on ablation outcomes, and a major future goal should be the performance of such studies. Those studies currently available primarily focus on supraventricular tachycardia and atrial flutter, which are already treated with high success rate with fluoroscopy. Some of those suggest improved outcomes, but most have shown clinical equivalence with fluoroscopy, the principal demonstrable benefits being reductions of fluoroscopy exposure and procedure time.[22-24]

With respect to procedure development, two areas seem promising for the development of imaging technologies in coming years in the field of pediatric and congenital ablation: the application of advanced imaging to the reduction and eventual elimination of fluoroscopy from ablation of common forms of SVT in the normal heart, and the real-time imaging of ablations lesions as they are created in complex ablations.

A major drawback of fluoroscopy often which is unfortunately often rationalized as a "necessary evil" is the exposure of patients to ionizing radiation. The importance of diagnostic and therapeutic procedures as a major source of radiation exposure has recently been made evident.[25] This exposure is of particular concern for children, who have growing, radiosensitive tissues, long life expectancy, and underlying arrhythmia problems that often have little or no lethal potential. Although exposure to ionizing radiation during fluoroscopy-based ablation procedures is low in comparison, for example, to radiotherapy, it is calculated to be of significant lifelong epidemiological consequence.[26] Additionally, it is also evident that both occupational exposure to fluoroscopy-generated radiation and the use of shielding to mitigate that exposure are associated with long-term health consequences for operators and staff.[27] In as-yet small and uncontrolled studies, several authors have now demonstrated the feasibility of safe and effective use of electroanatomical and ultrasound-based adjunctive techniques, initially designed to facilitate the performance of complex procedures, to perform SVT ablations in children with normal hearts.[28-30] Although this approach has a prolonged learning curve, and it is sometimes impossible to avoid the use of fluoroscopy altogether, further development of these techniques ultimately seems likely to displace the use of fluoroscopy as a primary modality to perform ablation procedures in many large patient groups.

For complex patients, the final link in the procedural chain "DIAGNOSE – NAVIGATE – ABLATE – EVALUATE" includes the assessment of lesion size and transmurality. A variety of real-time technologies have been proposed to determine whether an applied ablation lesion has modified the myocardium sufficiently to achieve the therapy goal. Recent developments in magnetic resonance imaging and catheter design compatible with high magnetic field strength suggest that MR-guided ablation may have the unique potential to allow real-time or near real-time imaging of ablation lesions as they are actually formed in the myocardium.[31] If this is brought to fruition, it will be of considerable value to complex procedures in congenital heart patients.

References

1. Walsh EP. Interventional electrophysiology in patients with congenital heart disease. *Circulation*. 2007;115:3224-3234.
2. Kugler JD, Danford DA, Houston K, Felix G. Radiofrequency catheter ablation for paroxysmal supraventricular tachycardia in children and adolescents without structural heart disease. *Am J Cardiol*. 1997;80:1438-1443.
3. Van Hare GF, Javitz H, Carmelli D, et al. Prospective assessment after pediatric cardiac ablation: demographics, medical profiles, and initial outcomes. *J Cardiovasc Electrophysiol*. 2004;15:759-770.
4. Collins KK, Love BA, Walsh EP, Saul JP, Epstein MR, Triedman JK. Location of acutely successful radiofrequency catheter ablation of intraatrial reentrant tachycardia in patients with congenital heart disease. *Am J Cardiol*. 2000;86:969-974.
5. Kannankeril PJ, Anderson ME, Rottman JN, Wathen MS, Fish FA. Frequency of late recurrence of intra-atrial reentry tachycardia after radiofrequency catheter ablation in patients with congenital heart disease. *Am J Cardiol*. 2003;92:879-881.
6. Cosio FG, Arribas F, Barbero JM, Kallmeyer C, Goicolea A. Validation of double-spike electrograms as markers of conduction delay or block in atrial flutter. *Am J Cardiol*. 1988;61:775-780.
7. Gepstein L, Hayam G, Ben-Haim SA. A novel method for nonfluoroscopic, catheter-based electroanatomical mapping of the heart: in vitro and in vivo results. *Circulation*. 1997;95:1611-1622.
8. Schilling R, Peters N, Davies D. Feasibility of a noncontact catheter for endocardial mapping of human ventricular tachycardia. *Circulation*. 1999;99:2543-2552.
9. Wittkampf FH, Wever EF, Derksen R, et al. LocaLisa: new technique for real-time 3-dimensional localization of regular intracardiac electrodes. *Circulation*. 1999;99:1312-1317.
10. Triedman JK, Alexander ME, Love BA, et al. Influence of patient factors and ablative technologies on outcomes of radiofrequency ablation of intra-atrial re-entrant tachycardia in patients with congenital heart disease. *J Am Coll Cardiol*. 2002;39:1827-1835.
11. Zeppenfeld K, Schalij MJ, Bartelings MM, et al. Catheter ablation of ventricular tachycardia after repair of congenital heart disease: electroanatomic identification of the critical right ventricular isthmus. *Circulation*. 2007;116:2241-2252.
12. Dravid SG, Hope B, McKinnie JJ. Intracardiac echocardiography in electrophysiology: a review of current applications in practice. *Echocardiography*. 2008;25:1172-1175.
13. Hsu SJ, Hubert JL, Smith SW, Trahey GE. Intracardiac echocardiography and acoustic radiation force impulse imaging of a dynamic ex vivo ovine heart model. *Ultrason Imaging*. 2008;30:63-77.
14. Forleo GB, Pappalardo A, Avella A, Visigalli L, Dello Russo A, Tondo C. Real-time integration of intracardiac echocardiography and 3D electroanatomical mapping to guide catheter ablation of isthmus-dependent atrial flutter in a patient with complete situs inversus and interruption of the inferior vena cava with azygos continuation. *J Interv Card Electrophysiol*. 2009;30(3):273-7. doi:doi: 10.1007/s10840-009-9427-2.
15. El Yaman MM, Asirvatham SJ, Kapa S, Barrett RA, Packer DL, Porter CB. Methods to access the surgically excluded cavotricuspid isthmus for complete ablation of typical atrial flutter in patients with congenital heart defects. *Heart Rhythm*. 2009;6:949-956.
16. Peichl P, Kautzner J, Gebauer R. Ablation of atrial tachycardias after correction of complex congenital heart diseases: utility of intracardiac echocardiography. *Europace*. 2009;11:48-53.
17. Packer DL, Johnson SB, Kolasa MW, Bunch TJ, Henz BD, Okumura Y. New generation of electro-anatomic mapping: full intracardiac ultrasound image integration. *Europace*. 2008;10(3):iii35-iii41.
18. Richmond L, Rajappan K, Voth E, et al. Validation of computed tomography image integration into the EnSite NavX mapping system to perform catheter ablation of atrial fibrillation. *J Cardiovasc Electrophysiol*. 2008;19:821-827.
19. Bertaglia E, Brandolino G, Zoppo F, Zerbo F, Pascotto P. Integration of three-dimensional left atrial magnetic resonance images into a real-time electroanatomic mapping system: validation of a registration method. *Pacing Clin Electrophysiol*. 2008;31:273-282.
20. Dong J, Dalal D, Scherr D, et al. Impact of heart rhythm status on registration accuracy of the left atrium for catheter ablation of atrial fibrillation. *J Cardiovasc Electrophysiol*. 2007;18:1269-1276.
21. Pflaumer A, Deisenhofer I, Hausleiter J, Zrenner B. Mapping and ablation of atypical flutter in congenital heart disease with a novel three-dimensional mapping system (Carto Merge). *Europace*. 2006;8:138-139.
22. Sporton SC, Earley MJ, Nathan AW, Schilling RJ. Electroanatomic versus fluoroscopic mapping for catheter ablation procedures: a prospective randomized study. *J Cardiovasc Electrophysiol*. 2004;15:310-315.
23. Schneider MA, Ndrepepa G, Dobran I, et al. LocaLisa catheter navigation reduces fluoroscopy time and dosage in ablation of atrial flutter: a prospective randomized study. *J Cardiovasc Electrophysiol*. 2003;14:587-590.
24. Kopelman HA, Prater SP, Tondato F, Chronos NA, Peters NS. Slow pathway catheter ablation of atrioventricular nodal re-entrant tachycardia guided by electroanatomical mapping: a randomized comparison to the conventional approach. *Europace*. 2003;5:171-174.
25. Fazel R, Krumholz HN, Wang Y, et al. Exposure to low-dose ionizing radiation from medical imaging procedures. *N Engl J Med*. 2009;361:849-857.
26. Beels L, Bacher K, de Wolf D, Werbrouck J, Thierens H. γ-H2AX foci as a biomarker for patient X-ray exposure in pediatric cardiac catheterization. *Circulation*. 2009;120:1903-1909.
27. Klein LW, Miller DL, Balter S, et al. Occupational health hazards in the interventional laboratory: time for a safer environment. *Heart Rhythm*. 2006;6:439-444.
28. Clark J, Bockoven JR, Lane J, Patel CR, Smith G. Use of three-dimensional catheter guidance and trans-esophageal echocardiography to eliminate fluoroscopy in catheter ablation of left-sided accessory pathways. *Pacing Clin Electrophysiol*. 2008;31:283-289.
29. Tuzcu V. A nonfluoroscopic approach for electrophysiology and catheter ablation procedures using a three-dimensional navigation system. *Pacing Clin Electrophysiol*. 2007;30:519-525.
30. Papagiannis J, Tsoutsinos A, Kirvassilis G, et al. Nonfluoroscopic catheter navigation for radiofrequency catheter ablation of supraventricular tachycardia in children. *Pacing Clin Electrophysiol*. 2006;29:971-978.
31. Knowles BR, Caulfield DR, Cooklin MR, et al. 3-D visualization of acute RF ablation lesions using MRI for the simultaneous determination of the patterns of necrosis and edema. *IEEE Trans Biomed Eng*. 2010;6:1467-1475.

Part III

Device Therapy

Substrate Assessment: Echocardiography, MRI, and CCT

13

Victoria Delgado, Matteo Bertini, Rutger J. van Bommel, C. Jan Willem Borleffs, Martin J. Schalij, and Jeroen J. Bax

Abstract

A large body of published evidence from multicenter controlled trials has demonstrated the clinical benefits of ICD and CRT in high-risk patients. However, to achieve the largest benefit, risk-stratification strategies and identification of candidates for these therapies should be optimized. In this regard, analysis of the cardiac substrate may provide important information. In candidates for ICD implantation, MRI and nuclear imaging provide meaningful insight in the arrhythmogenic substrate. In addition, selection of candidates to CRT should rely on an integrated approach including the assessment of LV dyssynchrony (latest activated segment), location and extent of myocardial scar tissue, and cardiac venous anatomy. As shown in numerous studies, patients with significant LV dyssynchrony and with an LV pacing lead positioned in or near the latest activated segments will benefit most from CRT implantation. Echocardiographic techniques provide useful information in this pathophysiological aspect. Furthermore, in patients with ischemic heart failure, positioning of the LV pacing lead in a segment with transmural scar reduces the CRT response. MRI and nuclear imaging are key imaging tools to evaluate the extension and location of scar tissue. Finally, the lack of suitable cardiac venous anatomy for transvenous LV lead implantation at the latest activated segments may be common in patients with prior myocardial infarction. In this subgroup of patients, MDCT may constitute a valuable imaging modality to anticipate the LV lead implantation strategy (transvenous or minimally invasive surgical approach). Therefore, a multimodality imaging approach, including echocardiographic techniques, MRI, nuclear imaging, and MDCT, may refine the risk-stratification and selection of candidates to these device therapies.

Keywords

Implantable cardioverter defibrillator • Cardiac resynchronization therapy • Device cardiac therapy • Imaging studies for cardiac device selection • Substrate assessment in cardiac imaging

Over the last decade, implantable cardioverter defibrillator (ICD) and cardiac resynchronization therapy (CRT) have become well-accepted, device-based therapies for high-risk cardiac patients. Landmark randomized clinical trials have demonstrated that ICD reduces by 30–54% the mortality risk of patients with heart failure and coronary artery disease, and CRT improves heart failure symptoms, clinical outcome, and left ventricular (LV) function of end-stage heart failure patients.[1-5] Based on this evidence, the American Heart Association/American College of Cardiology/Heart Rhythm Society guidelines have recently considered ICD and CRT a class I recommendation in selected high-risk patients.[6]

V. Delgado (✉)
Department of Cardiology, Leiden University Medical Center, Leiden, ZA, The Netherlands
e-mail: v.delgado@lumc.nl

A. Auricchio et al. (eds.), *Cardiac Imaging in Electrophysiology*,
DOI 10.1007/978-1-84882-486-7_13, © Springer-Verlag London Limited 2012

Despite these clinical benefits, several uncertainties confront implantable device therapy. In high-risk subpopulations, the clinical benefit of ICD is not consistent; only 35% of patients receive appropriate ICD shocks and an increased risk of heart failure has been reported due to unnecessary ventricular pacing.[7] Furthermore, up to 20–30% of heart failure patients do not show clinical improvement after CRT implantation, and 40% do not improve in LV function or show LV reverse remodeling.[8]

In order to optimize risk-stratification and identification of patients who will benefit from these device therapies, extensive research has been conducted, mainly in the field of multimodality cardiac imaging. Thus far, echocardiography, cardiac magnetic resonance imaging (MRI), nuclear imaging, and multidetector computed tomography (MDCT) have provided meaningful insight into the substrate on which ICD and CRT act. This chapter provides an overview of imaging modalities that may improve selection of patients who benefit from ICD and/or CRT.

13.1 Implantable Cardioverter Defibrillator

Sudden cardiac death accounts for nearly 50% of all cardiovascular mortality worldwide with an estimated annual incidence of 184,000–462,000 cases.[9] The World Health Organization defines sudden cardiac death as unexpected death within 1 h of symptom onset if witnessed or within 24 h of the person having been observed alive and symptom-free if unwitnessed.[10] In 80–90% of the cases, ventricular tachycardia and ventricular fibrillation are the causes of sudden cardiac death.[11] Among several structural cardiac diseases that have an increased risk for sudden death, coronary artery disease is the most frequent underlying cause, accounting for 65–70% of the cases.[11] Despite several interventions aiming to improve an early access to medical care and early advanced cardiopulmonary resuscitation, including external defibrillation, the mortality from life-threatening arrhythmias remains high.[9] Recent multicenter, randomized trials have demonstrated that ICD constitutes an effective strategy to prevent and treat sudden death.[12,13]

13.1.1 Secondary and Primary Prevention ICD Trials

Three major secondary prevention ICD trials are summarized in Table 13.1: the Antiarrhythmic vs. Implantable Defibrillators (AVID) trial, the Cardiac Arrest Study Hamburg (CASH), and the Canadian Implantable Defibrillator Study (CIDS).[14-16] In brief, the target study populations included patients resuscitated from ventricular fibrillation or ventricular tachycardia. Pooled analysis of these three secondary prevention ICD trials demonstrated 28% relative risk reduction in all-cause mortality and 50% relative risk reduction in arrhythmic mortality with ICD ($p=0.0006$ and $p<0.0001$, respectively) (Fig. 13.1).[23]

In addition, ten major primary prevention ICD trials have been conducted (Table 13.1, Fig. 13.2).[1,4,5,17,18,21,22,24] Initially, the target populations of these trials included patients with ischemic cardiomyopathy (LV ejection fraction ≤ 35%) and with inducible, non-suppressible ventricular tachyarrhythmias at the electrophysiological study. Thus, the Multicenter Automatic Defibrillator Implantation Trial (MADIT) showed a significant relative risk reduction of 59% in total mortality in the ICD arm after 27-month follow-up ($p=0.009$).[5] In addition, the study design of the Multicenter Unsustained Tachycardia Trial (MUSTT) provided further insight and indicated that patients with ischemic cardiomyopathy but without inducible life-threatening arrhythmias were also at a high risk for sudden death and may benefit from ICD therapy.[18] To test this hypothesis, the Multicenter Automatic Defibrillator Implantation Trial-II (MADIT-II) evaluated the beneficial effects of ICDs in 1,232 patients with prior myocardial infarction and LV ejection fraction ≤30% but without electrophysiological risk-stratification.[19] During 20-month follow-up, the superiority of ICD over antiarrhythmic medical treatment in preventing total mortality was demonstrated, with a relative risk reduction of 31% ($p=0.016$).[19] Finally, the benefits of ICD on survival in patients with nonischemic cardiomyopathy have been also evaluated (Table 13.1, Fig. 13.2).[1,2,20-22] Pooled data from 5 major randomized trials, including 1,854 patients, demonstrated that ICD was superior to medical therapy, reducing significantly the relative risk of all-cause mortality by 31% ($p=0.002$).[25]

Despite these encouraging results, two major trials did not demonstrate benefit of ICD for primary prevention in selected populations but provided valuable information on life-threatening arrhythmia mechanisms: the Coronary Artery Bypass Graft (CABG) Patch trial and the Defibrillator In Acute Myocardial Infarction Trial (DINAMIT).[4,17] The CABG-Patch trial randomized 1,055 patients undergoing CABG surgery to routine medical care vs. ICD.[17] The mortality rate in the ICD group was comparable to the medical care group (23% vs. 21%) with no difference in all-cause mortality at 32-month follow-up ($p=0.64$). In this study, the revascularization of jeopardized but viable myocardium probably reduced the risk for life-threatening arrhythmias, diminishing the power of the study to demonstrate the superiority of ICD over medical treatment in reducing all-cause mortality and sudden cardiac death.[17]

In addition, the DINAMIT trial evaluated the benefits of ICD in patients with recent myocardial infarction (within the first 40 days).[4] Patients randomized to ICD had similar total mortality rate as compared to patients randomized to standard medical therapy (7.5% vs. 6.9%, $p=0.66$).[4] Importantly, the annually arrhythmia-related deaths were lower in the ICD group than in the medical treatment group (1.5% vs.

Table 13.1 Major secondary and primary prevention ICD trials

Trial, year (Ref.#)	Randomization	No.	Population	Mean follow-up (months)	Main findings
Secondary prevention					
AVID (1997)[14]	ICD vs. Antiarrhythmic drugs (97% amiodarone)	1,016	Survived VT/VF/cardiac arrest; VT with syncope; VT with LVEF ≤40%	18	28% reduction in total mortality with ICD therapy (p=0.02)
CASH (2000)[15]	ICD vs. Antiarrhythmic drugs	288	Survived VT/VF/cardiac arrest	57	23% reduction in total mortality with ICD therapy (p=0.08)
CIDS (2000)[16]	ICD vs. Amiodarone	659	Survived VT/VF/cardiac arrest; VT with syncope; VT with LVEF ≤35%	35	20% reduction in total mortality with ICD therapy (p=0.14)
Primary prevention					
MADIT (1996)[5]	ICD vs. Antiarrhythmic drugs (74% amiodarone)	196	Prior MI; LVEF ≤35%; asymptomatic NSVT; NYHA class I-III; inducible VT refractory on EPS	27	54% reduction in total mortality with ICD therapy (p=0.009)
CABG-Patch (1997)[17]	CABG surgery plus ICD vs. CABG surgery plus conventional therapy	900	Patients scheduled for CABG; LVEF ≤35%; positive signal average ECG result	32	No reduction in total mortality with ICD therapy (p=0.64)
MUSTT (1999)[18]	EPS-guided therapy (antiarrhythmic or ICD) vs. conventional therapy	704	Prior MI; LVEF ≤40%; CAD; NSVT; inducible VT on EPS	39	58% reduction in total mortality with ICD therapy (p≤0.001)
MADIT II (2002)[19]	ICD vs. conventional therapy	1,232	Prior MI; LVEF ≤30%	20	Reduction in total mortality with ICD therapy (p=0.02)
DINAMIT (2004)[4]	ICD vs. conventional therapy	674	Recent MI (within 4–40 days), LVEF ≤35%; impaired cardiac autonomic modulation (heart rate variability)	39	No reduction in total mortality with ICD therapy (p=0.66)
CAT (2002)[20]	ICD vs. conventional therapy	104	NYHA class II-III, nonischemic heart failure; LVEF ≤30%; asymptomatic NSVT	66	No reduction in total mortality with ICD therapy (p=0.55)
AMIROVIT (2003)[21]	ICD vs. Amiodarone	103	NYHA class I-III, nonischemic heart failure; LVEF ≤35%; asymptomatic NSVT	36	No reduction in total mortality with ICD therapy (p=0.80)
DEFINITE (2004)[22]	ICD vs. conventional therapy	458	Nonischemic heart failure; LVEF ≤35%; NSVT or PVCs	29	Reduction in total mortality with ICD therapy (p=0.08)
COMPANION (2004)[2]	Conventional therapy Vs. CRT vs. CRT/ICD	1,520	NYHA class II-III ischemic and nonischemic heart failure; LVEF ≤35%; QRS≥120 ms	17	Reduction in total mortality with CRT alone (p=0.06)
SCD-HeFT (2005)[1]	ICD vs. conventional therapy	2,521	NYHA class II-III ischemic and nonischemic heart failure; LVEF ≤35%	45	Reduction in mortality with ICD therapy (p=0.007)

AMIOVIRT Amiodarone vs Implantable Defibrillator in Patients with Nonischemic Cardiomyopathy and Asymptomatic Nonsustained Ventricular Tachycardia, *AVID* Antiarrhythmics vs Implantable Defibrillators, *CABG* coronary artery bypass graft; *CASH* Cardiac Arrest Study Hamburg, *CAT* Cardiomyopathy Trial, *CIDS* Canadian Implantable Defibrillator Study, *CRT* cardiac resynchronization therapy-defibrillator, *DEFINITE* Prophylactic Defibrillator Implantation in Patients with Nonischemic Dilated Cardiomyopathy, *DINAMIT* Defibrillator in Acute Myocardial Infarction Trial, *ECG* electrocardiogram, *EPS* electrophysiological study, *ICD* implantable cardioverter defibrillator, *LVEF* left ventricular ejection fraction, *No* number of patients, *MADIT* Multicenter Automatic Defibrillator Trial, *MI* myocardial infarction, *MUSTT* Multicenter Unsustained Tachycardia Trial, *NSVT* nonsustained ventricular tachycardia, *NYHA* New York Heart Association, *PVC* premature ventricular complexes, *SCD-HeFT* Sudden Cardiac Death in Heart Failure Trial, *VF* ventricular fibrillation, *VT* ventricular tachycardia

3.5%).[4] This may indicate that the mechanisms of death in the ICD group were other than tachyarrhythmias, such as progression of LV dysfunction, recurrent heart failure events, bradyarrhythmias or electromechanical dissociation.[4]

The aggregate evidence of all these secondary and primary prevention trials provided meaningful information on patient risk-stratification and their results were implemented rapidly into the practice guidelines.[6] However, secondary analyses of these major primary prevention trials demonstrated that only 35% of patients received appropriate ICD (shock) therapy after 4-year follow-up.[1,7] This indicates that substantial heterogeneity exists among patients with depressed LV ejection fraction, and that not all patients may benefit from ICD implantation.

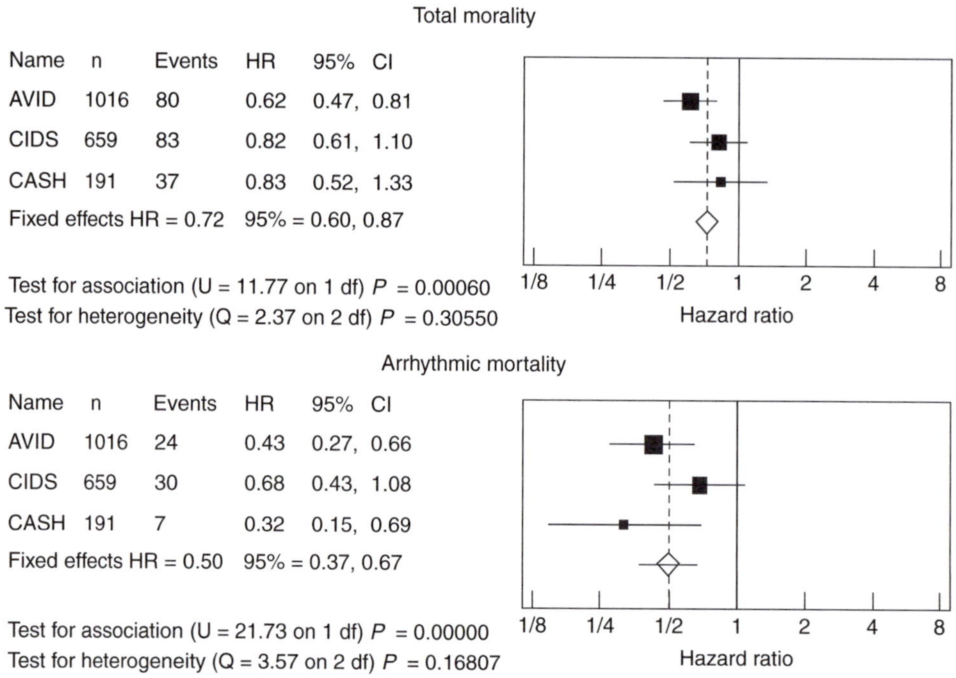

Fig. 13.1 Secondary prevention ICD trials: effects of ICD on total mortality and arrhythmic mortality. Pooling data from AVID, CIDS, and CASH trials demonstrated significant 28% relative reduction in total mortality and 50% relative reduction in arrhythmic mortality with ICD. *AVID* antiarrhythmics vs implantable defibrillators, *CABG* coronary artery bypass graft, *CASH* Cardiac Arrest Study Hamburg, *CIDS* Canadian Implantable Defibrillator Study, *ICD* implantable cardioverter defibrillator (Adapted from Connolly et al.[23])

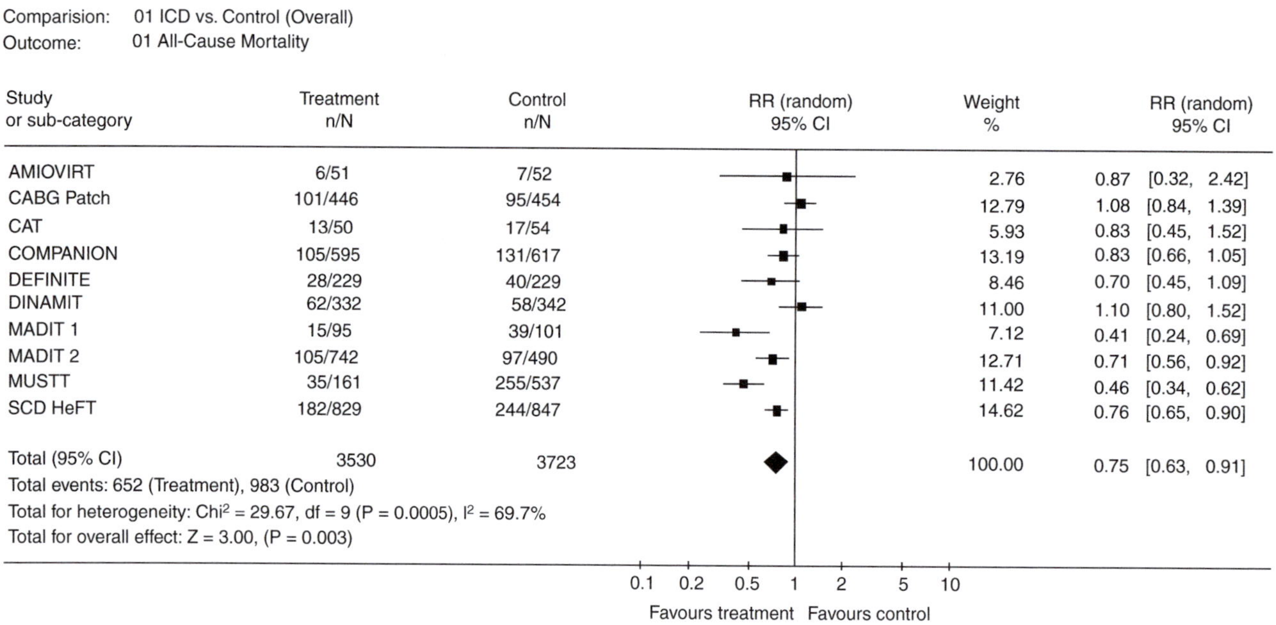

Fig. 13.2 Primary prevention ICD trials: effects of ICD on total mortality. Pooling data from primary prevention trials demonstrated a significant relative reduction in total mortality with ICD. Abbreviations: *ICD* implantable cardioverter defibrillator (Adapted from Nanthakumar et al.[24])

In order to optimize identification of patients who will benefit from ICD, several risk-stratification strategies based on the combination of clinical, echocardiographic, and electrophysiological parameters have been proposed.[26,27] However, detailed characterization of the arrhythmogenic substrate with sophisticated cardiac imaging techniques such as MRI or nuclear imaging may refine the risk-stratification of high-risk patients.

13.1.2 Identification and Selection of Candidates to ICD: Substrate Assessment

Structural cardiac abnormalities of the myocardium, coronary arteries, and cardiac innervation system constitute a substrate on which transient risk factors, such as myocardial ischemia, hemodynamic and electrolyte disturbances, electrophysiological abnormalities, and transient effect of drugs, may operate predisposing to the initiation of life-threatening arrhythmias.[11] The mechanisms of ventricular tachycardia and ventricular fibrillation rely on the presence of re-entry pathways at the border zone of scarred myocardial areas (i.e., due to an old myocardial infarction). The border zone of the scar tissue forms areas of viable myocardium with heterogeneous electrophysiological properties, such as slow conduction of the electrical breakthrough, dispersion of the repolarization wave, and higher arrhythmogenic response to several stimuli as compared to normal myocardium. In addition, the presence of increased sympathetic nervous system activity with elevated plasma levels of catecholamines interacting with the abnormal myocardium may enhance the arrhythmogenic response.[11] The electrophysiological study is an effective tool to detect the re-entrant circuits and identifies most of the patients at a high risk for sudden cardiac death.[11] In addition, current cardiac imaging modalities (mainly MRI and nuclear cardiac imaging) may be helpful to characterize the substrate at different levels by assessing the presence and extent of scar tissue, characterizing the border zone of the scar, and evaluating the response to sympathetic stimulation.

13.1.2.1 Magnetic Resonance Imaging

MRI is considered the gold standard for LV volumes and ejection fraction quantification. In addition, gadolinium contrast-enhanced MRI permits tissue characterization with high spatial resolution, providing information on the location, extent, and tissue heterogeneity of the myocardial scar.[28-30] These parameters have been all related to arrhythmogenic substrates. Bello et al. explored the role of contrast-enhanced MRI in 48 patients with coronary artery disease referred for an electrophysiological study.[28] The authors demonstrated the superiority of infarct surface area and mass assessed with contrast-enhanced MRI over LV ejection fraction to predict the occurrence of monomorphic ventricular tachycardia.[28] Subsequently, Yan et al. provided more insight into the prognostic stratification of post-myocardial infarction patients.[31] In 144 patients with prior myocardial infarction, contrast-enhanced MRI was performed. By using a computer-assisted semiautomatic algorithm, the total infarct size was quantified and divided into the core and peri-infarct regions based on signal-intensity thresholds (Figs. 13.3a and b).[31] The peri-infarct zone was identified as "gray zone" areas with different intensities of gray indicating the presence of highly heterogeneous tissue. After adjusting for age and LV ejection fraction, the percentage of normalized peri-infarct zone was independently related to all-cause mortality (hazard ratio, 1.42; $p=0.005$) and cardiovascular mortality (hazard ratio, 1.49; $p=0.01$).[31] Whether this tissue heterogeneity of the infarct zone was related to high risk for life-threatening arrhythmias was recently addressed in two studies.[29,30] In 47 post-myocardial infarction patients (>1 month), Schmidt et al. evaluated the infarct extent and tissue heterogeneity within the hyperenhanced region ("gray zone") by using contrast-enhanced MRI.[30] Afterward, patients underwent an electrophysiological study and ICD implantation for primary prevention of sudden cardiac death. Patients with inducible, monomorphic ventricular tachycardia had larger "gray zone" areas as compared to patients without inducible ventricular tachycardia (19 ± 8 g vs. 13 ± 9 g, $p=0.015$). The extent of "gray zone" areas was the only independent predictor of inducible ventricular tachycardia ($p=0.003$).[30] These results were extended by Roes et al. in a series of 91 patients with ischemic cardiomyopathy undergoing ICD implantation.[29] Patients were evaluated with contrast-enhanced MRI prior to ICD implantation, and the total infarct size and the extent of infarct "gray zone" were quantified. The extent of the infarct "gray zone" was the strongest predictor of the occurrence of spontaneous ventricular tachycardia with subsequent appropriate ICD shock therapy (hazard ratio, 1.49/each 10 g of "gray zone"; $p=0.04$).[29] In addition, a cutoff value of 16.7 g was used to separate patients with a large extent of infarct "gray zone" (>16.7 g) and patients with small extent (≤16.7 g). During follow-up, a higher percentage of patients with large extent infarct "gray zones" received appropriate ICD shock therapies than patients with small areas of "gray zone" (33% vs. 7%, $p=0.003$) (Fig. 13.3c).[29]

Finally, in nonischemic cardiomyopathy patients, the extent of scar tissue was also related to an increased risk of ventricular tachyarrhythmia.[32-34] In this subgroup of patients, hyperenhanced scar tissue has a patchy distribution, mainly located in the mid-wall, whereas the subendocardial disposition (typical for ischemic heart failure patients) is less frequently observed.[33] Interestingly, contrast-enhanced MRI may also be a useful tool to anticipate the interventional management of these patients. Bogun et al. investigated the value of contrast-enhanced MRI to guide the ablation of ventricular arrhythmias in patients with nonischemic cardiomyopathy.[32] In this study, the authors identified scar tissue in almost 50% of the patients. The contrast-enhanced MRI images and the location of the scar were integrated into an electroanatomical map, resulting in a valuable guide for catheter ablation (Fig. 13.4). Of note, catheter ablation was unsuccessful in all patients with isolated mid-wall fibrosis.[32]

13.1.2.2 Multidetector Computed Tomography

Beyond the assessment of coronary atherosclerotic burden, contrast-enhanced MDCT has demonstrated to be a valuable tool to evaluate the myocardial viability and the extent of scar tissue in a similar fashion to MRI.[35] Recent experimental studies have also demonstrated that contrast-enhanced MDCT may characterize the peri-infarct zone heterogeneity.[36] Schuleri et al. demonstrated an excellent correlation between contrast-enhanced MDCT, MRI, and histology in the quantification of the infarct size.[36] In addition, contrast-enhanced MDCT detected the peri-infarct zone (Fig. 13.5).[36] Additional studies to demonstrate the role of contrast-enhanced MDCT in the risk-stratification of patients at a high risk for life-threatening arrhythmias are warranted.

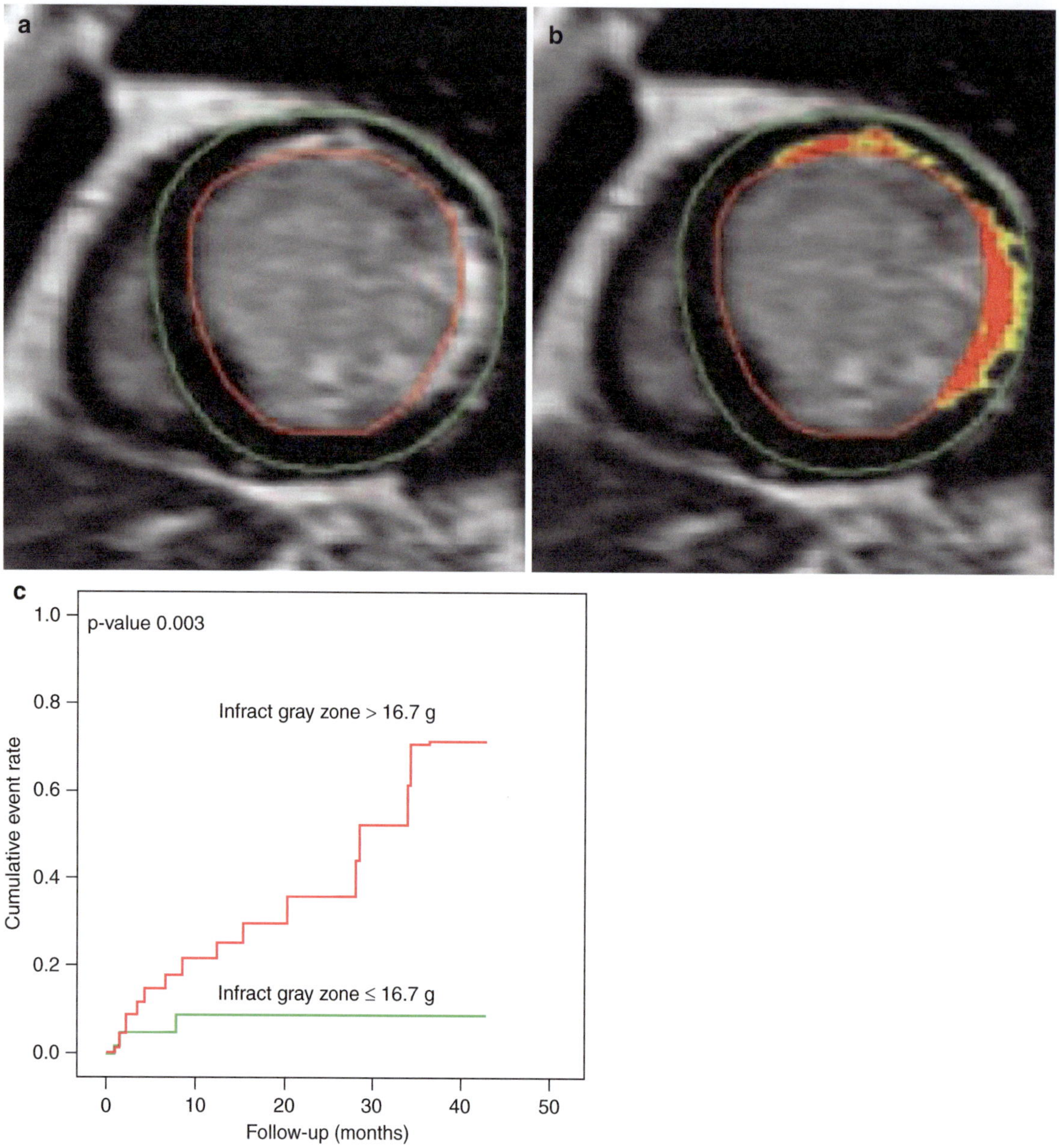

Fig. 13.3 Assessment of the infarct gray zone. From a short-axis contrast-enhanced MRI image of a patient with a previous MI, the endocardial (*red*) and epicardial (*green*) borders are outlined manually (**a**). Subsequently, the maximum signal intensity within the infarct region is determined and the infarct core is defined as myocardium with signal intensity ≥50% of the maximum signal intensity (*red area*) whereas the infarct gray zone is defined as myocardium with signal intensity ≥35% but with signal intensity <50% of the maximum signal intensity (*yellow area*) (**b**). Kaplan-Meier curve analysis shows the difference in appropriate ICD shock therapy when patients are divided according to the mean value of infarct gray zone (**c**)

13.1.2.3 Nuclear Imaging: ECG-Gated Single Photon Emission Computed Tomography (SPECT)

ECG-gated SPECT imaging provides comprehensive information on myocardial ischemia and viability, scar tissue, and LV ejection fraction. As such, this imaging technique may constitute an invaluable tool for the risk-stratification and clinical management of high-risk patients. Data from the CABG-Patch trial showed that coronary revascularization diminished the benefit of ICD, highlighting the role of jeopardized but viable myocardium as arrhythmogenic substrate.[17] The revascularization of this jeopardized myocardium may result in an increased electrical stability with a reduced arrhythmic death rate.

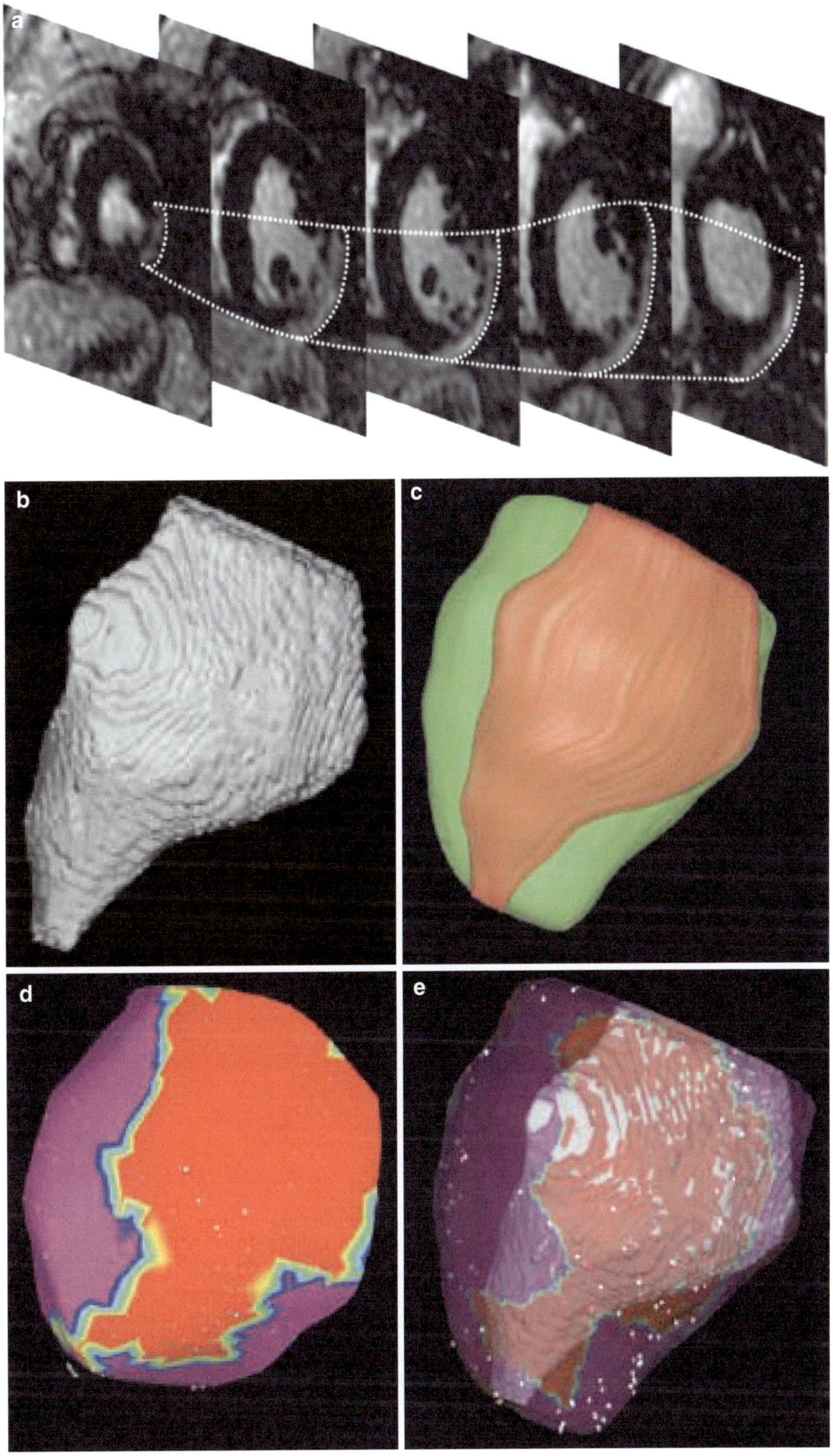

Fig. 13.4 Integration of 3-dimensional contrast-enhanced MRI within the electroanatomic map. From a stack of short-axis contrast-enhanced images showing the scar in the epicardial LV lateral wall (**a**), a 3-dimensional display of the scar is extracted (**b**) and integrated within the surrounding LV epicardium (**c** in *orange* is presented the scar extending from the LV apex to the base). At the epicardial voltage map (**d**), the scar tissue, coded in red, shows low voltage (<1.5 mV). Finally, the voltage map is merged with the 3-dimensional reconstruction of the scar seen on contrast-enhanced MRI (**e**) (Adapted from Bogun et al.[32])

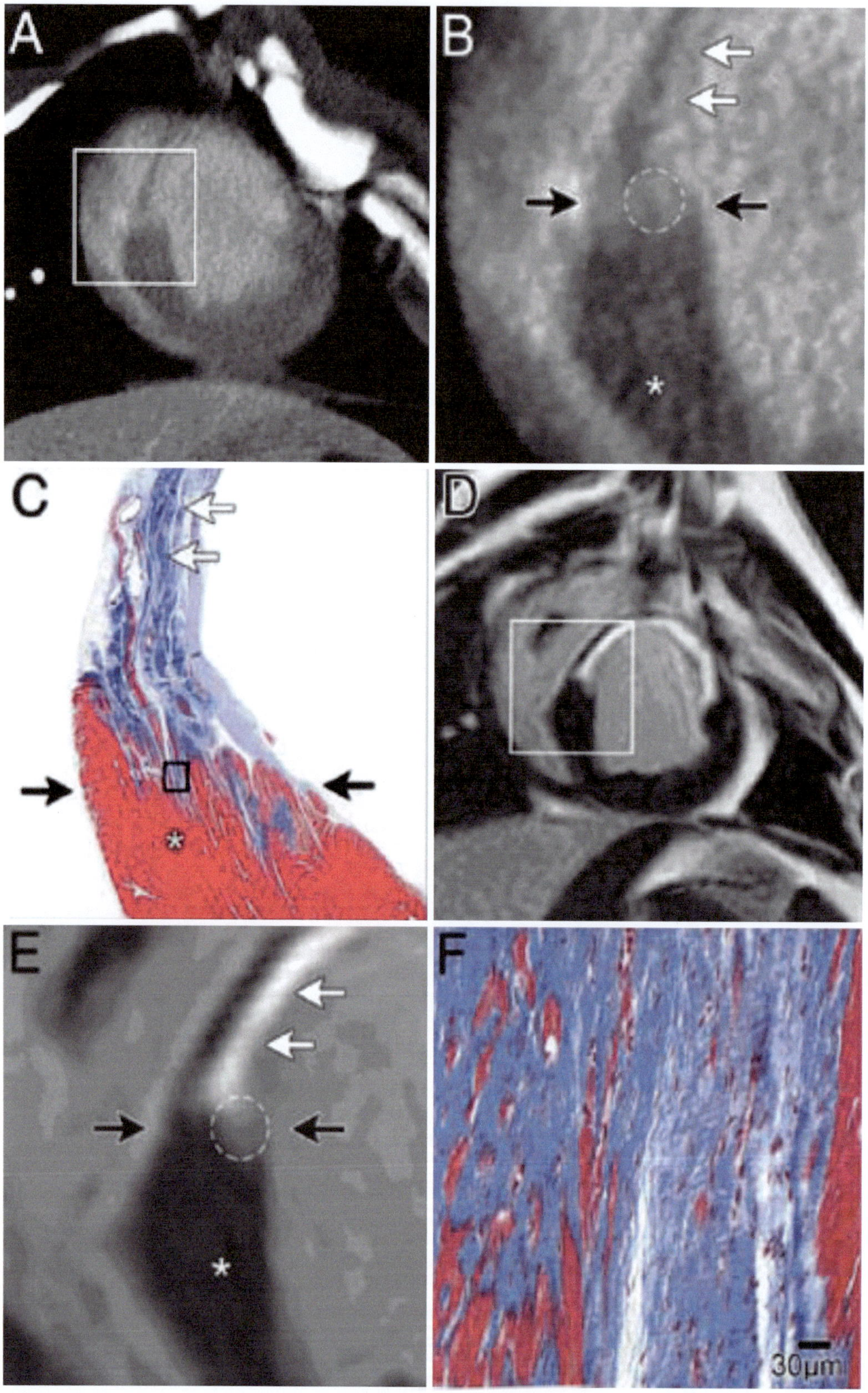

Fig. 13.5 Characterization of tissue heterogeneity in the peri-infarct zone by MDCT. The squares indicate the area zoomed in panels **B** and **D** Contrast-enhanced MDCT permits characterization of scar tissue and viable myocardium. The viable myocardium (*) shows lower attenuation values than enhanced scar tissue (*white arrows*) (**a** and **b**). The peri-infarct zone is visualized between two *black arrows* by intermediate signal intensity (*white circle*). Contrast-enhanced MRI characterizes the viable myocardium (*), contrast-enhanced infarct scar (*white arrows*), and the peri-infarct zone (*white circle*) (**d** and **e**). Masson trichrome stain depicts viable myocardium in *red* (*) from nonviable tissue in *blue*. At higher magnification, the densely packed collagenous extracellular matrix (*white arrows*) in the chronic infarct scar can be appreciated. Islands of viable myocytes (*red*) within the scar tissue are visualized, showing the heterogeneity of the peri-infarct zone (area between *black arrows*) (**c** and **f**) (Adapted from Schuleri et al.[36])

In a series of 5,183 consecutive patients, Hachamovitch et al. demonstrated the superiority of stress/rest SPECT imaging over the prescan likelihood information (including clinical, historical, and exercise data) to predict cardiac death.[37] The annualized event rate for cardiac death was 4.2% in patients having severely abnormal scans as compared to 0.5% in patients with normal scans ($p<0.001$).[37] More important, confirming the hypothesis of the CABG-Patch trial and the results of the MADIT-II trial, the study demonstrated that in patients with severely abnormal scans, early revascularization reduced the relative risk for cardiac death, and that revascularized patients had a significantly lower annualized death rate as compared to medically treated patients (1.3% vs. 4.6%, $p<0.001$).[37]

Recently, Borger van der Burg et al. evaluated the impact of viability, ischemia, scar tissue, and revascularization on clinical prognosis of survivors of sudden death.[38] In this study, a total of 153 consecutive survivors of sudden cardiac death with significant coronary artery disease underwent stress/rest SPECT to assess stress-inducible ischemia, viability, scar tissue, and LV ejection fraction, and patients were revascularized whenever possible ($n=73$). In addition, according to electrophysiological study results and LV ejection fraction, 112 received an ICD. During 3-year follow-up, 15 patients died and 42 patients had recurrent ventricular tachycardia. Patients with events had lower LV ejection fraction, more scar tissue, and less jeopardized myocardium and underwent coronary revascularization less frequently. The detection of jeopardized but viable myocardium in this study led to more aggressive therapy with coronary revascularization, if possible. Therefore, assessment of myocardial viability and coronary revascularization were beneficial parameters.[38] In addition, when patients with jeopardized myocardium were analyzed according to the treatment, patients who were treated medically had a higher risk of cardiac death as compared to revascularized patients (38% vs. 14%, $p<0.005$).[38]

Finally, from stress SPECT exams, valuable data on chronotropic response may be easily obtained, providing information on sympathetic autonomic function and adding incremental value to the risk-stratification of high-risk patients.[39] Azarbal et al. evaluated the presence of chronotropic incompetence in 10,021 patients undergoing stress SPECT. During a mean follow-up of 2 years, there were 93 cardiac deaths. The presence of severely abnormal scans (assessed with summed stress score) was the strongest predictor of cardiac death, but the presence of chronotropic incompetence provided substantial incremental value in the prediction of cardiac death.[39] Chronotropic incompetence may be a surrogate of autonomic dysfunction and, as demonstrated previously, this autonomic dysfunction leads to an increased mortality (similar to that observed in heart failure patients).[39] Therefore, the assessment of autonomic dysfunction may be of value to refine risk-stratification of these patients.

Nuclear imaging: innervation imaging. Cardiac innervation and denervation may play an important role in the arrhythmogenesis.[40,41] Cardiac sympathetic nerve fibers run over the epicardial surface and follow the coronary arteries. These fibers are affected by the oxygen deprivation more than the cardiomyocytes, and myocardial ischemia and infarction cause transient or permanent sympathetic denervation.[42] Regions of denervation have an altered response to sympathetic stimulation as compared to normal myocardium, and this electrophysiological heterogeneity within the myocardium may constitute a substrate for life-threatening arrhythmias. Furthermore, in heart failure syndrome, there is a hyperactivity of the sympathetic nervous system with an increase in plasma norepinephrine levels which may constitute an important trigger for the development of life-threatening arrhythmias.[11]

Specific nuclear imaging techniques enable evaluation of the cardiac autonomic nervous system. Cardiac innervation activity can be evaluated with positron emission tomography and carbon-11 metahydroxyephedrine, but also with SPECT imaging using metaiodobenzylguanidine (mIBG).[43] The mIBG, a norepinephrine analog labeled with 123-iodine (^{123}I), is taken up and stored in the presynaptic nerve endings. In heart failure patients, in whom a global cardiac sympathetic denervation can exist, ^{123}I-mIBG uptake is globally reduced; in addition, patients with previous infarction also show regional defects in cardiac innervation. ^{123}I-mIBG is used with both planar and SPECT imaging. As previously described, images are acquired at two different phases: Early planar and SPECT imaging are performed 10–20 min after tracer administration, whereas late planar and SPECT imaging are performed 3–4 h later. From the planar images, global myocardial ^{123}I-mIBG uptake can be assessed using the heart-to-mediastinum (H/M) ratio (Fig. 13.6). In addition, regional cardiac ^{123}I-mIBG uptake can be quantified with a summed score, and sympathetic denervated myocardial areas can be also quantified from the SPECT images.[43] Several small studies and one large, multicenter trial have demonstrated the role of ^{123}I-mIBG imaging to predict long-term prognosis of heart failure patients.[44-47] Recently, in 964 heart failure patients followed up for 2 years, the prospective open-label, multicenter ADMIRE-HF (AdreView Myocardial Imaging for Risk Evaluation in Heart Failure) trial confirmed the superiority of ^{123}I-mIBG imaging over LV ejection fraction or natriuretic peptide plasma levels to predict prognosis. Patients with an H/M ratio <1.2 tended to die more from heart failure progression, whereas arrhythmic events tended to occur more in patients with H/M ratio between 1.2 and 1.6. These findings indicate that the assessment of regional distribution of denervated myocardial areas may be of incremental value to the H/M ratio, by providing meaningful insight into the genesis of life-threatening arrhythmias. In this regard, several studies have demonstrated the presence of larger innervation than

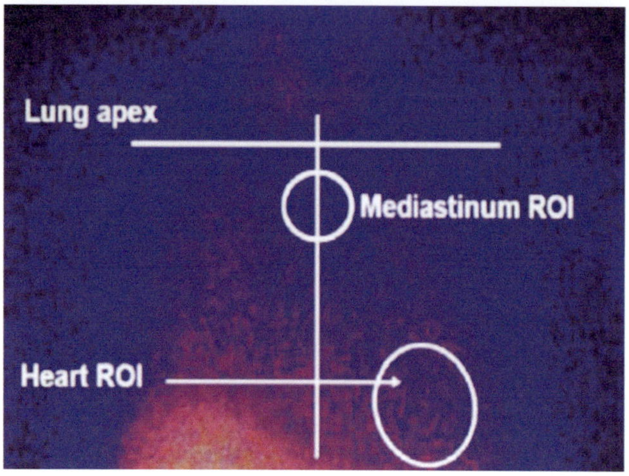

Fig. 13.6 Nuclear imaging: innervation imaging. Delayed planar *123I*-mIBG imaging displays global cardiac sympathetic innervation, expressing the heart-to-mediastinum (H/M) ratio. Regions of interest are manually drawn over the heart, and upper mediastinum and the mean cardiac counts per pixel are divided by the mean mediastinum counts per pixel. *ROI* region of interest (Adapted from [43])

perfusion defects in patients with prior myocardial infarction.[48-51] Arora et al. demonstrated in 17 heart failure patients with an ICD that a reduced late H/M ratio was associated with increased likelihood of an ICD shock during a mean follow-up of 14 ± 11 months.[48] In addition, patients with appropriate ICD shocks had more extensive ^{123}I-mIBG/perfusion (Tecnetium-99m sestamibi [^{99m}Tc]) mismatches on SPECT imaging. More recently, in a phase II multicenter study including 50 ischemic heart failure patients, the value of ^{123}I-mIBG-SPECT to predict the inducibility of ventricular arrhythmias during an electrophysiological study was assessed.[49] Patients with a positive electrophysiological study had significantly larger defects on ^{123}I-mIBG-SPECT but not different H/M ratio as compared to patients with a negative electrophysiological study.[49] Moreover, the ^{123}I-mIBG defect score was the only independent predictor for ventricular tachycardia inducibility during electrophysiological testing and a summed defect score of 37 was able to predict a positive electrophysiological study with a sensitivity of 77% and specificity of 75% (Fig. 13.7).[49] With this evidence, ^{123}I-mIBG imaging may be a valuable tool for identifying patients who will benefit from ICD implantation.

In summary, the assessment of arrhythmogenic myocardial substrate with current cardiac imaging modalities (contrast-enhanced MRI and nuclear imaging) may constitute a key step in the stratification of patients at risk for sudden death, providing valuable data on myocardial viability, scar

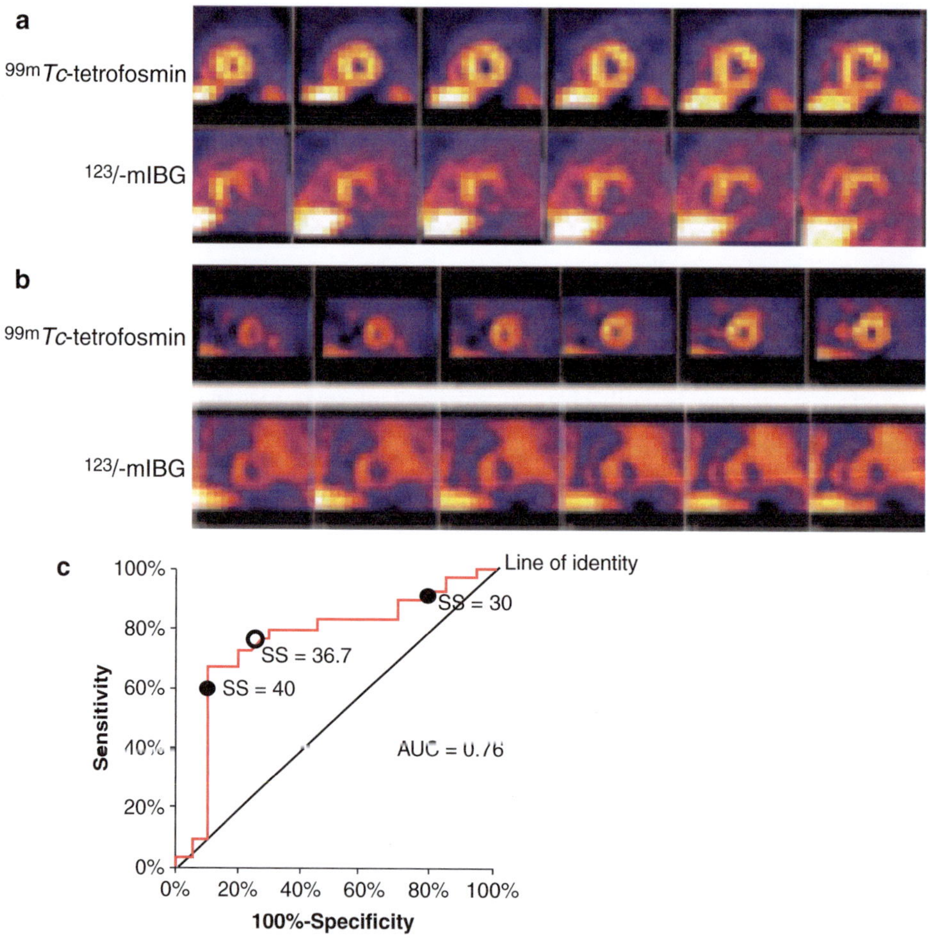

Fig. 13.7 Nuclear imaging: innervation imaging. (**a** and **b**) Present examples of late *123I*-mIBG and *99mTc*-tetrofosmin short-axis SPECT images in patients with negative and positive electrophysiological study, respectively. (**a**) Illustration of the example of a 56-year-old man with NYHA class II heart failure, LV ejection fraction 30%, and late H/M of 1.47. *99mTc*-tetrofosmin SPECT shows a small basal lateral wall infarction with summed score of 3. Late *123I*-mIBG SPECT shows large inferior and lateral wall defects, with summed score of 34. (**b**) Illustration of the example of a 72-year-old man, NYHA class III heart failure, LV ejection fraction 33%, late H/M of 1.41. *99mTc*-tetrofosmin SPECT shows a small apicolateral infarct with summed score of 5. *123I*-mIBG uptake is globally reduced (reflected by high lung and liver activity) and a large apical and inferolateral defect is shown; summed score was 37. (**c**) Late *123I*-mIBG SPECT receiver operating curve analysis where a summed defect score of 37 was able to predict a positive electrophysiological study with a sensitivity of 77% and specificity of 75% (area under the curve [*AUC*] = 0.76)

tissue, and cardiac innervation. The inclusion of such techniques in current risk-stratification strategies may optimize the selection of patients who will benefit from ICD implantation. Large, multicenter studies will be warranted in order to confirm the role of multimodality cardiac imaging in the risk-stratification for sudden cardiac death.

13.2 Cardiac Resynchronization Therapy

Heart failure is the most frequent cardiovascular diagnosis in developed countries with an estimated 550,000 new diagnoses each year in the United States.[52] Despite advances in diagnosis and medical therapy in the last few decades, the morbidity and mortality of heart failure patients remain high. Data from 8 randomized, multicenter trials, including 4,017 patients, have shown that CRT improves functional status and reduces all-cause mortality and heart failure hospitalizations in patients with end-stage, drug-refractory heart failure (Table 13.2).[2,3,53-57,59] In addition, available echocardiographic data of these trials demonstrated that CRT induces LV reverse remodeling, improvement in LV systolic function, and reduction in mitral regurgitation. Pooled data from 5 trials randomizing 2,371 patients to CRT (n=1,343) vs. medical therapy (n=1,028) have shown the superiority of CRT over medical treatment in reducing all-cause mortality with 29% of relative risk reduction.[60] These results led to the American Heart Association/American College of Cardiology/Heart Rhythm Society guidelines consider CRT as a class I indication in patients with end-stage heart failure (New York Heart Association functional class III-IV) with LV ejection fraction < or equal to 35% and a QRS complex duration > or equal to 120 ms.[6]

Despite the widely proven beneficial effects of CRT, 20–30% of the patients treated with CRT do not show clinical improvement, being considered "non-responder" patients.[8] Furthermore, when echocardiographic criteria of response are applied, the percentage of non-responder patients increases up to 40%.[8] This indicates the presence of a high heterogeneity among end-stage heart failure patients who fulfill current selection criteria for CRT device implantation.

Several factors may determine the response to CRT: LV dyssynchrony, extent and location of scarred tissue, and position of the LV lead.[61] Despite current guidelines define LV dyssynchrony by QRS complex duration, numerous studies have demonstrated the value of several parameters of mechanical LV dyssynchrony (based mostly on echocardiography imaging) to predict response to CRT, yielding high sensitivities and specificities (both 80–90%).[8] Finally, the extent and location of myocardial scar and the LV lead position have also shown to be determinants of CRT response.[8]

Table 13.2 Results of main clinical trials in CRT

Trial (Ref. #)	No.	Primary endpoints	Secondary endpoints	Main findings
MUSTIC-SR[53]	58	6 MWT	NYHA class, QoL, peak VO_2 LV volumes, MR Hospitalizations, mortality	Improvement in 6 MWT, NYHA class, QoL, peak VO_2 Reduction in LV volumes and MR Reduction in hospitalizations
PATH-CHF[54]	41	Peak VO_2 6 MWT	NYHA class QoL Hospitalizations	Improvement in NYHA class, QoL and 6 MWT Reduction in hospitalizations
MIRACLE[55]	453	NYHA class, 6 MWT, QoL	Peak VO_2 LVEDD, LVEF, MR Clinical composite response	Improvement in NYHA class, QoL and 6 MWT Reduction in LVEDD, MR, increase in LVEF
MIRACLE-ICD[56]	369	NYHA class, 6 MWT, QoL	Peak VO_2 LVEDD, LVEF, MR Clinical composite response	Improvement in NYHA class, QoL, peak VO_2
PATH-CHF II[57]	86	Peak VO_2 6 MWT	NYHA class QoL	Improvement in 6 MWT, QoL, peak VO_2
CONTAK-CD[58]	490	NYHA class, 6 MWT, QoL	LV volume, LVEF Composite of mortality, VT/VF, hospitalizations	Improvement in 6 MWT, NYHA class, QoL Reduction in LV volume, increase in LVEF
COMPANION[2]	1,520	All-cause mortality or hospitalization	All-cause mortality, cardiac mortality	Reduction in all-cause mortality or hospitalization
CARE-HF[3]	813	All-cause mortality or hospitalization	All-cause mortality, NYHA class, QoL	Reduction in all-cause mortality or hospitalization Improvement in NYHA class, QoL

CARE-HF Cardiac Resynchronization-Heart Failure, *CONTAK-CD* CONTAK-Cardiac Defibrillator, *COMPANION* Comparison of Medical Therapy, Pacing and Defibrillation in Heart Failure, *CRT* cardiac resynchronization therapy, *LV* left ventricular, *LVEDD* left ventricular end-diastolic dimension, *LVEF* left ventricular ejection fraction, *LVESV* left ventricular end-systolic volume, *MIRACLE* Multicenter InSync Randomized Clinical Evaluation, *MIRACLE-ICD* Multicenter InSync Implantable Cardioverter Defibrillator trial, *MR* mitral regurgitation, *MUSTIC* Multisite Simulation in Cardiomyopathies, *No* number of patients, *NYHA* New York Heart Association, *PATH-CHF* Pacing Therapies in Congestive Heart Failure trial, *QOL* quality-of-life score, *VF* ventricular fibrillation, VO_2 volume of oxygen, *VT* ventricular tachycardia, *6MWT* 6-min walk test

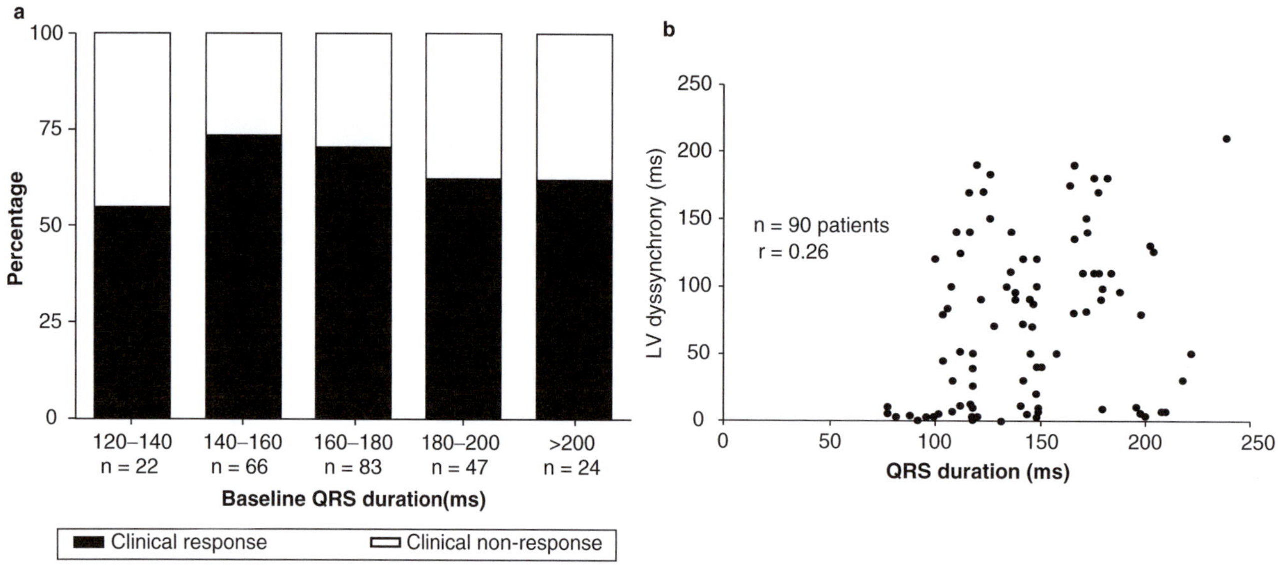

Fig. 13.8 Relation between QRS complex duration and clinical response to CRT (**a**) and LV dyssynchrony assessed with echocardiography (**b**). QRS complex duration is not a good predictor of clinical response to CRT as no differences in baseline QRS complex duration are observed between responders and non-responders (**a**). There is no relation between electrical dyssynchrony (reflected by the QRS complex duration) and LV mechanical dyssynchrony assessed with TDI echocardiography (**b**) (Adapted from Bax et al.[61] and Bleeker et al.[63])

Gadolinium contrast-enhanced MRI and nuclear imaging provide valuable information on myocardial viability and scar tissue whereas MDCT permits an exact characterization of the cardiac venous anatomy.[8] Therefore, noninvasive multimodality cardiac imaging may play a central role in patient selection for CRT. Whether LV dyssynchrony, LV scar, or LV lead position has the same value or not to predict response to CRT remains unknown.

13.2.1 LV Dyssynchrony Assessment

End-stage heart failure patients may show impaired electromechanical coupling, which may further impair LV performance.[6] The most common conduction abnormalities include prolonged atrio-ventricular conduction (first-degree atrio-ventricular block) and prolonged ventricular conduction (most common, left bundle branch block). Particularly, prolonged ventricular conduction may cause regional mechanical delay within the LV which in turn reduces LV systolic function, and causes mitral regurgitation and LV dilatation with further impairment of LV performance. CRT devices improve LV performance by restoring the synchronicity at three levels (atrio-ventricular, interventricular, and intraventricular) which, subsequently, increases LV filling time, reduces mitral regurgitation, and corrects septal dyskinesis.[8]

Based on current evidence, the assessment of LV dyssynchrony seems crucial to identify patients who may well respond to CRT.[62] Several studies have demonstrated that QRS complex duration fails to predict clinical or echocardiographic CRT response (Fig. 13.8a).[63,64] In addition, QRS complex duration is poorly related to LV dyssynchrony assessed with TDI, and 30% of patients with a QRS complex >150 ms do not show LV dyssynchrony whereas 30% of the patients with narrow QRS complex (<120 ms) may show LV dyssynchrony (Fig. 13.8b).[63,65] These observations indicate that LV electrical dyssynchrony is not synonymous to LV mechanical dyssynchrony, and several cardiac imaging modalities have been explored to search for LV mechanical dyssynchrony parameters that can predict the response to CRT.

13.2.1.1 Echocardiography

Various echocardiographic techniques have been proposed to quantify LV dyssynchrony (Table 13.3). One of the first echocardiographic LV dyssynchrony parameters was the measurement of the time delay between the peak systolic thickening of the septum and the peak systolic thickening of the posterior wall, the so-called septal-to-posterior wall motion delay, on *M-mode echocardiography* recordings (Fig. 13.9a).[66] Despite promising results in initial small series of patients, further studies involving larger series of patients tempered the predictive value of this LV dyssynchrony parameter and questioned the feasibility of this technique since a clear peak systolic excursion of the septal or the posterior wall may be not observed in up to 50% of patients (mostly patients with ischemic heart disease).[67] *Tissue Doppler imaging* (TDI) – an echocardiographic technique that measures the velocity of cardiac motion – has provided

Table 13.3 Echocardiographic studies on LV dyssynchrony assessment and predictive value of CRT response

Author (Ref. #)	No.	Measurement	Echocardiographic technique	LV dyssynchrony cutoff value	Sensitivity (%)	Specificity (%)
Pitzalis et al.[66]	20	Septal-to-posterior wall motion delay	M-mode	≥130 ms	100	63
Diaz-Infante et al. [67]	67	Septal-to-posterior wall motion delay	M-mode	≥130 ms	50	38
Penicka et al.[68]	49	Sum of LV and VV dyssynchrony (pulsed wave systolic velocities)	Pulsed wave TDI	>102 ms	96	77
Bax et al.[62]	85	Delay in peak systolic velocities (Four segments: basal septum, lateral, anterior, and inferior walls)	Color-coded TDI	≥65 ms	92	92
Yu et al.[69]	56	Standard deviation of time to peak systolic velocities (12 LV segments)	Color-coded TDI	≥34.4 ms	87	81
Dohi et al.[70]	38	Delay in peak radial strain (septal to posterior wall)	TDI-derived strain	≥130 ms	95	88
Delgado et al.[71]	161	Delay in peak radial strain (anteroseptal to posterior wall)	2-dimensional radial strain	≥130 ms	83	80
Marsan et al.[72]	57	Systolic dyssynchrony index: standard deviation of time to minimum volume (16 LV segments)	Real-time 3-dimensional echocardiography	≥6.4%	88	85
Van de Veire et al.[73]	60	Standard deviation of time to peak systolic velocities (12 LV segments)	Triplane TDI	>33 ms	90	83

LV left ventricular, *No* number of patients, *TDI* tissue Doppler imaging, *VV* interventricular

several parameters to quantify LV dyssynchrony by measuring the time to onset or peak systolic velocity in relation to the electrical activity (QRS complex).[8] There is limited evidence on the value of *pulsed wave TDI* to quantify LV dyssynchrony and to predict response to CRT.[68] With this technique, only one myocardial region can be interrogated at a time, precluding the measurement of two regions simultaneously. In contrast, *color-coded TDI* allows the assessment of two opposing LV walls simultaneously by post-processing the time-velocity tracings.[62,69] Numerous non-randomized, single-center studies have applied this technique to quantify LV dyssynchrony (Fig. 13.9b). For example, in 85 heart failure patients treated with CRT, LV dyssynchrony was assessed by measuring the time delay between the peak systolic myocardial velocities of the basal segments of four opposing LV walls (septal, lateral, anterior and inferior).[62] A cutoff value of ≥65 ms was highly predictive of both clinical (sensitivity/specificity 80% for both) and echocardiographic responses (sensitivity/specificity 92% for both).[62] In addition, Yu et al. used a 12-segment model to measure LV dyssynchrony.[69] A cutoff value of 34.4 ms predicted echocardiographic response to CRT yielding a sensitivity and specificity of 87% and 81%, respectively.[69] However, TDI-derived myocardial velocities or other derivate techniques (such as tissue tracking derived myocardial displacement) may be suboptimal approaches to measure dyssynchrony in ischemic heart failure patients since these techniques cannot differentiate active deformation (active contraction of viable myocardium) from passive motion (scar segments tethered by the adjacent viable segments). In this regard, *TDI-derived strain (rate) imaging* may provide more information since this imaging modality evaluates the active deformation (contraction) of the interrogated segments. Although several studies have shown that LV dyssynchrony can be quantified by TDI-derived strain (rate) imaging by measuring time to peak strain, its predictive value of response to CRT remains controversial.[70,74] Dohi et al. evaluated 38 heart failure patients treated with CRT and demonstrated that a time delay between peak TDI-derived strain of the septal and posterior wall ≥130 ms was predictive of acute improvement in stroke volume after CRT device implantation (sensitivity 95%, specificity 88%).[70]In contrast, Yu et al. demonstrated that LV dyssynchrony measurements based on longitudinal TDI-derived strain were not predictive of LV reverse remodeling at 3-month follow-up.[74]

In addition, *2-dimensional speckle tracking strain (rate) imaging* has emerged as a novel echocardiographic technique that enables multidirectional and angle-independent assessment of LV strain (Fig. 13.9c; see this chapter movies on the Springer Extras) [71,75] Initial studies have demonstrated that a time delay between peak radial strain of the (antero)septal and (postero)lateral walls ≥130 ms predicts echocardiographic response yielding a sensitivity of 83–89% and specificity of 80–83%.[71,75]Finally, *3-dimensional echocardiography* is a promising technique to evaluate LV dyssynchrony since, conceptually, it may provide a global measurement of the entire LV.[72,73,76] Initial studies have demonstrated the role of this technique by using direct volumetric analysis, with *real-time 3-dimensional echocardiography*, or by using *triplane TDI*.[72,73] The predictive

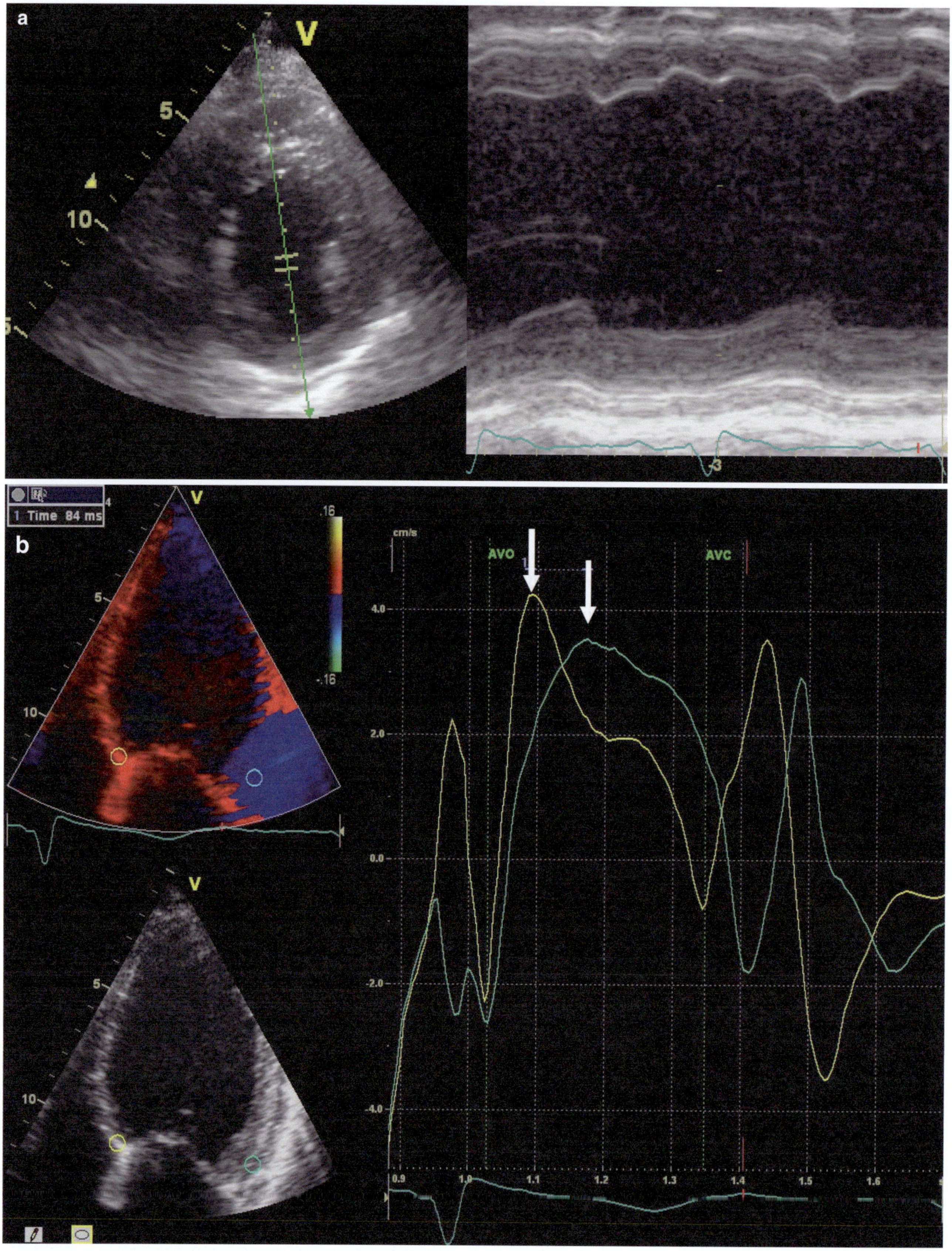

Fig. 13.9 LV dyssynchrony assessment with echocardiography. Several echocardiographic methods have been developed to quantify LV dyssynchrony (see text): septal-to-posterior wall motion delay on M-mode images (**a**), time delay between peak systolic velocities on TDI images (**b**), time delay between peak radial strain of the anteroseptal (*AS*) and postero(lateral) (*P*) walls (**c**), systolic dyssynchrony index (**d**) obtained from the standard deviation of time to minimum systolic volume of 16 sub-volumes (*left*) and color-coded polar maps showing the latest activated areas (*right*), the standard deviation of time to peak systolic velocities of 12 segments on triplane TDI images (**e**)

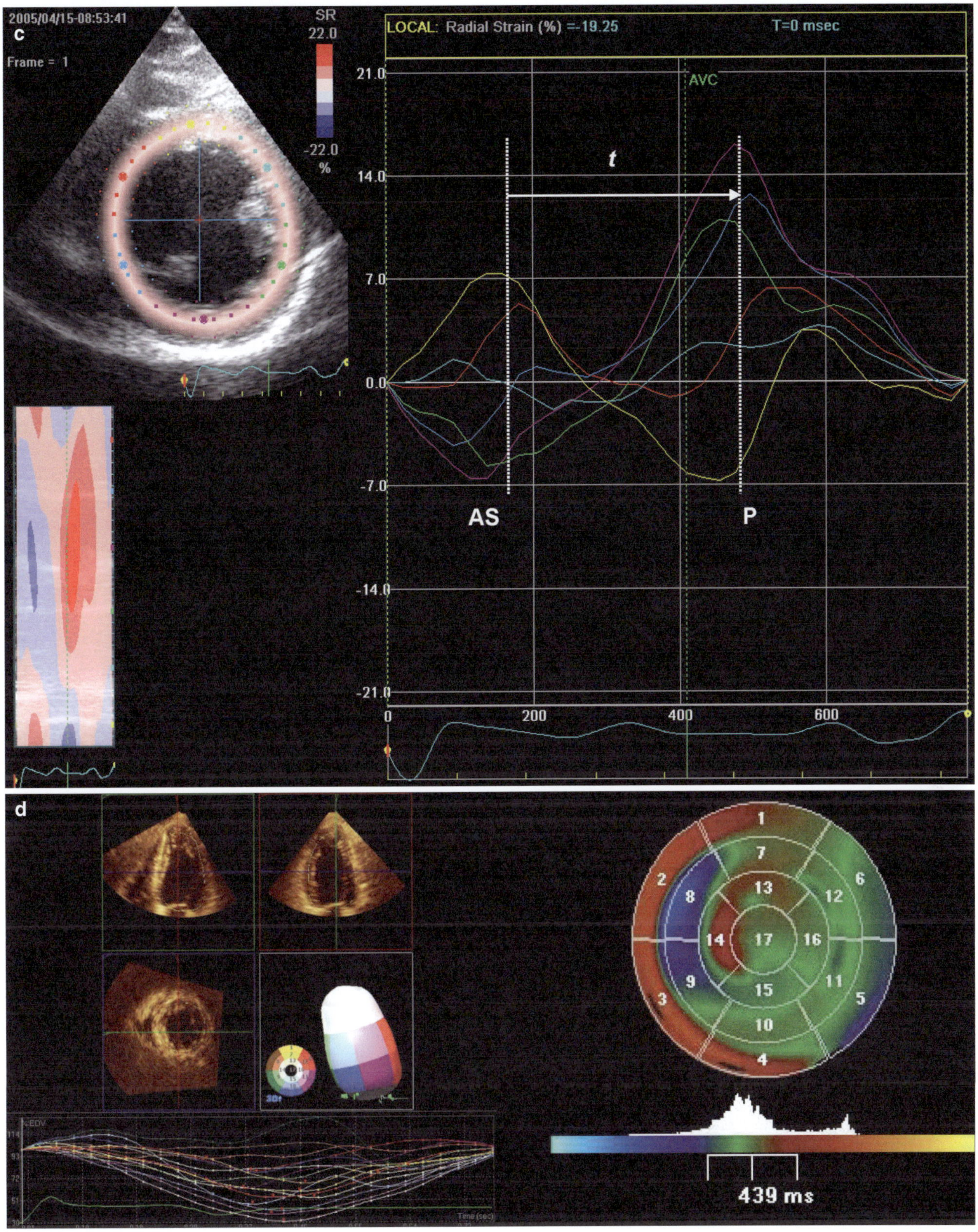

Fig. 13.9 (continued)

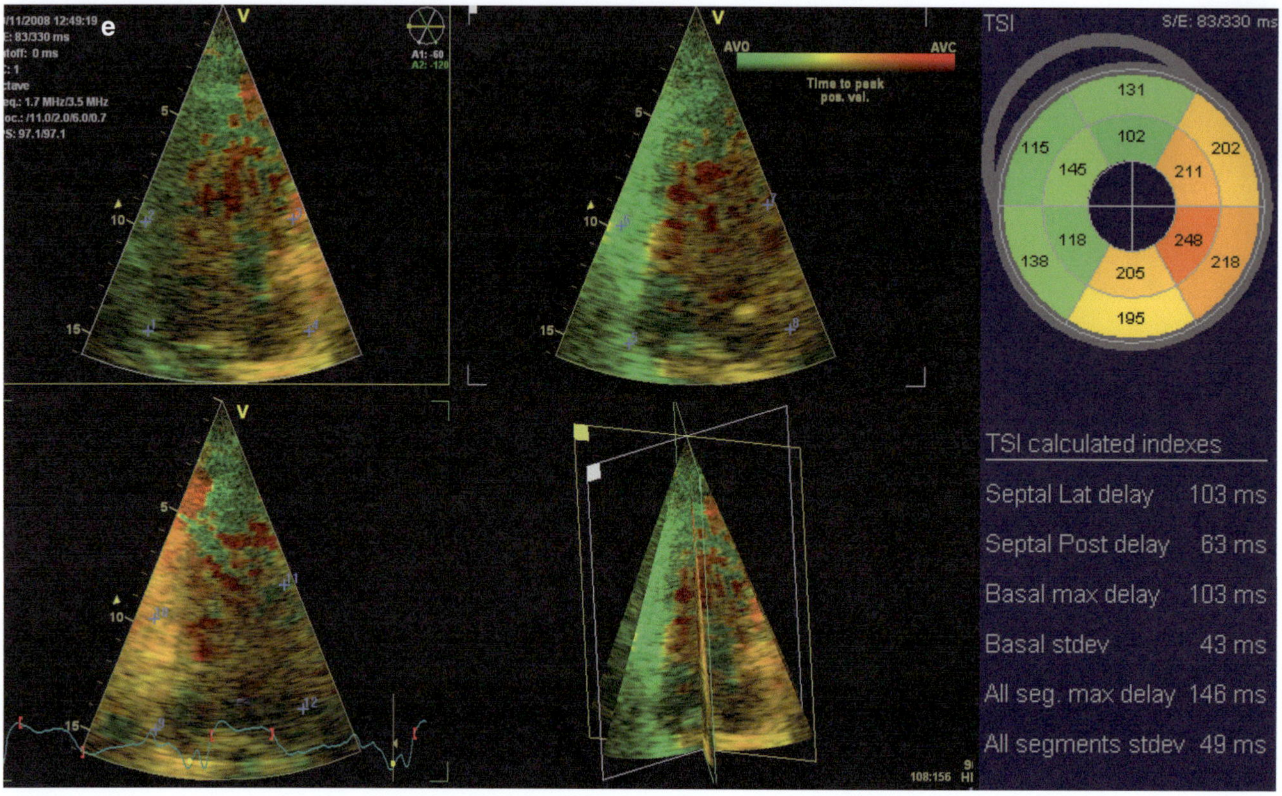

Fig. 13.9 (continued)

value of real-time 3-dimensional echocardiography for response to CRT has been recently demonstrated by Marsan et al.[72] With this technique, the systolic dyssynchrony index is calculated as the standard deviation of time to minimum systolic volume of the 16 sub-volumes in which the LV is divided (Fig. 13.9d). A cutoff value of 6.4% predicted acute LV reverse remodeling after CRT implantation with a sensitivity and specificity of 88% and 85% respectively.[72] In addition, van de Veire et al. demonstrated the value of triplane TDI to measure LV dyssynchrony and to predict clinical and echocardiographic responses to CRT.[73] From the triplane LV full volume dataset, time to peak systolic velocity was calculated in 12 segments and the time dispersion was derived as LV dyssynchrony index (Fig. 13.9e). A cutoff value of 33 ms predicted clinical (sensitivity/specificity of 89% and 82%) and LV reverse remodeling (sensitivity/specificity of 90% and 83%).[73]

Despite this plethora of echocardiographic LV dyssynchrony measurements, there is no consensus on which echocardiographic parameter should be used to identify patients who will respond to CRT. The Predictors Of Response to Cardiac Resynchronization Therapy (PROSPECT) trial demonstrated that no single echocardiographic measurement of LV dyssynchrony (based on M-mode, pulsed wave TDI, or color-coded TDI) was highly predictive of CRT response.[77] Beyond patient selection and technical issues that can explain these results, there are several pathophysiologic issues that were not considered in the PROSPECT trial such as the extent and location of LV scar tissue and the LV lead position, crucial determinants of CRT response as well. Therefore, a more integrated approach may optimize the selection of patients who will benefit from CRT.

13.2.1.2 Magnetic Resonance Imaging

The assessment of LV dyssynchrony based on MRI techniques has been shown feasible and provides reproducible data on 3-dimensional myocardial activation pattern. Currently, different approaches have been proposed to measure LV dyssynchrony and predict CRT response.[78] Tagged MRI enables 3-dimensional circumferential and longitudinal strain analysis, and several acquisition and post-processing algorithms have been developed to characterize the electromechanical activation sequence (Fig. 13.10a).[81,82] By measuring the time to onset or peak strain, a polar map can be obtained illustrating the LV mechanical activation pattern and identifying the latest activated areas.[82] In addition, with

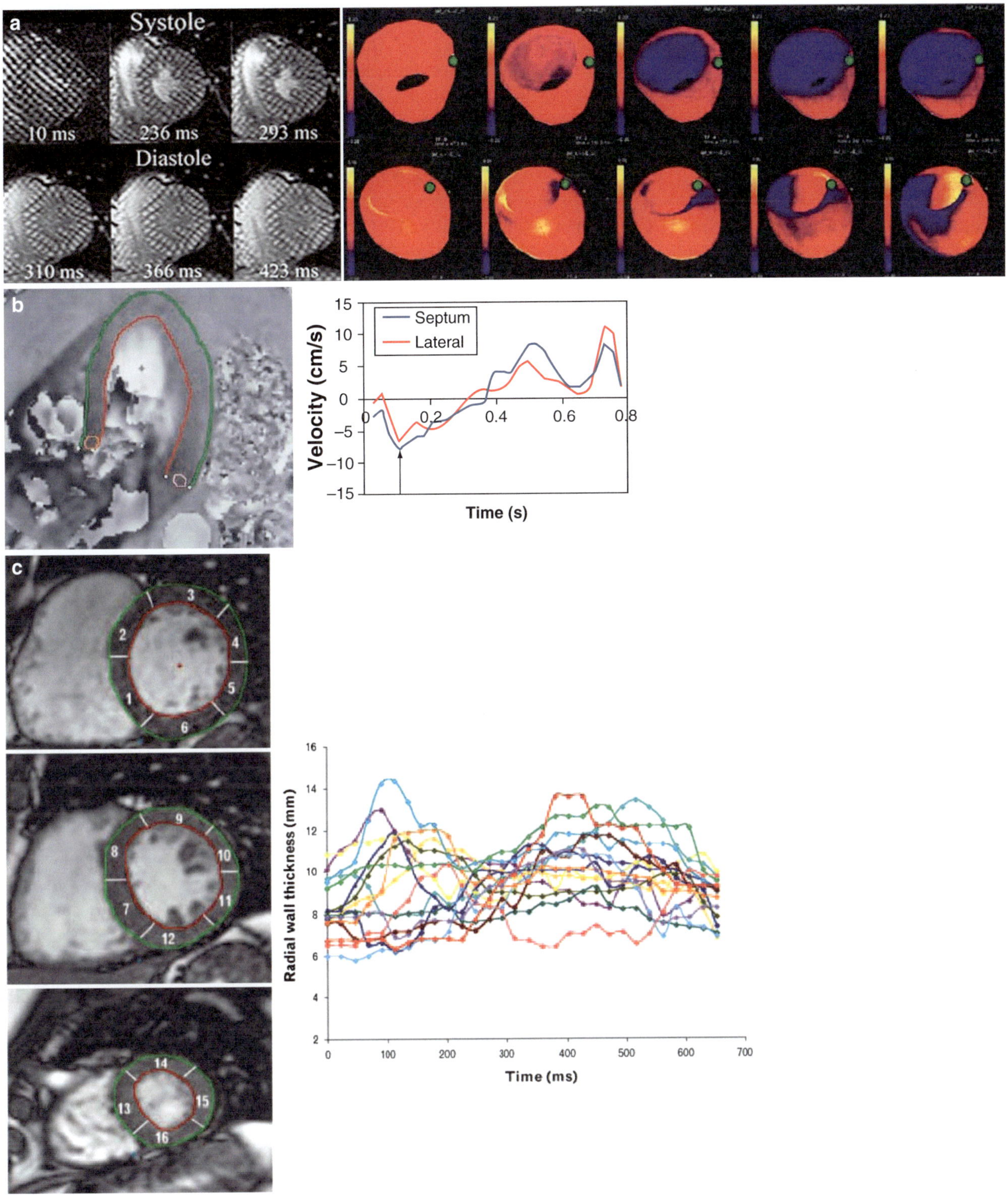

Fig. 13.10 LV dyssynchrony assessed with CMR techniques. Several CMR techniques have been proposed to measure LV dyssynchrony: tagged-CMR measures time to onset or peak deformation of myocardial segments and provides parametric images where the onset of the contraction and the most delayed segments can be readily identified (**a**); with velocity-encoded CMR, time delay in peak systolic velocities between basal septal and basal lateral segments can be quantified (**b**); regional wall motion analysis of LV short-axis CMR images yields time-radial wall thickness curves of 16 segments and the standard deviation of time to maximum wall thickness of 16 segments provides the LV dyssynchrony index (**c**) (Reproduced with permission from Lardo et al.[79]and Westenberg et al.[80])

the use of velocity-encoded MRI, Westenberg et al. assessed LV dyssynchrony in 20 heart failure patients undergoing CRT implantation.[80] From myocardial velocity tracings obtained with velocity-encoded MRI, time differences in peak systolic velocities between two opposing segments can be calculated and LV dyssynchrony can be quantified in a similar fashion to echocardiographic TDI (Fig. 13.10b).[80] The authors found an excellent agreement between velocity-encoded MRI and echocardiographic TDI in scoring the severity of LV mechanical dyssynchrony (95% of patients were classified identical).[80] Finally, LV dyssynchrony can be quantified with regional wall motion analysis from short-axis cine MRI (Fig. 13.10c).[83] By using a16-segment model, the time-wall thickness tracings can be obtained and the standard deviation of the time to peak wall thickness can be calculated as an indicator of LV dyssynchrony.[83]

13.2.1.3 Nuclear Imaging

ECG-gated SPECT imaging has been shown to be a useful tool to evaluate LV dyssynchrony.[84] A count-based method has been developed to measure amplitude (wall systolic thickening) and phase angle from the regional LV count changes along the cardiac cycle. The phase angle reflects timing of conduction within the cardiac cycle (0–360°) and this phase angle distribution is homogeneous when LV contraction is synchronous and heterogeneous when LV dyssynchrony exists.[84] In addition, the phase angle data can be displayed in histograms where the bandwidth can be measured as an indicator of the phase angle distribution. The value of the phase standard deviation and the bandwidth of the histogram to predict CRT response have been recently demonstrated in 40 heart failure patients treated with CRT.[85] A cutoff value of 72.5° for bandwidth and 19.6° for phase standard deviation predict LV reverse remodeling yielding a sensitivity and specificity of 83% and 81% respectively.[85]

However, as previously indicated, assessment of LV dyssynchrony alone may not be sufficient to identify patients who will or will not respond to CRT. Thus, patients with large areas of transmural scar or without suitable veins in the latest activated areas (where the LV pacing lead should be ideally placed) may show less benefit from CRT. Therefore, selection of patients who will benefit from CRT can be refined by integrating information on LV dyssynchrony, the latest activated area, extent and location of scar tissue, and venous anatomy.

13.2.2 LV Viability and Scar Tissue Assessment

Pacing the LV in a segment with transmural scar may reduce the effectiveness of CRT.[86] In addition, large areas of scar tissue may limit the ability of the myocardium to respond to CRT.[87,88] Echocardiography, MRI, and nuclear imaging provide information on the characteristics of the viable myocardium and on the location and extent of myocardial scar tissue.

13.2.2.1 Echocardiography

Low-dose dobutamine stress echocardiography is one of the most frequently used imaging techniques to evaluate myocardial viability.[89] Myocardial segments that show contractile reserve during increasing low-dose dobutamine infusion are considered viable.[90] In contrast, nonviable myocardial segments (transmural scar) do not show contractile reserve.[90] Few studies have used low-dose dobutamine echocardiography to assess scar and viability prior to CRT implantation.[91,92] In 31 patients with advanced heart failure undergoing CRT device implantation, Ypenburg et al. evaluated global myocardial viability with low-dose dobutamine echocardiography. In addition, regional myocardial viability was assessed at the segment targeted by the LV lead using 2-dimensional speckle tracking strain imaging.[92] The patients who responded to CRT had a significant increase in LV ejection fraction during low-dose dobutamine infusion (from 27%±7% to 40%±8%, $p<0.001$), whereas non-responders did not show significant improvement in LV ejection fraction. Furthermore, myocardial contractile reserve at the segment targeted by the LV lead was present only in responder patients as indicated by an increase in radial strain value (6±5% vs. −1±4%, $p=0.002$ responders vs. non-responders).[92]

Another echocardiographic technique useful to evaluate myocardial viability is myocardial contrast echocardiography. With the infusion of contrast agents consisting of microbubbles of a high-molecular-weight gas, encapsulated in a shell of lipids or proteins, myocardial perfusion can be assessed. Areas with normal perfusion appear as enhanced myocardium, whereas those areas with impaired perfusion appear as dark or patchy (Fig. 13.11). A perfusion score index can be derived from the summed segmental perfusion scores and the higher the perfusion score index, the larger the extent of viable myocardium. In 21 heart failure patients treated with CRT, Hummel et al. demonstrated the incremental value of the perfusion score index over LV dyssynchrony for predicting improvement in LV ejection fraction after CRT implantation.[93]

Magnetic resonance imaging. This 3-dimensional imaging technique permits a precise delineation of scar tissue with high spatial resolution yielding information concerning the exact location and transmurality of scar tissue, but also on the total extent of scar in the LV (the total scar burden). The gadolinium-based contrast agent diffuses within the scar tissue areas that have larger interstitial spaces between the collagen fibers as compared to regions of normal myocardium. In those areas, the outwash of gadolinium-based contrast agents is slower and with an inversion-recovery 3-dimensional gradient

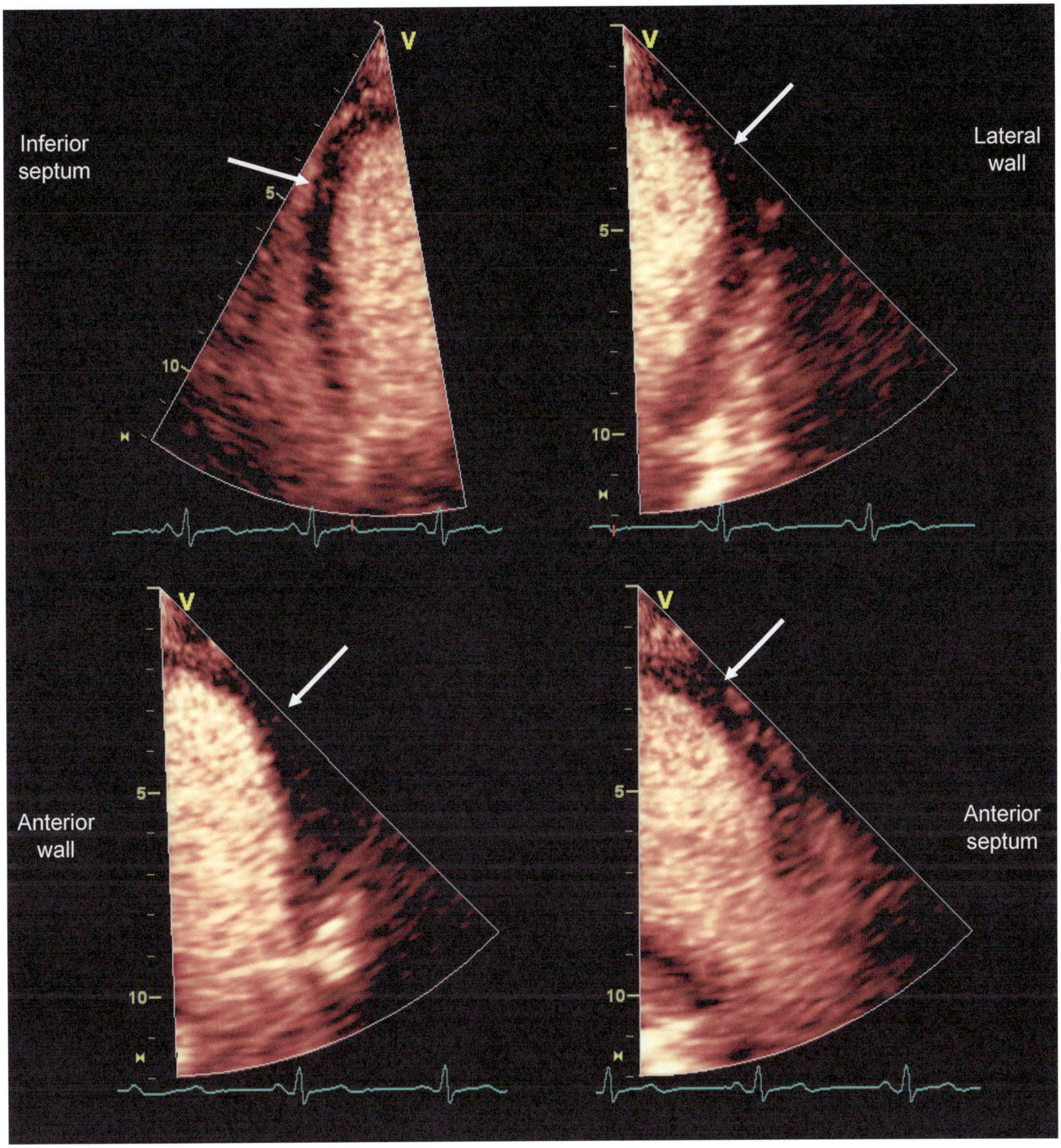

Fig. 13.11 Myocardial contrast echocardiography. The total scar burden can be evaluated with myocardial contrast echocardiography. A perfusion score index can be derived from the summed segmental perfusion scores and the higher the perfusion score index, the larger the extent of viable myocardium. The figure illustrates the example of a 46-year-old patient in NYHA class III with LV ejection fraction of 32% and large perfusion defects. After 6 months of CRT, the patient did not show LV reverse remodeling

echo sequence acquisition, the scarred areas appeared as hyperenhanced regions. The inversion time is usually determined with real-time plan scan to null normal myocardial signal (Fig. 13.12a). With the use of contrast-enhanced MRI, Bleeker et al. evaluated the presence and location of transmural scar in 40 patients with end-stage heart failure before CRT implantation.[86] In this study, patients with substantial baseline LV dyssynchrony (>65 ms assessed with TDI) and without transmural scar at posterolateral segments (preferred LV lead positioning) showed a significantly higher CRT response rate than patients with transmural posterolateral scar and/or absent LV dyssynchrony (95% vs. 11%, $p<0.05$).[86]

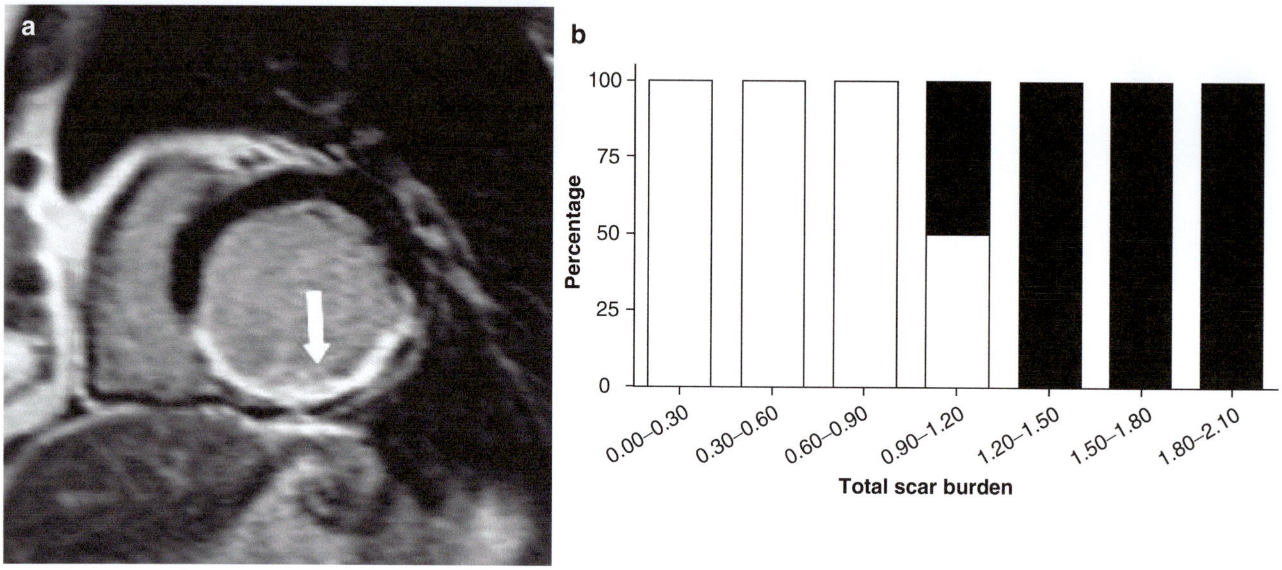

Fig. 13.12 Assessment of location and total scar burden with contrast-enhanced CMR. Positioning the LV pacing lead in a segment with transmural scar (**a**) may result in non-response to CRT. In addition, a larger amount of myocardial scarred tissue (total scar burden, **b**) increases the likelihood of non-response to CRT (Adapted Ypenburg et al.[94])

In addition, the influence of the total scar burden on LV reverse remodeling after CRT implantation was tested in 34 ischemic heart failure patients undergoing contrast-enhanced MRI prior CRT implantation.[94] The extent and transmurality of scar tissue were evaluated before CRT implantation. Using a 17-segment model with a 5-point hyperenhancement scale (from score 0=no hyperenhancement indicating no scar, to score 4=hyperenhancement>76% indicating transmural scar), the total scar burden was estimated. Linear regression analysis demonstrated a strong direct relation between the total scar burden and the relative change in LV end-systolic volume after CRT implantation ($r = -0.91$, $p < 0.05$) indicating that the higher scar burden is, the less reverse remodeling may be observed. In addition, based on the 5-point hyperenhancement scale, a total scar burden of >1.2 resulted in a complete functional non-response (Fig. 13.12b).[94]

Therefore, not only the presence of posterolateral transmural scar but also the total extent of scar tissue may limit the response to CRT. Contrast-enhanced MRI is an ideal technique to refine the selection of CRT candidates.

Nuclear imaging. Scintigraphic techniques, positron emission tomography (PET) or SPECT, provide comprehensive information on myocardial viability. Several radiolabeled tracers are used to evaluate the characteristics of the viable myocardium (thallium-201 for myocardial perfusion and cell membrane integrity assessment; ^{99m}Tc-tetrofosmin/sestamibi for myocardial perfusion, cell membrane integrity, and mitochondrial function; F^{18}-fluorodeoxyglucose (FDG) for glucose metabolism).[43] The role of these imaging techniques for prediction of CRT response has been evaluated in various studies. In a series of 20 advanced heart failure patients undergoing CRT implantation, the presence of non-viable myocardium as assessed with ^{99m}Tc-tetrofosmin ECG-gated SPECT was related to LV reverse remodeling at short-term follow-up.[95] Patients with significant perfusion defects on ECG-gated SPECT studies (tracer uptake<50%) did not show LV reverse remodeling or LV function improvement whereas patients with a normal myocardial perfusion pattern showed favorable echocardiographic response.[95] These results were extended by Ypenburg et al. in a recent study involving 51 ischemic heart failure patients with substantial LV dyssynchrony.[88] ^{99m}Tc-tetrofosmin ECG-gated SPECT data were displayed in a polar map and analyzed in a 17-segment model. Segmental tracer uptake was quantified and segments showing a tracer uptake≥75% were considered normal (viable) and segments showing a tracer uptake<50% were considered to have extensive scar tissue (transmural infarction). The total scar score was derived by summation of the segmental scores and the higher scores indicated more extensive scar tissue. In addition, the presence of transmural scar (tracer activity <50%) in the segments targeted by the LV lead was also assessed. The number of viable segments and the total scar burden were strongly related to LV reverse remodeling. Patients with ≥10 viable segments or a total scar score ≤15 had higher likelihood of showing favorable response to CRT.[88] Finally, patients with transmural scar tissue in the region of the LV pacing lead showed with less frequency significant LV reverse remodeling.

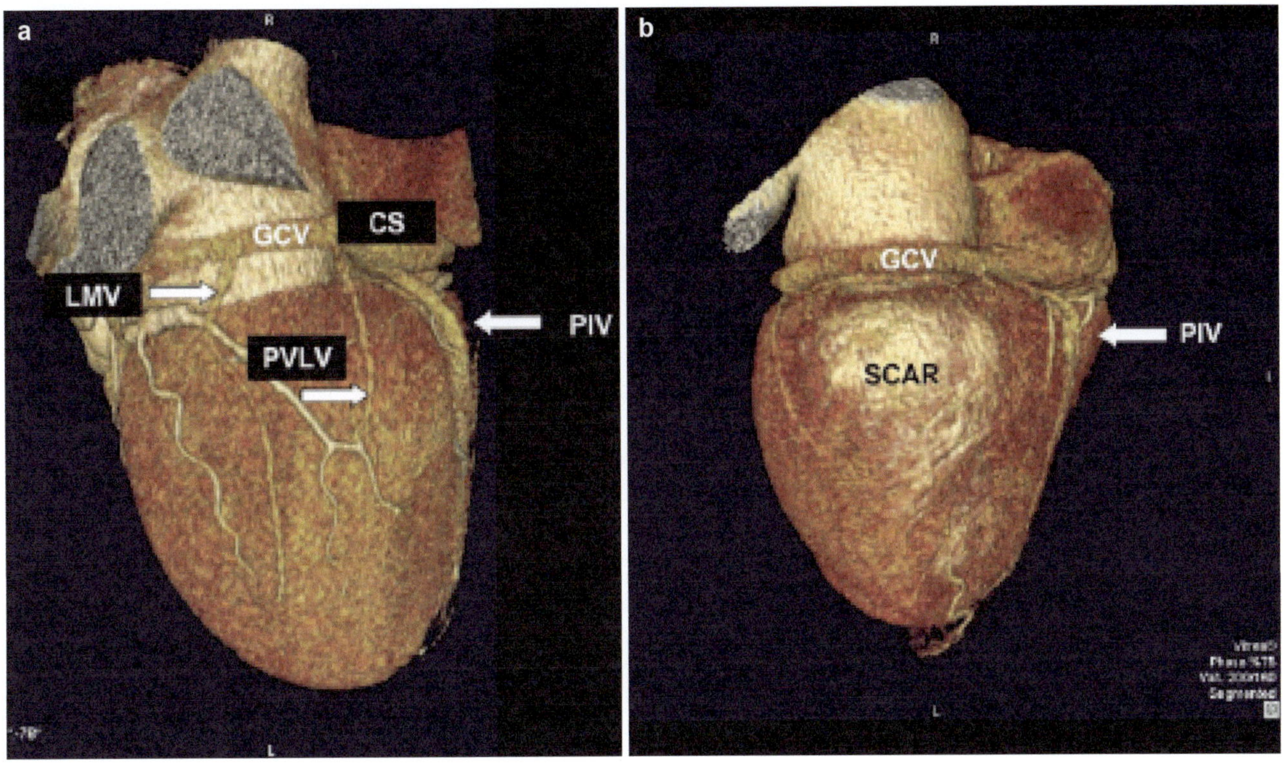

Fig. 13.13 Assessment of venous anatomy with MDCT. The coronary sinus and the tributaries can be evaluated with MDCT. (**a**) Left marginal vein (*LMV*) and posterior vein of the left ventricle (*PVLV*) are identified. In patients with prior myocardial infarction, suitable veins for LV lead implantation are less frequently observed and only minimal cardiac veins are identified (**b**). Abbreviations: *CS* coronary sinus, *GCV* great cardiac vein, *LMV* left marginal vein, *PIV* posterior interventricular vein, *PVLV* posterior vein of the left ventricle (Adapted from Van de Veire et al. [101])

13.2.3 LV Lead Position

Response to CRT is also influenced by the LV lead position and, indeed, initial studies showed that patients with the LV lead in the posterolateral regions had the best hemodynamic improvement.[96] In addition, several studies have demonstrated the lack of response to CRT when the LV pacing lead is positioned outside the latest activated segment.[97-99] Therefore, the identification of the latest activated areas and the assessment of the venous anatomy in those areas may be of importance to increase the CRT response rate. A minimally invasive surgical approach to implant an epicardial LV lead may be preferred when there are no suitable cardiac veins in the latest activated regions.

To identify the latest activated segments, several echocardiographic techniques may be of value.[97-99] Murphy et al. demonstrated the role of TDI echocardiography techniques to identify the latest activated segments and to guide the LV pacing lead positioning.[99] In 54 heart failure patients undergoing CRT implantation, the latest mechanical activated segments were evaluated with TDI. Patients with an LV lead positioned at the area of latest mechanical activation on TDI showed significant LV reverse remodeling after 6 months of CRT.[99] However, TDI does not distinguish between myocardial segments with active contraction or passively tethered by the adjacent and viable segments. In this regard, 2-dimensional strain (rate) imaging may provide more realistic information on the latest activated segments.[97,100] In 257 heart failure patients treated with CRT, the latest activated segments were identified with 2-dimensional speckle tracking radial strain imaging.[100] The LV lead position was assessed from chest X-ray. At 6-month follow-up, significant LV reverse remodeling was observed only in patients with the LV lead positioned at the latest activated segments.[100] More important, a concordant LV lead position was shown to be an independent predictor of hospitalization-free survival at long-term follow-up.[100]

To ensure concordant LV lead positioning at the latest site of activation, it may be important to assess the availability and distribution of cardiac veins in order to plan whether a transvenous approach is feasible, or minimally invasive surgery is preferred. MDCT or MRI techniques can noninvasively visualize venous anatomy.[101,102] Particularly, in patients with ischemic heart failure and previous lateral infarction, MDCT has demonstrated the absence of the left marginal vein (Fig. 13.13).[101] This finding has important clinical implications since the absence of (postero)lateral veins may preclude to target the latest activated segments when the LV lead is placed transvenously. In those cases, a surgical approach to implant the LV lead may be preferred.

13.3 Conclusions

A large body of published evidence from multicenter controlled trials has demonstrated the clinical benefits of ICD and CRT in high-risk patients. However, to achieve the largest benefit, risk-stratification strategies and identification of candidates for these therapies should be optimized. In this regard, analysis of the cardiac substrate may provide important information. In candidates to ICD implantation, MRI and nuclear imaging provide meaningful insight in the arrhythmogenic substrate. In addition, selection of candidates to CRT should rely on an integrated approach including the assessment of LV dyssynchrony (latest activated segment), location and extent of myocardial scar tissue, and cardiac venous anatomy. As shown in numerous studies, patients with significant LV dyssynchrony and with an LV pacing lead positioned in or near the latest activated segments will benefit most from CRT implantation. Echocardiographic techniques provide useful information in this pathophysiological aspect. Furthermore, in patients with ischemic heart failure, positioning of the LV pacing lead in a segment with transmural scar reduces the CRT response. MRI and nuclear imaging are key imaging tools to evaluate the extension and location of scar tissue. Finally, the lack of suitable cardiac venous anatomy for transvenous LV lead implantation at the latest activated segments may be common in patients with prior myocardial infarction. In this subgroup of patients, MDCT may constitute a valuable imaging modality to anticipate the LV lead implantation strategy (transvenous or minimally invasive surgical approach). Therefore, a multimodality imaging approach, including echocardiographic techniques, MRI, nuclear imaging, and MDCT, may refine the risk-stratification and selection of candidates to these device therapies.

References

1. Bardy GH, Lee KL, Mark DB, et al. Amiodarone or an implantable cardioverter-defibrillator for congestive heart failure. *N Engl J Med.* 2005;352:225 237.
2. Bristow MR, Saxon LA, Boehmer J, et al. Cardiac-resynchronization therapy with or without an implantable defibrillator in advanced chronic heart failure. *N Engl J Med.* 2004;350:2140-2150.
3. Cleland JG, Daubert JC, Erdmann E, et al. The effect of cardiac resynchronization on morbidity and mortality in heart failure. *N Engl J Med.* 2005;352:1539-1549.
4. Hohnloser SH, Kuck KH, Dorian P, et al. Prophylactic use of an implantable cardioverter-defibrillator after acute myocardial infarction. *N Engl J Med.* 2004;351:2481-2488.
5. Moss AJ, Hall WJ, Cannom DS, et al. Improved survival with an implanted defibrillator in patients with coronary disease at high risk for ventricular arrhythmia. Multicenter Automatic Defibrillator Implantation Trial Investigators. *N Engl J Med.* 1996;335:1933-1940.
6. Epstein AE, DiMarco JP, Ellenbogen KA, et al. ACC/AHA/HRS 2008 guidelines for device-based therapy of cardiac rhythm abnormalities: executive summary a report of the American College of Cardiology/American Heart Association Task Force on Practice Guidelines (Writing Committee to Revise the ACC/AHA/NASPE 2002 Guideline Update for Implantation of Cardiac Pacemakers and Antiarrhythmia Devices) developed in collaboration with the American Association for Thoracic Surgery and Society of Thoracic Surgeons. *J Am Coll Cardiol.* 2008;51:2085-2105.
7. Moss AJ, Greenberg H, Case RB, et al. Long-term clinical course of patients after termination of ventricular tachyarrhythmia by an implanted defibrillator. *Circulation.* 2004;110:3760-3765.
8. Bax JJ, Abraham T, Barold SS, et al. Cardiac resynchronization therapy: part 1 – issues before device implantation. *J Am Coll Cardiol.* 2005;46:2153-2167.
9. Goldberger JJ, Cain ME, Hohnloser SH, et al. American Heart Association/American College of Cardiology Foundation/Heart Rhythm Society scientific statement on noninvasive risk stratification techniques for identifying patients at risk for sudden cardiac death: a scientific statement from the American Heart Association Council on Clinical Cardiology Committee on Electrocardiography and Arrhythmias and Council on Epidemiology and Prevention. *Circulation.* 2008;118:1497-1518.
10. Chugh SS, Jui J, Gunson K, et al. Current burden of sudden cardiac death: multiple source surveillance versus retrospective death certificate-based review in a large U.S. community. *J Am Coll Cardiol.* 2004;44:1268-1275.
11. Zipes DP, Wellens HJ. Sudden cardiac death. *Circulation.* 1998;98: 2334-2351.
12. Goldberger Z, Lampert R. Implantable cardioverter-defibrillators: expanding indications and technologies. *JAMA.* 2006;295:809-818.
13. Goldenberg I, Moss AJ. Implantable device therapy. *Prog Cardiovasc Dis.* 2008;50:449-474.
14. The Antiarrhythmics versus Implantable Defibrillators (AVID) Investigators. A comparison of antiarrhythmic-drug therapy with implantable defibrillators in patients resuscitated from near-fatal ventricular arrhythmias. *N Engl J Med.* 1997;337:1576-1583.
15. Kuck KH, Cappato R, Siebels J, et al. Randomized comparison of antiarrhythmic drug therapy with implantable defibrillators in patients resuscitated from cardiac arrest: the Cardiac Arrest Study Hamburg (CASH). *Circulation.* 2000;102:748-754.
16. Connolly SJ, Gent M, Roberts RS, et al. Canadian implantable defibrillator study (CIDS): a randomized trial of the implantable cardioverter defibrillator against amiodarone. *Circulation.* 2000;101:1297-1302.
17. Bigger JT Jr. Prophylactic use of implanted cardiac defibrillators in patients at high risk for ventricular arrhythmias after coronary-artery bypass graft surgery. Coronary Artery Bypass Graft (CABG) patch trial investigators. *N Engl J Med.* 1997;337:1569-1575.
18. Buxton AE, Lee KL, Fisher JD, et al. A randomized study of the prevention of sudden death in patients with coronary artery disease. Multicenter Unsustained Tachycardia Trial Investigators. *N Engl J Med.* 1999;341:1882-1890.
19. Moss AJ, Zareba W, Hall WJ, et al. Prophylactic implantation of a defibrillator in patients with myocardial infarction and reduced ejection fraction. *N Engl J Med.* 2002;346:877-883.
20. Bansch D, Antz M, Boczor S, et al. Primary prevention of sudden cardiac death in idiopathic dilated cardiomyopathy: the Cardiomyopathy Trial (CAT). *Circulation.* 2002;105:1453-1458.
21. Strickberger SA, Hummel JD, Bartlett TG, et al. Amiodarone versus implantable cardioverter-defibrillator: randomized trial in

patients with nonischemic dilated cardiomyopathy and asymptomatic nonsustained ventricular tachycardia – AMIOVIRT. *J Am Coll Cardiol.* 2003;41:1707-1712.
22. Kadish A, Dyer A, Daubert JP, et al. Prophylactic defibrillator implantation in patients with nonischemic dilated cardiomyopathy. *N Engl J Med.* 2004;350:2151-2158.
23. Connolly SJ, Hallstrom AP, Cappato R, et al. Meta-analysis of the implantable cardioverter defibrillator secondary prevention trials. AVID, CASH and CIDS studies. Antiarrhythmics vs implantable defibrillator study. Cardiac Arrest Study Hamburg. Canadian Implantable Defibrillator Study. *Eur Heart J.* 2000;21:2071-2078.
24. Nanthakumar K, Epstein AE, Kay GN, et al. Prophylactic implantable cardioverter-defibrillator therapy in patients with left ventricular systolic dysfunction: a pooled analysis of 10 primary prevention trials. *J Am Coll Cardiol.* 2004;44:2166-2172.
25. Ezekowitz JA, Armstrong PW, McAlister FA. Implantable cardioverter defibrillators in primary and secondary prevention: a systematic review of randomized, controlled trials. *Ann Intern Med.* 2003;138:445-452.
26. Buxton AE, Lee KL, Hafley GE, et al. Limitations of ejection fraction for prediction of sudden death risk in patients with coronary artery disease: lessons from the MUSTT study. *J Am Coll Cardiol.* 2007;50:1150-1157.
27. Goldenberg I, Vyas AK, Hall WJ, et al. Risk stratification for primary implantation of a cardioverter-defibrillator in patients with ischemic left ventricular dysfunction. *J Am Coll Cardiol.* 2008;51: 288-296.
28. Bello D, Fieno DS, Kim RJ, et al. Infarct morphology identifies patients with substrate for sustained ventricular tachycardia. *J Am Coll Cardiol.* 2005;45:1104-1108.
29. Roes SD, Borleffs JW, van der Geest RJ, et al. Infarct tissue heterogeneity assessed with contrast-enhanced magnetic resonance imaging predicts spontaneous ventricular arrhythmia in patients with ischemic cardiomyopathy and implantable cardioverter-defibrillator. *Circ Cardiovasc Imaging.* 2009;2:183-190.
30. Schmidt A, Azevedo CF, Cheng A, et al. Infarct tissue heterogeneity by magnetic resonance imaging identifies enhanced cardiac arrhythmia susceptibility in patients with left ventricular dysfunction. *Circulation.* 2007;115:2006-2014.
31. Yan AT, Shayne AJ, Brown KA, et al. Characterization of the peri-infarct zone by contrast-enhanced cardiac magnetic resonance imaging is a powerful predictor of post-myocardial infarction mortality. *Circulation.* 2006;114:32-39.
32. Bogun FM, Desjardins B, Good E, et al. Delayed-enhanced magnetic resonance imaging in nonischemic cardiomyopathy: utility for identifying the ventricular arrhythmia substrate. *J Am Coll Cardiol.* 2009;53:1138-1145.
33. McCrohon JA, Moon JC, Prasad SK, et al. Differentiation of heart failure related to dilated cardiomyopathy and coronary artery disease using gadolinium-enhanced cardiovascular magnetic resonance. *Circulation.* 2003;108:54-59.
34. Nazarian S, Bluemke DA, Lardo AC, et al. Magnetic resonance assessment of the substrate for inducible ventricular tachycardia in nonischemic cardiomyopathy. *Circulation.* 2005;112:2821-2825.
35. Chang H, Georges R, Schuleri KH, et al. Prospective electrocardiogram-gated delayed enhanced multidetector computed tomography accurately quantifies infarct size and reduces radiation exposure. *J Am Coll Cardiol Imaging.* 2009;2:412-420.
36. Schuleri KH, Centola M, George RT, et al. Characterization of peri-infarct zone heterogeneity by contrast-enhanced multidetector computed tomography: a comparison with magnetic resonance imaging. *J Am Coll Cardiol.* 2009;53:1699-1707.
37. Hachamovitch R, Berman DS, Shaw LJ, et al. Incremental prognostic value of myocardial perfusion single photon emission computed tomography for the prediction of cardiac death: differential stratification for risk of cardiac death and myocardial infarction. *Circulation.* 1998;97:535-543.
38. Borger van der Burg AE, Bax JJ, Boersma E, et al. Impact of viability, ischemia, scar tissue, and revascularization on outcome after aborted sudden death. *Circulation.* 2003;108:1954-1959.
39. Azarbal B, Hayes SW, Lewin HC, et al. The incremental prognostic value of percentage of heart rate reserve achieved over myocardial perfusion single-photon emission computed tomography in the prediction of cardiac death and all-cause mortality: superiority over 85% of maximal age-predicted heart rate. *J Am Coll Cardiol.* 2004;44:423-430.
40. Barron HV, Lesh MD. Autonomic nervous system and sudden cardiac death. *J Am Coll Cardiol.* 1996;27:1053-1060.
41. Verrier RL, Antzelevitch C. Autonomic aspects of arrhythmogenesis: the enduring and the new. *Curr Opin Cardiol.* 2004;19:2-11.
42. Zipes DP. Influence of myocardial ischemia and infarction on autonomic innervation of heart. *Circulation.* 1990;82:1095-1105.
43. Bax JJ, Boogers MM, Schuijf JD. Nuclear imaging in heart failure. *Cardiol Clin.* 2009;27:265-276.
44. Agostini D, Verberne HJ, Burchert W, et al. I-123-mIBG myocardial imaging for assessment of risk for a major cardiac event in heart failure patients: insights from a retrospective European multicenter study. *Eur J Nucl Med Mol Imaging.* 2008;35:535-546.
45. Manrique A, Bernard M, Hitzel A, et al. Prognostic value of sympathetic innervation and cardiac asynchrony in dilated cardiomyopathy. *Eur J Nucl Med Mol Imaging.* 2008;35:2074-2081.
46. Merlet P, Valette H, Dubois-Rande JL, et al. Prognostic value of cardiac metaiodobenzylguanidine imaging in patients with heart failure. *J Nucl Med.* 1992;33:471-477.
47. Minardo JD, Tuli MM, Mock BH, et al. Scintigraphic and electrophysiological evidence of canine myocardial sympathetic denervation and reinnervation produced by myocardial infarction or phenol application. *Circulation.* 1988;78:1008-1019.
48. Arora R, Ferrick KJ, Nakata T, et al. I-123 MIBG imaging and heart rate variability analysis to predict the need for an implantable cardioverter defibrillator. *J Nucl Cardiol.* 2003;10:121-131.
49. Bax JJ, Kraft O, Buxton AE, et al. 123I-mIBG scintigraphy to predict inducibility of ventricular arrhythmias on cardiac electrophysiology testing: a prospective multicenter pilot study. *Circ Cardiovasc Imaging.* 2008;1:131-140.
50. Kioka H, Yamada T, Mine T, et al. Prediction of sudden death in patients with mild-to-moderate chronic heart failure by using cardiac iodine-123 metaiodobenzylguanidine imaging. *Heart.* 2007; 93:1213-1218.
51. Nagahara D, Nakata T, Hashimoto A, et al. Predicting the need for an implantable cardioverter defibrillator using cardiac metaiodobenzylguanidine activity together with plasma natriuretic peptide concentration or left ventricular function. *J Nucl Med.* 2008;49: 225-233.
52. McAlister FA, Ezekowitz J, Hooton N, et al. Cardiac resynchronization therapy for patients with left ventricular systolic dysfunction: a systematic review. *JAMA.* 2007;297:2502-2514.
53. Cazeau S, Leclercq C, Lavergne T, et al. Effects of multisite biventricular pacing in patients with heart failure and intraventricular conduction delay. *N Engl J Med.* 2001;344:873-880.
54. Auricchio A, Stellbrink C, Sack S, et al. Long-term clinical effect of hemodynamically optimized cardiac resynchronization therapy in patients with heart failure and ventricular conduction delay. *J Am Coll Cardiol.* 2002;39:2026-2033.
55. Abraham WT, Fisher WG, Smith AL, et al. Cardiac resynchronization in chronic heart failure. *N Engl J Med.* 2002;346:1845-1853.
56. Young JB, Abraham WT, Smith AL, et al. Combined cardiac resynchronization and implantable cardioversion defibrillation in advanced chronic heart failure: the MIRACLE ICD Trial. *JAMA.* 2003;289:2685-2694.
57. Auricchio A, Stellbrink C, Butter C, et al. Clinical efficacy of cardiac resynchronization therapy using left ventricular pacing in heart failure patients stratified by severity of ventricular conduction delay. *J Am Coll Cardiol.* 2003;42:2109-2116.

58. Higgins SL, Hummel JD, Niazi IK, et al. Cardiac resynchronization therapy for the treatment of heart failure in patients with intraventricular conduction delay and malignant ventricular tachyarrhythmias. *J Am Coll Cardiol*. 2003;42:1454-1459.
59. Lozano I, Bocchiardo M, Achtelik M, et al. Impact of biventricular pacing on mortality in a randomized crossover study of patients with heart failure and ventricular arrhythmias. *Pacing Clin Electrophysiol*. 2000;23:1711-1712.
60. Rivero-Ayerza M, Theuns DA, Garcia-Garcia HM, et al. Effects of cardiac resynchronization therapy on overall mortality and mode of death: a meta-analysis of randomized controlled trials. *Eur Heart J*. 2006;27:2682-2688.
61. Bax JJ, Gorcsan J III. Echocardiography and noninvasive imaging in cardiac resynchronization therapy: results of the PROSPECT (Predictors of Response to Cardiac Resynchronization Therapy) study in perspective. *J Am Coll Cardiol*. 2009;53:1933-1943.
62. Bax JJ, Bleeker GB, Marwick TH, et al. Left ventricular dyssynchrony predicts response and prognosis after cardiac resynchronization therapy. *J Am Coll Cardiol*. 2004;44:1834-1840.
63. Bleeker GB, Schalij MJ, Molhoek SG, et al. Relationship between QRS duration and left ventricular dyssynchrony in patients with end-stage heart failure. *J Cardiovasc Electrophysiol*. 2004;15:544-549.
64. Mollema SA, Bleeker GB, van der Wall EE, et al. Usefulness of QRS duration to predict response to cardiac resynchronization therapy in patients with end-stage heart failure. *Am J Cardiol*. 2007;100:1665-1670.
65. Ghio S, Constantin C, Klersy C, et al. Interventricular and intraventricular dyssynchrony are common in heart failure patients, regardless of QRS duration. *Eur Heart J*. 2004;25:571-578.
66. Pitzalis MV, Iacoviello M, Romito R, et al. Cardiac resynchronization therapy tailored by echocardiographic evaluation of ventricular asynchrony. *J Am Coll Cardiol*. 2002;40:1615-1622.
67. Díaz-Infante E, Sitges M, Vidal B, et al. Usefulness of ventricular dyssynchrony measured using M-mode echocardiography to predict response to resynchronization therapy. *Am J Cardiol*. 2007;100:84-89.
68. Penicka M, Bartunek J, De BB, et al. Improvement of left ventricular function after cardiac resynchronization therapy is predicted by tissue Doppler imaging echocardiography. *Circulation*. 2004;109:978-983.
69. Yu CM, Zhang Q, Fung JW, et al. A novel tool to assess systolic asynchrony and identify responders of cardiac resynchronization therapy by tissue synchronization imaging. *J Am Coll Cardiol*. 2005;45:677-684.
70. Dohi K, Suffoletto MS, Schwartzman D, et al. Utility of echocardiographic radial strain imaging to quantify left ventricular dyssynchrony and predict acute response to cardiac resynchronization therapy. *Am J Cardiol*. 2005;96:112-116.
71. Delgado V, Ypenburg C, Van Bommel RJ, et al. Assessment of left ventricular dyssynchrony by speckle tracking strain imaging comparison between longitudinal, circumferential, and radial strain in cardiac resynchronization therapy. *J Am Coll Cardiol*. 2008;51:1944-1952.
72. Marsan NA, Bleeker GB, Ypenburg C, et al. Real-time three-dimensional echocardiography as a novel approach to assess left ventricular and left atrium reverse remodeling and to predict response to cardiac resynchronization therapy. *Heart Rhythm*. 2008;5:1257-1264.
73. Van de Veire N, Yu CM, Jmone-Marsan N, et al. Triplane tissue Doppler imaging: a novel three dimensional imaging modality that predicts reverse left ventricular remodelling after cardiac resynchronisation therapy. *Heart*. 2008;94:e9.
74. Yu CM, Fung JW, Zhang Q, et al. Tissue Doppler imaging is superior to strain rate imaging and postsystolic shortening on the prediction of reverse remodeling in both ischemic and nonischemic heart failure after cardiac resynchronization therapy. *Circulation*. 2004;110:66-73.
75. Suffoletto MS, Dohi K, Cannesson M, et al. Novel speckle-tracking radial strain from routine black-and-white echocardiographic images to quantify dyssynchrony and predict response to cardiac resynchronization therapy. *Circulation*. 2006;113:960-968.
76. Kapetanakis S, Kearney MT, Siva A, et al. Real-time three-dimensional echocardiography: a novel technique to quantify global left ventricular mechanical dyssynchrony. *Circulation*. 2005;112:992-1000.
77. Chung ES, Leon AR, Tavazzi L, et al. Results of the predictors of response to CRT (PROSPECT) trial. *Circulation*. 2008;117:2608-2616.
78. Ypenburg C, Westenberg JJ, Bleeker GB, et al. Noninvasive imaging in cardiac resynchronization therapy – part 1: selection of patients. *Pacing Clin Electrophysiol*. 2008;31:1475-1499.
79. Lardo AC, Abraham TP, Kass P. Magnetic resonance imaging assessment of ventricular dyssynchrony: current and emerging concepts. *J Am Coll Cardiol*. 2005;46:2223-2228.
80. Westenberg JJ, Lamb HJ, van der Geest RJ, et al. Assessment of left ventricular dyssynchrony in patients with conduction delay and idiopathic dilated cardiomyopathy: head-to-head comparison between tissue Doppler imaging and velocity-encoded magnetic resonance imaging. *J Am Coll Cardiol*. 2006;47:2042-2048.
81. Helm RH, Leclercq C, Faris OP, et al. Cardiac dyssynchrony analysis using circumferential versus longitudinal strain: implications for assessing cardiac resynchronization. *Circulation*. 2005;111:2760-2767.
82. Zwanenburg JJ, Gotte MJ, Kuijer JP, et al. Timing of cardiac contraction in humans mapped by high-temporal-resolution MRI tagging: early onset and late peak of shortening in lateral wall. *Am J Physiol Heart Circ Physiol*. 2004;286:H1872-H1880.
83. Koos R, Neizel M, Schummers G, et al. Feasibility and initial experience of assessment of mechanical dyssynchrony using cardiovascular magnetic resonance and semi-automatic border detection. *J Cardiovasc Magn Reson*. 2008;10:49.
84. Henneman MM, van der Wall EE, Ypenburg C, et al. Nuclear imaging in cardiac resynchronization therapy. *J Nucl Med*. 2007;48:2001-2010.
85. Boogers MM, Van Kriekinge SD, Henneman MM, et al. Quantitative gated SPECT-derived phase analysis on gated myocardial perfusion SPECT detects left ventricular dyssynchrony and predicts response to cardiac resynchronization therapy. *J Nucl Med*. 2009;50:718-725.
86. Bleeker GB, Schalij MJ, van der Wall EE, et al. Postero-lateral scar tissue resulting in non-response to cardiac resynchronization therapy. *J Cardiovasc Electrophysiol*. 2006;17:899-901.
87. White JA, Yee R, Yuan X, et al. Delayed enhancement magnetic resonance imaging predicts response to cardiac resynchronization therapy in patients with intraventricular dyssynchrony. *J Am Coll Cardiol*. 2006;48:1953-1960.
88. Ypenburg C, Schalij MJ, Bleeker GB, et al. Impact of viability and scar tissue on response to cardiac resynchronization therapy in ischaemic heart failure patients. *Eur Heart J*. 2007;28:33-41.
89. Bax JJ. Assessment of myocardial viability in ischemic cardiomyopathy. *Heart Lung Circ*. 2005;14:S8-S13.
90. Sicari R, Nihoyannopoulos P, Evangelista A, et al. Stress echocardiography expert consensus statement: European Association of Echocardiography (EAE) (a registered branch of the ESC). *Eur J Echocardiogr*. 2008;9:415-437.
91. Da Costa A, Thevenin J, Roche F, et al. Prospective validation of stress echocardiography as an identifier of cardiac resynchronization therapy responders. *Heart Rhythm*. 2006;3:406-413.
92. Ypenburg C, Sieders A, Bleeker GB, et al. Myocardial contractile reserve predicts improvement in left ventricular function after cardiac resynchronization therapy. *Am Heart J*. 2007;154:1160-1165.

93. Hummel JP, Lindner JR, Belcik JT, et al. Extent of myocardial viability predicts response to biventricular pacing in ischemic cardiomyopathy. *Heart Rhythm*. 2005;2:1211-1217.
94. Ypenburg C, Roes SD, Bleeker GB, et al. Effect of total scar burden on contrast-enhanced magnetic resonance imaging on response to cardiac resynchronization therapy. *Am J Cardiol*. 2007;99: 657-660.
95. Sciagra R, Giaccardi M, Porciani MC, et al. Myocardial perfusion imaging using gated SPECT in heart failure patients undergoing cardiac resynchronization therapy. *J Nucl Med*. 2004;45:164-168.
96. Butter C, Schlegl M. Effect of epicardial or biventricular pacing to prolong QT interval and increase transmural dispersion of repolarization. *Circulation*. 2003;108:e27-e28.
97. Becker M, Franke A, Breithardt OA, et al. Impact of left ventricular lead position on the efficacy of cardiac resynchronisation therapy: a two-dimensional strain echocardiography study. *Heart*. 2007;93:1197-1203.
98. Becker M, Hoffmann R, Schmitz F, et al. Relation of optimal lead positioning as defined by three-dimensional echocardiography to long-term benefit of cardiac resynchronization. *Am J Cardiol*. 2007;100:1671-1676.
99. Murphy RT, Sigurdsson G, Mulamalla S, et al. Tissue synchronization imaging and optimal left ventricular pacing site in cardiac resynchronization therapy. *Am J Cardiol*. 2006;97:1615-1621.
100. Ypenburg C, Van Bommel RJ, Delgado V, et al. Optimal left ventricular lead position predicts reverse remodeling and survival after cardiac resynchronization therapy. *J Am Coll Cardiol*. 2008;52:1402-1409.
101. Van de Veire N, Schuijf JD, De Sutter J, et al. Non-invasive visualization of the cardiac venous system in coronary artery disease patients using 64-slice computed tomography. *J Am Coll Cardiol*. 2006;48:1832-1838.
102. van de Veire N, Marsan NA, Schuijf JD, et al. Noninvasive imaging of cardiac venous anatomy with 64-slice multi-slice computed tomography and noninvasive assessment of left ventricular dyssynchrony by 3-dimensional tissue synchronization imaging in patients with heart failure scheduled for cardiac resynchronization therapy. *Am J Cardiol*. 2008;101:1023-1029.

Dyssynchrony Evaluation: Echocardiography

14

John Gorcsan III

Abstract

Cardiac resynchronization therapy (CRT) has been proven to be a major benefit to severely symptomatic heart failure patients with widened QRS duration ≥ 120 ms, and reduced ejection fraction (EF) ≤ 35%. The prevailing evidence supports that multi-site pacing improves abnormalities of left ventricular (LV) mechanical activation, known as dyssynchrony, as the principal therapeutic effect of CRT. The most common pattern of dyssynchrony is represented by left bundle branch block (LBBB), which is characterized by early septal mechanical activation followed by delayed posterior and lateral wall activation. Although several variations in abnormalities of electrical and mechanical activation may exist, the electrical dispersion manifest by QRS widening is believed to be a surrogate for mechanical dyssynchrony. Current clinical guidelines based on selection criteria used in the CRT trials use QRS widening as a marker for dyssynchrony. Unfortunately, approximately one-third of patients receiving CRT do not seem to benefit using these routine clinical selection criteria. Since CRT implantation is expensive and associated with a small but significant procedural risk of complications, efforts have been focused on improvement in patient selection. The study of echocardiographic dyssynchrony has led to the important observation that there is a subset of patients with widened QRS duration that do not have significant mechanical dyssynchrony. The reasons for the disassociation of mechanical dyssynchrony from QRS widening are unclear, but most existing information indicates that the patients with QRS widening who lack significant mechanical dyssynchrony do not respond favorably to CRT and appear to have a worse prognosis.

Keywords

Echocardiography • Dyssynchrony evaluation in echocardiography • Doppler echocardiography • Speckle Tracking Echocardiography • Resynchronization Therapy in cardiology

14.1 Overview of Echocardiographic Dyssynchrony

14.1.1 Pathophysiological Basis of Dyssynchrony

Cardiac resynchronization therapy (CRT) has been proven to be a major benefit to severely symptomatic heart failure patients with widened QRS duration ≥ 120 ms, and reduced ejection fraction (EF) ≤ 35%. Several randomized clinical CRT trials have demonstrated significant improvements in

J. Gorcsan III
Department of Medicine, Cardiovascular Institute,
University of Pittsburgh, Pittsburgh, PA, USA
e-mail: gorcsanj@upmc.edu

A. Auricchio et al. (eds.), *Cardiac Imaging in Electrophysiology*,
DOI 10.1007/978-1-84882-486-7_14,

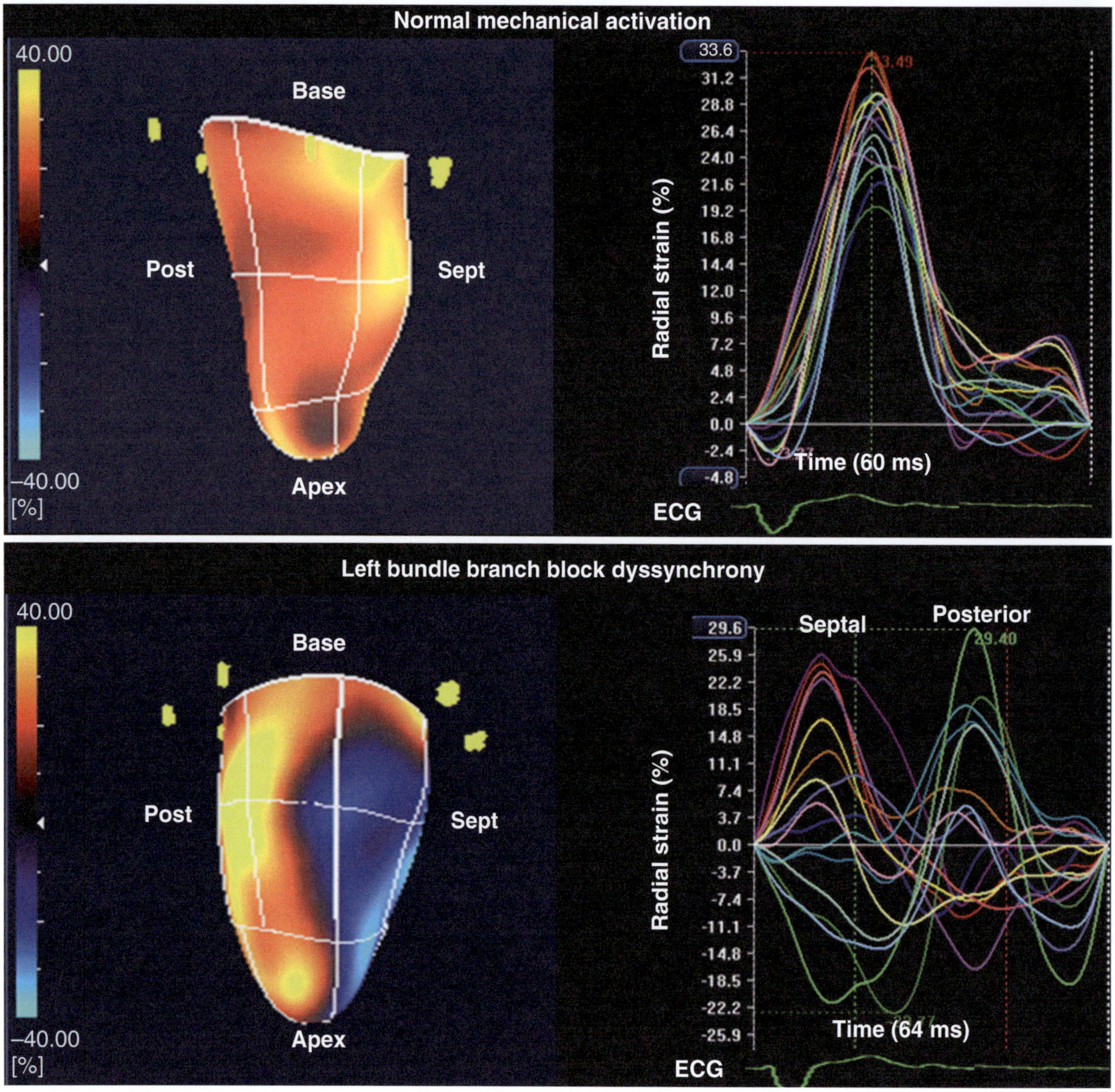

Fig. 14.1 Three-dimensional echocardiographic strain images and corresponding segmental time-strain curves from a normal subject (*top panels*) and a heart failure patient with left bundle branch block (*LBBB*) (*bottom panels*). The normal subject illustrates synchronous contraction with similar timing of peak strain curves, and the patient with LBBB demonstrates the typical dyssynchronous pattern of early septal contraction followed by delayed posterior segmental contraction

heart failure symptoms, ventricular function, and survival.[1-3] The prevailing evidence supports that multi-site pacing improves abnormalities of left ventricular (LV) mechanical activation, known as dyssynchrony, as the principal therapeutic effect of CRT. The most common pattern of dyssynchrony is represented by left bundle branch block (LBBB), which is characterized by early septal mechanical activation followed by delayed posterior and lateral wall activation (Fig. 14.1). Although several variations in abnormalities of electrical and mechanical activation may exist, the electrical dispersion manifest by QRS widening is believed to be a surrogate for mechanical dyssynchrony. Current clinical guidelines based on selection criteria used in the CRT trials use QRS widening as a marker for dyssynchrony. Unfortunately, approximately one-third of patients receiving CRT do not seem to benefit using these routine clinical selection criteria.[4-8] Since CRT implantation is expensive and associated with a small but significant procedural risk

Table 14.1 Dyssynchrony indices associated with response to cardiac resynchronization therapy

Index	Method	Cut-off (ms)	Calculation	Comments
Septal to posterior wall delay	M-mode	≥130	Time difference between peak inward motion of the septum and posterior wall	Largely affected by passive motion or tethering. Difficulties with segmental akinesis
Interventricular mechanical delay	Routine pulsed Doppler	≥40	Time difference between right ventricular ejection and left ventricular ejection	Widely available. Highly reproducible
Opposing wall delay, 2-sites	Color tissue Doppler peak velocity	≥65	Time difference between peak systolic velocities of opposing walls during ejection interval	Requires color TD equipment. Affected by passive motion and tethering
Maximum wall delay, 12-sites	Color tissue Doppler peak velocity	≥100	Maximum time difference between peak systolic velocities of 12 sites during ejection interval	More complete detection of longitudinal dyssynchrony. Affected by passive motion and tethering
Yu index	Color tissue Doppler, 12-segment standard deviation	≥32	Standard deviation in time from onset of QRS to peak velocities during ejection interval from 12 sites	Time-consuming but complete assessment of longitudinal dyssynchrony. Affected by passive motion and tethering
Delayed longitudinal contraction	Color tissue Doppler strain/strain rate	N/A	Delayed onset of contraction from basal sites	Technically demanding. Less affected by passive motion or tethering
Septal to posterior wall delay	Radial strain	≥130	Time difference between peak positive strain in septum to posterior wall	Requires specialized software for analysis. Reflects active contraction

of complications, efforts have been focused on improvement in patient selection. The study of echocardiographic dyssynchrony has led to the important observation that there is a subset of patients with widened QRS duration that do not have significant mechanical dyssynchrony.[4,5,7,9-12] (Table 14.1) The reasons for the disassociation of mechanical dyssynchrony from QRS widening are unclear, but most existing information indicates that the patients with QRS widening who lack significant mechanical dyssynchrony do not respond favorably to CRT and appear to have a worse prognosis.

14.1.2 Dyssynchrony and the PROSPECT Study

A large body of literature exists that supports the utility of echocardiographic dyssynchrony to predict response to CRT and improve patient selection. However, it remains controversial as to which echocardiographic dyssynchrony approach is most reproducible and clinically useful. The multicenter study of the predictors of responders to cardiac resynchronization therapy, known as the PROSPECT study, attempted to determine which echo-Doppler approach would be the most favorable predictor of response to CRT.[13] In this study of 426 patients with routine CRT indications from 53 international sites, M-mode, routine pulsed Doppler, and tissue Doppler methods were examined at baseline, and the outcome variables of clinical symptomatic response and reductions in end-systolic LV volumes assessed at 6 months.

Unfortunately, no single echocardiographic dyssynchrony measure was found to be highly predictive of CRT response. Furthermore, PROSPECT reported a low feasibility of M-mode and tissue Doppler dyssynchrony measures in many patients, and a high variability in their analysis. Since PROSPECT was the first prospective multicenter study of echocardiographic dyssynchrony, it concluded that routine clinical selection criteria have not yet been replaced.[13] Subsequently, several problems with the PROSPECT which most likely contributed to the variability in data have been identified including three different echocardiographic systems, three different off-line software programs used by three different echo core labs. Furthermore, the technical quality of many of the echocardiographic studies was limited, even for routine measures such as LV end-systolic volumes, which interfered with interpretation of the results. Furthermore, the follow-up period of 6 months may have been too short to demonstrate any significant outcome prediction in these patients. In the aftermath of PROSPECT, the clinical utility of echocardiographic dyssynchrony to predict response to CRT has become less clear. Further evidence suggested that response to CRT is multifactorial, including improvement in dyssynchrony, scar burden, LV lead positioning, atrio-ventricular delays, and LV-right ventricular interactions.[8,14-16] (Fig. 14.2). With the acknowledgment that echocardiographic assessment of mechanical dyssynchrony is complex and that its significance is not completely elucidated, the remainder of this chapter will review the body of data that support the potential usefulness of echocardiographic dyssynchrony and response to CRT. Since no echocardiographic or Doppler method is ideal, a multi-modality echocardiographic approach is prudent utilizing the three most supported methods of routine pulsed Doppler, tissue Doppler, and the newer approach of speckle tracking assessment of dyssynchrony.

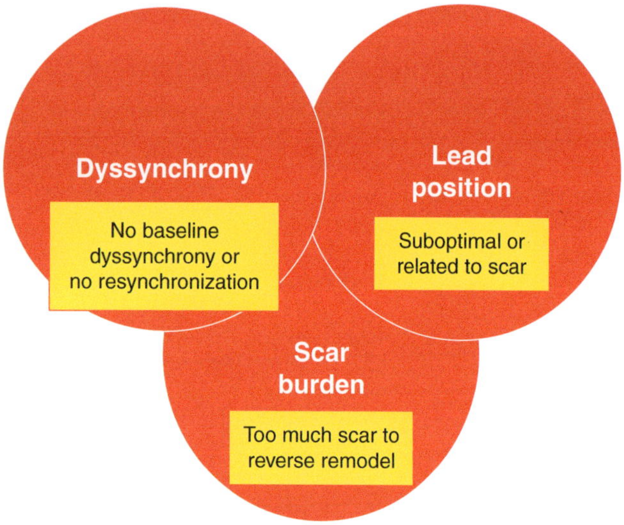

Fig. 14.2 Approximately one-third of patients who receive cardiac resynchronization therapy do not respond favorably. Although lack of dyssynchrony appears to play a role in non-response to cardiac resynchronization therapy, other important potential reasons for non-response exist

14.2 Routine Pulsed Doppler Methods

Routine pulsed Doppler echocardiography is widely available and highly reproducible to measure the timing and duration of cardiac blood flow events. Cazeau et al. introduced the use of routine pulsed Doppler to assess effects of cardiac dyssynchrony.[17] The most widely used measures include the onset of LV outflow, also known as the pre-ejection delay (PED), the difference between the right ventricular pre-ejection period and LV pre-ejection period, also known as the interventricular mechanical delay (IVMD), and mitral inflow filling time, usually expressed as a ratio to the R-R interval (FT/RR). The pathophysiological basis for delays in LV ejection is thought to be intraventricular dyssynchrony that result in increases in the LV pre-ejection period and IVMD. (Fig. 14.3). Although confounding variables for IVMD include effects of heart rate and right ventricular function, these simple measures have been shown to be predictive of patient outcomes following CRT. The commonly used cutoffs considered as significant dyssynchrony include: ≥140 ms for PED, ≥40 ms for IVMD, and ≤40% for FT/RR. Of these three routine pulsed Doppler indices, IVMD has been shown to be most predictive of response to CRT, in particular in patients with wider QRS durations. Achilli et al. reported the SCART study of 133 patients, where a positive response to CRT was predicted by IVMD > 44 ms with a sensitivity of 66% and a specificity of 55%.[18] Richardson et al. also reported that an IMVD > 50 ms was associated with patient survival following CRT as part of the CARE-HF trial analysis.[19] More recent data have shown that IVMD is associated with wider QRS duration and thus uncommon in patients with shorter QRS duration <130 ms. Accordingly, significant IVMD > 40 ms is likely the result of a more severe degree of dyssynchrony and is a potentially useful addition to measures of intraventricular dyssynchrony as discussed in detail subsequently.

14.3 Tissue Doppler Imaging

The largest body of published literature on echocardiographic dyssynchrony analysis to predict response to CRT is represented by tissue Doppler imaging.[4,5,7-10,12,20] The most frequent approach takes advantage of the favorable Doppler angle of incidence from the apical windows to measure LV longitudinal mechanical events. The most direct measurement utilizing tissue Doppler is velocity, and differences in time-to-peak velocity have been used widely as markers of dyssynchrony. The most straightforward approaches have been either the opposing wall delay, or the 12-site standard deviation, also known as the Yu Index (Fig. 14.4).[8,12,20] Tissue Doppler data are recorded from three standard apical views, including apical 4-chamber view, apical 2-chamber view, and apical long-axis view (Figs. 14.5 and 14.6). A summary of practical technical steps to help achieve reproducible longitudinal tissue Doppler velocity results is as follows:

- Utilize color-coded tissue Doppler imaging and off-line analysis.
- Align the left ventricle as vertical as possible within the ultrasound sector, to achieve the most favorable Doppler angle of incidence.
- Limit analysis to the timing of LV ejection, excluding post-systolic peaks. The event timing is usually extracted from routine Doppler tracings of the LV outflow tract.
- Use large regions of interest (7 × 15 mm) to enhance spatial averaging.
- Limit analysis to the basal and middle thirds of the LV, excluding the LV apex, because of an unfavorable Doppler angle of incidence and near-field noise.
- Adjust the region of interest within the segment to determine the most reproducible peak velocity signal, excluding short-duration high-frequency noise.

The cutoff values utilized commonly for significant dyssynchrony by tissue Doppler velocity include an opposing wall delay of ≥65 ms from a single view and a 12-site standard deviation of >32 ms. Other groups have been successful with post-processing approaches to tissue Doppler, including tissue Doppler strain and strain rate; however, the velocity data have the more robust signal to noise ratio.[10] The major disadvantage to tissue Doppler velocity assessment is that velocity cannot differentiate between the timing of active contraction, which is believed to be important for dyssynchrony analysis and passive motion, which is a confounding variable with ischemic heart disease and myocardial scar.

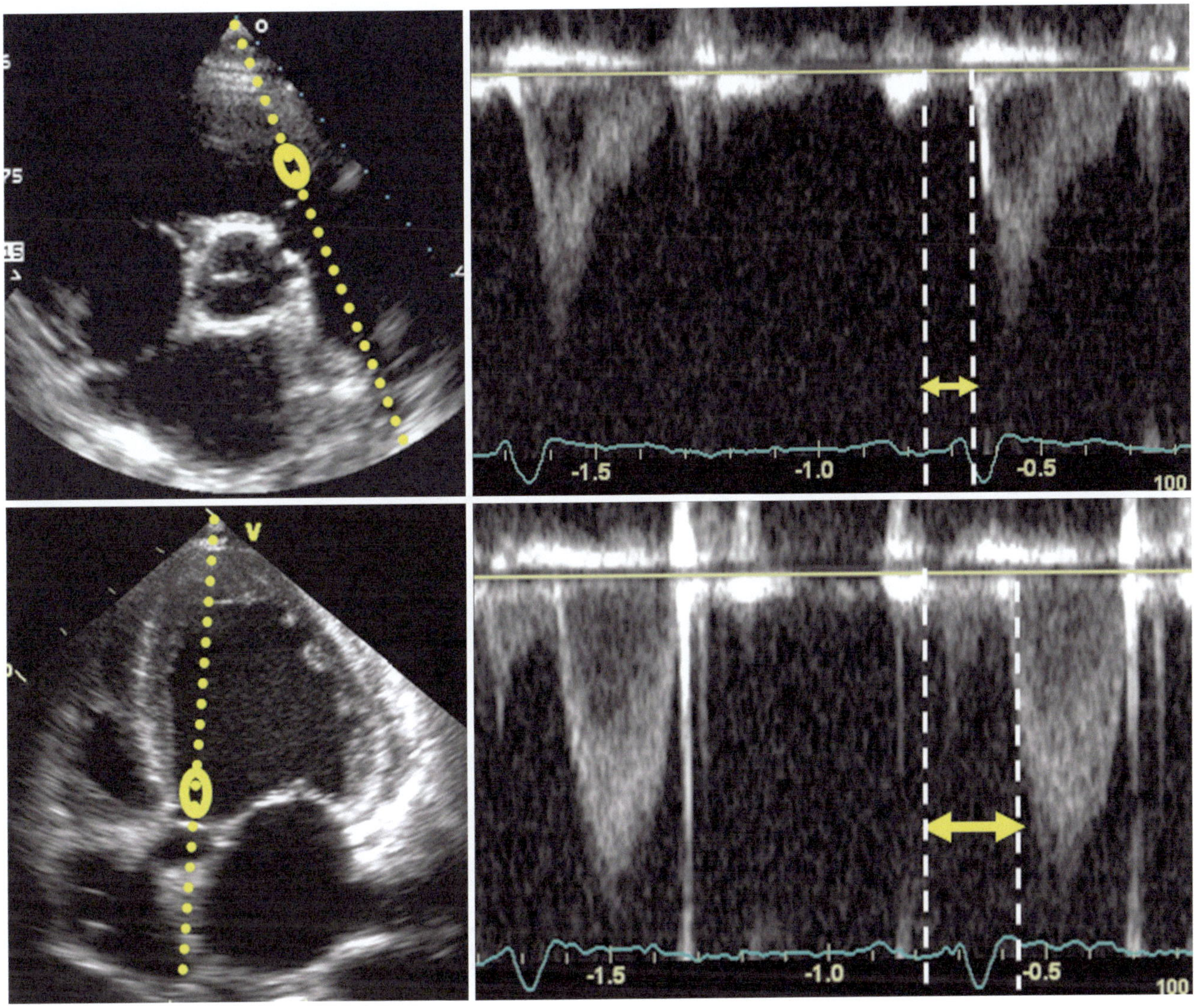

Fig. 14.3 Routine pulsed Doppler recordings demonstrating the calculation of interventricular mechanical delay (*IVMD*). The time difference between onset of right ventricular ejection and onset of left ventricular ejection is the IVMD. Lines in *left panels* indicate the location of pulsed Doppler sample volumes. *Arrow top-right* indicates time to onset of right ventricular ejection, *arrow bottom-right* indicates time to onset of left ventricular ejection

14.4 Speckle Tracking Echocardiography

14.4.1 Two-Dimensional Speckle Tracking

A recent advance in echocardiographic assessment of dyssynchrony is speckle tracking, which may be applied to routine gray scale digital images. The concept of speckle tracking is to use computer programming to recognize patterns of myocardial speckles within a region of interest from frame-to-frame.[9,11] Quantitative data is then extracted to determine myocardial strain as the relative change in distance or length. Three variations in two-dimensional strain have been described, including radial strain and circumferential strain from short-axis images and longitudinal strain from apical images. Radial strain from the mid-ventricular short-axis plane has been described to be advantageous to the alternate methods for dyssynchrony analysis, although data continue to emerge on these newer methods (Figs. 14.7 and 14.8). A summary of practical advice on achieving reproducible two-dimensional radial strain data from mid-LV short-axis images for dyssynchrony analysis with currently available equipment is as follows:

- Acquire high image quality data with frame rates 60–90 Hz. Frame rates too slow are excessively smoothed, and frame rates too fast are noisy.
- Carefully place the endocardial region of interest slightly within the LV cavity and place the epicardial region of

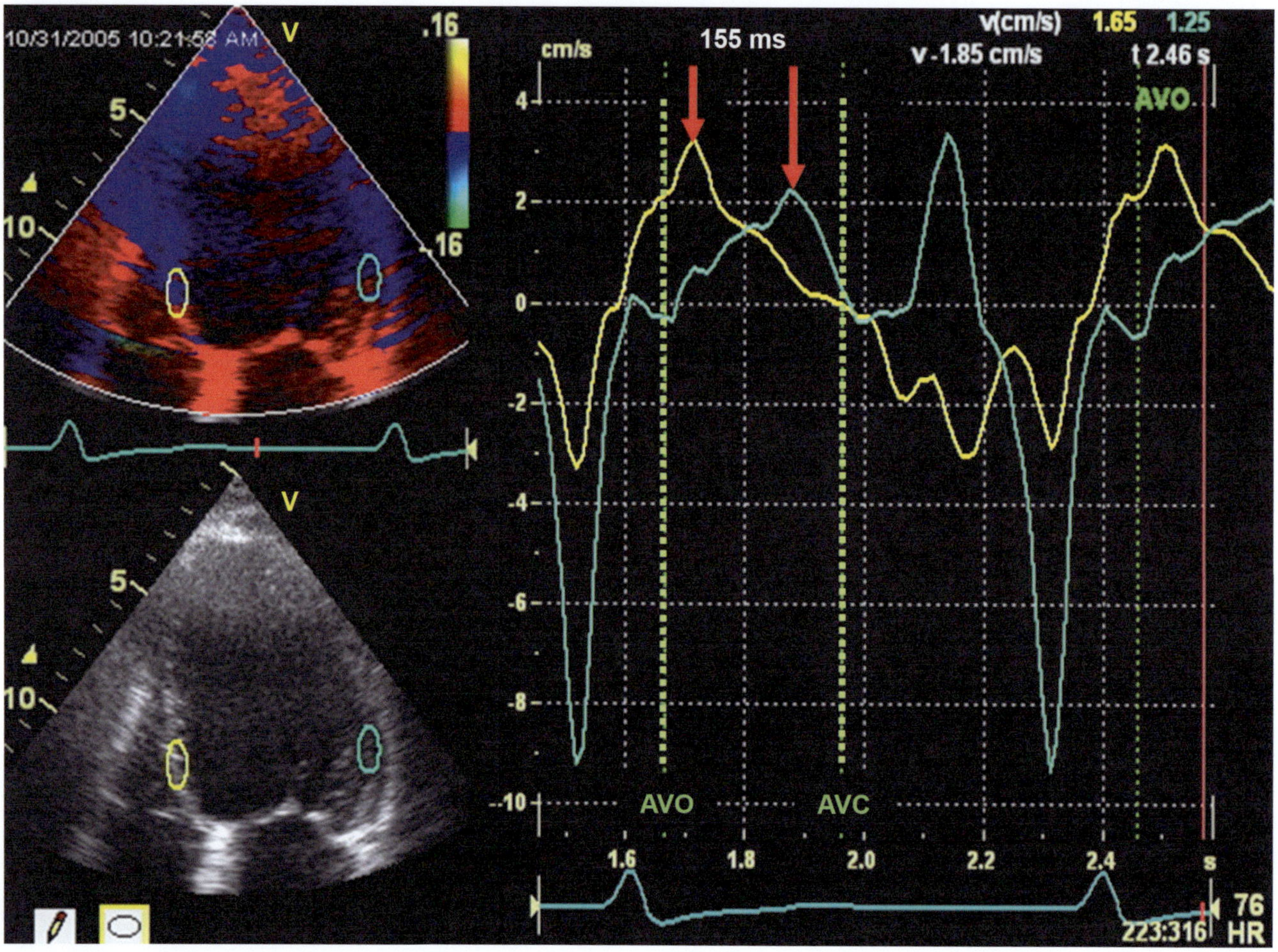

Fig. 14.4 An example determination of the tissue Doppler two-site opposing wall delay from the 4-chamber view from a patient with left bundle branch block. The opposing wall delay is assessed as the time difference between the septal peak velocity and the lateral wall peak velocity. Analysis of peak velocities is limited to the ejection interval between aortic valve opening (*AVO*) and aortic valve closure (*AVC*)

interest to match the epicardial contour. Care must be taken to adjust the septal region of interest to be slightly less wide than the free wall, consistent with the anatomical contour.

- Use the cine play feature to ensure visually that the region of interest is tracking the myocardial motion, in particular tracking the endocardium. Adjust and fine-tune the region of interest until satisfactory tracking is achieved.
- Determine the timing of peak positive strain (above the zero baseline), which represents thickening toward the LV cavity center, in septal segments and posterior and/or lateral segments. Note that the typical dyssynchrony pattern is early septal thickening and delayed free wall thickening.
- Analyze several beats and average values if beat-to-beat variability occurs.
- The time difference between anteroseptal to posterior wall peak strain has been shown to be predictive of response to CRT.

A variation of speckle tracking echocardiography is velocity vector imaging. This software utilizes routine digital DICOM images, and can extract quantitative data off-line.[16] Both longitudinal velocity data may be obtained from apical views for dyssynchrony analysis and radial strain from parasternal short-axis views. Although details for analysis of speckle tracking echocardiograms are related to specific manufacturer's software, similar results may be obtained.[21] Accordingly, speckle tracking echocardiography for dyssynchrony analysis appears promising for future applications.

14.4.2 Three-Dimensional Speckle Tracking

The most recent technological advance in speckle tracking echocardiography has been three-dimensional strain. A pyramid of echocardiographic data is acquired using a three-dimensional transducer, which allows myocardial dynamics to be tracked in three-dimensional space. This enables the complexities of three-dimensional dyssynchrony to be quantified and mechanical activation sequences displayed.

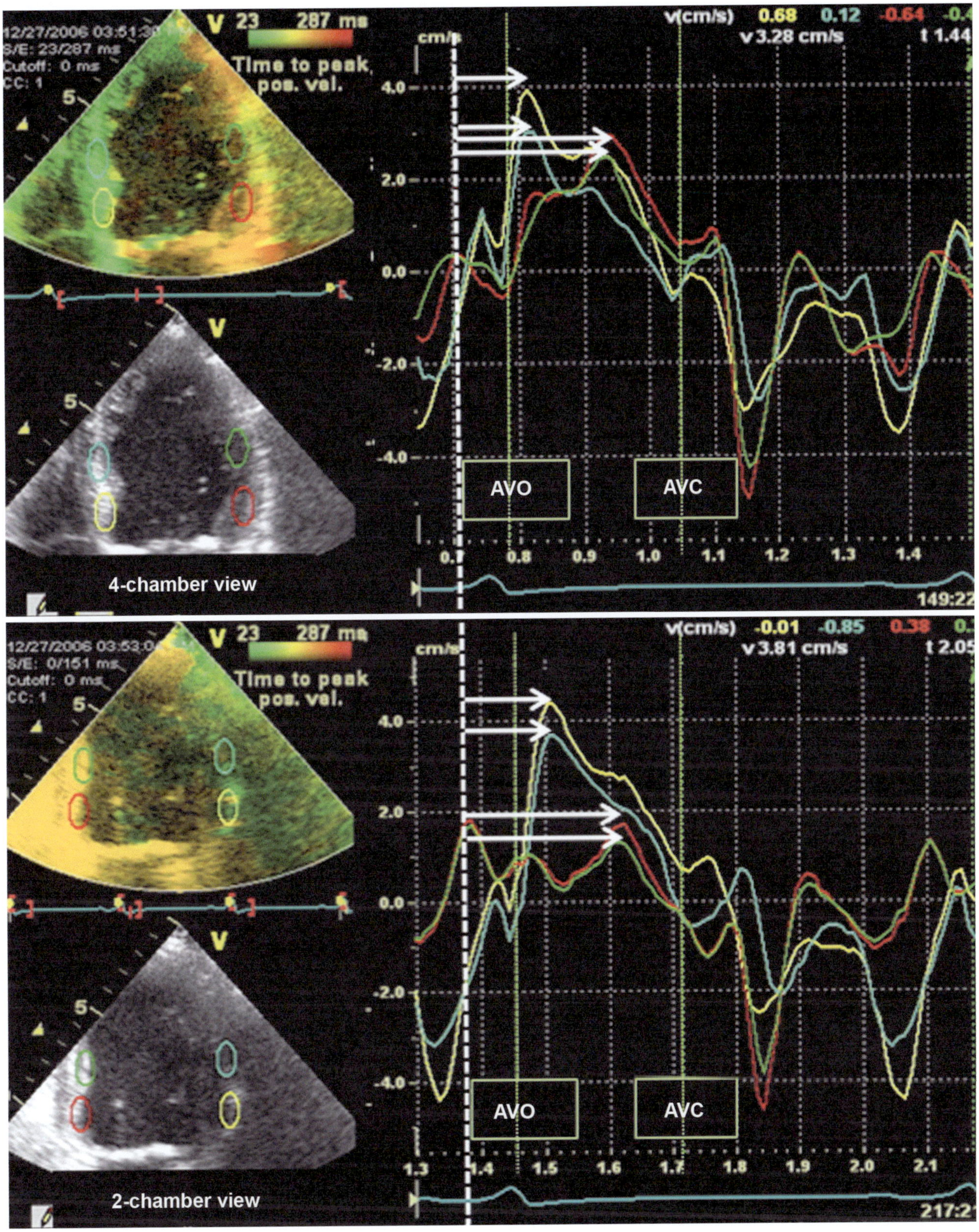

Fig. 14.5 A representative example of a patient with left bundle branch block and significant dyssynchrony. Time from onset of the electrocardiographic QRS complex to peak systolic velocities is shown from basal and mid segments using apical 4-chamber, apical 2-chamber, and apical long-axis views. Calculation of the standard deviation of the time-to-peak velocity from 12 sites is known as the Yu Index. Analysis of peak velocities is limited to the ejection interval between aortic valve opening (*AVO*) and aortic valve closure (*AVC*)

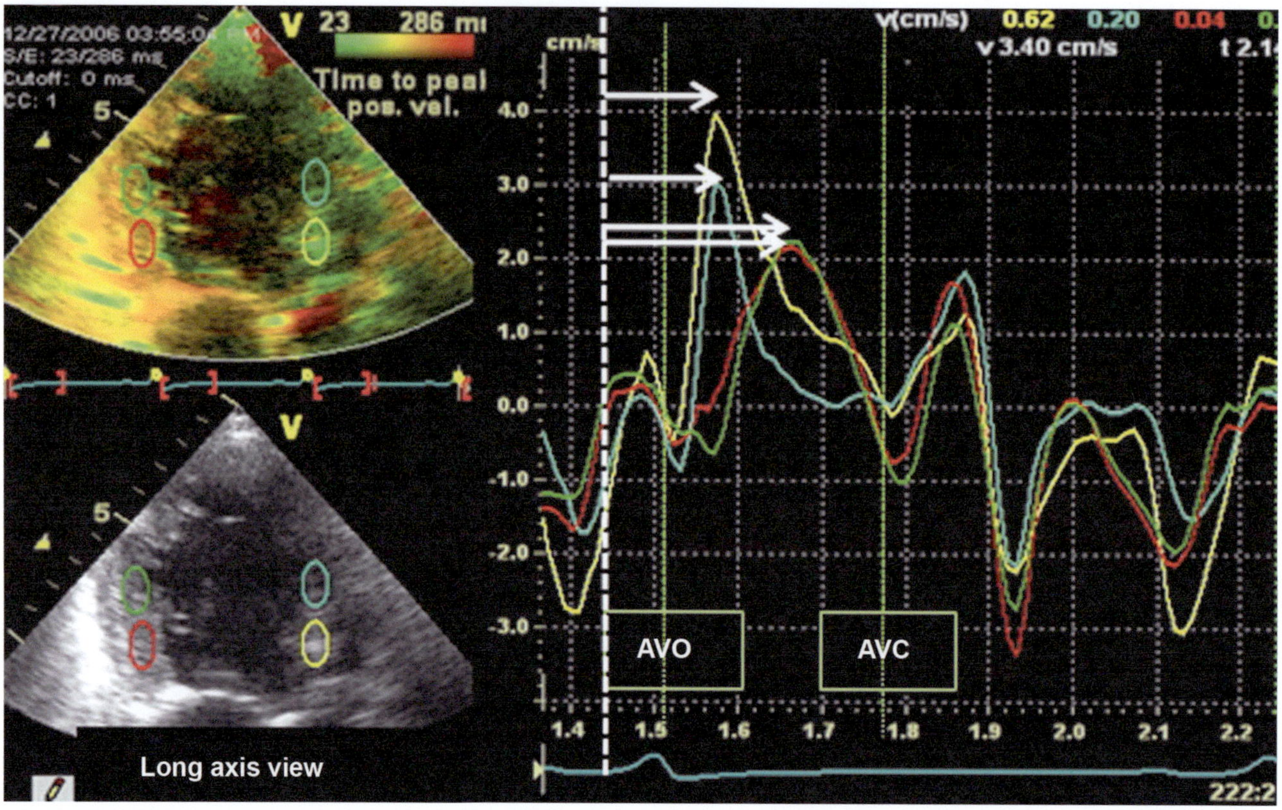

Fig. 14.5 (continued)

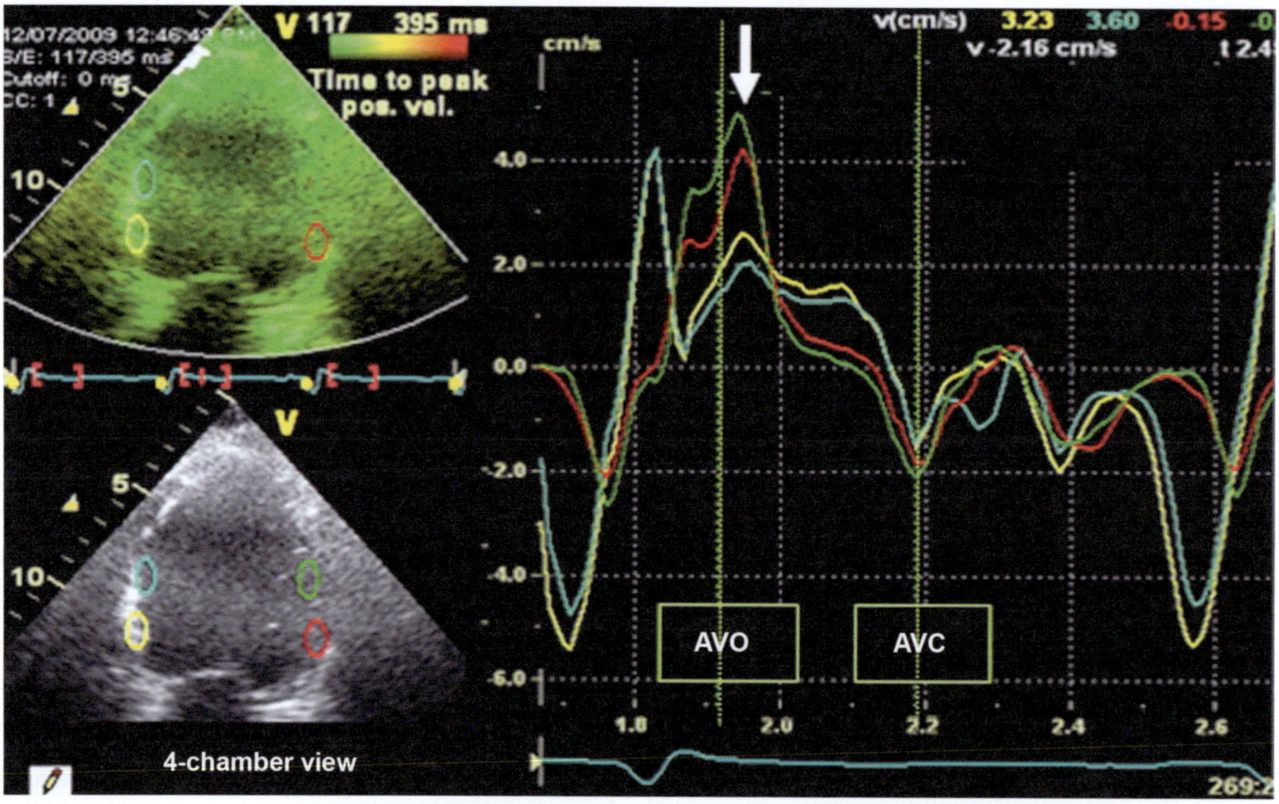

Fig. 14.6 A representative example of a patient with significant QRS widening 150 ms, but no significant dyssynchrony using tissue Doppler longitudinal velocities. Note that the time-to-peak velocities are similar in the three apical views. This patient was a non-responder to cardiac resynchronization therapy. Analysis of peak velocities is limited to the ejection interval between aortic valve opening (*AVO*) and aortic valve closure (*AVC*)

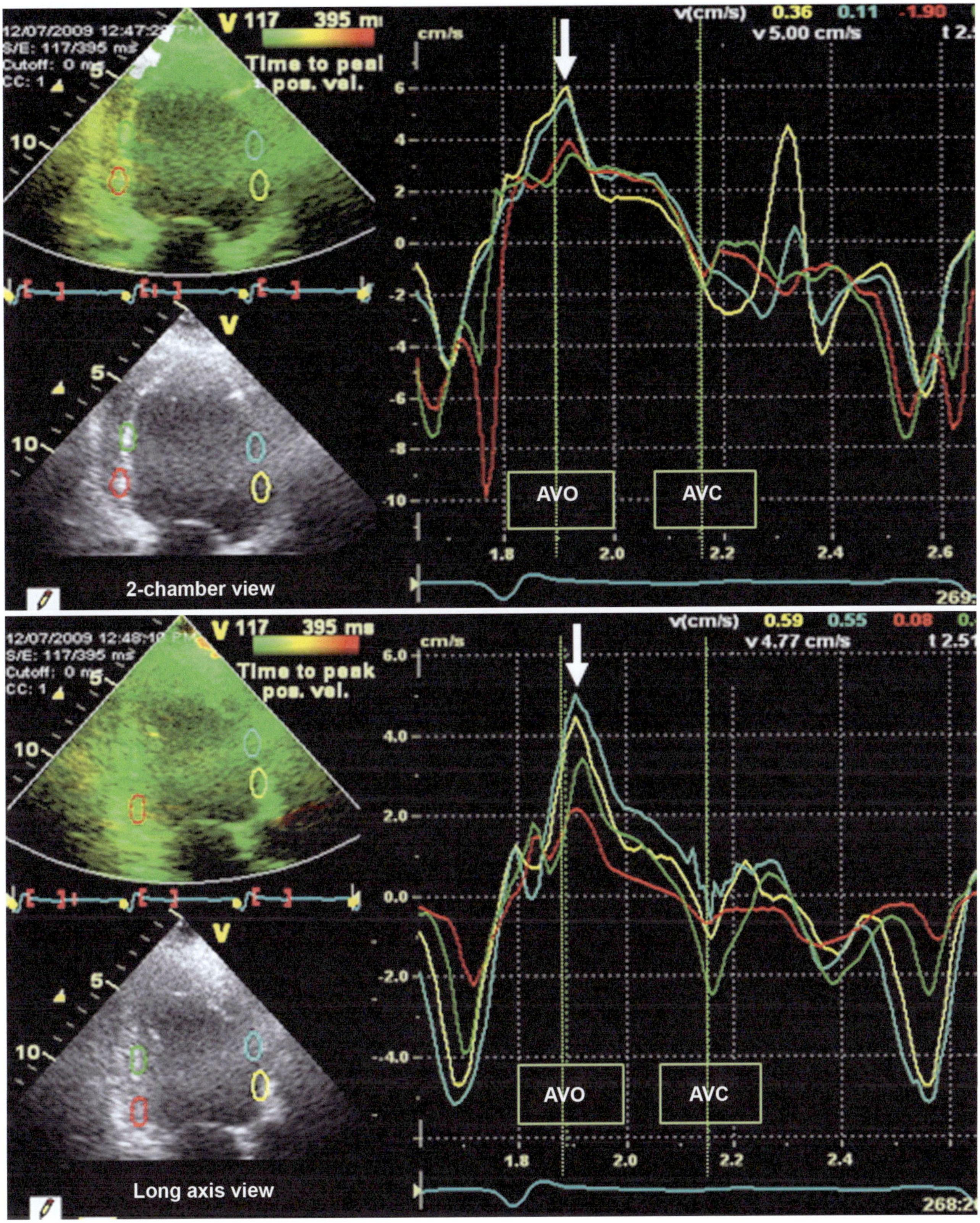

Fig. 14.6 (continued)

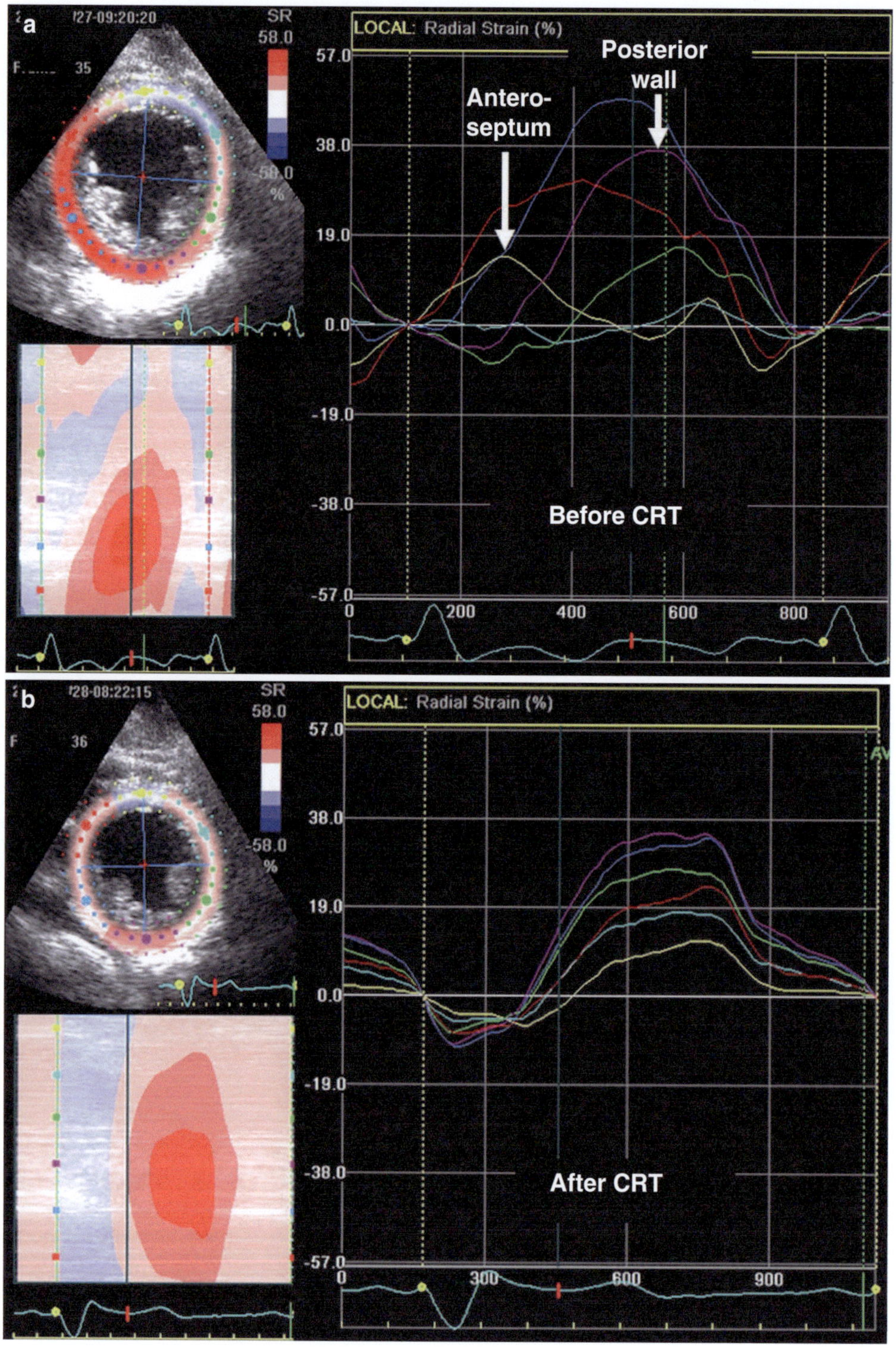

Fig. 14.7 (**a**) A mid-ventricular short-axis image with two-dimensional speckle tracking radial strain analysis in a heart failure patient with left bundle branch block before resynchronization therapy. Note the typical early contraction of the anteroseptum (*yellow line*) and delayed contraction of the posterior wall (*purple line*) and lateral wall (*green line*). (**b**) The same patient in (**a**) after biventricular pacing implantation demonstrating improvement in alignment of time-strain curves consistent with mechanical resynchronization

Formats used for strain display include three-dimensional color-coded LV strain and polar maps (Figs. 14.9 and 14.10). The initial clinical experience has demonstrated that the greatest degree of dyssynchrony in patients with wide QRS duration may be observed in the basal and mid levels of the left ventricle.[22] Furthermore, variations in site of latest activation may be observed in individual patients with LBBB or similar QRS duration, which suggests the potential for using this technology to assist with guiding optimal LV lead positioning.

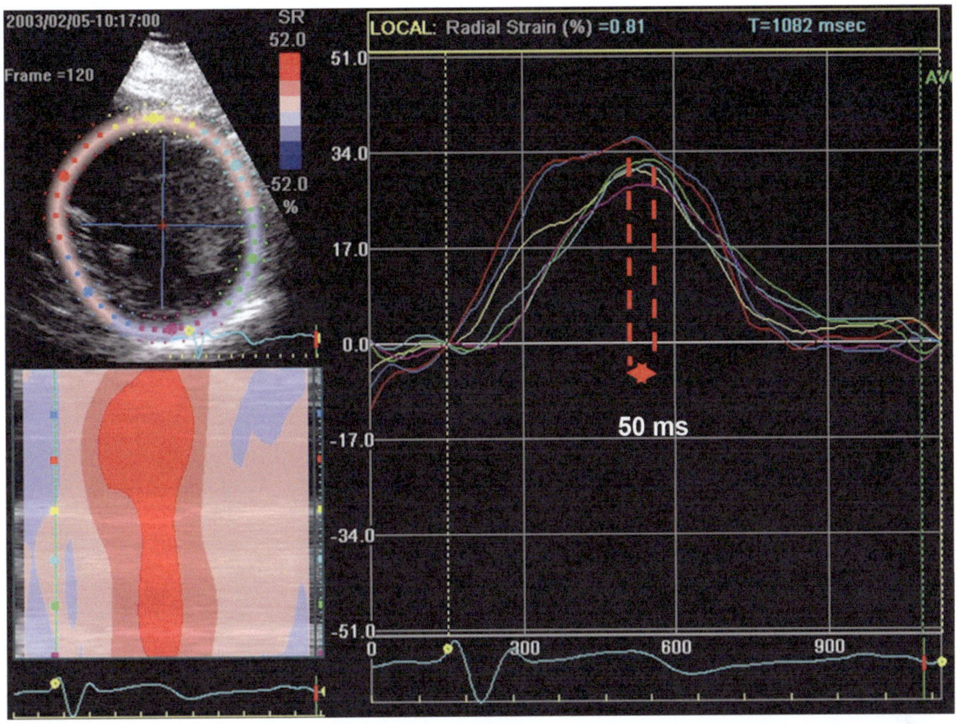

Fig. 14.8 A mid-ventricular short-axis image from a patient with left bundle branch block but paradoxically no significant dyssynchrony; 50 ms delay from anteroseptum to posterior wall. This patient was a non-responder to cardiac resynchronization therapy

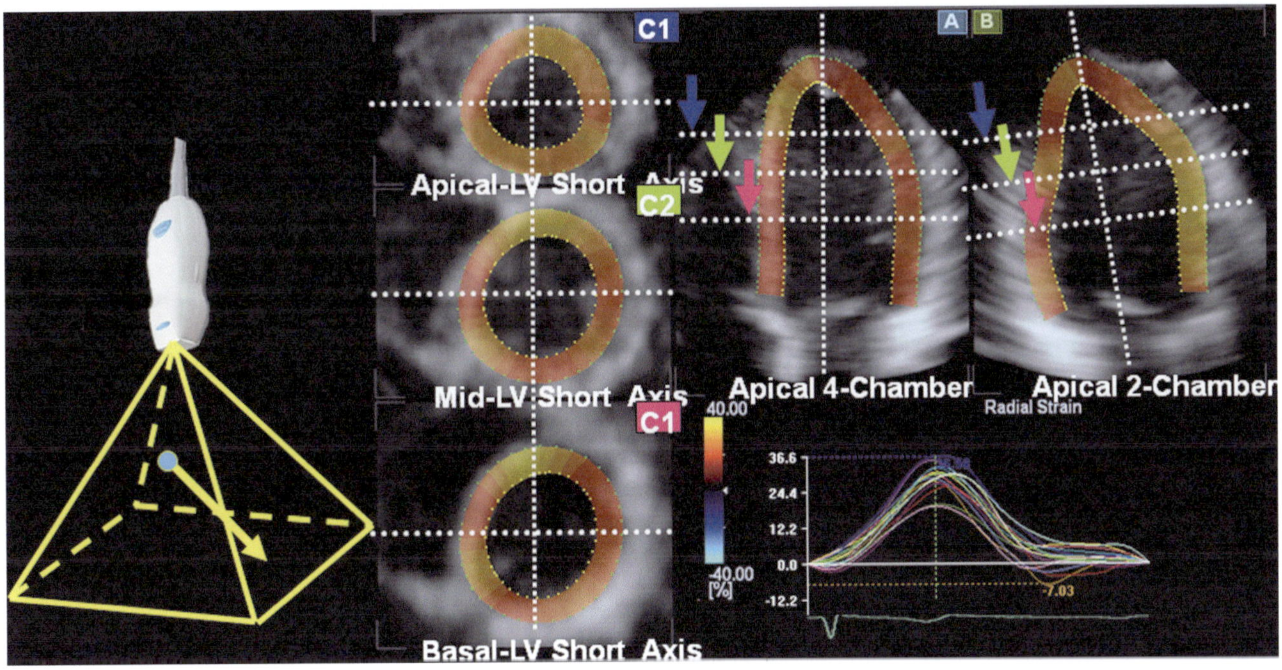

Fig. 14.9 Pyramidal ultrasound sector acquisition for three-dimensional speckle tracking analysis in a normal subject. Note similar alignment of segmental time-strain curves in *bottom right panel*

14.5 Dyssynchrony and Response to Resynchronization Therapy

When evaluating critically markers that predict response to CRT, the definitions of response have varied in the existing scientific literature. Clinical responses have included improvements in heart failure class, 6 min walk distance, quality of life, and a clinical composite score. The difficulties in utilizing these assessments are that there is inherent subjectivity to all of these measures, and the sample size must be sufficiently large to overcome the placebo effect associated with all clinical heart failure therapeutic trials. For example,

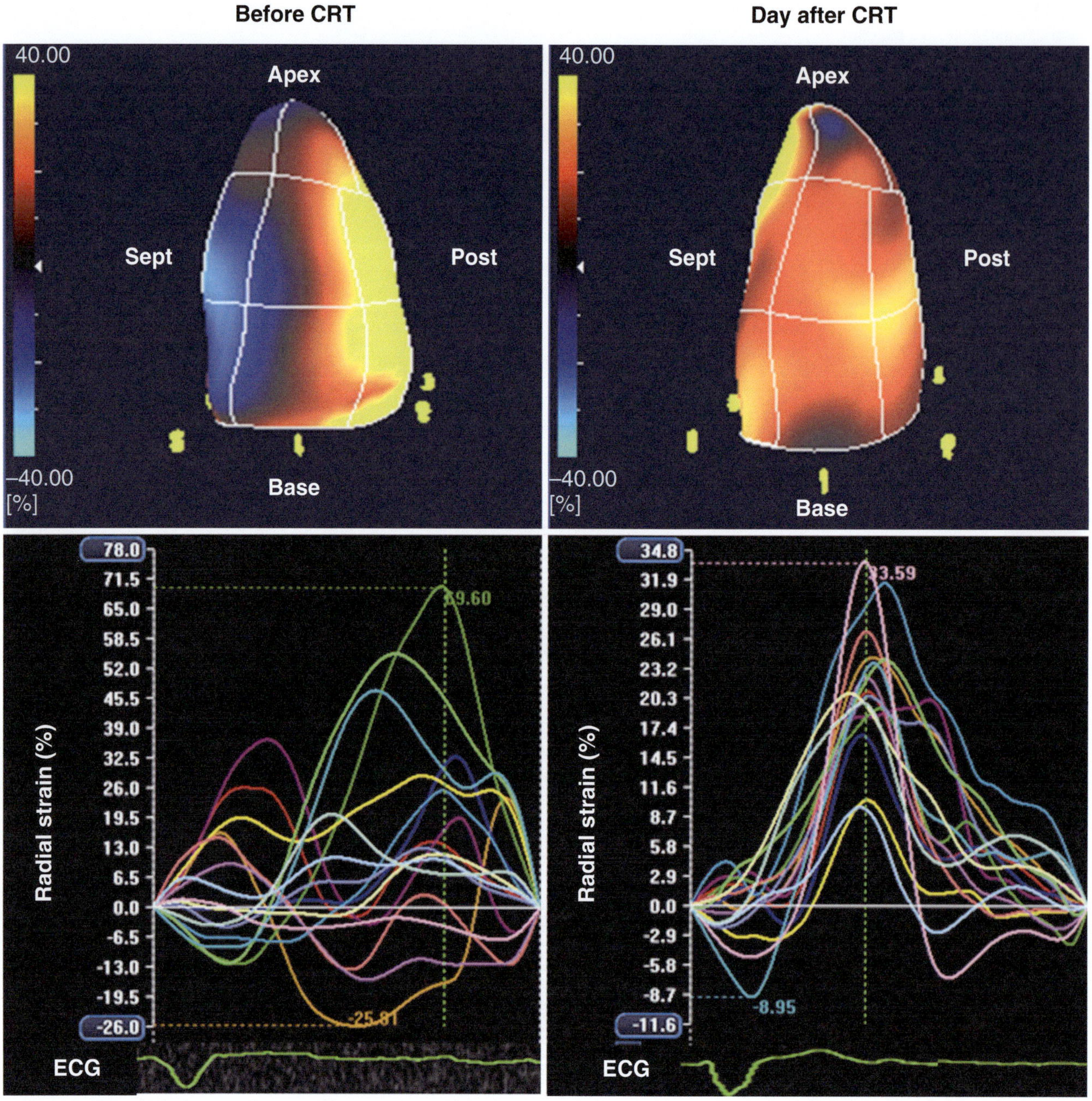

Fig. 14.10 Three-dimensional speckle tracking of a heart failure patients with left bundle branch block and dyssynchrony before (*left*) and the day after (*right*) cardiac resynchronization therapy (*CRT*). Note early septal and late posterior mechanical activation before and improvement in mechanical synchrony after therapy

the sample size was likely too small to overcome the limitations with the clinical end-point of the composite score in the PROSPECT study.[13] Alternate widely utilized markers of response to CRT have been LV functional or LV reverse remodeling measures, such as relative or absolute increase in EF or relative decreases in end-systolic volume. These measures have been useful because they are more objective than clinical assessments alone. Dyssynchrony measures of opposing wall delay ≥65 ms, alone or more powerfully in combination with speckle tracking radial strain anteroseptal to posterior wall delay >130 ms were associated with a relative EF response to CRT (Fig. 14.11).[9] In particular, the absence of both significant longitudinal and radial dyssynchrony was predictive of a lack of EF response to CRT. The lack of reverse remodeling determined by at least a 10% reduction in end-systolic volume following CRT has been associated with a poor prognosis. More recently, echocardiographic dyssynchrony measures have been shown to be associated with long-term survival. Patients with dyssynchrony characterized by aYu Index with a cutoff of≥32 ms

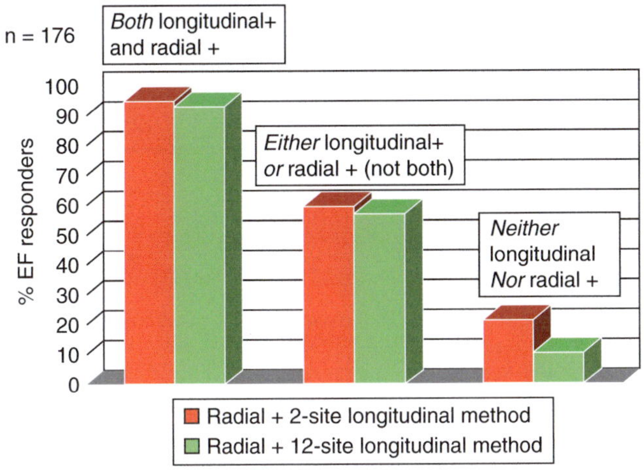

Fig. 14.11 Results of a combined dyssynchrony assessment of longitudinal velocity by tissue Doppler and radial strain by speckle tracking. When longitudinal and radial dyssynchrony results were similar, ejection fraction response (*EF*) to cardiac resynchronization therapy was predicted

or speckle tracking radial strain anteroseptal to posterior wall delay ≥ 130 ms were both associated with a more favorable survival free from heart transplant or ventricular assist device implantation after CRT (Fig. 14.12).[23] The clinical implications of these observations is not entirely clear, but they add evidence to the notion that echocardiographic dyssynchrony is predictive of patient outcome following CRT.

14.6 Borderline QRS Duration

Current clinical criteria for CRT include patients with New York Heart Association functional class III or IV (on optimal pharmacological therapy), EF ≤ 35%, and QRS duration of ≥ 120 ms. The existing evidence does not currently support the use of echocardiographic dyssynchrony information to withhold CRT to patients who meet these criteria. However, clinical scenarios may exist where patients are considered as borderline candidates, usually because of a borderline QRS duration, where echocardiographic dyssynchrony information may be used as an adjunct to assist with clinical decision making. The CARE-HF trial, which demonstrated a significant survival benefit to CRT, required patients with QRS duration of 120–150 ms to have two out of three echocardiographic dyssynchrony indices to be present: IVMD > 40 ms, pre-ejection delay > 140 ms, and/or a posterior lateral wall delay.[3] These data indirectly support that potential utility of echocardiographic dyssynchrony for patient selection for CRT with lesser degrees of QRS widening. More recent data have shown that patients with QRS duration between 100 and 130 ms have less prevalent dyssynchrony as compared to those with wide QRS > 130 ms.[24] In a series of 201 CRT patients, a smaller proportion of borderline QRS patients (53%) were EF responders compared with 75% with widened QRS ($p < 0.05$). Differences were observed in the ability of dyssynchrony to predict EF response to CRT.[24] Specifically, IVMD ≥ 40 ms and opposing wall delay ≥ 65 ms were predictive of EF response in the wide QRS group, but not the borderline QRS group. Speckle tracking radial dyssynchrony ≥ 130 ms, however, was predictive of EF response in both wide QRS patients and borderline QRS patients and associated with reverse remodeling demonstrated by significant reductions in end-systolic volume. (Fig. 14.13). These data suggest that speckle tracking radial strain may be useful to assist with patient selection for CRT in those individuals who are otherwise considered borderline candidates.

14.7 Future Applications of Echocardiographic Dyssynchrony

Potential future clinical applications of echocardiographic dyssynchrony information include guidance of LV lead positioning, discussed elsewhere in detail, and selection of CRT for patients with narrow QRS duration who have mechanical dyssynchrony. Previous single-center studies have shown that a subset of heart failure patients may have mechanical dyssynchrony with QRS duration < 120 ms, and have suggested that these patients may benefit from CRT. (Fig. 14.14) The first and presently the only randomized clinical trial of CRT in patients with narrow QRS and echocardiographic dyssynchrony was the RethinQ trial.[25] Patients were selected for randomization if they had New York Heart Association Class III heart failure, EF ≤ 35%, QRS <130 ms, and evidence of mechanical dyssynchrony. Dyssynchrony in RethinQ was defined as a tissue Doppler septal to lateral wall cutoff of ≥ 65 ms from either apical 4-chamber views or apical long-axis views, or M-mode septal to posterior wall delay ≥ 130 ms. More than 90% of randomized patients had dyssynchrony by tissue Doppler criteria.[25] This trial did not show a therapeutic effect of CRT on the primary end-point of peak myocardial oxygen consumption at 6 months. However, positive effects of CRT were observed on the secondary end-points of improvement in NYHA functional class, 6 min walk distance in the non-ischemic subgroup, and peak myocardial oxygen consumption in a pre-specified subgroup of patients with borderline QRS duration between 120 and 130 ms. Although RethinQ was reported as a negative randomized trial, the positive aspects of this trial have led other investigators to interpret these results as inconclusive. It is unclear presently if the type or degree of dyssynchrony may be refined in this narrow QRS population to predict response to CRT. Larger randomized clinical trials are underway to explore more definitively if patients with mechanical dyssynchrony and narrow QRS may benefit from CRT and the potential pivotal role that echocardiography will play in their selection for therapy.

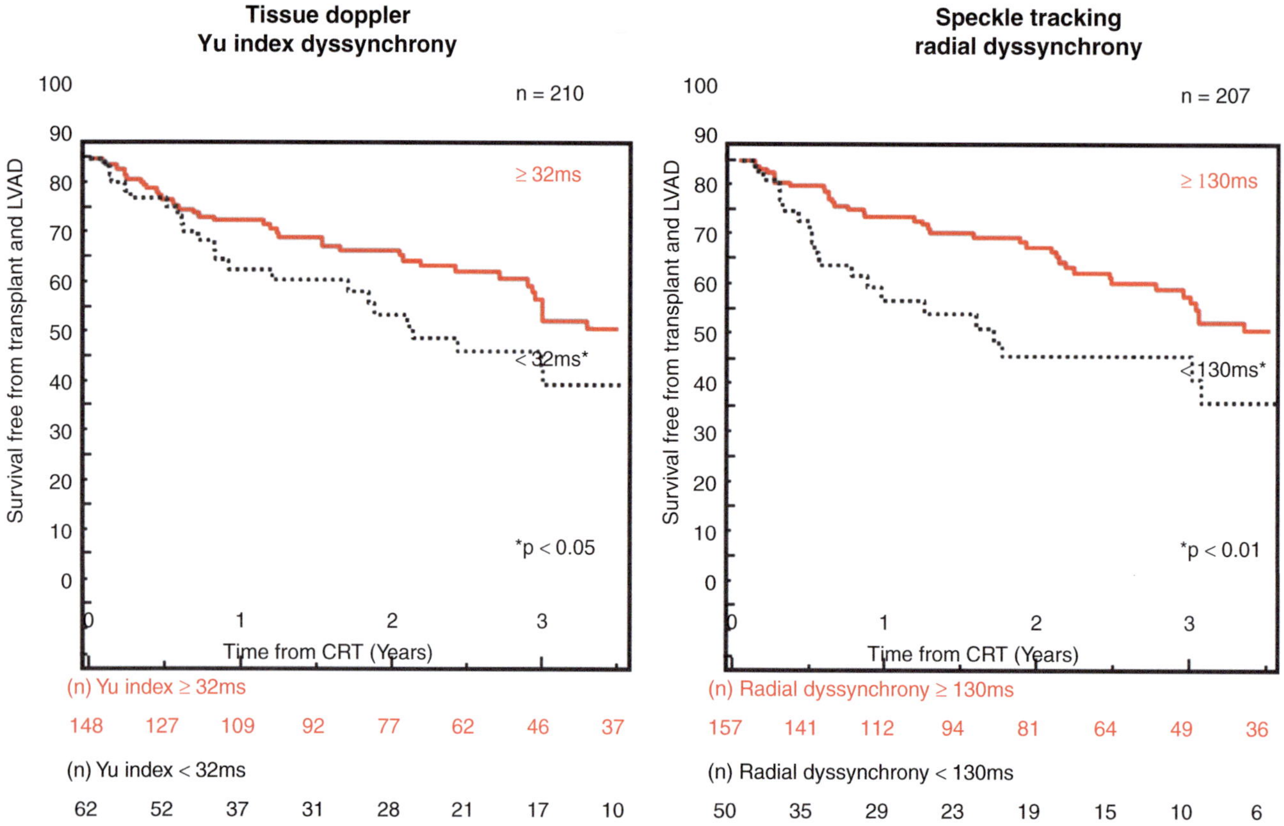

Fig. 14.12 Kaplin–Meier plots of survival free from heart transplantation or left ventricular assist device (*LVAD*) implantation following cardiac resynchronization therapy (*CRT*). Patients with significant dyssynchrony by tissue Doppler Yu Index ≥ 32 ms or speckle tracking radial dyssynchrony ≥ 130 ms had significantly more favorable event-free survival following CRT

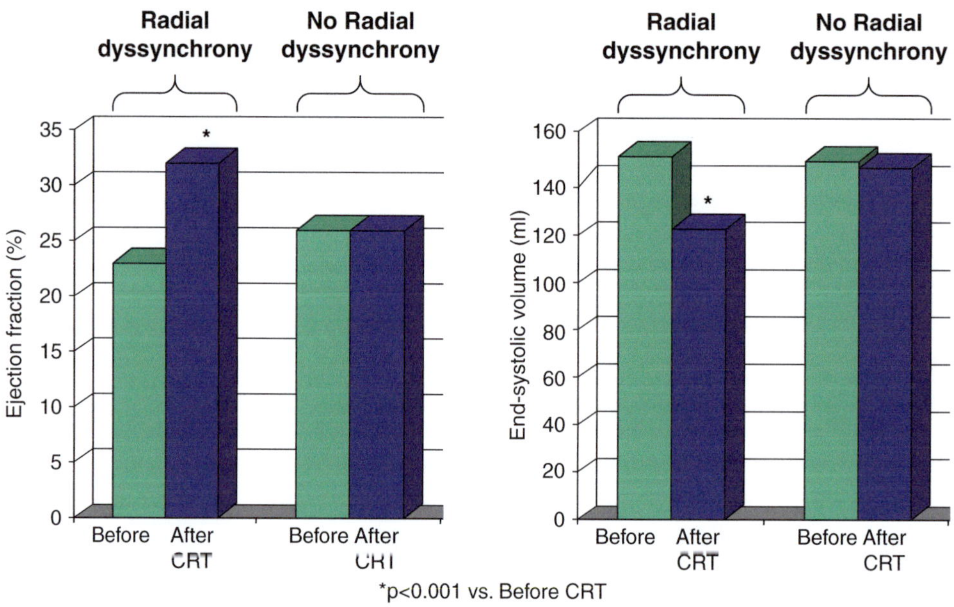

Fig. 14.13 Bar graphs of group mean data from 72 patients with borderline QRS duration (100–130 ms) demonstrating ejection fraction and end-systolic volume responses to cardiac resynchronization therapy (*CRT*). The 36 patients with significant radial dyssynchrony ≥ 130 ms by speckle tracking had a more favorable response than the 35 without significant radial dyssynchrony

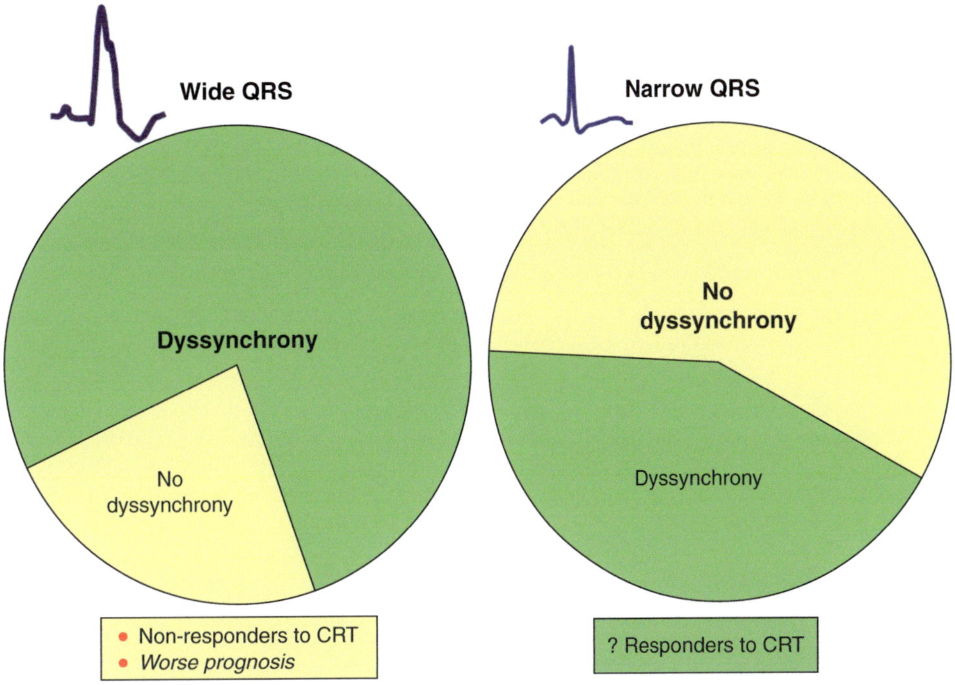

Fig. 14.14 Pie graphs illustrating the relationship of QRS width, estimated prevalence of dyssynchrony, and clinical utility of dyssynchrony analysis

References

1. Abraham WT, Fisher WG, Smith AL, et al. Cardiac resynchronization in chronic heart failure. *N Engl J Med*. 2002;346:1845-1853.
2. Bristow MR, Saxon LA, Boehmer J, et al. Cardiac-resynchronization therapy with or without an implantable defibrillator in advanced chronic heart failure. *N Engl J Med*. 2004;350:2140-2150.
3. Cleland JG, Daubert JC, Erdmann E, et al. The effect of cardiac resynchronization on morbidity and mortality in heart failure. *N Engl J Med*. 2005;52:1539-1549.
4. Bax JJ, Abraham T, Barold SS, et al. Cardiac resynchronization therapy: part 1 – issues before device implantation. *J Am Coll Cardiol*. 2005;46:2153-2167.
5. Bax JJ, Bleeker GB, Marwick TH, et al. Left ventricular dyssynchrony predicts response and prognosis after cardiac resynchronization therapy. *J Am Coll Cardiol*. 2004;44:1834-1840.
6. Bilchick KC, Dimaano V, Wu KC, et al. Cardiac magnetic resonance assessment of dyssynchrony and myocardial scar predicts function class improvement following cardiac resynchronization therapy. *JACC Cardiovasc Imaging*. 2008;1:561-568.
7. Gorcsan J 3rd, Abraham T, Agler DA, et al. Echocardiography for cardiac resynchronization therapy: recommendations for performance and reporting – a report from the American Society of Echocardiography Dyssynchrony Writing Group endorsed by the Heart Rhythm Society. *J Am Soc Echocardiogr*. 2008;21:191-213.
8. Yu CM, Wing-Hong Fung J, Zhang Q, Sanderson JE. Understanding nonresponders of cardiac resynchronization therapy – current and future perspectives. *J Cardiovasc Electrophysiol*. 2005;16:1117-1124.
9. Gorcsan J 3rd, Tanabe M, Bleeker GB, et al. Combined longitudinal and radial dyssynchrony predicts ventricular response after resynchronization therapy. *J Am Coll Cardiol*. 2007;50:1476-1483.
10. Sogaard P, Egeblad H, Kim WY, et al. Tissue Doppler imaging predicts improved systolic performance and reversed left ventricular remodeling during long-term cardiac resynchronization therapy. *J Am Coll Cardiol*. 2002;40:723-730.
11. Suffoletto MS, Dohi K, Cannesson M, Saba S, Gorcsan J 3rd. Novel speckle-tracking radial strain from routine black-and-white echocardiographic images to quantify dyssynchrony and predict response to cardiac resynchronization therapy. *Circulation*. 2006;113:960-968.
12. Yu CM, Fung WH, Lin H, Zhang Q, Sanderson JE, Lau CP. Predictors of left ventricular reverse remodeling after cardiac resynchronization therapy for heart failure secondary to idiopathic dilated or ischemic cardiomyopathy. *Am J Cardiol*. 2003;91:684-688.
13. Chung ES, Leon AR, Tavazzi L, et al. Results of the predictors of response to CRT (PROSPECT) trial. *Circulation*. 2008;117:2608-2616.
14. Adelstein EC, Saba S. Scar burden by myocardial perfusion imaging predicts echocardiographic response to cardiac resynchronization therapy in ischemic cardiomyopathy. *Am Heart J*. 2007;153:105-112.
15. Bleeker GB, Kaandorp TA, Lamb HJ, et al. Effect of posterolateral scar tissue on clinical and echocardiographic improvement after cardiac resynchronization therapy. *Circulation*. 2006;113:969-976.
16. Ypenburg C, Sieders A, Bleeker GB, et al. Myocardial contractile reserve predicts improvement in left ventricular function after cardiac resynchronization therapy. *Am Heart J*. 2007;154:1160-1165.
17. Cazeau S, Gras D, Lazarus A, Ritter P, Mugica J. Multisite stimulation for correction of cardiac asynchrony. *Heart*. 2000;84:579-581.
18. Achilli A, Peraldo C, Sassara M, et al. Prediction of response to cardiac resynchronization therapy: the selection of candidates for CRT (SCART) study. *Pacing Clin Electrophysiol*. 2006;29(Suppl 2):S11-S19.
19. Richardson M, Freemantle N, Calvert MJ, Cleland JG, Tavazzi L. Predictors and treatment response with cardiac resynchronization therapy in patients with heart failure characterized by dyssynchrony: a pre-defined analysis from the CARE-HF trial. *Eur Heart J*. 2007;28:1827-1834.
20. Yu CM, Chau E, Sanderson JE, et al. Tissue Doppler echocardiographic evidence of reverse remodeling and improved synchronicity by simultaneously delaying regional contraction after biventricular pacing therapy in heart failure. *Circulation*. 2002;105:438-445.
21. Tanaka H, Hara H, Saba S, Gorcsan J 3rd. Prediction of response to cardiac resynchronization therapy by speckle tracking

echocardiography using different software approaches. *J Am Soc Echocardiogr*. 2009;22:677-684.

22. Tanaka H, Hara H, Samir S, Gorcsan J. Usefulness of three-dimensional speckle tracking strain to quantify dyssynchrony and the site of latest mechanical activation. *Am J Cardiol*. 2010;105:235-242.
23. Gorcsan J, Oyenuga O, Habib PJ, Tanaka H, Adelstein EC, Hara H, McNamara DM, Saba S. Relationship of echocardiographic dyssynchrony to long-term survival after cardiac resynchronization therapy. *Circulation*. 2010;122:1910–1918.
24. Oyenuga O, Hara H, Tanaka H, et al. Usefulness of echocardiographic dyssynchrony in patients with borderline QRS duration to assist with selection for cardiac resynchronization therapy. *JACC Cardiovasc Imaging*. 2010;3(2):132-140.
25. Beshai JF, Grimm RA, Nagueh SF, et al. Cardiac-resynchronization therapy in heart failure with narrow QRS complexes. *N Engl J Med*. 2007;357:2461-2471.

Dyssynchrony Evaluation: MRI and CCT

15

Kenneth Bilchick, Eliot McVeigh, and Albert Lardo

Abstract

Cardiac resynchronization therapy (CRT) has been shown to improve outcomes in clinical trials for appropriate selected patients with heart failure; however, there is still a significant nonresponse rate when the electrocardiographic QRS duration is used to identify mechanical dyssynchrony. From a physiological standpoint, the appropriate cardiac substrate for CRT is most appropriately identified by assessment of mechanical dyssynchrony and myocardial scar burden, but QRS duration has poor sensitivity and specificity in this regard. For this reason, cardiac imaging has been evaluated in order to improve CRT candidate selection, but results using echocardiography have been disappointing. In contrast, clinical studies show great promise for CMR in this regard. As the gold standard for assessment of cardiac scar and strain, cardiac magnetic resonance (CMR) offers clear advantages over other imaging modalities for assessment of the cardiac substrate for resynchronization. In particular, CMR is regarded as the best imaging modality for assessment of circumferential strain, which corresponds to the primary orientation of cardiac myofibers. This chapter reviews important principles of CMR dyssynchrony and scar imaging, then discusses in detail clinical and technical aspects of myocardial tissue tagging, displacement encoding with stimulated echoes (DENSE), strain-encoded MR (SENC), velocity-encoded MR, and contour tracking methods. The role of CMR in patients with pacemakers and defibrillators is also discussed. The potential role of cardiac CT (CCT) for dyssynchrony evaluation is also discussed. Cardiac CT provides excellent assessment of coronary venous anatomy, while methods for assessment of dyssynchrony and scar based on CCT are being developed.

Keywords

Cardiac dyssynchrony evaluation • Dyssynchrony evaluation • MRI in dyssynchrony evaluation • CCT in dyssynchrony evaluation

15.1 Rationale for CMR and CCT for Cardiac Dyssynchrony Evaluation

15.1.1 Epidemiology of Heart Failure and Dyssynchrony

Nearly five million Americans have heart failure[1]; more than 500,000 are diagnosed each year[2] and 2.5 million require hospitalization for their disease.[3] Overall 5-year survival in patients diagnosed with heart failure is approximately 25% in men and 38% in women,[4] with an estimated annual mortality

A. Lardo (✉)
Department of Medicine and Biomedical Engineering, Heart and Vascular Institute, Johns Hopkins University, Baltimore, MD, USA
e-mail: al@jhmi.edu

A. Auricchio et al. (eds.), *Cardiac Imaging in Electrophysiology*,
DOI 10.1007/978-1-84882-486-7_15, © Springer-Verlag London Limited 2012

rate of 300,000.[2] In addition to the traditional predictors of mortality, including age, systolic dysfunction, and functional class, QRS duration (QRSd) has an incremental prognostic value, with a 3-year mortality risk of 20%, 36%, and 58% in patients with a QRSd of less than 120 ms, between 120 and 160 ms, and greater than 160 ms, respectively,[5] with left bundle branch block (LBBB) carrying a worse prognosis than right bundle branch block (RBBB).[6]

15.1.2 Definitions: Electrical and Mechanical Dyssynchrony

Electrical dyssynchrony merely refers to a prolonged conduction time in the ventricles resulting in a prolonged QRS duration. LV electrical conduction proceeds from the AV node and His bundle to the right and left bundle branches. A prolonged QRS duration may be due to some combination of bundle branch block (LBBB or RBBB) and intramyocardial conduction delay (myocyte-to-myocyte conduction). Mechanical dyssynchrony refers to mechanical discoordination in the hearts of patients with heart disease, usually associated with simultaneous contraction and stretch in different regions of the left ventricle, as well as delays in time to peak contraction from one region of the left ventricle to another. Mechanical dyssynchrony is more important than electrical dyssynchrony for identification of appropriate CRT candidates because it is the mechanical abnormality rather than the electrical abnormality that causes symptoms, and CRT corrects the mechanical abnormality.

Typical hemodynamic effects of CRT, generated by simultaneous right ventricular (RV) and left ventricular (LV) free wall stimulation (biventricular pacing) with a shortened atrioventricular (AV) delay, on heart failure with LBBB have been described.[7] Clinical trials have established the benefit of CRT. For example, in the MIRACLE,[8] COMPANION,[9] CARE-HF trials,[10] and other trials, CRT resulted in improvements in morbidity and/or mortality in patients with New York Heart Association class III-IV heart failure and a prolonged QRSd.

15.1.3 The Clinical Imperative for Accurate Imaging in CRT Candidates

Despite the impressive results reported by the clinical trials, there remains a significant nonresponse rate (approximately 30–35%) if patients are selected by current guideline-recommended non imaging based clinical criteria, which are: (1) ejection fraction 35% or less; (2) class III or IV heart failure symptoms; and (3) QRSd greater than or equal to 120 ms or frequent dependence on ventricular pacing.[11] Echocardiography-based criteria for CRT selection designed to improve this response rate have looked promising in small single-center studies, but unfortunately the recent multicenter blinded PROSPECT[12] and RethinQ[13] trials showed that echocardiography adds little predictive value for CRT response, highlighting the need for alternative imaging modalities such as cardiac magnetic resonance imaging (CMR) and cardiac CT to identify CRT responders.

15.2 Cardiac Strain Assessment for Mechanical Dyssynchrony

15.2.1 Mechanical Dyssynchrony

Mechanical dyssynchrony within the left ventricles is termed intraventricular dyssynchrony, while mechanical dyssynchrony between the ventricles is termed interventricular dyssynchrony. Although interventricular delay has been used to help set intervals for CRT, recent studies have not found a significant relationship between its basal value and chronic clinical CRT response.[14] Instead, recent studies for CRT selection have been based almost entirely on evaluation of intraventricular dyssynchrony, which is illustrated in Fig. 15.1 and the accompanying movie (Movie 1 in Chap. 15 folder on Springer Extras) based on a now classic MR-tagging-based mechanical activation study.[15] With LBBB-induced LV dyssynchrony, the LV septum often shortens up to 10% prior to ejection, has minimal subsequent systolic shortening, and undergoes late systolic stretch. The lateral wall is pre-stretched up to 15% in early systole and then undergoes systolic shortening. Of note, a distinction may also be made between timing-based dyssynchrony, in which regional timing delays in peak velocity or strain are measured, and discoordination-based dyssynchrony, in which the extent of simultaneous regional stress and contraction is measured.

15.2.2 Strain Versus Velocity

One challenge in imaging patients for CRT has been quantitatively characterizing this mechanical dyssynchrony. Although cardiac motion can be indexed by tissue velocity, tissue velocity is influenced not only by myocardial contraction but also by the rigid body translation and rotation of the heart. Although tissue displacement and velocity describe myocardial motion, they are influenced by myocardial contraction as well as the rigid body translation and rotation of the heart and by tethering to other tissue. Contractile function alone is more precisely quantified by strain, a tensor quantity that accounts for the complete three-dimensional complex deformation of the myocardium, quantifies the change in shape of regions of the myocardium, and directly assesses myocardial contraction and thickening. Strain may

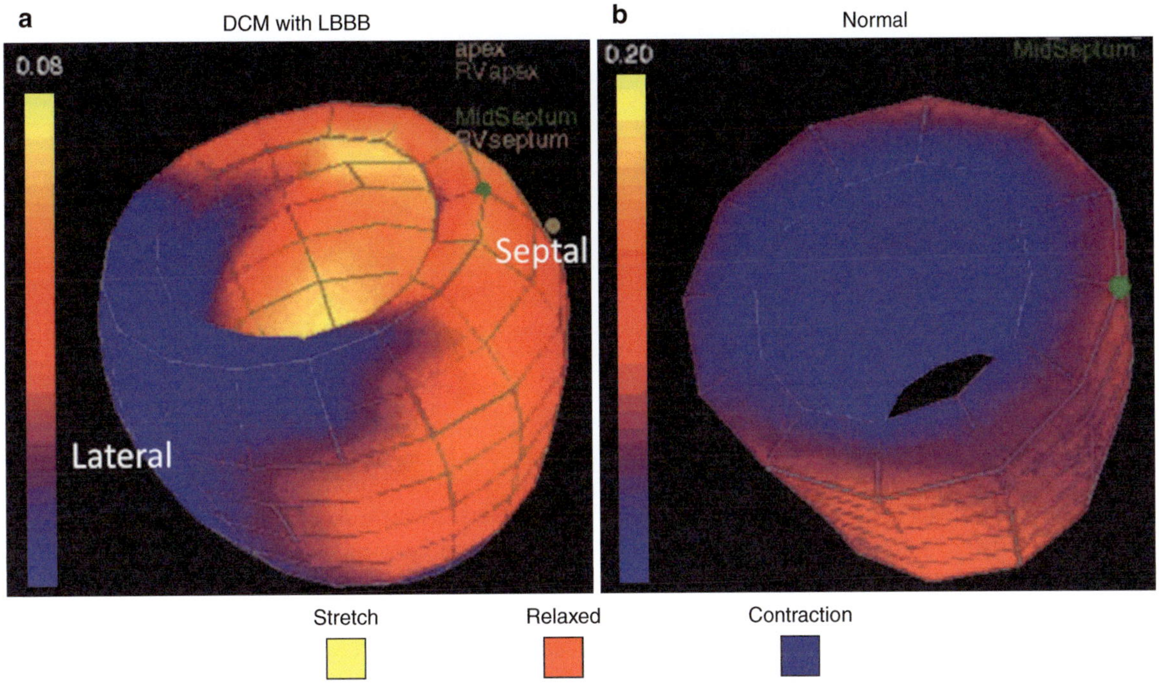

Fig. 15.1 Mechanical dyssynchrony in cardiomyopathy with LBBB – CMR strain data depicting contraction (*blue*) and stretch (*yellow*) obtained from a canine with heart failure and LBBB (**a**) compared to a normal canine (**b**). (**a**) Shows the late lateral wall contraction with simultaneous septal stretch characteristic of heart failure with LBBB. See also Movie 1 on Springer Extras. (From Curry et al.,[15] with permission)

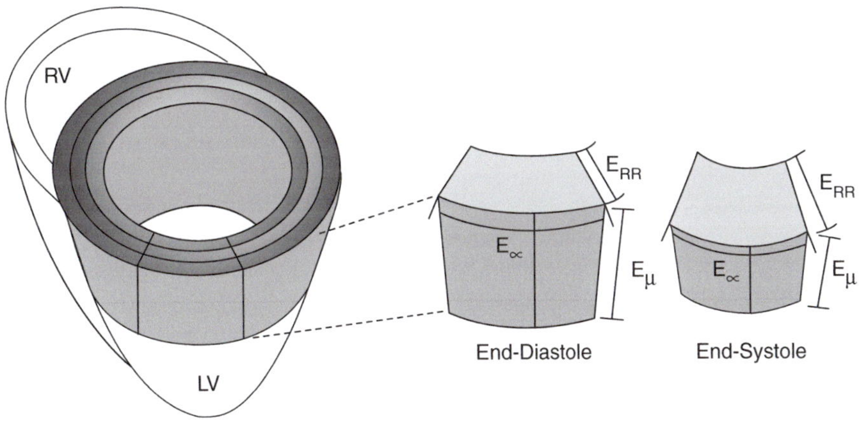

Fig. 15.2 Cardiac strain – Standard components of the strain tensor for the left ventricle are circumferential (E_{cc}), radial (E_{rr}), and longitudinal (E_{ll}) strain

be calculated as Lagrangian strain, in which material coordinates are used to characterize the deformation, or Eulerian strain, in which spatial coordinates are used. As shown in Fig. 15.2, the components of strain that are commonly used to describe the heart are E_{rr}, E_{cc}, and E_{ll}, which describe radial thickening, circumferential shortening, and longitudinal shortening, respectively. Of note, E_{rr} is thus positive during systole because the myocardium thickens in the radial direction, while E_{cc}, and E_{ll} are negative during systole because the myocardium shortens in these directions. Typical values of end-systolic E_{rr}, E_{cc}, and E_{ll} for normal volunteers are approximately 0.35, −0.20, and −0.15 in the mid-ventricle, respectively.

15.2.3 Rationale for Dyssynchrony Based on Circumferential Strain

As shown in Fig. 15.3, circumferential strain may be a superior measure for dyssynchrony compared to radial or longitudinal strain because the predominant orientation of cardiac myofibers is circumferential.[16] A study comparing LBBB HF animals before and after resynchronization found that circumferential dyssynchrony indices have twice the dynamic range and less than half the intrasubject variance than longitudinal parameters, although the longitudinal dyssynchrony parameters in this study were based on extrapolated strain values from a spline function.[17] We confirmed the superiority

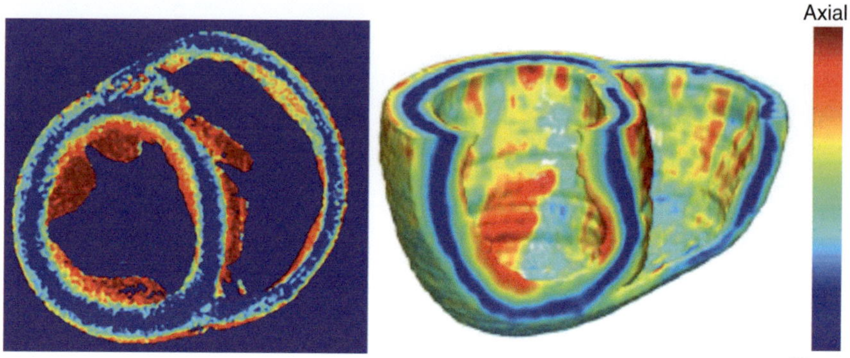

Fig. 15.3 Predominance of circumferential myofiber orientation in the heart – Diffusion magnetic resonance imaging has shown that circumferentially oriented cardiac myofibers are dominant. (From Helm et al.,[16] with permission)

of circumferential strain in another study with canines having heart failure with or without LBBB.

Circumferential contraction directly results in radial inward myocardial motion, and therefore, radial strain could presumably be a surrogate measure of circumferential strain. Yet it is known that regional segmental radial strain is subject to some degree of tethering, such that radial strain in a given segment is somewhat dependent on radial strain in adjacent segments. This suggests another potential advantage for circumferential strain, as it does not have this limitation. In a clinical study, circumferential strain measured with myocardial tagging and indexed by Fourier-transform–based dyssynchrony index with circumferential strain was shown to be highly predictive of functional class improvement following CRT.[18] The controversy surrounding this question is highlighted by a recent study suggesting that radial strain as measured by speckle tracking was better for indexing dyssynchrony than circumferential strain,[19] although the accuracy of speckle tracking for measuring circumferential strain has not been established

15.3 Basic MR Concepts Related to Strain and Dyssynchrony Imaging

A basic description of some key MR imaging concepts is in order prior to a discussion of the MR pulse sequences and analysis methods that have been used for assessment cardiac dyssynchrony. The following sections discuss how MR images are generated, various protocols for cardiac MR, the importance of ECG gating, how contrast between tissue types is formed, and the importance of k-space and Fourier transformation in the generation of images. These concepts are integral to an understanding of the role of MR for dyssynchrony imaging.

15.3.1 Magnetization and RF Excitation

Magnetic resonance imaging is possible because the human body is 95% water, and each water molecule contains two hydrogen atoms, or protons. The protons are really very small magnets, each with a very small magnetic field, called a magnetic moment. When an external magnetic field (B_0) is applied, the magnetic moments of the protons will precess around the axis of the magnetic field at a frequency equal to 42.6 MHz/T multiplied by B_0. With the typical 1.5 T superconducting magnet present in most scanners, protons exposed to the magnetic field will thus precess at a frequency of 64 MHz.

To produce an MR image, four different hardware components are required. First, the superconducting magnetic is necessary to produce the baseline magnetization in the cranial-caudal, or z direction. Second, a transmitting coil emits a radiofrequency (RF) pulse that generates a second, much smaller magnetic field, B_1 (on the order of 10–20 μT and oriented in an orthogonal direction) that changes the magnetization of the protons. When the RF pulse is stopped, the magnetization undergoes relaxation as it returns to the baseline orientation. Third, the changing magnetization induces an electrical current in a receiving coil, usually a phased-array coil positioned close to the body. The signal received by the coil is called the echo. Finally, three pairs of gradient coils are used to generate magnetic field gradients in the x, y, and z directions, sometimes referred to as the frequency encoding, phase encoding, and slice-select directions, or alternatively as the P, R, and S directions. These gradients help to localize the signal based on small spatial gradients of the magnetic field. This process is then repeated 128–512 times to form the image.

15.3.2 MR Pulse Sequences

MR pulse sequences are defined by the relationship between the RF pulses and the detection of the signal generated by relaxing protons, referred to as the echo. The RF pulse in the orthogonal direction causes some of the magnetic moment to be translated from the longitudinal (z) direction to the transverse (xy), or orthogonal plane, resulting in transverse magnetization. The resulting angle of the

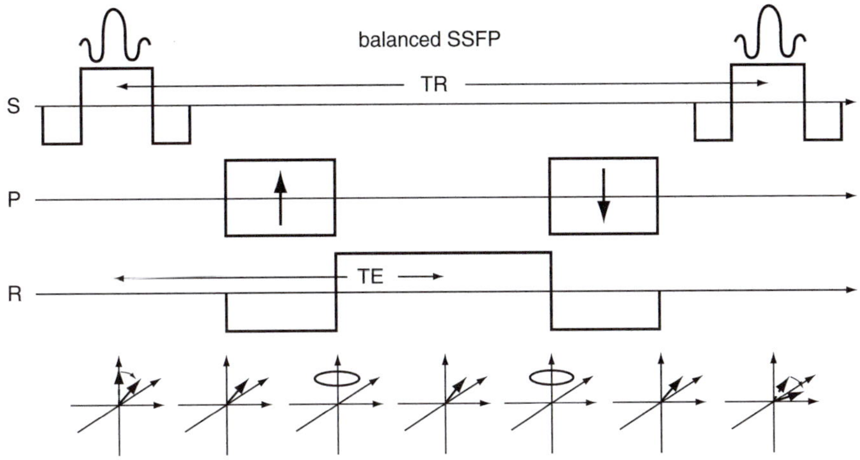

Fig. 15.4 Steady-State Free Precession (*SSFP*) pulse sequence – The pulse sequence for balanced SSFP is shown. See text for explanation of P, R, and S axes. See also Movie 2 on Springer Extras for clinical examples. (From Scheffler and Lehnhardt,[20] with permission)

magnetization is called the flip angle. The time between RF pulses is referred to as the relaxation time, or TR, while the time between the RF pulse and the echo is referred to as the echo time, or TE. As protons relax, first the coherence of the precessing magnetic moment is lost, which is called T2 relaxation. Second, the transverse magnetization is lost and longitudinal magnetization returns as protons return to their original magnetization. This is called T1 relaxation. The pulse sequence also defines the gradients used for spatial localization.

The two main types of MR images are gradient echo and spin echo images. For dyssynchrony and functional imaging, gradient echo imaging and a variant called steady-state free precession (SSFP)[20] are used for the most part. With gradient echo imaging, an RF pulse results in transverse magnetization with a flip angle less than 90° causing a coherent transverse magnetization. A gradient in the *x* direction results in accelerated, or forced, dephasing after the RF pulse, while a second *x* gradient results in accelerated rephasing. The gradient echo is then formed when the phase is maximally coherent. The process is then repeated a number of times. As a result, gradient echo imaging is fast at the cost of some signal relative to noise, and is ideal for assessment of cardiac function.

For cine functional images, SSFP or balanced SSFP is usually used and has largely replaced other types of cine gradient echo sequences (such as spoiled gradient echo sequences) at most centers. Examples of SSFP imaging in patients with and without mechanical dyssynchrony are shown in Movies 2A and 2B on Springer Extras, respectively. Figure 15.4 shows an example of a pulse sequence diagram for balanced SSFP, with *P*, *R*, and *S* axes as defined above.[20] RF pulses are applied throughout the SSFP sequence at very tight intervals (TR less than 4–5 ms with segmented k-space filling) to keep the magnetization in a steady state. SSFP offers images with high spatial resolution, as well as very good temporal resolution on the order of 40–50 ms, but is more sensitive to field inhomogeneities and artifacts.

15.3.3 Electrocardiographic Gating

Electrocardiographic (ECG) gating, the mechanism by which MR or CT data acquisition is linked to the phase of the cardiac cycle, is very important for acquisition of high-quality MR or CT images. The two main types of gating are prospective triggering and retrospective gating. With prospective triggering, MR data acquisition begins at a preset time interval following the R wave, depending on the phase of the cardiac cycle being imaged. With prospective gating, the acquisition window is usually about 90% of the R-R interval, in order to account for small beat-to-beat differences in the R-R interval. With retrospective gating, both the data and the ECG tracing are acquired continuously and throughout the entire cardiac cycle. After the acquisition has been completed, the data are then retrospectively sorted based on the echo timing relative to the R wave, such that image reconstruction is more complex than with prospective triggering. Patients with arrhythmias, particularly atrial fibrillation and frequent premature beats (which are not uncommon in patients with heart failure and dyssynchrony), present a challenge for dyssynchrony imaging. For these arrhythmias, prospective triggering or real-time sequences may improve image quality. Prospective gating can reduce radiation exposure in cardiac CT by reducing scan times.[21]

15.3.4 Myocardial Signal and Contrast Agents

MR pulse sequences may be either T1-weighted or T2-weighted. T1 relaxation times are typically much longer than T2 relaxation times. Normally myocardial tissue such as myocardium have relatively short T1 and T2 times, while fluids such as blood have relatively long T1 and T2 times. Relaxation times for fat are also distinct from blood and myocardium. These differences permit contrast of different types of tissue in the body, which may be expressed by a quantity called the CNR, or contrast-to-noise ratio.

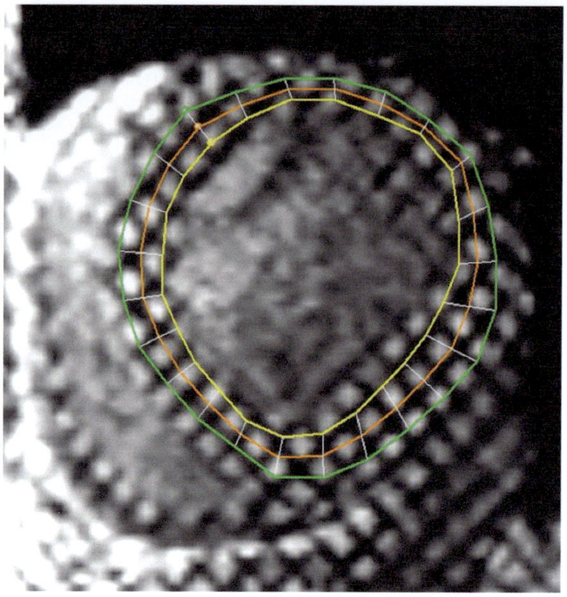

Fig. 15.5 Automatic tracking of myocardial grid tags – A typical appearance of myocardial grid tags is shown along with the result of automatic tracking. The epicardial, midmyocardial, and endomyocardial contours are represented by the colors *green*, *orange*, and *yellow*, respectively. See also Movie 3 on Springer Extras and reference Young et al.[26]

Gadolinium chelates are effective contrast agents for cardiac MR imaging. Gadolinium has the effect of shortening T1 relaxation, resulting in increased T1 signal. Because gadolinium preferentially accumulates in areas of myocardial scar, late gadolinium enhancement is typically used to image myocardial scar. As discussed later, scar imaging holds important prognostic significance of patients with heart failure being considered for CRT.

15.4 Myocardial Tissue Tagging

15.4.1 Technical Aspects

The most widely used MR tissue tracking technique is myocardial tissue tagging (MR-MT). Developed in the 1980s by Zerhouni and colleagues,[22] tagging is typically performed using breath-hold segmented ECG-gated acquisitions.[23] Like breath-hold cine MRI, tagged cine images are generally acquired using an ECG-gated, segmented method, requiring 12–16 heartbeats during suspended respiration. Spoiled gradient echo imaging has most commonly been used for signal generation; however, recent studies have demonstrated better tag contrast and reduced tag fading using SSFP.[24,25]

As shown in Fig. 15.5, radiofrequency and gradient pulses result in saturation of the tissue magnetization (tagging) in a stripe or grid pattern.[27] A movie of myocardial tagging in a patient with mechanical dyssynchrony is also provided (Movie 3 on Springer Extras). These pulses are applied with detection of the ECG R wave, which occurs before the onset of systolic contraction. After application of the tagging pulses, signal nulling in a stripe or grid pattern is evident, because no contraction has occurred between application of the tags and acquisition of the images. The figure shows that a movie of the tagged images is created by combining frames from sequential TR intervals. Although tagging pulses can be added to any imaging sequence, the sequence must provide good contrast between tagged and non-tagged myocardium with adequate temporal and spatial resolution.

Myocardial strain can then be calculated from the deformation of the stripe or grid pattern in subsequent phases because the saturated tissue moves coherently as the heart contracts. Essentially, a Fourier transformation permits conversion between the actual tagged image and k-space, with the latter containing all the frequency information used to calculate strains. The k-space generated from a Fourier transform of the image with vertical tag stripes is oriented in the horizontal direction, while the k-space orientation is vertical for the horizontal tag stripes. As time progresses during the cardiac cycle, the magnetization recovers toward equilibrium due to spin-lattice, or T1, relaxation. As a result, the saturated tissue does not retain diminished signal intensity indefinitely. Because the T1 of the heart at a magnetic field strength of 1.5 T is 850 ms, the tags remain adequately saturated throughout systole and into early diastole.

The processing and analysis of tagged MR images involves three stages.[28] First, the left ventricular myocardium must be identified on two-dimensional images. Second, the tag lines must be identified. These first two steps are accomplished through manual or semiautomatic computer-assisted detection of the epicardial border, endocardial border, and tag lines,[26] although semiautomatic techniques generally require some extent of manual correction and either technique is usually quite time consuming. Third, strain must be estimated from the tag lines and contours. Initially finite element methods, global polynomial fitting, and model-free stochastic estimation approaches were used, but these were limited by lack of automation and dependence on interpolation. With these methods, detection of tag lines[29] was time intensive and impractical for routine use. The Harmonic Phase, or HARP method, was developed at Johns Hopkins in the late 1990s to address these problems and has become the standard method for tagging anlaysis to date.

The HARP method, as illustrated in Fig. 15.6, is based on the use of SPAMM tag patterns, which modulates the underlying images and produces an array of spectral peaks through Fourier analysis containing information about tissue motion. The initial single-shot HARP image analysis technique was based on reconstructing synthetic tag lines, and calculating displacement fields and Eulerian strain for a single time-frame in the cardiac cycle. This method uses a

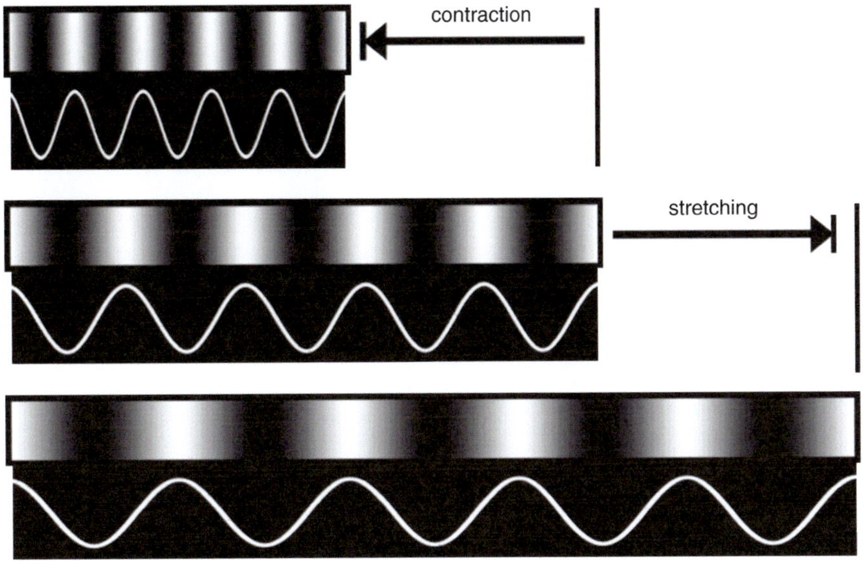

Fig. 15.6 Harmonic phase analysis – This illustration shows the relation between local strain and tag frequency. The contraction of a tagged fiber increases the tagging frequency (density of tag lines) as shown by the top fiber. Stretching causes a reduction in local frequency. See also reference Zerhouni et al.[22]

bandpass filter to isolate the lowest harmonic frequency, or HARP frequency, in a certain tag direction. Using an inverse Fourier transform of the bandpass region, a harmonic phase image containing a detailed picture of myocardial motion is created. The harmonic phase image contains a characteristic harmonic phase and HARP angle that are material properties of the tagged tissue, and this permits tracking of tagged lines as a set of points having the same harmonic phase. The net result is that Lagrangian circumferential and radial strain can be determined using this method. Tracking points on concentric circles within the myocardium yields radial strain, while tracking points along a ray intersecting the center of the left ventricle results in radial strain. Twist and torsion associated with tag deformation is performed automatically. Circumferential strain, E_{cc}, is an accurately estimated element of the strain tensor from tagged images and most commonly reported component of strain. Of note, HARP analysis applies significant filters to the raw MRI data, and the spatial resolution of strain is significantly lower than the spatial resolution of the original tagged image.[30] For some applications, this may be an important limitation.

15.4.2 MRI-Based Measures of Dyssynchrony

Myocardial tagging was employed in some very important early studies of dyssynchrony physiology. These studies demonstrated the distinction between electrical/mechanical dyssynchrony,[31] the favorable dynamic range for circumferential strain,[17] and the "sweet spot" for left ventricular pacing for maximal synchrony and cardiac function.[32] We recently reported the results of a study applying this technique to patients with heart failure referred for CRT.[18] Using the HARP method for rapid analysis of MR-MT data, we determined circumferential strain for each of 24 segments of several short-axis left ventricular slices during systole and early diastole; incorporated the data into a multi-slice, multiphase, multisegment circumferential dyssynchrony index called CURE (circumferential uniformity ratio estimate);[31] evaluated CURE for prediction of improvement in function class after CRT; and compared the results of MR-MT-based CURE to the TDI septal-lateral delay. Determination of CURE is from regional circumferential strain and is shown in Fig. 15.7. In this series, CURE predicted improvement in function class after CRT with 90–95% accuracy depending on whether the left ventricular scar burden was also included. Furthermore, TDI and CURE offered discordant dyssynchrony assessments in 30% of heart failure patients, and the TDI actually indicated dyssynchrony in 44% of normal patients, who all had a normal CURE.[18]

15.5 SENC and Fast-SENC

15.5.1 Technical Aspects

Strain-encoded magnetic resonance imaging, or SENC-MR, is a variation on myocardial tagging that was first reported in 2001.[34] SENC uses a similar SPAMM pulse sequence, but differs from myocardial tagging with respect to the orientation of the tagged surface relative to short-axis images. With traditional myocardial tagging, the planes of saturated magnetization are oriented orthogonal to the image plane; however, with SENC, the tag planes are initially oriented parallel to the imaging plane.

The chief strength of this pulse sequence is that measurement of longitudinal strain is possible from measurement of

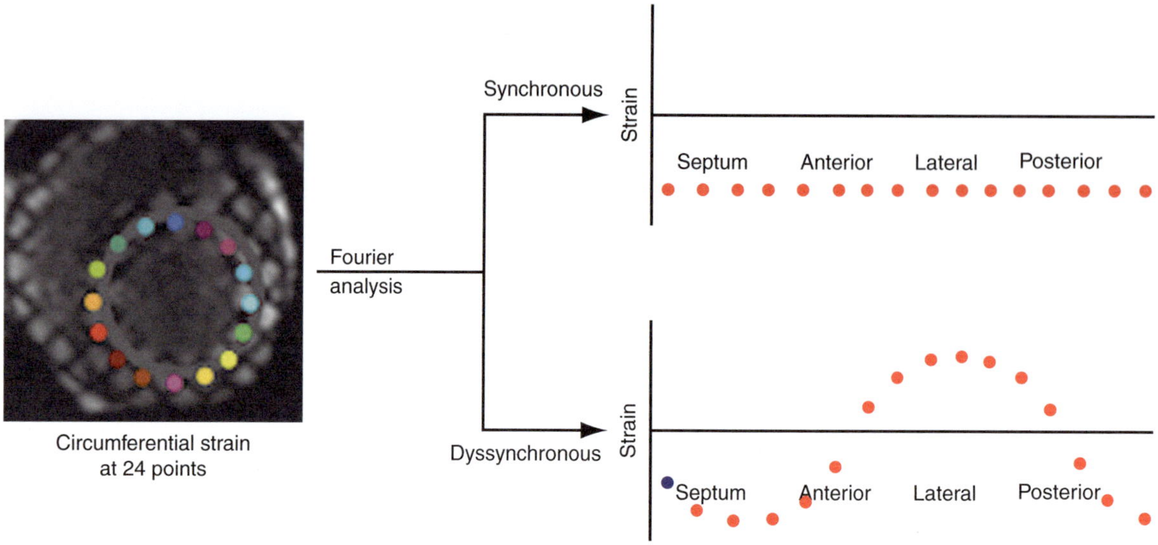

Fig. 15.7 Derivation of CURE (Circumferential Uniformity Ratio Estimate) – The derivation of CURE for circumferential mechanical dyssynchrony is shown. CURE (0–1; 1 = synchrony; 2 = dyssynchrony) is a ratio measure based on the zero- and first-order Fourier transform terms for regional strain at a specific time point based on a short-axis left ventricular slice. Simultaneous regional stretch and contraction, as is typically found in heart failure with LBBB and is evident in (**a**) of Fig. 15.1, will cause the measure to be lower. The CURE may then be averaged over multiple time points to obtain the overall measure for that slice. See also reference Bilchick et al.[33]

two short-axis left ventricular MR slices, as long as the slice thickness is several times larger than the tag period. The SENC images yielding longitudinal strain may then be combined with standard short-axis tagged images, the latter yielding circumferential and radial strain. With standard myocardial tagging, long-axis images would be required to determine longitudinal strain. Furthermore, only six regional strains per long-axis level are usually obtained with the usual three long-axis views obtained with myocardial tagging. With SENC, determination of regional longitudinal strain at any point along the circumference of the left ventricle is possible. However, the density of points along the long axis of the left ventricle for a specific circumferential location for which longitudinal strain may be determined with SENC is limited, as the slice thickness must be at least four times the tagging period. Of note, longitudinal strain may also be determined from MR short-axis images using DENSE, and this technique is currently under development.

The SENC pulse sequence employs a magnetic field (tagging) gradient oriented in the slice-selection direction and placed between the two RF pulses at end-diastole, with a crusher in place at the end to spoil all transverse spins. A tuning (demodulation) gradient is then applied during the refocusing lobe of the slice-selection gradient, which allows image acquisition at a specific tuning frequency (i.e., ω, the local frequency of the spatial sinusoidal tag pattern). Deformation of the heart throughout the cardiac cycle changes the local frequency of the tagging pattern, and this changes regional signal intensity of the images acquired using a specific tuning frequency. As with standard myocardial tagging, the tag pattern during myocardial contraction undergoes both compression and tilting, corresponding to the deformation and displacement undergone by the left ventricle. The effect on the tagging pattern may be represented by the baseline tag frequency (w) and the overall tag frequency during systole. The latter frequency has a component (v) in the z (slice-select) direction representing longitudinal strain and another component (ω) representing the local modulation frequency of the myocardium at a specific position. Two tuning frequencies (ω), both a high frequency and low frequency, must be applied. There is then a shift in v so that it is greater than the original tag frequency during contraction, but is less than the original tag frequency during relaxation. Longitudinal strain is then calculated in standard fashion based on the shift in tag frequency compared to the baseline tag frequency.

Recently, Osman has developed a modification of the SENC pulse sequence called fast-SENC, which enables scan durations as short as a single heartbeat.[33] Fast-SENC images are shown in Fig. 15.8 and Movie 4 on Springer Extras. The rationale for fast-SENC is that faster acquisition times would make the technique useful for applications such as stress testing and provide an MR technique useful for patients with limited capacity for breath-holds or arrhythmias. Fast-SENC achieves accelerated SENC images by combining three techniques. First, localized excitation helps achieve a reduced field of view (FOV) without foldover artifacts. This reduces the size of the sampled matrix in k-space, which in turn reduces

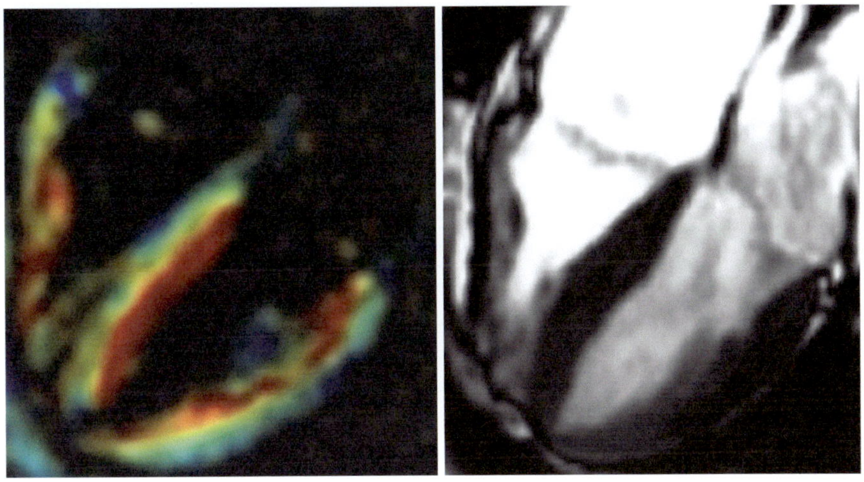

Fig. 15.8 Fast-SENC and SSFP – Typical images obtained with the fast-SENC technique are shown on the *left*, with a corresponding SSFP image on the *right*. See also Movie 4 on Springer Extras

the scan time. Second, interleaved tuning refers to the fact that the two images obtained with different tuning frequencies can be obtained in a single acquisition. Third, spiral imaging is employed for rapid acquisition. This takes advantage of the fact that most of the information from SENC images is localized in the center of K space and provides both short echo times (TEs) and reduced sensitivity to flow artifacts.

With fast-SENC, cine SENC images may be acquired in as little as 1 heartbeat, (although more cardiac cycles may be used to improve the signal-to-noise ratio), while 12 heartbeats (for a breath-hold lasting approximately 10 s) are usually used with SENC. Initial application was on a 1.5 T scanner, although implementation on a 3.0 T scanner to improve SNR has recently been described.[35] A typical frequency configuration on a 1.5 T scanner would be a high-tuning frequency of 0.4 mm^{-1} and a low-tuning frequency of 0.3 mm^{-1}. Typical inplane resolution is 4.0 $mm^2 \times 4.0$ mm^2, which is somewhat lower resolution than the 4.0 $mm^2 \times$ 4.0 mm^2 used in standard SENC.

15.5.2 Clinical Studies

In an animal study, SENC has been shown to identify regions of reduced contractility in dogs with myocardial infarction, with areas of reduced strain corresponding to regions with delayed hyperenhancement.[33] Peak systolic and diastolic strains acquired by fast-SENC performed at 3 T have been found to correlated closely with myocardial tagging in patients with heart failure and normal volunteers.[35] In this study, quantitative analysis was significantly faster with fast-SENC as compared to tagging. Another recent study of SENC at 1.5 T confirmed that characterization of regional heterogeneity of strain with SENC correlated closely with strain results obtained from myocardial tissue tagging.[36]

15.6 Velocity-Encoded MR Imaging

15.6.1 Technical Aspects

A second method for measuring intramyocardial function is velocity-encoded phase contrast MRI. This technique works by measuring instantaneous myocardial tissue velocities at successive cardiac phases throughout the cardiac cycle. Instantaneous velocity is measured by creating transverse magnetization, applying bipolar velocity-encoding gradients, and detecting phase shifts that are linearly proportional to velocity.[37] The successive instantaneous velocities can be interpreted in a manner analogous to tissue Doppler echocardiography, or can be used to estimate displacements, strains, and strain rates.[38] Advantages of this technique include pixel-wise spatial resolution and inherent quantification of velocity and strain rate without the need for tag detection. However, a disadvantage of this technique is that errors in the velocity measurements are cumulatively propagated into estimates of displacement as each volume of tissue is tracked through time.[39,40]

15.6.2 Clinical Studies

Longitudinal myocardial velocities derived from magnetic resonance velocity-encoded (VENC) MR were shown to correlate with longitudinal velocity by tissue Doppler imaging in a small study with ten normal volunteers and ten patients with dyssynchrony,[41] although the velocities with MR were noted to be consistently higher than the corresponding velocities obtained with tissue Doppler imaging. Both techniques also agreed with respect to time-to-peak velocity, and reproducibility of the MR velocity measurements appeared reasonable. Similar results were found in two other studies with about 30 patients each (composed of normal volunteers and patients with heart failure).[42,43] In these studies, VENC-MR results for peak systolic longitudinal velocity, time-to-peak velocity, and delay in time-to-peak velocities between the septal and lateral walls (SLD) agreed with those from tissue Doppler imaging. Of note, one of these studies, estimation of filling pressures as measured by E/E′, also showed significant agreement. Although these studies established agreement between tissue Doppler and VENC-MR techniques, none of them demonstrated clinical significance for CRT response. As noted earlier, the PROSPECT study showed very limited value for tissue Doppler measurements with respect to CRT response, so the fact that the VENC-MR correlated with another poorly predictive technique does not appear to be clinically important.

In another study, velocity-encoded (VENC) MR was compared with pulsed-wave echocardiography for the quantification of interventricular mechanical dyssynchrony (IVMD).[44] In this study, VENC MR was used to measure the timing between the onset of aortic and pulmonary blood flow in 45 heart failure patients with or without left bundle branch block. Interobserver and intraobserver agreements were very good for the VENC measurements, and there was a strong correlation between the two modalities for the quantification of IVMD. Although this study once again established agreement between the two modalities, clinical significance was not demonstrated. Of note, IVMD measurements have historically had limited predictability for CRT response as compared to intraventricular dyssynchrony measurements. Furthermore, as discussed above, measurement of circumferential versus longitudinal dyssynchrony may be more physiologic, and dyssynchrony parameters based on strain are preferable to those based on velocity.

15.7 SSFP Contour Tracking Methods

15.7.1 Technical Aspects

Due to the specialized expertise required for acquisition and analysis of cardiac strain with certain MR strain imaging techniques, there has been interest in performing strain measurements based on tracking of contours using steady-state free procession (SSFP) MR images. (See later discussion for further details on SSFP.) Cine white blood imaging using SSFP also has high temporal resolution with excellent definition at the myocardial/blood pool interface, permitting high-quality contour detection. As shown in Fig. 15.9 (and Movie 5 on Springer Extras), these cine images are usually acquired for any patient undergoing CMR and can be analyzed post-hoc even if strain imaging was not intended based on the original study. This technique may also facilitate strain imaging in patients with pacemakers and defibrillators, as well as other patients with artifacts present that may preclude strain analysis with other techniques. Several software packages are now available for this analysis, and validation studies are presently underway.

15.7.2 Clinical Studies

In a recently published European study, investigators performed SSFP imaging in patients with heart failure undergoing CRT.[45] A short-axis stack of approximately 8 LV slices was obtained, then radial wall motion was determined for 6 segments in each slice at 20 time points. The radial wall motion was then fitted to an empirical sine wave function to account for the cyclical nature of myocardial motion. The chosen dyssynchrony parameter was the CMR-TSI, which was calculated as the standard deviation of the segmental phase shifts for each segment, which essentially reflects the time delay to peak contraction and bears some similarity to the Ts-SD validated by Yu and colleagues. (The Ts-SD is calculated as the standard deviation in time-to-peak longitudinal velocity in six basal and six mid-cavity segments from tissue Doppler long-axis left ventricular images.) Intraobserver and interobserver variability for this technique was satisfactory. In a cohort of 77 heart failure patients, the authors found that CMR-TSI > 110 ms was an independent predictor of cardiovascular death with a hazard-ratio of 3.8 ($p<0.001$). These results run counter to the major findings in the body of the dyssynchrony literature indicating the increasing mechanical dyssynchrony is associated with a better response to resynchronization. The patients in the group with the greater CMR-TSI (increased dyssynchrony) had more severe disease, more severe left ventricular systolic dysfunction, more left ventricular dilatation, and an increased likelihood of ischemic cardiomyopathy. These factors likely account for the poorer prognosis in this group. The authors did not perform any regression analysis to account for these differences between groups, so the significance of the CMR-TSI for risk-stratification has not really been established.

Other studies comparing SSFP tissue tracking for circumferential and radial strain to myocardial tagging are presently in progress. It is possible that this technique, when studied in an appropriate population with appropriate dyssynchrony parameters, may prove to be useful in the future.

15.8 Displacement Encoding with Stimulated Echoes (DENSE)

15.8.1 Technical Aspects

Although myocardial tagging is a good technique, it has some limitations. For example, HARP analysis applies significant filters to the raw CMR data resulting in a significantly reduced spatial resolution,[30] and there are only a limited number of tags that can fit in certain walls of thin dilated hearts. The

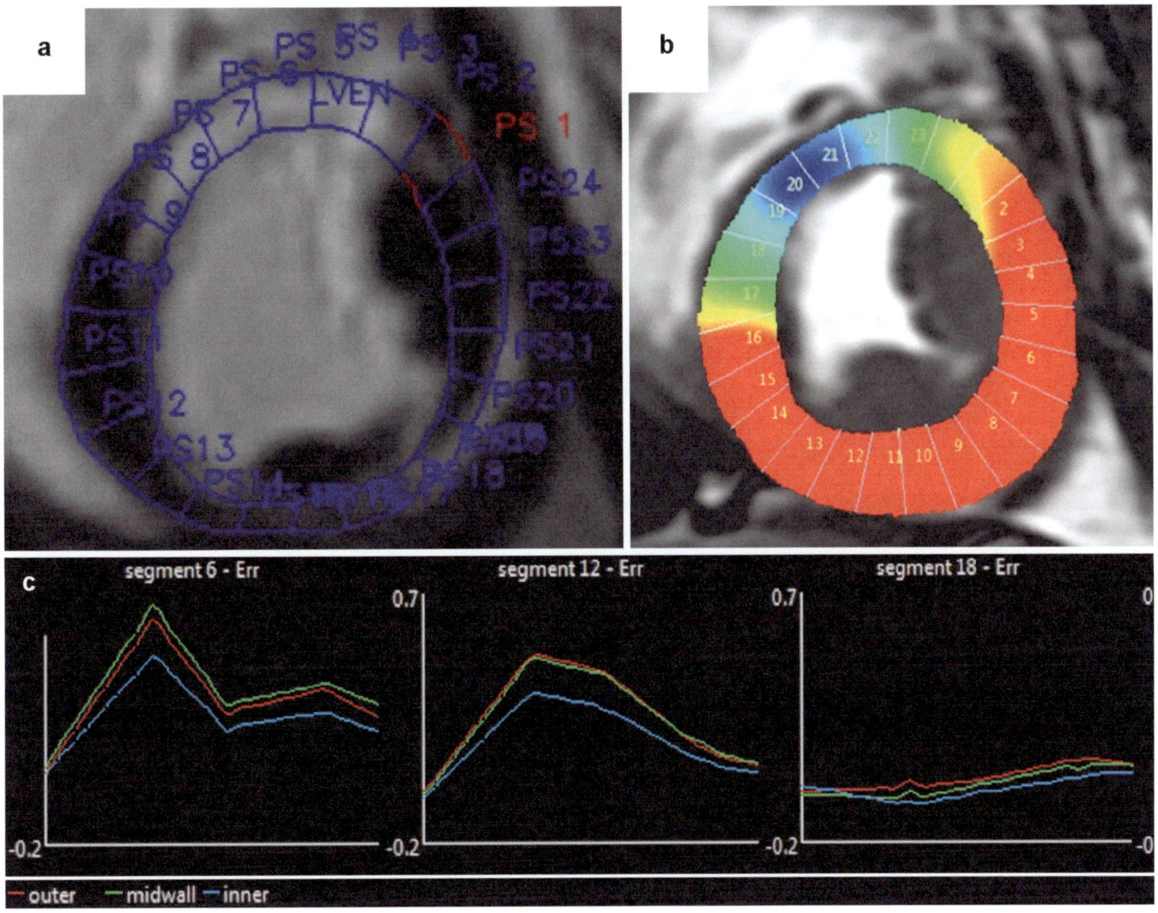

Fig. 15.9 Tissue tracking for cardiac strain – Tracking of myocardial tissue contours from short-axis SSFP images is demonstrated in a patient with anteroseptal myocardial scar. (**a**) Short-axis delayed enhanced MRI image of myocardial scar divided in 24 segments. (**b**) Color-coded radial strain plot generated from a non-tagged MRI cine showing contraction (*red*) in viable myocardium and simultaneous stretching (*blue*) in the anteroseptal myocardium. (**c**) Plots of strain versus time for select myocardial segments show that tissue tracking results correlate well with the viability data. See also Movie 5 on Springer Extras

DENSE technique[46-48] has high accuracy, high spatial and temporal resolution, inherent tissue tracking without tag detection, and straightforward strain analysis.

In a manner similar to conventional myocardial tagging, DENSE tags the signal upon detection of the R wave at end-diastole and samples the displacement-encoded signal later in the cardiac cycle, thereby avoiding the error accumulation problem inherent to velocity-encoded imaging. In addition, as shown in Fig. 15.10 (and Movie 6 on Springer Extras), instead of encoding displacement information into the amplitude of the signal like myocardial tagging, the displacement information is encoded into the phase of the signal. Thus, DENSE has the property that displacement relative to the end-diastolic position, not instantaneous velocity, is measured in the signal phase. Also, because displacement is measured using the phase, pixel-wise spatial resolution and inherent tissue tracking are achieved. Although DENSE and MR-MT are based on similar principles, the hallmark of DENSE has been much higher spatial and temporal resolution realized through prospective pulse sequence design rather than filtering the raw data acquired using conventional tagging sequences.

15.8.2 Clinical Studies

Initial implementations of DENSE focused primarily on high spatial resolution, and image acquisition was limited to a single cardiac phase. Cine DENSE data acquisition[50] and analysis[48] methods were subsequently developed, such that 2D displacement-encoded images with spatial resolution of $2.5 \times 2.5 \times 8$ mm^3 and temporal resolution of 34 ms of a single slice can now be achieved in a single 13 – 19 heartbeat breathhold.[49] It is also possible to perform 3D tissue tracking with DENSE, albeit at the cost of increased data acquisition time. Using motion phantoms, for displacements and velocities representative of the human heart, the displacement accuracy of cine DENSE is approximately 0.25 mm.[47] Strain analysis of cine DENSE has also been validated in normal human subjects versus conventional tagging.[46]

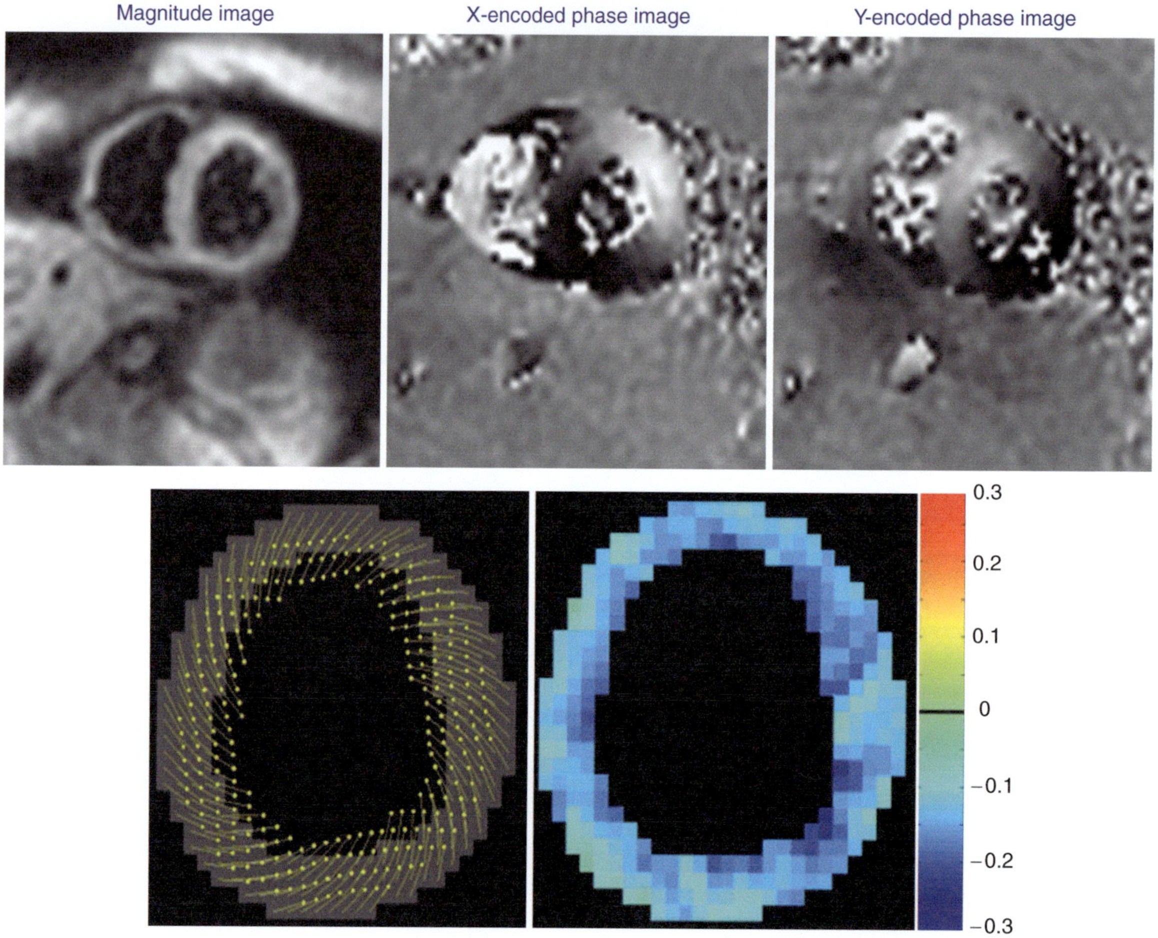

Fig. 15.10 Displacement Encoding with Stimulated Echoes (DENSE) – This figure demonstrates analysis of strain data using DENSE for a synchronous heart. The magnitude and phase images are shown in the *top row*. With DENSE, strain is directly encoded in the phase of the MR signal, which decreases the amount of post-processing required compared to myocardial tagging. In the *bottom row*, strain may be depicted in terms of a vector-based movie or a color-coded movie. See also Movie 6 on Springer Extras and reference Zhong et al.[49]

15.9 MR Scar Imaging with Late Gadolinium Enhancement

15.9.1 Technical Aspects

Delayed contrast-enhanced MR, also referred to as late gadolinium enhancement (LGE), is used to assess areas of infarct. Typically, a gadolinium contrast injection of 0.1–0.2 mmol/kg is given intravenously and the myocardium imaged 10–30 min later. Due to delayed washout of gadolinium from infracted myocardium and other areas with scar, fibrosis and infarct appear bright with LGE (as shown in Fig. 15.11), as gadolinium shortens T1 and T2 times, as described above. Typically, an ECG-triggered inversion recovery sequence is used, with the inversion pulse at 150–200 ms after the R wave so that imaging occurs during diastole. Either a spoiled gradient echo or steady-state free precession sequence is used. Usually, a short-axis stack and three long-axis images are performed. Analysis of regional transmurality and extent of scar, as well as percent LV scar volume, may then be performed using standard software.[53]

15.9.2 Clinical Studies

Although mechanical intraventricular dyssynchrony has been shown to be useful for identifying CRT candidates, 30–40% of patients with mechanical dyssynchrony confirmed by echocardiographic metrics actually do not respond to CRT.[12] This may be due to inaccuracies (false positives) in the assessment of the mechanical dyssynchrony by echocardiography or characteristics of the underlying complex myocardial substrate such as myocardial scar burden/distribution. Evidence suggests that patients with significant posterolateral scar[52] or greater than 15% overall left ventricular total scar are less likely to respond to CRT.[54] Precise assessment of LV fibrosis volume and location of scar represents an important advantage of cardiac magnetic resonance imaging over echocardiography.

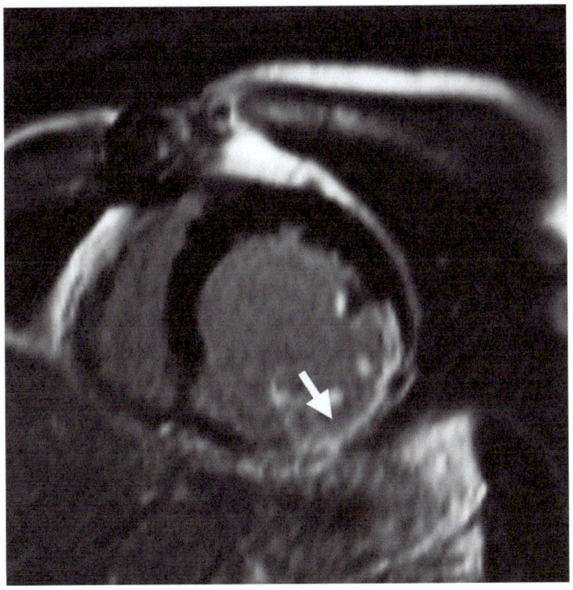

Fig. 15.11 MR Late Gadolinium Enhanced Scar Imaging – A postero-lateral infarct (*arrow*) is evident with LGE imaging. As shown in Truong et al.,[51] This predicts a decreased response rate to CRT. (From Bleeker et al.,[52] with permission)

15.10 MR Imaging in Patients with Pacemakers and Defibrillators

15.10.1 Clinical Importance

Traditionally, MR imaging of patients with pacemakers and defibrillators has not been performed due to concern about the potential effects of the magnetic field on the devices and leads. However, many patients with pacemakers and defibrillators have a clinical indication for noncardiac MR imaging. Furthermore, the ability to perform CMR in patients with non-CRT defibrillators or pacemakers in order to assess mechanical dyssynchrony and scar could be very useful in patients with devices referred for CRT upgrades. An example of how adequate images may be obtained even with artifact from the device is shown in Fig. 15.12. Upgrades have become an increasing component of referrals for CRT, as compared to patients referred for new CRT implants.

15.10.2 Testing of MR Safety in Patients with Cardiac Devices

Cardiac MR is not presently approved for routine use in patients with pacemakers and defibrillators, although recent safety data is promising. In an animal study, Roguin et al. tested pacemakers and defibrillators manufactured before and after the year 2000 for lead heating, device function, force acting on the device, and image distortion at 1.5 T, and found that devices in the later group had stable function before and after the MR study without significant force or heating effects. Nazarian et al. then reported a series of 58 patients with pacemakers or defibrillators who had MR imaging (including 13 cardiac viability studies) performed with appropriate precautions.[55] The clinical question was answered in all nonthoracic studies and 93% of thoracic studies, with no inappropriate activation or inhibition of pacing and stable lead parameters before and after the study.

A recent published statement from the FDA notes that while there have been some studies suggesting the safety of MR for patients with pacemaker and defibrillators,[56] there is still not enough data to justify routine use, although the benefits may outweigh the risks in selected patients. At this time, MR-conditional pacemaker systems are now clinically available, while implantable cardioverter defibrillator systems designed for the MR environment are in the advanced stages of development. In addition to overcoming the safety issue, application of MR for dyssynchrony imaging in patients with preexisting pacemaker and defibrillators must also overcome possible cardiac artifacts due to the device and leads.

15.11 Cardiac Computed Tomography (CCT) for Cardiac Resynchronization Therapy

There has been interest in the use of multidetector cardiac computed tomography (MDCT) for patients with heart failure referred for resynchronization because it has the potential to provide many different types of information relevant to these patients. Like other imaging modalities, CCT can provide some information about LV contractile function and mechanical dyssynchrony, although MR is still the gold standard. In addition, CCT can provide a detailed three-dimensional image of the coronary veins, as discussed below. Third, CCT can provide some approximation to the scar information available with MR late gadolinium enhancement. Finally, since the CRT procedure is performed with fluoroscopy, the potential integration of CCT information with the fluoroscopy in real time is another possible application of CCT in these patients.[51]

15.11.1 Coronary Vein Imaging

The ability of MDCT to provide detailed information about the coronary venous anatomy is the distinct advantage of this imaging modality compared to others such as CMR. While it is possible to image the coronary veins with CMR, the detail provided by MDCT is better.[57] There have now been several studies showing the accuracy of MDCT, in particular 64-slice MDCT, for coronary venography.[58-62] Several of these demonstrated accuracy of MDCT compared to invasive coronary sinus venography obtained during the CRT procedure.

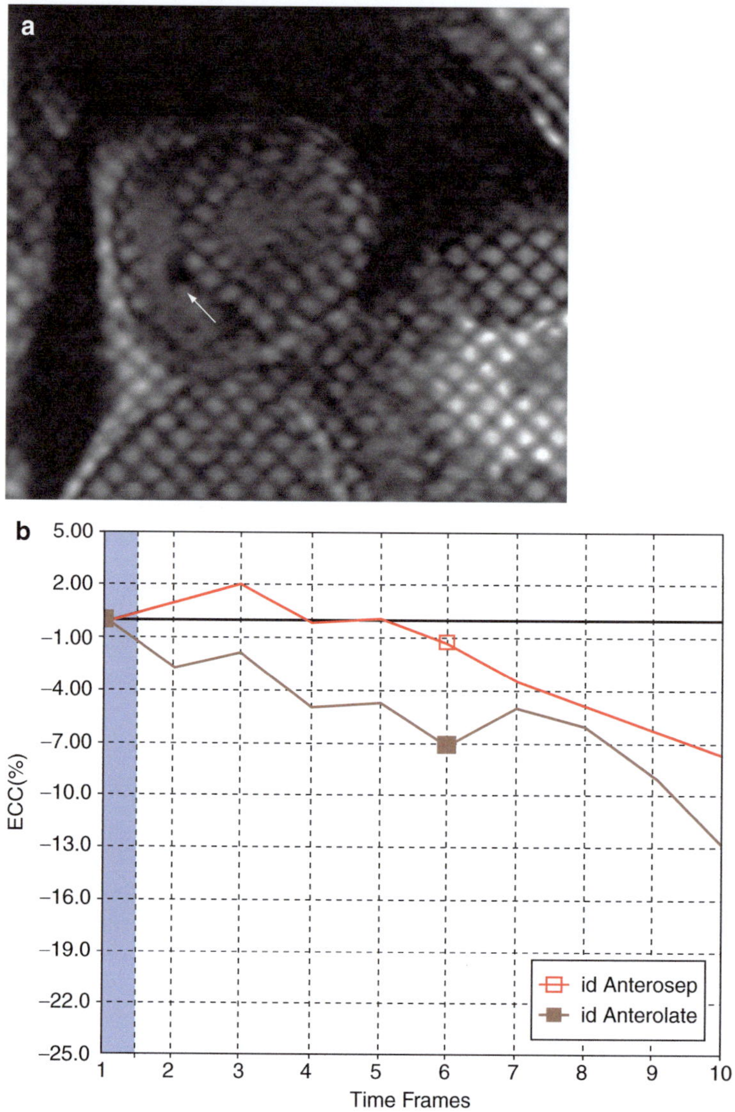

Fig. 15.12 Myocardial tagging in a patient with CRT-D – (**a**) Mid-myocardial, mid-systole short-axis MRI tagging image in a patient with a CRT device. Tag quality is good and sufficient for strain analysis. The right ventricular lead is shown by the *small arrow*. (**b**) Strain versus frame/time plots of two myocardial segments

A CT protocol similar to that used for CT coronary arteriography has been shown to be effective for coronary venography[61]; however, modifications to the arterial protocol have been suggested.[51] These include using a test bolus rather than bolus tracking to calculate contrast agent transit time. Also, an empiric 5–10 s delay for image acquisition may be added to the time of peak contrast opacification of the ascending aorta to facilitate acquisition during the venous rather than the arterial phase. An example of high-quality CT venous imaging is shown in Fig. 15.13.

Classification and nomenclature of the coronary sinus branches may be based on names for specific veins or by the region of LV myocardium drained, as recently proposed.[63] Regarding the specific coronary sinus branches, the recent CT literature refers to three veins of particular interest, in order of proximal to distal takeoff from the main coronary sinus: the posterior interventricular vein (PIV), the posterior vein of the left ventricle (PVLV), and the left marginal vein (LMV).[61] The coronary sinus beyond the PVLV is termed the greater cardiac vein (GCV) and eventually gives off the anterior interventricular vein. Pacing leads in anterior branches are less likely to generate effective resynchronization.

One goal of this body of literature has been to determine the effect of ischemic heart disease on the likelihood of patency of the coronary veins used for the resynchronization lead. In one study, the PIV was present in nearly all patients, regardless of whether they were normal patients, had coronary artery disease (CAD), or had prior myocardial infarction (MI). However, the PVLV was present in only 82–84% of patients with prior MI or CAD, compared to 96% of controls. The LMV was much less likely to be present with prior myocardial infarction (27%) as compared to patients with CAD without prior MI or controls (61–71%).[61] Among patients with myocardial infarction, patients with Q wave infarctions were noted to have fewer branches than those with non-Q wave infarctions. Of note, about 20% of patients with CAD or prior MI

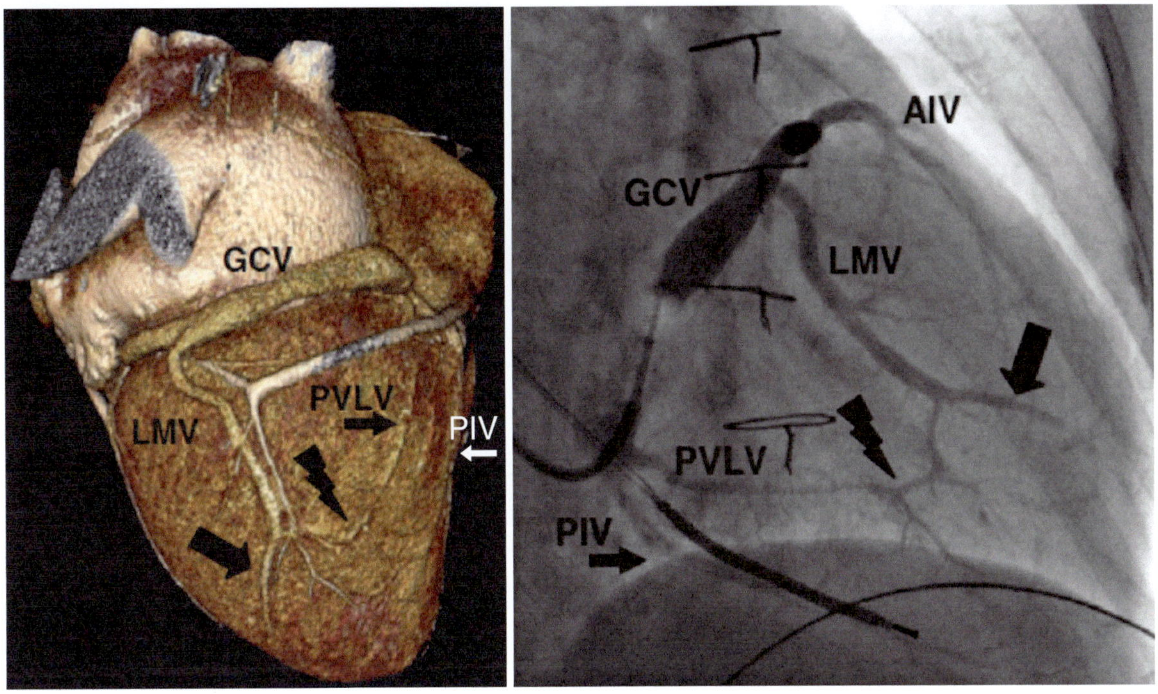

Fig. 15.13 CT venography of coronary sinus – Excellent correlation between CT venography and invasive findings is demonstrated. (From van de Veire et al.,[61] with permission)

had an absent PVLV and LMV, indicating that the PIV would be the only posterolateral branch available for the resynchronization lead in these patients. Of note, a previous invasive coronary venography study of 86 patients referred for ICD implantation found a PVLV present in 55% and LMV present in 83%, while the PIV was present in all but one patient.[64]

15.11.2 CCT-Based Mechanical Dyssynchrony

Although CCT does not offer the full cardiac strain assessment capabilities of CMR, inferences regarding dyssynchrony and cardiac function can be made based on wall thickening and wall motion. Truong et al. recently evaluated the ability of CCT to characterize mechanical dyssynchrony in 27 patients with HF and LVEF less than 35% compared to 11 control patients.[62] A dyssynchrony index based on the standard deviation to maximal wall thickening or maximal wall motion in six short-axis cardiac segments was compared to dyssynchrony characterized by the maximal difference in thickening or wall motion between opposing segments (Fig. 15.14). The authors were able to demonstrate acceptable intraobserver and interobserver variability for the standard deviation in six segments in time to maximal thickening (the best of the parameters) with this technique. Although they showed that this measure was significantly greater in heart failure patients compared to controls, this metric was not effective for discrimination between the narrow-QRS and wide-QRS HF groups. Although QRS duration does not correlate well with CRT response, it is uncommon with patients with QRS durations less than 110–120 ms to have significant mechanical dyssynchrony based on the RETHINQ study,[13] such that the failure of a dyssynchrony metric to discriminate between patients with wide- and narrow-QRS complexes may indicate decreased potential to predict CRT response.

As with MR/SSFP-based contour tracking methods, the use of wall thickening to characterize mechanical dyssynchrony depends on a radial-based measurement and is suboptimal compared to CMR techniques offering direct characterization of circumferential, longitudinal, and radial strain. However, improved dyssynchrony characterization may be possible with improvements in temporal resolution available with multisegment reconstruction algorithms.[65] In addition, evaluation of better methods to characterize mechanical dyssynchrony based on information from the CT may be possible.

15.11.3 Scar Assessment

Since posterolateral scar and overall increased volume of myocardial scar as assessed by CMR are poor prognostic indicators for CRT response,[54,66] an accurate CCT-based assessment of scar would be desirable. Although CMR is the gold standard for scar assessment, CT has been investigated by Dickfeld and others for scar characterization in

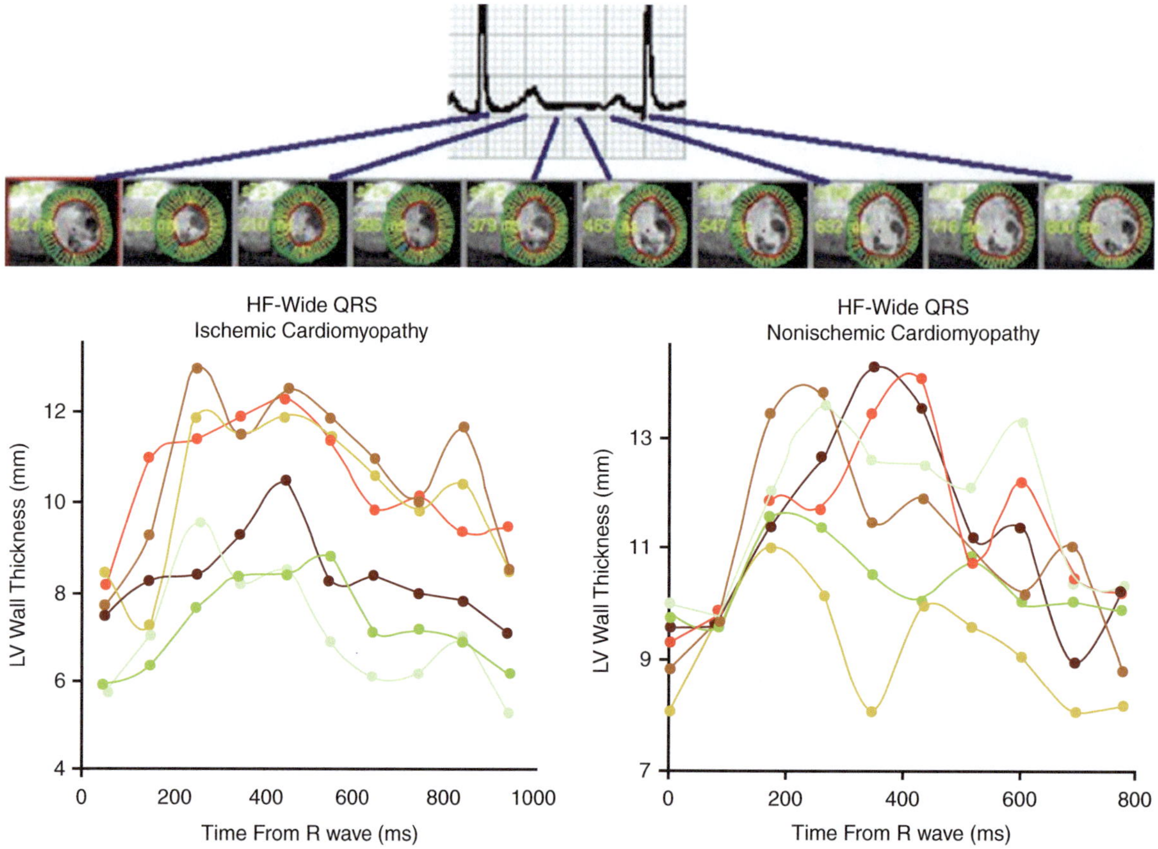

Fig. 15.14 Evaluation of dyssynchrony with CCT – Mechanical data can be determined from CCT based on changes in wall thickness during the cardiac cycle, although MR is still the gold standard. (From Truong et al.,[51] with permission)

patients referred for ventricular tachycardia ablation.[67-69] Because definition of scar is somewhat limited with CCT as compared to CMR, metabolic PET imaging is often used in combination with CCT to provide optimal scar imaging. An example of CCT scar imaging is shown in Fig. 15.15. Using this technique, a good correlation has been demonstrated between voltage mapping at the time of ablation and the CCT scar map. In addition, CT/PET scar mapping was also found to identify metabolically active channels within the myocardial scar not detected by voltage mapping.[68] However, CCT without PET mapping has not yet been demonstrated to identify scar accurately and consistently compared to CMR.

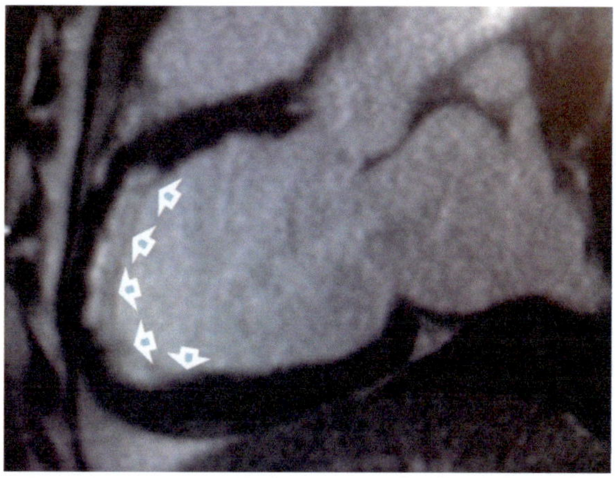

Fig. 15.15 CT Scar imaging – An example of anteroapical scar from CCT imaging is shown

15.11.4 Future of CCT for Dyssynchrony

Due to its potential to image scar, mechanical function, and the coronary veins, as well as ability to integrate with fluoroscopy and confirmed safety in patients with preexisting pacemakers and non-CRT defibrillators, CCT may find further applications in patients referred for CRT in the future. However, at this time, CCT has still not been demonstrated to provide the degree of accuracy of scar and mechanical assessment provided by CMR. Although spatial resolution is quite good, protocols to improve temporal resolution are still under evaluation. In addition, CCT exposes patients to a significant dose of radiation, as well as intravenous contrast, which may be problematic for patients with renal dysfunction. As a result, further investigation is required to determine if CCT will eventually prove useful in patients referred for CRT.

References

1. Roger VL, Weston SA, Redfield MM, et al. Trends in heart failure incidence and survival in a community-based population. *JAMA*. 2004;292(3):344-350.
2. Levy D, Kenchaiah S, Larson MG, et al. Long-term trends in the incidence of and survival with heart failure. *N Engl J Med*. 2002; 347(18):1397-1402.
3. Haldeman GA, Croft JB, Giles WH, Rashidee A. Hospitalization of patients with heart failure: National Hospital Discharge Survey, 1985 to 1995. *Am Heart J*. 1999;137(2):352-360.
4. Ho KK, Pinsky JL, Kannel WB, Levy D. The epidemiology of heart failure: the Framingham study. *J Am Coll Cardiol*. 1993;22 (4 Suppl A):6A-13A.
5. Shamim W, Francis DP, Yousufuddin M, et al. Intraventricular conduction delay: a prognostic marker in chronic heart failure. *Int J Cardiol*. 1999;70(2):171-178.
6. Baldasseroni S, Opasich C, Gorini M, et al. Left bundle-branch block is associated with increased 1-year sudden and total mortality rate in 5517 outpatients with congestive heart failure: a report from the Italian network on congestive heart failure. *Am Heart J*. 2002;143(3):398-405.
7. Bilchick KC, Helm RH, Kass DA. Physiology of biventricular pacing. *Curr Cardiol Rep*. 2007;9(5):358-365.
8. Abraham WT, Fisher WG, Smith AL, et al. Cardiac resynchronization in chronic heart failure. *N Engl J Med*. 2002;346(24):1845-1853.
9. Bristow MR, Saxon LA, Boehmer J, et al. Cardiac-resynchronization therapy with or without an implantable defibrillator in advanced chronic heart failure. *N Engl J Med*. 2004;350(21):2140-2150.
10. Cleland JG, Daubert JC, Erdmann E, et al. The effect of cardiac resynchronization on morbidity and mortality in heart failure. *N Engl J Med*. 2005;352(15):1539-1549.
11. Epstein AE, DiMarco JP, Ellenbogen KA, et al. ACC/AHA/HRS 2008 Guidelines for Device-Based Therapy of Cardiac Rhythm Abnormalities: executive summary a report of the American College of Cardiology/American Heart Association Task Force on practice guidelines (Writing Committee to Revise the ACC/AHA/NASPE 2002 Guideline Update for Implantation of Cardiac Pacemakers and Antiarrhythmia Devices) developed in collaboration with the American Association for Thoracic Surgery and Society of Thoracic Surgeons. *J Am Coll Cardiol*. 2008;51(21):2085-2105.
12. Chung ES, Leon AR, Tavazzi L, et al. Results of the predictors of response to CRT (PROSPECT) trial. *Circulation*. 2008;117(20): 2608-2616.
13. Beshai JF, Grimm RA, Nagueh SF, et al. Cardiac-resynchronization therapy in heart failure with narrow QRS complexes. *N Engl J Med*. 2007;357(24):2461-2471.
14. Bordachar P, Lafitte S, Reuter S, et al. Echocardiographic parameters of ventricular dyssynchrony validation in patients with heart failure using sequential biventricular pacing. *J Am Coll Cardiol*. 2004;44(11):2157-2165.
15. Curry CW, Nelson GS, Wyman BT, et al. Mechanical dyssynchrony in dilated cardiomyopathy with intraventricular conduction delay as depicted by 3D tagged magnetic resonance imaging. *Circulation*. 2000;101(1):E2.
16. Helm PA, Younes L, Beg MF, et al. Evidence of structural remodeling in the dyssynchronous failing heart. *Circ Res*. 2006;98(1):125-132.
17. Helm RH, Leclercq C, Faris OP, et al. Cardiac dyssynchrony analysis using circumferential versus longitudinal strain: implications for assessing cardiac resynchronization. *Circulation*. 2005;111(21): 2760-2767.
18. Bilchick KC, Dimaano V, Wu KC, et al. Cardiac magnetic resonance assessment of dyssynchrony and myocardial scar predicts function class improvement following cardiac resynchronization therapy. *JACC Cardiovasc Imaging*. 2008;1(5):561-568.
19. Delgado V, Ypenburg C, Van Bommel RJ, et al. Assessment of left ventricular dyssynchrony by speckle tracking strain imaging comparison between longitudinal, circumferential, and radial strain in cardiac resynchronization therapy. *J Am Coll Cardiol*. 2008;51(20): 1944-1952.
20. Scheffler K, Lehnhardt S. Principles and applications of balanced SSFP techniques. *Eur Radiol*. 2003;13(11):2409-2418.
21. Shuman WP, Branch KR, May JM, et al. Prospective versus retrospective ECG gating for 64-detector CT of the coronary arteries: comparison of image quality and patient radiation dose. *Radiology*. 2008;248(2):431-437.
22. Zerhouni EA, Parish DM, Rogers WJ, Yang A, Shapiro EP. Human heart: tagging with MR imaging – a method for noninvasive assessment of myocardial motion. *Radiology*. 1988;169(1):59-63.
23. McVeigh ER, Atalar E. Cardiac tagging with breath-hold cine MRI. *Magn Reson Med*. 1992;28(2):318-327.
24. Zwanenburg JJ, Kuijer JP, Marcus JT, Heethaar RM. Steady-state free precession with myocardial tagging: CSPAMM in a single breathhold. *Magn Reson Med*. 2003;49(4):722-730.
25. Herzka DA, Guttman MA, McVeigh ER. Myocardial tagging with SSFP. *Magn Reson Med*. 2003;49(2):329-340.
26. Young AA, Imai H, Chang CN, Axel L. Two-dimensional left ventricular deformation during systole using magnetic resonance imaging with spatial modulation of magnetization [erratum appears in Circulation 1994 Sep;90(3):1584]. *Circulation*. 1994;89(2): 740-752.
27. McVeigh ER. MRI of myocardial function: motion tracking techniques. *Magn Reson Imaging*. 1996;14(2):137-150.
28. Osman NF, Kerwin WS, McVeigh ER, Prince JL. Cardiac motion tracking using CINE harmonic phase (HARP) magnetic resonance imaging. *Magn Reson Med*. 1999;42(6):1048-1060.
29. Bundy JM, Lorenz CH. TAGASIST: a post-processing and analysis tools package for tagged magnetic resonance imaging. *Comput Med Imaging Graph*. 1997;21(4):225-232.
30. Garot J, Bluemke DA, Osman NF, et al. Fast determination of regional myocardial strain fields from tagged cardiac images using harmonic phase MRI. *Circulation*. 2000;101(9):981-988.
31. Leclercq C, Faris O, Tunin R, et al. Systolic improvement and mechanical resynchronization does not require electrical synchrony in the dilated failing heart with left bundle-branch block. *Circulation*. 2002;106(14):1760-1763.
32. Helm RH, Byrne M, Helm PA, et al. Three-dimensional mapping of optimal left ventricular pacing site for cardiac resynchronization. *Circulation*. 2007;115(8):953-961.
33. Pan L, Stuber M, Kraitchman DL, Fritzges DL, Gilson WD, Osman NF. Real-time imaging of regional myocardial function using fast-SENC. *Magn Reson Med*. 2006;55(2):386-395.
34. Osman NF, Sampath S, Atalar E, Prince JL. Imaging longitudinal cardiac strain on short-axis images using strain-encoded MRI. *Magn Reson Med*. 2001;46(2):324-334.
35. Korosoglou G, Youssef AA, Ibrahim EH, Bilchick KC, Lardo AC, Osman NF. Real-time fast strain-encoded magnetic resonance imaging to evaluate regional myocardial function at 3.0 Tesla: comparison to conventional tagging. *J Magn Reson Imaging*. 2008;27(5):1012-1018.
36. Neizel M, Lossnitzer D, Korosoglou G, et al. Strain-encoded (SENC) magnetic resonance imaging to evaluate regional heterogeneity of myocardial strain in healthy volunteers: comparison with conventional tagging. *J Magn Reson Imaging*. 2009;29(1):99-105.
37. Pelc NJ, Drangova M, Pelc LR, et al. Tracking of cyclic motion with phase-contrast cine MR velocity data. *J Magn Reson Imaging*. 1995;5(3):339-345.

38. Zhu Y, Drangova M, Pelc NJ. Estimation of deformation gradient and strain from cine-PC velocity data. *IEEE Trans Med Imaging*. 1997;16(6):840-851.
39. Constable RT, Rath KM, Sinusas AJ, Gore JC. Development and evaluation of tracking algorithms for cardiac wall motion analysis using phase velocity MR imaging. *Magn Reson Med*. 1994;32(1):33-42.
40. Lingamneni A, Hardy PA, Powell KA, Pelc NJ, White RD. Validation of cine phase-contrast MR imaging for motion analysis. *J Magn Reson Imaging*. 1995;5(3):331-338.
41. Delfino JG, Bhasin M, Cole R, et al. Comparison of myocardial velocities obtained with magnetic resonance phase velocity mapping and tissue Doppler imaging in normal subjects and patients with left ventricular dyssynchrony. *J Magn Reson Imaging*. 2006;24(2):304-311.
42. Marsan NA, Westenberg JJ, Tops LF, et al. Comparison between tissue Doppler imaging and velocity-encoded magnetic resonance imaging for measurement of myocardial velocities, assessment of left ventricular dyssynchrony, and estimation of left ventricular filling pressures in patients with ischemic cardiomyopathy. *Am J Cardiol*. 2008;102(10):1366-1372.
43. Westenberg JJ, Lamb HJ, van der Geest RJ, et al. Assessment of left ventricular dyssynchrony in patients with conduction delay and idiopathic dilated cardiomyopathy: head-to-head comparison between tissue Doppler imaging and velocity-encoded magnetic resonance imaging. *J Am Coll Cardiol*. 2006;47(10):2042-2048.
44. Muellerleile K, Baholli L, Groth M, et al. Interventricular mechanical dyssynchrony: quantification with velocity-encoded MR imaging. *Radiology*. 2009;253(2):364-371.
45. Chalil S, Stegemann B, Muhyaldeen S, et al. Intraventricular dyssynchrony predicts mortality and morbidity after cardiac resynchronization therapy: a study using cardiovascular magnetic resonance tissue synchronization imaging. *J Am Coll Cardiol*. 2007;50(3):243-252.
46. Kim D, Gilson WD, Kramer CM, Epstein FH. Myocardial tissue tracking with two-dimensional cine displacement-encoded MR imaging: development and initial evaluation. *Radiology*. 2004; 230(3):862-871.
47. Spottiswoode BS, Zhong X, Hess AT, et al. Tracking myocardial motion from cine DENSE images using spatiotemporal phase unwrapping and temporal fitting. *IEEE Trans Med Imaging*. 2007;26(1):15-30.
48. Spottiswoode BS, Zhong X, Lorenz CH, Mayosi BM, Meintjes EM, Epstein FH. Motion-guided segmentation for cine DENSE MRI. *Med Image Anal*. 2009;13(1):105-115. Epub 2008 Jul 9.
49. Zhong X, Spottiswoode BS, Meyer CH, Epstein FH. Two-dimensional spiral cine DENSE. *Proc Intl Soc Magn Reson Med*. 2007;15:756.
50. Zhong X, Spottiswoode BS, Cowart EA, Gilson WD, Epstein FH. Selective suppression of artifact-generating echoes in cine DENSE using through-plane dephasing. *Magn Reson Med*. 2006;56(5): 1126-1131.
51. Truong QA, Hoffmann U, Singh JP. Potential uses of computed tomography for management of heart failure patients with dyssynchrony. *Crit Pathw Cardiol*. 2008;7(3):185-190.
52. Bleeker GB, Kaandorp TA, Lamb HJ, et al. Effect of posterolateral scar tissue on clinical and echocardiographic improvement after cardiac resynchronization therapy. *Circulation*. 2006;113(7):969-976.
53. Schmidt A, Azevedo CF, Cheng A, et al. Infarct tissue heterogeneity by magnetic resonance imaging identifies enhanced cardiac arrhythmia susceptibility in patients with left ventricular dysfunction. *Circulation*. 2007;115(15):2006-2014.
54. White JA, Yee R, Yuan X, et al. Delayed enhancement magnetic resonance imaging predicts response to cardiac resynchronization therapy in patients with intraventricular dyssynchrony. *J Am Coll Cardiol*. 2006;48(10):1953-1960.
55. Nazarian S, Roguin A, Zviman MM, et al. Clinical utility and safety of a protocol for noncardiac and cardiac magnetic resonance imaging of patients with permanent pacemakers and implantable-cardioverter defibrillators at 1.5 Tesla. *Circulation*. 2006;114(12):1277-1284.
56. Faris OP, Shein M. Food and Drug Administration perspective: magnetic resonance imaging of pacemaker and implantable cardioverter-defibrillator patients. *Circulation*. 2006;114(12):1232-1233.
57. van de Veire NR, Schuijf JD, Bleeker GB, Schalij MJ, Bax JJ. Magnetic resonance imaging and computed tomography in assessing cardiac veins and scar tissue. *Europace*. 2008;10 (suppl 3):iii110-iii113.
58. Jongbloed MR, Lamb HJ, Bax JJ, et al. Noninvasive visualization of the cardiac venous system using multislice computed tomography. *J Am Coll Cardiol*. 2005;45(5):749-753.
59. Tada H, Kurosaki K, Naito S, et al. Three-dimensional visualization of the coronary venous system using multidetector row computed tomography. *Circ J*. 2005;69(2):165-170.
60. Muhlenbruch G, Koos R, Wildberger JE, Gunther RW, Mahnken AH. Imaging of the cardiac venous system: comparison of MDCT and conventional angiography. *AJR Am J Roentgenol*. 2005;185(5): 1252-1257.
61. van de Veire NR, Schuijf JD, De SJ, et al. Non-invasive visualization of the cardiac venous system in coronary artery disease patients using 64-slice computed tomography. *J Am Coll Cardiol*. 2006; 48(9):1832-1838.
62. Truong QA, Singh JP, Cannon CP, et al. Quantitative analysis of intraventricular dyssynchrony using wall thickness by multidetector computed tomography. *JACC Cardiovasc Imaging*. 2008;1(6): 772-781.
63. Singh JP, Houser S, Heist EK, Ruskin JN. The coronary venous anatomy: a segmental approach to aid cardiac resynchronization therapy. *J Am Coll Cardiol*. 2005;46(1):68-74.
64. Meisel E, Pfeiffer D, Engelmann L, et al. Investigation of coronary venous anatomy by retrograde venography in patients with malignant ventricular tachycardia. *Circulation*. 2001;104(4):442-447.
65. Flohr TG, Raupach R, Bruder H. Cardiac CT: how much can temporal resolution, spatial resolution, and volume coverage be improved? *J Cardiovasc Comput Tomogr*. 2009;3(3):143-152.
66. Bleeker GB, Schalij MJ, van der Wall EE, Bax JJ. Postero-lateral scar tissue resulting in non-response to cardiac resynchronization therapy. *J Cardiovasc Electrophysiol*. 2006;17(8):899-901.
67. Dickfeld T, Kocher C. The role of integrated PET-CT scar maps for guiding ventricular tachycardia ablations. *Curr Cardiol Rep*. 2008;10(2):149-157.
68. Dickfeld T, Lei P, Dilsizian V, et al. Integration of three-dimensional scar maps for ventricular tachycardia ablation with positron emission tomography-computed tomography. *JACC Cardiovasc Imaging*. 2008;1(1):73-82.
69. Tian J, Smith MF, Chinnadurai P et al. Clinical application of PET/CT fusion imaging for three-dimensional myocardial scar and left ventricular anatomy during ventricular tachycardia ablation. *J Cardiovasc Electrophysiol*. 2009;20(6): 597-604.

16 Nuclear Imaging for Device Therapy

Amparo L. Figueroa and Ahmed Tawakol

Abstract

Nuclear imaging procedures have grown since the first gamma camera was invented in the late 1960s, and have now become an essential part of clinical cardiology practice. Major clinical applications of SPECT and PET in cardiology continue to be in the diagnosis, prognostic evaluation, and risk stratification of patients with known or suspected coronary artery disease. Additionally, both SPECT and PET remain gold standard in determining myocardial viability using radiotracers of blood flow and glucose metabolism. Recent developments in cardiology research suggest that nuclear imaging may prove to be an important clinical tool for evaluating patients with cardiac electrophysiologic pathologies. Although still in the investigational stage, nuclear imaging techniques such as gated SPECT MPI with phase analysis have been shown clinically useful in selecting appropriate candidates for CRT. Cardiac neurotransmission imaging with SPECT using 123I-MIBG has also been shown to predict fatal cardiac outcomes and may be useful in combination with other clinical parameters for selecting patients who would greatly benefit from ICD. Studies also demonstrated the potential clinical utility of 123I-MIBG in assessing CRT response through improvement in cardiac adrenergic activity. Additionally, PET has provided metabolic characterization of myocardial scar that has been shown valuable in guiding ablation of scar-related ventricular tachycardia. In this chapter, the potential role of nuclear imaging for device therapy is discussed.

Keywords

Nuclear imaging for device therapy • SPECT in cardiology • PET in cardiology • Electrophysiologic pathologies • Cardiac neuronal imaging

16.1 Historical Perspectives

Nuclear imaging relies on the detection of photons emitted through the principle of radiodecay. Historically, radioactivity was originally observed through decaying products in purified ores of radium and uranium. Hevesy conducted early plant circulation experiments involving radioactive isotope of lead in the 1920s which earned him a Nobel peace prize.[1] Subsequent development of atomic reactors and accelerators in the 1940s paved the way to the production of artificial radionuclides including ^{99m}Tc, C-11, N-13, F-18, ^{131}I.[1-4] By the late 1940s, scintillator counters consisting of NaI(Tl) and CsI(Tl) crystals were developed to allow for effective photon detection.[1] Shortly thereafter, Cassen, a UCLA physicist developed a rectilinear scanner, the first nuclear imaging device using a NaI (Tl) probe to obtain a single analog tomographic scan.[1,5] The first gamma camera was soon developed by Anger at the Lawrence Radiation Laboratory in the period 1950s–1960s[1,6] and made dynamic imaging of the cardiac distribution of radioactivity possible.[7,8] With the first commercially available thallium-201 (^{201}Tl) introduced in 1976, myocardial perfusion scintigraphy with two-dimensional planar imaging was shown

A.L. Figueroa (✉)
Department of Radiology, Massachusetts General Hospital, Boston, MA, USA
e-mail: alfigueroa@partners.org

A. Auricchio et al. (eds.), *Cardiac Imaging in Electrophysiology*,
DOI 10.1007/978-1-84882-486-7_16,

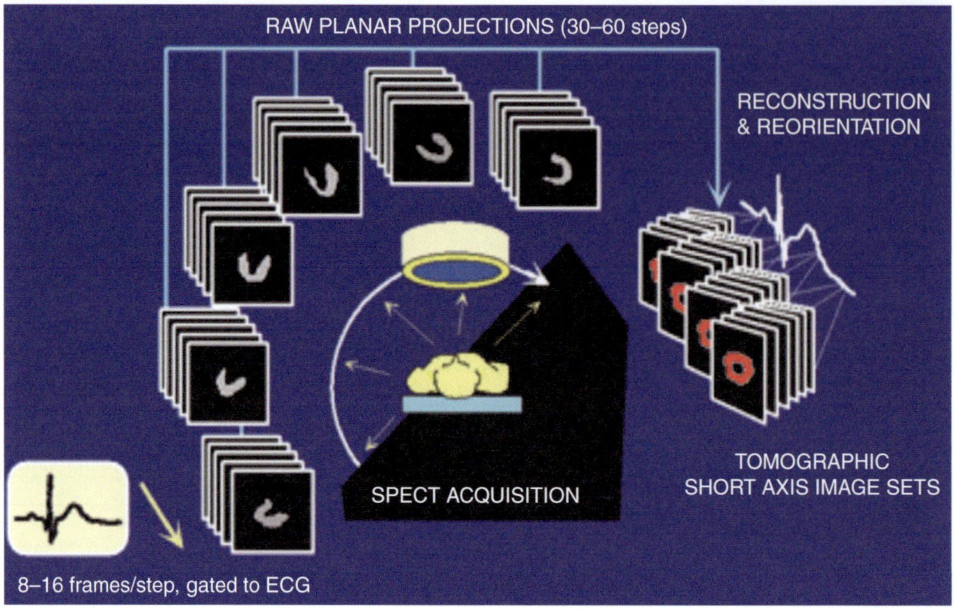

Fig. 16.1 SPECT acquisition Single Photon Emission Tomography (SPECT) acquisition is performed using a gamma camera mounted on a gantry which is then rotated around the patient. Detector orbits can be orbital or elliptical. A common approach to image acquisition is called the step and shoot method. There, the detector stops at pre-defined intervals to acquire the images, which can be obtained while gated to the ECG. After the orbit is complete, the raw planar images are reconstructed to produce the tomographic image sets for interpretation (From Fuster et al.[20])

useful for detection of coronary artery stenosis and eventually in the risk stratification of patients.[7,8] In mid-1980s, single photon emission computed tomography (SPECT) using planar gamma camera was invented.[1,9] The approval for use of technetium-99 m (^{99m}Tc)-sestamibi in the United States in 1990 followed shortly by another ^{99m}Tc-based agent, ^{99m}Tc-tetrofosmin, made it possible to obtain images from different parts of the cardiac cycle (gated SPECT) owing to their high myocardial count rates. Gated SPECT was routinely applied in the clinics in the late 1990s.[7,8]

The earliest work on positron imaging using coincidence-counting techniques was done by Brownell and Sweet using a positron probe[2,10,11], while Aronow later used a positron scanner for brain studies.[2,12] Thereafter, the first positron emission tomograph initially intended for use in brain imaging was built by a group of researchers led by Hoffman, Phelps, and Ter Porgassan at Washington University in the early 1970s[2,13,14] and was shortly introduced as a new imaging modality in cardiology.[2,15] In recent years, hybrid imaging devices such as PET/CT and SPECT/CT have become available allowing direct image fusion to localize radioactivity within a specific anatomic site.[16,17]

16.1.1 Imaging Mechanism and Current Techniques

16.1.1.1 SPECT

The scintillation camera detectors used by SPECT rotate 180° around the patient in a semicircular or elliptical fashion to collect sets of transaxial tomographic images.[1,7] The three-dimensional (3D) reconstruction of myocardial radioactivity from the two-dimensional projections produces a sequence of slices of images in the short axis, vertical long axis, and horizontal long axis orientations.[7] Electrocardiographically (ECG) gated single photon emission computed tomography (SPECT) myocardial perfusion imaging (MPI) is the most commonly utilized diagnostic test for determining the presence of coronary artery stenosis. It is primarily useful in the early evaluation of patients suspected of coronary artery (CAD) and in the risk stratification and prognostic evaluation of patients with established CAD.[18] Gated SPECT images are obtained in 8–16 phases of the cardiac cycle through the process of electrocardiographic (ECG) triggering (gating)[19] which allows evaluation of global and regional left ventricular function (Fig. 16.1).[18,21,22] The most commonly used radiotracers are thallium (Tl-201) and the two technetium-99 m (Tc-99 m) labeled compounds, sestamibi (Cardiolite) and tetrofosmin (Myoview).[18,19]

In patients who can exercise adequately, exercise stress protocols are typically employed with SPECT MPI while[19,23] for patients who cannot achieve an adequate level of exercise (85% of maximal predicted HR),[19,23] pharmacologic stress testing is used.[19,24] Coronary vasodilators, adenosine or dipyridamole, are the preferred pharmacologic stress agents for SPECT MPI.[19] In addition to the perfusion information, the SPECT images provide the size, shape, wall motion, and overall function of the left ventricle (LV), transient ischemic dilation (TID) of the LV, right ventricular (RV) myocardial uptake pattern, RV size, and abnormalities of lung uptake or other abnormal extracardiac activity.[7,25,26] SPECT MPI can additionally be used to assess myocardial viability.[7,27]

SPECT has also been performed with equilibrium radionuclide angiography (multiple-gated blood pool imaging) where a blood-pool tracer, usually ^{99m}Tc-labeled red blood cells, is used.[7,28,29] Resting images of the blood-pool tracer within the cardiac chambers are obtained at a constant heart rate by EKG gating through multiple cycles.[29] Methods of acquisition and processing have been shown to be similar to that of SPECT MPI.[7,28] In multiple-gated blood-pool scanning, images are

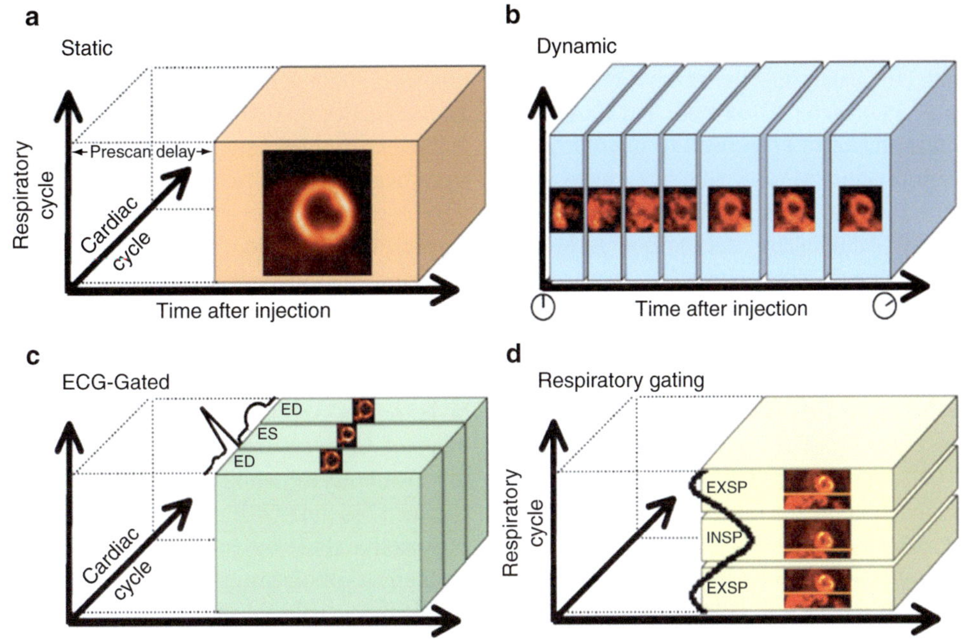

Fig. 16.2 Coincidental photons are captured by the gamma camera and are continuously recorded along with information about time after start of acquisition, electrocardiographic (ECG) data, and optionally breathing data. Reconstruction of images is performed afterwards. (**A**) High-count static images can be reconstructed by summing all information after a pre-defined delay time after tracer injection (pre-scan delay). (**B**) Dynamic imaging sequences,used for tracer kinetic analysis, are obtained by serial temporal reconstruction at different times after injection. (**C**) ECG-gated images, done to assess ventricular function are obtained at multiple phases of the cardiac cycle. (**D**) Respiratory gated images, used to suppress artifacts from respiratory motion, can be obtained at different phases of the breathing cycle. ED = end diastole; ES = end systole; EXSP = expiratory phase; INSP = inspiratory phase; PET = positron emission tomography. PET acquisition (From Bengel et al.[33])

taken at different views or angles to avoid tissue overlap and thereby enhance evaluation of regional function.[7,30]

16.1.1.2 PET

PET imaging involves the use of radionuclide agents that decay by positron emission.[15,31] The positron (β+particle) that is ejected from the nucleus collides with a nearby electron (β−particle) which leads to an annihilation event[32] during which the positron and electron masses are converted into energy and a pair of gamma (γ) rays[2]. The total energy is conserved with each gamma ray having an energy of 511 kiloelectron volt (keV) with the two gamma rays traveling in opposite directions (180° apart).[32] These gamma rays are detected by using a pair of collinearly aligned detectors which are installed in a ringlike pattern, and this allows measurement of radioactivity within the volume of interest at a series of angle and radial distances. This angular information is used to re-construct tomographic images of regional distribution of the positron-emitting radionuclide[2,32](Fig. 16.2).

PET, with its diverse utility and function, has become a valuable imaging tool over the years. PET has provided assessment of the relative distributions of regional myocardial blood flow, contractile function and has also provided information on regional functional processes (i.e., biochemical changes, substrate flow, and neuronal and membrane receptor function). Other tracers, although still in the experimental phase, can target and monitor specific gene expression, cell trafficking, and provide insight about molecular processes and cardiac innervation.[2,18,19] Recent introduction of multimodality imaging such as PET/CT and PET/MR has allowed merging of functional information (i.e., myocardial blood flow, vascular inflammation, or apoptosis) with structural information (i.e., coronary calcifications, arterial plaques, or regional or global left ventricular (LV) contractile function) which has provided further understanding of the pathophysiology of atherosclerosis[2,19,34,35] and can potentially help gain deeper insight about other cardiovascular pathologies. Over the years, PET's major clinical application in cardiology remains in the detection, prognostic evaluation, and risk stratification of patients with Coronary Artery Disease. Additionally, PET remains the gold standard for assessing myocardial viability (using radiotracers of blood flow and glucose metabolism) in the clinical setting.[19] The most commonly used tracers for visual assessment of regional myocardial tracer distribution (qualitative myocardial blood flow) under rest and stress conditions are rubidium-82 and N-13 ammonia.[2,36-41]

C-11 palmitate is the first radiopharmaceutical used for PET measurement of regional cardiac metabolism. C-11 acetate is converted to C-11 acetyl CoA in the mitochondria and enters the TCA (tricarboxylic acid) cycle, where it interacts with the TCA cycle intermediates and exits from the myocardium in the form of C-11-CO_2.[3] C-11 allows measurement of myocardial oxygen consumption providing parameters of oxidative metabolism as well as blood flow.[2,42-46] PET using Flourine-18 (^{18}F) Fluorodeoxyglucose (FDG) has been used for myocardial viability imaging.[47] Like D-glucose, FDG is transported into the myocardium and is trapped in the cell because of the very low levels of the enzyme, glucose-6-phosphatase responsible for catalyzing the reverse reaction.[47,48] PET with Oxygen-15 Oxygen (^{15}O-oxygen) and Carbon-11 acetate (^{11}C-acetate) has also been used to measure myocardial oxygen consumption (MVO_2), with ^{11}C-acetate as the most commonly used and most accurate method of measuring MVO_2 noninvasively. Additionally, fluoromisonidazole has also been used to measure tissue hypoxia.[47-52]

16.1.2 Nuclear Imaging in Cardiac Resynchronization Therapy

Cardiac resynchronization therapy (CRT), which has become an important therapeutic option for patients with refractory heart failure, is typically employed in patients with NYHA class II-IV heart failure, depressed LV function, sinus rhythm, and an intraventricular conduction defect (IVCD). Despite these selection criteria, several trials show that a large portion of patients fail to improve after CRT.[53] Accordingly, other factors to aid in the selection of patients that are likely to respond to CRT have been sought. One such factor, the presence of LV dyssynchrony, which has been shown in several trials to identify patients likely to benefit from CRT, is most often measured using echocardiography (ECHO).[54] However, when these ECHO approaches have been applied in a large multicenter setting, as was done in the PROSPECT trial, the value of using ECHO information in the selection of patients for CRT is rather modest,[55] possibly a result of the relatively large intra- and inter-reader variability in ECHO measurements. Accordingly, better methods for measuring cardiac dyssynchrony are desired.

Nuclear imaging techniques have long been investigated for evaluation of dyssynchrony, the most commonly employed nuclear method being gated blood-pool ventriculography scanning (GPBS) with phase analysis.[53,56] During GPBS phase analysis, a first-, second-, or third-harmonic Fourier analysis of the blood-pool-time-versus-radioactivity curve is typically performed (Fig. 16.3). To accomplish this, the time-activity curve from each pixel is analyzed as a series of cosine waves, each having a frequency, amplitude, and phase shift relative to the R-wave. As a result, several characteristics of cardiac contraction are derived, including the timing of ventricular contraction (from phase angle data), and relative ventricular synchrony (from standard deviation of ventricular phase data).[53,56,58]

Henneman et al. observed that two indices of phase analysis, histogram bandwidth, and phase standard deviation (measures degree of dyssynchrony) correlate well with ECHO measures of LV dyssynchrony (tissue Doppler imaging, or TDI).[59] In their study, they reported correlation coefficients of 0.89 and 0.80 between TDI and histogram bandwidth and phase SD measures of dyssynchrony, respectively ($P<0.001$ for both). Furthermore, phase analysis measures can be largely automatically measured, and have relatively low inter- and intra-observer variability,[60] making it an attractive modality for clinical use.

Several studies have been performed to evaluate cardiac dyssynchrony with phase analysis of radionuclide angioscintigraphy.[53,56,61-64] In an early pilot study of patients with idiopathic dilated cardiomyopathy and IVCD, phase analysis was employed to evaluate the effect of biventricular pacing on contractile synchrony and left ventricular ejection fraction (LVEF).[56] This study demonstrated improvement of interventricular synchrony and an increase in LVEF during biventricular pacing in 9 out of 13 patients. The improvement of interventricular dyssynchrony correlated with the increase in LVEF while no significant correlation was found between baseline QRS duration and LVEF improvement. A prospective study by Toussant et al.[64] investigated the long-term effects of biventricular pacing on ventricular electromechanical resynchronization as assessed with radionuclide angioscintigraphy phase analysis on 34 patients. All patients had NYHA class III-IV, LVEF <40, and a wide QRS (>150 ms) with a left bundle branch block (LBBB) pattern. Radionuclide angioscintigraphy was performed before CRT, 8 days after implantation, and every 6 months thereafter with follow-up duration of 20 ± 7 months. Results demonstrated that biventricular pacing reduced electrical and mechanical dyssynchrony and improved clinical status (Fig. 16.4).

While the above studies suggest utility of GPBS phase analysis to improve selection among patients that meet standard ACC/AHA criteria for CRT, studies using GPBS phase analysis are underway to identify patients that do not meet classical criteria but which might none-the-less benefit from CRT. One recent study shows that patients with mild-to-moderate LV dysfunction (EF 36–50%) exhibit significant degrees of dyssynchrony compared to normal controls.[65] Another recent study demonstrates that approximately 29% of patients with LVEF 35–50% and QRS duration <120 ms manifest mechanical dyssynchrony.[66] Such provocative observations suggest that GPBS phase analysis might also prove useful for identifying patients who might benefit from CRT despite not meeting current standard criteria.

Studies suggest that the presence of scar tissue in the region of the LV pacing lead (usually in the posterolateral wall) may limit response to CRT.[64,65] Gated SPECT using ^{99m}Tc-tetrofosmin was used by Ypenburg et al.[67] to evaluate the presence of scar tissue before CRT and found out that patients without scar tissue in the LV pacing target region showed improvement in the NYHA class, quality of life, 6 min walk test, LV volumes, and LVEF at 6 month follow-up whereas no improvement was observed in patients with scar tissue. Additionally, extent of viable myocardium was positively related to decreased LV volumes and improved LVEF. Other studies reported similar findings.[68,69] Furthermore, using FDG imaging, studies showed that

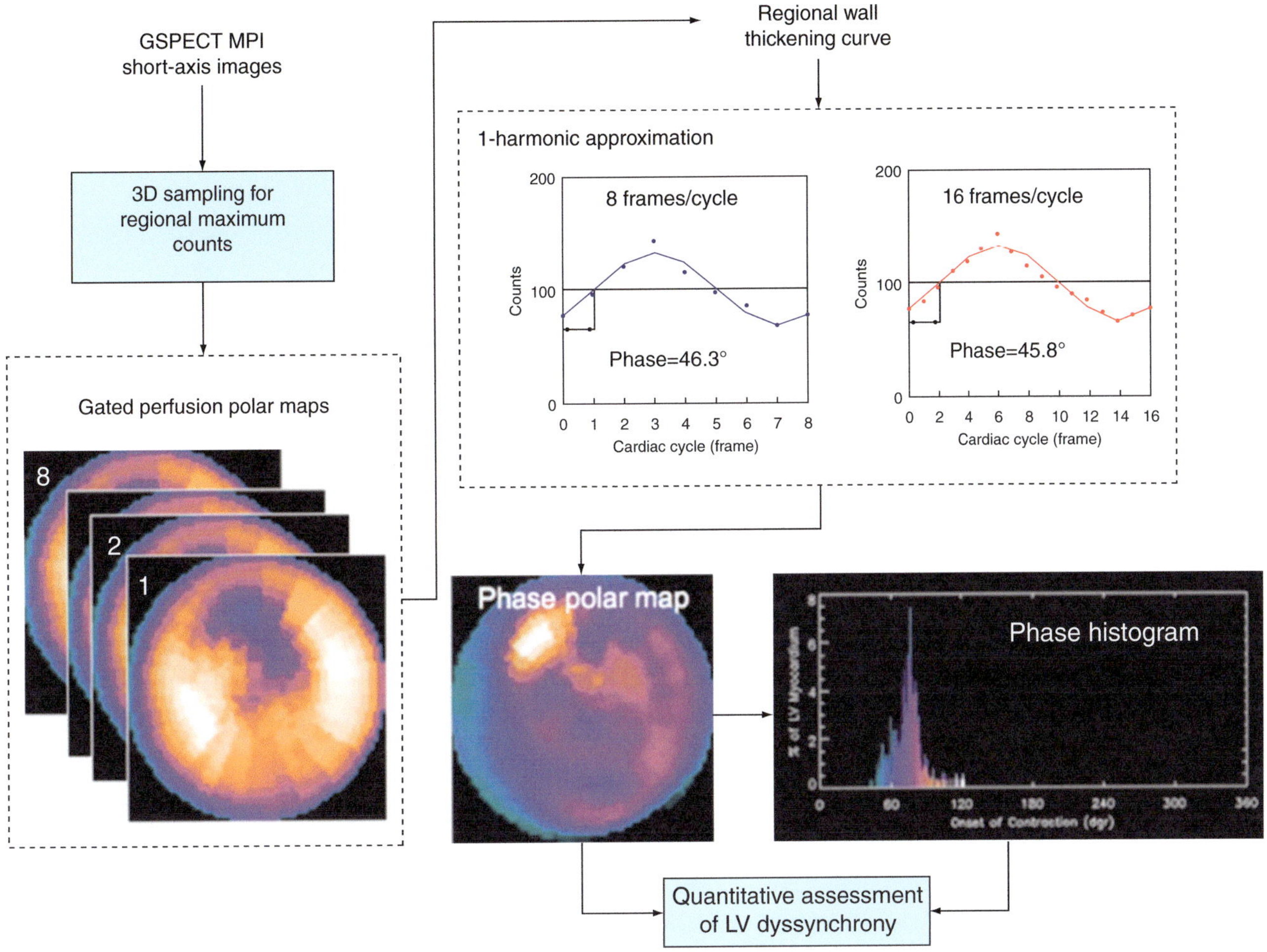

Fig. 16.3 Demonstrating the derivation of SPECT myocardial perfusion imaging to derive information about myocardial perfusion as well as phase analysis to assess LV dyssynchrony. The points in the sinograms represent regional wall thickening data. The first-harmonic approximation for 8 or 16 frames/cycle is shown as *solid lines*. The phase polar map shows a significant phase delay (bright region) at the anterior and apical wall. The location of the phase delay matches well with the perfusion defect shown in the perfusion polar map (Adapted from Chen et al.,[57] with permission)

clinical response to CRT could be predicted with a sensitivity of 74% and a specificity of 87% when substantial viability was present. On the other hand, extensive scar tissue predicted a non-response to CRT with a sensitivity of 83% and a specificity of 74%.[54,70,71]

Nuclear imaging techniques have been used to study the mechanisms involved in the benefits seen with CRT. Several studies[54,72-76] have investigated the effect of CRT on global MVO_2 using PET with ^{11}C-acetate. The results of these studies showed no significant change in global MVO_2 despite a significant improvement in LV systolic function. Moreover, several other studies examined the effect of CRT on myocardial blood flow (MBF), none of which showed significant difference in global MBF with CRT compared with no CRT.[53,70,72-75,77-80] The effect of CRT on myocardial energetics had also been investigated with ^{18}FDG-PET. Such studies have demonstrated that CRT leads to more homogeneous glucose metabolism.[53,70,71]

16.1.2.1 Cardiac Neuronal Imaging

Cardiac neural autonomic function plays an important role in cardiac electrophysiology. Several studies have demonstrated that abnormal cardiac sympathetic function is associated with worsening heart failure and increased cardiac mortality. Systemic catecholamine levels increase in advanced heart failure. At the level of the cardiac sympathetic nerve terminal, there is an initial increase in receptor-mediated NE uptake (uptake-1 mechanism). However, with advancing heart failure, there is eventual downregulation of uptake-1-related proteins and loss of neurons. This neuronal deregulation has proven to be amenable to imaging (Fig. 16.5).

Several radiotracers have been developed to characterize cardiac sympathetic function. ^{123}I metaiodobenzylguanidine (MIBG) has been used extensively in Europe and Japan for cardiac imaging. This agent is an analog of norepinephrine that diffuses into the synaptic space where it accumulates in the pre-synaptic nerve terminal by way of the uptake-1

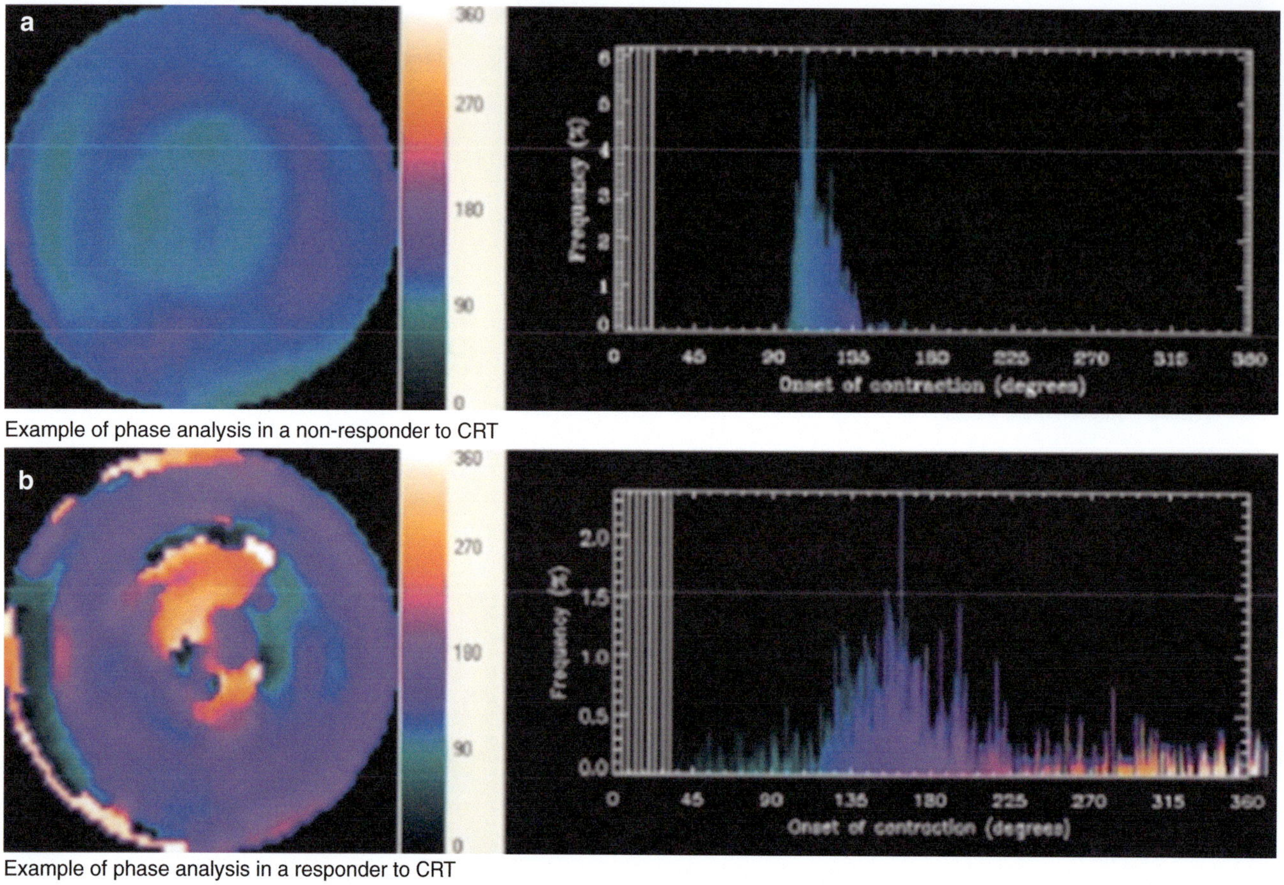

Fig. 16.4 Example images from two patients with NYHA functional class III, depressed left ventricular ejection fraction (LVEF) (<35%), and prolonged QRS duration (>120 ms) who underwent CRT. Phase analyses are displayed for a non-responder (**a**) and a responder (**b**) to CRT. LV dyssynchrony with phase analysis was absent in the non-responder, but present in the responder. At 6 month after CRT, the non-responder experienced worsened CHF (NYHA functional class increased from III to IV), whereas the responder improved (NYHA functional class decreased from III to II), while the change in LVEF post-CRT was minimal for both patients (non-responder: from 32% to 33%; responder: from 27% to 33%) (Adapted from Chen et al.,[57] with permission)

pathway (a protein-mediated, energy-dependent pathway). This agent is imaged using SPECT, during which MIBG uptake is typically assessed as a heart to mediastinal ratio (HMR). MIBG washout is also sometimes assessed when planar imaging techniques are employed, and may reflect turnover of catecholamines.

In advanced heart failure, the HMR is decreased and the washout is increased. Several studies have shown that a decreased HMR in patients with heart failure is associated with a poor prognosis.[82,83] One study of 414 patients that underwent MIBG imaging found HMR to be a more powerful predictor of overall cardiac death than NYHA class, the presence of previous myocardial infarction, and age.[84] Further, Agostini et al. performed 123I-mIBG imaging on 290 patients with heart failure and adjudicated events over the subsequent 24 months.[85] They demonstrated HMR and LVEF as the only significant predictors of cardiac events. Moreover, they demonstrated that a normal HMR can be used to identify patients with low risk of cardiac event within patients with low LVEF.

The risk stratification enabled by MIBG imaging raises the possibility that MIBG imaging can be used to select patients for ICD implantation. Nagahara et al. showed that MIBG activity combined with plasma brain natriuretic peptide (BNP) and LVEF has been shown to be associated with fatal cardiac events and may be useful in the selection of patients who would benefit from an implantable cardioverter defibrillators (ICD). The study prospectively followed up 54 ICD-treated patients after assessment of MIBG activity, plasma BNP, and LVEF. The results showed that a low level of MIBG activity, a high plasma BNP, and an LVEF of <50% can predict fatal outcome, either cardiac death or an ICD shock precipitated by lethal arrhythmias.[86] Multicenter, randomized prospective trials are needed to further clarify the utility of using MIBG imaging for selection of patients for ICD therapy.

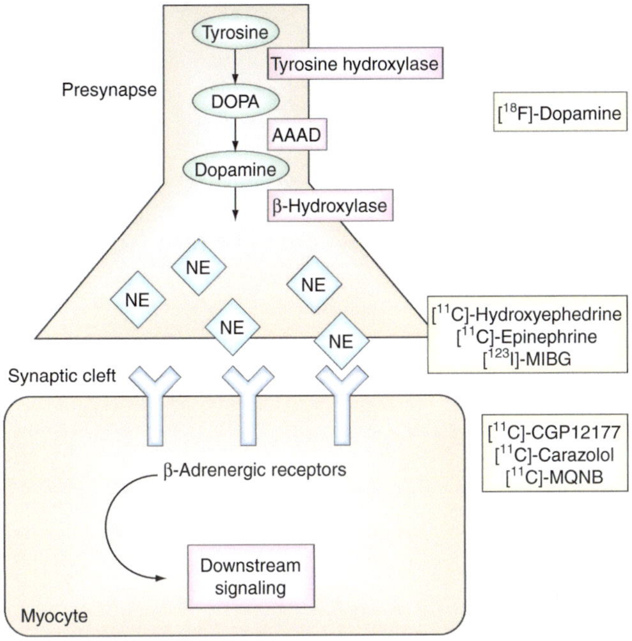

Fig. 16.5 Cardiac neurotransmission. Presynaptic production, release, and reuptake of norepinephrine enable the use of several imaging tracers to assess cardiac neuronal function. Norepinephrine is synthesized from tyrosine via DOPA and dopamine through the enzymatic actions of tyrosine hydroxylase, AAAD, and -hydroxylase. Radiopharmaceuticals (shown in yellow boxes) bind to the adrenergic receptors on the postsynaptic myocyte. *AAAD* aromatic L-amino acid decarboxylase, *DOPA* dihydroxyphenylalanine, *MIBG* meta-iodobenzylguanide, *MQNB* methylquinuclidinyl benzilate, *NE* norepinephrine (From Chun et al.,[81] with permission)

MIBG imaging may also be useful in the evaluation of other electrophysiologic disorders. It is appreciated that neuronal system activity plays an important role for the prognosis of patients with atrial fibrillation (AF). Akutsu et al. recently investigated the role of MIBG imaging toward evaluating risk of vascular events in 69 patients with paroxysmal AF who did not have structural heart disease.[87] They found that after adjustment for age, left atrial dimension, and left ventricular function, HMR was an independent predictor of vascular events with a hazard ratio of 4.1 [95% confidence interval (CI): 1.3–12.6, $p=0.014$].

MIBG has also been used to assess changes in cardiac adrenergic activity after CRT therapy. In one such study, CRT responders defined as those patients who had ≥5% absolute increase in LVEF + improvement in ≥1 NYHA class + absence of heart failure hospitalization were shown to have lower MIBG washout at follow-up compared to non-responders, moderately correlating with LVEF improvement at follow-up but not at baseline.[88] Similar results were obtained in another study which reported a decrease in MIBG washout after 6 months of CRT, together with an improvement in NYHA classification, reduction in QRS width, improvement in echo parameters such as decrease LV end-diastolic and systolic diameter, septal to lateral delay, and decreased BNP levels.[89]

Cardiac sympathetic imaging can also be accomplished using PET, which provides greater quantitation and better spatial resolution compared to SPECT. Further, current neuronal tracers used with PET, such as C11-hydroxyephedrine (HED), have better specificity for sympathetic nerve terminals compared to MIBG. As seen with MIBG, global sympathetic denervation has been demonstrated with HED PET.[90,91] Further in a study of 46 NYHA class II-III CHF patients (mean LVEF 35% ± 8%) who had undergone HED imaging, reduced HED retention predicted death or need for heart transplantation over the subsequent 5 years.[92]

16.1.2.2 Characterization of Scar Prior to VT Ablation

Another potential role for nuclear imaging methods in electrophysiology is in the metabolic detection and characterization of myocardial scar prior to VT ablation. Current clinical mapping systems used for VT ablation have a limited ability to detect and characterize intramyocardial or epicardial scar. A recent study by Fahmy et al. evaluated feasibility and accuracy of PET/CT integration with voltage-based electroanatomical map. The investigators demonstrated that PET/CT images could be well-registered with electroanatomical maps, with a mean surface registration error of 5.1 ± 2.1 mm. Conventional voltage-based scar set at 0.5 mV (of the eight different thresholds evaluated) was found to have the greatest total percent error when compared to biological PET-based scar which correlated best with voltage threshold of 0.9 mV (Fig. 16.6). Based on these results, the authors suggested that there may be a need for a re-adjustment of the conventional voltage-based scar for a more accurate scar localization prior to VT ablation. Moreover, they proposed that in view of the better accuracy of PET/CT in defining scar, the need for acquiring detailed voltage maps may be obviated with this technique.[93] In a similar study, Tian et al. studied ten patients scheduled for VT ablation who underwent contrast-enhanced computed tomography and Rubidium-82 perfusion/F-18 Fluorodeoxyglucose metabolic PET imaging. Data from PET/CT reconstruction allowed 3D display of myocardial scar and border zone and has provided a more detailed scar anatomy, with CT providing additional anatomic changes. LV scar area from FDG PET 3D reconstruction correlated well with voltage-defined scar.[94] The above results are promising and may prove useful for accurate localization and ablation of ventricular tachycardia (VT).

16.1.3 Future Directions

Using current selection criteria, CRT is unsuccessful in about 20–30% of patients.[53] The presence of LV dyssynchrony and tissue viability has been shown to be important for predicting

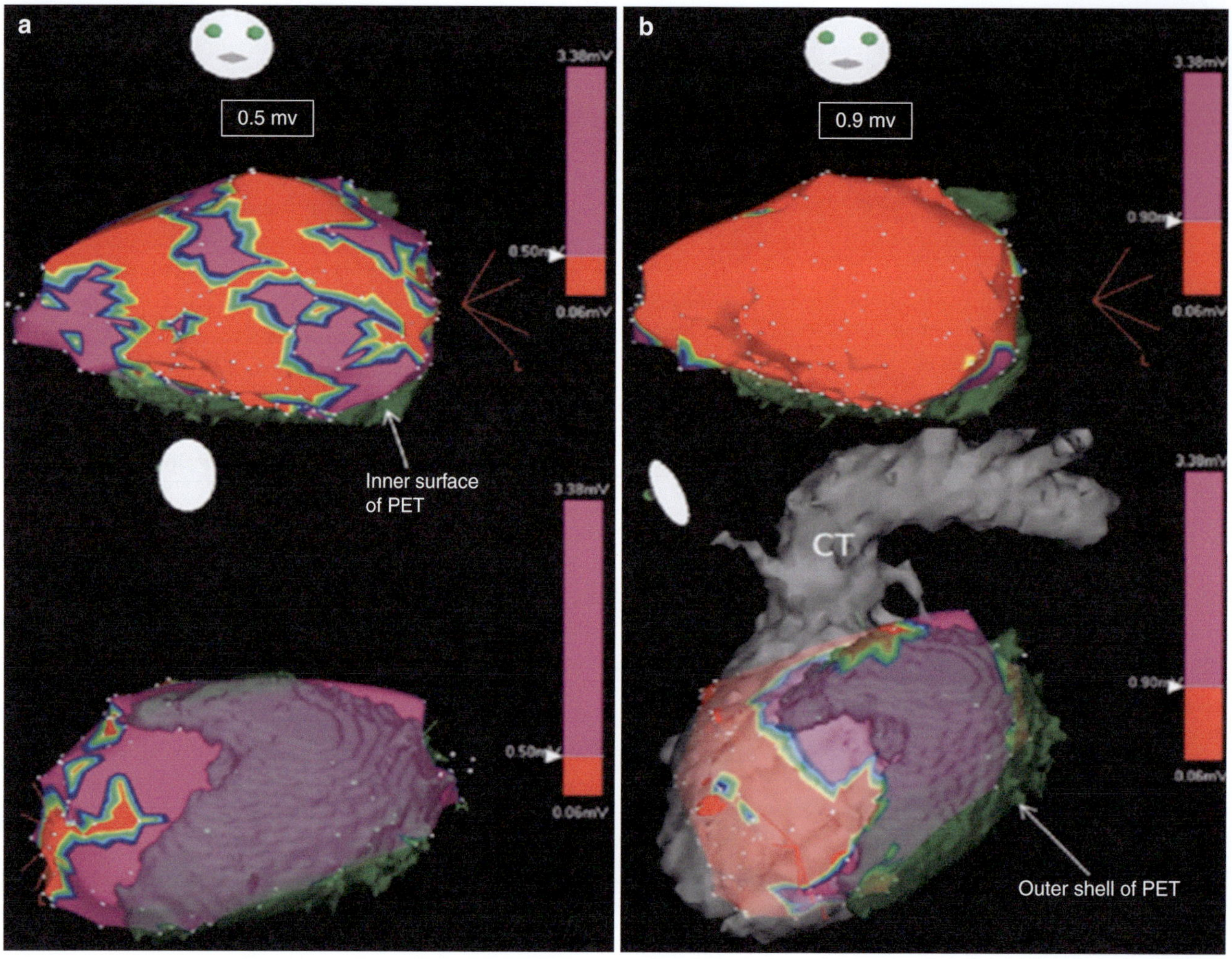

Fig. 16.6 Integration of PET/CT with electroanatomical mapping in a patient with a large scar. (**a**) Slight left anterior oblique and right post-oblique to demonstrate scar areas at threshold 0.5 mV. (**b**) Same views at threshold 0.9 mV. Note that the latter electroanatomical map correlates better with PET (From Fahmy et al.,[93] with permission)

a response to CRT, and nuclear imaging techniques such as gated blood-pool ventriculography and phase analysis can provide such clinical information and may be considered to be a promising alternative or may even prove superior to echocardiography. The results of large prospective trials should clarify the role of GPBS phase analysis for the selection of patients for device therapy.

Scintigraphic imaging for managing patients with risk for arrhythmic events is rapidly changing. Cardiac MIBG activity in combination with other clinical parameters such as BNP and LVEF may prove clinically useful in the risk-stratification and selection of patients for ICD. The development of highly sensitive and specific neuronal PET tracers has the potential to substantially change the approach for the selection of patients for ICD implantation. Newer agents looking at parasympathetic cardiac activity may provide better understanding of the sympatho-vagal balance in cardiac disease. Additionally, novel PET approaches for use with VT scar mapping systems may prove clinically valuable (Table 16.1).

16.2 Summary

Nuclear imaging techniques are rapidly emerging as important tools for use in selecting patients for device therapy. Further investigation, however, is needed in order to elucidate the true clinical value of these techniques. As such, large prospective trials are essential to document and validate the clinical utility of these nuclear imaging techniques. Additionally, the development of novel molecular PET tracers targeting important electrophysiologic pathways may lead to a greater understanding of relevant electrophysiologic mechanisms.

Table 16.1 PET tracers with reported human application

Tracer	Half-life, min	Myocardial uptake mechanism
Rb-82	1.3	Mechanism Flow
N-13 Ammonia	10	Flow
F-18 Deoxyglucose	110	Glucose metabolism
F-18 Misonidazole	110	Hypoxia
$H_2{}^{15}O$	2	Free diffusion (perfusion)
^{62}Cu-PTSM	9.7	Intracellular binding (perfusion)
^{18}F-BMS747158	110	Mitochondrial binding (perfusion)
^{11}C-acetate	20	Krebs cycle flux, oxidative metabolism
^{11}C-palmitate	20	Fatty acid metabolism
^{11}C-glucose	20	Glucose metabolism
^{18}F-FTHA	110	Fatty acid uptake
^{11}C-hydroxyephedrine	20	Sympathetic neuronal catecholamine uptake
^{11}C-epinephrine	20	Sympathetic neuronal catecholamine storage
^{11}C-phenylephrine	20	Sympathetic neuronal catecholamine turnover
^{18}F-fluorodopamine	110	Sympathetic neuronal uptake and function
^{11}C-CGP12177	20	β-adrenergic receptor density
^{11}C-CGP12388	20	β-adrenergic receptor density
^{11}C-GB67	20	α1-adrenergic receptor density
^{11}C-MQNB	20	Muscarinic receptor density

Adapted From Bengel et al.,[33] with permission
FTHA fluorothioheptadecanoic acid, *MQNB* methylquinuclidinyl benzilate, *PET* positron emission tomography, *PTSM* pyruvaldehyde bis(N4-methylthiosemicarbazone)

References

1. Williams LE. Anniversary paper: nuclear medicine: fifty years and still counting. *Med Phys*. 2008;35:3020-3028.
2. Schwaiger M, Ziegler S. Positron emission tomography. In: Topol E, ed. *Textbook of Cardiovascular Medicine*, vol. 55. 3rd ed. Philadelphia: Lippincott Williams & Wilkins; 2007:932-947.
3. Buchanan J, Hastings A. The use of isotopically marked carbon in the study of intermediary metabolism. *Physiol Rev*. 1946;26:120-155.
4. Tobias C, Lawrence J, Roughton F. The elimination of carbon monoxide from human body with reference to the possible conversion of carbon monoxide to carbon dioxide. *Am J Physiol*. 1945;145:253-263.
5. Cassen B, Curtis L, Reed C, Libby R. Instrumentation for 131I use in medical studies. *Nucleonics*. 1951;9:46-50.
6. Anger HO. Scintillation camera. *Rev Sci Instrum*. 1958;29:27-33.
7. Berman D, Hachamovitch R, Shaw LJ, et al. Nuclear cardiology. In: Fuster V, O'Rourke RA, Walsh RA, Poole-Wilson P, eds. *Hurst's The Heart*. 12th ed. China: Mcgraw-Hill Companies, Inc; 2008:19.
8. Germano G, Berman DS. Quantification of ventricular function. In: Germano G, Berman DS, eds. *Clinical Gated Cardiac SPECT*. 2nd ed. Oxford: Blackwell Publishing; 2006:93-137.
9. Jaszczak RJ, Tsui BMW. Single photon emission computed tomography (SPECT). In: Szabo WZ, Buchanan JW, eds. *Principles of Nuclear Medicine*. Philadelphia: WB Saunders; 1995:317-341.
10. Brownell G, Sweet W. Localization of brain tumors with positron-emitters. *Nucleonics*. 1953;11:40-45.
11. Schlyer DJ. PET tracers and radiochemistry. *Ann Acad Med Singapore*. 2004;33:146-154.
12. Aronow S. Positron scanning. In: Hine G, ed. *Instrumentation in Nuclear Medicine*. New York: Academic; 1967:461-483.
13. Phelps M, Hoffman E, Mullani N, Ter-Pogossian M. Application of annihilation coincidence detection to transaxial reconstruction tomography. *J Nucl Med*. 1975;16:210-224.
14. Ter Pogossian M, Phelps M, Hoffman E. A positron emission transaxial tomograph for nuclear medicine imaging (PETT). *Radiology*. 1975;114:89-98.
15. Hoffman EJ, Phelps ME, Weiss ES, Welch MJ, et al. Transaxial tomographic imaging of canine myocardium with 11C-palmitic acid. *J Nucl Med*. 1977;18:57-61.
16. Blankespoor SC et al. Attenuation correction of SPECT using x-ray CT on an emission-transmission CT system: myocardial perfusion assessment. *IEEE Trans Nucl Sci*. 1996;43:2263-2274.
17. Beyer T, Townsend DW, Brun T, Kinahan PE, et al. A combined PET/CT scanner for clinical oncology. *J Nucl Med*. 2000;41:1369-1379.
18. Cerqueria MD. Imaging techniques in nuclear cardiology. In: Topol E, ed. *Textbook of Cardiovascular Medicine*, vol. 53. 3rd ed. Philadelphia: Lippincott Williams & Wilkins; 2007:885-896.
19. Schelbert HR. Positron emission tomography for the noninvasive study and quantitation of myocardial blood flow and metabolism in cardiovascular disease. In: Fuster V, O'Rourke RA, Walsh RA, Poole-Wilson P, eds. *Hurst's The Heart*. 12th ed. China: Mcgraw-Hill Companies, Inc; 2008:23.
20. Fuster V, O'Rourke RA, Walsh RA, Poole-Wilson P. *Hurst's The Heart*. 12th edition. New York: The McGraw-Hill Companies, Inc. 2007. http://www.accessmedicine.com.
21. Germano G, Kiat H, Kavanagh PB, et al. Automatic quantification of ejection fraction from gated myocardial perfusion SPECT. *J Nucl Med*. 1995;36:2138-2147.
22. Bateman TM, Berman DS, Heller GV, et al. American Society of Nuclear Cardiology position statement on electrocardiographic gating of myocardial perfusion SPECT scintigrams. *J Nucl Cardiol*. 1999;6:470-471.
23. Klocke FJ, Baird MG, Bateman TM, et al. ACC/AHA/ASNC guidelines for the clinical use of cardiac radionuclide imaging: a report of the American 1995 guidelines for the clinical use of radionuclide imaging. *Circulation*. 2003;108:1404-1418.
24. Verani MS, Mahmarian JJ, Hixson JB, et al. Diagnosis of coronary artery disease by controlled coronary vasodilation with adenosine and thallium-201 scintigraphy in patients unable to exercise. *Circulation*. 1990;82:80-87.
25. Weiss AT, Berman DS, Lew AS, et al. Transient ischemic dilation of the left ventricle on stress thallium-201 scintigraphy: a marker of severe and extensive coronary artery disease. *J Am Coll Cardiol*. 1987;9:752-759.
26. Mazzanti M, Germano G, Kiat H, et al. Identification of severe and extensive coronary artery disease by automatic measurement of transient ischemic dilation of the left ventricle in dual-isotope myocardial perfusion SPECT. *J Am Coll Cardiol*. 1996;27:1612-1620.
27. Berman DS, Germano G. Interpretation and reporting of gated myocardial perfusion SPECT. In: Germano G, Berman DS, eds. *Clinical Gated Cardiac SPECT*. 2nd ed. Oxford: Blackwell Publishing; 2006:139-171.
28. Van Kriekinge SD, Berman DS, Germano G. Automatic quantification of left ventricular ejection fraction from gated blood pool SPECT. *J Nucl Cardiol*. 1999;6:498-506.
29. Nishimura Rick A, Gibbons Raymond J, Glockner James F, Tajik AJ. Noninvasive cardiac imaging: echocardiography, nuclear cardiology, and MRI/CT imaging. In: Fauci AS, Braunwald E, Kasper DL, Hauser SL, Longo DL, Jameson JL, Loscalzo J, eds. *Harrison's Principles of Internal Medicine*; chap 222; 17e: http://www.accessmedicine.com/content.aspx?aID=2887731.

30. Boogers MM, Chen J, Bax J. Myocardial perfusion single photon emission computed tomography for the assessment of mechanical dyssynchrony. *Curr Opin Cardiol.* 2008;23:431-439.
31. Valk PE, Bailey DL, Townsend DW, Maisey MN. *Positron Emmission Tomography: Basic Science and Clinical Practice.* London: Springer; 2002.
32. Di Filippo FP. Instrumentation and principles of imaging: PET. In: Di Carli MF, Lipton MJ, eds. *Cardiac PET and PET/CT Imaging,* vol. 1. New York: Springer; 2007:3-18.
33. Bengel FM, Higuchi T, Javadi MS, Lautamaki R. Cardiac positron emmission tomography. *J Am Coll Cardiol.* 2009;54:1-15.
34. Hoffmann U, Butler J, Tawakol A. Evaluating high-risk, vulnerable plaques with integrated PET/CT. In: Di Carli MF, Lipton MJ, eds. *Cardiac PET and PET/CT Imaging,* vol. 23. New York: Springer; 2007:347-359.
35. Hyafil F, Fayad ZA. Evaluating vulnerable atherosclerotic plaque with MRI. In: Di Carli MF, Lipton MJ, eds. *Cardiac PET and PET/CT Imaging.* New York: Springer; 2007;24:360-359-372.
36. Gould K. Assessment of coronary stenoses with myocardial perfusion imaging during pharmacologic coronary vasodilation. *Am J Cardiol.* 1978;42:761-768.
37. Go RT, Marwick TH, MacIntyre WJ, et al. A prospective comparison of rubidium-82 PET and thallium-201 SPECT myocardial perfusion imaging utilizing a single dipyridamole stress in the diagnosis of coronary artery disease [see comments]. *J Nucl Med.* 1990;31:1899-1905.
38. Stewart RE, Schwaiger M, Molina E, et al. Comparison of rubidium-82 positron emission tomography and thallium-201 SPECT imaging for detection of coronary artery disease. *Am J Cardiol.* 1991;67:1303-1310.
39. Tamaki N, Yonekura Y, Senda M, et al. Myocardial positron computed tomography with 13N-ammonia at rest and during exercise. *Eur J Nucl Med.* 1985;11:246-251.
40. Gould KL, Goldstein RA, Mullani NA, et al. Noninvasive assessment of coronary stenoses by myocardial perfusion imaging during pharmacologic coronary vasodilation. VIII. Clinical feasibility of positron cardiac imaging without a cyclotron using generator-produced rubidium-82. *J Am Coll Cardiol.* 1986;7:775-789.
41. Demer LL, Gould KL, Goldstein RA, et al. Assessment of coronary artery disease severity by positron emission tomography. Comparison with quantitative arteriography in 193 patients. *Circulation.* 1989;79:825-835.
42. Buxton DB, Schwaiger M, Nguyen A, et al. Radiolabeled acetate as a tracer of myocardial tricarboxylic acid cycle flux. *Circ Res.* 1988;63:628-634.
43. Brown MA, Marshall DR, Sobel BE, Bergmann SR. Delineation of myocardial oxygen utilization with carbon-11 labeled acetate. *Circulation.* 1987;76:687-696.
44. Beanlands RS, Nahmias C, Gordon E, et al. The effects of beta(1)-blockade on oxidative metabolism and the metabolic cost of ventricular work in patients with left ventricular dysfunction: a double-blind, placebo controlled, positron emission tomography study. *Circulation.* 2000;102:2070-2075.
45. Buck A, Wolpers HG, Hutchins GD, et al. Effect of carbon-11-acetate recirculation on estimates of myocardial oxygen consumption by PET. *J Nucl Med.* 1991;32:1950-1957.
46. Gropler RJ, Siegel BA, Geltman EM. Myocardial uptake of carbon-11-acetate as an indirect estimate of regional myocardial blood flow. *J Nucl Med.* 1991;32:245-251.
47. Machac J. Radiopharmaceuticals for clinical cardiac PET imaging. In: Di Carli MF, Lipton MJ, eds. *Cardiac PET and PET/CT Imaging.* New York: Springer; 2007:73-82.
48. Gallagher BM, Ansari A, Atkins H, et al. Radiopharmaceuticals XXVII. 18F-labeled 2-deoxy-2-fluoro-D-glucose as a radiopharmaceutical for measuring regional myocardial glucose metabolism in vivo: tissue distribution and imaging studies in animals. *J Nucl Med.* 1977;18:990-996.
49. Shelton ME, Dence CS, Hwang DR, Welch MJ, Bergman SR. Myocardial kinetics of fluorine-18 misonidazole: a marker of hypoxic myocardium. *J Nucl Med.* 1989;30:351-358.
50. Nunn A, Linder K, Strauss HW. Nitroimidazoles and imaging hypoxia. *Eur J Nucl Med.* 1995;22:265-280.
51. Dearling JLJ, Lewis JS, Mullen GED, Welch MJ, Blower MJ. Copper bis (thio-semicarbazone) complexes as hypoxia imaging agents: structure-activity relationships. *J Biol Inorg Chem.* 2002;7:249-259.
52. Fujibayashi Y, Taniuchi H, Yonekura Y, Ohtani H, Konishi J, Yokoyama A. Copper-62-ATSM: a new hypoxia imaging agent with high membrane permeability and low redox potential. *J Nucl Med.* 1997;38:1155-1160.
53. Henneman M, van der Wall EE, Ypenburg C, et al. Nuclear imaging in cardiac resynchronization therapy. *J Nucl Med.* 2007;48: 2001-2010.
54. Bax JJ, Bleeker GB, Marwick TH, et al. Left ventricular dyssynchrony predicts response and prognosis after cardiac resynchronization therapy. *J Am Coll Cardiol.* 2004;44:1834-1840.
55. Chung ES, Leon AR, Tavazzi L, et al. Results of predictors of response to CRT (PROSPECT) trial. *Circulation.* 2008;117(20): 2608-2616.
56. Botvinick EH. Scintigraphic blood pool and phase image analysis: the optimal tool for the evaluation of resynchronization therapy. *J Nucl Cardiol.* 2003;10:424-428.
57. Chen J, Bax JJ, Henneman MM, Boogers MJ, Garcia EV. Is nuclear imaging a viable alternative technique to assess dyssynchrony? *Europace.* 2008;10:iii101-iii105.
58. Chen J, Garcia E, Lerakis S, et al. Left ventricular mechanical dyssynchrony as assessed by phase analysis of ECG-gated SPECT myocardial perfusion imaging. *Echocardiography.* 2008;25: 1186-1194.
59. Henneman MM, Chen J, Ypenburg C, et al. Phase analysis of gated myocardial perfusion SPECT compared to tissue Doppler imaging for the assessment of left ventricular dyssynchrony. *J Am Coll Cardiol.* 2007;49:1708-1714.
60. Trimble MA, Velazquez EJ, Adams GL, et al. Repeatability and reproducibility of phase analysis of gated single-photon emission computed tomography myocardial perfusion imaging used to quantify cardiac dyssynchrony. *Nucl Med Commun.* 2008;29(4): 374-381.
61. Kerwin WF, Botvinick EH, O'Connell JW, et al. Ventricular contraction abnormalities in dilated cardiomyopathy: effect of biventricular pacing to correct interventricular dyssynchrony. *J Am Coll Cardiol.* 2000;35:1221-1227.
62. Fauchier L, Marie O, Casset-Senon D, et al. Interventricular and intraventricular dyssynchrony in idiopathic dilated cardiomyopathy: prognostic study with Fourier phase analysis of radionuclide angioscintigraphy. *J Am Coll Cardiol.* 2002;40:2022-2030.
63. Fauchier L, Marie O, Casset-Senon D, et al. Reliability of QRS duration and morphology on surface electrocardiogram to identify ventricular dyssynchrony in patients with idiopathic dilated cardiomyopathy. *Am J Cardiol.* 2003;92:341-344.
64. Toussaint JF, Lavergne T, Kerrou K, et al. Basal asynchrony and resynchronization with biventricular phasing predict long-term improvement of LV function in heart failure patients. *PACE.* 2003; 26:1815-1823.
65. Trimble MA, Borges-Neto S, Honeycutt EF, et al. Evaluation of mechanical dyssynchrony using phase analysis of gated SPECT imaging in patients with left ventricular dysfunction. *J Nucl Cardiol.* 2008;15:663-670.
66. Atchley AE, Trimble MA, Samad Z, et al. Use of phase analysis of gated SPECT perfusion imaging to quantify dyssynchrony in patients with mild-to-moderate left ventricular dysfunction. *Nucl Cardiol.* 2009;16(6):888-894. Epub 2009 Aug 19.
67. Ypenburg C, Schalij MJ, Bleeker GB, et al. Impact of viability and scar tissue on response to cardiac resynchronization therapy in ischemic heart failure patients. *Eur Heart J.* 2007;28:33-41.

68. Sciagra R, Giaccardi M, Porciani MC, et al. Myocardial perfusion imaging using gated SPECT in heart failure patients undergoing cardiac resynchronization therapy. *J Nucl Med*. 2004;45:164-168.
69. Adelstein EC, Saba S. Scar burden by myocardial perfusion imaging predicts echocardiographic response to cardiac resynchronization therapy in ischemic cardiomyopathy. *Am Heart J*. 2007;153: 105-112.
70. Neri G, Zanco P, Zanon F, Buchberger R. Effect of biventricular pacing on metabolism and perfusion in patients affected by dilated cardiomyopathy and left bundle branch block: evaluation by positron emission tomography. *Europace*. 2003;5:111-115.
71. Nowak B, Sinha AM, Schaefer WM, et al. Cardiac resynchronization therapy homogenizes myocardial glucose metabolism and perfusion in dilated cardiomyopathy and left bundle branch block. *J Am Coll Cardiol*. 2003;41:1523-1528.
72. Sundell J, Engblom E, Koitinen J, et al. The effects of cardiac resynchronization therapy on left ventricular function, myocardial energetics, and metabolic reserve in patients with dilated cardiomyopathy and heart failure. *J Am Coll Cardiol*. 2004;43: 1027-1033.
73. Braunschweig F, Sorensen J, von Bibra H, et al. Effects of biventricular pacing on myocardial blood flow and oxygen consumption using carbon-11 acetate positron emission tomography in patients with heart failure. *Am J Cardiol*. 2003;92:95-99.
74. Lindner O, Vogt J, Kammeier A, et al. Effect of cardiac resynchronization therapy on global and regional oxygen consumption and myocardial blood flow in patients with non-ischemic and ischemic cardiomyopathy. *Eur Heart J*. 2005;26(1):2670-2676.
75. Ukkonen H, Beanlands RS, Burwash IG, et al. Effect of cardiac resynchronization on myocardial efficiency and regional oxidative metabolism. *Circulation*. 2003;107:28-31.
76. Lindner O, Sorensen J, Vogt J, et al. Cardiac efficiency and oxygen consumption measured with ^{11}C-acetate PET after long-term cardiac resynchronization therapy. *J Nucl Med*. 2006;47:378-383.
77. Lindner O, Vogt J, Kammeier A, et al. Cardiac resynchronization therapy: effects on myocardial perfusion at rest, after vasodilatation and oxygen consumption. *Nuklearmedizin*. 2006;45:10-14.
78. Nowak B, Stellbrink C, Sinha AM, et al. Effects of cardiac resynchronization therapy on myocardial blood flow measured by oxygen-15 water positron emission tomography in idiopathic-dilated cardiomyopathy and left bundle branch block. *Am J Cardiol*. 2004;93:496-499.
79. Nielsen JC, Bottcher M, Jensen HK, Nielsen TT, Pedersen AK, Mortensen PT. Regional myocardial perfusion during chronic biventricular pacing and after acute change of the pacing mode in patients with congestive heart failure and bundle branch block treated with an atrioventricular sequential biventricular pacemaker. *Eur J Heart Fail*. 2003;5:179-186.
80. Knaapen P, van Campen LM, de Cock CC, et al. Effects of cardiac resynchronization therapy on myocardial perfusion reserve. *Circulation*. 2004;110:646-651.
81. Chun HJ et al. Intracellular and extracellular targets of molecular imaging in the myocardium. *Nat Clin Pract Cardiovasc Med*. 2008;5(suppl 2):S33-S41. doi:10.1038/ncpcardio1161
82. Anastasiou-Nana MI, Terrovitis JV, Athanasoulis T, et al. Prognostic value of iodine-123-metaiodobenzylguanidine myocardial uptake and heart rate variability in chronic congestive heart failure secondary to ischemic or idiopathic dilated cardiomyopathy. *Am J Cardiol*. 2005;96(3):427-431.
83. Somsen GA, Verberne HJ, Fleury E, Righetti A. Normal values and within-subject variability of cardiac I-123 MIBG scintigraphy in healthy individuals: implications for clinical studies. *J Nucl Cardiol*. 2004;11(2):126-133.
84. Nakata T, Miyamoto K, Doi A, et al. Cardiac death prediction and impaired cardiac sympathetic innervation assessed by MIBG in patients with failing and nonfailing hearts. *J Nucl Cardiol*. 1998; 5:579-590.
85. Agostini D, Verberne HJ, Burchert W, et al. I-123- m IBG myocardial imaging for assessment of risk for a major cardiac event in heart failure patients: insights from a retrospective European multicenter study. *Eur J Nucl Med Mol Imaging*. 2008;35:535-546.
86. Nagahara D, Nakata T, Hashimoto A, et al. Predicting the need for an implantable cardioverter defibrillator using cardiac metaiodobenzylguanidine activity together with plasma natriuretic peptide concentration or left ventricular function. *J Nucl Med*. 2008;49:225-233.
87. Akutsu Y, Kaneko K, Kodama Y, et al. Significance of cardiac sympathetic nervous system abnormality for predicting vascular events in patients with idiopathic paroxysmal atrial fibrillation. *Eur J Nucl Med Mol Imaging*. 2009;37(4):742-9. Epub 2009 Dec 17.
88. Burri H, Sunthorn H, Somsen A, et al. Improvement in cardiac sympathetic nerve activity in responders to resynchronization therapy. *Europace*. 2008;10:374-378.
89. Erol-Yilmaz A, Verbene HJ, Schrama TA, et al. Cardiac resynchronization induces favorable neurohumoral changes. *Pacing Clin Electrophysiol*. 2005;28:304-310.
90. Bengel FM, Schwaiger M. Assessment of cardiac sympathetic neuronal function using PET imaging. *J Nucl Cardiol*. 2004;11: 603-616.
91. Bengel FM, Permanetter B, Ungerer M, Nekolla S, Schwaiger M. Relationship between altered sympathetic innervation, oxidative metabolism and contractile function in the cardiomyopathic human heart; a non-invasive study using positron emission tomography. *Eur Heart J*. 2001;22:1594-1600.
92. Pietila M, Malminiemi K, Ukkonen H, et al. Reduced myocardial carbon-11 hydroxyephedrine retention is associated with poor prognosis in chronic heart failure. *Eur J Nucl Med*. 2001;28:373-376.
93. Fahmy TS, Wazni OM, Jaber WA, et al. Integration of positron emission tomography/computed tomography with electroanatomical mapping: a novel approach for ablation of scar-related ventricular tachycardia. *Heart Rhythm*. 2008;5:1538-1545.
94. Tian J, Smith MF, Chinnadurai PJ, et al. Clinical application of PET/CT fusion imaging for three-dimensional myocardial scar a left ventricular anatomy during ventricular tachycardia ablation. *J Cardiovasc Electrophysiol*. 2009;20:597-604.

Post-CRT Imaging Based Optimization

17

Veronica Lea J. Dimaano, Aurélio C. Pinheiro,
and Theodore P. Abraham

Abstract

The utility of CRT optimization remains to be confined to patients whose response to the treatment is either suboptimal or negligible. There are several techniques, mostly echocardiographic, to optimize device settings but none has been universally accepted as gold standard. Methods differ in recording techniques and may therefore vary significantly in performance. Several studies have employed various techniques in interrogating AV interval with or without concomitant V-V interval optimization. Optimization was performed at varying time points. Even the definition for favorable response to CRT varied. Optimization of AV and V-V intervals is patient-specific and optimal values change over time. This may relate to time-related LV reverse remodeling or to progression of disease. The effects of inter-atrial conduction defect, right atrial pacing, and exercise on AV delay add complexities to performance of AV delay optimization. At the very least, it has to be ensured that the programmed AV delay should neither produce fusion of mitral E and A waves nor truncation of the latter. Despite evidences of hemodynamic benefits, the utility of V-V interval optimization remains controversial. Evidence for the incremental long-term benefits on mortality and morbidity is still lacking. Three large clinical trials reported conflicting results in terms of clinical and echocardiographic improvement. Differences in the study designs could have contributed to the conflicting results. It is important to remember, however, that simultaneous biventricular pacing is not always synonymous with suboptimal V-V interval, nor should sequential pacing be routinely regarded as ideal.

Keywords

Cardiac resynchronization therapy • Imaging post CRT • Echocardiography • CRT optimization • AV delay optimization

Cardiac resynchronization therapy (CRT) has been shown to improve cardiac structure and function, quality of life (QOL), and exercise capacity; alleviate symptoms; and lower mortality and morbidity in patients with moderate-to-severe heart failure and intraventricular conduction defect.[1-5] The effectiveness of CRT per se depends on the adequacy of mechanical synchrony achieved at atrio-ventricular (AV), interventricular, and intraventricular levels. The clinical and echocardiographic response to CRT may vary significantly among patients. Like any effective drug or device therapy for heart failure patients, response to therapy can be heterogeneous. Dubious response to CRT may be attributed to the interplay of factors that relate to suboptimal medical therapy; inappropriate lead placement and suboptimal programming of AV timing; and occurrence of atrial tachyarrhythmias that may result in intermittent or complete loss of CRT. Addressing these issues during follow-up evaluation of CRT patients may impact on their response to therapy.[6]

T.P. Abraham (✉)
Division of Cardiology, Department of Mediciine,
Johns Hopkins University, Baltimore, MD, USA
e-mail: tabraha3@jhmi.edu

A. Auricchio et al. (eds.), *Cardiac Imaging in Electrophysiology*,
DOI 10.1007/978-1-84882-486-7_17,

17.1 Optimizing CRT

There are a considerable number of patients who have minimal benefit or no improvement at all from CRT and are considered as "non-responders." A recent study comprehensively evaluated 75 ambulatory recipients of CRT plus defibrillator (CRT-D) devices for a minimum duration of 6 months who persistently experienced heart failure symptoms. The investigators reported a high percentage (47%) of suboptimal AV delay programming detected in this cohort of clinical non-responders.[6] While the European Society of Cardiology recommended performance of AV and ventriculo-ventricular (V-V) interval optimization to maintain atrial-synchronous permanent biventricular pacing,[7] there is still no consensus for routine performance of AV and V-V delay optimization after CRT. The current use of CRT optimization is limited and is still considered as an *option* in an attempt to convert patients judged to be "non-responders" to "responders."[8]

17.2 Echocardiography and Optimal CRT

Currently, echocardiography is the modality most widely used in optimizing AV and V-V intervals in CRT.[8] Echocardiographic measurements are safe to acquire, easily repeatable, and are accurate when compared to invasive hemodynamic measurements. Optimizing pacemaker settings can be incorporated into a follow-up study allowing simultaneous assessment of reverse remodeling, mitral regurgitation (MR), and left ventricular (LV) function. However, it requires a certain level of expertise among those acquiring and interpreting the studies. Moreover, optimizing pacemaker settings using some echocardiographic techniques may require a significant length of time which, in turn, may impact on the patient, the consistency in image acquisition, and the availability of the personnel performing the optimization.

17.3 Atrio-Ventricular Interval Optimization

During AV sequential pacing, the effect of AV interval on LV filling impacts on LV systolic performance by modulating preload. Inadequate programming of the atrial to *left* ventricular delay has the potential to curtail the beneficial effects of CRT. While optimizing AV delay may benefit "under-responders," it cannot convert a *true* "non-responder" to a "responder."[9] AV delay optimization has been shown to improve the acute hemodynamic response to CRT and may improve long-term clinical outcomes. The optimal AV delay in CRT exhibits great variability from patient to patient.[10-12] Moreover, recent studies have suggested that the optimal AV delay changes over time.[12-16]

17.3.1 Atrio-Ventricular Relationships

In the setting of a prolonged AV conduction, atrial depolarization occurs relatively early in diastole resulting in the superimposition of atrial contraction on the early LV filling phase, thereby compromising atrial contribution to cardiac output (CO). Following atrial contraction, the mitral valve remains open as a result of delayed LV contraction. As LV diastolic pressure exceeds left atrial (LA) pressure during atrial relaxation, diastolic MR ensues. This condition decreases preload and reduces LV end-diastolic pressure (LVEDP) at the onset of LV systole with subsequent reduction in LV rate of pressure rise (LV dP/dt_{max}) and ultimately, CO.[10] In the setting of a short AV delay, on the other hand, the LA contribution to LV filling is compromised by premature mitral valve closure that results from early LV contraction. Thus, an optimal AV relationship is achieved by an AV interval that provides the longest LV filling duration that allows maximal LA contribution to LV filling, eliminates diastolic MR, and shortens isovolumic contraction phase, thus resulting in greater stroke volume (SV) and CO.[17]

Translated echocardiographically, prolonged AV conduction is characterized by the fusion of the transmitral Doppler E and A wave and the appearance of MR signals during late diastole (reduced LV filling time). Unphysiologically short AV delay is characterized by a distinct separation of the Doppler E wave and a truncated A wave (increased LV filling time) (Fig. 17.1). The optimal AV delay is the shortest AV delay that will allow maximal EA duration without causing abbreviation of transmitral Doppler A wave. The onset of rise of LV pressure should coincide with the end of atrial contraction.

17.3.2 Echocardiography-Based Atrio-Ventricular Delay Optimization

There are many ways of optimizing the AV delay. Doppler echocardiographic methods include analysis of mitral inflow and LV outflow (LVOT) profile using pulsed-Doppler, and

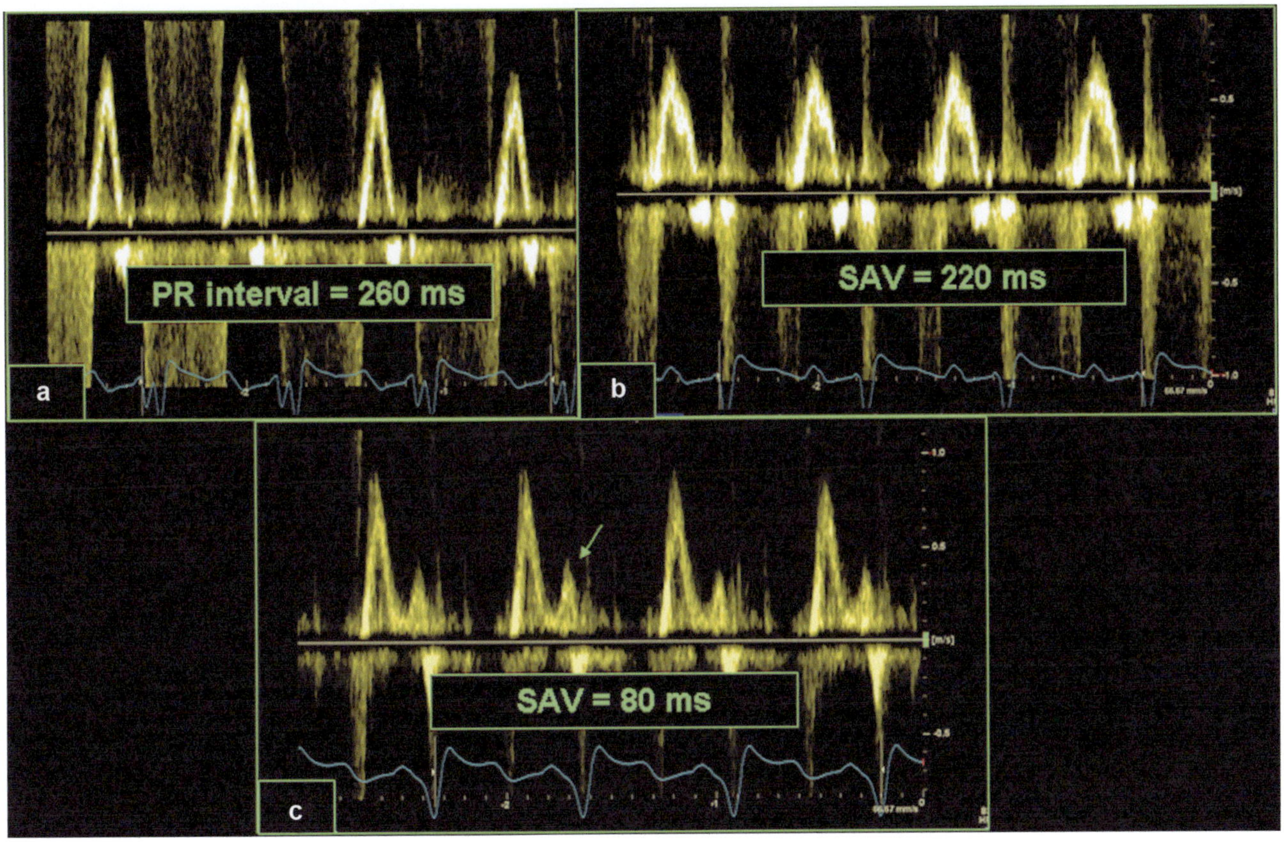

Fig. 17.1 Effects of atrio-ventricular dyssynchrony on diastolic filling. During (**a**) prolonged atrio-ventricular conduction, or (**b**) programmed long AV delay, LV filling is compromised by a relative delay in mitral valve closure and opening, causing superimposition of early and late diastolic filling and is reflected by fused mitral Doppler E and A waves. (**c**) With a programmed short AV delay, the contribution of atrial kick to LV filling is abbreviated by the early closure of the mitral valve causing truncation of A wave (*green arrow*). *SAV-sensed AV delay*

trans-aortic flow and MR velocity profiles using continuous wave (CW)-Doppler techniques. Using these flow velocity profiles, optimal AV delay can be calculated from temporal parameters measured during specific short and long AV delays (calculation-based) or estimated by a single hemodynamic parameter measured against a series of practicable AV intervals (dynamic AV interval-based). A wide array of optimization techniques has been used in several small studies, but none has been universally considered as the gold standard. The results may vary according to recording techniques and therefore may differ substantially in performance.[18]

17.3.2.1 Calculation-Based Method of AV Delay Optimization

Optimization of AV delay by calculation-based method is performed by pulsed-Doppler interrogation of mitral inflow. It aims to maximize LV filling time without compromising the atrial kick. Techniques that employ analysis of mitral inflow velocity profiles generally vary in terms of the location of the sample volume. In most protocols, Doppler waveforms are recorded by sampling at the tips of the mitral leaflets where the flow velocity profile reflects the diastolic function and filling parameters rather than volume. Conversely, sample volume placed at the level of the mitral annulus reflects the diastolic filling volume rather than function. Normally, E wave velocity is higher at the tips of the leaflets, thereby relatively increasing the prominence of the A wave when waveforms are recorded by sampling at the mitral annulus.[19] Calculation-based method involves measurements of several time intervals during a programmed short and long AV delays. The measurements rely on the A wave configuration. Because it entails recordings at only two AV intervals, the time required for performing optimization is generally less.

Ritter's technique has been the most popular calculation-based technique of optimizing AV delay after CRT. According to Ritter, the most ideal AV interval is that setting that will

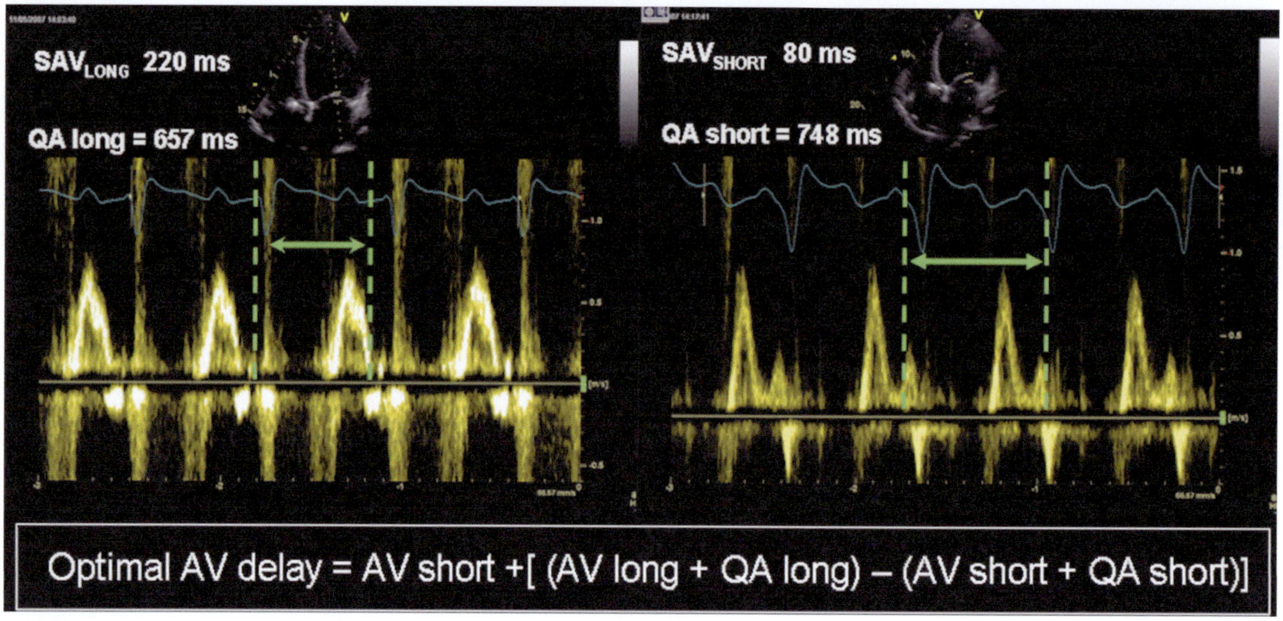

Fig. 17.2 Ritter's formula for calculating optimal AV delay. *Left and right panels* represent mitral inflow velocity profiles during respective long and short sensed AV (SAV) delays. QA is measured from the QRS onset to the end of mitral A wave. In this example, optimal AV delay using the Ritter formula is calculated at 150 ms (149 ms)

allow mitral valve closure, as a result of ventricular contraction, to coincide with the completion of the atrial transport represented by the end of the mitral A wave. Ritter used the QRS onset on a simultaneously recorded EKG signals to time the onset of ventricular contraction during a programmed long (usually 200 ms) and short (usually 60 ms) AV delays. The programmed long AV delay has to be reduced by the difference between the time intervals from the QRS onset to the end of the mitral A wave during the programmed short and long AV delays in order to determine the optimal AV delay[20] (Fig. 17.2).

Ritter's method has originally been used to optimize AV delays in dual-chamber pacemakers where patients have AV block and normal LV function. MIRACLE and INSYNCH III trials have used the technique to set the AV delay settings in their respective studies.[21] In a study that evaluated 40 patients within 24-h post-CRT device implantation, the optimal AV delay obtained by aortic velocity-time integral (VTI) method was significantly longer than that calculated by Ritter's method.[22] In two other studies, the optimized AV delay by this technique showed poor correlation with the invasive LV d*P*/d*t* measurements and may therefore not represent the maximum achievable hemodynamic benefit.[18,23] There are some limitations to the application of this technique. In heart failure, elevated LVEDP promotes immediate closure of the mitral valve. It may be difficult to ascertain A wave abbreviation. In the presence of a short or normal PR interval, biventricular pacing can be compromised during long AV delay setting.[9]

Several techniques have been derived from Ritter's method. Ishikawa used the first component of the first heart sound (S1) using a phonocardiogram or, alternately, the end of the diastolic MR velocity signals. The optimal AV delay is a slightly prolonged AV delay less the duration of diastolic MR at that AV delay setting. A slightly prolonged AV delay ensures full biventricular pacing during the procedure even in patients with intact AV conduction. Like Ritter's technique, *Ishikawa's technique* was used initially applied to optimizing dual-chamber pacemaker settings.[24] In a study that evaluated five heart failure patients who received biventricular pacemakers, optimal AV delay predicted by Ishikawa's technique yielded the highest echocardiographic CO and longest LV filling time than when the AV delay was set 25 ms earlier and later than the optimal AV delay. In addition, regular evaluation of the optimal AV delay every 3–6 months over a 2-year-follow-up period was associated with improvement of the patients' functional class.[25] However, in the absence of a control group on top of a modification in the medical management, such improvement may be attributed to the concerting effect of CRT itself and the medical management.

Instead of diastolic MR, Meluzin used the onset of high velocity (systolic) component of MR to index the onset of

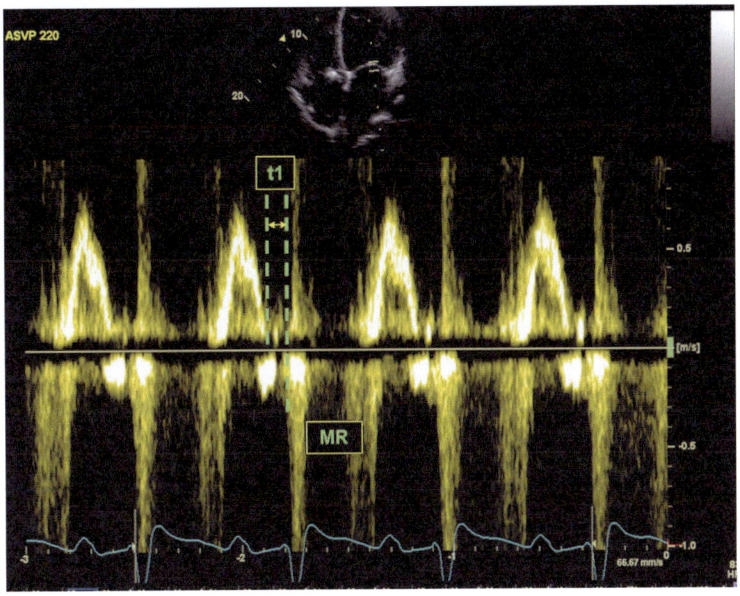

Fig. 17.3 Meluzin technique. In a apical 4-chamber view, pulsed-Doppler interrogation of the mitral inflow is performed with sample volume placed at the tips of the mitral leaflets to record early and late diastolic filling and mitral regurgitation flows. The testing long AV delay by which the recording was made has to be shortened by t1 (*yellow double arrow*) to determine the optimal AV delay. In this figure, testing long AV delay was 220 ms during an atrial-sensed, ventricular paced mode (ASVP 220). T1 measured from the end of fused mitral E and A waves to the onset of the high velocity signal of mitral regurgitation (MR) was 82 ms. Thus, optimal AV interval was set at 140 ms (138 ms)

ventricular contraction. The time interval between the end of the A wave (representing the late diastolic transmitral flow at atrial contraction) and the onset of the systolic (high velocity) component of MR measured during a programmed long AV interval represents the time that the testing long interval must be shortened to achieve the optimal AV delay (Fig. 17.3). *Meluzin's technique* predicted the AV delay that yielded the maximum CO (thermodilution technique) in 78% of 18 CRT patients optimized 3 months post-device implantation. The technique is simple. In contrast to the Ritter's and Ishikawa's methods, it only requires a single echo recording and a single measurement to calculate the optimal AV delay.[26] It however requires the presence of systolic MR which is a limitation to the technique. Ismer et al. used the ventricular pacing stimulus to mark the onset of ventricular contraction.

Ismer's technique requires the use of bipolar esophageal leads to record LA electrogram. Simultaneous recording of transmitral flow, LA electrogram, and real-time pacemaker sensed-event markers allows determination of the components of the optimal AV delay.[27] According to Ismer, the hemodynamically optimal AV delay can be calculated as the sum of the inter atrial conduction interval and the LA electromechanical action reduced by the latency of mitral valve closure induced by ventricular stimulation.[27] Ismer's method was used to determine the optimal AV delay in 11 CRT patients. Echocardiographic LV ejection fraction (LVEF), measured at three programmed AV delays (optimal AV delay, optimal AV delay + 50 ms, and optimal AV delay – 50 ms), was maximal with optimal AV delay.[28] The same group has employed this technique to compare optimal AV delay at rest and during submaximal exercise in 20 CRT-D patients.[29]

17.3.2.2 Dynamic AV Interval-Based Optimization of AV Delay (Iterative Method)

Dynamic AV interval-based method of optimizing AV delay or the iterative method determines the ideal AV delay by successive estimation of a single hemodynamic index from a range of programmed AV delays commencing from the longest AV interval that allows full capture of biventricular pacing. Alternately, the longest AV delay is set 30 ms less of the measured interval during an atrial-sensed and ventricular-sensed setting. Iterative method requires recordings at more than two programmed delays and entails long procedural time. Moreover, consistency in the sample volume location where pulsed-Doppler waveforms are recorded is essential. Variability and errors may arise from these two limiting factors.

Unlike the calculation-based method, the application of iterative method is not limited to optimizing diastolic filling time. In addition to LV filling volume, iterative method can also be used to optimize markers of systolic function such as SV and CO, Doppler-derived LV dP/dt_{max}, and myocardial performance index (MPI). These hemodynamic indices of systolic performance can also be used in optimizing V-V delay as well. Thus, iterative method allows the assessment of relative changes in a referenced hemodynamic parameter when combined AV and V-V delay optimization procedure is performed.

Diastolic filling time (DFT) is the time interval from the onset of the mitral E wave to the end of A wave. The AV interval that produces the complete separation of the Doppler E and A waves without causing A wave truncation is considered optimal (Fig. 17.4). *Mitral inflow* VTI is an index of LV filling volume. In the absence of MR or shunt abnormalities, mitral inflow VTI serves as a surrogate for SV assuming a

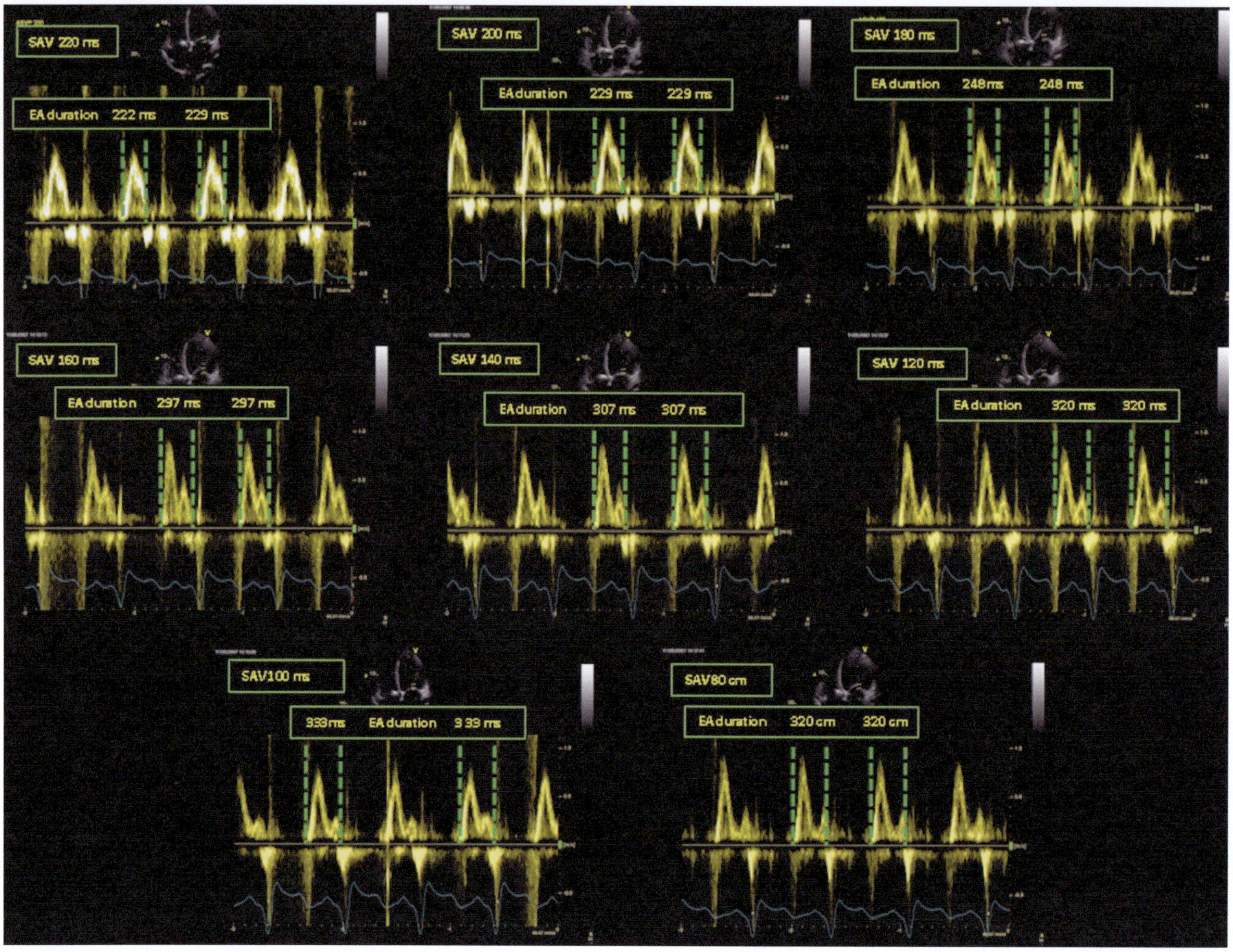

Fig. 17.4 Iterative method using diastolic filling time technique. Diastolic filling time measured from the onset of mitral E to the end of mitral A is repeatedly measured during progressive uniform reduction in the programmed sensed AV delay (SAV). As the AV delay shortens, DFT increases. DFT starts to decrease once a significant interruption in the mitral A wave is reached. In this example, DFT was maximal at the AV delay of 100 ms. Close examination revealed beginning A wave truncation that did not improve by increasing the AV delay by 10 ms (not shown). Thus, optimal AV delay was 120 ms

constant cross-sectional area of the mitral annulus. Functional MR is very common among heart failure patients. Thus, the LV filling volume cannot reflect the actual SV as the flow across the regurgitant orifice will generally be higher.[30] The programmed AV interval that yields the maximum mitral inflow VTI is considered optimal (Fig. 17.5).

In a series of 30 patients, 4 techniques of AV optimization were compared against invasive LV dP/dt_{max}. Mitral inflow VTI, DFT, and aortic VTI predicted optimal AV delay in 97%, 67%, and 43% of patients respectively. Ritter method failed to predict optimal AV delay.[18] In this study, V-V delay optimization was likewise performed. In addition more than half of the patients have moderate-to-severe MR. The extent to which the presence of MR and the optimization of the V-V delay affected the results is not known. Neither invasive SV nor CO measurement was reported. Moreover, aortic VTI was collectively represented by both pulsed-Doppler recordings at the LVOT and CW Doppler recordings across the aortic valve.

Improvement in the hemodynamic indices of systolic function is a function of both AV and intraventricular synchrony. Thus, these markers can be used to optimize both AV and V-V interval. Assessment of SV and CO can be performed by pulsed-Doppler interrogation of the LVOT or by CW Doppler recording of flow across the aortic valve. Trans-aortic VTI technique (*aortic VTI technique*) is more reproducible than measuring LVOT VTI (*LVOT VTI technique*) (Fig. 17.6). It is however influenced by the morphology of the aortic valve and may not be accurate in the subset of CRT patients with concomitant aortic valve disease.

A randomized, prospective, single-blind trial consisting of 40 CRT patients compared aortic VTI-guided AV delay optimization ($n=20$) using AV intervals of 60–200 ms (119 ± 34 ms) to an empirically programmed AV interval of 120 ms ($n=20$)

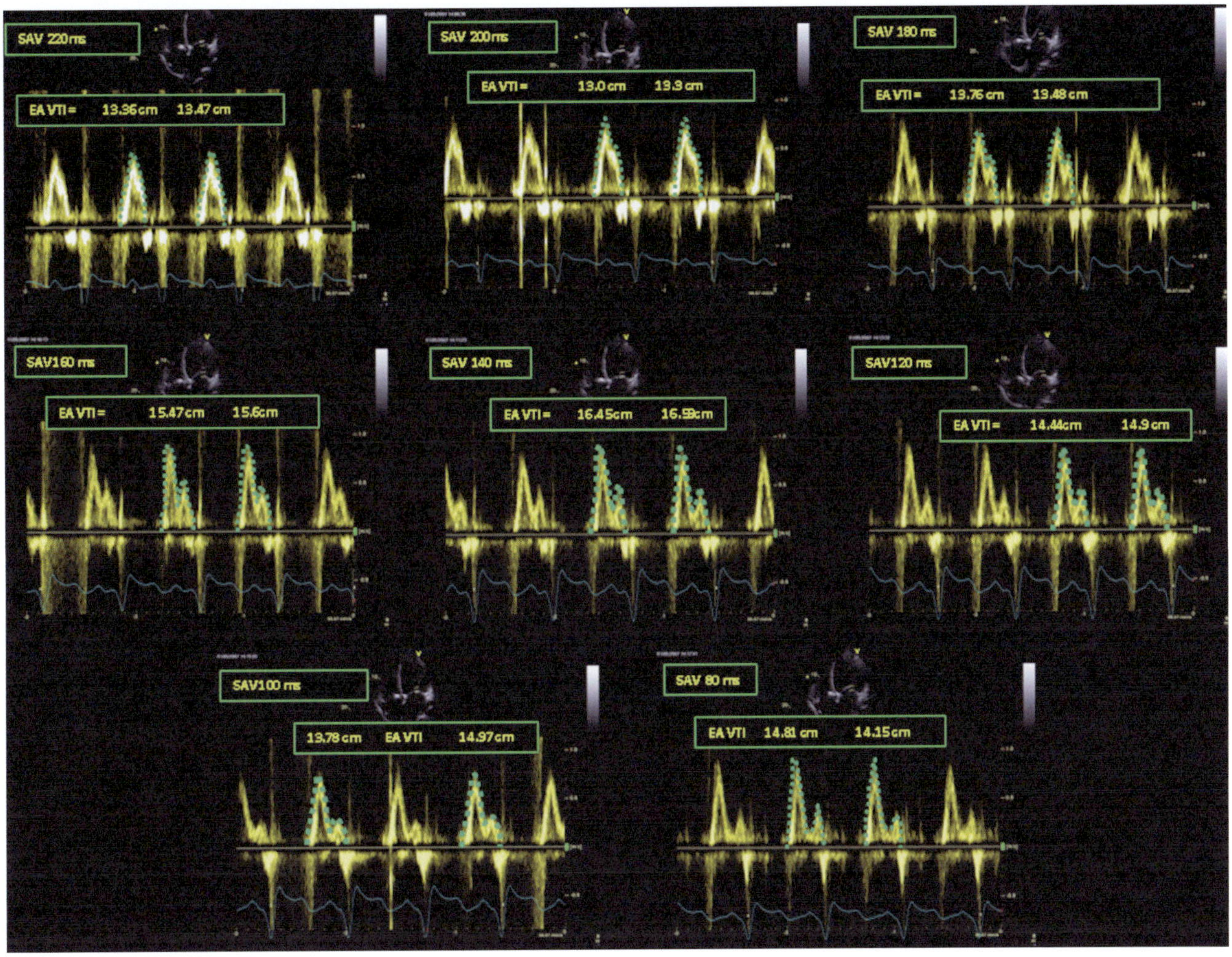

Fig. 17.5 Iterative method using mitral inflow (EA) VTI technique. Representative images of mitral inflow velocity profiles recorded during progressive uniform reduction in the programmed sensed AV delay (SAV). Shortening of AV delay increases EA VTI until the optimal AV delay is achieved. In this example, optimal SAV is 140 ms where EA VTI is maximal

in VDD pacing mode at the initiation of CRT. After initiation of CRT, optimized AV delay resulted in a significant improvement in both the aortic VTI (4.0±1.7 cm vs. 1.8±3.6 cm, $p<0.02$) and LVEF (7.8±6.2% vs 3.4±4.4%, $p<0.02$) compared to the empirically programmed AV delay. At 3-month follow-up, there was greater improvement in the NYHA class and QOL in the optimized group. Both groups achieved improvement in the objective 6 min walk test (6MHWT) to a comparable degree.[31] The same group of investigators further compared aortic VTI with Ritter's technique. The intra-observer reproducibility was excellent for both aortic VTI ($r=0.99$) and Ritter's technique ($r=0.98$ and 0.99 at long and short AV interval). Aortic VTI-guided optimized AV delay resulted in a greater improvement in LV SV (as estimated change in aortic VTI from baseline) compared to optimized AV delay calculated by Ritter's formula in 36 of the 40 patients. Both techniques resulted in a comparable improvement in the diastolic filling. There was, however, no correlation in the AV delay values predicted by the two techniques ($r=0.03$)[22] In two studies, aortic VTI technique showed better correlation with invasive LV dP/dt_{max} compared with the Ritter formula in terms of sensed and paced AV delay optimization[23] or combined AV and V-V delay optimization.[18]

The chronic effects of AV delay optimization guided by LVOT VTI (sensed and paced) performed at 31±8 weeks after initiation of CRT were investigated in 33 patients. AV delay optimization resulted in a bidirectional magnitude of change from the pre-programmed AV delay at implantation. After a mean follow-up of 43 days, a significant increase from baseline in the 6MHWT (from 449±16 m to 475±17 m, $p<0.05$) and reduction in the NT pro-BNP level (from 3192±765 ng/L to 2593±675 ng/L) were reported in the absence of improvement in the LVEF and diastolic filling. These benefits were significantly greater in the subgroup of patients with greater corrections in the AV delay.[32] In the absence of a control group, it is however difficult to attribute the observed improvements solely to the beneficial effect of AV delay optimization.

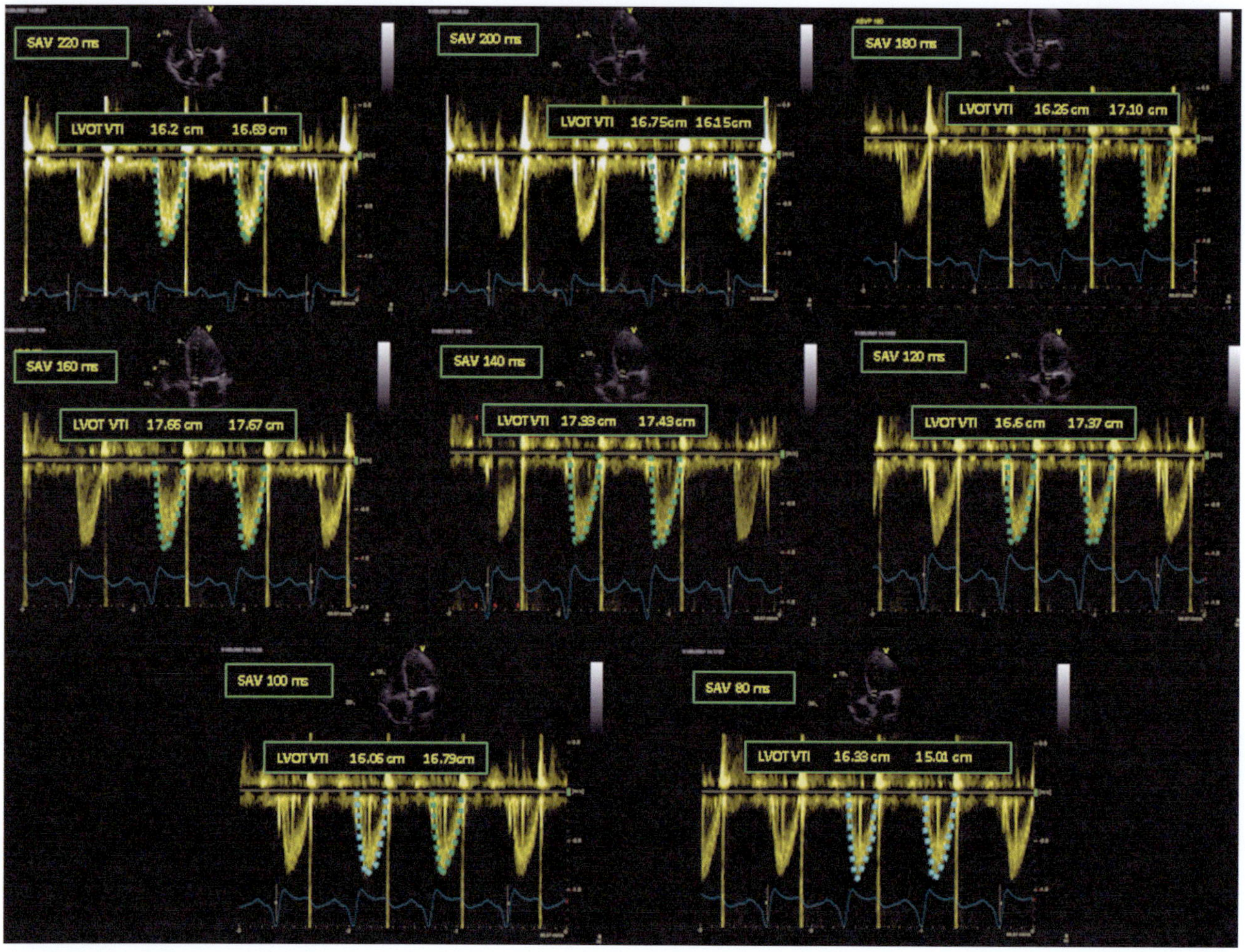

Fig. 17.6 Iterative method using LV outflow (LVOT) VTI technique. Representative images of LVOT velocity profiles recorded during progressive uniform reduction in the programmed sensed AV delay (SAV). In this example, repeated measurements of LVOT VTI across a range of SAV's estimated the optimal SAV at 160 ms

Doppler-derived LV dP/dt$_{max}$ – Optimizing AV synchrony has been shown to increase invasive LV d*P*/d*t*$_{max}$ at the time of device implantation.[10,33] LV d*P*/d*t*$_{max}$ derived from invasive LV pressure measurements has been used as the gold standard for validating various noninvasive AV and V-V delay optimization techniques.[18,23] An approach to program optimal AV delay based on this invasive method is to evaluate the response in the LV dP/dt max determined by CW Doppler measurements of the time of acceleration in the MR jet velocity (Fig. 17.7).

A recent study allocated 41 consecutive CRT patients on a 2:1 basis to optimization using Doppler-derived LV d*P*/d*t*$_{max}$ or empirical 120 ms AV delay 3 months post-device implantation. A wide range of optimal AV delay was observed with majority of patients falling between 80 and 120 ms. Intra- and inter-observer repeatability were excellent ($r = 0.99$ and 0.98, respectively). The results demonstrated the superiority of Doppler-derived d*P*/d*t*$_{max}$ technique over a fixed AV delay of 120 ms in terms of improvement in the functional classification and echocardiographically determined LVEF at 6-month follow-up.[34] Nevertheless, this technique requires the presence of a well-defined MR jet-envelope by CW Doppler and may not be applicable in all patients.

MPI is considered a measure of global cardiac function. It is correlated with LV dP/dt and the clinical severity of heart failure. MPI is calculated by dividing the total isovolumic time by the ejection time.[35,36] Ejection time is the interval between the start and the end of the aortic flow. Total isovolumic time is the difference between the interval from the mitral A terminus to the onset of the subsequent mitral E wave and the ejection time[37] (Fig. 17.8).

Postoperative optimization of AV and VV intervals using MPI has been used in two non-randomized small studies that did not include clinical outcome. Both did not include any control group and used baseline non-CRT measurements for comparison. Selection of optimal AV/V-V delay combination was based on minimum MPI. Optimized intervals improved the LV mechanical efficiency by reducing the isovolumic phases of the cardiac cycle.[14,38]

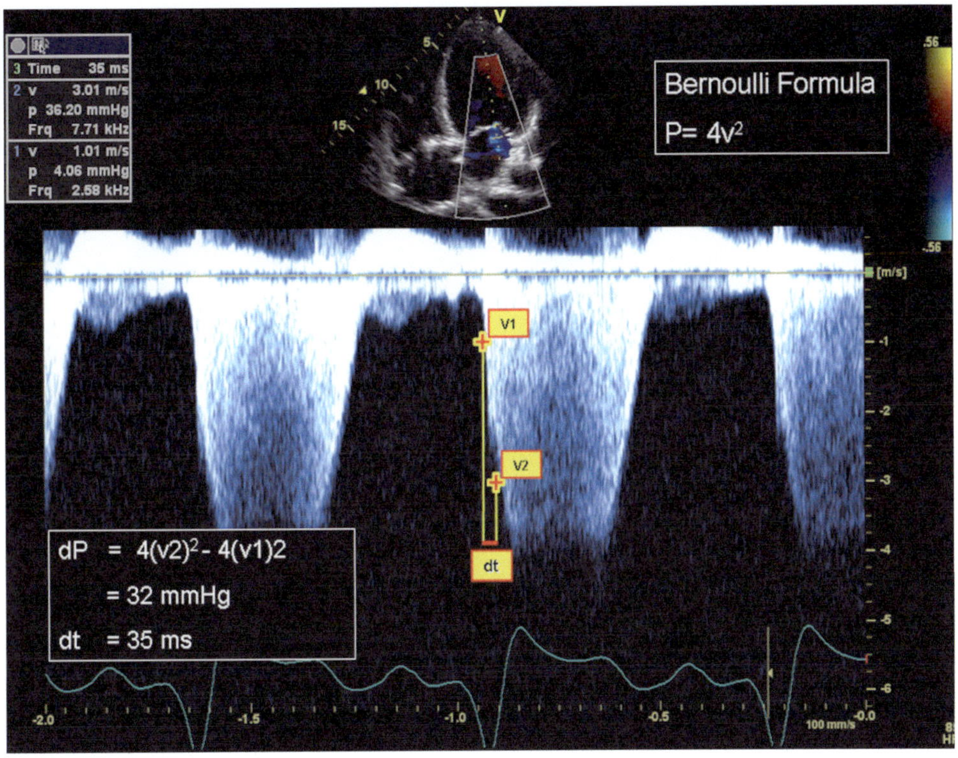

Fig. 17.7 Doppler-derived rate of LV pressure rise (LV d*P*/d*t*max). The MR CW Doppler velocity curve reflects pressure gradient between LV and left atrium during systole. The LV d*P*/d*t* can be estimated by measuring the time (d*t*, *red line*) for the MR velocity to increase from 1 (V1) to 3 (V2) mmHg. Thus LV d*P*/d*t* in this example is 32 mmHg divided by 0.035 s

In the absence of consensus for routine performance of AV delay optimization after CRT, the American Society of Echocardiography has proposed a simplified Doppler screening protocol using mitral inflow velocity analysis. Accordingly, performance of AV delay optimization is not required in the presence of a clearly identified and separated E and A waves provided that the termination of the A wave occurs at least 40 ms before QRS onset or mitral valve closure click.[39] In addition, presence of stage one LV diastolic filling pattern may obviate the need for optimization.[30,40]

17.3.3 Optimal AV Delay with Right Atrial Pacing or During Exercise

AV interval interrogation has been routinely done at rest and most of the published literatures on AV delay optimization were based on intrinsic atrial conduction. Data on the effect of exercise on optimized AV delay reported conflicting results.[29,41] Thus, the utility of rate-adaptive AV delay remains controversial.

Echocardiographic methods of programming the optimal AV interval in patients that require DDD pacing during CRT have not been studied extensively. In contrast to the VDD mode where both atria are intrinsically activated, pacing the right atrial appendage leads to a delayed electromechanical activation of the left atrium.[42] Compensating for such delay entails longer paced AV interval at the expense of reduced LV filling duration. There are reports that optimal AV delay during right atrial pacing is longer than the optimized sensed AV interval.[23,29,43] However, there is no general consensus as to what value the default offset between sensed and paced AV intervals should be, as parameter will depend on intra-atrial and inter-atrial conduction delay as well as right atrial lead position.[44,45] Thus, optimizing paced AV delay is deemed necessary in patients who are expected to require atrial pacing. When optimizing AV delay in DDD mode, mitral inflow technique is, however, inferior to aortic VTI technique.[23]

17.4 Ventriculo-Ventricular Delay Optimization

Technically, the benefit of acute and long-term CRT to heart failure patients is dependent on the restoration of synchronous ventricular contraction. Post-CRT mechanical ventricular synchrony can be compromised by suboptimal LV lead position, abnormal global activation, and regional conduction delays across infarcted myocardium. Compensating for these delays is a function of tailored sequence of ventricular stimulation.[21,46,47]

Tailored sequence of ventricular stimulation, in combination with optimal atrio-ventricular coupling, produces an ideal coalition of the activation wave fronts derived from RV and LV pacing, and spontaneous AV conduction.[48,49] Hence, the LV area that is simultaneously depolarized is substantially increased, and LV synchrony is improved.

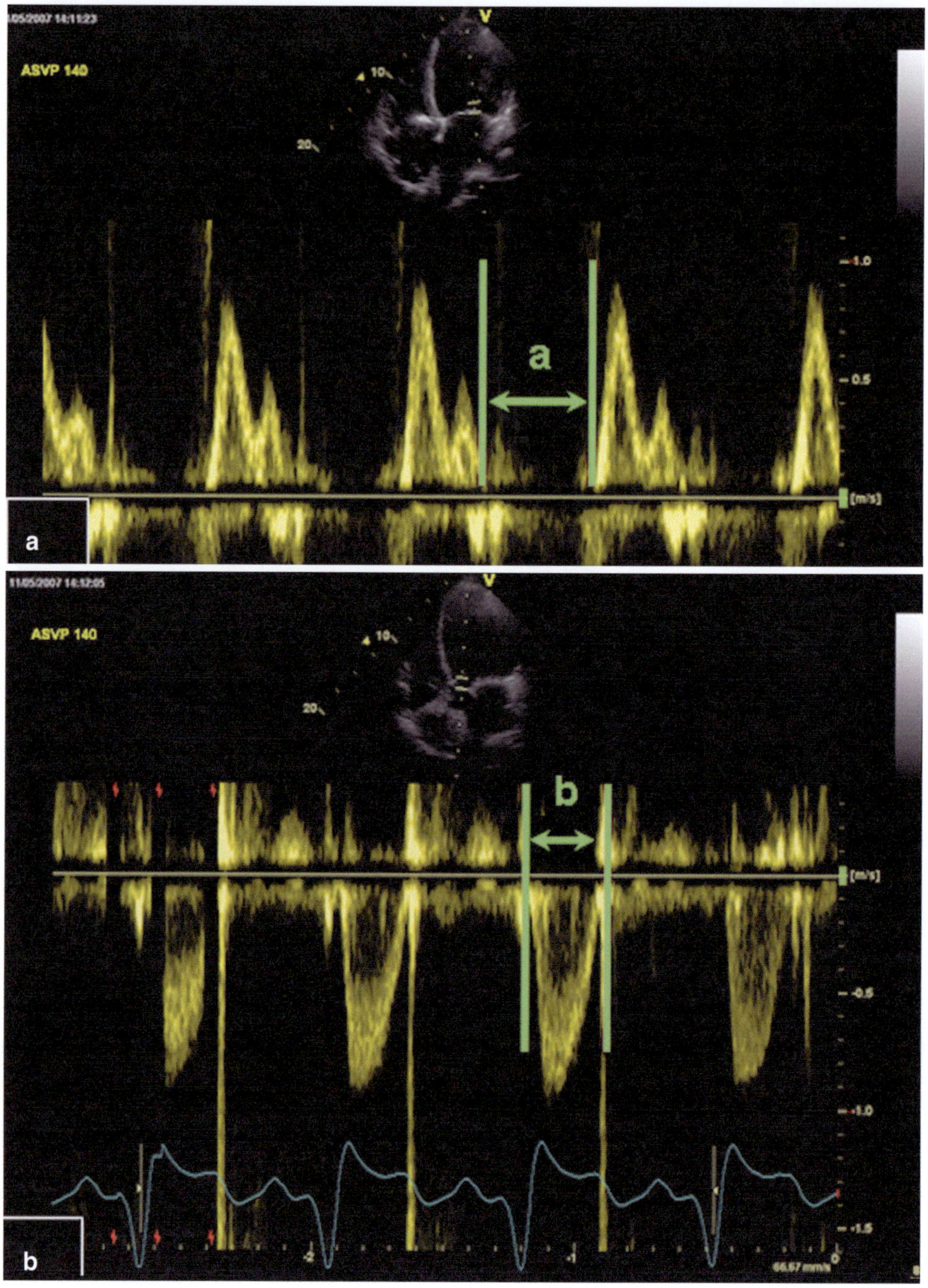

$$\text{MPI} = \frac{a-b}{b}$$

Fig. 17.8 Myocardial performance index (MPI). MPI is calculated from several temporal measurements derived from both pulsed-Doppler mitral inflow (**a**) and LVOT flow (**b**) velocity profiles. It is the ratio of the total isovolumic time to ejection time. Ejection time (**b**) is measured from the onset to the end of the LVOT velocity envelope. The time interval from the mitral A terminus to the onset of the succeeding mitral E wave (**a**) minus the ejection time is the total isovolumic time

Contemporary CRT devices permit programming of AV interval and sequential stimulation of the ventricles. Interrogation of optimal V-V delay proceeds in such a way that the optimal AV delay obtained during simultaneous biventricular pacing is used across all sequence of ventricular stimulations. It is important that the optimized AV delay be maintained by the right atrial-LV channels throughout the series of V-V interval interrogation.

17.4.1 Echocardiography-Based Techniques of V-V Delay Optimization

Echocardiographic V-V interval programming is guided by the same techniques used in AV delay optimization, more commonly with aortic or LVOT VTI technique. Evaluation of the extent of residual LV dyssynchrony after V-V delay programming requires the application of more sophisticated echocardiographic techniques such as Tissue Doppler Imaging (TDI), Tissue Synchronization Imaging, and strain echocardiography.[47] It has to be noted, however, that the role of these echocardiographic techniques in selecting patients for CRT has been challenged by the result of PROSPECT trial. Significant intra- and inter-observer variability posed important limitations to the use of several Doppler-derived indices of mechanical dyssynchrony in predicting response to CRT. Accordingly, the use of these echocardiographic dyssynchrony indices cannot be recommended for CRT candidate selection and device function optimization.[50]

17.4.1.1 Impact of Optimized V-V Interval on Echocardiographic and Clinical Outcome

Two small, non-randomized studies optimized V-V delay using LVOT VTI and assessed residual intraventricular and interventricular dyssynchrony after CRT. In a study 41 patients that underwent VV-delay interrogation including uni-ventricular pacing immediately post-device implantation, simultaneous biventricular pacing was the optimal V-V setting in 15% of the patients. AV delay optimized-nominal V-V delay setting resulted in increased CO, decreased MR, and improved echocardiographic ventricular dyssynchrony indices. Incremental benefits were observed when V-V interval was optimized. Such changes translated to improvement of NYHA class, QOL, exercise capacity with increased LVEF, and reduced LV end-systolic (LVESV) and end-diastolic volumes (LVEDV) at 3-month follow-up.[51] In another study consisting of 20 patients whose pacemaker settings were optimized at prehospital discharge, improvement in CO, MR, and ventricular dyssynchrony translated to improved NYHA class and LVEF at 6-month follow-up. There was a 10% non-response to CRT based on a 25% cut-off in terms of LVEF and LV end-diastolic diameter (LVEDD) change from baseline. Optimal V-V interval at simultaneous biventricular pacing was noted in three patients.[52]

A subset of patients (*n*-34) enrolled in a multi-center, non-randomized, observational 3 months trial ($n = 189$) underwent V-V interval optimization guided by aortic VTI technique at two time points: prehospital discharge and at 3-month follow-up. Optimal V-V delay with LV pre-excitation was reported in only 38% at prehospital discharge and 53% at 3-month follow-up. With V-V delay optimization, SV and aortic VTI increased by 23% and 18% at prehospital discharge and 3-month follow-up respectively. SV and VTI were found to be optimal within a small range of V-V delays and changes in these V-V delays were minimal or none at all in 79% at 3-month follow-up. NYHA class and distance covered at 6MHWT at 3-month follow-up were comparable between groups of patients programmed on simultaneous biventricular pacing or periodic AV and V-V optimization.[53] Using a follow-up period of 12-months, another study that included 37 patients performed serial AV/V-V optimization within 48 h after device implantation, at 6 month follow-up, and in less than half of the patients ($n = 14$) at 12-month follow-up. At implant, the optimal V-V delay was achieved by simultaneous biventricular pacing in four patients. A non-concordance was found in the optimal V-V delay between each assessment. Increased aortic VTI associated with V-V delay optimization at all three time points translated to minimal but significant increase in LVEF at 6 months (from a median of 25% at baseline to 28%, $p < 0.002$). Median LVEF at 12 months was 35% compared with 26% at baseline, ($p < 0.14$). CRT response based on a minimum 15% reduction in LVESV was observed in 20 of 37 and 10 of 14 patients at 6- and 12-month follow-up respectively. AV and V-V delay were comparable between responders and non-responders.[15]

There are two large multi-center trials that investigated incremental benefits of optimized biventricular pacing on clinical outcomes. *In-Synch III* is a non-randomized trial that evaluated 359 patients. LVOT VTI-guided pre-discharged V-V interval optimization yielded 8.6% increase in stroke volume. Clinical outcomes at 6 months were compared against the treatment arm of MIRACLE trial. There was no significant difference in the effect of optimized sequential and simultaneous CRT on NYHA functional class and QOL. The In-Synch III cohort experienced greater improvement in the 6MHWT from baseline to 6 months compared to the MIRACLE simultaneous CRT treatment group with or without adjusting for differences in the baseline 6MHW, beta-blocker use, QRS duration, gender, and LVEDD between the two cohorts. Moreover, significant improvement in SV with sequential biventricular pacing has been consistently observed among a subset of patients with prior history of myocardial infarction.[21] A single-blind randomized trial (*RHYTHYM II trial*) allocated 121 CRT patients at prehospital discharge to simultaneous or LVOT VTI-guided optimized biventricular pacing on a 1:3 sampling method. Response to CRT was assessed on the basis of clinical (NYHA functional class improvement by at least one class)[54] or echocardiographic (at least 5% increase in LVEF and/or at least 10% decrease in LVESV)[12] during a follow-up period of 6 months. Optimal V-V setting is simultaneous biventricular stimulation in 30% of patients. Contradictory to the In-Synch III results, optimizing V-V delay failed to confer additional clinical benefits over simultaneous biventricular pacing in terms of NYHA functional class, QOL, and 6MHWT at 6-month follow-up. It also failed to promote additional reverse LV remodeling and improve the proportion of echocardiographic responders.[12,54] It has to be recognized however that simultaneous biventricular pacing is a dichotomous variable. In both studies, V-V delay interrogation was performed only on the group randomized to optimized V-V setting. As such, the proportion of patient randomized to simultaneous biventricular pacing whose setting is actually optimal cannot be determined and the extent to which it influenced the results is not known.[12]

The DECREASE-HF trial is a randomized double-blind, 3-arm trial designed to show treatment equivalence of sequential biventricular and LV pacing to simultaneous biventricular pacing in terms of echocardiographic response at 6 months follow-up. Patients were randomly assigned without testing the hemodynamic response to each modality. Timing of sequential biventricular pacing was programmed on the basis of baseline intrinsic conduction and may have not represented the most hemodynamically optimal V-V setting. Like RHYTHM II, sequential biventricular pacing failed to show superiority over simultaneous biventricular pacing in terms of echocardiographic improvement at 6-month follow-up when compared to baseline pre-CRT values.[46]

17.4.2 Nuclear Imaging-Based V-V Delay Optimization

Radionuclide ventriculography has been used in the assessment of mechanical dyssynchrony and measurement of LVEF in patients with dilated cardiomyopathy.[55,56] Compared to echocardiography, nuclear ventriculography has lower temporal resolution. One study ($N=27$) has utilized radionuclide ventriculography to optimize LVEF in selecting optimal V-V interval. Simultaneous biventricular LV pacing yielded maximal LVEF in 33% of patients. A relative increase in LVEF by 18% during sequential pacing was found in the remaining patients. As with echocardiography, there was great heterogeneity in individual response to V-V delay programming. Improvement in LVEF, however, was not to accompanied by improvement in intraventricular dyssynchrony.[57]

17.5 Summary

The utility of CRT optimization remains confined to patients whose response to the treatment is either suboptimal or negligible. There are several techniques, mostly echocardiographic, to optimize device settings but none has been universally accepted as gold standard. Methods differ in recording techniques and may therefore vary significantly in performance.

Several studies have employed various techniques in interrogating AV interval with or without concomitant V-V interval optimization. Optimization was performed at varying time points. Even the definition for favorable response to CRT varied. Optimization of AV and V-V intervals is patient-specific and optimal values change over time. This may relate to time-related LV reverse remodeling or to progression of disease. The effects of inter-atrial conduction defect, right atrial pacing, and exercise on AV delay add complexities to performance of AV delay optimization. At the very least, it has to be ensured that the programmed AV delay should neither produce fusion of mitral E and A waves nor truncation of the latter.

Despite evidences of hemodynamic benefits, the utility of V-V interval optimization remains controversial. Evidence for the incremental long-term benefits on mortality and morbidity is still lacking. Three large clinical trials reported conflicting results in terms of clinical and echocardiographic improvement. Differences in the study designs could have contributed to the conflicting results. It is important to remember, however, that simultaneous biventricular pacing is not always synonymous with suboptimal V-V interval, nor should sequential pacing be routinely regarded as ideal.

References

1. Abraham WT, Fisher WG, Smith AL, et al. Cardiac resynchronization in chronic heart failure. *N Engl J Med.* 2002;346(24):1845-1853.
2. Bristow MR, Saxon LA, Boehmer J, et al. Cardiac-resynchronization therapy with or without an implantable defibrillator in advanced chronic heart failure. *N Engl J Med.* 2004;350(21):2140-2150.
3. Cleland JG, Daubert JC, Erdmann E, et al. Longer-term effects of cardiac resynchronization therapy on mortality in heart failure [the CArdiac REsynchronization-Heart Failure (CARE-HF) trial extension phase]. *Eur Heart J.* 2006;27(16):1928-1932.
4. Cleland JG, Daubert JC, Erdmann E, et al. The effect of cardiac resynchronization on morbidity and mortality in heart failure. *N Engl J Med.* 2005;352(15):1539-1549.
5. Linde C, Leclercq C, Rex S, et al. Long-term benefits of biventricular pacing in congestive heart failure: results from the MUltisite STimulation in cardiomyopathy (MUSTIC) study. *J Am Coll Cardiol.* 2002;40(1):111-118.
6. Mullens W, Grimm RA, Verga T, et al. Insights from a cardiac resynchronization optimization clinic as part of a heart failure disease management program. *J Am Coll Cardiol.* 2009;53(9):765-773.
7. Authors/Task Force Members, Vardas PE, Auricchio A, et al. Guidelines for cardiac pacing and cardiac resynchronization therapy: the Task Force for Cardiac Pacing and Cardiac Resynchronization Therapy of the European Society of Cardiology. Developed in Collaboration with the European Heart Rhythm Association. *Eur Heart J.* 2007;28(18):2256-2295.
8. Gras D, Gupta MS, Boulogne E, Guzzo L, Abraham WT. Optimization of AV and VV delays in the real-world CRT patient population: an international survey on current clinical practice. *Pacing Clin Electrophysiol.* 2009;32(suppl 1):S236-S239.
9. Barold SS, Ilercil A, Herweg B. Echocardiographic optimization of the atrioventricular and interventricular intervals during cardiac resynchronization. *Europace.* 2008;10(suppl 3):iii88-iii95.
10. Auricchio A, Stellbrink C, Block M, et al. Effect of pacing chamber and atrioventricular delay on acute systolic function of paced patients with congestive heart failure. The Pacing Therapies for Congestive Heart Failure Study Group. The Guidant Congestive Heart Failure Research Group. *Circulation.* 1999;99(23):2993-3001.
11. Auricchio A, Ding J, Spinelli JC, et al. Cardiac resynchronization therapy restores optimal atrioventricular mechanical timing in heart failure patients with ventricular conduction delay. *J Am Coll Cardiol.* 2002;7:1163-1169.

12. Boriani G, Biffi M, Muller CP, et al. A prospective randomized evaluation of VV delay optimization in CRT-D recipients: echocardiographic observations from the RHYTHM II ICD study. *Pacing Clin Electrophysiol*. 2009;32(suppl 1):S120-S125.
13. O'Donnell D, Nadurata V, Hamer A, Kertes P, Mohamed U. Long-term variations in optimal programming of cardiac resynchronization therapy devices. *Pacing Clin Electrophysiol*. 2005;28(suppl 1):S24-S26.
14. Porciani MC, Dondina C, Macioce R, et al. Temporal variation in optimal atrioventricular and interventricular delay during cardiac resynchronization therapy. *J Card Fail*. 2006;12(9):715-719.
15. Valzania C, Biffi M, Martignani C, et al. Cardiac resynchronization therapy: variations in echo-guided optimized atrioventricular and interventricular delays during follow-up. *Echocardiography*. 2007;24(9):933-939.
16. Zhang Q, Fung JW, Chan YS, et al. The role of repeating optimization of atrioventricular interval during interim and long-term follow-up after cardiac resynchronization therapy. *Int J Cardiol*. 2008;124(2):211-217.
17. Bax JJ, Abraham T, Barold SS, et al. Cardiac resynchronization therapy: Part 2 – issues during and after device implantation and unresolved questions. *J Am Coll Cardiol*. 2005;46(12):2168-2182.
18. Jansen AH, Bracke FA, van Dantzig JM, et al. Correlation of echo-Doppler optimization of atrioventricular delay in cardiac resynchronization therapy with invasive hemodynamics in patients with heart failure secondary to ischemic or idiopathic dilated cardiomyopathy. *Am J Cardiol*. 2006;97(4):552-557.
19. Jaffe WM, Dewhurst TA, Otto CM, Pearlman AS. Influence of Doppler sample volume location on ventricular filling velocities. *Am J Cardiol*. 1991;68(5):550-552.
20. Bleeker GB, Yu CM, Nihoyannopoulos P, et al. Optimal use of echocardiography in cardiac resynchronisation therapy. *Heart*. 2007;93(11):1339-1350.
21. Leon AR, Abraham WT, Brozena S, et al. Cardiac resynchronization with sequential biventricular pacing for the treatment of moderate-to-severe heart failure. *J Am Coll Cardiol*. 2005;46(12):2298-2304.
22. Kerlan JE, Sawhney NS, Waggoner AD, et al. Prospective comparison of echocardiographic atrioventricular delay optimization methods for cardiac resynchronization therapy. *Heart Rhythm*. 2006; 3(2):148-154.
23. Gold MR, Niazi I, Giudici M, et al. A prospective comparison of AV delay programming methods for hemodynamic optimization during cardiac resynchronization therapy. *J Cardiovasc Electrophysiol*. 2007;18(5):490-496.
24. Ishikawa T, Sumita S, Kimura K, et al. Prediction of optimal atrioventricular delay in patients with implanted DDD pacemakers. *Pacing Clin Electrophysiol*. 1999;22(9):1365-1371.
25. Inoue N, Ishikawa T, Sumita S, et al. Long-term follow-up of atrioventricular delay optimization in patients with biventricular pacing. *Circ J*. 2005;69(2):201-204.
26. Meluzin J, Novak M, Mullerova J, et al. A fast and simple echocardiographic method of determination of the optimal atrioventricular delay in patients after biventricular stimulation. *Pacing Clin Electrophysiol*. 2004;27(1):58-64.
27. Ismer B, von Knorre GH, Voß W, Körber T. Definition of the optimal atrioventricular delay by simultaneous measurement of electrocardiographic and Doppler-echocardiographic parameters. *Prog Biomed Res*. 2002;7(2):116-120.
28. Melzer C, Borges AC, Knebel F, et al. Echocardiographic AV-interval optimization in patients with reduced left ventricular function. *Cardiovasc Ultrasound*. 2004;2:30.
29. Melzer C, Bondke H, Korber T, Nienaber CA, Baumann G, Ismer B. Should we use the rate-adaptive AV delay in cardiac resynchronization therapy-pacing? *Europace*. 2008;10(1):53-58.
30. Quinones MA, Otto CM, Stoddard M, Waggoner A, Zoghbi WA, Doppler Quantification Task Force of the Nomenclature and Standards Committee of the American Society of Echocardiography. Recommendations for quantification of Doppler echocardiography: a report from the Doppler Quantification Task Force of the Nomenclature and Standards Committee of the American Society of Echocardiography. *J Am Soc Echocardiogr*. 2002;15(2): 167-184.
31. Sawhney NS, Waggoner AD, Garhwal S, Chawla MK, Osborn J, Faddis MN. Randomized prospective trial of atrioventricular delay programming for cardiac resynchronization therapy. *Heart Rhythm*. 2004;1(5):562-567.
32. Hardt SE, Yazdi SH, Bauer A, et al. Immediate and chronic effects of AV-delay optimization in patients with cardiac resynchronization therapy. *Int J Cardiol*. 2007;115(3):318-325.
33. Kass DA, Chen CH, Curry C, et al. Improved left ventricular mechanics from acute VDD pacing in patients with dilated cardiomyopathy and ventricular conduction delay. *Circulation*. 1999; 99(12):1567-1573.
34. Morales MA, Startari U, Panchetti L, Rossi A, Piacenti M. Atrioventricular delay optimization by doppler-derived left ventricular dP/dt improves 6-month outcome of resynchronized patients. *Pacing Clin Electrophysiol*. 2006;29(6):564-568.
35. Tei C, Nishimura RA, Seward JB, Tajik AJ. Noninvasive Doppler-derived myocardial performance index: correlation with simultaneous measurements of cardiac catheterization measurements. *J Am Soc Echocardiogr*. 1997;10(2):169-178.
36. Tei C, Ling LH, Hodge DO, et al. New index of combined systolic and diastolic myocardial performance: a simple and reproducible measure of cardiac function – a study in normals and dilated cardiomyopathy. *J Cardiol*. 1995;26(6):357-366.
37. Porciani MC, Dondina C, Macioce R, et al. Echocardiographic examination of atrioventricular and interventricular delay optimization in cardiac resynchronization therapy. *Am J Cardiol*. 2005;95(9): 1108-1110.
38. Stockburger M, Fateh-Moghadam S, Nitardy A, Langreck H, Haverkamp W, Dietz R. Optimization of cardiac resynchronization guided by Doppler echocardiography: haemodynamic improvement and intraindividual variability with different pacing configurations and atrioventricular delays. *Europace*. 2006;8(10):881-886.
39. Gorcsan J 3rd, Abraham T, Agler DA, et al. Echocardiography for cardiac resynchronization therapy: recommendations for performance and reporting – a report from the American Society of Echocardiography Dyssynchrony Writing Group endorsed by the Heart Rhythm Society. *J Am Soc Echocardiogr*. 2008;21(3): 191-213.
40. Kedia N, Ng K, Apperson-Hansen C, et al. Usefulness of atrioventricular delay optimization using Doppler assessment of mitral inflow in patients undergoing cardiac resynchronization therapy. *Am J Cardiol*. 2006;98(6):780-785.
41. Scharf C, Li P, Muntwyler J, et al. Rate-dependent AV delay optimization in cardiac resynchronization therapy. *Pacing Clin Electrophysiol*. 2005;28(4):279-284.
42. Cha YM, Nishimura RA, Hayes DL. Difference in mechanical atrioventricular delay between atrial sensing and atrial pacing modes in patients with hypertrophic and dilated cardiomyopathy: an electrical hemodynamic catheterization study. *J Interv Card Electrophysiol*. 2002;6(2):133-140.
43. Bernheim A, Ammann P, Sticherling C, et al. Right atrial pacing impairs cardiac function during resynchronization therapy: acute effects of DDD pacing compared to VDD pacing. *J Am Coll Cardiol*. 2005;45(9):1482-1487.
44. Burri H, Sunthorn H, Shah D, Lerch R. Optimization of device programming for cardiac resynchronization therapy. *Pacing Clin Electrophysiol*. 2006;29(12):1416-1425.
45. Porciani MC, Sabini A, Colella A, et al. Interatrial septum pacing avoids the adverse effect of interatrial delay in biventricular pacing: an echo-Doppler evaluation. *Europace*. 2002;4(3):317-324.

46. Rao RK, Kumar UN, Schafer J, Viloria E, De Lurgio D, Foster E. Reduced ventricular volumes and improved systolic function with cardiac resynchronization therapy: a randomized trial comparing simultaneous biventricular pacing, sequential biventricular pacing, and left ventricular pacing. *Circulation*. 2007;115(16):2136-2144.
47. Sogaard P, Egeblad H, Pedersen AK, et al. Sequential versus simultaneous biventricular resynchronization for severe heart failure: evaluation by tissue Doppler imaging. *Circulation*. 2002;106(16): 2078-2084.
48. Verbeek XA, Auricchio A, Yu Y, et al. Tailoring cardiac resynchronization therapy using interventricular asynchrony. Validation of a simple model. *Am J Physiol Heart Circ Physiol*. 2006;290(3):H968-H977.
49. Vernooy K, Cornelussen RN, Verbeek XA, et al. Cardiac resynchronization therapy cures dyssynchronopathy in canine left bundle-branch block hearts. *Eur Heart J*. 2007;28(17):2148-2155.
50. Chung ES, Leon AR, Tavazzi L, et al. Results of the predictors of response to CRT (PROSPECT) trial. *Circulation*. 2008;117(20): 2608-2616.
51. Bordachar P, Lafitte S, Reuter S, et al. Echocardiographic parameters of ventricular dyssynchrony validation in patients with heart failure using sequential biventricular pacing. *J Am Coll Cardiol*. 2004;44(11):2157-2165.
52. Vanderheyden M, De Backer T, Rivero-Ayerza M, et al. Tailored echocardiographic interventricular delay programming further optimizes left ventricular performance after cardiac resynchronization therapy. *Heart Rhythm*. 2005;2(10):1066-1072.
53. Mortensen PT, Sogaard P, Mansour H, et al. Sequential biventricular pacing: evaluation of safety and efficacy. *Pacing Clin Electrophysiol*. 2004;27(3):339-345.
54. Boriani G, Muller CP, Seidl KH, et al. Randomized comparison of simultaneous biventricular stimulation versus optimized interventricular delay in cardiac resynchronization therapy. The Resynchronization for the HemodYnamic Treatment for Heart Failure Management II implantable cardioverter defibrillator (RHYTHM II ICD) study. *Am Heart J*. 2006;151(5):1050-1058.
55. Fauchier L, Marie O, Casset-Senon D, Babuty D, Cosnay P, Fauchier JP. Interventricular and intraventricular dyssynchrony in idiopathic dilated cardiomyopathy: a prognostic study with Fourier phase analysis of radionuclide angioscintigraphy. *J Am Coll Cardiol*. 2002;40(11):2022-2030.
56. Kerwin WF, Botvinick EH, O'Connell JW, et al. Ventricular contraction abnormalities in dilated cardiomyopathy: effect of biventricular pacing to correct interventricular dyssynchrony. *J Am Coll Cardiol*. 2000;35(5):1221-1227.
57. Burri H, Sunthorn H, Somsen A, et al. Optimizing sequential biventricular pacing using radionuclide ventriculography. *Heart Rhythm*. 2005;2(9):960-965.

18 Sudden Cardiac Risk Assessment

Stephan Danik and Jagmeet Singh

Abstract

For the purposes of this discussion, the focus on the use of imaging as a means for risk assessment for sudden cardiac death will be limited to patients with structural heart disease with an emphasis on the detection and characterization of scar. The concept that high-resolution imaging can detect the hallmarks of electrical instability assumes that there are specific characteristics of scar and the surrounding surviving myocardium that are indeed "arrhythmogenic." While most of the work has focused on the detection of this at-risk tissue in patients with coronary artery disease, this chapter will also highlight the data presently available in other myopathies, including hypertrophic cardiomyopathy, arrhythmogenic right ventricular dysplasia, and sarcoidosis. While the pathological process of these entities differs considerably, each of them shares a common endpoint of the development of scar and/or fibrosis. In addition, because the majority of the work to date has focused on using cardiac magnetic resonance imaging to risk-stratify patients for ventricular tachyarrhythmias, much of this review will highlight the use of this modality and the data presently available.

Keywords

Sudden cardiac death • Imaging in assessment for sudden cardiac death • Hypertrophic cardiomyopathy • Arrhythmogenic right ventricular dysplasia • Sarcoidosis • Cardiac magnetic resonance imaging

Sudden cardiac death is the leading cause of cardiovascular mortality accounting for approximately 400,000 lives each year in the USA.[1,2] Patients with a history of coronary artery disease, previous myocardial infarction, and depressed left ventricular function are at increased risk for ventricular arrhythmias that are responsible for the vast majority of this syndrome.[3] As a result, many of these patients with an ejection fraction ≤35% receive prophylactic placement of an implantable cardioverter-defibrillator (ICD). Closer examination of the two trials[4,4a,5] responsible for this current standard of care found that only slightly more than 20% of implanted ICDs had fired appropriately during 4 years of follow-up,[6] subjecting the majority of patients to the morbidity of having implanted hardware without receiving benefit. While ejection fraction is a strong predictor of ventricular tachyarrhythmias, it is relatively nonspecific, and other clinical characteristics are needed to help appropriately risk-stratify these patients.[7]

As a result, the search for a diagnostic test that could serve as a noninvasive means of identifying patients at risk for sudden cardiac death has rapidly evolved. Traditional methods, such as exercise treadmill testing and myocardial perfusion imaging, provide important prognostic information for patients at risk for cardiovascular morbidity and mortality. While medical therapy improves outcomes in patients with structural heart disease, the use of ICDs has dramatically

S. Danik (✉)
Experimental Electrophysiology Laboratory,
Department of Medicine, Massachusetts General Hospital,
Boston, MA, USA
e-mail: sdanik@partners.org

A. Auricchio et al. (eds.), *Cardiac Imaging in Electrophysiology*,
DOI 10.1007/978-1-84882-486-7_18,

altered the treatment of primary and secondary prevention of sudden cardiac death.

Most of the experimental work that has elucidated the mechanism of scar-related ventricular tachycardia has been performed in models of myocardial infarction from the occlusion of an epicardial coronary artery. It is important to review the pathological process of scar-related ventricular tachyarrhythmias as it has allowed to launch the means to accurately detect its hallmarks to identify the high-risk patient.

18.1 The Substrate of Sudden Cardiac Death

18.1.1 Mechanism of Ventricular Tachycardia

In order to find a more specific way of assessing who is at risk for sudden death, an understanding of the basic mechanism of ventricular tachycardia in the patient with coronary artery disease is critical. Detailed works in animal and human pathological specimens have shown that both a substrate and a trigger are required for the initiation and propagation of the reentry circuit.[8-11] The initiation may be due to multiple factors such as ischemia, changes in autonomic tone, and neurohormonal and metabolic influences.[12] The anatomic substrate is scar tissue around which propagation exhibits slow conduction, thus allowing for maintenance of reentry circuits. The surviving myocardial fibers surrounding and interspersed within the infarcted region provide the electrical heterogeneity that is required for differences in conduction velocity, functional block, and alterations in cell-to-cell coupling.[13-19]

In explanted human hearts, high-resolution mapping revealed that the mechanism of slow conduction that is critical for reentry occurred along a complex network of surviving myocytes interspersed in and around the scar[20,21] (Fig. 18.1). These findings helped to complement earlier work in animal models that demonstrated the presence of the circuit in the "border zone" of tissue made of scar and surviving myocardium.[22,23]

18.1.2 Viable Myocardium, Hibernating Myocardium, and Cardiovascular Prognosis

At about the same time that the relevance of these surviving fibers was determined in the context of reentrant circuits, the role of viable myocardium (also referred to as jeopardized myocardium) was emerging as a risk factor for death in patients with a history of myocardial infarction. In patients with an area of previously infarcted myocardium, the tissue is often a heterogeneous mixture of both scar and noncontractile but still surviving myofibers.

This dysfunctional but viable tissue in the setting of chronic coronary disease has been classified as hibernating myocardium.[24] In patients with viable myocardium, the ability to restore blood flow through revascularization improves survival as compared to those who are treated medically[25-28] (Fig. 18.2). This mortality benefit occurs even without an improvement in left ventricular function.[29] However, revascularization is not possible in many patients, especially those with a prior coronary artery bypass grafting or diabetics with poor target vessels. Despite areas of viable myocardium, these patients are treated medically with the subsequent risk of higher mortality. Not surprisingly, in small, nonrandomized studies, sudden cardiac death was the leading cause of mortality in patients with hibernating myocardium who were treated medically.[30-32]

18.2 Imaging the Substrate

18.2.1 The Use of Magnetic Resonance Imaging to Determine Viable Myocardium

Cardiac magnetic resonance imaging (MRI) in the assessment of myocardial dysfunction has become a powerful tool for the assessment and quantification of myocardial tissue characteristics as well as cardiac function.[33,34] Its ability to distinguish living from chronically infarcted tissue with great detail is done so with a technique known as delayed-enhancement contrast MRI (DE-MRI).[35-37] Studies have validated its use in the assessment of viable myocardium[38,39]; it can quantify the extent of transmural scar,[40] and it can distinguish viability from scar in the different myocardial layers.[41,42] There are recent studies that have utilized this technique for general cardiovascular outcomes,[43,44] and none to date in large scale for VT/VF.

18.2.2 Image Integration to Characterize Scar in Patients with Ventricular Tachycardia

Image integration of scar in patients with ventricular arrhythmias using multiple imaging modalities has been performed during electrophysiologic study in patients with ventricular tachycardia. Because imaging is performed in patients who have had a prior arrhythmogenic event, the information derived from visualization of the substrate has yielded important insights when considering mapping and ablating patients who have had VT. Initial studies in the porcine model of healed myocardial infarction demonstrated the feasibility of aligning pre-procedural imaging (in this case MRI) with real-time electroanatomic mapping to aid in mapping and ablation of the heart.[45] This early work suggested that DE-MRI could begin to visualize the substrate that had been

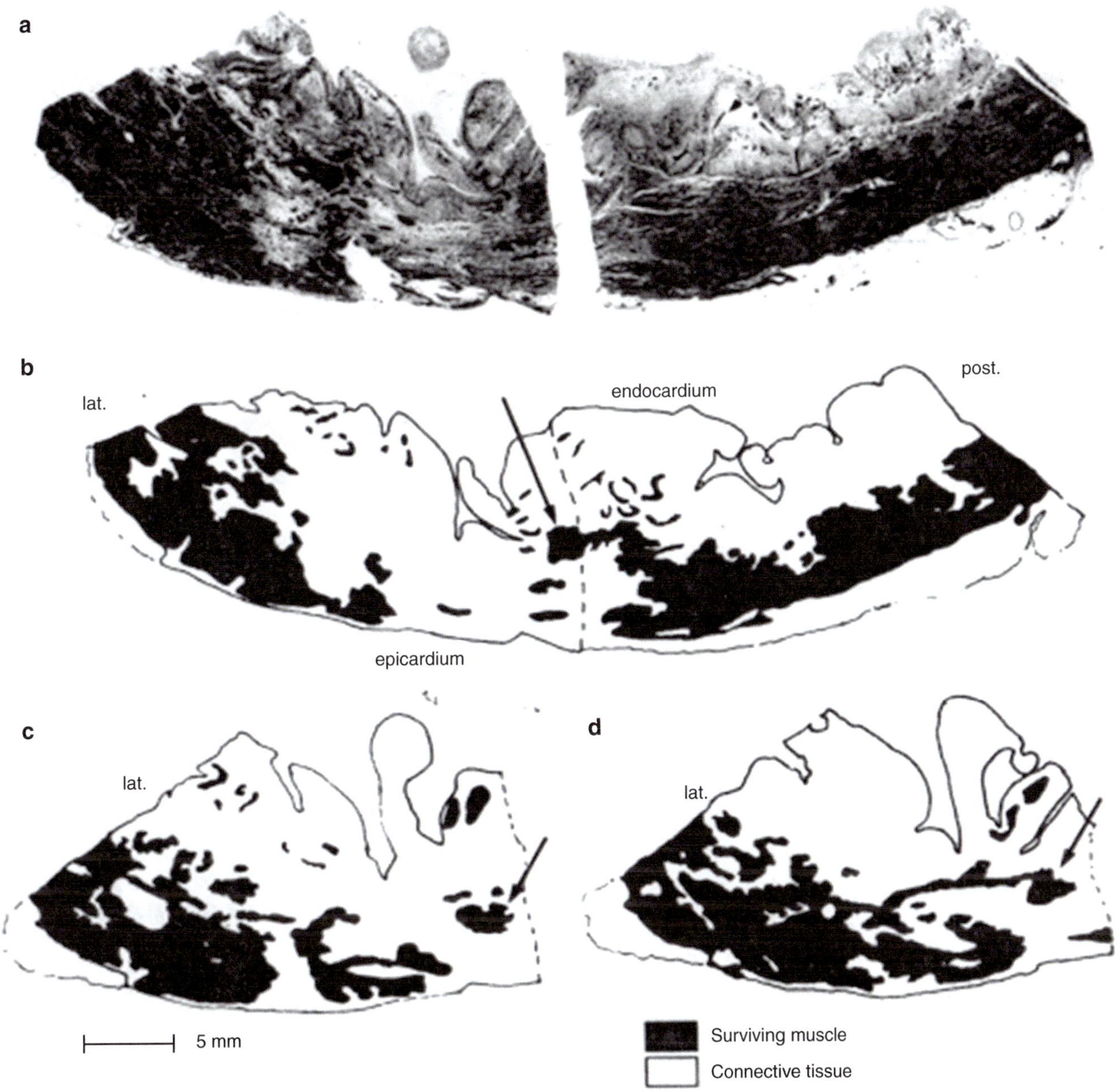

Fig. 18.1 (**a**) Photomicrograph of sections of Langendorff-perfused human heart. The patient had a history of myocardial infarction, coronary artery disease, and ventricular tachycardia. *Dark areas* mark surviving cardiac tissue; *light areas* point to fibrotic and fatty tissue. (**b**) Schematic drawings of the section. (**c** and **d**) Schematic drawings of sections beneath **b**. *Arrow* notes continuous patch of viable tissue coursing through lateral and posterior wall (From de Bakker et al.,[20] with permission)

described in the earlier experimental models of reentry (Fig. 18.3).

Both MRI and CT have since been used to characterize the tissue to be targeted for ablation. DE-MRI was used in characterize and aid in the localization of the critical site for successful catheter ablation of ventricular arrhythmias in 29 patients with nonischemic cardiomyopathy.[46] Importantly, when pre-procedural imaging identified scar in the epicardium or midmyocardium without extension to the endocardium, endocardial ablation was unsuccessful. These findings are important as these patients are more likely than those with coronary disease to have arrhythmias that can only be successfully ablated from the epicardial surface.

Integration of positron emission tomography (PET) with voltage mapping has also been described to aid in the ablation of ventricular tachycardia. The potential ability of PET-CT to provide both biological and anatomical characteristics of the scar and interdigitating fibers that can be combined with the electrical properties of the surviving tissue has been performed in patients with coronary artery disease

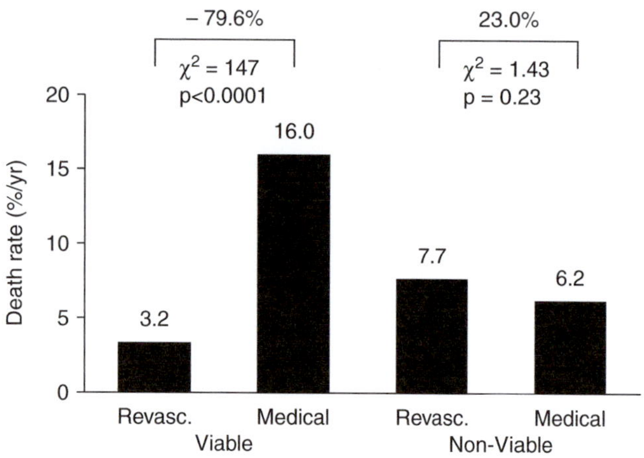

Fig. 18.2 Death rates for patients with and without myocardial viability treated by revascularization or medical therapy. There is 79.6% reduction in mortality for patients with viability treated by revascularization ($p<0.0001$). In patients without myocardial viability, there was no significant difference in mortality with revascularization versus medical therapy (From Allman et al., [25] with permission)

undergoing ablation of ventricular tachycardia. Reports indicate that the delineation of such tissue is markedly enhanced with this technology. In addition, areas of low voltage due to poor catheter contact can be quickly confirmed as such if the areas represent metabolically normal tissue. Indeed, these studies have shown that successful integration of these imaging modalities can be reliably performed when performing an anatomical approach for patients with scar-related VT (substrate modification).[47,48]

18.3 What Are the Critical Components?

18.3.1 Imaging of Scar and Ventricular Tachyarrhythmias: Is Quantification of Scar Enough?

The concept that the presence of scar and/or fibrosis represents a potential substrate for ventricular tachyarrhythmias has led investigators to determine whether prognostic information can be obtained in a variety of pathological processes, including hypertrophic cardiomyopathy, and ischemic and nonischemic cardiomyopathies. For patients with a history of myocardial infarction, prior data has suggested that total cardiac enzyme release is a predictor of cardiovascular outcomes.[49] The degree of tissue necrosis may therefore be reflected by the extent of scarred myocardium. It has been able to provide prognostic information for patients treated with percutaneous coronary intervention during an acute ST segment elevation myocardial infarction. DE-MRI performed 1 week post intervention ($n=122$) was able to be a better predictor of subsequent death, myocardial infarction, or heart failure.[50]

The extent of scar tissue, as determined by stress testing with myocardial perfusion (technetium) imaging, has been demonstrated to be an important predictor of death and ventricular arrhythmias.[51] In 349 patients with ischemic cardiomyopathy and depressed ejection fraction, patients who were more likely to die or undergo cardiac transplantation were found to have a greater degree of quantifiable scar than those who did not reach this endpoint.[52] In addition, this study

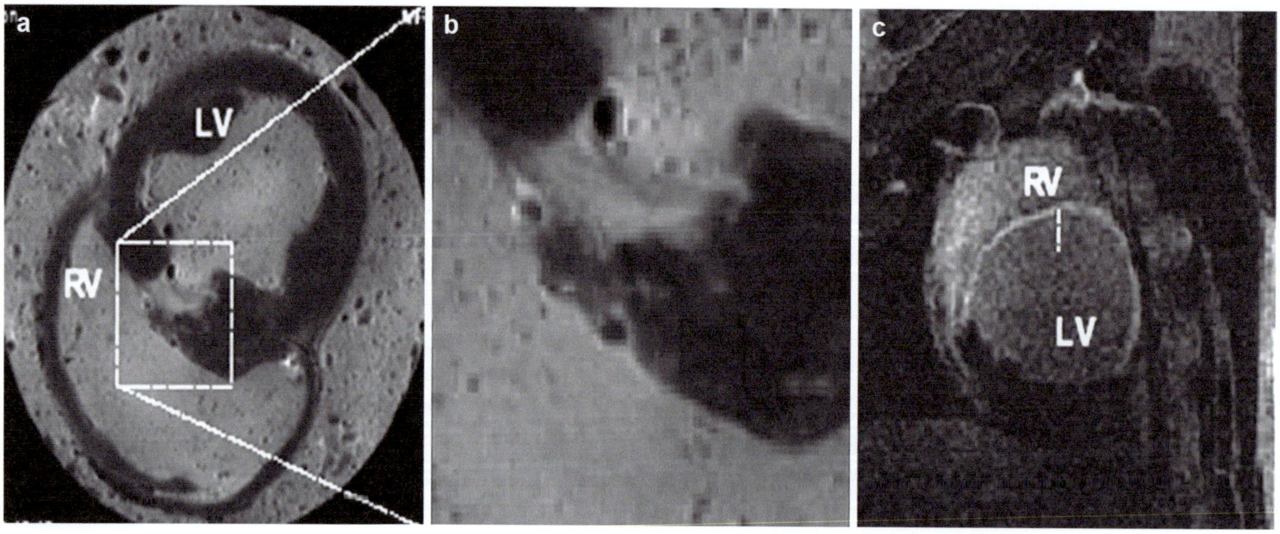

Fig. 18.3 MR image of explanted and in vivo porcine infarcts. IV gadolinium was infused in a chronic infarcted pig. The explanted heart was imaged (**a** and **b**) using MRI surface coils (1.5 T MRI). Alternatively, IV gadolinium was infused and an in vivo MRI was performed (**c**) using surface coils. In (**a**), the infarct can be seen involving the interventricular septum, and the magnification (**b**) of this region is suggestive of interdigitation of normal and abnormal tissue. In (**c**), the infarcted tissue is hyperenhanced (Courtesy of Vivek Reddy, MD)

found that women with scar were also more likely to be predictors of cardiac events. Similarly, DE-MRI performed in 231 patients with a prior healed myocardial infarction demonstrated that infarct size was a better predictor of long-term mortality than left ventricular ejection fraction.[53]

While the ability to accurately quantify scar in patients with coronary artery disease may be useful to accurately reflect the size of a prior myocardial infarction, it is not clear if this information will be useful to predict who is likely to develop ventricular tachyarrhythmias as opposed to mortality due to congestive heart failure or myocardial infarction. Efforts to answer this question have grown with the hope of providing a noninvasive measure to prognosticate cardiovascular outcomes, especially as it relates to arrhythmogenesis.

Infarct mass, as determined by DE-MRI, was shown to be a better predictor than left ventricular ejection fraction of inducible monomorphic ventricular tachycardia at electrophysiology study in patients ($n=48$) with a history of coronary artery disease.[54]

18.3.2 DETERMINE Study

Based on the data suggesting that infarct size may be an important predictor of future arrhythmogenic events, the Defibrillators to Reduce Risk by Magnetic Resonance Imaging Evaluation (DETERMINE) Trial (trial (http://clinicaltrials.gov/ct2/show/NCT00487279) was designed with the hope of answering this question. The goal of the study was to test the hypothesis that patients with an infarct size of ≥10% randomized to ICD and medical therapy will have improved survival as compared to those randomized to medical therapy alone.[55] DE-MRI would have been performed in patients with coronary disease and an ejection fraction of >35% and less than 50% (or patients with an ejection fraction of 30–35% and NYHA class I heart failure without a history of ventricular tachyarrhythmias). The primary endpoint was to have been death from any cause. Unfortunately, to reach the target randomization, approximately 10,000 patients would have to been screened with DE-MRI. Due to slow enrollment, the study was recently halted.

18.3.3 Fibrosis/Scar in Patients with Nonischemic Cardiomyopathy

Patients with a dilated cardiomyopathy (without obstructive coronary artery disease) are also at risk for the development of scar and fibrosis, thus creating the substrate for reentrant ventricular arrhythmias. While it has been known that fibrosis and scar can be present in myopathic hearts without coronary disease,[56] the ability to detect these changes with noninvasive imaging has only recently been realized. One of the initial studies[57] demonstrated that 28% patients with a nonischemic cardiomyopathy were found to have delayed enhancement; the pattern was patchy in nature and distinct in territory as compared to those with coronary artery disease.

Further studies have sought to use this finding to prognosticate this subgroup of patients. Of 101 patients with a history of nonischemic dilated cardiomyopathy, 35% were found to have midwall fibrosis as detected by DE-MRI.[58] Patients with fibrosis were more likely than those without to reach the primary endpoint of mortality or hospitalization for cardiovascular causes. In addition, although there were few patients with sudden cardiac death or ventricular tachyarrhythmias, those with fibrosis were more likely to reach this secondary endpoint as well.

To determine which characteristics of scar may play a role in the inducibility of sustained monomorphic ventricular tachycardia during electrophysiologic study, one group[59] performed DE-MRI in 26 patients with moderate-to-severe left ventricular dysfunction without obstructive coronary artery disease. Five of the 26 patients were inducible; the mean cycle length of the tachycardias was 290 ms. Inducible patients were more likely to have scar distribution that involved 26–75% of the wall thickness, suggesting that a certain degree of transmural necrosis may be necessary for the maintenance of reentry in patients without a history of significant coronary disease. This cohort was limited in the number of patients, and it was noted that inducible patients were more likely to have a lower ejection fraction and fewer segments without hyperenhancement of myocardial tissue.

Another group[60] performed DE-MRI in 65 patients with nonischemic cardiomyopathy who subsequently underwent implantation of an ICD for primary prevention of sudden cardiac death. Delayed enhancement was detected in 42% of these patients. At 17 months, patients with fibrosis were significantly more likely to reach a combined endpoint of hospitalization for heart failure, appropriate ICD discharge, or cardiac death. While 4 of 27 patients with delayed enhancement had an appropriate ICD discharge, it is important to note that 3 out of 38 patients without delayed enhancement did receive an appropriate ICD discharge. The latter finding may be due to the mechanism of ventricular tachycardia being independent of the presence of scar or because the degree of fibrosis could not be detected by DE-MRI. Regardless, this finding is an important consideration when making clinical decisions in such patients.

The largest study to date that evaluated the prognostic capability of DE-MRI included 857 patients with both ischemic ($n=642$) and nonischemic cardiomyopathy($n=215$) with the primary endpoint being all-cause mortality or cardiac transplantation.[61] Patients with a known etiology (hypertrophic cardiomyopathy, sarcoidosis, amyloidosis) were excluded. At a median follow-up of 4.4 years, 29% of patients reached this endpoint underscoring the fact that these patients were selected for referral for MRI. Most of these patients had a history of coronary artery disease (75%) with mild-to-moderate left ventricular dysfunction. The scar index independently predicted death or transplantation regardless of whether the patients had coronary disease. This finding was true even in patients with an ejection fraction of less than 30%. Interestingly, even in patients with preserved ejection fraction, the presence of delayed enhancement significantly increased the likelihood of death or progression to transplantation.

However, while the presence of scar results in a potential substrate, the actual mixture of viable and nonviable tissue in and around the infracted tissue may be a more important predictor.[43]

18.4 Characterizing the Border Zone

The ability to characterize scar and the surrounding border zone in patients with a prior myocardial infarction has been studied using both DE-MRI as well as PET CT. Furthermore, initial reports seem to yield promising results regarding prognostic information in potentially high-risk patients.

Fernandes and colleagues[62] characterized the anatomical and mechanical properties of left ventricular wall segments that contain different degrees of scar tissue and the location of these segments from the interface between infarcted and noninfarcted myocardial tissue. All 46 patients underwent electrophysiologic testing before implantation of a cardioverter-defibrillator for primary prevention of sudden cardiac death. Patients with inducible monomorphic ventricular tachycardia during electrophysiology study were more likely to have a greater number of infracted and border zone segments as compared to those who were noninducible. In addition, inducible patients were more likely to have border zones segments with greater systolic contractility as compared to noninducible patients. These findings suggested that "enhanced" border zone function as determined by DE-MRI may be a marker of inducible monomorphic ventricular tachycardia. This characterization of the mixture of viable and nonviable tissue (tissue heterogeneity) and how it relates to arrhythmogenesis further strengthened the potential prognostic capabilities of DE-MRI.[63]

More recently, Roes and colleagues[64] examined whether infarct tissue heterogeneity could be used to predict spontaneous ventricular tachyarrhythmias. Ninety one patients with a prior myocardial infarction underwent DE-MRI before placement of an implantable cardioverter-defibrillator (ICD). After a median follow-up of 8.5 months, 18 patients had received appropriate ICD therapy. Infarct tissue heterogeneity was a better predictor of appropriate ICD therapy as compared to left ventricular function, volume, and total infarct size.

Specific characteristics of the interface between scar and surviving tissue are likely a crucial determinant of whether a substrate exists; while DE-MRI has traditionally been used to characterize this area (Fig. 18.4), computed tomography with delayed enhancement has also been used given the number of patients who already have ICDs (Fig. 18.5).

There are important technical considerations when assessing the characteristics of the peri-infarct territory known as the border zone. This mixture of viable and nonviable tissue does alter signal intensity as partial volume effects produce intermediate signal intensities along the border zone.[65] The quality of the study is paramount; proper adjustment of inversion time (T1) is crucial in interpreting the images.[66] Manually adjusting the T1 to null signal from normal myocardium is necessary in each patient on an individual basis to optimize signal intensity.[67] This method is necessary to maximize the difference between infracted and noninfracted tissue. As the field rapidly evolves, measures of infarct tissue heterogeneity continue to grow.

18.5 Arrhythmogenic Right Ventricular Cardiomyopathy/Dysplasia (ARVC/D)

Arrhythmogenic right ventricular cardiomyopathy/dysplasia is a cardiomyopathy with a progressive replacement of right ventricular myocytes with adipose and fibrous tissue; the left ventricle can also be affected. The incidence has been estimated at 1:5000[68]; it can exist in a sporadic as well as an inherited fashion, most commonly in an autosomal dominant fashion. There is mounting evidence that genetic mutations of desmosomal proteins located at the intercalated disk are altered and susceptible to stress and damage ultimately leading to myocyte death.[69] Sudden cardiac death is common, and up to half of patients die before the age of 35.[70] The diagnosis has been aided by criteria established by the Task Force of cardiomyopathies.[71]

18.5.1 The Impact of Imaging for Diagnosing ARVC/D

The use of MRI to aid in the diagnosis of arrhythmogenic right ventricular dysplasia has been widely adopted. While the original task force recommendations include that the diagnosis of fibrofatty replacement of myocardium be made by endomyocardial biopsy (a major criteria), this determination has been largely replaced by noninvasive imaging, either by DE-MRI or CT.[72,73] This practice has become more acceptable not only due to the potential complications due to biopsy of the right ventricle, but also because of the patchy nature of the fibrofatty replacement of tissue of the disease which may result in a relatively low sensitivity as a means of diagnosis.

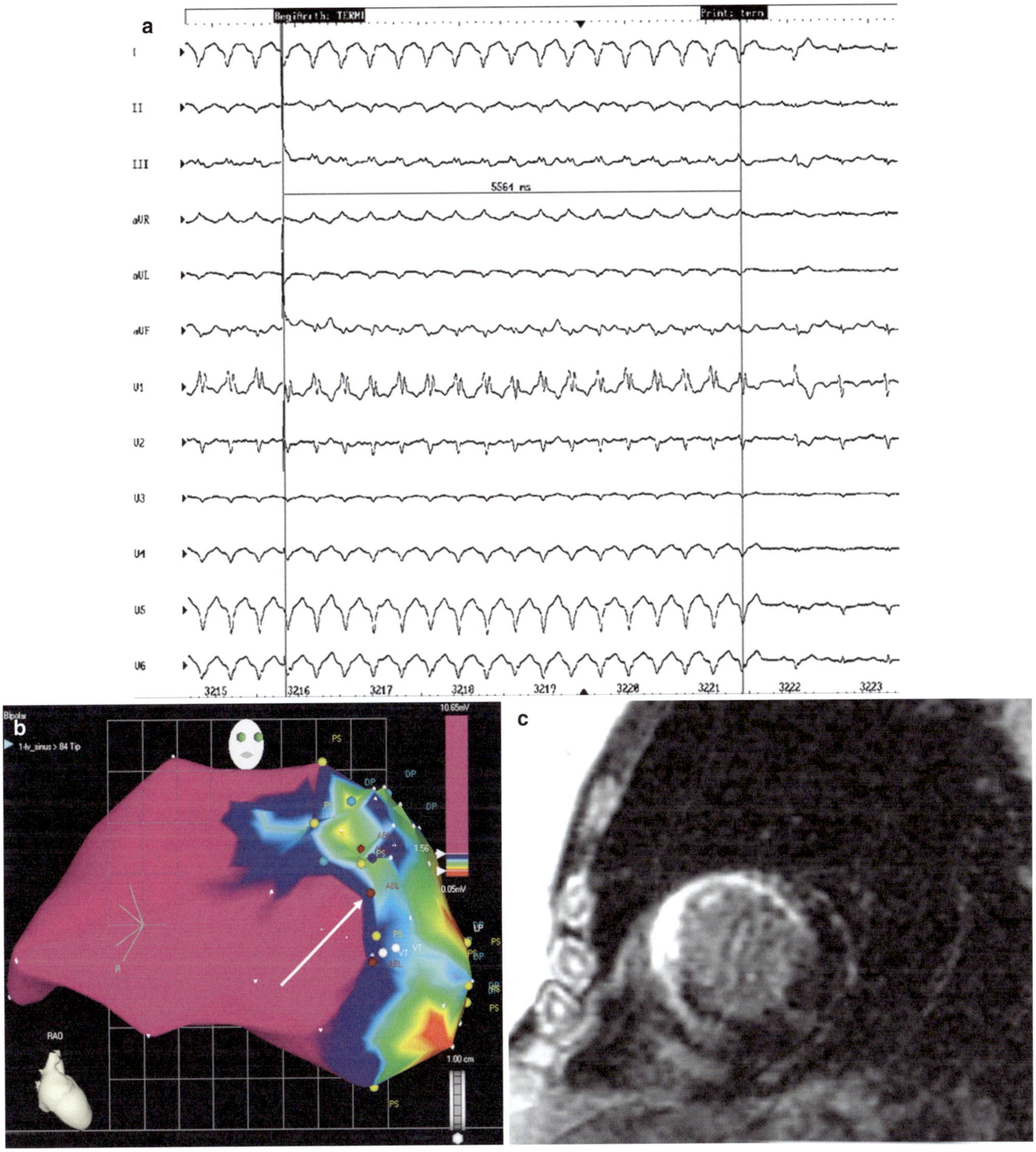

Fig. 18.4 Electroanatomic voltage map created during electrophysiologic study and radiofrequency ablation of a 61-year-old female with a prior history of myocardial infarction with sustained monomorphic ventricular tachycardia. Successful termination of tachycardia (**a**) with application of radiofrequency energy at the site marked with *white arrow* on electroanatomical map (**b**). The site is located at the border zone of scar and healthy tissue (*purple* is surviving, healthy tissue with normal amplitude while *red* is scar with the colors in between representing a mixture of viable and nonviable tissue. Cardiac MRI with scar extending anteriorly and septally (**c**). In the distal septal region, the scar is surrounded by viable tissue; it is this region where the tachycardia was ablated (Courtesy of Vivek Reddy, MD)

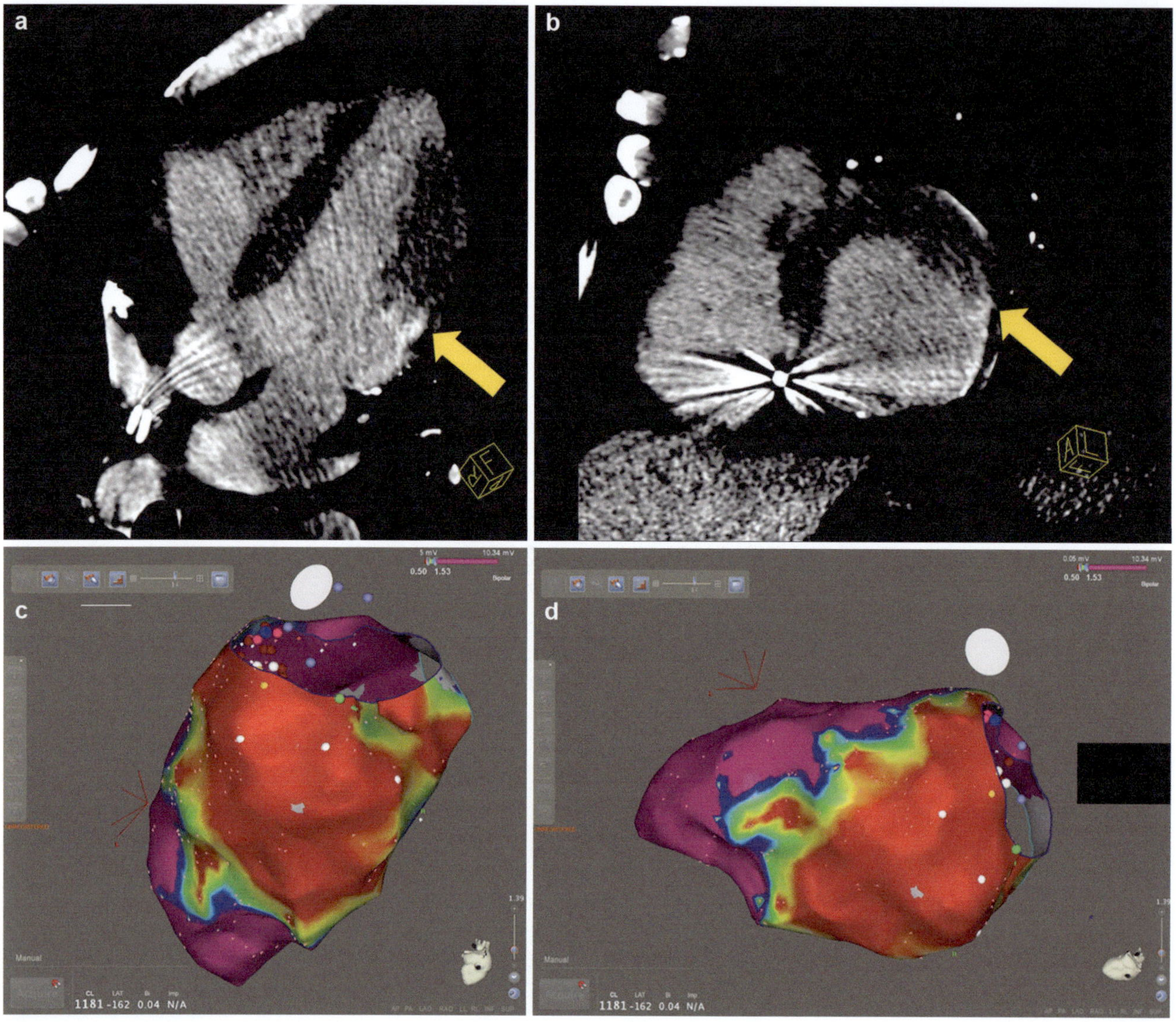

Fig. 18.5 Pre procedural computed tomography imaging (**a** and **b**) with delayed enhancement of a 59-year-old male with a prior history of myocardial infarction and ICD who subsequently developed sustained monomorphic ventricular tachycardia demonstrated scar and thinning in the lateral wall with areas of intersection of scar and surviving tissue in the inferolateral aspect toward the base of the heart. Voltage mapping (**c** and **d**) demonstrated a large inferolateral scar in the basal segment of the heart. Successful ablation of the tachycardia was performed in this region

18.5.1.1 Has Imaging Overstated the Incidence?

There is limited data as to whether DE-MRI can be used to determine which patients are at risk for malignant tachyarrhythmias. While several studies have attempted to determine whether the presence of scar in patients with HOCM may help stratify those at risk for ventricular tachycardia, a thorough evaluation has not yet been undertaken in patients with ARVC/D. It was reported that six of eight patients suspected of having ARVC/D who had the presence of delayed enhancement were found to have inducible sustained monomorphic ventricular tachycardia during electrophysiologic testing.[74]

A unique study reported the prognosis of 64 patients with or without ARVC/D all of whom presented with nonischemic scar-related VT of right ventricular origin.[75] A diagnosis of ARVC/D was made based on the task force criteria. Scar was determined either by DE-MRI or contrast echocardiography; further details were not provided. There was no difference in the recurrence of VT in patients diagnosed with or without ARVC/D based on the task force criteria.

It is important to note that the potential concern for the overdiagnosis if only fat is present on DE-MRI. While the presence of fibrosis should raise concern, fat replacement of the right ventricle has been found to be present in patients without ARVD/C, and likely represents a distinct entity of uncertain significance.[76] The presence of fat replacement in the anterior wall of the apex is probably not an abnormal finding. Prior data has suggested that the presence of significant amounts of fat infiltration of the right ventricle occurs in greater than half of normal hearts in elderly patients.[77,78] DE-MRI has been used to show that patients with marked fat deposition without other morphological features of ARVD/C is a distinct clinical entity that must be treated as such to avoid unnecessary referral for ICD implantation.[79] The presence of fibrosis in the form of delayed enhancement appears to improve the diagnostic accuracy of DE-MRI for ARVD/C.[80]

18.5.2 Hypertrophic Cardiomyopathy

18.5.2.1 Present Components for Risk Stratification

The mechanism of sudden cardiac death in patients with hypertrophic cardiomyopathy has been shown to be mainly due to ventricular tachycardia and/or fibrillation.[81] The traditional risk factors for assessment of sudden cardiac death include family history of sudden death, history of syncope, nonsustained ventricular tachycardia on Holter, abnormal blood pressure response during exercise stress testing, and left ventricular hypertrophy of 30 mm or greater.[82,83] Unfortunately, most data suggests that many patients experience an event with only one risk factor. Indeed, data with long-term follow-up suggests that appropriate ICD discharges are just as likely in patients with one risk factor as they are in those with 3 or more risk factors.[84] As a result, recommendations on prophylactic implantation of a defibrillator are difficult in patients who may have only one such risk factor.

Triggers are thought to include abnormal reflex control of the peripheral vasculature leading to inappropriate vasodilatation, especially during or immediately after exercise.[85,86] This abnormal response is thought to account for the observation that a substantial portion of patients with hypertrophic cardiomyopathy experience sudden cardiac death with exertion.[82]

18.5.2.2 Scarring in HOCM

The pattern of scarring in hypertrophic cardiomyopathy is distinct in that it typically does not occur in the territory of the epicardial coronary arteries. There are indeed typical patterns of fibrosis seen on DE-MRI. Recent data has suggested that the presence of scar and fibrosis in patients can be reliably detected using delayed enhanced MRI. In addition, data suggests that there is good histological correlation of the scar with areas of delayed enhancement.[87] More importantly, the use of this imaging modality may have prognostic significance in these patients.[88]

Several studies have sought to show a correlation between scar and fibrosis as detected by DE-MRI and the presence of nonsustained ventricular tachycardia as detected by Holter monitoring. One such investigation[89] found that delayed enhancement was present in 41% of 177 patients with hypertrophic cardiomyopathy. Although this finding was an independent predictor of NSVT on Holter, there was no difference in the extent of tissue with delayed enhancement in patients with or without NSVT. Another study ($n=47$) found that delayed enhancement was present in nearly all patients with NSVT on Holter, but that it was also present in 60% of patients without NSVT. It also found that the degree of tissue with delayed enhancement was not different in patients with or without NSVT.[90] Importantly, while patients with delayed enhancement may be more likely to have NSVT, the lack of delayed enhancement as seen by MRI does not preclude the presence of NSVT by Holter.[91,92] As a result, whether this imaging modality will prove to have a negative predictive value that is clinically acceptable during risk-stratification of patients has yet to be determined. Most of these studies have found that the presence of fibrosis at the RV insertion site of the septum is "classic" for hypertrophic cardiomyopathy.[93]

The largest study published to date included 424 patients who underwent DE-MRI.[94] While 56% were found to have delayed enhancement, there was no relationship to symptoms or functional class as compared to those without this finding. Patients with delayed enhancement were more likely to have NSVT on Holter than those without it (27 vs. 8.5%). Many of the patients had undergone genotyping as well. Gene "positive" patients were more likely to have areas of delayed enhancement as well. Importantly, sudden cardiac death occurred in four patients, and an additional four patients received appropriate ICD therapy; all eight patients had tissue with delayed enhancement (Fig. 18.6).

It is important that the absence of NSVT on Holter in children is by no means reassuring. Indeed, certain genotypes have been shown to have higher rates of sudden death, but with extensive myocardial disarray and less myocardial hypertrophy and fibrosis at autopsy. In addition, these patients tend to be much younger.[95] Myocardial disarray may act as a substrate for ventricular arrhythmias due to the alteration of the location and extent of gap junctions which are critical for cell coupling and impulse propagation.

Because it is known that sudden death can occur in patients without significant left ventricular hypertrophy,[96] it is still not known whether these patients would demonstrate a significant amount of scarring on DE-MRI that would warrant concern. While it is believed that the substrate is due to myocardial disarray as well as fibrosis, it is not yet clear which of the two is the more important component in the maintenance of ventricular tachyarrhythmias. There can be significant amounts of disorganized myocardial architecture in myocardial tissue that is not hypertrophied.[97] In addition, it is not yet certain of the relationship between myocardial disarray and fibrosis that is detected during imaging.

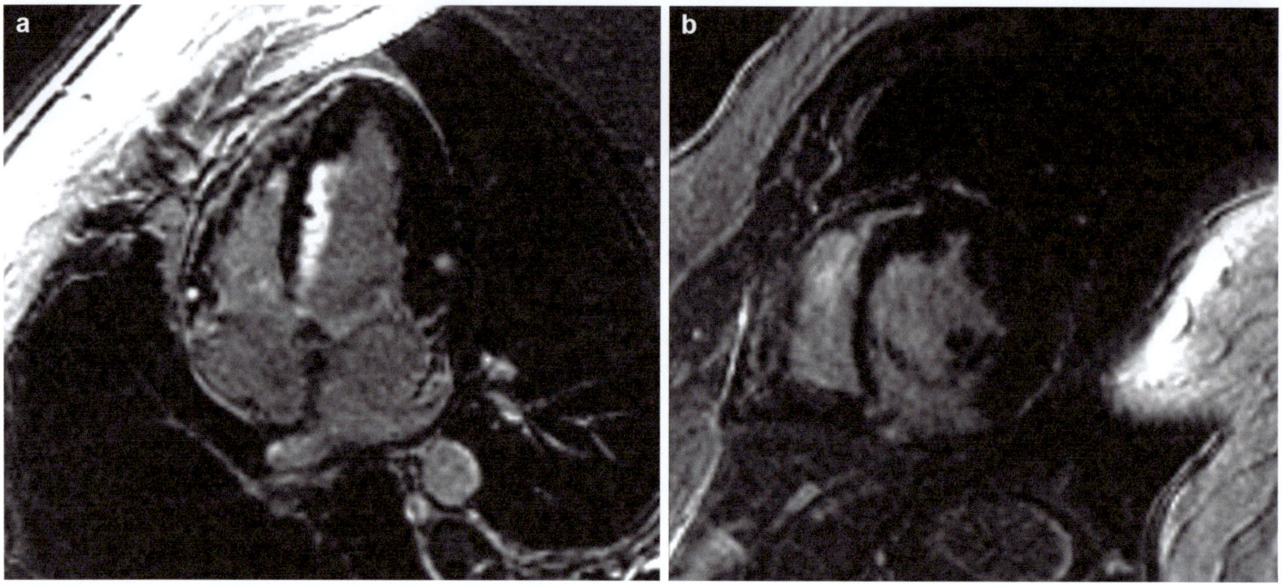

Fig. 18.6 DE-MRI images of a 52-year-old male with a known history of apical hypertrophic cardiomyopathy who had none of the traditional risk factors for sudden cardiac death. Maximal wall thickness was 16 mm. While on a routine business trip, he collapsed at the airport. Within minutes, a passerby placed an automated external defibrillator which revealed ventricular fibrillation. He was successfully defibrillated. Delayed enhancement was seen in a subepicardial distribution in the basal to mid anteroseptal wall

It has been suggested that the scar and fibrosis detected by DE-MRI in patients with hypertrophic cardiomyopathy may be directly related to the presence and degree of small intramural coronary arteriole dysplasia which is the result of pressure necrosis due to left ventricular hypertrophy and dynamic outflow tract gradients.[98] These findings also correlated with the occurrence of ventricular tachycardia as detected by Holter monitoring.

While there is no general consensus as to how to use this data, many clinicians are using DE-MRI as a "tie breaker" in risk-stratifying their patients for those who have one or more of the established features in the original guidelines. Others, however, suggest that the finding of scar alone is enough to warrant concern.

18.5.3 Sarcoidosis

18.5.3.1 Background

Sarcoidosis is a multisystem disease characterized by noncaseating granulomas. While it most commonly affects the lungs and lymph nodes, other organs and tissues, including the heart, liver, eyes, skin, and spleen, can be involved as well. While the exact etiology has not yet been determined, it is believed that the granulomas may be an immunologic response to an antigenic trigger.[99] In the USA, the annual incidence has been estimated at 10.9 per 100,000 in whites and up to 35.5 per 100,000 in African Americans.[100]

Autopsy studies have shown cardiac involvement in at least 25% of patients with sarcoidosis in the USA[101] and is responsible for up to a quarter of the deaths.[102] The presence of noncaseating granulomas can be found anywhere in the heart, but especially in the left ventricular free wall and interventricular septum; conduction abnormalities are common, including complete heart block.[103] Ventricular arrhythmias are not uncommon, and sudden cardiac death is responsible for up to 65% of cases of mortality due to cardiac sarcoid.[104,105] DE-MRI has been shown to correlate with cardiac involvement with findings of delayed enhancement suggesting fibrogranulomatous replacement and inflammation.[106,107] When delayed enhancement is found on MRI, it is typical in the basal aspects of the septum and the free wall of the left ventricle.[108]

18.5.3.2 Present Use of Imaging

There is limited data on the prognostic utility of DE-MRI in predicting future cardiac events. One of the only studies to do so[109] prospectively evaluated 81 consecutive patients with biopsy-proven extracardiac sarcoidosis with DE-MRI for evaluation of cardiac involvement and followed them for approximately 2 years. Delayed enhancement was present in 26% of patients, representing a twofold higher identification than the consensus criteria for the diagnosis of cardiac sarcoidosis developed by the Japanese Ministry of Health and Welfare (JMH).[110] All patients with hyperenhancement underwent coronary angiography that excluded the presence of obstructive coronary disease. While there was no difference in LV volumes, patients with hyperenhancement had significantly lower, albeit only mildly so, ejection fractions (median = 45 vs. 57% respectively) as compared to patients without evidence delayed enhancement. Patients with delayed enhancement were nine times more likely to

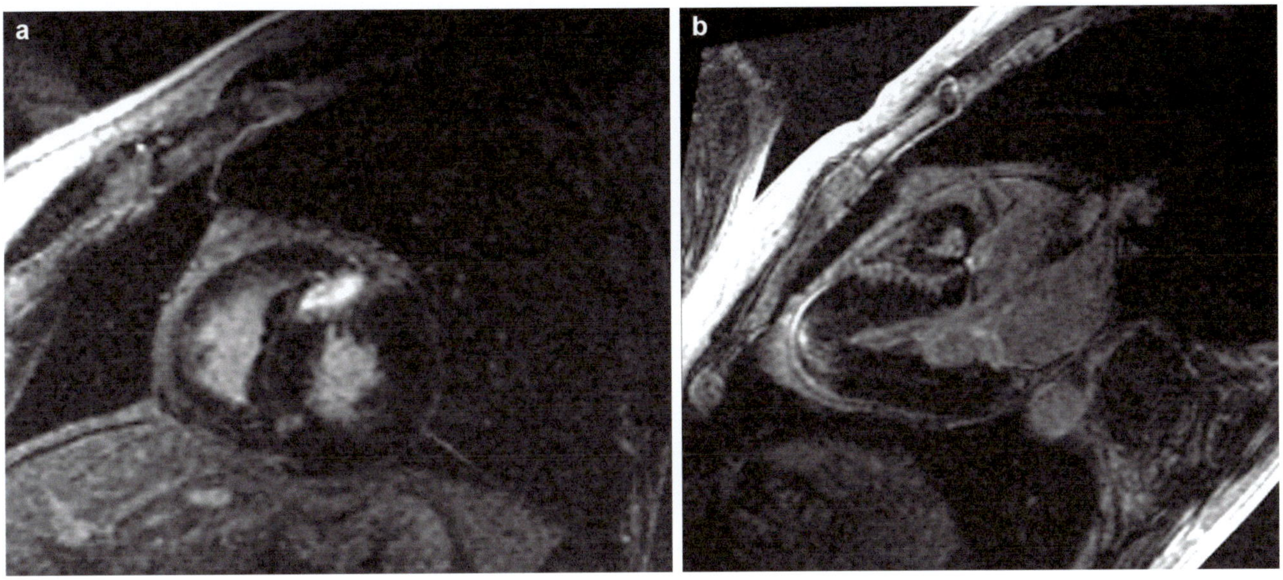

Fig. 18.7 DE-MRI images of a 55-year-old female with a history of ventricular tachycardia. Workup included a normal cardiac catheterization, mildly dilated left ventricle with an ejection fraction of 50%, and a biopsy of the right ventricle which was unrevealing. (**a** and **b**) Demonstrate scar in the septum and inferior wall at the midventricular levels. A diagnosis of sarcoidosis was eventually made, and she received an implantable defibrillator due to sustained ventricular tachycardia. She also developed high degree AV block

reach a combined endpoint of death, defibrillator therapy, or requirement of pacemaker implantation as compared to those without this finding. In addition, four patients with delayed enhancement reached the endpoint of cardiac death as opposed to one patient without it (Fig. 18.7).

18.6 Left Ventricular Non-compaction

Isolated left ventricular non-compaction (LVNC) is a cardiomyopathy characterized by persistent prominent ventricular trabeculations with deep intertrabecular recesses (Fig. 18.8) resulting from a defect in embryogenesis.[111] It is considered to be a rare cause of cardiomyopathy, although an important one in the pediatric population.[112] There is very little data about the natural course of this entity. Initially thought to have a poor prognosis, more recent data suggests otherwise.[113] These inconsistencies are not surprising due to the lack of long-term follow-up and the small numbers cited in the literature. In addition, most reports have been cited in patients who presented with symptoms. A registry of 105 adult patients in France diagnosed by echocardiography found that over 2.3 years of follow-up, severe heart failure occurred in a third of patients, ventricular arrhythmias occurred in 7, embolic events occurred in 9, and death in 12 patients.[114]

Magnetic Resonance Imaging is now considered to be the gold standard for the diagnosis of this entity. It can conclusively make the diagnosis when results of other imaging modalities are not definitive. Of a total of 763 patients referred for DE-MRI for the further characterization of a cardiomyopathy, 42 patients were diagnosed with LVNC.

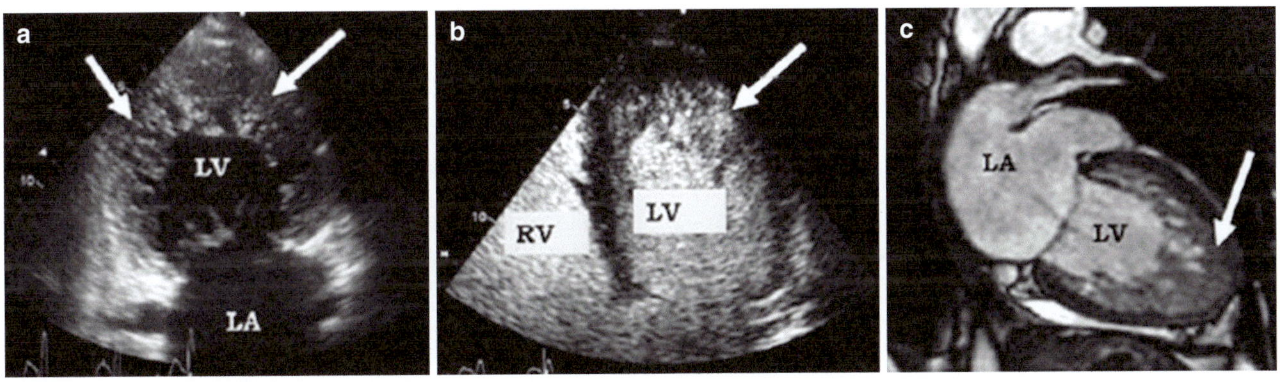

Fig. 18.8 Various imaging modalities used to demonstrated LVNC with the *arrows* highlighting the trabeculations. Transthoracic echocardiography (**a**), contrast echocardiography (**b**), DE-MRI (**c**), and 3D echocardiography (**d**). The explanted heart is shown in (**e**)

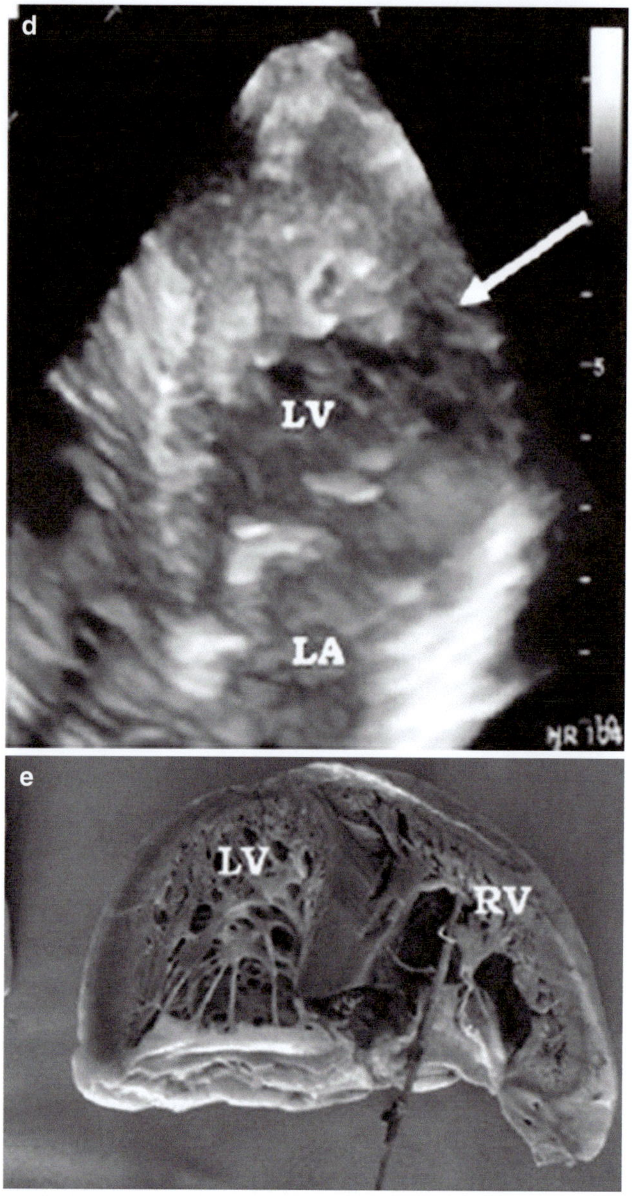

Fig. 18.8 (continued)

Echocardiography had made the diagnosis in only 10% of these patients. Half the patients with LVNC had presented with dyspnea, and a further 14% had evidence of thromboembolic phenomenon (pulmonary embolism, stroke, and brachial artery embolism).[115] The understanding of this disease is still in its infancy, and long-term data with prognostic indicators is still lacking.

18.7 Future Directions

To date, the ability of DE-MRI to image the substrate has focused on detecting areas of scar as well as the interface of surviving and infarcted tissue (gray zone, areas of interdigitation). Newer methods, such as diffusion spectrum MRI tractography (DSI tractography), have been developed to view the myofiber architecture in normal and diseased hearts.[116] This technique is able to visualize fibers at the microstructural level; the ability to do so is based on the preferential diffusion of water along the muscle fiber. This imaging modality has shown that in the excised rodent heart, the normal architecture consists of a highly organized pattern consisting of tracts of crossing helical myofibers. In contrast, 3 weeks after the induction of myocardial infarction, DSI tractography is able to show profound changes including the loss of myofibers as well as the replacement of the normal trajectory of these fibers with a meshwork network of orthogonally directed fibers.[117] These areas of disrupted architecture represent the mechanical remodeling that occur which likely have profound electrophysiological consequences. While not yet available in the clinical setting, the field is advancing rapidly to achieve this goal.

18.8 Conclusion

The search to image the substrate for sudden cardiac death has rapidly evolved. The use of noninvasive imaging, particularly DE-MRI, to quantify and characterize scar in a variety of myopathic substrates continues to move forward with the ultimate goal to be able to more effectively predict which scar is more likely to give rise to ventricular tachyarrhythmias. Whether this assessment of scar will replace, or be used in conjunction with, ejection fraction as the test of choice remains to be seen.

References

1. American Heart Association (online September 2004) Cardiovascular disease statistics. Accessed May 23, 2005.
2. Zheng ZJ et al. Sudden cardiac death in the United States, 1989 to 1998. *Circulation*. 2001;104:2158-2163.
3. Zipes DP, Wellens HJ. Sudden cardiac death. *Circulation*. 1998;98:2334-2351.
4. Moss AJ, Zareba W, Jackson Hall W, et al. Prophylactic implantation of a defibrillator in patients with myocardial infarction and reduced ejection fraction. *N Engl J Med*. 2002;346:877-883.

4a. Spirito P, Seidman CE, McKenna WJ, Maron BJ. The management of hypertrophic cardiomyopathy. *N Engl J Med*. 1997;336:775-785.

5. Brady G, Lee KL, Mark DB, et al. Amiodarone or an implantable cardioverter-defibrillator for congestive heart failure. *N Engl J Med*. 2005;352:225-237.
6. McClellan MB, Tunis SR. Medicare coverage of ICDs. *N Engl J Med*. 2005;352:222-224.
7. Gehl A, Haas D, Fuster V. Primary prophylaxis with the implantable cardioverter-defibrillator: the need for improved risk stratification. *JAMA*. 2005;294(8):958-960.
8. Josephson ME, Horowitz LN, Farshidi A, Kastor JA. Recurrent sustained ventricular tachycardia. Mechanisms. *Circulation*. 1978; 57:431-440.

9. Wit AL, Allessie MA, Bonke FI, et al. Electrophysiologic mapping to determine the mechanism of experimental ventricular tachycardia initiated by premature impulses. Experimental approach and initial results demonstrating reentrant excitation. *Am J Cardiol.* 1982;49(1):166-185.
10. El Sherif N, Scherlag BJ, Lazzara R. Electroide cather recording during malignant ventricular arrythmia following experimental acute myocardial ischemia. Evidence for re-entry due to conduction delay and block in ischemic myocardium. *Circulation.* 1975;51(6):1003-1014.
11. de Bakker JM, van Capelle FJ, Janse MJ, et al. Reentry as a cause of ventricular tachycardia in patients with chronic ischemic heart disease: electrophysiologic and anatomic correlation. *Circulation.* 1988;77(3):589-606.
12. Akhtar M, Garan H, Lehmann MH, Troup PJ. Sudden cardiac death: management of high-risk patients. *Ann Intern Med.* 1991; 114(6):499-512.
13. Josephson ME, Almendral JM, Buxton AE, Marchlinski FE. Mechanisms of ventricular tachycardia. *Circulation.* 1987;75(4 pt 2):III41-III47.
14. Garan H, Fallon JT, Rosenthal S, Ruskin JN. Endocardial, intramural, and epicardial activation patterns during sustained monomorphic ventricular tachycardia in late canine myocardial infarction. *Circ Res.* 1987;60(6):879-896.
15. Peters NS, Coromilas J, Severs NJ, Wit AL. Disturbed connexin43 gap junction distribution correlates with the location of reentrant circuits in the epicardial border zone of healing canine infarcts that cause ventricular tachycardia. *Circulation.* 1997;95(4):988-996.
16. Yao JA, Hussain W, Patel P, Peters NS, Boyden PA, Wit AL. Remodeling of gap junctional channel function in epicardial border zone of healing canine infarcts. *Circ Res.* 2003;92(4):437-443.
17. Stevenson WG, Friedman PL, Sager PT, et al. Exploring postinfarction reentrant ventricular tachycardia with entrainment mapping. *J Am Coll Cardiol.* 1997;29(6):1180-1189.
18. Fenoglio JJ Jr, Pham TD, Harken AH, Horowitz LN, Josephson ME, Al W. Recurrent sustained ventricular tachycardia: structure and ultrastructure of subendocardial regions in which tachycardia originates. *Circulation.* 1983;68(3):518-533.
19. Soejima K, Suzuki M, Maisel WH, et al. Catheter ablation in patients with multiple and unstable ventricular tachycardias after myocardial infarction: short ablation lines guided by reentry circuit isthmuses and sinus rhythm mapping. *Circulation.* 2001;104(6): 664-669.
20. de Bakker JM, Coronel R, Tasseron S, et al. Ventricular tachycardia in the infracted, Langendoff-perfused human heart: role of the arrangement of surviving cardiac fibers. *J Am Coll Cardiol.* 1990;15(7):1594-1607.
21. de Bakker JM, van Capelle FJ, Janse MJ, et al. Slow conduction in the infracted human heart. 'Zigzag' course of activation. *Circulation.* 1993;88(3):915-926.
22. Dillon SM, Allessie AM, Ursell PC, Wit AL. Influences of anisotropic tissue structure on reentrant circuits in the epicardial border zone of subacute canine infarcts. *Circ Res.* 1988;63(1):182-206.
23. Wit AL, Dillon SM, Coromilas J, Saltman AE, Waldecker B. Anisotropic reentry in the epicardial border zone of myocardial infarcts. *Ann N Y Acad Sci.* 1990;591:86-108.
24. Rahimtoola SH. The hibernating myocardium. *Am Heart J.* 1989;117:211-221.
25. Allman KC, Shaw LJ, Hachamovitch R, Udelson JE. Myocardial viability testing and impact of revascularization on prognosis in patients with coronary artery disease and left ventricular dysfunction: a meta analysis. *J Am Coll Cardiol.* 2002;39:1151-1158.
26. Toda K, Mackenzie K, Mehra MR, et al. Revascularization in severe ventricular dysfunction (15%<OR=LVEF<OR=30%): a comparison of bypass grafting and percutaneous intervention. *Ann Thorac Surg.* 2002;74(6):2082-2087.
27. Bell MR, Gersh BJ, Schaff HV, et al. Effect of completeness of revascularization on long-term outcome of patients with three-vessel disease undergoing coronary artery bypass surgery: a report from the Coronary Artery Surgery Study (CASS) Registry. *Circulation.* 1992;86:446-457.
28. Jones EL, Craver JM, Guyton RA, et al. Importance of complete revascularization in performance of the coronary bypass operation. *Am J Cardiol.* 1983;51:7-12.
29. Samady H, Elefteriades JA, Abbott BG, et al. Failure to improve left ventricular function after coronary revascularization for ischemic cardiomyopathy is not associated with worse outcome. *Circulation.* 1999;100:1298-1304.
30. Pagano D, Lewis ME, Townend JN, et al. Coronary revascularization for postischemic heart failure: how myocardial viability affects survival. *Heart.* 1990;82:684-688.
31. Podio V, Spinnler MT, Bertuccio G, et al. Prognosis of hibernating myocardium is independent of recovery of function: evidence from a routine based follow up study. *Nucl Med Commun.* 2002;23:933-942.
32. Chaudhry FA, Tauke JT, Alessandrini RS, et al. Prognostic implications of myocardial contractile reserve in patients with coronary artery disease and left ventricular dysfunction. *J Am Coll Cardiol.* 1999;34:730-738.
33. Pearlman JD, Laham RJ, Simons M. Coronary angiogenesis: detection in vivo with MR imaging sensitive to collateral neocirculation-preliminary study in pigs. *Radiology.* 2000;214:801-807.
34. Poon M, Fuster V, Fayad Z. Cardiac magnetic resonance imaging: a "one-stop-shop" evaluation of myocardial dysfunction. *Curr Opin Cardiol.* 2002;17:663-670.
35. Simonetti OP, Kim RJ, Fieno DS, et al. An improved MR imaging technique for the visualization of myocardial infarction. *Radiology.* 2001;218(1):215-223.
36. Kim RJ, Fieno DS, Parrish TB, et al. Relationship of MRI delayed contrast enhancement to irreversible injury, infarct age, and contractile function. *Circulation.* 1999;100:1992-2002.
37. Gupta A, Lee VS, Chung YC, Babb JS, Simonetti OP. Myocardial infarction: optimization of inversion times at delayed contrast-enhanced MR imaging. *Radiology.* 2004;233:921-926.
38. Ramani K, Judd RM, Holly TA, et al. Contrast magnetic resonance imaging in the assessment of myocardial viability in patients with stable coronary artery disease and left ventricular dysfunction. *Circulation.* 1998;98:268-294.
39. Kim RJ, Wu E, Rafael A, et al. The use of contrast-enhanced MRI to identify reversible myocardial dysfunction. *N Engl J Med.* 2006;343:1445-1453.
40. Wu E, Judd RM, Vargas JD, et al. Visualisation of presence, location, and transmural extent of healed Q-wave and non-Q-wave myocardial infarction. *Lancet.* 2001;357:21-28.
41. Schinkel AF, Bax JJ, Poldermans D. Clinical assessment of myocardial hibernation. *Heart.* 2005;91:111-117.
42. Judd RM, Wagner A, Rehwald WG, Albert T, Kim RJ. Technology insight: assessment of myocardial viability by delayed-enhancement magnetic resonance imaging. *Nat Clin Pract Cardiovasc Med.* 2005;2(3):150-158.
43. Kwong RY, Chan AK, Brown KA, et al. Impact of unrecognized myocardial scar detected by cardiac magnetic resonance imaging on event-free survival in patients with signs and symptoms of coronary artery disease. *Circulation.* 2006;113:2733-2743.
44. Yan AT, Shayne AJ, Brown KA, et al. Characterization of the peri-infarct zone by contrast-enhanced cardiac magnetic resonance imaging is a powerful predictor of post-myocardial infarction mortality. *Circulation.* 2006;114:32-39.
45. Reddy VY, Malchano ZJ, Holmvang G, et al. Integration of cardiac magnetic resonance imaging with three-dimensional electroanatomic mapping to guide left ventricular catheter manipulation: feasibility in a porcine model of healed myocardial infarction. *J Am Coll Cardiol.* 2004;44(11):2202-2213.

46. Bogun F, Desjrdins B, Good E, et al. Delayed-enhanced magnetic resonance imaging in nonischemic cardiomyopathy utility for identifying the ventricular arrhythmia substrate. *J Am Coll Cardiol.* 2009;53:1138-1145.
47. Fahmy TS, Wazni OM, Jaber WA, et al. Integration of positron emission tomography/computed tomography with electroanatomical mapping: a novel approach for ablation of scar-related ventricular tachycardia. *Heart Rhythm.* 2008;5(11):1538-1545.
48. Tian J, Smith MF, Chinnadurai P, et al. Clinical application of PET/CT fusion imaging for three-dimensional myocardial scar and left ventricular anatomy during ventricular tachycardia ablation. *J Cardiovasc Electrophysiol.* 2009;20:597-604.
49. Thompson PL, Fletcher EE, Katavatis V. Enzymatic indices of myocardial necrosis: influence on short- and long-term prognosis after myocardial infarction. *Circulation.* 1979;59:113-119.
50. Wu E, Ortiz JT, Tejedor P, et al. Infarct size by contrast enhanced cardiac magnetic resonance is a stronger predictor of outcomes than left ventricular ejection fraction or end-systolic volume index: prospective cohort study. *Heart.* 2008;94(6):730-736.
51. van der Burg AE, Bax JJ, Boersma E, Pauwels EK, van der Wall EE, Schalij MJ. Impact of viability, ischemia, scar tissue, and revascularization on outcome after aborted sudden death. *Circulation.* 2003;108(16):1954-1959.
52. Kwon DH, Smedira NG, Rodriguez ER, et al. Cardiac magnetic resonance detection of myocardial scarring in hypertrophic cardiomyopathy: correlation with histopathology and prevalence of ventricular tachycardia. *J Am Coll Cardiol.* 2009;54(3):242-249.
53. Roes SD, Kelle S, Kaandorp TA, et al. Comparison of myocardial infarct size assessed with contrast-enhanced magnetic resonance imaging and left ventricular function and volumes to predict mortality in patients with healed myocardial infarction. *Am J Cardiol.* 2007;100:930-936.
54. Bello D, Fieno DS, Kim RJ, et al. Infarct morphology identifies patients with substrate for sustained ventricular tachycardia. *J Am Coll Cardiol.* 2005;45:1104-1108.
55. Kadish A, Bello D, Finn JP, et al. Rationale and Design for the Defibrillators to Reduce Risk by Magnetic Resonance Imaging Evaluation (DETERMINE) trial. *J Cardiovasc Electrophysiol.* 2009;20(9):982-987.
56. de Leeuw N, Ruiter DJ, Balk AH. Histopathologic findings in explanted heart tissue from patients with end-stage idiopathic dilated cardiomyopathy. *Transpl Int.* 2001;14:299-306.
57. McCrohon J, Moon J, Prasad S, et al. Differentiation of heart failure related to dilated cardiomyopathy and coronary artery disease using gadolinium-enhanced cardiovascular magnetic resonance. *Circulation.* 2003;108:54-59.
58. Assomull RG, Prasad SK, Lyne J, et al. Cardiovascular magnetic resonance, fibrosis, and prognosis in dilated cardiomyopathy. *J Am Coll Cardiol.* 2006;48:1977-1985.
59. Nazarian S, Bluemke DA, Lardo AC, et al. Magnetic resonance assessment of the substrate for inducible ventricular tachycardia in nonischemic cardiomyopathy. *Circulation.* 2005;112:2821-2825.
60. Wu KC, Weiss RG, Thiemann DR, et al. Late gadolinium enhancement by cardiovascular magnetic resonanceheralds an adverse prognosis in nonischemic cardiomyopathy. *J Am Coll Cardiol.* 2008;51:2414-2421.
61. Cheong BY, Muthupillai R, Wilson JM, et al. Prognostic significance of delayed-enhancement magnetic resonance imaging: survival of 857 patients with and without left ventricular dysfunction. *Circulation.* 2009;120(21):2069-2076.
62. Fernandes VR, Wu KC, Rosen BD, et al. Enhanced infarct border zone function and altered mechanical activation predict inducibility of monomorphic ventricular tachycardia in patients with ischemic cardiomyopathy. *Radiology.* 2007;245(3):712-719.
63. Schmidt A, Azevedo CF, Cheng A, et al. Infarct tissue heterogeneity by magnetic resonance imaging identifies enhanced cardiac arrhythmia susceptibility in patients with left ventricular dysfunction. *Circulation.* 2007;115(15):2006-2014.
64. Roes SD, Borleffs CJ, van der Geest RJ, et al. Infarct tissue heterogeneity assessed with contrast-enhanced MRI predicts spontaneous ventricular arrhythmia in patients with ischemic cardiomyopathy and implantable cardioverter-defibrillator. *Circ Cardiovasc Imaging.* 2009;2(3):183-190.
65. Klocke FJ, Wu E, Lee DC. "Shades of gray" in cardiac magnetic resonance images of infarcted myocardium. Can they tell us what we'd like them to? *Circulation.* 2006;114:8-10.
66. Wagner A, Mahrholdt H, Thomson L, et al. Effects of time, dose and inversion time for acute myocardial infarct size measurements based on magnetic resonance imaging – delayed contrast enhancement. *J Am Coll Cardiol.* 2006;47:2027-2033.
67. Weinmann HJ, Laniado M, Mutzel W. Pharmacokinetics of Gd-DTPA/dimeglumine after intravenous injection into healthy volunteers. *Physiol Chem Phys Med NMR.* 1984;16:167-172.
68. Frances RJ. Arrhythmogenic right ventricular dysplasia/cardiomyopathy. A review and update. *Int J Cardiol.* 2006;110:279-287.
69. Norman M, Simpson M, Mogensen J, et al. Novel mutation in desmoplakin causes arrhythmogenic left ventricular cardiomyopathy. *Circulation.* 2005;112(5):636-642.
70. Hamilton R. Arrhythmogenic right ventricular cardiomyopathy. *PACE.* 2009;32:S44-S51.
71. McKenna WJ, Thiene G, Nava A, et al. Diagnosis of arrhythmogenic right ventricular dysplasia/cardiomyopathy. Task Force of the Working Group Myocardial and Pericardial Disease of the European Society of Cardiology and of the Scientific Council on Cardiomyopathies of the International Society and Federation of Cardiology. *Br Heart J.* 1994;71(3):215-218.
72. Jain A, Tandri H, Calkins H, Bluemke DA. Role of cardiovascular magnetic resonance imaging in arrhythmogenic right ventricular dysplasia. *J Cardiovasc Magn Reson.* 2008;10(32):1-14.
73. Tandri H, Bomma C, Calkins H, Bluemke DA. Magnetic resonance and computed tomography imaging of arrhythmogenic right ventricular dysplasia. *J Magn Reson Imaging.* 2004;19(6):848-858.
74. Tandri H, Saranathan M, Rodriguez ER, et al. Noninvasive detection of myocardial fibrosis in arrhythmogenic right ventricular cardiomyopathy using delayed-enhancement magnetic resonance imaging. *J Am Coll Cardiol.* 2005;45(1):98-103.
75. Wijnmaalen AP, Schalij MJ, Bootsma M, et al. Patients with scar-related right ventricular tachycardia: determinants of long-term outcome. *J Cardiovasc Electrophysiol.* 2009;20(10):1119-1127.
76. Burke AP, Farb A, Tashko G, Virmani R. Arrhythmogenic right ventricular cardiomyopathy and fatty replacement of the right ventricular myocardium: are they different diseases? *Circulation.* 1998;97(16):1571-1580.
77. Fontaliran F, Fontaine G, Fillette F, Aouate P, Chomette G, Grosgogeat Y. Nosologic frontiers of arrhythmogenic dysplasia: quantitative variations of normal adipose tissue of the right heart ventricle. *Arch Mal Coeur Vaiss.* 1991;84:33-38.
78. Shirani J, Berezowski K, Roberts WC. Quantitative measurement of normal and excessive (coradiposum) subepicardial adipose tissue, its clinical significance, and its effect on electrocardiographic QRS voltage. *Am J Cardiol.* 1995;76:414-418.
79. Macedo R, Prakasa K, Tichnell C, et al. Marked lipomatous infiltration of the right ventricle: MRI findings in relation to arrhythmogenic right ventricular dysplasia. *AJR Am J Roentgenol.* 2007;188(5):W423-W427.
80. Pfluger HB, Phrommintikul A, Mariani JA, Cherayath JG, Taylor AJ. Utility of myocardial fibrosis and fatty infiltration detected by cardiac magnetic resonance imaging in the diagnosis of arrhythmogenic right ventricular dysplasia – a single centre experience. *Heart Lung Circ.* 2008;17(6):478-483.
81. Maron BJ, Shen WK, Link MS, et al. Efficacy of implantable cardioverter-defibrillators for the prevention of sudden death in

patients with hypertrophic cardiomyopathy. *N Engl J Med*. 2000; 342:365-373.

82. Spirito P, Bellone P, Harris KM, et al. Magnitude of left ventricular hypertrophy and risk of sudden death in hypertrophic cardiomyopathy. *N Engl J Med*. 2000;342:1778-1785.
83. Maron BJ. Hypertrophic cardiomyopathy: a systematic review. *JAMA*. 2002;287:1308-1320.
84. Maron B, Spirito P, Shen WK, Haas TS, Formisano F, et al. Implantable cardioverter-defibrillators and prevention of sudden cardiac death in hypertrophic cardiomyopathy. *JAMA*. 2007;298(4):405-412.
85. Frenneaux MP, Counihan PJ, Caforio AL, et al. Abnormal blood pressure response during exercise in hypertrophic cardiomyopathy. *Circulation*. 1990;82:1995-2002.
86. Sadoul N, Prasad K, Elliott PM, et al. Prospective prognostic assessment of blood pressure response during exercise in patients with hypertrophic cardiomyopathy. *Circulation*. 1997;96:2987-2991.
87. Moon JC, Reed E, Sheppard MN, et al. The histologic basis of late gadolinium enhancement cardiovascular magnetic resonance in hypertrophic cardiomyopathy. *J Am Coll Cardiol*. 2004;43:2260-2264.
88. Teraoka K, Hirano M, Ookubo H, et al. Delayed contrast enhancement of MRI in hypertrophic cardiomyopathy. *Magn Reson Imaging*. 2004;22(6):901.
89. Adabag AS, Maron BJ, Appelbaum E, et al. Occurrence and frequency of arrhythmias in hypertrophic cardiomyopathy in relation to delayed enhancement on cardiovascular magnetic resonance. *J Am Coll Cardiol*. 2008;51:1369-1374.
90. Dimitrow PPKP, Vliengenthart R, Pasowicz M, et al. Late hyperenhancement in gadolinium magnetic resonance imaging: comparison of hypertrophic cardiomyopathy patients with and without nonsustained ventricular tachycardia. *Int J Cardiovasc Imaging*. 2008;24(1):85-87.
91. Paya E, Marin F, Gonzalez J, Gimeno JR, Feliu E, et al. Variables associated with contrast-enhanced cardiovascular magnetic resonance in hypertrophic cardiomyopathy: clinical implications. *J Card Fail*. 2008;14:414-419.
92. Dumont C, Monserrat L, Soler R, et al. Clinical significance of late gadolinium enhancement on cardiovascular magnetic resonance in patients with hypertrophiccardiomyopathy. *Rev Esp Cardiol*. 2007; 60(1):15-23.
93. Suk T, Edwards C, Hart H, Christiansen JP. Myocardial scar detected by contrast-enhanced cardiac magnetic resonance imaging is associated with ventricular tachycardia in hypertrophic cardiomyopathy patients. *Heart Lung Circ*. 2008;17(5):370-374.
94. Rubinshtein R, Glockner JF, Ommen SR, et al. Characteristics and clinical significance of late gadolinium enhancement by contrast-enhanced magnetic resonance imaging in patients with hypertrophic cardiomyopathy. *Circ Heart Fail*. 2010;3(1):51-58.
95. Varnava AM, Elliott PM, Baboonian C, Davison F, Davies MJ, McKenna WJ. Hypertrophic cardiomyopathy: histopathological features of sudden death in cardiac troponin T disease. *Circulation*. 2001;104(12):1380-1384.
96. Maron BJ, Kragel AH, Roberts WC. Sudden death in hypertrophic cardiomyopathy with normal left ventricular mass. *Br Heart J*. 1990;63:308-310.
97. Maron BJ, Wolfson JK, Roberts WC. Relation between extent of cardiac muscle cell disorganization and left ventricular wall thickness in hypertrophic cardiomyopathy. *Am J Cardiol*. 1992;70: 785-790.
98. Kwon DH, Halley CM, Carrigan TP, et al. Extent of left ventricular scar predicts outcomes in ischemic cardiomyopathy patients with significantly reduced systolic function: a delayed hyperenhancement cardiac magnetic resonance study. *JACC Cardiovasc Imaging*. 2009;2(1):34-44.
99. Kim JS, Judson MA, Donnino R, et al. Cardiac sarcoidosis. *Am Heart J*. 2009;157:9-21.
100. Rybicki BA, Major M, Popovich J Jr, et al. Racial differences in sarcoidosis incidence: a 5 year study in a health maintenance organization. *Am J Epidemiol*. 1997;145:234-241.
101. Iwai K, Sekiguti M, Hosoda Y, et al. Racial difference in cardiac sarcoidosis incidence observed at autopsy. *Sarcoidosis*. 1994;11:26-31.
102. Silverman KJ, Hutchins GM, Bulkley BH. Cardiac sarcoid: a clinicopathologic study of 84 unselected patients with systemic sarcoidosis. *Circulation*. 1978;58:1204-1211.
103. Roberts WC, McAllister HA, Ferrans VJ. Sarcoidosis of the heart: a clinicopathologic study of 35 necropsy patients and review of 78 previously described necropsy patients. *Am J Med*. 1977;63:86-108.
104. Fleming HA, Bailey SM. Sarcoid heart disease. *J R Coll Physicians London*. 1981;15:245-246. 249–523.
105. Fleming HA, Bailey SA. The prognosis of sarcoid heart disease in the United Kingdom. *Ann N Y Acad Sci*. 1986;465:543-550.
106. Vignaux O, Dhôte R, Duboc D, et al. Clinical significance of myocardial magnetic resonance abnormalities in patients with sarcoidosis. A 1 year follow-up study. *Chest*. 2002;122:1895-1901.
107. Cheong BYC, Muthupillai R, Nemeth M, et al. The utility of delayed-enhancement magnetic resonance imaging for identifying nonischemic myocardial fibrosis in asymptomatic patients with biopsy-proven systemic sarcoidosis. *Sarcoidosis Vasc Diffuse Lung Dis*. 2009;26:39-46.
108. Smedema JP, Snoep G, Van Kroonenburgh MP, et al. Evaluation of the accuracy of gadolinium-enhanced cardiovascular magnetic resonance in the diagnosis of cardiac sarcoidosis. *J Am Coll Cardiol*. 2005;45:1683-1690.
109. Patel MR, Cawley PJ, Heitner JF, et al. Detection of myocardial damage in patients with sarcoidosis. *Circulation*. 2009;120(20): 1969-1977.
110. Hiraga H, Yuwa K, Hiroe M. *Guideline for the Diagnosis of Cardiac Sarcoidosis: Study Report on Diffuse Pulmonary Disease [in Japanese]*. Tokyo: The Japanese Ministry of Health and Welfare.1993:23-24.
111. Elliott P, Andersson B, Arbustini E, et al. Classification of the cardiomyopathies: a position statement from the European Society of Cardiology Working Group on Myocardial and Pericardial Diseases. *Eur Heart J*. 2008;29:270-276.
112. Nugent AW, Daubeney PE, Chondros P, et al. The epidemiology of childhood cardiomyopathy in Australia. *N Engl J Med*. 2003; 248:1639-1646.
113. Murphy RT, Thaman R, Blanes JG, et al. Natural history and familial characteristics of isolated left ventricular non-compaction. *Eur Heart J*. 2005;26:1144-1145.
114. Habib G, Charron P, Eicher JC, et al. Isolated left ventricular non-compaction in adults: clinical and echocardiographic features in 105 patients. Results from a French registry. *Eur J Hear Fail*. 2010;13(2):177-185. Epub 2010 Dec 29.
115. Yousef ZR, Foley PW, Khadjooi K, et al. Left ventricular non-compaction: clinical features and cardiovascular magnetic resonance imaging. *BMC Cardiovasc Disord*. 2009;9:37.
116. Sosnovik DE, Wang R, Dai G, et al. Diffusion spectrum MRI tractography reveals the presence of a complex network of residual myofibers in infarcted myocardium. *Circ Cardiovasc Imaging*. 2009;2:206-212.
117. Sosnovik DE, Wang R, Dai G, Reese TG, Wedeen VJ. Diffusion MR tractography of the heart. *J Cardiovasc Magn Reson*. 2009; 11:47.

Part IV

Advances in Navigation and Integration

MRI/CCT Fusion into Fluoroscopic Imaging

19

Nassir F. Marrouche and Gaston R. Vergara

Abstract

The combination of an EAM system and a CCT/cMRI is able to provide real-time information with regard to catheter position; however, this approach requires registration of two different models with different appearance. Furthermore, EAM systems are dedicated, expensive, and require operator familiarity with its function and display modalities. Registration of CCT/cMRI with fluoroscopy-based systems has several benefits. Advantages related to fluoroscopy-based systems are its widespread use and the familiarity of its images to most electrophysiologists. Some of the disadvantages are related to the low soft-tissue resolution, images that are 2D projections, and high exposure to ionizing radiation from X-rays. CCT and cMRI have the capability to provide very high quality and resolution imaging which can be processed to generate 3D renderings or models of the cardiac chamber selected in great detail. Disadvantages are related to the use of contrast to delineate the endocardial surface of the cardiac chambers, the inability to reliably image patients with implantable cardiac defibrillators (cMRI), and exposure to ionizing radiation (CCT), among others. Potential pitfalls related to this newer methodology are related to difficulty performing an accurate registration and potentially double exposure to ionizing radiation (if CCT is used).

Keywords

MRI/CCT fusion • Fluoroscopic imaging and MRI/CCT • Electroanatomical mapping • Cardiac magnetic resonance imaging • Cardiac computerized tomography

Integration of cardiac Magnetic Resonance Imaging (cMRI) and Cardiac Computerized Tomography (CCT) images with a non-fluoroscopy-based electroanatomical mapping (EAM) system is a significant step in assisting with navigation and mapping of complex arrhythmias, since it allows for precise catheter monitoring in a real-time three-dimensional manner during ablation. Integration typically consists in fusing two images: a pre-acquired CCT or cMRI image with an electroanatomical map.

EAM systems record continuously and in real time the position of a moving catheter in the cardiac chamber. By assigning points in space and integrating these points, interpolating the surface in between them, a 3D shell of such chamber can be created. This information is then integrated with electrophysiological data collected and a map of the electrical activity of a cardiac chamber can be created. This, however, has limitations such as the need for a dedicated mapping system, some requiring special catheters, and limiting the anatomic information to those areas that can be reached by the catheter. Newer mapping systems can combine this electroanatomical information provided by the catheter with a 3D rendering from a CCT or cMRI of the cardiac chamber of interest allowing combination of both

N.F. Marrouche (✉)
Division of Cardiology, Comprehensive Arrhythmia Research & Management Center,
University of Utah Health Sciences Center,
Salt Lake City, UT, USA
e-mail: nassir.marrouche@hsc.utah.edu

A. Auricchio et al. (eds.), *Cardiac Imaging in Electrophysiology*,
DOI 10.1007/978-1-84882-486-7_19,

data sets for more accurate spatial resolution.[1] In this case, the CCT or cMRI provides accurate, three-dimensional anatomical information, while the electroanatomical mapping system provides real-time information about catheter position and electrical activity within such chamber.

Integration of CCT and cMRI with fluoroscopy has the potential advantage to combine high spatial resolution and anatomical detail from 3D reconstructions from CCT/cMRI with a familiar 2D fluoroscopic image able to provide real-time information without the need for an expensive and dedicated EAM system.[2-6]

19.1 Cardiac Computerized Tomography and Cardiac Magnetic Resonance Registration into a 3D Electroanatomical Mapping System

Integration of CCT/cMRI with 3D EAM system consists in combining a high-resolution image from either a CCT or cMRI, which provides a high level of anatomic detail, with a 3D shell of the cardiac chamber of interest obtained in real time from the 3D EAM system, with the ultimate goal to be able to have high quality anatomical data combined with real-time information to aid in catheter navigation.

The process of registration CCT/cMRI images into a 3D EAM system consists of three basic steps: (1) image acquisition from CCT/cMRI, (2) image segmentation, and (3) registration of both systems.

1. Images *acquired* from CCT/cMRI are uploaded as raw files into the EAM system.
2. *Segmentation* of these images consists in identifying and separating the component or chamber of interest from those other structures visualized in the image. This is a semi-automated process which requires isolation of the structure of interest, annotation of pertinent associated structures, and finally manual adjustments to the model.
3. The final step consists in *registration* of both image systems (CCT/cMRI and EAM system). This can be performed by matching anatomical landmarks or *fiduciary points* between both systems[2,3] or by *registration of chamber surface*, in which a 3D shell of the chamber of interest acquired with EAM system is combined with the image acquired and segmented from the CCT/cMRI. Disadvantages and limitations of registration based on *fiduciary points* are: changes in chamber size between time of acquisition of CCT/cMRI and time of EAM, and differences in position and chamber size due to suboptimal gating at the time of image acquisition.[1] Limitations related to surface registration consist in either poor distribution of the point sample collected during mapping or collection of points in highly mobile parts of the cardiac chamber of interest (such as the left atrial anterior wall) which may lead to indentations and inaccuracies in the 3D shell generated.[5]

19.2 Integration of Cardiac Computerized Tomography and Cardiac Magnetic Resonance with Fluoroscopy

The process of registration of CCT or cMR images into a fluoroscopy system consists in acquiring, segmenting, and registering into the fluoroscopy system the images of the cardiac chamber of interest.

19.2.1 Image Acquiring

The process of cMRI/CCT image acquisition is essentially the same regardless of the final system combined for catheter navigation. Gating for image acquisition is a critical step to obtain quality images. For CCT, the optimal portion of the cardiac cycle length (CL) for image acquisition was 70–80% of the RR interval for patients in normal sinus rhythm and 45% for those in atrial fibrillation.[7,8] This last portion of the CL is chosen due to shorter and variable RR intervals in this last group. For cMRI acquisition of atrial images has been gated at the end of systole, to coincide with atrial diastole.[6] cMRI is not only ECG gated but also respiratory navigated, this allows for tracking of the diaphragm and cardiac structures with respiratory motion.[9]

Optimal visual resolution of the blood–myocardium interface is achieved by using radio-opaque contrast during CCT imaging. This presents all the drawbacks related to iodinated contrasted agents. cMRI is more flexible and allows for visualization of the blood pool and the myocardium with or without the use of contrast. Gadolinium-based contrast agents have been used for endoluminal imaging and delineation of the endocardial surface[8]; with cMR angiography, the images can then be segmented and reconstructed from this data set. Alternatively, cMRI allows for imaging of the endocardial–blood interface without contrast with the use of special MRI acquisition sequences.[6] The advantage of this methodology is that inhomogeneous contrast distribution leading to low chamber image resolution is avoided.

19.2.2 Segmentation

The segmentation process is essentially the same for integration with fluoroscopy than that used for EAM systems. However, segmentation is processed in separate workstations and then exported to the fluoroscopy system.[5,10]

19.2.3 Registration

The approach to registration of CCT/cMRI images with fluoroscopy-based systems is different than that used for EAM systems. The first step involves calibration of the fluoroscopy system; for this, a catheter with a known interelectrode distance, which can be measured, is imaged on two orthogonal planes. The next step consists in introducing the 3D model of the segmented chamber from the cMR images. After calibration, the catheter position from the fluoroscopic images is projected in the 3D model obtained from cMR images.[6] With this technique, the alignment errors for a heart-cast model experiment were negligible (0.37–0.6 mm), suggesting that the alignment errors seen between the 3D reconstruction from the cMRI and patients angiographic views of the cardiac chambers were related to cardiac motion or changes in chamber size between imaging and ablation procedure.[6]

Alternatively, a catheter can be introduced from the superior vena cava distally into the coronary sinus. This catheter is left at the time of fluoroscopy imaging and is then used to align those images with the corresponding 3D model of segmented structures from CCT images, the superior vena cava and the coronary sinus (CS).[7] Using catheter motion images on fluoroscopy, the CS position is captured and registered during diastole. The CS position is then aligned with that recorded during CCT imaging and used to assist in the registration process.

19.3 Pitfalls of CCT and cMRI Registration with Fluoroscopy

There could be several potential issues that affect accurate image registration between two systems:

The optimal portion of the cardiac cycle length for image acquisition has not been completely established.[11] Studies suggest 45% of CL during AF and 70–80% of the CL during sinus rhythm. However, as newer multidetector CT scanners able to image the atrium in a single cycle became available, this would potentially be less of an issue.

Respiratory navigated image acquisition is also important. Respiration and diaphragmatic movement causes changes in cardiac position and may affect atrial size as it produces changes venous return.[12]

CCT/cMR images are usually acquired hours to days prior to the actual electrophysiological study. Changes in the patient volume status and different cardiac loading conditions may affect cardiac chamber size. During the procedure, differences in rhythm, and pressure exerted by the catheters against the cardiac walls may also affect chamber geometry and size.

19.4 Summary

The combination of an EAM system and a CCT/cMRI is able to provide real-time information with regard to catheter position; however, this approach requires registration of two different models with different appearance. Furthermore, EAM systems are dedicated, expensive, and require operator familiarity with its function and display modalities.

Registration of CCT/cMRI with fluoroscopy-based systems has several benefits. Advantages related to fluoroscopy-based systems are its widespread use and the familiarity of its images to most electrophysiologists. Some of the disadvantages are related to the low soft-tissue resolution, images that are 2D projections, and high exposure to ionizing radiation from X-rays.

CCT and cMRI have the capability to provide very high quality and resolution imaging which can be processed to generate 3D renderings or models of the cardiac chamber selected in great detail. Disadvantages are related to the use of contrast to delineate the endocardial surface of the cardiac chambers, the inability to reliably image patients with implantable cardiac defibrillators (cMRI), and exposure to ionizing radiation (CCT), among others. Potential pitfalls related to this newer methodology are related to difficulty in performing an accurate registration and potentially double exposure to ionizing radiation (if CCT is used).

References

1. Verma A, Marrouche N, Natale A. Novel method to integrate three-dimensional computed tomographic images of the left atrium with real-time electroanatomic mapping. *J Cardiovasc Electrophysiol*. 2004;15(8):968. PubMed PMID: 15333101.
2. Hajnal JV, Saeed N, Soar EJ, Oatridge A, Young IR, Bydder GM. A registration and interpolation procedure for subvoxel matching of serially acquired MR images. *J Comput Assist Tomogr*. 1995; 19(2):289-296. PubMed PMID: 7890857.
3. Solomon SB, White P Jr, Wiener CM, Orens JB, Wang KP. Three-dimensional CT-guided bronchoscopy with a real-time electromagnetic position sensor: a comparison of two image registration methods. *Chest*. 2000;118(6):1783-1787. PubMed PMID: 11115473.
4. Kistler PM, Rajappan K, Jahngir M, et al. The impact of CT image integration into an electroanatomic mapping system on clinical outcomes of catheter ablation of atrial fibrillation. *J Cardiovasc Electrophysiol*. 2006;17(10):1093-1101. PubMed PMID: 16989651.
5. Fahmy TS, Mlcochova H, Wazni OM, et al. Intracardiac echo-guided image integration: optimizing strategies for registration. *J Cardiovasc Electrophysiol*. 2007;18(3):276-282. Epub 2007 Feb 2. PubMed PMID: 17284265.
6. Ector J, De Buck S, Adams J, et al. Cardiac three-dimensional magnetic resonance imaging and fluoroscopy merging: a new approach for electroanatomic mapping to assist catheter ablation. *Circulation*. 2005;112(24):3769-3776. Epub 2005 Dec 5. PubMed PMID: 16330683.

7. Sra J, Krum D, Malloy A, et al. Registration of three-dimensional left atrial computed tomographic images with projection images obtained using fluoroscopy. *Circulation*. 2005;112(24):3763-3768. PubMed PMID: 16344405.
8. Gutiérrez LF, Silva R, Ozturk C, et al. Technology preview: X-ray fused with magnetic resonance during invasive cardiovascular procedures. *Catheter Cardiovasc Interv*. 2007;70(6):773-782. PubMed PMID: 18022851; PubMed Central PMCID: PMC2677196.
9. Vergara GR, Marrouche NF. Tailored management of atrial fibrillation using a LGE-MRI based model: from the clinic to the electrophysiology laboratory. *J Cardiovasc Electrophysiol*. 2010;22(4):481-487. doi:10.1111/j.1540-8167.2010.01941.x.. Epub 2010 Nov 2. PubMed PMID: 21044212.
10. Sra J, Krum D, Belanger B, Vaillant R. Registration of three-dimensional left atrial computed tomographic images with fluoroscopy. *Heart Rhythm*. 2005;2(9):1020. PubMed PMID: 16171764.
11. Kistler PM, Earley MJ, Harris S, et al. Validation of three-dimensional cardiac image ntegration: use of integrated CT image into electroanatomic mapping system to perform catheter ablation of atrial fibrillation. *J Cardiovasc Electrophysiol*. 2006;17(4):341-348. PubMed PMID: 16643352.
12. Noseworthy PA, Malchano ZJ, Ahmed J, Holmvang G, Ruskin JN, Reddy VY. The impact of respiration on left atrial and pulmonary venous anatomy: implications for image-guided intervention. *Heart Rhythm*. 2005;2(11):1173-1178. PubMed PMID: 16253905.

Magnetic Navigation: Description of Technique, Advantages, and Technical Issues

20

Sabine Ernst

Abstract

Magnetic navigation has been established as one of the first remote navigation systems in invasive cardiology in recent years. Mainly in the field of electrophysiology, but also in interventional coronary procedures, this new platform has proven its benefit in complex procedures. This chapter reviews the baseline concept of the system and analyzes the technical progress and the magnetic tools available up to date.

Keywords

Magnetic navigation • Magnetic catheter ablation • 3D mapping systems and magnetic navigation • Remote-controlled ablation of atrial fibrillation

Since more than 8 years, the magnetic navigation system (Stereotaxis Inc.) has been introduced into clinical practice and has meanwhile established its place in advanced arrhythmia and interventional procedures.[1-4] This chapter reviews the technique of remote navigation and the progress made in recent years, but also discusses the shortcomings and limitations of the currently available system.

20.1 Baseline Concept

Small magnets are integrated in the tip of a very soft ablation catheter that can be moved by using a well-defined outer magnetic field (0.08 T).[3,4] By changing the orientation of the outer magnets, the small magnets in the catheter tip arrange parallel to the outer magnetic field lines. The combination with a mechanical motor drive allows operating the magnetic mapping and ablation catheter fully remote-controlled.[2]

S. Ernst
Royal Brompton and Harefield Hospital, London, UK
e-mail: sernst1708@aol.com, s.ernst@imperial.ac.uk, s.ernst@rbht.nhs.uk

20.2 Available Ablation Catheters

The first-generation ablation catheter equipped with a single magnet (1M) consisted of a 4-mm tip and a single ring electrode (2 mm) allowing single bipolar recordings only. It was mostly used in supraventricular tachycardia (SVT) ablations.[2,5-8] The second-generation catheter had a total of three magnets embedded (3M) in the distal shaft, but still only two electrodes.[9] Since all embedded magnets will parallel out in the outer magnetic field, the ability of this catheter to respond to more acute vectors was much improved. Further development led to the implementation of two additional ring electrodes to allow two bipolar recordings (3M quad). Subsequently, the catheter tip was extended to a total length of 8 mm (3M 8 mm) to achieve larger lesion formation. And finally two different irrigated tip catheters have become available recently: a CARTO-sensor (Biosense Webster) equipped one and a gold-tipped version (Trignum Flux, BIOTRONIK).

20.3 Compatibility with 3D Mapping Systems

Initially, only the LocaLisa system (Medtronic, Minneapolis, MN) and the NAVx mapping system (St. Jude Medical, St. Pauls, MN) were compatible with the magnetic navigation

A. Auricchio et al. (eds.), *Cardiac Imaging in Electrophysiology*,
DOI 10.1007/978-1-84882-486-7_20,

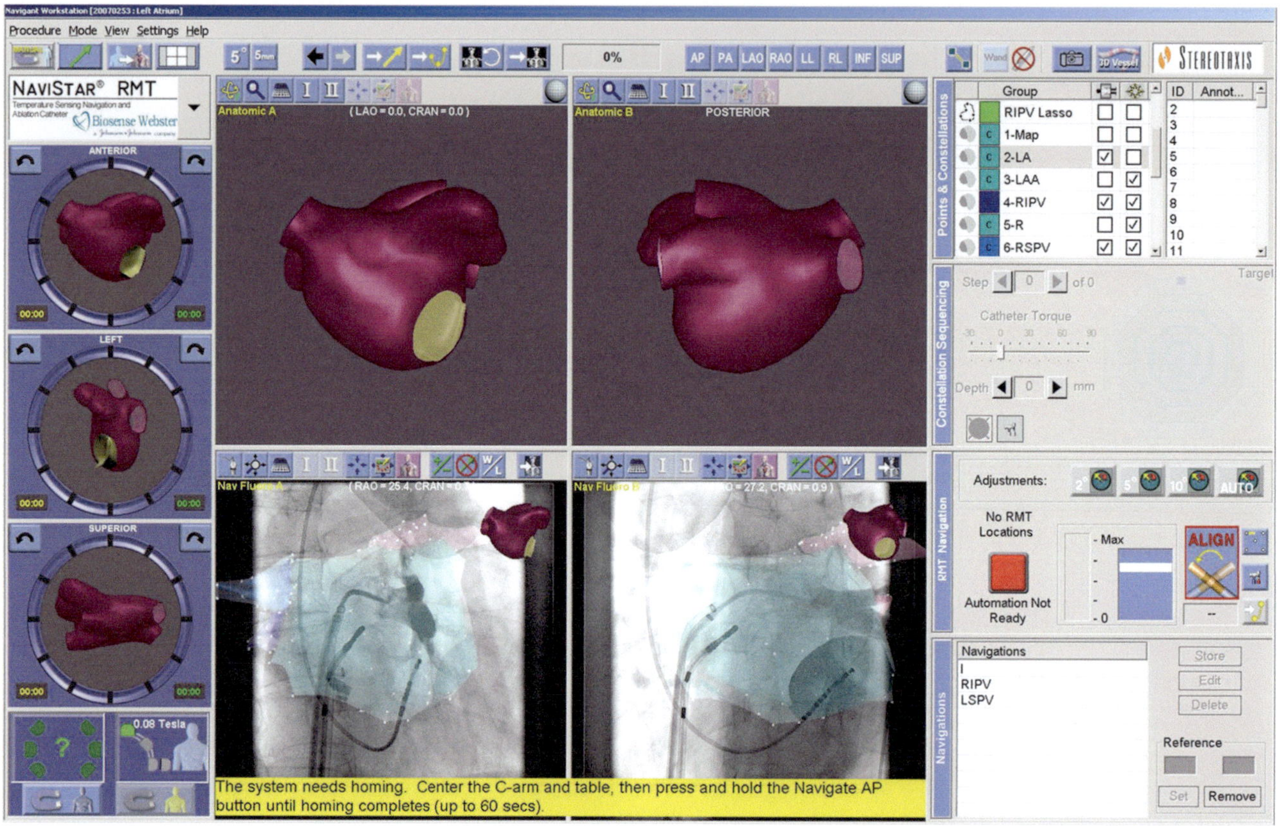

Fig. 20.1 Example of the Navigant workstation with CARTO maps of both right (*RA*) and left atrium (*LA*) displayed in the corresponding fluoroscopy reference pictures

system. Both work on the same principle which is able to locate all catheters positioned in the thorax between the condensator-like patches.

Subsequently, a fully integrated 3D electroanatomical CARTO system (CARTO RMT, Biosense Webster, Brussels, Belgium) was introduced that integrates its ultralow electromagnetic field (10^{-12} T) with the permanent magnetic field of the navigation system (0.08 T). In addition, cross-talk between the fluoroscopy system, the magnetic navigation system, and the CARTO RMT system is enabled, such that the information is displayed superimposed on the fluoroscopic reference pictures (Fig. 20.1).

20.4 Tachycardia Substrates

After the first report on remote-controlled catheter ablation of AV nodal reentrant tachycardia, most of the initial reports have dealt with supraventricular tachycardia substrates. However, several reports have focused on ventricular arrhythmias with excellent success rates.[10-13] Theoretically, the soft catheter shaft should reduce the risk of mechanical block during mapping of the left ventricular septum.

20.5 Remote-Controlled Ablation of Atrial Fibrillation

Despite initial reports of a single center on their experience using the solid 4-mm tip, the majority of groups have not yet touched on this most interesting, but also most challenging substrate.[14-17] The urgently awaited irrigated tip catheter will certain encourage more centers to apply their AF ablation strategy remote-controlled. So far, mostly mapping and ablation in re-procedures have been published. However, switching to a handheld irrigated catheter certainly reduces the effect on radiation exposure for the investigator.

20.6 Irrigated Tip AF Ablation

Recently, a magnetically enable irrigated tip catheter that also incorporates a CARTO location sensor has been introduced.[18] Therefore, the operator can now perform both mapping and ablation totally remote-controlled without the need to convert to conventional techniques. Certainly, the operator has to apply different strategies than the direct transfer of conventional electrode positions, since the limited contact force might make lesion formation otherwise very difficult (Fig. 20.2).

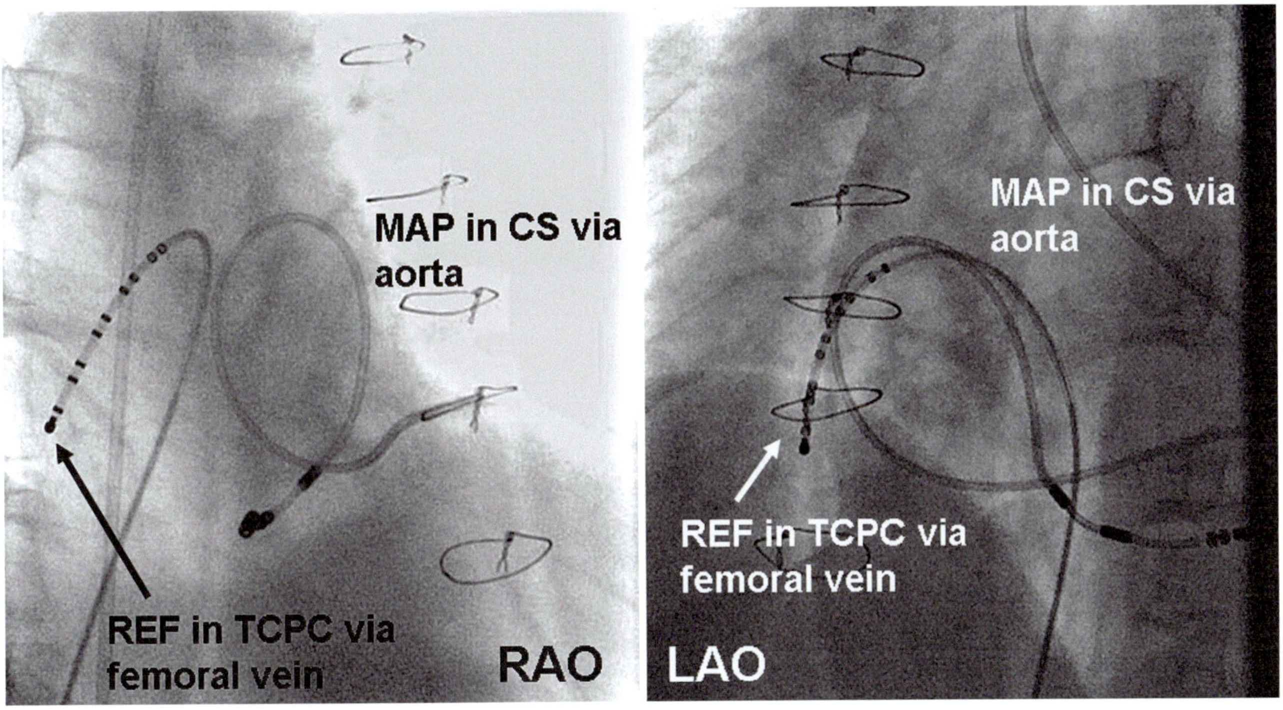

Fig. 20.2 Unusual catheter positioning required to improve stability when addressing difficult-to-reach sites. The figure depicts the position of the magnetic catheter inside the coronary sinus of a patient after total cavo-pulmonary connection advanced retrogradely across the aortic valve and through a double inlet left ventricle

20.7 ACHD Ablation Using Magnetic Navigation

One of the recently published achievements is the introduction of magnetic navigation to the challenging group of arrhythmia patients with surgically corrected congenital heart disease. While not only the understanding of the underlying anatomy is a challenge of its own right, accessibility of the target chamber might prove to require technically difficult access such as trans-baffle punctures.[19] Using the soft magnetic catheter, the target chamber can be reached atraumatically, e.g., across the aortic valve and retrograde via the corresponding AV valve.[20] By integration of previously acquired 3D imaging using magnetic resonance or computer tomography, the whole anatomy including large vessels can visualized on the fluoroscopy reference screens (Fig. 20.3).

20.8 Further Advantages of Remote Magnetic Navigation

One of the most obvious advantages of a remote-controlled catheter navigation system is the reduced fluoroscopy exposure for the investigator. Once all catheters are inserted and positioned in the specific locations, mapping and subsequent ablation will be performed from the control room without further exposure to scattered radiation. However, if this comes at the expense of increased radiation exposure for the patients, this remote position of the investigator would not be very beneficial since it would demonstrate the poor orientation of the investigator. Because of the outer magnetic field's permanent nature (it is formed by magnetic metals called "rare earthers" that have magnetic properties), the position of the magnetic catheter is very stable and will only change when the field vector is altered. Therefore, the catheter position is precise and good contact is ascertained by little beat-to-beat variation of the distal bipolar signal. Once accustomed to being "remote" and after fulfillment of the learning curve, reduction of the overall fluoroscopy exposure can be demonstrated.

20.9 Other Magnetic Tools

Generally speaking, any device that is equipped with magnets will attempt to align parallel to the outer magnetic field direction. Clinically established tools are magnetically enabled guidewires that have been used to address complex tortuous vessels and complex coronary artery disease.[1,21] Following the same concept, left ventricular lead placement guided by magnetically steering into all CS side branches has been reported.[22] In selected cases, the operator might even omit to visualize the CS side branches using contrast injection, but would instead enter in candidate vessels

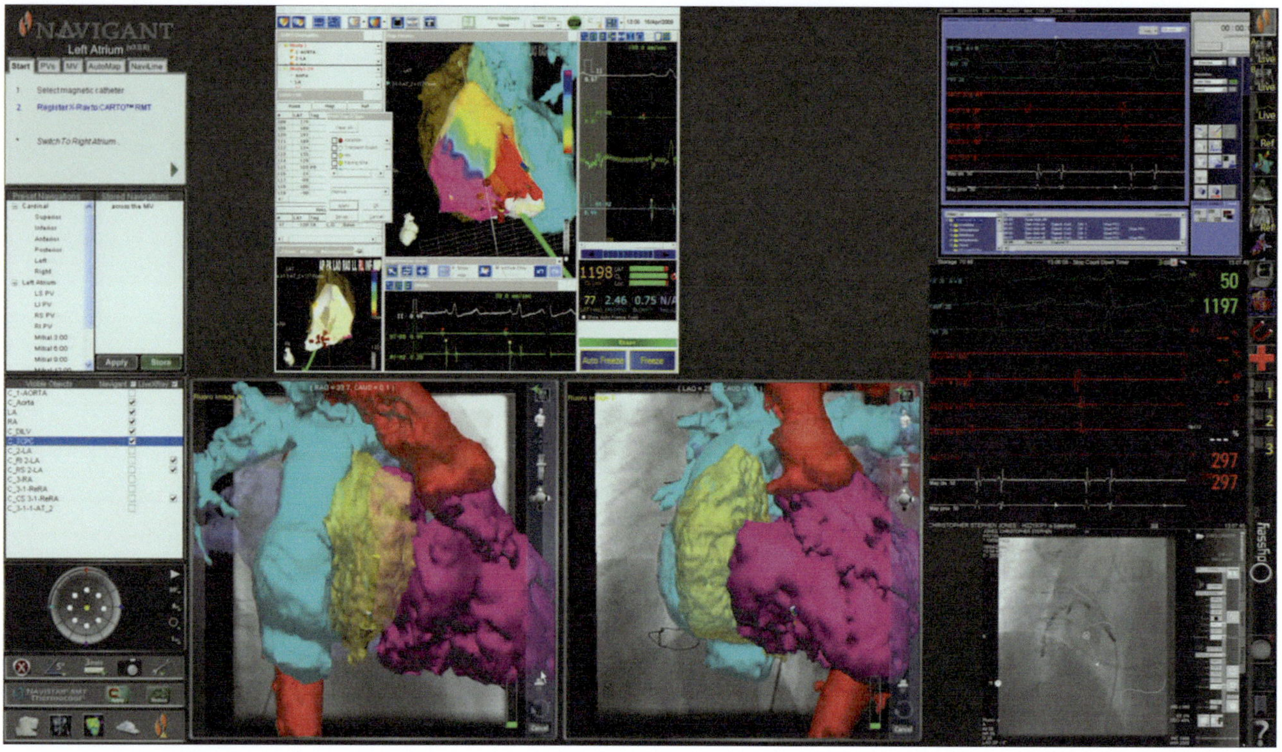

Fig. 20.3 Image integration of previously acquired 3D magnetic resonance imaging and superimposition on the magnetic navigation workstation Odyssey (Stereotaxis Inc.)

"blindly" guided only by the appropriate magnetic field vector. This allows avoiding the use of any long guidance sheath, which on removal might dislodge the LV lead (at least in the hands of non-experienced operators).

20.10 Magnetic Navigation in Patients with Implantable Devices

Since the magnetic field strength of the system amounts only to 0.08 T, any cardiac procedure can be carried out safely when the magnets are in "park" position. Patients with implanted devices convert into "magnet mode" (e.g., VVI pacing 100 bpm), once the magnets are positioned in "navigate" position. By changing the direction of the magnetic field, the device might revert to non-inhibited mode depending which field direction is applied. Therefore, in a pacing dependent patient, a temporary pacing lead has to be safely positioned before the magnets are placed in "navigate" position. In order to avoid possibly disturbing action of the implanted devices, the output of the capturing lead should be reduced to the minimum in order to not to interfere with the EP study.

20.11 Conclusion

The magnetic navigation system is, more than 6 years after its introduction into clinical practice, able to successfully perform all varieties of tachycardia substrates including ventricular tachycardia, AF, and complex arrhythmia ablation in patients with congenital heart disease.

References

1. Atmakuri SR, Lev EI, Alviar C, et al. Initial experience with a magnetic navigation system for percutaneous coronary intervention in complex coronary artery lesions. *J Am Coll Cardiol*. 2006;47(3): 515-521.
2. Ernst S, Ouyang F, Linder C, et al. Initial experience with remote catheter ablation using a novel magnetic navigation system: magnetic remote catheter ablation. *Circulation*. 2004;109(12):1472-1475.
3. Faddis MN, Blume W, Finney J, et al. Novel, magnetically guided catheter for endocardial mapping and radiofrequency catheter ablation. *Circulation*. 2002;106(23):2980-2985.
4. Faddis MN, Lindsay BD. Magnetic catheter manipulation. *Coron Artery Dis*. 2003;14(1):25-27.

5. Chun JK, Ernst S, Matthews S, et al. Remote-controlled catheter ablation of accessory pathways: results from the magnetic laboratory. *Eur Heart J*. 2007;28(2):190-195.
6. Ernst S, Hachiya H, Chun JK, Ouyang F. Remote catheter ablation of parahisian accessory pathways using a novel magnetic navigation system – a report of two cases. *J Cardiovasc Electrophysiol*. 2005;16(6):659-662.
7. Ernst S, Ouyang F, Linder C, et al. Modulation of the slow pathway in the presence of a persistent left superior caval vein using the novel magnetic navigation system Niobe. *Europace*. 2004;6(1):10-14.
8. Kerzner R, Sanchez JM, Osborn JL, et al. Radiofrequency ablation of atrioventricular nodal reentrant tachycardia using a novel magnetic guidance system compared with a conventional approach. *Heart Rhythm*. 2006;3(3):261-267.
9. Thornton AS, Janse P, Theuns DA, Scholten MF, Jordaens LJ. Magnetic navigation in AV nodal re-entrant tachycardia study: early results of ablation with one- and three-magnet catheters. *Europace*. 2006;8(4):225-230.
10. Aryana A, d'Avila A, Heist EK, et al. Remote magnetic navigation to guide endocardial and epicardial catheter mapping of scar-related ventricular tachycardia. *Circulation*. 2007;115(10):1191-1200.
11. Burkhardt JD, Saliba WI, Schweikert RA, Cummings J, Natale A. Remote magnetic navigation to map and ablate left coronary cusp ventricular tachycardia. *J Cardiovasc Electrophysiol*. 2006;17(10):1142-1144.
12. Thornton AS, Jordaens LJ. Remote magnetic navigation for mapping and ablating right ventricular outflow tract tachycardia. *Heart Rhythm*. 2006;3(6):691-696.
13. Thornton AS, Res J, Mekel JM, Jordaens LJ. Use of advanced mapping and remote magnetic navigation to ablate left ventricular fascicular tachycardia. *Pacing Clin Electrophysiol*. 2006;29(6):685-688.
14. Di Biase L, Fahmy TS, Patel D, et al. Remote magnetic navigation: human experience in pulmonary vein ablation. *J Am Coll Cardiol*. 2007;50(9):868-874.
15. Katsiyiannis WT, Melby DP, Matelski JL, Ervin VL, Laverence KL, Gornick CC. Feasibility and safety of remote-controlled magnetic navigation for ablation of atrial fibrillation. *Am J Cardiol*. 2008;102(12):1674-1676.
16. Pappone C, Augello G, Gugliotta F, Santinelli V. Robotic and magnetic navigation for atrial fibrillation ablation. How and why? *Expert Rev Med Devices*. 2007;4(6):885-894.
17. Pappone C, Vicedomini G, Manguso F, et al. Robotic magnetic navigation for atrial fibrillation ablation. *J Am Coll Cardiol*. 2006;47(7):1390-1400.
18. Haghjoo M, Hindricks G, Bode K, Piorkowski C, Bollmann A, Arya A. Initial clinical experience with the new irrigated tip magnetic catheter for ablation of scar-related sustained ventricular tachycardia: a small case series. *J Cardiovasc Electrophysiol*. 2009;20(8):935-939.
19. Schwagten B, Cuypers J, Szili-Torok T. The magnetic navigation system allows avoidance of puncturing a baffle during ablation of a postincisional macroreentrant tachycardia. *Cardiol Young*. 2009;19(2):216-219.
20. Wu J, Pflaumer A, Deisenhofer I, et al. Mapping of intraatrial reentrant tachycardias by remote magnetic navigation in patients with d-transposition of the great arteries after mustard or senning procedure. *J Cardiovasc Electrophysiol*. 2008;19(11):1153-1159.
21. Tsuchida K, Garcia-Garcia HM, van der Giessen WJ, et al. Guidewire navigation in coronary artery stenoses using a novel magnetic navigation system: first clinical experience. *Catheter Cardiovasc Interv*. 2006;67(3):356-363.
22. Thornton AS, Maximo Rivero-Ayerza M, Jordaens LJ. Magnetic assisted navigation in electrophysiology and cardiac resynchronisation: a review. *Indian Pacing Electrophysiol J*. 2006;6(4):202-213.

Magnetic and Robotic Navigation

21

Bruce D. Lindsay and Oussama Wazni

Abstract

The advent of ablation procedures for supraventricular and ventricular arrhythmias accelerated the development of technology to facilitate catheter manipulation and imaging. Deflectable catheters, electroanatomic imaging, and intracardiac echocardiography represent technologies that have had a high impact on the field. The early work on catheter-based ablation of arrhythmias focused on elimination of accessory pathways, dual AV nodal physiology, right atrial tachycardias, or typical isthmus-dependent right atrial flutter. These arrhythmias demand significant skills for catheter manipulation, but the ability to direct a catheter to the target rarely proves difficult for an experienced electrophysiologist. Two technologies developed to meet these objectives include the magnetic navigation system designed by Stereotaxis, Inc., and a robotic-controlled catheter system manufactured by Hansen Medical. While neither is approved specifically for ablation of atrial fibrillation or ventricular tachycardia, the impetus to develop these technologies is to use them for complex ablation procedures. The potential utility of these remote navigation systems has not been fully developed, but they have improved. Prospective studies are not available to compare these technologies or to demonstrate definitively that either system shortens procedure time, improves outcomes of ablation, or improves the safety profile of these and other complex ablation procedures. Nonetheless, preliminary evidence offers promise that these goals could be achieved.

Keywords

Robotic navigation in cardiology • Magnetic navigation in cardiology • Left atrial mapping and ablation • Ablation procedures • Electrophysiology and robotic navigation

The advent of ablation procedures for supraventricular and ventricular arrhythmias accelerated the development of technology to facilitate catheter manipulation and imaging. Deflectable catheters, electroanatomic imaging, and intracardiac echocardiography represent technologies that have had a high impact on the field. The early work on catheter-based ablation of arrhythmias focused on elimination of accessory pathways, dual AV nodal physiology, right atrial tachycardias, or typical isthmus-dependent right atrial flutter. These arrhythmias demand significant skills for catheter manipulation, but the ability to direct a catheter to the target rarely proves difficult for an experienced electrophysiologist.

As electrophysiologists began to ablate more complex arrhythmias, the limitations of standard deflectable catheters placed greater demands on the skills of the electrophysiologists. In most cases, catheter deflection is bound by a fixed radius of curvature in a single plane. Side-to-side catheter motion is determined by transmission of torque, which may be limited by tortuosity of vessels and the orientation of the catheter within the cardiac chamber. Once the shaft of the catheter has undergone one or more curvatures imposed by

B.D. Lindsay (✉)
Department of Cardiovascular Medicine,
Cleveland Clinic, Cleveland, OH, USA
e-mail: lindsab@ccf.org

A. Auricchio et al. (eds.), *Cardiac Imaging in Electrophysiology*,
DOI 10.1007/978-1-84882-486-7_21,

anatomy, the response of the distal catheter becomes less predictable when torque is applied to the handle. In addition, stable endocardial contact at the tip of the catheter may be compromised by cardiac or respiratory motion. These limitations are particularly important for ablation of complex arrhythmias such as ventricular tachycardia, atypical atrial flutter, or atrial fibrillation, which depend on accurate control of the catheter for focal or linear lesions. Imprecise control may result in gaps within a linear lesion, excessive force resulting in perforation, or inadequate contact which affects lesion formation.

Radiation exposure to patients and physicians and the orthopedic consequences of prolonged ablation procedures raise additional concerns about safety and long-term consequences. Radiation-induced skin injury is a well-documented complication of complex procedures that depend on X-ray imaging.[1-4] While adherence to good technique can reduce the risk of this complication, it does not eliminate the risk entirely. Longer procedures require prolonged periods of wearing lead aprons, which puts additional stress and strain on the shoulders and spinal column.[5] Accordingly, electrophysiologists are at an increased risk of spinal injuries,[6] that may cause prolonged absence from work, surgery, or in career changes.

Advances in technology offer potential solutions for many of these issues. The objectives of developing new technologies to facilitate complex ablation procedures include:

1. Precise and stable catheter navigation
2. 3D imaging systems to track the position of the catheter and guide effective ablation of tissue
3. Improved delivery of energy for more effective delivery of ablative lesions
4. Assessment of effective lesion formation
5. Reduced radiation exposure
6. Shorter procedures
7. Cost-effective

While new technologies generally increase the cost of a procedure when they are introduced, the costs may be justified if they improve outcomes. For example, it is common for patients to require a second ablation procedure to eliminate atrial fibrillation. If advances in technology reduced the need for a second or third ablation procedure, the additional cost incurred during the initial ablation might be justified. The concept of remote catheter navigation is appealing for the operator because these systems may reduce radiation exposure to the physician and the risk of developing orthopedic problems related to prolonged use of protective lead aprons during protracted cases. They also may facilitate analysis of intracardiac electrograms and 3D images because the catheter navigation and analysis can be performed from a work station where the operator is seated. Other key objectives are more accurate navigation to a specified target, improved lesion formation, and a reduction in the risk of perforation.

Two technologies developed to meet these objectives include the magnetic navigation system designed by Stereotaxis, Inc., and a robotic-controlled catheter system manufactured by Hansen Medical. While neither is approved specifically for ablation of atrial fibrillation or ventricular tachycardia, the impetus to develop these technologies is to use them for complex ablation procedures. The potential utility of these remote navigation systems has not been fully developed, but they have improved. Prospective studies are not available to compare these technologies or to demonstrate definitively that either system shortens procedure time, improves outcomes of ablation, or improves the safety profile of these and other complex ablation procedures. Nonetheless, preliminary evidence offers promise that these goals could be achieved.

21.1 Magnetic Navigation

21.1.1 Principles of Magnetic Navigation

The fundamental concept underlying magnetic navigation is that the catheter aligns itself with a magnetic field vector. This principle can be illustrated by holding a compass in one hand and a large magnet in the other. As the magnet is moved toward the compass, the needle of the compass will align itself with the magnet. The magnetic catheter navigations system approved for human use (Stereotaxis, Inc.) does not pull the catheter forward. In order to advance or retract the catheter, a mechanical device is attached to the catheter near the vascular access site and is controlled by the operator at the workstation.

The magnetic navigation system developed by Stereotaxis, Inc., uses two large fixed magnets located on either side of the patient that rotate to generate a composite field vector. The maximum field strength is in the range of 0.08 T, which is less than one tenth of an MRI scanner. A computer workstation located in the control room allows the operator to select a field vector that adjusts the positions of the magnets. The catheter aligns with the vector provided that there are no anatomic barriers. Once the vector is selected, the operator advances the catheter, which tends to move along the line of the field vector to the target.

The degree to which the catheter deflects depends on the strength of the magnetic field, the angle of the field vector relative to the shaft of the catheter, the length of the magnet, and the ferromagnetic materials that comprise the magnet. The maximal force on the catheter occurs when the field is perpendicular to the catheter, and the force decreases to zero when the catheter is parallel to the field vector. As a result, the catheter aligns with the magnetic field because this is the path of least resistance. The size and number of magnets within the catheter determine the response to the magnets and the flexibility of the distal portion of the catheter.

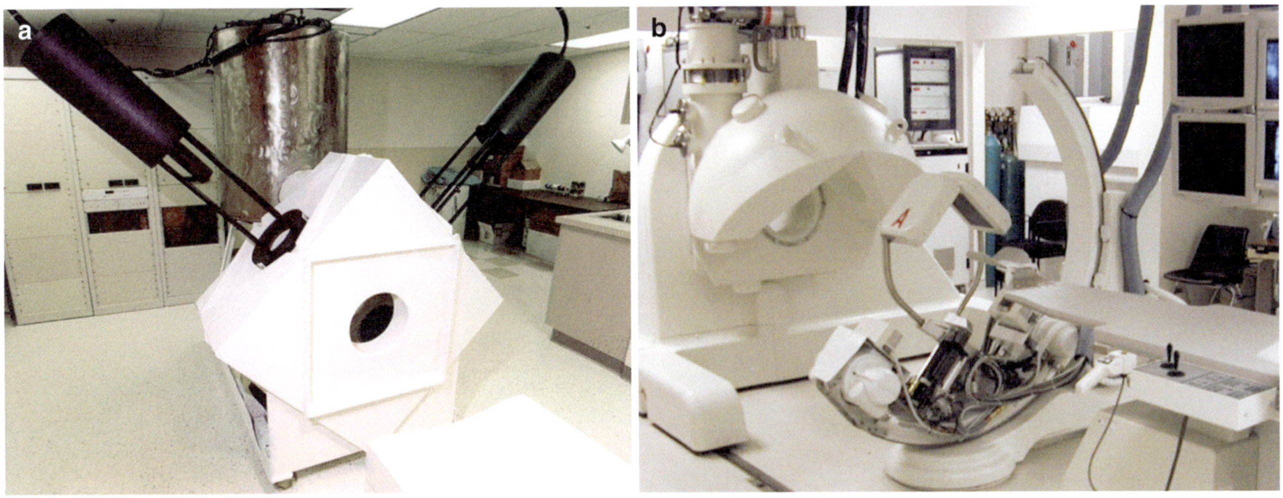

Fig. 21.1 (**a**) Photograph of the initial prototype used for early proof of concept in animals. The core of the magnet was relatively small because it was designed for neurosurgical procedures. A large tank of liquid nitrogen is positioned behind the core of the magnet. An orthogonal array of three electromagnets was built into the core. Flat panel monitors were used to avoid image deterioration caused by the interaction of the magnetic field with a cathode ray tube monitor. (**b**) The second prototype of the magnetic guidance system was designed for cardiac mapping in humans. The core of the magnet was considerably larger. A helium compressor was used to cool the magnet, so the large tank of liquid nitrogen was eliminated. Flat plates were positioned over the patient within the core of the magnet for fluoroscopic imaging

21.1.2 Early Development

The concept of magnetic navigation had its roots in neurosurgery. The initial work by Rogers Ritter examined the possibility that a flexible implement could be introduced through the skull via a burr hole and guided to a target deep in the brain using a magnetic field to steer the implement.[7] The first prototype, which was used for the initial cardiac experiments in relatively small dogs or pigs, is shown in Fig. 21.1a. It was developed for neurosurgery and consisted of an orthogonal array of electromagnets that were cooled by liquid nitrogen. The bore of the magnet was small, which limited the size of animals used for magnetic cardiac navigation and was not large enough for a human torso. The expense and engineering support required for this system was impractical for commercial release. The software used to direct the magnetic field was not optimal, and there were delays of 10–15 s whenever the field vector was changed.

Although the initial prototype was used successfully for proof of concept in animals, a larger magnet was required to encompass the torso of a human and higher fidelity fluoroscopy was mandatory. The second prototype, shown in Fig. 21.1b, met these requirements, but it had other characteristics that were not attractive for clinical use. The electromagnets were cooled by liquid helium which had to be compressed by pumps that created substantial background noise. The sound generated by the helium pumps made communication difficult. The patient was inserted into the donut of the magnet, which could induce claustrophobia, imposed an additional constraint on monitoring and communication, and limited the size of the patient. Nonetheless, the initial human trials were conducted successfully without any significant complications.

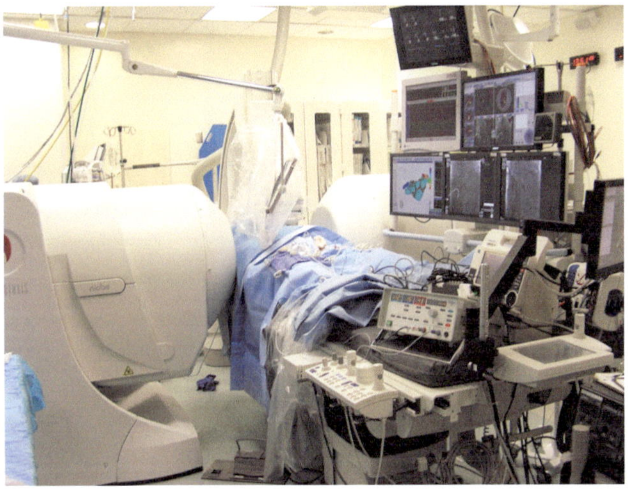

Fig. 21.2 The Niobe II system was approved for commercial release. It uses fixed magnets that are housed within the encasements positioned on either side of the patient

Figure 21.2 illustrates the third design, which replaced electromagnets and the need for helium to cool the system. This design, which is currently in use (Niobe II, Stereotaxis, Inc.), employs large fixed magnets, housed on either side of the patient, which the operator controls by a computer interface. The system rotates the magnets quietly and rapidly to adjust the field vector in about 1–2 s. There is no barrier to communication. The system accommodates relatively large patients without inducing claustrophobia, but patients who

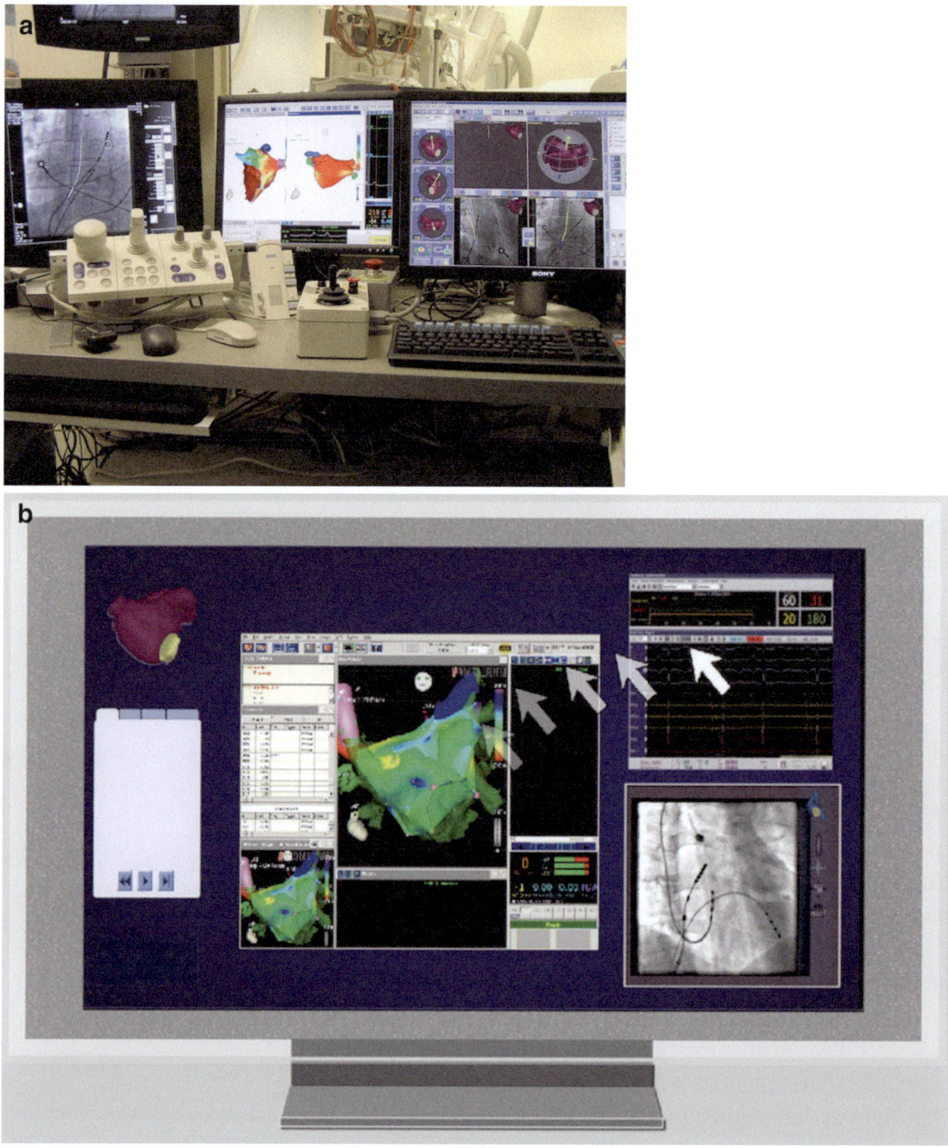

Fig. 21.3 (**a**) Computer interface with X-ray images, the CARTO electroanatomic image, and the Stereotaxis images are shown. The control box for the catheter advancer system is designated by an *arrow*. (**b**) An updated display, which is controlled by a single keyboard and mouse, integrates X-ray images, the CARTO image, and the Stereotaxis image on a single large plasma screen

are broad shouldered or exceed approximately 280 lb may be not fit well between the magnets. To date, rotational fluoroscopy is employed, but bi-planar fluoroscopy cannot be used with the magnets in place. There have been substantial improvements in catheter and software design to improve catheter navigation. Figure 21.3 shows the workstation where the physician sits while operating the system. New competing technologies offer single screen displays to reduce clutter at the workstation.

The initial animal studies tested the feasibility of magnetic navigation and identified areas for further development.[8] Three questions were evaluated regarding the safety and efficacy of the catheter. First, would electrical interference induced by the magnetic field limit the fidelity of recording standard surface ECGs or intracardiac electrograms? Second, would the magnetic catheter apply excessive force that could result in injury or perforation? Third, would the catheter meet demands for accurate navigation, intracardiac recording, and radiofrequency ablation of cardiac tissue in vivo? The initial studies showed that the magnetic field induced an electrical potential that was superimposed on the ECG recorded on the body surface and distorted the T-wave predominantly. It is attributed to a potential generated by blood flow in the aorta by the magnetic field. The temporal distribution of the interfering signal component does not compromise analysis of the P-wave or QRS morphology. Intracardiac recordings with conventional filter settings were qualitatively unaffected by the magnetic field. Intracardiac signals can be recorded without substantial effect on the signals.

The maximal endocardial contact force exerted by the magnetic catheter was determined to be less than the maximal force applied by a conventional catheter, and attempts in animals to intentionally perforate the heart with the magnetic catheter did not result in endocardial injury or perforation. The third objective of the study confirmed accurate navigation

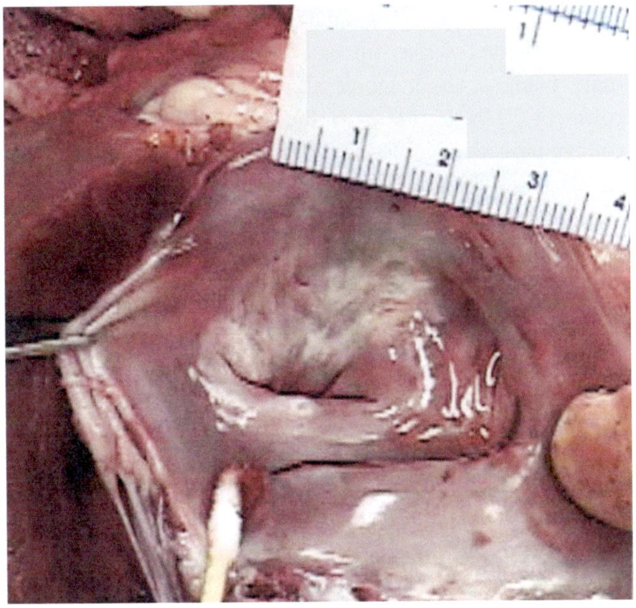

Fig. 21.4 Canine pulmonary vein ostium appearance 3 months after catheter ablation. A typical pulmonary vein ostium is pictured 3 months after remote-controlled pulmonary vein isolation. The figure demonstrates prominent endocardial fibrosis of the peri-ostial region of the right superior pulmonary vein after a segmental ablation strategy. No pulmonary vein stenosis was present (From Greenberg et al.,[9] with permission)

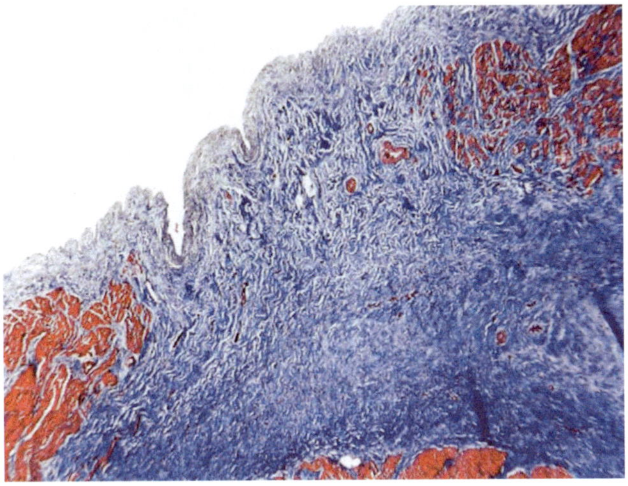

Fig. 21.5 Histologic appearance of catheter ablation lesion. Microscopic examination of the pulmonary vein ostium RF ablation lesion 3 months after catheter ablation shows full thickness fibrosis with replacement of the muscle bundles (trichrome stain at 40×) (From Greenberg et al.,[9] with permission)

of the catheter to all four cardiac chambers. In addition, navigation to the pulmonary veins was achieved by transseptal and retrograde approaches.

From its inception, the ultimate goal of cardiac magnetic navigation was to facilitate ablation of atrial fibrillation. A second series of animal experiments focused on electrical isolation of pulmonary veins in seven canines using an ostial segmental ablation strategy.[9] A remote control mechanism was used to advance or retract the catheter. The procedures were performed using an irrigated catheter. All targeted pulmonary veins were successfully isolated without evidence of stenosis during long-term follow-up. Figure 21.4 demonstrates the appearance of the pulmonary vein ostium 3 months after ablation, and Fig. 21.5 shows the histological appearance of transmural ablation lesions when the animal was sacrificed 3 months later. The conclusion of this study was that electrical isolation of the pulmonary veins was feasible with a magnetically guided irrigated catheter. The force of catheter contact appeared to be adequate and the irrigation technology induced transmural lesions without charring or other complications.

21.1.3 Clinical Trials

The first clinical trial in humans, performed at Washington University in St. Louis, was designed to demonstrate the equivalence of magnetic catheter navigation with conventional 4-mm tip catheters.[10] The primary objectives were to test the system for intracardiac navigation, pacing, and recording. A secondary endpoint was measurement of stimulation thresholds in the right atrium and right ventricle. Lessons learned early in the study led to modification of the catheter to improve its response to navigation. Electrograms and pacing thresholds were comparable to a conventional catheter. In addition, the magnetic catheter was used to ablate supraventricular tachycardia successfully in each of the seven patients in whom this was attempted. One limitation of this trial was that the catheter advancer system had not been approved for use in humans, so the catheter was advanced or retracted manually.

Subsequent clinical trials were performed at other centers in the United States, Europe, and other international sites as the system became available. Although a major long-term objective for remote magnetic navigation is ablation of complex arrhythmias that are technically challenging, clinical trials for ablation of common supraventricular arrhythmias provide experience for more complex problems. These initial trials confirmed the safety and efficacy of the magnetic navigation system, developed operator experience, and identified areas for improvement in the design of the system. The studies by Ernst, Thornton, and Kerzner confirmed that remote navigation could be used safely and with a high degree of success for ablation of AV nodal reentry.[11-13]

Chun reported the success rate for three successive catheter designs for the ablation of accessory pathways in patients with atrioventricular reentry tachycardia using remote magnetic navigation.[14] The improved success rate with successive models (67%, 85%, and 92%) was attributable to the enhanced responsiveness in the third-generation catheter and experience gained with using the system. This investigation is characteristic of an evolving technology.

In Chun's study,[14] patient fluoroscopy exposure was reduced by the magnetic navigation system compared to ablations performed with a conventional catheter. One reason is that the catheter can be redirected to a previously visited location using stored coordinates. The catheters are so flexible that they can be navigated without live fluoroscopy, and can be guided using only 3D imaging systems in conjunction with the magnetic navigation system, which further reduces the fluoroscopy dose.

Chun observed that retrograde mapping of the mitral annulus can be performed successfully, but it was more difficult to position the catheter with this approach than with transseptal navigation.[14] This observation is concordant with the author's experience. The soft shaft of the catheter tends to prolapse into the left ventricular apex rather than moving the tip forward as the operator tries to advance it. This observation has implications for the design of catheters intended primarily for left ventricular mapping, and could also be an important consideration when using a retrograde approach to access left atrial sites that are difficult to reach from the transseptal route.

A multicenter trial reported by Wood on behalf of the Stereotaxis Heart Study Investigators found an overall success rate of 91%, which did not differ significantly from patients who underwent ablation with a conventional catheter.[15] The authors found that the number of lesions delivered to achieve success was less with magnetic navigation and the fluoroscopy time was shorter, which has been a consistent finding in several trials. Another study by Davis et al. found that typical AV nodal reentry could be ablated in all 16 patients who were studied. The time of onset to a junctional tachycardia was less with magnetic navigation compared with conventional catheters used in a control group, and magnetic navigation was associated with less temperature variation.[16] These observations suggest that magnetic navigation catheter provides excellent contact along the right side of the septum, and other reports using magnetic navigation of parahisian pathways attests to the accuracy of small incremental changes that can be achieved with this system.[17,18]

Several conclusions can be drawn from these initial clinical trials for ablation of AV nodal reentry and accessory pathways. The risk of perforation associated with this system is extremely low. The composite data demonstrate a high degree of success despite the fact that it is an evolving technology with characteristics that were new to the investigators. It is clear that the soft, flexible magnetically controlled catheters provide adequate tissue contact for the ablation of cardiac tissue.

The experience with magnetic navigation for ablation of atrial fibrillation is more limited. Pappone evaluated the magnetic guidance system for wide circumferential ablation in 40 patients with paroxysmal or chronic atrial fibrillation.[19] A nonirrigated 4-mm tip catheter was employed for this study using a target temperature of 65° and a maximum power of 50 W. Figure 21.6 illustrates changes in left atrial voltage induced by ablation of left atrial tissue and the distribution of lesions used in a representative case. As illustrated in Fig. 21.7, procedural times decreased as the investigators gained experience with the system. The median procedure time was 192 min in the first 12 patients and 148 min in the last 28. There was no reduction in the overall procedure time of the last 28 patients compared to controls in whom a conventional catheter was used.

DiBiase et al. reported significant limitations when they used a remote magnetic navigation of a 4-mm tip catheter for ablation of atrial fibrillation.[20] In their experience, the catheter could be accurately navigated to the target sites with reduced fluoroscopy exposure, but they did not observe a reduction in total procedure time. Two particularly important observations were that pulmonary vein antrum isolation was not reliably achieved with a 4-mm tip catheter and there was a high incidence of charring during tissue ablation. Experience with remote navigation is required to assess tissue contact, but the incidence of charring reported by DiBiase suggests that adequate contact was not a problem. A reasonable conclusion is that the 4-mm tip catheter, which was designed for ablation of supraventricular tachycardia, should not be promoted for ablation of atrial fibrillation.

Katsiyniannis et al. reported a more favorable experience when they used the 4-mm tip catheter in 20 patients who had predominantly paroxysmal atrial fibrillation.[21] They were able to isolate all four pulmonary veins in all patients and 80% of the patients were free of atrial fibrillation at 1 year. Their experience was that procedural times were shorter in a control group who underwent ablation with a conventional catheter, but the fluoroscopy time was substantially shorter with magnetic navigation compared with conventional catheter navigation (58.6 ± 21 min vs 19.5 ± 9.8 min, $P < 0.001$)

Based on the results of pulmonary vein isolation in animals,[9] an irrigated magnetic navigation catheter should address the concerns raised by DiBiase. Charring was not observed in those studies and consistent pulmonary vein isolation was achieved. A magnetic irrigated catheter has been released for human use, but no multicenter, randomized, controlled trials have been performed to compare magnetic navigation to conventional catheters for ablation of atrial fibrillation.

No major trials have been conducted to evaluate the magnetic navigation system for ablation of ventricular tachycardia. Reports by Thornton have demonstrated its use for right ventricular outflow ventricular tachycardia and left ventricular fascicular ventricular tachycardia,[22,23] and

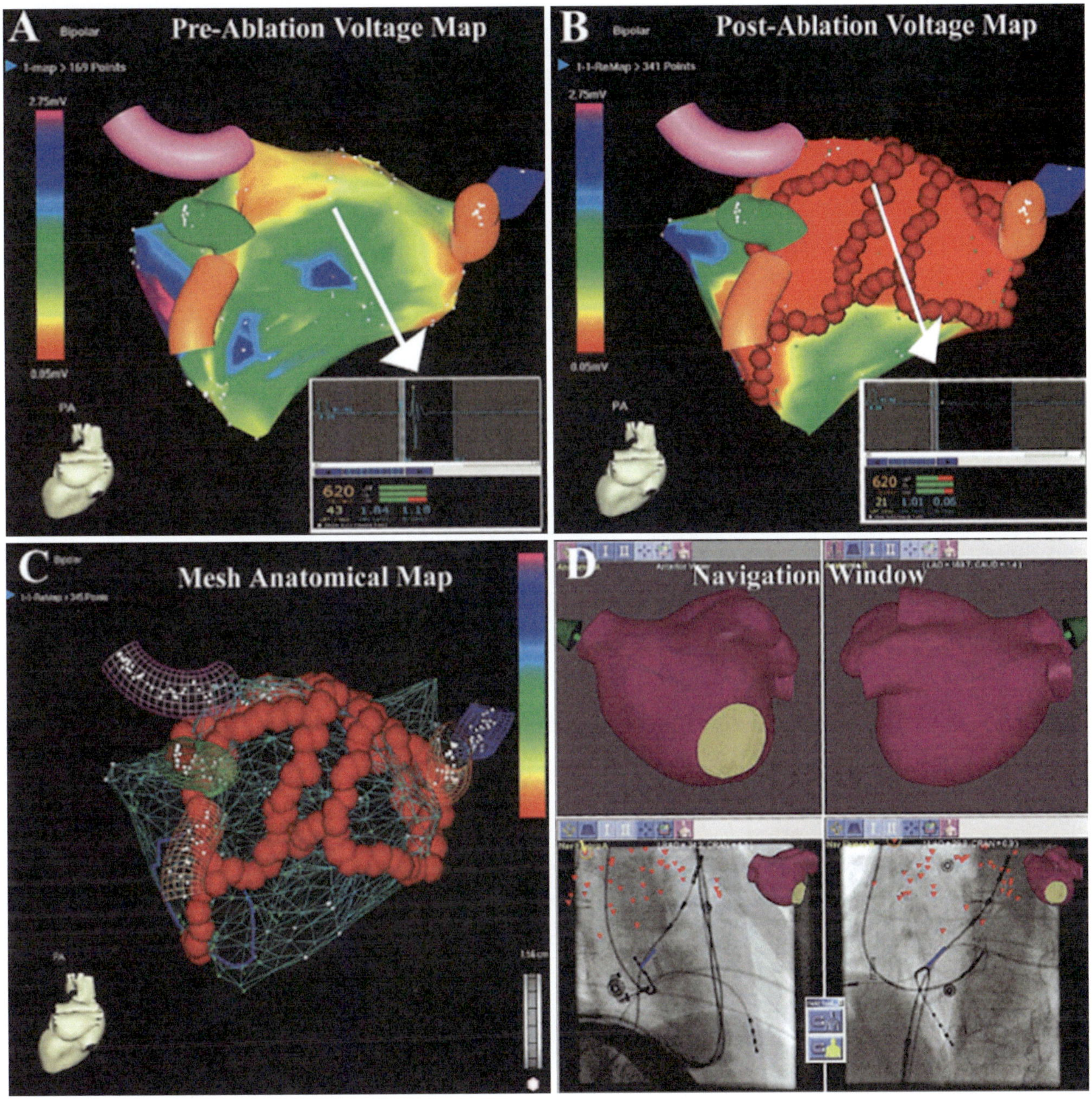

Fig. 21.6 (**a**) Posteroanterior view showing color-coded preablation electroanatomical voltage map of the left atrium and pulmonary veins using a CARTO RMT integration system. (**b**) Post-ablation electroanatomical map with encircled lesions around all targeted pulmonary veins and the mitral isthmus line. Note atrial electrograms before (**a**) (*arrow*) and after ablation (**b**). No potentials were recorded inside encircled lesions (*arrow*). The mitral isthmus ablation line is evident. (**c**) A mesh anatomical map is shown. (**d**) *Red points* depicted on fluoroscopic views indicate ablation points sent from the CARTO RMT integration system. *Blue line* shown on both fluoroscopic images represents the base for target control navigation on Navigant (From Pappone et al.,[19] with permission)

Burkhardt reported a case in which magnetic navigation was used to ablate a ventricular arrhythmia arising from the left aortic cusp.[24] Reddy used magnetic navigation with excellent results in 24 patients with ventricular tachycardia related to scar from myocardial infarction, dilated cardiomyopathy, arrhythmogenic right ventricular cardiomyopathy, hypertrophic cardiomyopathy, or sarcoidosis.[25] The system was used successfully to ablate 75 of 77 ventricular tachycardias, including a few that required percutaneous epicardial access. Both transseptal and retrograde aortic approaches were used for mapping the left ventricular endocardium.

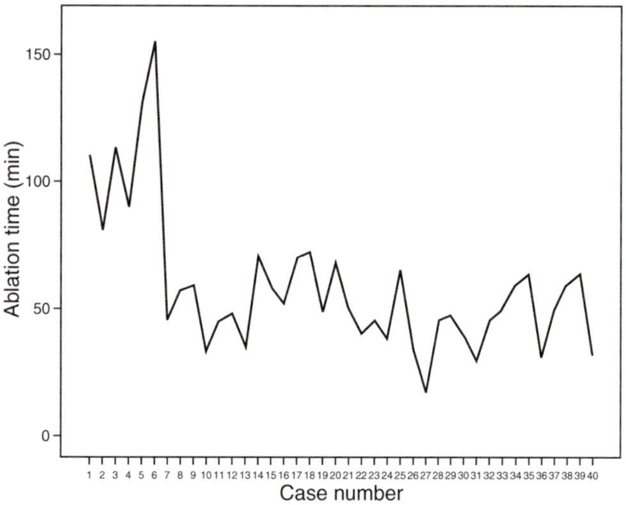

Fig. 21.7 Graph showing the learning curve of remote magnetic navigation for atrial fibrillation ablation. Sequence of the first 40 procedures performed at Hospital San Raffaele. Note that the ablation time curve stabilizes after the first 12 cases. In the first and third patient, radiofrequency ablation was completed manually (From Pappone et al.,[19] with permission)

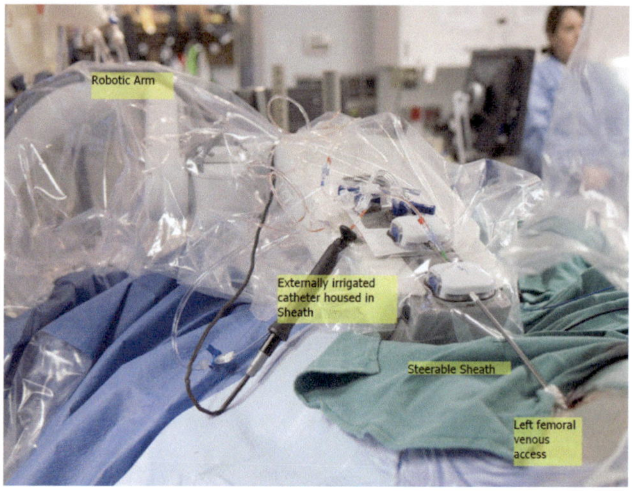

Fig. 21.8 The robotic arm is mounted over the patient and the sheaths are inserted via the femoral vein

21.1.4 Overview of Magnetic Navigation

There have been significant improvements in the system since introduction of the initial prototype. The Niobe II model, which features fixed magnets that rotate in response to the computer interface, has performed reliably. Successive improvements in catheter design have made the catheters more responsive to the operator, and software improvements facilitate accurate navigation to the target. An irrigated catheter offers potential advantages for ablation of atrial fibrillation and ventricular tachycardia. The system performs well in all four chambers of the heart, but it can be difficult to navigate beneath the right inferior pulmonary vein, especially in patients with a relatively small left atrium.

In patients who require left ventricular mapping, the catheter is easy to maneuver along the septum where conventional catheter manipulation can be difficult. The only limitation in the left ventricle is that when the inferior wall is targeted near the mitral annulus, the soft shaft of the catheter may prolapse into the apex as the catheter is advanced if the patient has a large apical aneurysm. Future catheter designs may need to modify the stiffness of the shaft and the transition point between the shaft and the highly flexible distal portion of the catheter.

When electroanatomic mapping is required, magnetic navigation is compatible with either the Carto or NavX systems. Special anesthesia equipment, designed to be compatible with MR imaging, is needed if general anesthesia is used during the procedure. The staff must be educated in safety procedures when working around large magnets because instruments or other metallic equipment may be drawn forcefully to the magnets if appropriate precautions are not used.

The main limitations are that large patients may not fit between the magnets, and it is not approved for use with pacemakers or ICDs. Although the magnetic field strength of the system is less than one tenth of an MRI system, there may be a risk of damaging components of pacemakers or ICDs if they are positioned between the magnets, and exposure of leads to the magnetic field induces a current that could cause tissue injury. To date, the author has not observed any change in pacemaker or ICD pacing or sensing thresholds after exposure to the system, but the number of patients studied is not sufficient to provide broad safety guidelines. This system is not approved for use in patients with implantable pacemakers or defibrillators and further study is needed to verify the safety of magnetic navigation in patients with these devices.

21.2 Robotic Navigation

21.2.1 Principles of Robotic Navigation

The Hansen Robotic system shown in Figs. 21.8 and 21.9 is approved for cardiac navigation and mapping in humans. The robotic system (Sensei™ Robotic Catheter System, Hansen Medical) consists of a physician workstation (PWS) that includes an instinctive motion controller (IMC) – otherwise simply the remote catheter manipulator, and a setup joint that is mounted on table at the patient's side. The remote catheter manipulator directly controls a robotic hollow catheter (Artisan™ catheter) consisting of an internal steerable guide sheath and an outer sheath comprising the steerable

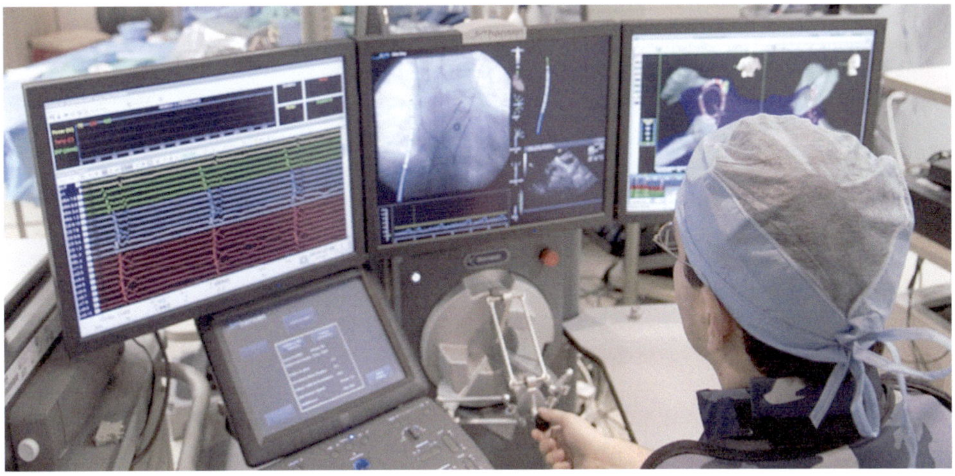

Fig. 21.9 The operator controls the robotic arm from a console at the workstation

sheath system (SSS Artisan Catheter). This robotic hollow catheter is in reality a hollow robotically steered sheath that can house any mapping or ablation catheter. The physician workstation is placed at a remote location from the patient's table. The control is provided via a master and slave electromechanical system: The operator movements at the IMC are updated constantly with resultant motion of the catheter. Using this system, catheter control is performed in three dimensions.

21.2.2 Left Atrial Mapping and Ablation

The Hansen steerable sheath system is not approved for ablation of atrial fibrillation, but there is considerable experience with left atrial mapping and ablation. The following discussion is based on experience acquired at the Cleveland Clinic.[26] The steerable sheath system is inserted via a 14 Fr sheath in the right femoral vein and advanced manually into the right atrium under fluoroscopic visualization. When patients undergo ablation of atrial fibrillation, two transseptal punctures are performed manually under guidance by intracardiac echocardiography. A long sheath is introduced through the first transseptal puncture and a circular mapping catheter is introduced through the sheath into the left atrium. Systemic anticoagulation is initiated just prior to the first transseptal puncture, using intravenous heparin with a target ACT of approximately 400 s. A wire is introduced into the left atrium through the second transseptal puncture. The robotically guided ablation catheter housed in the steerable sheath is then driven into the left atrium alongside the wire. With this configuration, the final setup before initiation of ablation includes the following: A robotically controlled steerable sheath housing the ablation catheter is operated by the physician at the console and a circular mapping catheter is positioned manually by a second physician at the procedure table.

Radiofrequency power is set at 25–40 W with a maximum temperature limit of 45°C with irrigation using heparinized saline infusion (2,000 IU/L) at a rate of 30 cm^3/min via the Cool Flow pump (Biosense Webster, Diamond Bar, CA, USA). At the end of the procedure, systemic anticoagulation is discontinued and partially reversed with intravenous protamine prior to removal of vascular sheaths.

21.2.3 Early Development and Clinical Studies

Preliminary studies in animals demonstrated the feasibility and safety of cardiac navigation and the safety of transseptal puncture in animals with preliminary evidence that procedure time and radiation exposure might be reduced.[27,28]

Reddy has systematically evaluated robotic navigation for ablation of atrial fibrillation in animals and humans.[29] CT scans were performed and segmented data sets were manually converted into 3D surface reconstruction, which were combined with electroanatomic mapping using custom imaging software (Fig. 21.10). The registered images were used to guide robotic navigation using either an 8-mm or 3.5-mm irrigated tip catheter. During the preclinical phase, mapping and ablation was performed in nine pigs. Left and right pulmonary veins were successfully targeted. The authors felt that the catheter was stable and responded well during navigation. Postmortem studies showed that lesions were distributed around the pulmonary vein ostia and there was no evidence of perforation.

The second phase of the study reported by Reddy included nine patients who underwent pulmonary vein isolation for atrial fibrillation.[29] They were able to manipulate the catheter into the pulmonary veins and left atrial appendage without difficulty. Intracardiac echocardiography was performed to verify accurate positioning of the catheter at these sites. An irrigated catheter was used to perform extra-ostial pulmonary vein isolation in all nine patients. The lesion set was

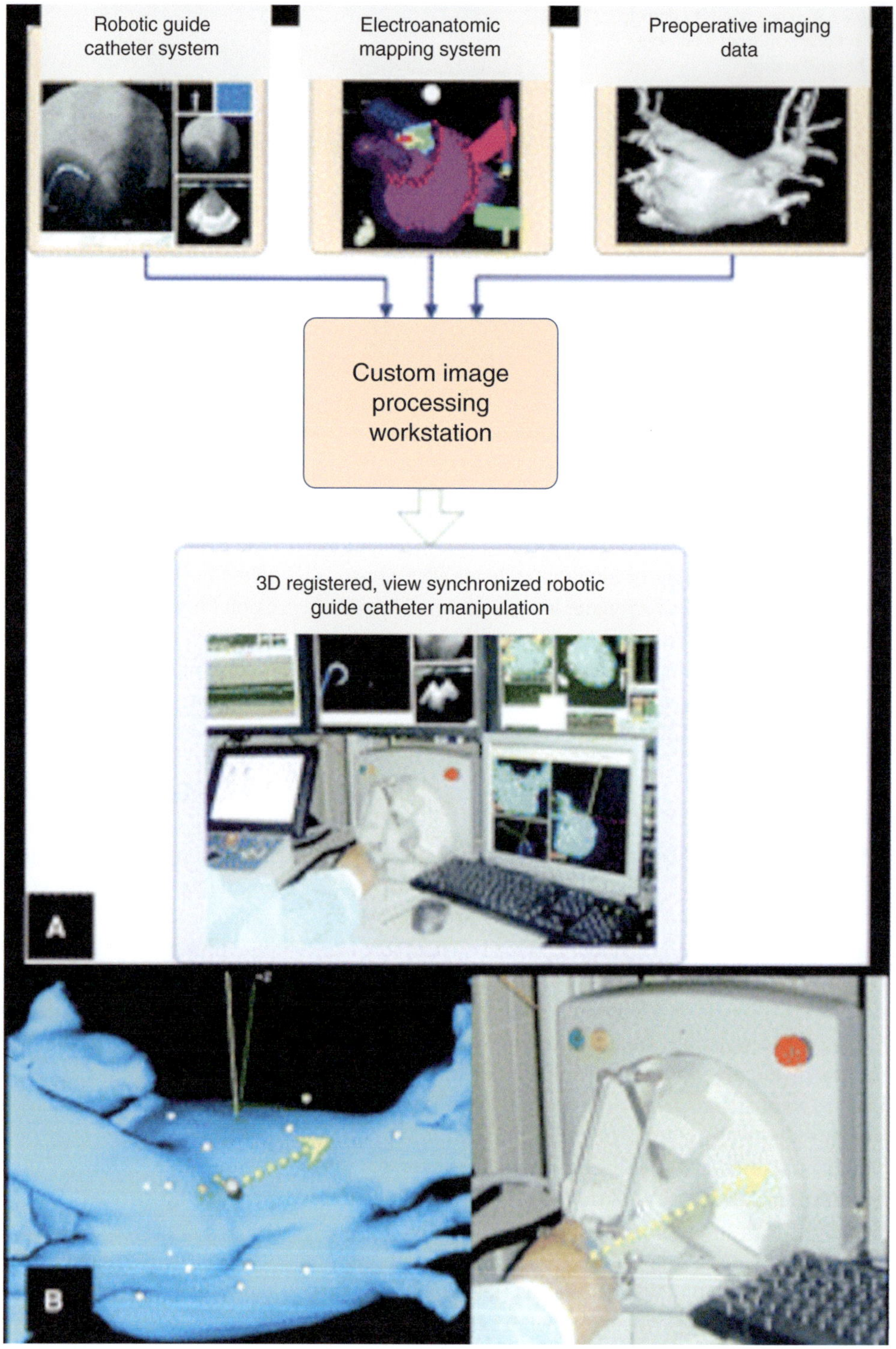

Fig. 21.10 View-synchronized robotic mapping. (**a**) The custom image processing workstation integrated the navigation, mapping, and imaging information. The 3D CT image was imported and registered to the real-time electroanatomic mapping system. Synchronizing the 3D imaging view to that of the RNS made intuitive cardiac mapping feasible. (**b**) Movement of the catheter to the right on this posterior oblique view of the synchronized 3D CT image is accomplished by moving the 3D joystick in the same direction (*yellow arrows*). Accordingly, 3D view synchronization provides a facile means for cardiac mapping regardless of the perspective from which the chamber is viewed (From Reddy et al.,[29] with permission)

extended on the posterior left atrial wall to include the pulmonary vein antrum. Figure 21.11 shows an integrated image demonstrating precise placement of ablation lesions during isolation of the pulmonary veins. Low-amplitude fractionated signals were targeted during atrial fibrillation. In three patients with chronic atrial fibrillation, the lesion sets included a cavotricuspid line, left atrial roof line, and a line between the mitral annulus and the left pulmonary veins. No acute or chronic complications were observed. The overall procedure time was 338 ± 89 min. To put the relatively long procedure times in perspective, the experience with the system was limited and the intent was to evaluate the capabilities of the system as opposed to assessing the shortest possible time to complete the procedure.

An initial report by Saliba included 40 patients in whom the procedure was performed in Europe between January and

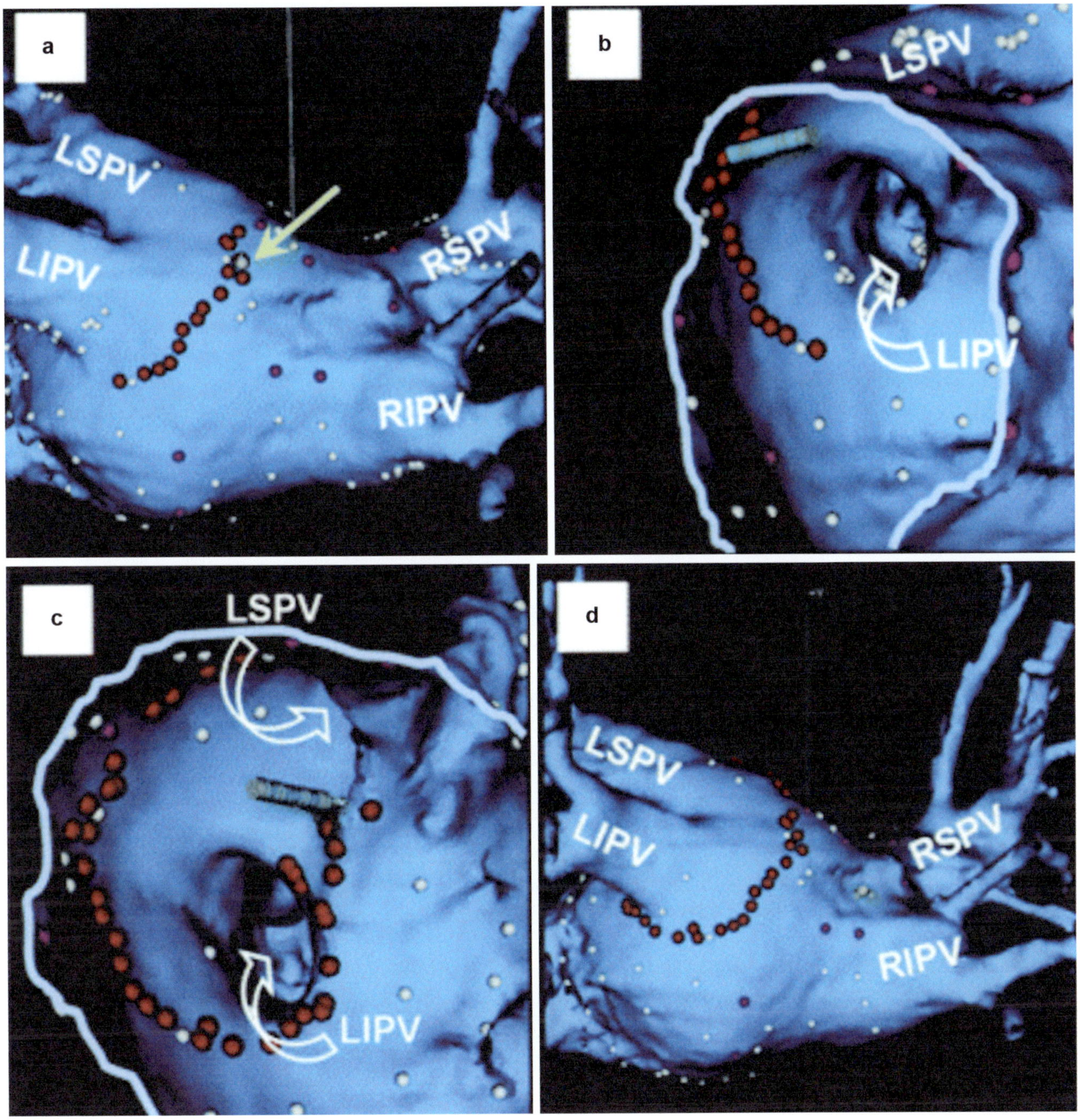

Fig. 21.11 View-synchronized catheter ablation. In this patient with paroxysmal AF, external posterior views (**a**, **d**) and endoluminal views (**b**, **c**) of the left-sided pulmonary veins are shown. The irrigated ablation catheter is shown during ablation along the posterior LA wall (**a**, **b**) and at the ridge between the left superior pulmonary vein and the left atrial appendage (**c**); this latter lesion completed extra-ostial electrical isolation of these left-sided pulmonary veins (also shown in **d**). In (**a**), the *arrow* indicates the tip of the ablation catheter. *LSPV* indicates left superior PV, *RSPV* right superior PV, *LIPV* left inferior PV, and *RIPV* right inferior PV (From Reddy et al.,[29] with permission)

July 2005. The total procedure time was 163 ± 88 min, ablation time 89.6 ± 43 min, and fluoroscopy time 64 ± 33 min. The average radiation exposure at the procedure table is 149 vs. 13 microseiverts at the physician workstation ($P < 0.05$). Acute isolation of all pulmonary veins was achieved in all patients. Two patients undergoing pulmonary vein isolation developed pericardial tamponade requiring pericardiocentesis. The procedure was aborted in one patient after manual transseptal puncture. However, procedural success was similar to the manual approach.[30]

Several other groups have evaluated the Hansen robotic navigation system with encouraging results. Kantner et al. compared robotic navigation to conventional manual navigation in patients who underwent pulmonary vein antral isolation for atrial fibrillation and found that robotic navigation reduced the cumulative time of radiofrequency energy delivery and fluoroscopy time.[31] This study also noted a decrease in procedure time. Although Schmidt found that robotic navigation reduced fluoroscopy time, it did not reduce the procedural time.[32] Steven et al. compared robotic navigation to manual navigation in patients who required ablation of common right atrial flutter.[33] In their experience, robotic navigation reduced fluoroscopy time and cumulative time for applications of radiofrequency energy, but the total procedural time was longer. In aggregate, these studies suggest that robotic navigation provides stable tissue contact for energy delivery and reduces the time of radiofrequency application. While the reduction in fluoroscopy time probably reflects accurate navigation, the total procedural time has not been consistently reduced. It is likely that this goal will be achieved with greater operator experience and technical improvements.

If robotic navigation provides better contact with the tissue during ablation, one must question whether energy delivery should be modified to reduce the risk of steam pops or collateral damage. DiBiase evaluated the relationship between lesion formation and pressure in animals using the robotic navigation system and Intellisense force sensor.[34] Intellisense provides an indirect measure of force based on variations of the resistance to forward motion as measured from the proximal as opposed to the distal portion of the catheter. Therefore, the force is an estimate of the actual contact with tissue. He found that 30 W lesions were more likely to be transmural at >40 g pressure compared with <30 g pressure, but the risk of steam pops and crater formation were greater with the higher pressure.[34] The investigators found that transmural lesions could be achieved with 20–30 g of pressure at 40 W with reduced risk of steam pops. This study was not designed to assess whether applications with this power and degree of pressure have an associated risk of esophageal injury. It is a difficult problem to study because animal models do not replicate the anatomy of humans. The unresolved questions regarding force, power of energy, and duration of the application are applicable to manual navigation, magnetic navigation, and robotic navigation. Other work by Cummings demonstrated that in animals, there is a substantial temperature gradient between the esophageal lumen and the external wall of the esophagus during applications of radiofrequency energy to the posterior wall of the left atrium.[35] Moreover, in humans, the power of the application did not correlate well with temperature recorded from the esophageal lumen.[36] The distance between the esophagus and the atrium, displacement of the posterior wall of the atrium toward the esophagus during catheter navigation, and the vascular supply of the esophagus are among the other variables that are difficult to control.

An extensive single center report, based on experience acquired at the Cleveland Clinic from July 2007 to October 2008, included 71 patients with AF (31 paroxysmal) who underwent attempted ablation with the Hansen system.[26] This is the largest single center report and reflects a learning curve. In this study, 67 patients underwent pulmonary vein antral isolation with the Hansen system. Four patients were lost to follow-up. In the remaining 63 patients, 49 (78%) had no recurrence of arrhythmia off antiarrhythmic drugs after 161 ± 90 days of follow-up.

The experience gained from this series will be reviewed in detail because it may be of value to those who are not familiar with the Hansen system. The mean left atrial instrumentation time was 171 ± 51 min. In 70 patients, left atrial instrumentation was achieved with the Hansen system, but in one patient, the robotic sheath could not be advanced across a thick septum.

One procedure was aborted intraoperatively due to a vascular complication with excessive femoral bleeding at the Hansen sheath site after the robotically performed transseptal puncture was completed. This complication occurred before ablation in left atrium. In one patient, the procedure was converted to the manual approach because the Hansen system malfunctioned, and the procedure was terminated in another because of tamponade prior to ablation. In the remaining 67 procedures, the pulmonary veins were isolated acutely with the Hansen system. Further left atrial ablation was performed on the posterior wall in all patients. In three (4%) of the 67 patients, the right inferior pulmonary veins were isolated manually because of difficulty reaching the target with the Hansen system. Ablation in the posterior wall was limited due to rise in esophageal temperature in 22 patients.

21.2.4 Avoiding Complications

The experience gained from the Cleveland Clinic experience helps to identify ways to avoid complications.[26] The robotic sheath is inherently stiff because of the steering mechanisms. The introducer sheath is large (14 Fr) which increases the risk of complications, which includes the risk of venous dissection if the steerable sheath is inadvertently driven into the wall of the vein. In order to minimize the risk of vascular complications, it is advisable to obtain venous access under ultrasound guidance. First, an 8 Fr sheath is inserted and then upsized sequentially to an 11 Fr and then a 14 Fr sheath. A long 30-cm 14 Fr sheath is positioned with the end at the level of the liver. This is used to insert the steerable sheath into the right atrium leading with the flexible ablation catheter by at least 10 cm. At this level, the ablation catheter is withdrawn into the steerable sheath with only the distal electrodes protruding. When this technique was employed, there were no further vascular complications

Cardiac tamponade during ablation procedures often is caused by perforation because of improper catheter movement or a steam pop during an application of radiofrequency energy. There was only one steam pop phenomenon leading to cardiac tamponade early in our experience. Following this episode, we altered our power output to a maximum of 35 W in most areas in the left atrium. Due to stability afforded by the robotic nature of the system, the power delivered could be lowered with the same resultant attenuation of myocardial signal.

The steerable sheath did not respond appropriately despite numerous realignments in one patient. The operator noticed uncontrolled rebound ("fling") of the catheter with steering. The patient became hemodynamically unstable with sinus tachycardia and hypotension. Intracardiac echocardiography and fluoroscopy confirmed the presence of tamponade which was quickly treated with pericardiocentesis. On further investigation and discussion with the manufacturer, it became apparent that certain catheters with flat wire deflection mechanisms, such as the current version of the CoolPath Ablation Catheter manufactured by St. Jude, are not compatible with the Hansen system, thereby rendering the system a semi-open rather than fully open platform. Subsequent to this event, Hansen Medical issued a field notification regarding this issue to all Hansen users. The flat wire mechanism allows the catheter to only bend along a 2D plane described by the flat surface of the wire without causing tension. When the robotic sheath moves in a direction that is not along the 2D plane of the flat wire, tension is created resulting in an uncontrolled rebound rapid correction of the system. This rarely happens with manual manipulation of the catheter alone because the operator will rotate the catheter appropriately to achieve the desired position.

Sometimes it is difficult to cross a thick septum. This was encountered in one patient who later on developed tamponade after attempted transseptal puncture. The operator realized that the system was not responding appropriately and had advanced anteriorly instead of the intended posterior trajectory even after several realignments. Subsequent analysis revealed that one of the pull wires of the robotic sheath had broken. As with all new technologies, vigilance is required to detect erratic or unanticipated performance. In such cases, the system should be abandoned and the manufacturer should be notified so that a root cause analysis can determine the reason.

In order to reduce the risk of a misdirected transseptal puncture, one solution is to perform a manual transseptal puncture in the plane of the left pulmonary veins under guidance by intracardiac echocardiography. This provides a visual landmark for transseptal puncture with the Hansen system. Next, fluoroscopy and intracardiac echocardiography is used to guide the Hansen system housing the ablation catheter, which is driven bluntly alongside this wire into the left atrium. Currently, there is no robotically driven transseptal needle available for use in the United States; however, there is a specially designed Hansen steerable sheath dilator over which the steerable sheath can be advanced into the left atrium. Our recommendation is to perform the transseptal puncture manually and then advance the system into the left atrium over this wire. In some cases, the dilator may not be stiff enough to adequately support the advancement of the steerable sheath into the left atrium without buckling at the level of the septum.

Ablation on the posterior wall of the left atrium is associated with the risk of injury to the esophagus or para-esophageal nerves that control gastric motility. The same risks exist with the robotic navigation system and can be even greater if lower energy is not employed because of the stable contact afforded by the steering sheath. In one patient, ablation was limited on the posterior wall because of a rapid rise in temperature even with powers as low as 5 W. Therefore, if posterior wall ablation is planned using the Hansen system temperatures, the esophageal temperature should be monitored closely. Currently, we limit power in the posterior wall according to a predetermined maximal change in esophageal temperature recorded in the esophagus. RF delivery is terminated when luminal esophageal temperature reaches 38°C. There were no cases of atrial esophageal fistulas; however, there was one case of gastroparesis. These observations emphasize the need to limit power while ablating on the posterior wall of the left atrium.

21.2.5 Conclusions Drawn from the Cleveland Clinic Experience

While the robotic system affords greater stability at the tissue interface, complications that occur with the manual approach can also occur with the robotic system. The same features that provide catheter stability are associated with stiffness and rigidity of the sheath that must be taken into consideration by the operator. As with any evolving technology, it is incumbent on the operator to be cognizant of potential complications and to avoid maneuvers that could increase these risks.

In the future, design improvements that address the following issues will add to the safety profile of the system:

1. Robotic transseptal puncture with a dedicated transseptal needle that can be robotically driven.
2. Decrease of the steerable sheath size may be possible as smaller profile catheters become available.
3. Development of more advanced catheters with accurate measurement of force exerted by sensors at the catheter tip.
4. Optimal energy titration while using the system still needs to be determined given the stability afforded.

21.2.6 Overview of Robotic Navigation

This evolving technology provides accurate catheter navigation and contact for ablation of tissue in the atria or along the mitral or tricuspid annuli. The risk of perforation by an ablation catheter that is steered by a relatively stiff guiding sheath appears to be diminished by technology that provides feedback about the relative degree of force at the tip of the catheter. It does not require special anesthesia equipment and can be used in patients with pacemakers or implantable cardioverter defibrillators. The relatively large supporting sheath (14 Fr) may increase the risk of bleeding when it is removed from the body, but this potential problem has not been clinically significant in the early clinical reports. The supporting and steering sheaths are designed for use in the atria. They are not long enough to be used in the left ventricle. Moreover, careful study would be required to assess the safety of introducing a stiff sheath into the left ventricle. It is likely that this will be accomplished at some point in the future, but the current system is not designed for this purpose.

21.3 Comparison of Magnetic and Robotic Navigation

No direct comparison of magnetic and robotic navigation has been performed. A valid comparison would require a randomized multicenter trial performed by physicians who are experienced with both systems and have no financial ties to either manufacturer through consulting fees, grant support, or other incentives. Individual assessments, including the authors', may be influenced by personal experience that is greater with one system compared with the other.

Both systems require experience to master use of the software for navigation, and there is undoubtedly a learning curve that affects procedural time. The duration of the learning curve probably depends on how frequently the system is used. It takes longer to gain proficiency if the system is used only once a week. The repetition of daily use is likely to shorten the time needed to take full advantage of the technology. For most physicians, this is likely to be a minimum of 20–30 cases. Although the systems may facilitate catheter navigation, this technology is no substitute for other fundamental skills. Experience, analytical ability, and judgment will continue to be the defining attributes of excellence, but the ability to maneuver a catheter to a specified target may be reduced to a common denominator by these and other innovative advances in technology.

Magnetic and robotic navigation can be used to direct the catheter to a target with a high degree of precision. Sometimes the catheter may not move as expected because it meets resistance from contact with the endocardium. One component of the learning curve is to recognize when the catheter must be retracted so that it can respond to the control mechanism. It may need to be advanced after it shifts position in order to achieve optimal tissue contact. The Stereotaxis system estimates tissue contact by the difference between the selected field vector and the position of the catheter. The assumption is that if the catheter is not fully aligned with the field, then it must be in contact with the tissue. One of the limitations of this estimate is that it provides less information about force perpendicular to the endocardium. In that case, the contact can be estimated fluoroscopically by deflection of the catheter tip. The catheter is so flexible that the risk of perforation is negligible. In contrast, Hansen's robotic system depends on relatively stiff sheaths to direct a standard ablation catheter. An accurate feedback mechanism is particularly important to assess tissue contact and reduce the risk of perforation. Both systems have the potential to improve tissue contact, which may change the energy requirements for ablation of tissue. This is an important consideration on the posterior wall of the left atrium, which is relatively thin and in close proximity to other structures such as the esophagus.

There are three distinct differences between the systems. Hansen's robotic system is portable, so it can be moved to the room where it is required. This provides flexibility for programs that have multiple rooms; however, a practical limitation is that the calibration may need to be adjusted if the system is moved. For that reason, it is generally left in one lab for day-to-day usage. In contrast, the large magnets required for Stereotaxis are not portable, they consume considerable space within the laboratory, and the floor of the laboratory requires additional structural support to bear the weight of the magnets. The second major difference is that specially designed catheters are required for magnetic navigation. In principle, any eight French ablation catheter can be inserted through the steering sheath developed by Hansen for robotic navigation; however, as noted in the preceding discussion, caution should be exercised with new catheters until the response to deflection has been analyzed. The third difference is that while the Stereotaxis system has been used safely in all four chambers of the heart, the Hansen system is designed for use in the atria.

We face an era in which the cost of technology will come under increasing scrutiny. Neither company can claim that their system reduces cost compared to conventional ablation technology. The argument that cost may be reduced by shorter procedure times has not been realized yet. Even if the time to complete an ablation is reduced, the overall procedure time may not be effected substantially because the time to set up equipment or perform the diagnostic portion of the study is not shorter. The remaining arguments by the manufacturers may center on the comparative costs of their systems. It is clear that the initial cost of Stereotaxis is substantially greater, but the incremental costs of the disposable guiding catheters manufactured by Hansen accrue over time.

The expenses related to these systems may be comparable at 5 years. Differences in contractual agreements between the manufacturers and the hospital make these comparisons even more difficult.

Expensive technologies should have a significant impact on efficiency, reduce the risk of complications, and improve long-term outcomes. Even if the time to complete an ablation is reduced, the overall procedure time may not be affected substantially because the time to set up equipment or perform the diagnostic portion of the study is unchanged. The argument that cost may be reduced by shorter procedure times has not been realized yet. It is likely that radiation exposure to the patient will be reduced by an experienced operator. Neither company can claim that their system reduces cost compared to conventional ablation technology, nor are there controlled studies to show that they reduce the risk of complications or improve long-term outcomes.

References

1. Shope TB. Radiation-induced skin injuries from fluoroscopy. *Radiographics*. 1996;16(5):1195-1199.
2. Wong L, Rehm J. Images in clinical medicine; radiation injury from a fluoroscopic procedure. *N Engl J Med*. 2004;350:e23.
3. Vano E, Arranz L, Sastre JM, et al. Dosimetric and radiation protection considerations based on some cases of patient skin injuries in interventional cardiology. *Br J Radiol*. 1998;71:510-516.
4. Hasimoto I, Sedo H, Inatsugi K, Hideki N, Arase S. Severe radiation-induced injury after cardiac catheter ablation: a case requiring free anterolateral thigh flap and vastus lateralis muscle flap reconstruction on the upper arm. *J Plast Reconstr Aesthet Surg*. 2008;61:704-708.
5. Fadl Y, Ellenbogen KA, Grub RL Jr, Khoo-Summers L, Lindsay BD. A review of spinal injuries in the invasive cardiologist: part I. Biomechanics and pain generation. *Pacing Clin Electrophysiol*. 2007;30:1012-1019.
6. Fadl Y, Ellenbogen KA, Grub RL Jr, Khoo-Summers L, Lindsay BD. A review of spinal injuries in the invasive cardiologist: part II: prevention and treatment. *Pacing Clin Electrophysiol*. 2007;30: 1149-1157.
7. McNeil RG, Ritter RC, Wang B, et al. Functional design features and initial performance characteristics of a magnetic-implant guidance system for stereotactic neurosurgery. *IEEE Trans Biomed Eng*. 1995;42(8):793-801.
8. Faddis MN, Blume W, Finney J, et al. Novel, magnetically guided catheter for endocardial mapping and radiofrequency catheter ablation. *Circulation*. 2002;106:2980-2985.
9. Greenberg S, Blume W, Faddis M, et al. Remote controlled magnetically guided pulmonary vein isolation in canines. *Heart Rhythm*. 2006;3:71-76.
10. Faddis MN, Chen J, Osborn J, Talcott M, Cain ME, Lindsay BD. Magnetic guidance system for cardiac electrophysiology: a prospective trial of safety and efficacy in humans. *J Am Coll Cardiol*. 2003;42:1952-1958.
11. Ernst S, Ouyang F, Linder C, et al. Initial experience with remote catheter ablation using a novel magnetic navigation system: magnetic remote catheter ablation. *Circulation*. 2004;109:1472-1475.
12. Kerzner R, Sanchez JM, Osborn JL, et al. Radiofrequency ablation of atrioventricular nodal reentrant tachycardia using a novel magnetic guidance system compared to a conventional approach. *Heart Rhythm*. 2006;3:261-267.
13. Thornton AS, Janse P, Dominic AKJ, et al. Magnetic navigation in AV nodal reentrant tachycardia study: early results of ablation with 1-, and 3-magnet catheters. *Europace*. 2006;8:225-230.
14. Chun J, Ernst S, Mathews S, et al. Remote-controlled catheter ablation of accessory pathways: results from the magnetic laboratory. *Eur Heart J*. 2007;28:190-195.
15. Wood MA, Orlov M, Ramaswamy K, Haffajee C, Ellenbogen K, The Stereotaxis Heart Study Investigators. Remote magnetic versus manual catheter navigation for ablation of supraventricular tachycardias. *Pacing Clin Electrophysiol*. 2008;31:1313-1321.
16. Davis DR, Tang ASL, Gollop MH, Lemery R, Green MS, Birnie DH. Remote magnetic navigation-assisted catheter ablation enhances catheter stability and ablation success with lower catheter temperatures. *PACE*. 2008;31:893-898.
17. Ernst S, Hachiy H, Chun JKR, Ouyang F. Remote catheter ablation of parahisian accessory pathways using a novel magnetic navigation system – a report of two cases. *J Cardiovasc Electrophysiol*. 2005;16:659-662.
18. Davis DR, Tang AS, Birnie DH, Gollob MH. Successful ablation of a concealed parahisian accessory pathway using remote magnetic navigation system following failure by conventional methods. *J Interv Card Electrophysiol*. 2006;16:149-151.
19. Pappone C, Vicedomini G, Manguso F, et al. Robotic magnetic navigation for atrial fibrillation ablation. *J Am Coll Cardiol*. 2006;47:1390-1400.
20. Di Biase LD, Fahmy TS, Patel D, et al. Remote magnetic navigation: initial human experience in pulmonary vein ablation. *J Am Coll Cardiol*. 2007;50:868-874.
21. Katsiyiannis WT, Melby DP, Matelski JL, Ervin VL, Laverence KL, Gornick CC. Feasibility and safety of remote-controlled magnetic navigation for ablation of atrial fibrillation. *Am J Cardiol*. 2008;102:1674-1676.
22. Thornton AS, Jordaens LJ. Remote magnetic navigation for mapping and ablating right ventricular outflow tract tachycardia. *Heart Rhythm*. 2006;3:691-696.
23. Thornton AS, Res J, Mekel JM, Jordaens LJ. Use of advanced mapping and remote magnetic navigation to ablate left ventricular fascicular tachycardia. *Pacing Clin Electrophysiol*. 2006;29:685-688.
24. Burkhardt JD, Saliba WI, Schweikert RA, Cummings J, Natale A. Remote magnetic navigation to map and ablate left coronary cusp ventricular tachycardia. *J Cardiovasc Electrophysiol*. 2006;17: 1142-1144.
25. Aryana A, d'Avila A, Heist K, et al. Remote magnetic navigation to guide endocardial and epicardial catheter mapping of scar-related ventricular tachycardia. *Circulation*. 2007;115:1191-1200.
26. Wazni OM, Barrett C, Martin DO, et al. Experience with the Hansen robotic system for atrial fibrillation ablation-lessons learned and techniques modified: Hansen in the real world. *J Cardiovasc Electrophysiol*. 2009;20:1193-1196.
27. Al-Ahmad A, Grossman JD, Wang PJ. Early experience with a computerized robotically controlled catheter system. *J Interv Card Electrophysiol*. 2005;3:199-202.
28. Saliba W, Cummings JE, Oh S, et al. Novel robotic catheter remote control system: feasibility and safety of transseptal puncture and endocardial catheter navigation. *J Cardiovasc Electrophysiol*. 2006;10:1102-1105.
29. Reddy VY, Neuzil P, Malchano ZJ, et al. View-synchronized robotic image guided therapy for atrial fibrillation ablation. Experimental validation and clinical feasibility. *Circulation*. 2007; 115:2705-2714.
30. Saliba W, Reddy VY, Wazni O, et al. Atrial fibrillation ablation using a robotic catheter remote control system: initial human experience and long-term follow-up results. *J Am Coll Cardiol*. 2008;51: 2407-2411.

31. Kautzner J, Peichl P, Cihak R, Wichterle D, Mlcochova H. Early experience with robotic navigation for catheter ablation of paroxysmal atrial fibrillation. *Pacing Clin Electrophysiol*. 2009;32: S163-S166.
32. Schmidt B, Tilz RR, Neven K, Julian Chun KA, Ouyang F. Remote robotic navigation and electroanatomical mapping for ablation of atrial fibrillation: considerations for navigation and impact on procedural outcome. *Circ Arrhythm Electrophysiol*. 2009;2:120-128.
33. Steven D, Rostock T, Servatius H, et al. Robotic versus conventional ablation for common-type atrial flutter: a prospective randomized trial to evaluate the effectiveness of remote catheter navigation. *Heart Rhythm*. 2008;5:1556-1560.
34. Di Biase L, Natale A, Barrett C, et al. Relationship between catheter forces, lesion characteristics, "popping", and char formation: experience with the robotic navigation system. *J Cardiovasc Electrophysiol*. 2009;20:436-440.
35. Cummings JE, Barrett CD, Litwak KN, et al. Esophageal luminal temperature measurement underestimates esophageal tissue temperature during radiofrequency ablation within the canine left atrium. *J Cardiovasc Electrophysiol*. 2008;19:641-644.
36. Cummings JE, Schweikert RA, Saliba WI, et al. Assessment of temperature, proximity, and course of the esophagus during radiofrequency ablation within the left atrium. *Circulation*. 2005; 112:459-464.

Index

A. Auricchio et al. (eds.), *Cardiac Imaging in Electrophysiology*,
DOI 10.1007/978-1-84882-486-7, © Springer-Verlag London Limited 2012

MIX
Papier aus verantwortungsvollen Quellen
Paper from responsible sources
FSC® C105338

If you have any concerns about our products,
you can contact us on
ProductSafety@springernature.com

In case Publisher is established outside the EU,
the EU authorized representative is:
Springer Nature Customer Service Center GmbH
Europaplatz 3, 69115 Heidelberg, Germany

Printed by Libri Plureos GmbH
in Hamburg, Germany